W9-ACU-354

THE GEOGRAPHY OF
MODERN AFRICA

Second Edition   *Fully Revised*

THE GEOGRAPHY OF MODERN AFRICA

by William A. Hance

Columbia University Press

New York

Copyright © 1964, 1975 by Columbia University Press
First Edition, 1964
Second Edition, 1975
Printed in the United States of America

Library of Congress Cataloging in Publication Data

Hance, William Adams, 1916–
    The geography of modern Africa.

    Bibliography: p.
    Includes index.
    1. Africa—Description and travel—1951–
I. Title.
DT12.2.H28 1975   916   75-2329
ISBN 0-231-03869-0
9 8 7 6 5 4 3 2

To Margie

Preface

Africa is an enormous, complex, unpredictable, and fascinating continent. It has been changing with sometimes startling rapidity. This contributes to its fascination, but greatly increases the task of reporting on it. Assessments such as are attempted here require constant revision as changes occur and as knowledge of the continent is broadened and deepened. Nonetheless, many of the problems and conditions affecting economic development are more or less timeless and others will persist for many years.

It is impossible to know the continent as well as one should to write a book such as this. The author's preparation has involved studying Africa since 1941, teaching courses about the continent since 1949, and ten field trips. All of the major countries and most of the lesser ones have been visited at one time or another, and return visits to many of them have proved valuable in assessing the growth and development of various economic sectors and regions. The debt to others who have studied the continent and its parts is enormous, however, and is acknowledged with gratitude.

Columbia University
February 1975

Many people have been generous in the help extended in preparation of both editions of this book. Academic colleagues, government officials, persons in business and industry, and citizens of African, European, Asian, and American origin have gone beyond the call of duty in attempting to supply me material requested. Their personal kindnesses and unfailing cooperation are deeply appreciated.

I have enjoyed working with my friends at the Columbia University Press—including Leslie Bialler, Manuscript Editor; Audrey Smith, of the production department; William F. Bernhardt, Assistant Executive Editor; and with the map designer for both editions, Vaughn Gray. Each of them is an expert in his or her own speciality, and their contributions are greatly appreciated. Sincere thanks go also to Dennis Grastorf, book designer, and Bronwen McLaughlin for her portion of the typing. Most especially, I express my abiding thankfulness for my wife who joyfully helped me in so many ways to bring this and other projects to fruition.

WILLIAM A. HANCE

Contents

Maps

Tables

Charts

Countries

Anglo-Egyptian Sudan	Sudan
Basutoland	Lesotho
Bechuanaland	Botswana
Belgian Congo	Zaïre
Congo (Kinshasa)	Zaïre
Fernando Po [a]	Ile Macias Nguema
French Somaliland	French Territory of Afars and Issas
French Soudan [b]	Mali
Gold Coast	Ghana
Madagascar	Malagasy Republic
Moyen Congo [c]	Congo (Brazzaville)
Northern Rhodesia	Zambia
Nyasaland	Malawi
Portuguese Guinea	Guinea (Bissau)
Rio de Oro	Spanish Sahara
Ruanda	Rwanda
South-West Africa	South-West Africa (de facto)
	Namibia (de jure per UN)
Southern Rhodesia	Rhodesia
Tanganyika	mainland portion of United Republic of Tanzania
Ubangi-Shari [c]	Central African Republic
United Arab Republic	Egypt
Urundi	Burundi

Lakes and rivers

Albert, L.	Sese Seko Mobutu, L.
Congo R.	Zaïre R.
Edward, L.	Idi Amin, L.
Leopold II, L.	Mayi Ndombe, L.
Nyasa, L.	Malawi, L.
Salisbury, L.	Opeta, L.
Stanley Pool	Malebo Pool

[a] part of Equatorial Guinea

[b] part of French West Africa

[c] part of French Equatorial Africa

Towns and cities (by country)

Algeria

Bône	Annaba
Bougie	Bejaia
Philippeville	Skikda

Botswana

Gaberones	Gaborone
Lobatsi	Lobatse
Ootsi	Ootse
Tsabong	Tshabong

Burundi

Usumbura	Bujumbura

Chad

Ft. Archambault	Sarh
Ft. Lamy	N'Djaména

Equatorial Guinea

Santa Isabel	Malabo

Gambia

Bathurst	Banjul

Malawi

Cholo	Thyolo
Fort Johnston	Mangoche
Mlanje	Mulanje
Port Herald	Nsanja

Mauritania

Fort Gouraud	F'Derik
Port Etienne	Nouadhibou

Morocco

Mazagan	El Jadida
Mogodor	Essaouira
Port Lyautey	Kenitra

Zaïre

Albertville	Kalemie
Bakwanga	Mbuji-Mayi
Banningville	Bandundu
Baudouinville	Moba
Coquilhatville	Mbandaka
Costermansville	Bukavu
Elisabethville	Lubumbashi
Jadotville	Likasi
Léopoldville	Kinshasa
Luluabourg	Kananga
Moerbeke	Kwilu Ngongo
Ponthierville	Ubundu
Port Francqui	Ilebo
Stanleyville	Kisangani
Thysville	Mbanza-Ngungu

Zambia

Abercorn	Mbala
Bancroft	Chililabombwe
Broken Hill	Kabwe
Ft. Jameson	Chipata
Ft. Rosebery	Mansa

Part One
INTRODUCTION

Main line of the Kenya-Uganda portion of the East African Railway on the east side of the Great Rift Valley about 40 miles (64km) from Nairobi
Mt. Longonot, an extinct volcano lying within the Rift Valley, is seen in the background.

The objective of this book is to assemble and analyze material that will contribute to a better understanding of the African economic scene. Specifically, it focuses upon two main questions: what is the present state of economic development? and what are the potentialities for development in the coming decades? Generalizing about a continent as vast as Africa is a hazardous undertaking. Hence the bulk of this study is devoted to analyses of African regions and subregions. Only the first four and final chapters are devoted to the continent as a whole.

The word "modern" appears in the title and it reflects two emphases, one on current economic conditions and the second on the modern exchange economy. This means that certain elements—such as historical background, ethnic subdivisions, and patterns of subsistence agriculture—receive less attention. Nor is the physical environment treated in detail; the attempt is to include only those aspects which have pertinence to the major objectives. Even with these deliberate exclusions, the task of covering those subjects which have been chosen for attention is a challenging one, made more difficult by the enormous diversity in the physical, eth-nographic, social, economic, and political realms.

The writings about Africa have been increasing at an exponential rate in recent years, and now no one can possibly read more than a fraction of what is available. But despite the proliferation of documentation there are gaps of considerable importance: statistical data are of highly variable quality and completeness from country to country; demographic information is often suspect; much of the area remains very poorly known scientifically; and securing comparable and up-to-date information for each of the political units is sometimes a frustrating exercise.

In this introductory section attention is given first to the economic setting. This is followed by chapters on the physical background, the population and peoples of Africa, and the political setting. Succeeding sections deal with major regions of Africa and their component parts, starting with the northern tier of countries and proceeding somewhat circuitously to southern Africa and the Indian Ocean islands. The tables giving comparative data for the continent use the same progression, which seems preferable to an alphabetical listing.

CHAPTER ONE

The Economic Setting

There is tremendous diversity in the economic position of individual African countries. There is also great variation and contrast from region to region within most countries. By some economic criteria, the Republic of South Africa is an exception, but certain generalizations may be made in the economic sphere that will have wide validity on the continental scale.

UNDERDEVELOPMENT

By almost any measure, Africa rates low on the scale of economic development. Traditional systems of cultivation and grazing, characterized by low production per man and low productivity per acre, prevail over most of the continent. The dominant activities are extractive; energy production and manufacturing output rate poorly. Social and physical infrastructures often remain grossly inadequate despite the very real improvements that have been made. There are shortages of skilled workers, entrepreneurs, and managers. Rates of literacy and levels of educational achievement remain low; incidence of disease, infant mortality, and death rates are high. Standards of housing and sanitation often leave much to be desired. Diets do not usually lack calories, but protein intake is insufficient and malnutrition is common.

The average annual gross national product (GNP) per capita in 1971 was about $223.[1] Libya, with only a little over 2 million people and an immense value of oil exports, was the only country with a per capita GNP

[1] International Bank for Reconstruction and Development, *World Bank Atlas 1973* (Washington, D.C., 1973).

above $1,000 ($1,770 in 1972); in 1971 only four other countries with 6.7 percent of the total population had per capita GNPs above $500 (Map 1); some eleven countries with 21.9 percent of the population had per capita GNPs under $100.

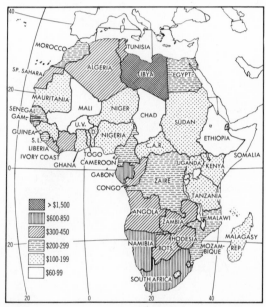

Map 1. Gross national product per capita, African countries, 1972

The total estimated GNP for fifty-three political units of Africa, including island appurtenances, was $79.6 billion in 1971; this was about the same GNP as for the four Fenno-Scandian countries, whose total population is only 6 percent of Africa's. Excluding South Africa with 23 percent and the five northern African countries with 27 percent of the African GNP, the remaining forty-seven African countries had a total GNP about the same as Switzerland and Austria combined, whose populations were 4.4 percent as large.

In 1972 Africa accounted for only 4.5 percent of world imports and 4.7 percent of world exports, while its population represented about 9.7 percent of the world total. Map 2 shows the per capita exports of African countries in 1972; for most countries it was below $50 and only four countries had exports above $150 per person.

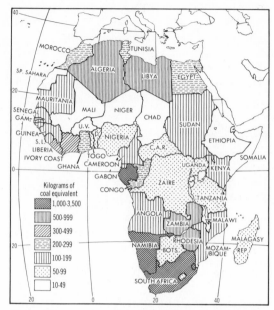

Map 3. Energy consumption per capita, African countries, 1974

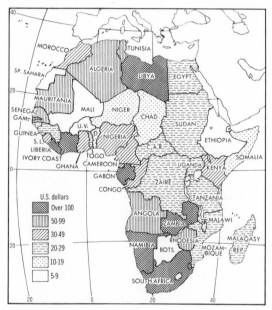

Map 2. Per capita exports, African countries, 1972

Energy consumption per capita stands at only about 18.3 percent of the world average (Map 3); only South Africa and Libya consume above the world average, while thirty-four countries consume less than 10 percent and twenty-three of these below 5 percent of the world average.

Per capita national budgetary expenditures are also very low; most for tropical African countries were under $50 in 1973–74 or 1975 (Map 4).

Since World War II, however, there has been considerable evidence of improvement in a variety of these measures. Many farmers

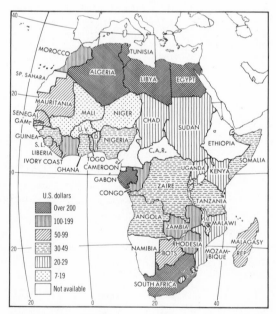

Map 4. National budgetary expenditures per capita, 1973–74 or 1975

have added cash crops to their traditional farming systems and measurable advances have been made in some regions and in some farming practices. Africa has increased its share of world production of tea, coffee, rubber, peanuts, bananas, sugar, and roundwood; it has maintained its dominant position in exports of cocoa, cloves, vanilla, sisal, pyrethrum, and gum arabic (Table 1). Since 1960, however, the value of agricultural exports from the less developed countries (LDCs) has risen more slowly than the value of such exports from the developed countries (DCs).

In the mineral realm Africa has greatly improved its relative position in output of gold, manganese, antimony, iron ore, columbium, petroleum, and natural gas; its leadership continues in the output of cobalt, diamonds, and gold.

A considerable expansion in industrial output may also be noted, though percentage figures are misleading because of the low level of development in most countries. Consumption of energy for Africa as a whole increased from 1.15 percent of the world total in 1953 to 1.8 percent in 1972; electric energy produced increased 472 percent during this period.

Notable advances have been made in modernizing and extending the transport infrastructure of the continent, as is reflected by the following increases for the items noted in the two decades to 1972: rail-ton-mileage, 237 percent; air passengers carried, 207 percent; goods loaded and unloaded in international trade, 673 percent.

Striking advances have also been made in the provision of social services, as measured by increases in the numbers attending school, the representation of colleges, universities, and technical institutions, the number of hospitals and clinics, the ratio of medical personnel to population, and the average life-span.

But the record is by no means entirely favorable, and current trends in some sections leave much to be desired. There are, of course, striking contrasts in the rates of advance from country to country, so there are

notable exceptions to the generalizations made above. For example, in the period 1960–71 nine countries with 11.7 percent of the total population of Africa had a decline or no increase in their per capita GNPs, seven additional countries had rates of increase under 1 percent per annum, and only three countries (Libya, Mauritania, and Gabon) had rates exceeding the 5 percent goal of the UN development decade. In the second half of the period 1960–71 twenty-three of fifty-four countries had lower rates of increase in per capita GNP than in the first half. Africa's share of production declined in the period 1961–72 for a considerable number of commodities, including a dozen minerals, palm products, olive oil, wine, manioc, wool, and citrus fruit. And in the social sector, it has frequently not proved possible to sustain rates of improvement comparable to the rate of population growth.

There are numerous explanations for the relatively less favorable rates of development witnessed in the 1960s and early 1970s, while in 1973 a new problem arose for most countries with the marked increases in petroleum prices, which are likely to more than offset the total aid received by African countries. There are longer-term reasons for the unsatisfactory growth: unfavorable changes in the terms of trade for some countries; a temporary decrease in administrative efficiency associated with the move from colonial to independent status in many countries; serious political disruptions in such countries as Algeria, Nigeria, Zaïre, Rwanda, and Burundi; a reduction in the rate of capital investment and a flight of capital from some countries; and some mistakes in development strategy, including rather costly experimentation with unworkable or at least unwise approaches. Economists have sometimes contributed to these mistakes.

Agriculture, for example, fell out of favor as a sector for priority, and industry was given particular attention. In fact, manufac-

Table 1. African share of world production of selected commodities, as percentage of world production [a]

MINERALS

Metals	1938	1961	1972	Nonmetals	1938	1961	1972
Gold	40	52	80 [b]	Diamonds	99	96	85
Cobalt	87	77 [b]	72	Phosphates	33	26	26
Vanadium	48	31 [b]	42	Vermiculite	...	26	33 [c]
Platinum group	9	30	30	Talc	...	...	32
Manganese	22	22	34	Asbestos	16	14	10
Chromite	35	35	33	Graphite	7	4	5 [c]
Antimony	4	22	26	Mica	6	2	5 [c]
Titanium	...	...	22	Fluorspar	...	...	5 [c]
Copper	18	22	20	Barite	—	4	4 [c]
Uranium oxide	—	15 [c]	18 [c]	Potash	—	—	3
Beryllium	...	40	16	Gypsum	...	...	2 [c]
Iron	4	3	10	Magnesite	...	...	1
Tin	12	10	9	Diatomite	—	...	1 [c]
Columbium-tantalum	—	...	9	Feldspar	...	...	1 [c]
Arsenic	...	...	8	*Energy minerals*			
Lead	3	8	6	Petroleum	—	1	11
Bauxite	—	7	6	Natural gas	—	—	2
Zinc	1	7	5	Coal	1	2	3
Cadmium	—	6	4				
Nickel	1	1	4				
Silver	—	5	3				

AGRICULTURAL, FORESTRY, AND FISHERY PRODUCTS

	1950–58, Average	1961	1972		1950–58, Average	1961	1972
Vegetable oils, oil seeds				*Fibers*			
Palm kernels	83	89	63	Sisal	68	49	52
Palm oil	81	74	53	Cotton	12	8	10
Peanuts	25 [d]	30	31	Wool	9	7	2
Cottonseed	9 [d]	7	10	Others: raffia, piassava, punga, kenaf			
Castor beans	...	...	9				
Olive oil	9 [d]	13	8	*Other foods*			
Copra	4 [d]	3	5	Dates	...	...	42
Others: cashew nuts, sesame seeds, tung oil, sunflower seeds				Bananas	5 [d]	5	12
				Cane sugar	5 [d]	8	9
				Citrus fruit	7 [d]	9	6
Stimulants and spices							
Cloves	c.99	c.99	c.99	*Other industrial products*			
Vanilla	...	69	84 [c]	Pyrethrum	...	c.90	c.90
Cocoa beans	64	74	70	Rubber	3 [d]	7	7
Coffee	16	22	28	Others: quinine, perfume essences			
Tea	3 [d]	5	10				
Wine	10	9	6	*Forest products*			
Tobacco	4	5	4	Gum arabic	c.90	c.90	c.90
Others: chat, pepper, cinnamon, ginger, kola nuts				Roundwood	7	...	12
				Others: sawnwood, wattle, cork			

AGRICULTURAL, FORESTRY, AND FISHERY PRODUCTS

Grains and roots	1950–58, Average	1961	1972	Animals and fish	1950–58, Average	1961	1972
Yams	...	...	98	Fish catch	7	6	5
Taro	...	...	83	Meat	...	...	3
Manioc	51	...	44	Milk	...	...	2
Millet	...	...	23				
Sorghum	...	...	22		*Percent of world population*		
Corn	6 [d]	5	4	Camels	...	...	70
Wheat	...	2	2	Goats	...	...	30
Rice	2	2	2	Sheep	12	13	14
				Cattle	...	13	14
				Chickens	...	...	8

··· = not available   — = none

SOURCES: U.S. Department of Agriculture, commodity publications; U.S. Department of Interior, Bureau of Mines, *Minerals Yearbook 1970;* U.N. *Statistical Yearbook, 1962, 1972;* U.N. Food and Agriculture Organization, *Production Yearbook 1972.*

[a] Some commodities are listed for which data are not available but for which Africa is known to have significant production.

[b] Excluding USSR and China

[c] 1970

[d] Average for 1948/49–1952/53

[e] Non-communist world

turing has not proved to be the panacea, primarily because of the limitations imposed by the small size of the individual national markets. Although there is now much wider recognition that agriculture is deserving of greater attention, government policies in many countries continue to be urban-biased to the detriment of agriculture and rural advance in general.

In any case, it is now realized that development is a very much more difficult process than had been thought. But there is no reason for disillusion; there is reason to seek more effective ways to promote the development of African countries.

SUBSISTENCE AND EXCHANGE ECONOMIES

Africa is marked by the coexistence, often the juxtaposition, of heterogenous economies. Most of it is in transition from a subsistence to an exchange economy, but there are relatively few areas which, and peoples who, have not been affected in some way by modern economic forces. About three-fifths of the total cultivated area of tropical Africa is still devoted to subsistence production, which occupies over half of the adult population.

Two prototype extremes between which most African economies lie may be recognized.[2] In both extremes there is a modern and a traditional economy. In the first, commercialization has been brought about chiefly by the transformation of part or parts of the traditional economy, usually by peasant production of export crops; there is relatively little foreign investment in large-scale enterprise, and the movement of workers to the modern economy remains relatively small. Falling largely into this category

[2] See UN, *Economic Survey of Africa since 1950* (1959), p. 12.

Cocoa drying on a small farm in Cameroon
In some areas, commercialization has been brought about
chiefly by transformation of the traditional economy,
usually by peasant production of export crops.

would be such countries as Gambia, Mali,
Upper Volta, Niger, Chad, Cameroon, the
Central African Republic, Uganda, and Tan-
zania.

The second prototype is characterized by
an exchange economy brought about largely
by foreign capital and enterprise, operating
with a high capitalization and advanced tech-

A tea plantation in Thyolo District, Malawi
In other areas, an exchange economy has been brought
about largely by foreign capital and enterprise.

niques, mainly in mining and agricultural
enterprises; it depends heavily on non-
African and foreign capital, and there is a
relatively large flow of workers from the traditional to the modern economy. South
Africa, Rhodesia, Gabon, Zambia, Zaïre, and
Kenya fall into this group.

The transition to a modern economy
frequently involves the development of economic distortions and social and political
tensions. Traditional systems of tenure and
land use may impede the most economic disposition of the land. Tribalism often directly
opposes modernist tendencies, leading to
some of the most dynamic of African political problems. The migration of labor from
traditional areas may have adverse effects on
the subsistence agricultural system and contribute to the problems associated with urbanization. In countries of the first prototype, movement into the modern economy is
likely to result in transformation of the traditional economy; in countries of the second
prototype, the result is more likely to be
disruption and disintegration of previous
forms.

THE PREFERENCE FOR A
SOCIALIST ECONOMY

A high proportion of African countries
opted at least nominally for a socialist economic policy after independence. Considerable efforts have been made to enunciate a
specifically African Socialism, or a more local
form such as "Arab Socialism," "Nkrumahist
Socialism," "Tanzanian Socialism," or "Zambian Socialism." Factors that have attracted
support for a socialist approach include the
claim that it prevents exploitation both
among citizens within a nation and by outside powers, and the concept that development will be more successful under the direction of the state than under a free
enterprise system.

The very great dependence of some countries on foreign-owned and -controlled en-

terprise, which is seen as sustaining a neocolonialist relationship, contributes to the desire to bring resources under national control. This is particularly true for mineral bodies, considered to be wasting resources that must be exploited for the benefit of the nation. Thus there have been nationalizations of petroleum and mining companies in, e.g. Algeria, Libya, Mauritania, Sierra Leone, Ghana, Togo, Zaïre, Zambia. The attitude toward use of the land often closely parallels that toward mineral resources, which means that foreign-owned plantations and estates are usually not considered appropriate for agricultural development.

No African country, however, has achieved a purely socialist economy; nor have the various definitions of local socialism differed significantly from existing forms in other parts of the world. The vast bulk of national economies are mixed, with no greater proportion of state-run enterprises than exists in the United States or in Western European countries. And even the most Marxist-oriented governments continue to stress their desire for increased foreign investment.

The main explanations for the gap between the stated goals of achieving socialism and the actual achievements are that the government bureaucracies are not adequate to sustain a pervasive socialism; the goals have been overambitious, not understood by many people, and not accepted by many others, hence it has proved very difficult to win mass support for socialization (particularly of agriculture and marketing); the inability of almost all countries to generate enough local capital to meet the needs of a socialist bureaucracy; and the observed and experienced failure of several efforts at socialization, which have adversely affected political stability in several countries.

Examples of failures to achieve the degree of socialism desired by governments include collapse of the efforts in Guinea and Tunisia to nationalize the agricultural and commer-

cial sectors of their economies; the poor record of state farms initiated in Ghana, Algeria, and several other countries; and the lack of acceptance of the concepts of collective fields in Mali or of village collectives in Tanzania. On the other hand, the nationalizations of several mining enterprises may be interpreted as successful examples, inviting comparable actions by other nations.

There is considerable evidence that those nations which have been most open to foreign enterprise have made substantially better records in economic growth; witness for example the achievements of such countries as Morocco, Libya, Ivory Coast, Kenya, Zambia, Swaziland, and South Africa. Some of these nations have, however, been criticized for not achieving a better record in development;[3] that is, the benefits of a rapid statistical growth have not accrued sufficiently to the betterment of the populace. They have instead, it is claimed, been absorbed by a swollen bureaucracy, dissipated on unproductive showpieces, or dispensed in outright graft. Misallocation of available funds is reprehensible, but it would be a mistake to assume that it is confined to free enterprise economies; indeed the reverse seems to be nearer the truth, because there are greater opportunities for misappropriation.

THE DOMINANCE OF EXTRACTIVE ACTIVITY

The importance of primary production is a dominant feature of the modern economy of Africa (Table 2). While most of the food crop production is consumed locally, much of the agricultural output and the vast bulk of minerals produced are destined for shipment outside the continent. At present there

[3] See, for example, Robert W. Clower et al., *Growth without Development: An Economic Survey of Liberia* (Evanston, Ill.: Northwestern University Press, 1955). But see also, Louis P. Beleky, "The Development of Liberia," *Journal of Modern African Studies*, 11, no. 1 (March 1973): 43–60.

Table 2. Exports of Africa and of tropical Africa by commodity realm as percentages of the total value of exports, 1961, 1972

Product realm	Africa [a]		Tropical Africa [b]	
	1961	*1972*	*1961*	*1972*
Mineral	35.9	56.3	28.6	45.9
Vegetal	45.2	25.7	61.2	42.1
Animal	5.9	4.2	3.2	2.7
Manufactures	4.4	6.3	1.0	1.9
Unidentified	8.6	7.5	6.0	7.4
Total	100.0	100.0	100.0	100.0

SOURCES: National trade statistics; 1961 data from first edition.

[a] Includes all countries except Botswana and Atlantic Ocean Islands, including rough estimates for Rhodesia, in 1972.

[b] Excludes northern tier, South Africa, South-West Africa, and Rhodesia (1972 only).

is relatively little exchange among African countries and domestic markets are not fully developed, but both of these characteristics are changing with some rapidity.

The dominance of extractive exports means that African countries may suffer severely in periods of declining prices for raw materials. In the period 1969–72, for example, the movement of export prices was unfavorable for many African countries; in the period 1972–74, however, notable increases in the price of many agricultural and mineral commodities reversed this trend. The impact of oil imports is likely, however, to have a serious effect on the overall balance of trade and payments of most countries. The marked shift from 1961 to 1972 from vegetal to mineral exports is accounted for mainly by the very large exports of oil from Libya, Nigeria, and Algeria in the latter year. If the value of oil exports from these three countries is extracted, the share of mineral exports in 1972 is reduced to 40.4 percent for Africa and to 32.1 percent for tropical Africa, while vegetal exports rise to 35.1 and 52.9 percent respectively.

The importance of extractive activity may also be indicated by examining the percentage of the gross domestic product (GDP) by industrial origin. The Economic Commission for Africa estimates for 42 independent African countries that the 1971 contributions to GDP were as follows: agriculture 31.9 percent, mining 11.6 percent, and manufacturing 11.0 percent.[4]

AGRICULTURE

As would be expected, agriculture is the main sector of economic activity. In tropical Africa about 80 percent of the population is rural and over 65 percent is engaged in agriculture. There is great variation in institutional agricultural forms. The largest proportion of commercial crops is produced in modified traditional systems. Plantations are important in the production of rubber in Liberia, Nigeria, and Zaïre; palm products in Ivory Coast, Cameroon, and Zaïre; tea in East Africa and Malawi; sisal in Kenya, Tanzania, Angola, Mozambique, and Madagascar; coconuts in Mozambique; and sugar in many countries. European-owned farms and estates account for a high percentage of commercial agricultural output in Rhodesia and South Africa, and continue to be important in Morocco, Kenya, Angola, and Swaziland. Collective farms have replaced European holdings in Algeria, while a large part of Sudanese cotton is produced under a somewhat unusual partnership arrangement.

Traditional systems of cultivation have great diversity. Four main types may be delineated: (1) shifting cultivation, in which virgin land or well-regenerated secondary bush is cleared for cropping; this type is found mainly in sudanic and savanna areas where clearance by fire is possible; (2) rotational bush fallow, which involves a few years of cropping followed by a longer period during which the field reverts to natural woody bush or grass, but where woodland per se is

[4] ECA, *Survey of Economic Conditions in Africa, 1972* (New York, 1973), p. 52.

not allowed to regenerate (see chapter 10, pp. 210–12); (3) rotation planted fallow, which differs from bush fallow only in the selection and planting of the fallow cover; and (4) mixed farming, where stock rearing is combined in various ways with cultivation. This last system is rare in tropical Africa.

Three other traditional systems are locally important but are usually complementary to one of the main types. They are: (1) permanent cultivation, under which the use of manures, compost, and household waste permits cropping each year; many groups maintain permanent cultivation in small kitchen gardens, and it is practiced on all or most of the cropped area only where pressure of population has required more intensive use, such as in crowded mountain refuges or in the close-settled zones around the emirate cities of the sudan-sahel belt, or in areas where unusually rich soils exist; (2) tree cultivation, which is a second type of permanent cultivation and is typically associated with rotational bush fallow systems; and (3) floodland and irrigation cultivation. In the Sahara, the great Nile oases, and in parts of Madagascar irrigation agriculture is all-important; elsewhere it is characteristically associated with, and a relatively minor part of, the total cropping pattern. Floodland cultivation involves the growing of catch crops after floodwaters have receded from seasonally inundated lands.

Since a large part of Africa is semiarid, pastoral activities occupy a vast expanse. In 1972 Africa had an estimated 158 million cattle and 140 million sheep, or about 14 percent of the world total for both. Tropical African pastoralists are strongly dependent on their herds of cattle and goats, but often prefer not to sell cattle unless it is necessary; commercialization is therefore poorly developed. Indigenous livestock are typically very poor milkers and have a low dressed weight as well. Livestock products constitute a very small part of total exports, though hides and skins do figure among the commodities shipped from most countries. The fact that the tsetse fly prevents the keeping of all bovines except resistant dwarf cattle in about half of tropical Africa helps to explain why most African countries are net importers of meat.

Africa's total agricultural production increased more rapidly than the world average from prewar years to 1961–62, but its per capita output actually decreased slightly during this period. Since then the record has been less favorable, while production of foodstuffs has experienced a lower than average increase in production in the entire postwar period. Imports of foodstuffs have tended to rise in most countries and many cities are now heavily dependent on foreign goods. These characteristics suggest an undesirable neglect of agriculture, a certain imbalance in crop emphases, and a partial failure to achieve more desirable output levels of both subsistence and commercial products.

FISHING

A second primary activity, fishing, is well developed commercially only in South Africa, South-West Africa (Namibia), Angola, and Morocco, all of which have relatively valuable exports of fish and other products of the sea. Modern vessels operate out of other ports, particularly those engaged in the well-commercialized tuna fisheries whose products are sold primarily in extracontinental markets. Fishing in the lakes of eastern Africa is of some significance, while elsewhere lagoons, rivers, swamps, and seasonally inundated areas provide fish for local areas plus some shipment to internal markets. A number of territories have promoted the establishment of fish ponds in which *tilapia*, in particular, are raised to produce a welcome addition to the characteristically protein-deficient diets.

FORESTRY

Although Africa is estimated to have 27 percent of the total forest area of the world, its share of the world production of roundwood is estimated at only 12 percent of the total, but the output of roundwood, sawn wood, plywood, and veneer has increased at a very favorable rate in the postwar period. A large amount of unrecorded cutting is consumed as fuel, mainly for cooking. Only in a few countries—including the Ivory Coast, Ghana, Nigeria, Cameroon, Gabon, Congo, and Swaziland—does forestry make more than a small contribution to the money economy. Still a net importer of forest products, Africa does have great potentialities for an intensified utilization of its forests. At present, less than a seventh of the valuable forests are under exploitation.

MINING

Among extractive activities, mining ranks after agriculture in value of production. Employing about 1.1 million people, including perhaps 125,000 non-Africans, mining is the most highly organized and capitalized sector of the modern economy, of which it has sometimes been a decisive initiator. But the mining industry is very unevenly distributed. The five leading mineral exporting countries—South Africa, Libya, Nigeria, Algeria, and Zambia—accounted for about 80 percent of the value of all mineral exports from Africa in 1972 and the top ten countries for 93.4 percent. While certain nations are critically dependent on mineral exports, others remain largely or entirely agricultural. In 1973, fifteen countries had a single mineral as their leading export, nine other states had minerals ranking among the top three exports, and a total of twenty-eight countries had mineral exports valued at 5 percent or more of their exports. At the same time, twenty-two countries had less than 5 percent of their exports in minerals. Mineral exports as a whole grew from 35.9 percent of total African exports in 1961 to 56.3 percent in 1972; price increases for oil and gold will undoubtedly increase the share of minerals as these two exports accounted for an estimated 36.3 percent of total continental exports in 1972.

There is a marked concentration by mineral in the value of exports, with the top five—petroleum, gold, copper, diamonds, and iron—accounting for about 85 percent of the total value of mineral exports and the top ten for 90 percent (see Table 3). The position of the continent in world production of specified minerals is given in Table 1.

The most significant impacts of mining operations for African producers are as sources of foreign exchange earnings and of government revenues. Mining is usually capital-intensive and provides less direct employment than other activities. Only a few countries, most especially South Africa, have an important segment of manufacturing based on minerals.

Notable trends in the mining industry in recent decades include production of an ever-increasing variety of minerals; a greater interest in lower-metallic-content ores; a marked intensification of exploration and a wider dispersion in the areas receiving attention; increased processing of minerals before export; construction of pipelines, new mineral railways and ports, and of special mechanical loading gear at shipping points; installation of several large-scale electric stations to serve mining-metallurgical complexes; the imposition of higher tax and royalty rates; acquisition by nationalization of partial or full ownership of mining, which has occurred in all the North African countries, plus Sierra Leone, Guinea, Ghana, Togo, Zaïre, Tanzania, and Zambia in tropical Africa, and which has usually involved compensatory payments and retention of the former concessionaires as managers; more rigid requirements regarding the localization of staff and provision of training facilities; the formation of multinational companies for the development of major bodies; and increased participation by consuming companies in the production of certain minerals, including iron ore, bauxite, manganese, and copper. Despite an apparently accelerating trend toward greater local participation or nationalization, almost all countries continue to seek foreign investment in exploration and exploitation of new mineral bodies.

Despite the favorable future for minerals in Africa, mining cannot be expected to solve the problems of African development and particularly the need to generate employment. With a few notable exceptions, agriculture will be the sector requiring prime government attention for the next several decades.

NONEXTRACTIVE ACTIVITIES

Nonextractive activities are rather poorly represented in most of Africa. Much of manufacturing is still closely associated with extractive pursuits, being concerned with the primary processing of mineral, agricultural, forestry, and fishery products. Recent decades have, however, seen a notable increase in the representation of establishments catering to the domestic national markets.

South Africa accounts for perhaps two-fifths of the manufactured output of Africa. The Maghrebian countries (Morocco, Algeria, and Tunisia) and Egypt have a large number of persons employed in handicraft industries and in a fairly broad range of modern manufacturing sectors. In tropical Africa, Rhodesia is the only country having at least minimal development in all major divisions of manufacturing. Nigeria, Zaïre, and Kenya rate relatively favorably. Most other countries have only a rudimentary industrial development.

Manufacturing in Africa is like mining in the importance of non-African capital, skill, and management. With the exceptions of a few large forestry plants, large-scale mineral smelters and refineries, and chemical plants in the north, industry outside South Africa is composed mainly of small plants using relatively simple manufacturing processes.

Tourism is a rapidly advancing sector in a number of countries. It contributes importantly to their balance of payment earnings, and since it is a labor-intensive activity it provides significant employment opportunities. Morocco and Tunisia in the north, East

Africa, and South Africa have seen particularly notable increases in the number of visitors, but tourism is important to the economies of many other countries, including Egypt, Gambia, Ivory Coast, Ethiopia, Rhodesia, Mozambique, the Seychelles, and Mauritius. Heavy investments have been made in the provision of facilities, but many countries do not yet possess a sufficiently developed infrastructure to attract large numbers of tourists, while excessive visa requirements and currency restrictions discourage many visitors.

Public administration and defense, trading, transportation, and domestic service figure importantly among other nonextractive activities. They account for a substantial percentage of wage employees in many countries, and their share of government consumption to the GDP has increased from an estimated 13 percent in 1960 to about 18 percent in 1973.

THE IMPORTANCE OF FOREIGN TRADE IN AFRICAN ECONOMIES

A high percentage of Africans, particularly in tropical Africa, are still involved primarily with subsistence production and have relatively little exchange or contact with communities outside their own, but the money economies of African countries are more involved in foreign trade than are those of most other developing countries. Africa exports somewhat over a quarter of its GNP. And using Spearman's rank correlation, the correlations between per capita exports, per capita GNP, per capita government expenditures, and per capita energy consumption were all found to be highly significant (see Maps 1–4). Indeed, for the sixteen poorest independent countries, exports accounted for only 14 percent of GDPs in 1971 as compared to an average of 25.6 percent for the continent as a whole, while the growth rates in per capita GNP were highest in those countries which had high exports in relation to GDPs.

The heavy focus of many African economies on foreign trade is reflected in the structure of their transport and urban systems. The vast bulk of African imports and exports are with extracontinental markets (less than 7 percent is with other African countries). South Africa has the highest total value of intracontinental trade; the three East African countries, aided by a partial customs union, have the highest percentage of their exports going to other African countries.

The dependence on exports of primary products involves serious problems, including vulnerability to wide fluctuations in the volume and value of sales, the difficulty of finding adequate markets for a variety of commodities, and the impediments to planning for orderly growth when revenues are heavily dependent on trade. Some of these problems are discussed below. It would be unwise to conclude from what has just been written, however, that African nations should not continue to strive to increase their exports and their competitiveness in world markets. Export production imparts a generative impulse to domestic enterprise; it has an important multiplier effect.

It is sometimes proposed that Africa cannot hope to be truly independent so long as it remains heavily dependent on international trade; self-reliance is suggested as the appropriate alternative. In fact, self-reliance is a counsel of despair if it implies self-subsis-

Table 3. Main exports of Africa and tropical Africa by percent of total value, 1961 and 1972 [a]

		Africa				Tropical Africa		
Rank in 1972	Commodity	1972 % of total	1972 Cumulative	1961 % of total	Commodity	1972 % of total	1972 Cumulative	1961 % of total
1	Crude petroleum	28.4		4.7	Crude petroleum	22.3		···
2	Gold	7.9	36.3	10.3	Copper	13.5	35.8	13.8
3	Copper	7.1	43.4	7.5	Coffee	10.8	46.6	9.3
4	Coffee	4.9	48.3	4.7	Cocoa	8.3	54.9	10.8
5	Cotton	4.4	52.7	7.3	Cotton	5.4	60.3	7.0
6	Cocoa	3.8	56.5	5.4	Iron ore	3.8	64.1	1.4
7	Animal products	3.2	59.7	···	Wood products	3.4	67.5	4.5
8	Diamonds	2.7	62.4	3.8	Peanut products	3.0	70.5	7.2
9	Iron ore	1.8	64.2	1.4	Diamonds	2.6	73.1	4.2
10	Textiles/clothing	1.8	66.0	···	Sugar	2.1	75.2	2.9
11	Wood products	1.6	67.6	2.3+	Animal products	2.0	77.2	1.0+
12	Peanut products	1.4	69.0	4.0	Palm products	1.2	78.4	4.4
13	Vegetables	1.3	70.3	···	Tea	1.1	79.6	1.1
14	Phosphates	1.2	71.5	1.5	Rubber	0.8	80.3	2.0
15	Sugar	1.0	72.5	1.9	Petroleum products	0.8	81.1	···
16	Fish products	1.0	73.5	1.5	Cobalt	0.7	81.8	···
17	Tobacco	0.6	74.1	1.9	Fish products	0.7	82.5	···
18	Palm products	0.5	74.6	2.2	Tin	0.6	83.1	1.1
19	Tea	0.5	75.1	···	Cloves	0.6	83.7	···
20	Olive oil	0.5	75.6	···	Cashew nuts	0.6	84.3	···

··· = less than 1.0 percent

SOURCE: Compiled from national trade statistics.

[a] See notes at end of Table 2.

tence; no nation in the world—large or small, developed or developing, free enterprise or state directed—is an economic island. Inward-looking growth is just not possible for countries as small as those in Africa, and even the larger and richer countries such as Nigeria, Zaïre, and South Africa must rely on exports to stimulate growth. It has been very common for African countries to adopt policies of import substitution, particularly in industry. While there are opportunities for such substitution, industrialization by the replacement of imports is often an unsatisfactory and temporary engine of growth. It impedes competition, creates a high-cost manufacturing sector, conserves little if any foreign exchange, does not reduce the dependence as was intended, reduces the attention that can be given to agriculture, contributes to the characteristic urban bias of African government policies, and stimulates smuggling, which is a prevalent phenomenon.

THE "ONE-PRODUCT" CHARACTER
OF AFRICAN ECONOMIES

Just as there is emphasis upon the extractive end of the economic spectrum, so there tends to be heavy dependence on a limited number of commodities as far as exports are concerned. Chart 1 shows the small number of products required to account for the bulk of exports from African countries. This "one-product" character has numerous actual and potential disadvantages. It makes the economy of an individual country excessively vulnerable to fluctuations in prices and in the size of market for its staple products—fluctuations very largely or completely beyond its control.

The degree of fluctuation in unit value appears to be larger for African commodities than for the world as a whole. Chart 2 shows the changing unit values for exports from selected countries in the period 1963–73, while Chart 3 illustrates the price fluctuations for specific commodities from 1960 to 1974. The latter gives average annual prices except for 1974 and hence conceals some rather wide swings within individual years.

Reductions in price and volume will cause reductions not only in export earnings but also in revenues to government, characteristically heavily dependent on export and import taxes. This can in turn significantly affect development expenditures. "One-product" emphases may create further difficulties in pursuing an effective labor market policy, cause serious unemployment, and lead to political unrest. They may also have physical disadvantages, including soil deterioration, increased insect and disease attack, and neglect of proper dietal standards. While declining prices and volumes of sales are the major threats to "one-product" economies, unusually high prices are a mixed blessing (see chapter 20).

It is not just individual countries that are characterized by dependence upon a limited number of commodities; the continent, or a large part of it, is similarly dependent. In 1972 only five commodities accounted for 52.7 percent of total exports by value from Africa and for 60.3 percent of exports from tropical Africa (Table 3), while the top twenty commodities accounted for 75.7 percent of African and 84.4 percent of tropical African exports. The five leading minerals accounted for 47.9 percent of African and 42.9 percent of tropical African exports in the same year; five leading items in the vegetable realm accounted for 16.1 percent of African and 30.9 percent of tropical African exports. While much lip service has been paid to the necessity to broaden the range of national exports, the record reveals greater concentration in exports in twelve countries and greater diversity in only eight countries over the period 1961–72.

Although "one-product" emphases undoubtedly have disadvantages, it does not necessarily follow that economic strength is directly correlated with degree of diversification. Madagascar and Mozambique, for ex-

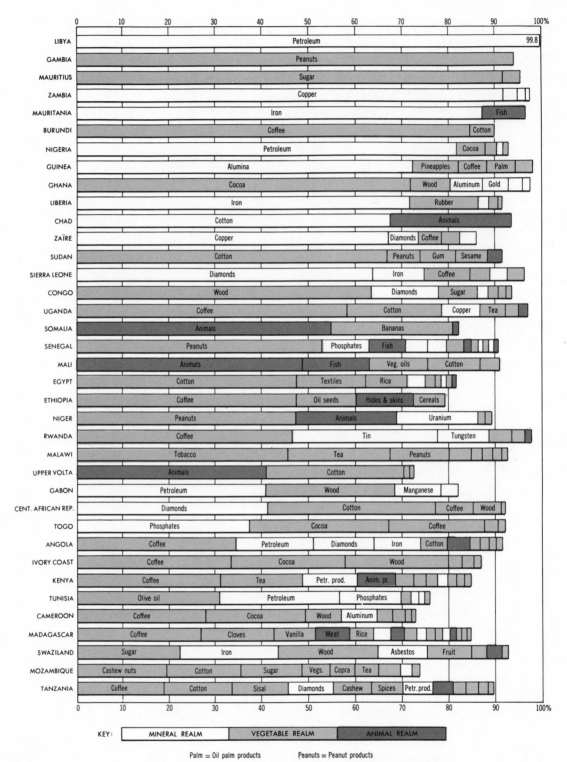

Chart 1. *Leading export commodities from selected African countries as percentages of total exports by value, nearest year to 1972*

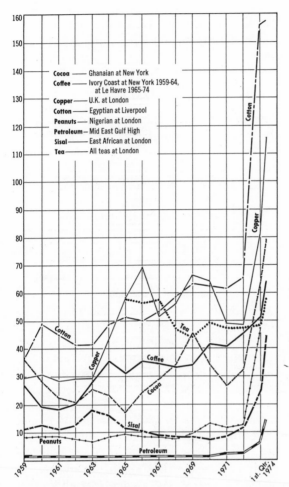

Chart 2. Price fluctuations for selected commodities, 1959–first quarter 1974
Wholesale price quotations in $U.S. per 100 lbs except for petroleum, in $U.S. per barrel. SOURCE: International Monetary Fund, *International Financial Statistics,* various issues to mid-1974.

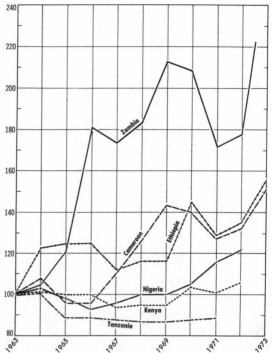

Chart 3. Export price indexes for selected African countries, 1963 (100) to 1973
SOURCE: International Monetary Fund, *International Financial Statistics,* June 1974.

ample, have an unusual variety of exports but have experienced lower growth rates than such strongly one-product economies as Libya, Mauritania, or Zambia.

Concern over the fragility of one-product economies has led to a variety of measures and programs designed to reduce the impact of their undesirable features. On the one hand, these efforts have been applied to the major commodities involved, with the goal of reducing price fluctuations and evening out

returns to the primary producer and to the country. On the national scene, there has been introduction of production quotas, controlled marketing, some control by taxation, and fixed purchase prices. On the international level, there have been efforts to promote commodity arrangements designed to stabilize raw material prices and to seek compensatory payments to the LDCs to offset deteriorating terms of trade. The example of price increases imposed by OPEC (Organization of Petroleum Exporting Countries) has suggested possible similar action by producers of other raw materials.

On the other hand, efforts have been made to diversify the economies of African countries by introducing new cash crops, producing import-substitute crops, building up domestic markets, developing mineral production, attracting new industries, promoting tourism, and so on. While these measures may be beneficial to individual countries, some of them may actually increase the difficulties for the continent as a whole; a new crop in one area, for example, may already be in surplus production in others.

Attention has also been given to the upgrading of African produce with the goal of commanding higher prices for what is sold on world markets. All these and other measures, including some involving concerted moves with other LDCs, may be expected to receive continuing attention.

While there have been strong tendencies to date to overproduce and thus to depress prices, the longer run may reveal shortages developing in a variety of commodities, especially minerals, if we are to accept the predictions in *The Limits to Growth* and similar publications.[5] Indeed, if one acceded to the arguments of the exponentialists it would be wise for Africa to eschew further economic growth, conserve material resources, discourage most forms of industrial development, prohibit further urban development, and rigidly control population growth. African elites have not accepted the necessity for these drastic measures.

THE INADEQUATE INFRA-STRUCTURE

The basic public services of African countries are in no case adequately developed. This is true for the social infrastructure (education, health services, etc.) and the physi-

[5] See, for example, "Blueprint for Survival," *The Ecologist* (January 3, 1972); Barry Commoner, *Science and Survival* (New York: Viking Press, 1966) and *The Closing Circle: Nature, Man and Technology* (New York: Knopf, 1971); Ann H. Ehrlich and Paul R. Ehrlich, *Population, Resources and Environment: Issues in Human Ecology* (San Francisco: W. H. Freeman, 1972); D. H. Meadows et al, *Limits to Growth* (New York: Universe Books, March 1972). But see also critiques of the major themes of these works in such sources as *The Economist* (March 11, 1972); Karl Kaysen, "The Computer that Printed W*O*L*F*," *Foreign Affairs,* July 1972, pp. 660–68; H. S. D. Cole et al, *Thinking about the Future: A Critique of the Limits to Growth* (London: Chatto and Windus, 1973).

cal infrastructure (transportation, energy, water supply, communications, waste disposal, etc.), both of which are essential for economic development.

EDUCATION

The single most important explanation for the relatively undeveloped position of most of Africa is the low level of educational achievement. The accomplishments of the relatively few educated persons in various parts of Africa give some notion of the really dramatic potentialities that will unfold as the educated African population broadens and deepens. Unfortunately, technical, professional, and experienced managerial personnel are not trained overnight and it will be some decades before shortages in these and other fields are no longer inhibiting. Furthermore, it is not sufficient just to have a limited elite at the top; if the mass of the population is to make an effective contribution toward development, a significant percentage of the population must be educated. Until a sizable number of people are asking themselves the question "how can I do this job better than I have been doing it?" it is not likely that effective progress will be made against the deadweights of inertia, ignorance, apathy, and inhibiting tradition.

At present, there is still, in most countries, a low percentage of school-age children actually attending school and a very rapid drop-off in numbers passing from primary to secondary and higher levels. Maps 5 and 6 indicate the comparative position of African countries as measured by educational attendance. Map 5 shows the percentage of primary school students to the total population; Map 6 gives secondary school attendance as a percent of primary school enrollment and thus reveals the rapid decline noted above. The drop-off from the secondary level to higher education is even more striking; in most countries higher educational institutions have registrations less than 5 percent of those in secondary schools and only two

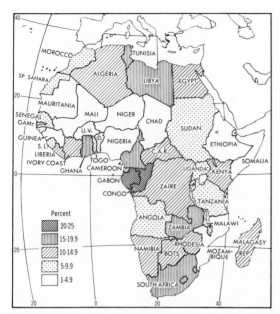

Map 5. *Primary school attendance as a percentage of total population of African countries, early 1970s*

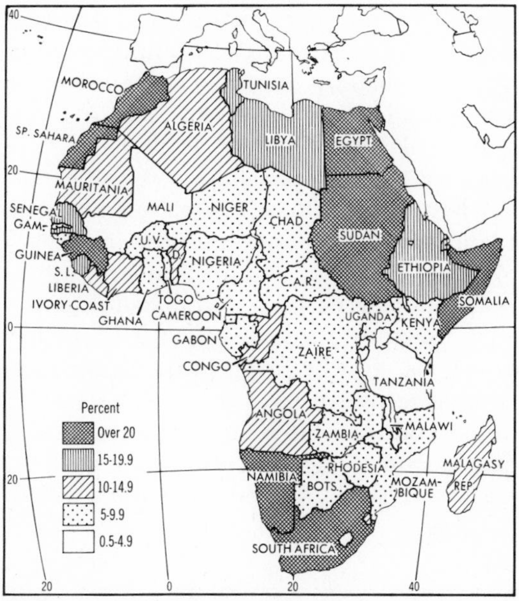

Map 6. *Secondary school attendance as a percentage of primary school attendance in African countries, early 1970s*

countries have registrations over 8 percent as high.

There have been notable improvements in the last several decades in the provision of education, which often receives the highest single allotment in national budgets, but most countries cannot move so rapidly as would be desired to universal education because of the lack of funds and of qualified teachers. Indeed, a few countries have had to postpone stated goals, in part due to the now very rapidly increasing populations. There are also needs for more technical training, for syllabi and readings related specifically to Africa, and for upgrading the quality of instruction at all levels.

One of the most encouraging characteristics of the educational scene is the intense dedication of many students; it holds great hope for Africa. Evidence of the interest in education is the considerable increase in numbers of colleges and universities in recent years (Map 7). It should be noted that many Africans are also attending higher educational institutions on other continents.

Map 7. *African colleges and universities, 1974*
A university is situated on Réunion, not shown on map.

HEALTH SERVICES

The health and medical infrastructure of Africa has undergone significant improvement in recent years but it remains inadequate to the massive tasks of improving nutrition, combatting disease, and bettering the health standards of the people. The high incidence of disease, high death rates, and short average life spans testify to the needs; they have direct impacts on development in several ways, including reduced ability to work, high rates of absenteeism, and an increased dependency load on the adult population. Some experts also hold that further reductions must be made in the death rates before effective progress can be expected in fertility control. Map 8 gives one measure of the inadequate health services in showing the ratio of doctors to population in African countries.

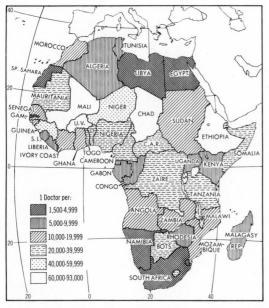

Map 8. The ratio of doctors to population in African countries, early 1970s

TRANSPORTATION

The provision of transport is fundamental to all other forms of development. Marked improvements have been made in postwar years in this field, which has received much larger allocations than before the war, but

Table 4. Africa's percentage share in world population, transport, and trade, 1961, 1971 and 1972

	1961	1971	1972
Population	8.6	9.6	9.6
Rail net ton kilometers	1.8	2.0	1.9
Registered motor vehicles	2.0	2.0	1.9
Aviation passenger kilometers	2.0	2.1	2.4
Goods moved in shipping	5.9	9.3	8.7
Imports	5.6	4.6	4.0
Exports	5.4	4.4	4.3

SOURCE: U.N. *Statistical Yearbook, 1962, 1972, 1973.*

needs continue to be substantial. Table 4 shows the share of Africa in world population and in selected key items related to transport.

Almost nowhere in tropical Africa is there a fully integrated transport complex; transport has to a considerable degree developed without proper planning. Political considerations have too often outweighed economics to the detriment of some of the transport agencies, individual territories, and particular regions, which are forced to use longer, higher-cost routes than might have been necessary. The goal should be integrated development not only on a national but on an international scale.

PORTS. Africa has the shortest coastline in relation to its area of any of the continents and most of the coastline is notoriously poor in good natural harbors. Morocco and Libya, the Guinea Gulf countries of West and Equatorial Africa, and Tanzania, Somalia, and Ethiopia on the east coast have been particularly handicapped by the absence of desirable sites for ports. River mouths and estuaries are frequently plagued by shifting sands and offshore bars. Lighterage ports are still important in several countries and vessels must sometimes be worked in completely open roadsteads.

Great progress has been made, however, in improving the port situation in Africa. No significant surf ports remain in use; modern installations have replaced the slow, expensive, inefficient, and dangerous system of loading and unloading vessels lying offshore with surf boats that land at inadequate piers or directly on the beach. Unprotected general cargo piers have also largely disappeared. Since World War II thirteen entirely new ports have been constructed in eleven countries, several existing ports have received new deepwater facilities where none existed before, and large-scale extensions and improvements have been made at practically all major ports. At present, twenty-eight of the thirty-one countries fronting on the sea have at least one modern port. It is, therefore, no longer accurate to generalize that African development is handicapped by difficulties of sea-land contacts. Nonetheless some national subregions are inadequately served, while periodic congestion and projected increased tonnages require extensions to a number of major ports such as Casablanca, Monrovia, Lagos, Douala, Matadi, Dar es Salaam, Mombasa, and Port Sudan.

Other important improvements affecting ocean transport include construction of several offshore petroleum loading posts in producing nations; greatly improved facilities for landing petroleum (and, incidentally, for delivering it in bulk to inland destinations, thus substantially reducing the need to handle drums by hand); special facilities for loading palm oil by pipe in a few ports; mechanical ore-loading gear capable of loading large ore carriers in a matter of hours, installed at some eighteen sites; the use of much more mechanical equipment on shore, including forklift trucks, straddle trucks, and mobile cranes; and the beginnings of containerization in a number of ports.

The relative importance of African ports as measured by tonnages handled is shown in Map 9. Allowance should be made for the very large shipments of petroleum from points in Libya, Nigeria, Algeria, and Tunisia and for large tonnages of mineral loadings from other ports in assessing the comparative importance of specific points.

Bales of long-staple cotton awaiting loading on the quay at Port Sudan
New and modernized ports have greatly improved the transport picture in Africa in recent decades.

WATERWAYS. Because of Africa's plateau character and the high seasonality of precipitation away from its equatorial core, navigable waterways have only limited importance in most African areas. Other handicaps to river transport are the floating vegetation known as "sudd" on the Upper Nile, water hyacinth (*Salvinia auriculata*)—which has required control measures on several rivers and lakes—and low-hanging vegetation on minor streams. In the north, only the Nile is of importance. In West Africa, the lower Niger is significant for bulk shipments; the upper Niger and Senegal rivers are only locally important; the Gambia, which might have been a major routeway, has had circumscribed use because of preferences given to use of the national land routes in surrounding Senegal; and tidal creeks are of some value in the countries from Guinea (Bissau) to Liberia, while coastal lagoons are generally of decreasing significance in Ivory Coast, Dahomey, and Nigeria.

In middle Africa, the estuaries and coastal indentations of Cameroon and Gabon are important in floating logs to shipping points.

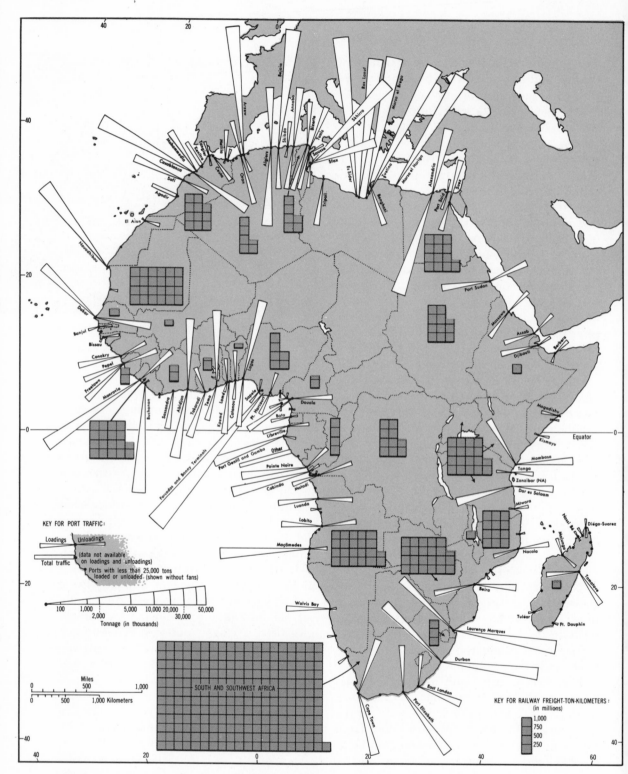

Map 9. *Goods loaded and unloaded at African ports, nearest year to 1972; freight net-ton-kilometers on African railways, c. 1971*

The Zaïre (Congo) system is the most important inland waterway of tropical Africa and the natural feeding system of the Matadi–Kinshasa and Brazzaville–Pointe Noire axes. The East African lakes, particularly Lake Victoria, are of some importance in inland navigation. The value of the long, narrow, rift-valley lakes can easily be exaggerated, however, for they are to a considerable degree impediments to transport rather than aids. Lake Tanganyika, for example, is the only break in a transcontinental rail route from Lobito to Dar es Salaam. A few of the rivers flowing into the Indian Ocean—such as the Zambezi, the Rufiji, and the Tana—are navigable for short stretches, but none are important routeways. In South Africa, inland waterways are almost entirely lacking.

The further development of inland waterways is not a crucial need in improvement of the African transport infrastructure, though some interesting possibilities present themselves. The blasting of rock shelves can sometimes extend the navigability of streams, as it has below Bangui on the Ubangui. Lock systems may some day permit the elimination of short transshipments on the Niger and the Zaïre. Lakes created by hydroelectric and irrigation dams may become locally significant as waterways, in some cases opening up possibilities for the carriage of bulk timber that would not otherwise be practicable.

RAILWAYS. The railway map of Africa is notable chiefly for its emptiness (Map 10). The Maghreb does have a skeletal system and South Africa has a reasonable network. But in most areas, single-track routes stretch inland with few branch lines and few connecting links with other lines. Lobito is the only point on the west coast that is connected by rail to the east coast. It would be a mistake to conclude, however, that a denser network and numerous interconnections would be desirable, because the level and direction of traffic movements would not begin to justify the heavy capital expenditures required.

Rail-lines are most efficient in the moderate to long-distance movement of bulk traffic with a reasonably steady year round flow. Thus lines constructed to move minerals or large tonnages of agricultural produce —particularly if shipment of the latter can be spread out over some months (e.g. peanuts and cotton)—have generally been successful. A good many of the railways built on the faith that they would stimulate sufficient traffic to make them viable have required heavy subsidization on a permanent basis.

Since World War II, entirely new rail lines have been constructed in Mauritania, Guinea, Liberia, southern Angola, Swaziland, northern Mozambique into Malawi, and southern Tanzania. The first five were for new mineral exploitations and have been entirely viable; the northern Mozambique line carries a relatively small tonnage of mixed agricultural products plus miscellaneous imports, while the Tanzanian line from Mtwara, built to serve the ill-fated Groundnut Scheme, was taken up because of the very low traffic generated. Other new mineral lines will probably be constructed in Liberia, Ivory Coast, Gabon, and South Africa.

Extensions to existing lines have been made in about sixteen countries, while branch lines have been abandoned in five countries. There have also been marked improvements in the efficiency of many African railways measured by the number of employees in relation to traffic carried, while tracking and alignments, power equipment and rolling stock, control systems, and other features have been modernized.

Despite a prevalent myth that Africa needs many more rail lines, relatively few new lines are, in fact, required or would be economically viable. Additional myths hold that the present pattern, presumably because of its colonial origin, is ill fitted to national needs and that it is particularly deficient in cross-boundary connections. An objective analysis would reveal that most paying lines are

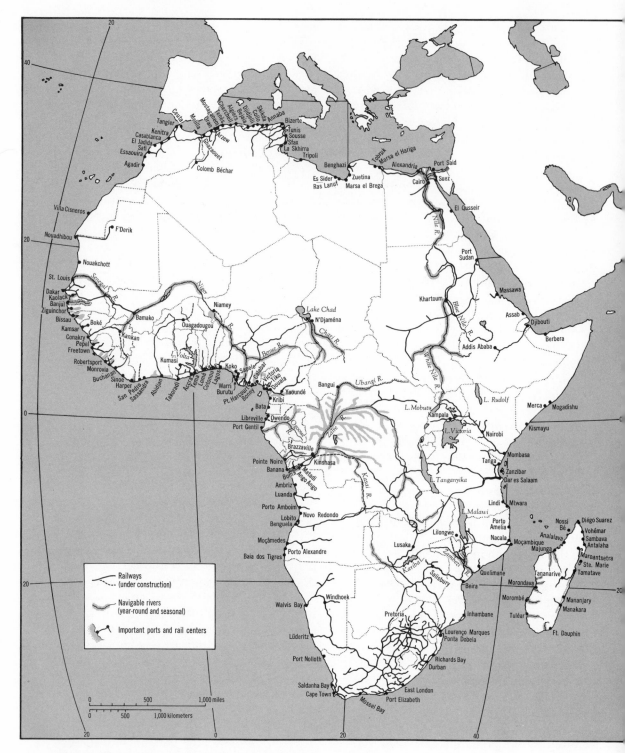

Map 10. Ports, rail lines, and navigable waterways in Africa

needed as much by independent governments as they were by their predecessors, and that road and shipping services can in most cases competently handle the present level of intercountry trade at substantially lower investment and operating costs.

ROADS. Africa has less than 5 percent of the world's paved highways and streets and 10 percent of its total road mileage. Roads are inadequately mapped, and information regarding their negotiability is frequently unavailable. Roads depicted on some maps as "main" turn out to be mere tracks or to represent planned networks. The aptness of the term "road" for some routeways is questionable: the unimproved tracks in Ethiopia have been described as the only roads where pedestrians overtake vehicles.

Roads in the tropical rainy areas are more likely to be built for permanent service because the year-round rainfall requires it, but they may still be periodically or even chronically impassable or so rough that they can be negotiated only by four-wheel-drive or other powerful vehicles. Laterite or murram, having a tendency to harden upon exposure, is often satisfactory when traffic is light, but it tends to corrugate or break down with heavier use. In areas that are alternately wet and dry, only the roads of greatest importance are likely to be open in the rainy season, and in the dry season it is sometimes more practical and more comfortable to drive alongside the road than in the rutted route. Wadis that are bone dry in the dry period may become raging torrents in the rainy season, but building bridges across them is often not justified at the present level of economic activity. Sometimes *radiers,* cement pavements across a stream, spread out the flow to permit fording except during floods. Washouts are a regular occurrence, even on permanent roads.

The life expectancy of vehicles operating on cross-country roads in Africa is greatly reduced by the pounding they take and the fine dust affecting moving parts and brake

A road in the Central African Republic near the Sudan border
The need for new and improved roads continues to be a major requirement for economic development. The road shown here was classified as one of the "seven excellent main roads" of the country in literature distributed by the government.

linings. Many vehicles appear to be held together with hairpins and baling wire and are operated without adequate safety inspections; the sides of some main roads are littered with the debris of abandoned cars and trucks.

There have been impressive improvements in road construction and extension in recent decades in most parts of Africa though the needs remain very great. Roads are far more flexible than rail lines; they can be constructed more quickly and at varying standards according to the need, can have sharper curves and steeper gradients, and require less capital per mile. They can be constructed as money becomes available, whereas investment in railroads is characterized by lumpiness, usually requiring heavy borrowing. A significant mileage of feeder roads has also been built on a self-help basis with very little expenditure. One of the most important assets of roads is the possibility of door-to-door service, which helps to explain their competitiveness over railways for short-haul, non-bulk traffic. Another advantage is that most vehicles will be purchased by private entrepreneurs or individuals, whereas the government must usually provide the engines and rolling stock for a railroad and remains responsible for their maintenance. Operation of privately owned trucks, buses, and "bush taxis" has provided investment opportunities and employment for many thousands of Africans, as have the closely linked filling stations and garages. Many countries have regulations reserving these activities to their own nationals.

AIR SERVICES. The major significance of airlines is their annihilation of time and distance. They play an important role in the movement of people, sometimes aiding internal political cohesion and administrative efficiency, contributing to international intercourse both among countries on the continent and on an intercontinental level, and greatly increasing the potentialities for tourism. They are rather insignificant in the volume of freight handled, but may save weeks or months in the delivery of a crucial piece of equipment, spare parts, or newspapers and periodicals. They are used to transport some perishable goods, such as carcasses to cities in the tsetse-ridden rainy tropics or fruits and vegetables on an off-season basis to Europe, and for the movement of high-value products such as diamonds and gold.

Africa now accounts for only about 2.4 percent of world air traffic as measured by passenger-kilometers flown, but passenger movements have been increasing about 11 to 12 percent per annum in the past decade. Many nations have a strong urge to establish national lines, but may be too small and too poor to justify the necessary investment; a number of existing lines have required excessive subsidization. There is often good reason to join with other countries as has been done by many of the francophone nations in Air Afrique or by the partners of East African Airways. It is usually beneficial also to contract with more experienced airlines for help in management, operations, maintenance, and training.

The present air route map of Africa, as compared with those of earlier years, shows a considerably greater intensity on the main connections, a notable filling-in by coverage of local services, and much-improved east–west links as contrasted with the predominance of north–south flights when the major termini were the capitals of the colonial powers. Some distortions from the previous pattern are occasioned by political policies; these particularly affect South African Airways, which is not permitted to overfly numerous African countries.

Internal routes served on a regularly scheduled basis in mid-1973 are depicted on Map 11, while intracontinental cross-border flights are shown on Map 12; there are also, of course, a large number of connections from Africa to other continents, most especially Europe. Some areas, such as East Africa and South Africa, also have well developed charter services; in the former area they handle more passengers than the regular services on internal flights.

Some countries appear to have an excessive coverage in relation to the level of their economies, including Madagascar, Gabon, Angola, and possibly Ethiopia, but the very poorly developed surface transport in these countries helps to explain the intensity. In 1973 a total of 550 points on the continent were regularly served with the national totals as shown below.

PROBLEMS OF TRANSPORTATION. The development of transportation in Africa is obviously faced with many problems and continuing needs. Some of the physical difficulties have been touched upon; others

	Number of airports served						
	1	2–4	5–9	10–19	20–29	30–39	40–49
Number of countries	15	11	10	13	3	2	3

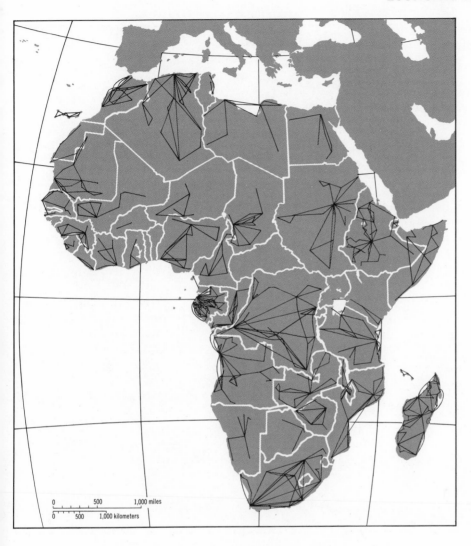

Map 11. Internal air routes in Africa in mid-1973

are noted in chapter 2. Size and distance must also be stressed; the continent is 3.24 times the size of the United States; tropical Africa alone is about 2.31 times as large. To provide an integrated transport system for so vast a region is obviously a staggering task. Many of the productive areas are far from the coast; to meet their needs, long routes traversing low-productive regions with low traffic potential must be constructed. Distance also means time. Therefore, perishable crops cannot be grown for export in some favored areas, and air freight can be no more than a minor solution to this problem.

African transport has also been plagued with problems of equipment and personnel. The cost of maintaining roads, rails, ports, and airlines is often very high. It becomes disproportionately high on short rail and air lines, especially where there are a variety of locomotives or airplanes in use. A considerable portion of an African locomotive's life is spent in the repair shops, because of more rapid deterioration under tropical conditions, poor water and fuel consumed, out-

Map 12. External air routes in Africa exclusive of intercontinental routes, mid-1973

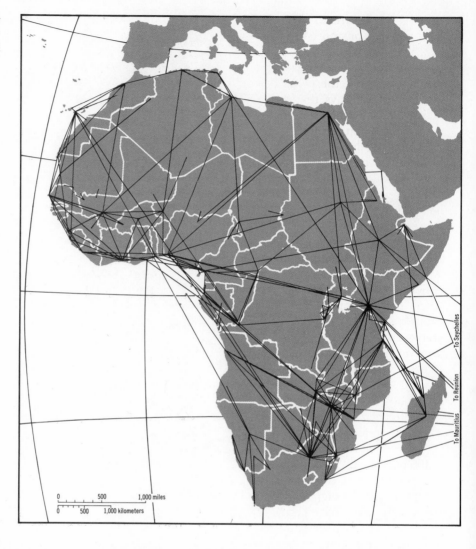

To Seychelles

To Réunion

To Mauritius

0   500   1,000 miles
0   500   1,000 kilometers

moded repair facilities, and inadequately trained engineers. Maintenance difficulties help to explain the switch to diesel engines on numerous African lines; dieselization may result in savings of up to 40 or 50 percent in operating and maintenance costs. Personnel problems are apparent at both the managerial level, which has sometimes been unimaginative and shortsighted, and at the middle and lower levels, where inadequate training and lack of interest are often apparent. In South Africa and Rhodesia certain jobs have been reserved for whites, but it has often been the less qualified whites who are

employed. In some cases there has been too rapid Africanization, with posts being assumed by inadequately prepared persons. Many of the difficulties noted will, of course, be removed in time. Nor should the shortcomings be permitted to obscure the very real achievements that have often been made under most difficult conditions.

The precarious earnings position of some of the railways, plus the importance of others as sources of governmental income, helps to explain a number of the transport policies adopted by African territories. Preoccupation with railways sometimes resulted in a failure to develop an adequate road network or, more often, an overall plan in which each means of transport would be used where it was most practical. In a number of cases, the policy was so weighted in favor of railways that road building was deliberately obstructed or made to follow devious routes that minimized possible competition with railways. Laws and licensing arrangements have been used to keep trucks from diverting traffic from the railroads. In some areas long-standing agreements have bound important users to specific transport routes.

Closely interwoven with social, economic, and financial factors influencing transport in Africa are strong forces operating to favor national ports, roads, and railways even when extranational facilities are cheaper, shorter, or less congested. The techniques employed are diverse and often difficult to uncover. Construction contracts may include clauses requiring shipment by vessels of a specific flag; contracts for purchase of African goods may prescribe the routes of shipment to be followed. There are understandable reasons for the desire to focus upon national routes—to support the territorial economy, to assist sometimes shakily financed transport companies, to save on expenditure of foreign exchange. But as a result some areas have had to pay excessively for their transport services, thus in some degree inhibiting their development.

The shortcomings of African transport are, then, numerous and weighty. The density of routes is typically low, the quality is often inferior, there are shortages of equipment, and there is inadequate planning and coordination within and among the individual countries. The difficulties of ameliorating and eliminating these shortcomings are formidable. Despite the difficulties, however, considerable progress has been achieved; witness the new ports and rail lines in numerous countries and the new roads in every country. Very large expenditures have been made on transport, which usually receives

one of the two largest allocations in national and developmental budgets.

But a glance at any modern map of transport routes in Africa reveals the continuing need. Landlocked nations are often poorly served; remote regions in many countries are neglected and many regions are not adequately served to permit their operating within the modern exchange economy. As many as six or seven transshipments are necessary in moving freight between overseas areas and such places as Rwanda, Burundi, Chad, southern Sudan, or parts of Zambia and Zaïre. Congestion is still present in many ports. Heavy expenditures must obviously continue to be made in the transport field, on which all basic development must in the last resort depend.

ENERGY

The provision of energy is a second major sector of the physical infrastructure. The resource position of Africa in energy raw materials is briefly summarized in chapter 2. The keynotes of Africa's comparative position in production and consumption are shown in Table 5. The most dramatic change has been the notable increase in output of oil, particularly from Libya, Nigeria,

Table 5. African energy production and consumption, percentages of world totals in selected years, 1948–1973

Production	1948	1961	1971	1973
Coal	1.9	2.3	3.0	3.2
Petroleum	.4	2.1	11.5	10.6
Natural Gas	—	···	0.3	0.4
Electric energy	1.6	1.8	1.8	1.8 [b]
Total	1.2 [a]	1.6	6.0	5.8 [b]
Consumption				
Total	1.5 [a]	1.4	1.7	1.8 [b]
Per capita	20.1 [a]	22.1	17.5	18.3 [b]

— = none   ··· = not available

SOURCE: U.N. Statistical Yearbook, 1962, 1972, 1973; Monthly Bulletin of Statistics, January 1975

[a] 1953
[b] 1972

Table 6. Production of petroleum in Africa,
1961, 1972, 1973 (in million metric tons)

	1961	1972	1973
Libya	.7	106.4	104.8
Nigeria	2.3	90.8	101.8
Algeria	15.7	50.5	49.6
Egypt	3.8	10.6	8.3
Angola	.1	7.1	8.3
Gabon	.8	6.3	7.6
Tunisia		4.0	3.9
Congo	.1	.3	2.1
Morocco	.1	<.1	<.1
Total	23.6	276.1	286.5

SOURCE: U.N. *Monthly Bulletin of Statistics*

and Algeria (Table 6); oil shipments from these three countries accounted for no less than 26.7 percent of total African exports by value in 1972. Their share has increased substantially since then with the multiple increases in oil prices in 1973 and 1974. Natural gas is also rapidly becoming an important export as liquefying plants, LNG tankers, and one pipeline are constructed to meet contracts for large-scale deliveries to Western Europe and the United States.

For the continent as a whole solid fuels still accounted in 1972 for an estimated 47.8 percent of total consumption, but this reflects the anomalous position of South Africa, which produces the vast bulk of the continent's coal and absorbs an estimated 53.2 percent of the total energy consumption of the continent. Petroleum accounted for approximately 41.8 percent of the energy consumed in Africa in 1972. Most countries have a high dependence on imported petroleum, the largest portion of which is used in transport, particularly for automotive vehicles but also for rail transport, which has largely been converted from coal and wood to diesel fuel; oil is also the main fuel in carboelectric stations except in southern Africa.

The much higher prices for petroleum after 1973 have put the economies of many African countries in considerable jeopardy. The already precarious balances of payments are seriously affected and it is often very difficult to offset the increased import bill with additional exports. Transport and electricity costs, already very high in remote areas, are markedly increased, which reduces the ability to market some produce. Tourist visits have declined and air services have been curtailed. The severity of the indirect impact will depend in considerable part on how seriously the economies of the DCs are affected, which will influence the level of raw material exports, the numbers of tourists going to Africa, and probably the level of aid extended.

High petroleum prices are likely to bring new interest in the exploitation of African power resources, particularly the immense hydroelectric potential of the continent. In the longer run they may induce several countries to convert their railways from diesel fuel to electricity or wood. High prices have also induced an accelerated search for oil and gas; in 1974 exploration was proceeding in some twenty-two countries in addition to the nine which were producing.

In 1973 the Organization of Arab Petroleum Exporting Countries (OAPEC) announced a boycott on oil shipments to white-ruled areas in southern Africa. This does not appear to have affected South Africa seriously, and the subsequent revolt in Portugal, followed by actual or prospective granting of independence to its territories, removed them from the boycott. The position of Rhodesia, already under UN mandatory sanctions, may be further weakened if independent Mozambique adopts these sanctions.

Electricity consumption represents about 9.8 percent of the energy used in Africa. South Africa consumes about 57 percent of the total. In most tropical African countries only the larger towns and cities have electricity laid on and in most of these only a fraction of the households are tied to the mains.

The position with respect to electricity consumption varies in relation to mining and industrial development; where these activities are important a few large plants, particularly those refining minerals, are likely to account for a high percentage of total consumption. Grid systems are, however, being extended to rural areas in a few favored countries, including Ghana, parts of Nigeria, Kenya, and Uganda.

Hydroelectricity provides a very small share of South African production but about 77 percent of the total electricity produced in tropical Africa and 50.7 percent of the total in north Africa; it accounts for the bulk of electricity in nineteen countries and more than half of the total in five more. Distinctions may be made among hydroelectric plants with reference to their associated consumption patterns:

(1) *Those producing for general supply.* Most plants in this category are small, although there are a few notable exceptions such as the multipurpose installations in the Maghreb, the Saad el Ali station in Egypt, Kainji on the Niger in Nigeria, and Cabora Bassa on the Zambezi in Mozambique, the bulk of whose output is transmitted to the South African grid system.

(2) *Those catering to the needs of existing mining-metallurgical complexes.* Major examples are the installations in Shaba, Zaïre, Kariba on the Zambezi serving Zambia and Rhodesia, and Kafue in Zambia. Inga II and III on the Zaïre are also to be used to provide electricity to the expanding mining-metal installations in Shaba; this involves unusually long-distance transmission—1,131 miles (1,820 km).

(3) *Those designed to attract low-cost power oriented industries such as electro-metallurgical or electro-chemical plants.* There are only a few examples of such installations in Africa: the Volta Dam in Ghana supplying an aluminum smelter at Tema, the station at Edéa in Cameroon also serving an aluminum smelter, and the Owen Falls plant in Uganda, much of whose output has been consumed in a copper refinery at Jinja.

OTHER FEATURES OF THE
INFRASTRUCTURE

With only a few exceptions, notably in the developed parts of South Africa, the continent ranks poorly in other aspects of the physical infrastructure: piped water supply, sewage disposal, piped cooking gas, and communications, including radio, television, telephone, telegraph, and mail services. Progress is being made in most of these services, albeit rather unevenly, but much remains to be done before most countries have modern, adequate, and widespread facilities.

SHORTAGES OF CAPITAL

Another major characteristic of African economies is the inadequacy of capital resources. The situation regarding sources of capital varies markedly from country to country with respect to both public and private funds. It is clear that the formidable character of African development problems requires that capital funds be secured from a variety of internal and external, public and private sources. Governments alone (excepting Libya, with its huge revenues from petroleum) are incapable of commanding amounts adequate to achieve desired expansion rates; economies are not likely to grow unless the private sectors in agriculture, trade, industry, and mining are advancing. Indeed, private investment is needed to pay for government expenses, which face marked increases to meet rising social requirements.

Two-thirds to three-fourths of government revenues usually come from indirect taxes—tariffs, export taxes, and excise taxes. A few countries—for example South Africa, Rhodesia, and Kenya—have greater dependence on direct taxes, while countries with a heavy reliance on mining secure revenues from royalties or, for those which have equity holdings, from earnings of the operations. Several countries have also secured revenues from marketing boards; originally intended to level out receipts to producers, these boards sometimes acquired large amounts of funds, which were then tapped by government to finance development.

It is one thing to secure revenues and another to expend them wisely. While the record is relatively good in most African countries, a number of features are less satisfactory, including an early penchant for showpiece construction; subsidization of noneconomic investments in housing, airlines, merchant marines, and even luxury hotels; expenditures for military establishments beyond any apparent needs; and payment of government salaries whose high level was derived from that of colonial officials (which tends to create an excessively wide gap between the government elite and the bulk of the populace). Graft has also reached unconscionable levels in more than a few countries. Much broader issues with respect to the suitability of government allocations are related to the weights given to the several productive sectors and to rural- or urban-biased investments.

Almost all African countries continue to depend on a flow of outside capital to assist in their development. Development budgets as contrasted with recurrent budgets are frequently dependent on loans and grants from developed countries for from half to all of the available funds. Kamarck estimates the flow of capital funds to Africa south of the Sahara as follows: $6 billion in the years up to World War II; $6 billion from private sources and $10 billion from public funds in the period 1945–60; and a net $2.5–3 billion of private capital and about $10 billion in public capital in the decade of the 1960s.[6] In the period 1969–71 net receipts of official development assistance to thirty-four tropi-

cal African countries from DAC countries averaged $758 million per annum while those from multilateral agencies averaged $448 million; the comparable figures for the five North African countries were $314 million and $59 million.[7] Most funds at concessional rates now come from official sources, with international institutions—particularly the World Bank Group and the European Economic Community (EEC)—being of increasing importance. At present the flow of funds, both private and public, to Africa is well below the desired level, particularly in view of the impact of increased petroleum prices on the balance of payments positions of most countries.

There has been much criticism of foreign investment in Africa, and the investor who pursues his profits without sensitivity to the needs of the host country is deserving of criticism, but there is mutual benefit in most of the private investment that has occurred. African countries have gained through new production capabilities, diversification of their economies, creation of jobs, stimulation of some forward and backward linkages, increased government revenues, improvements to their infrastructures, and the transfer of technical and managerial know-how.

There are, of course, problems associated with foreign investment: the introduction of industries inappropriate to the raw material and market conditions of a nation; what is sometimes seen as an excessive reliance on foreign capital and personnel (which presumably reduces the economic and hence the political independence of African countries); and the amassing of external indebtedness beyond the means of the nation to service it.

Steps can be taken to preclude or minimize these problems or to assure that they will be reduced in the future. These include the more careful preparation of feasibility

[6] Andrew M. Kamarck, *The Economics of African Development* (2nd edition, revised) (New York: Praeger Publishers, 1971), pp. 252–53.

[7] Economic Commission for Africa, *Survey of Economic Conditions in Africa,* 1972.

studies, which should be aided by the establishment of a center for the exchange of information and experience on industrial projects under UNIDO; the restriction of foreign investments to specific sectors; provision for joint ventures with local capital; contract specifications with respect to employment, training of nationals, and gradual reduction of expatriate personnel; provision for the renegotiation of agreements in the event of unusual profitability or for some sliding scale of taxation depending on profit levels; and renegotiation of debt-servicing schedules with longer-term, lower-interest loans. But until such time as there is sufficient local entrepreneurial and technical capacity and greater availability of indigenous capital, African countries would be well advised to maintain their present interest in the attraction of foreign enterprise.

The profit motive associated with foreign investment in manufacturing has several little-appreciated safety features not found in state enterprises: industrial types will be carefully selected because the investment is not likely to be made unless the company foresees a profitable future, and any loss that does occur is at the expense of the investor. The experience of Ghana under Nkrumah is a revealing one in this respect, as most of the state industrial enterprises turned out to be costly mistakes.

Aid programs play an important role in providing capital funds to Africa and also in technical assistance. France has been the largest single source of aid to Africa in postwar years, the bulk of its funds going to its former territories. In 1973 Africa was allotted about $220 million from the French Cooperation Budget. France has been attempting since 1963 to reduce its emphasis on bilateral aid and in more recent years to get away from the more paternalistic aspects of its relationships with African countries. In 1972 there were 9,143 French technical assistants in Africa, about 28 percent of whom were in the Ivory Coast.

Official United States development aid to Africa totaled $1.1 billion in the four years to mid-1973, but the yearly amount was down sharply in 1973 because of reduced shipments of food under PL480. While the U.S. is the world's largest aid-granting country its record leaves much to be desired: the percentage of GNP devoted to aid has declined from 2.78 in 1949 to 0.31 in 1970 and 0.22 percent in 1973–74, the lowest ratio of any DAC nation except Italy; Congress appears increasingly reluctant to provide funds to multilateral agencies such as the International Development Association (IDA), which provides low-interest, long-term loans to the poorest countries; and allocations tend to be too rigidly tied to projects and to procurement from U.S. sources.

U.S. aid has gone to 36 African countries in recent years, but a limited number receive the bulk of assistance. The three largest recipients have been Ethiopia, Morocco, and Tunisia; other important countries are Liberia, Ghana, Nigeria, Zaïre, Kenya, and Tanzania. Generous aid has been extended to nations suffering from drought in 1973 and 1974 and the U.S. will probably participate in longer term aid to a number of Francophone West African countries who were severely affected.

The United Kingdom ranks after France in aid extended to Africa. As would be expected, the bulk of its assistance, which has tended to increase slowly in recent years, goes to English-speaking countries, much of it through the Commonwealth Development Corporation (CDC). Other West European countries—particularly West Germany and Sweden, plus Japan and Canada—are also important in bilateral aid programs in Africa.

Israel began a program of foreign aid in Africa in the late 1950s in part to leapfrog the political isolation imposed by its Arab neighbors. Much of Israeli aid was in the form of technical assistance, for which it acquired an excellent reputation. Its position was eroded seriously beginning in 1972 as a determined and well-financed Arab campaign attempted to ostracize that country. After the Arab-Israeli War of 1973 many African countries joined the half-dozen that had previously broken diplomatic relations with Israel. African countries have not been entirely satisfied with the Arab response to their support and several have complained that they have not been given special treatment in the sale of oil. The Arab nations, and other producing nations, have said that they would not adopt a dual pricing system because of the difficulty of preventing fraudulent shipments; several have, however, promised to set up aid programs to assist African countries, as has Iran. Just how large these programs will be is not yet known.

The USSR began extending aid to Africa in 1958, concentrating its efforts in ensuing years on particular countries. Enthusiasm soon turned to disillusion, however, and aid levels declined. Major recipients of Soviet aid in recent years have been Egypt (particularly for the Saad el Ali, or High Dam, and military equipment), Algeria, and Somalia. China entered the field in 1964; its early efforts were supportive of subversion, which led to the expulsion of Chinese from more than one country; in more recent years its relations, like those of the USSR, have been "correct" and it has been commended for certain features of its programs, including the ratio of grants and of soft loans and the practice of having its technical experts adopt the same standards as experts in the host countries. Chinese aid is generally on a small scale; a major exception is its financing of and technical assistance to the Tazara Railway.

Multilateral assistance programs are increasingly important to Africa. In 1973 the World Bank Group's International Bank for Reconstruction and Development (IBRD) and IDA became the number one source of aid to Africa, and the Group has taken an increasing role in stimulating and coordinating aid efforts. In the early years the IBRD extended loans mainly for infrastructural needs; more recently it has attempted to assist agricultural development, and new policies call for focusing on the poorest 40 percent of countries, on providing low-cost housing and site and service schemes, and on improving health. The EEC's Economic Development Fund (FED) ranks second among multilateral agencies in aid extended to Africa (see chapter 4). Other such agencies are the International Monetary Fund (IMF), the African Development Bank (ADB), and the African Development Fund, affiliated with the ADB but having subscriptions from non-African countries as contrasted with the exclusively African nature of the Bank itself. Not all "aid" should really be dignified by that name. Much is in the form of loans, some at relatively high rates of interest; some is extended for questionable political, military, or cultural reasons; and some of the funds are tied to the purchase of machinery and equipment of dubious utility to Africa. Nonetheless both multilateral aid, which is more strictly economic, and bilateral aid are very important to Africa and are likely to remain so for some years.

UNEMPLOYMENT

Increasing unemployment, particularly in the main urban communities, is causing concern in most African countries. There are a variety of explanations for this phenomenon: the so-called population explosion; the propensity to opt for low-labor-intensive, high-capital-intensive techniques in African industries; improving labor productivity; the demise of some artisanal activities as mass-produced articles replace their output; the gap, often extremely wide, between rural and urban incomes and the urban bias in provision of social services, both of which encourage an exodus from the country often far in excess of the number of jobs being created; the reaction against farming and manual labor in general on the part of many "school leavers;" and pressure to leave rural areas because of their ecological constraints or the shortage of available land.[8]

A variety of proposals have been made to ameliorate this problem, many of which are subject to serious questioning:

(1) Reducing the numbers attending school at all levels. While some reorientation of curricular content may be desirable, this is a nonsolution in the long run, because a nation's prime asset is its people, and the single most important explanation for African underdevelopment is the average low educational-achievement level.

(2) Adoption of labor-intensive methods or "intermediate technology" in manufacturing and construction. Since the modern sector usually employs only a small percent of the population, the impact of this proposal would be minimal. Cost of the finished product would be increased to the disadvantage of domestic consumers and competitiveness on local and eventually international markets.

(3) Promote population control. This is unquestionably desirable, but it cannot have any measurable impact on unemployment for several decades because those people who will enter the job market during this period are already born or will be before any decrease in fertility rates may be expected to occur.

(4) Prohibiting migration to the cities and forcibly returning unemployed persons to their rural communities. This suggestion is objectionable on moral grounds, and all efforts along these lines have proved totally impractical.

[8] See Carl Eicher et al., "Employment Generation in African Agriculture," Research Paper No. 9, Institute of International Agriculture, Michigan State University, East Lansing, July 1970; and Carl Eicher, "Tackling Africa's Employment Problems," *Africa Report,* January 1971, pp. 30–33.

(5) Attempting to maintain and increase the numbers employed in agriculture. This prescription is economically sound but it will require reorientation away from urban-biased policies and a range of improvements in rural areas; the fact that rural populations will continue to increase for at least three or four decades suggests the importance of giving greater attention to the agricultural sector in practically all African countries. Despite much lip service paid to focusing expenditures on agriculture, only a few countries have in fact unequivocally adopted this policy.

DISTRIBUTIONAL ASPECTS OF ECONOMIC ACTIVITY

A number of features of the distributional pattern of economic activity are important to an understanding of the continent's economy. A crude visual impression of the relative importance of various regions and nodal points can be obtained from two maps: Maps 13, which shows the source areas of exports (as has been noted, the export economies of Africa represent a high percentage of the total money economies, particularly in tropical Africa), and Map 21, which depicts the cities and towns of Africa, which are the main commercial, administrative, and industrial centers of the continent.

THE ISLANDIC PATTERN OF ECONOMIC ACTIVITY

Map 13 displays a pattern of productive "islands" often set in vast seas of emptiness where activity is either nonexistent or is concerned almost entirely with subsistence production. There is no solid frontier of the sort that characterized the economic opening of North America or Australia. There is great unevenness in the size and distribution of these "islands" and the individual nodes are often either completely separated from each

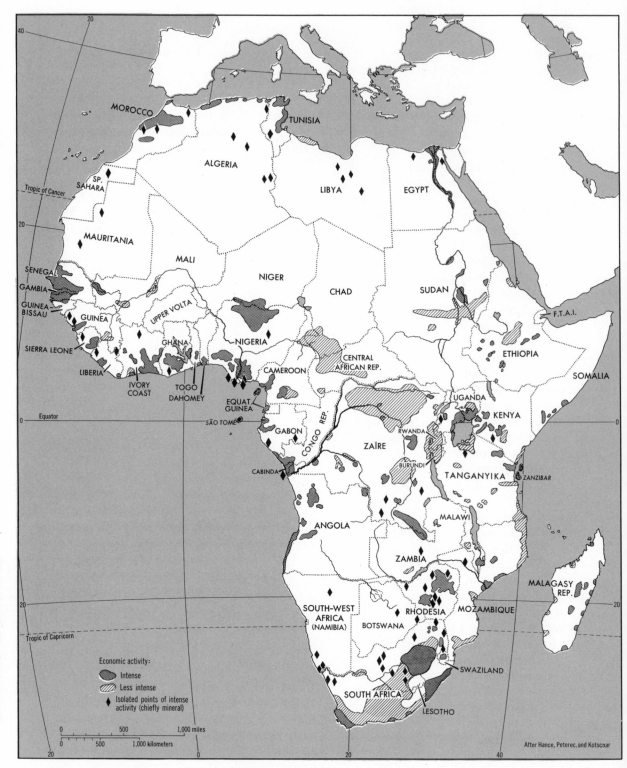

Map 13. *Main regions of commercial production in Africa*

Economic activity:

◖ Intense

▨ Less intense

◆ Isolated points of intense activity (chiefly mineral)

MOROCCO
TUNISIA
SP. SAHARA
ALGERIA
LIBYA
EGYPT
MAURITANIA
MALI
NIGER
CHAD
SUDAN
SENEGAL
GAMBIA
GUINEA-BISSAU
GUINEA
UPPER VOLTA
GHANA
NIGERIA
SIERRA LEONE
LIBERIA
IVORY COAST
TOGO
DAHOMEY
EQUAT. GUINEA
SÃO TOMÉ
CAMEROON
CENTRAL AFRICAN REP.
F.T.A.I.
ETHIOPIA
SOMALIA
UGANDA
KENYA
GABON
CONGO REP.
ZAÏRE
RWANDA
BURUNDI
TANGANYIKA
ZANZIBAR
CABINDA
ANGOLA
MALAWI
ZAMBIA
MALAGASY REP.
SOUTH-WEST AFRICA (NAMIBIA)
BOTSWANA
RHODESIA
MOZAMBIQUE
SWAZILAND
SOUTH AFRICA
LESOTHO

Tropic of Cancer
Equator
Tropic of Capricorn

0   500   1,000 miles
0   500   1,000 kilometers

After Hance, Peterec, and Kotscnar

other or joined by only very tenuous links. These "islands" may be classified into a limited number of categories.

The coastal or peripheral "islands" provide one of the dominant patterns of important economic areas, a pattern heavily reinforced by the high percentage of important cities that are coastal points. Conversely, possibly the most striking feature in the entire pattern is the emptiness of the vast, remote areas. Perhaps the most important explanation for this group of "islands" is the factor of accessibility—the relative ease of moving goods to export points. But climate is also an important factor. Subtropical coastal zones in the Maghreb and South Africa produce such characteristic exports as citrus and deciduous fruit, grapes and wine, olives and olive oil, and cork. Tropical rainy climate areas occur along the Guinea Gulf Coast, providing areas ecologically suitable for the production of, for example, cocoa, robusta coffee, bananas, oil palm products, timber, and rubber. Along and near the east coast and on the Indian Ocean islands one finds concentrations devoted to bananas, sisal, coconuts, cashew nuts, cloves, sugar, and vanilla. The historical fact of earlier contact may be credited for some of the significance of coastal areas. The cool Canaries and Benguela Currents help to explain exports of fish from the northwest and southwest coastal points. And finally, the division of Africa into numerous political units led to efforts by each metropole to produce the various tropical products from one or more of its own territories, a factor helping to account for the pattern along the coasts of West Africa, including the way in which empty areas often coincide with boundary lines.

A second category, the highland "islands," is explained by a variety of factors: better soils including rich volcanics; ameliorated temperature and humidity; lower incidence of insects and disease; attractiveness for white settlement; higher population densi-

ties; and ecological suitability for the production of high value crops such as arabica coffee, tea, tobacco, and pyrethrum.

A few areas can be classified as irrigation "islands," where the intensity permitted through irrigation is the prime factor in their importance. Often, too, these areas have excellent azonal soils. The Nile floodplain and delta, irrigated areas in north and south Africa, and the Gezira and other schemes in Sudan are important examples of irrigation "islands."

Mineral output is usually characterized by high productivity in a limited area. Mineral "islands" may consist of only one mine, though several are fairly extensive and produce a variety of minerals or large amounts of oil and natural gas. While the mining of relatively low value minerals such as iron ore, phosphates, and bauxite is favored by closeness to the coast, and it is typically the more accessible bodies of these minerals which have been opened up, much of African mineral production comes from interior locations. This is disadvantageous to the mining industries themselves, but it has ben-

Removal of overburden in phosphate mine, Togo
Most mineral output occurs in highly concentrated "islands" of production such as this mine in Togo, which accounts for one-third to one-half of Togolese exports by value.

efited the regions involved by justifying far better transport services than would otherwise be expected and by providing funds for social and economic development which certainly could not have come so quickly from agricultural improvements.

There are a few producing areas of importance that do not fit into the four classifications given above: southeast Sierra Leone, the peanut and cotton areas of northern Nigeria, the cotton belt of Cameroon, Chad, and the Central African Republic, the vegetable producing areas of the Zaïre Basin, and the belt of gum arabic, grain, peanut, and animal product exports in central Sudan. A complex of factors explains their importance, among which the availability of transport plays a decisive role. The unusual dispersion of production in the Zaïre Basin reflects the availability of an extended network of inland waterways. The rather large areas of sparse production in Chad and the Central African Republic is explained in part by administrative requirements regarding the planting of cotton by peasant cultivators.

PERCEIVED MALDISTRIBUTIONS OF ECONOMIC ACTIVITY

A number of criticisms have been made of the existing distributional patterns of economic activity in Africa. Objections have been raised regarding each of the following:

(1) The "islandic" concentrations of economic activity. These are said to reflect the regional dualism of the modern and the traditional, largely subsistent areas—a dualism that, it is argued, must be replaced by unification if nations are to avoid extreme inequalities, the underutilization of resources, inefficient location of industrial activity, political instability, and the constriction of national consumer markets. But the pattern also reflects the stage of economic development

and the relatively short historical period during which many regions have participated in the money economy.

With most of the continent classified as underdeveloped it is not to be expected that much attention could have been given to the most backward regions. There are examples of such efforts, however, including the Zande Scheme in southern Sudan, which was conceived as "an experiment for the social emergence of indigenous races in remote areas," the Gonja Scheme in northern Ghana, the Mokwa Scheme in a sparsely populated part of the middle belt of Nigeria, and the *paysannat indigène* program in pre-independence Zaïre.

Today, there is an increasing trend toward planning for individual regions as well as formulation of national and sectoral plans as, for example, in Senegal, Ivory Coast, Congo, Tanzania, and Madagascar, while the division of Nigeria into twelve states with a considerable degree of autonomy and shared returns from government revenues assures greater attention to some less developed regions.

Closely related to the subject of uneven distribution of economic activity in rural areas is the question of appropriate strategies for development of rural areas. The difficulties of dealing with thousands of individual farmers and with the inhibiting features of communal tenure, archaic technologies, and other characteristics of traditional agriculture led many countries to focus development efforts on major schemes, sometimes to the point of "projectitis." The high per capita costs and the limited areal extent of most of these projects obviously reduced the attention that could be given to the general populace. And a high percentage of these projects were not successful.

Spontaneous spreading out and coalescing of productive zones may be expected, but spreading out is difficult in many areas because of ethnic barriers or because ecological conditions deteriorate.

(2) The excessive concentrations of people in some rural areas. This may result in pressure on the land and a consequent decline in carrying capacity. Measures to offset the difficulties vary from the application of conservation techniques to the resettlement of parts of the population. Very few resettlement schemes have been successful, however, and this type of maldistribution requires increased attention, particularly as population growth rates are increasing.

(3) Inappropriate policies with respect to emphases given to rural vs. urban development. There is increasing concurrence that agriculture has not received appropriate priority and that this has contributed to an excessive migration from rural areas. As has already been suggested, correction of this situation will require a multipronged program to reduce the urban bias and generate satisfactory development in rural regions.

(4) An excessive concentration in the urban hierarchy on one (the primate) or a very few urban centers. Growth in these centers proceeds by internal momentum and by competitive and cumulative gains over their rivals. We still know relatively little about the phenomenon of primacy and it is not even certain that there is excessive concentration in the main cities given the present level of development, though some experts claim that the primate cities are parasites on the rest of their countries.

The promotion of growth poles is often proposed as one way to correct or ameliorate the maldistributions of existing distributional patterns, the thesis being that progress in the less advanced areas will be achieved through diffusion of new ideas from them. There is little agreement, however, regarding the type of growth pole that should be promoted. Various "experts" have called for focusing attention on: (a) the second node in a country; (b) a series of medium-sized nodes (a policy adopted, for example, by Kenya and Tanzania); (c) a hierarchical system of nodes (some have suggested that these be determined in part by reference to an idealized geometrical pattern); (d) existing incipient nodes; and (e) the agglomeration of dispersed dwellings into villages, which would become growth poles through the clustering of services in them, improving their transport linkages, and placing new industries in them.

Unfortunately, very little is known regarding the deliberate formation of growth poles in developing countries and the experiences of regional development in laggard areas of Western Europe have not been sufficiently clear cut or successful to provide obvious models. Nor do we know how significant centrifugal forces and the trickling down effect may be and whether the spontaneous rise of lesser centers will eventually lead to greater equality. Present prescriptions are usually dispensed on faith, but the high investment that may be involved in creating growth poles suggests that most nations should be wary of adopting programs without very careful study. As Johnson concludes, "It is much too early to know whether the model builders with their abstract conceptualization of areal design can be helpful to the planners in underdeveloped countries." [9]

(5) The division of the continent into an excessive number of national units. Kamarck writes that "probably the greatest contribution the political sector could make to economic development is in the creation of larger political and, therefore, economic units." [10] African successes and failures at achieving greater political and economic cooperation are discussed in chapter 4.

[9] E. A. J. Johnson, *The Organization of Space in Developing Countries* (Cambridge: Harvard University Press, 1970), p. 150.

[10] Kamarck, *Economics of African Development,* p. 304.

CHAPTER TWO

The Physical Assets and Limitations

The physical environment assumes a prominent role in African economic development—more persistent and pervasive than in developed areas. This chapter gives a brief overview of the natural factors affecting the present and potential economies of Africa; it is confined to presenting essential background for the succeeding discussions of individual areas, which provide additional data regarding the physical environment.

SIZE, SHAPE, AND LOCATION

Africa, with a total area of 11.7 million square miles (30.3 million sq km), ranks second only to Asia among the world's continents. It is 10 percent larger than non-Soviet Asia, about 1.7 times the size of South America, 3.24 times the size of the United States, and over 6 times as large as Europe west of the Soviet Union. Its huge size helps to account for the variety of resources and environments that characterize it; such diversity is usually considered advantageous, but one must look more closely before carrying this point very far. Large size also tends to mean massiveness, which makes for remoteness of interior portions and high cost of transport to the sea. It means great distances between points, which may involve great distances between raw material sources, between raw materials and markets, and between individual markets, all of which increases the cost of transport and the difficulty of integrating national economies. Large size also means continentality, which has a direct effect on climate because of the unequal heating and cooling of land and water masses.

In the analysis of size we almost unavoidably referred to shape when referring to massiveness. Theoretically, a large continent could be so cut up and penetrated by inland seas that massive would be a poor descriptive word. This is certainly not so for Africa; it is the most massive of all continents, having the shortest coastline in relation to its area of any. The most striking contrast is with Europe, whose gulfs and seas greatly increase the accessibility of the area to land-sea contact and permit the deep penetration of moderating marine influences. Africa, has usually rather straight coasts and there are no large indentations improving accessibility, though it might be considered an advantage that the great bulge of West Africa occurs where contact is easier with the developed world of Europe and North America. Only Asia has a larger interior area remote from the sea. In some measure, Africa's massiveness must help to explain the large number of landlocked countries—fourteen, many more than on any other continent. The more intimate details of the shape as revealed by features of the coastline also compare unfavorably with other continents. There are very few estuaries open to navigation, and some that do exist are plagued by offshore bars, shifting channels, shallow depths, and narrow turns. There are also few protected bays and first-class natural harbors (less than a dozen), and some of these are little used because of their location with respect to unproductive hinterlands.

Africa is the southward extension of the great Afro-Eurasian land mass. Africa south of the Sahara, surrounded by seas and the great "sea" of the Sahara, has no marked

Map 14. Globe focused on Africa
Relief Map Copyright Aero Service Corporation

geographic orientation toward any other continent (Map 14). It stands in a position of at least partial isolation. Northern Africa has not shared this isolation. Separated from Europe by only very narrow bands of water, open to navigation on the great, inland Mediterranean Sea, and tied by a land bridge to the Near East (now cut by the Suez Canal and by political confrontation), it participated in and contributed to the advance of civilization over many centuries. It has always had more important ties with Europe and Asia than with Black Africa (the lack of contact between north and south must not be exaggerated however—witness the historic dhow traffic along the east coast and the movement of peoples along the Nile and later by caravan across the Sahara). That the bulk of Africa was the last of the continental areas to be "opened up" to modern development and that the vast bulk of this development has occurred in the present century may be attributed in considerable measure to its isolation, explained by the continent's location, the difficulty of access to it from the sea, and its massive character. It must be noted, however, that numerous ethnic

groups in Black Africa had achieved high levels of craft production, artistic achievement, and politically sophisticated forms long before the arrival of the Europeans: the overworked appellation of "the dark continent" reflects the ethnocentrism of outside observers.

LANDFORMS

The great diversity of landforms makes generalization difficult; one must study them on a regional basis for an adequate understanding. Nonetheless, some major points have wide validity. Africa is frequently called the "plateau continent" (see frontispiece); it is highest in the Ethiopian massifs, and then in eastern and southern Africa, including a broad belt extending nearly across Angola, somewhat lower in the uplifted Zaïre (Congo) Basin, and lower again in West Africa, where the term elevated plains is more appropriate except for isolated regions such as the Jos and Bamenda Plateaus. Most of the plateau is composed of basement complex of metamorphosed rock. Much volcanic action has resulted in massive outpourings of lava as in Ethiopia and Kenya, while eastern Africa is rent by the two great rift valleys. Here too are the greatest volcanic cones— sometimes intimately associated with the rifts, sometimes lying at a considerable distance to one side—as Mounts Elgon, Kenya, and Kilimanjaro. Sedimentary formations are found: (1) on the uplifted plateau basins overlying the basement complex; (2) over much of the Sahara; (3) in usually rather confined and often separated coastal plains and basins; and (4) in the folded mountains of the Maghreb, a part of the great Alpine mountain system.

The landform patterns of Africa have several disadvantages related to economic development. First, they often make overland movement difficult. Coastal zones averaging only twenty miles (32 km) in width are typically backed by steep scarps or scarp zones, which make road and rail construction difficult and costly. The profiles of many African railroads reflect this adverse topographic pattern (Chart 4). A concomitant of the plateau character of the continent is a characteristic stream profile having rapids close to the coast. With few exceptions, such as the Nile, the Senegal, and the Niger, rivers have not provided access to the interior, and river valleys, unlike those of Europe and North America, have not been the natural routeways for road and rail penetration. The fact that the Zaïre was first traced from upstream rather than by following its course from the mouth illustrates the extreme inhospitality of river routes descending across the plateau edges.

But landform problems are not confined to coastal scarp zones. In eastern Africa, from Ethiopia to Mozambique, the rift zones provide some of the most difficult topographic barriers in the world. Also in East Africa there tends to be a series of antithetic scarps, so that what has been laboriously gained in mounting the steep, seaward-facing slopes is gradually lost on the plateau surface and a repetition of scarps must be overcome. Examination of the frontispiece will also reveal that the plateau is divided into several high basins with elevated rims often more like mountains which must be negotiated before reaching the flatter plateau surfaces at somewhat lower elevations. Elsewhere the country is highly incised and with a high percentage of the surface in slopes, as in Rwanda and Burundi and adjacent areas or in upland Madagascar.

Several large regions are also characterized by the grain of the country running at right angles to the main routeways: examples are the scarps and rift systems of eastern Africa and the parallel valleys of southwest Zaïre, which make rail construction so costly that there is still no rail connection between the highly productive mineral-producing province of Shaba and the rail line from Kinshasa to the national port of Matadi. The impact of difficult landforms on transport should not be exaggerated, however. They were more significant as obstructions to early penetration. There are now an increasing number of good to excellent ports, which offset the disadvantages mentioned in the previous section; and once roads and railways have been constructed across landform barriers they are no longer

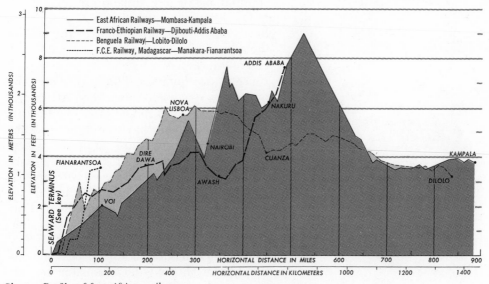

Chart 4. Profiles of four African railways

so inhibiting. Nor is it always topography that makes rail and road building expensive, while it is often easier in the long run to overcome the landform handicaps than it is to meet the problems occasioned by other physical conditions.

The landform patterns have varying impacts on land use. They influence climate, causing excessive precipitation on some slopes but providing orographic rainfall in some regions that would otherwise be empty (e.g., the Aïr and Tibesti Massifs), condemning some protected and leeward areas to greater aridity (e.g., the floors of the rift valleys, the leeward slopes of the great volcanoes, and the plateaus of South Africa), and moderating through elevation the high temperatures of tropical lands, helping to make them ecologically suitable for a wide range of crops including a number of high value. Some regions with steep slopes are highly vulnerable to erosion (e.g., Lesotho and upland Madagascar), though the plastic-

ity of some soils permits their use on surprisingly steep slopes if reasonable precautions are taken (e.g., in several of the highland zones of eastern Africa). At the other extreme, some areas of extreme flatness are inundated for months after the rainy season, though this may permit the off-season cultivation of crops reliant on the moisture stored in the soil. Elsewhere the generally flat to gently rolling topography is advantageous to agriculture and is actually or potentially conducive to mechanization of operations.

CLIMATE

While landform features were more important in impeding penetration in earlier years, climate is the most important physical factor retarding present economic development. Africa lies astride the equator, with Tunis 2,400 miles (3,860 km) to the north and Cape Town 2,600 miles (4,180 km) to the south. About 90 percent of the continent may be classified as tropical, the highest percentage for any continent, and there is a certain symmetry in the pattern of climate, vegetation, and soil regions, and hence, to a considerable degree, in land use regions.

The location of the continent also means that temperatures are high over the bulk of the area. This being so, it follows that the element of climate which is of most significance is precipitation (Map 15). There are numerous systems for classifying African climates, three of which are shown in Map 16. It should be noted that boundaries dividing the several regions are arbitrary except where topographic features make for sharp distinctions; elsewhere, as for example in West Africa or in the Sudan, one region grades into another imperceptibly, and annual differences would show marked alterations in the arbitrary lines selected. Map 17, containing climographs for selected stations, is helpful in showing the seasonal patterns of temperature and precipitation on the continent.

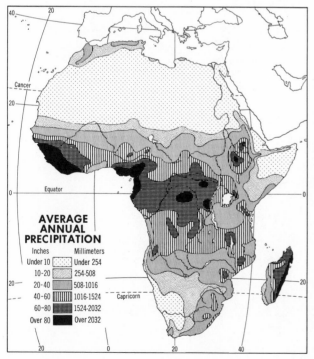

AVERAGE ANNUAL PRECIPITATION

Inches	Millimeters
Under 10	Under 254
10-20	254-508
20-40	508-1016
40-60	1016-1524
60-80	1524-2032
Over 80	Over 2032

Map 15. Average annual precipitation, Africa

Map 16. African climatic regions according to three systems

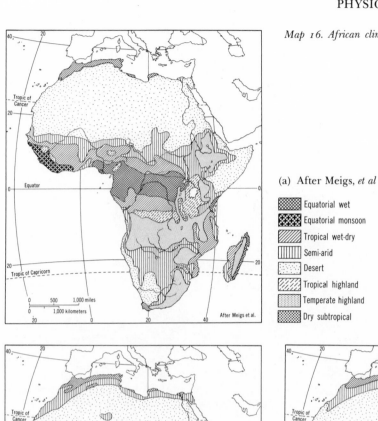

(a) After Meigs, *et al*

- Equatorial wet
- Equatorial monsoon
- Tropical wet-dry
- Semi-arid
- Desert
- Tropical highland
- Temperate highland
- Dry subtropical

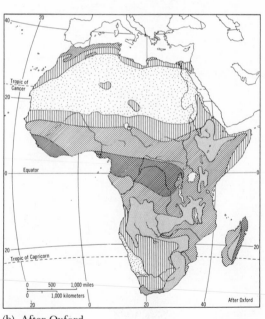

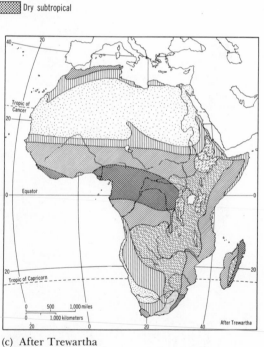

(b) After Oxford

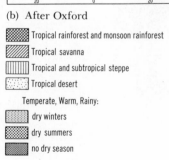

- Tropical rainforest and monsoon rainforest
- Tropical savanna
- Tropical and subtropical steppe
- Tropical desert

Temperate, Warm, Rainy:
- dry winters
- dry summers
- no dry season

(c) After Trewartha

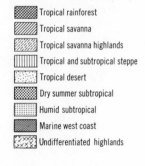

- Tropical rainforest
- Tropical savanna
- Tropical savanna highlands
- Tropical and subtropical steppe
- Tropical desert
- Dry summer subtropical
- Humid subtropical
- Marine west coast
- Undifferentiated highlands

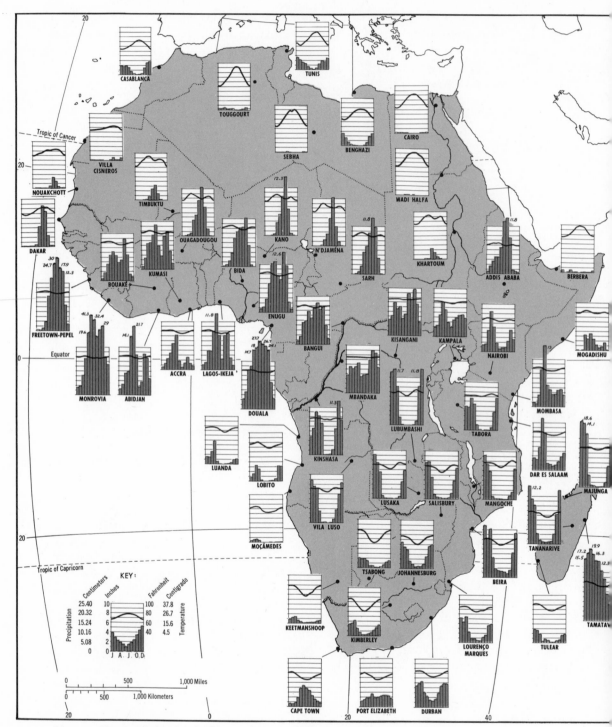

Map 17. Climographs for selected African stations

As far as human utility is concerned, the climatic pattern of Africa means that a very large proportion has low productivity. On a continental basis Africa ranks first in the extent of dry climates, possessing about a third of the arid lands of the world and having the highest percentage of arid lands of any continent except Australia. Meigs [1] estimates the position as follows:

	Million square miles	Million square kilometers	Percent of total area
Hyperarid	1.76	4.56	15.0
Arid	2.82	7.30	24.1
Semiarid	2.35	6.09	20.1
Total arid	6.93	17.95	59.2

Worthington [2] states that precipitation is scanty in about 75 percent of sub-Saharan Africa, and that in over half of Africa water is *the* principal physical factor limiting advance. Not only is there inadequate precipitation in these regions, but they suffer also from great fluctuations within individual seasons and from year to year.

Moving from the arid and semiarid areas toward the equator, we find that the savanna climate type occupies about 20 percent of the continent. Characterized by a long rainy season and a short dry season, climate is not so strong a limiting factor as in the arid areas. But regions in this climate type are again plagued by uncertainty, and the torrential character of much of the precipitation is another disadvantage. And new problems arise in these zones—the presence of the tsetse fly and poorer soils.

About 8 percent of Africa has a tropical rainy climate, marked by a ten- to twelve-month rainy season. Much of this area might be said to be too wet for optimum utility, at least under present known techniques. Its constantly high temperatures and high rainfall do present climatic conditions of great potential productivity, however.

Much of Africa's climatically more favorable land falls in the highland climate regions. Precipitation in these areas may have the characteristics of any of the other types depending upon exposure, distance from the equator and from the sea, and specific landform features. Temperatures are moderated by elevation. There are some fairly extensive areas with highland climates, such as the Ethopian Massif or the highveld south of the Zaïre Basin, but many are small, irregular, and separated from one another by other climate types. Additional disadvantages stem from the difficult landforms that must be expected in the highland areas and from the fact that many of the tropical highlands suffer from remoteness of location as, for example, those of eastern Zaïre or of southwestern Tanganyika.

The regions of subtropical climate are relatively small, and their utilizable portions are further restricted by difficult landform conditions.

About 92 percent of the continent may be said to suffer from one or another climatic disability—surely one of the most important facts concerning Africa. It presents a picture, as far as water is concerned, of plenty where it cannot be used and of paucity where it is most needed. It should also be noted that the climate is often considered undesirable from the standpoint of human health and energy, though there is little scientific evidence to support this hypothesis. In fact, Lee states that "apart from a reduced desire for activity, there seem to be no permanent deleterious *direct* effects in healthy persons living under natural conditions in the tropics." [3] We do know, however, that it is very difficult to acclimatize quality middle-latitude livestock to the high temperatures of tropical lands, one of the many fac-

[1] Peveril Meigs, "Arid and Semi-arid Climatic Types of the World," in *Proceedings, Seventeenth International Geographic Congress* (Washington, 1952), p. 137.

[2] Edgar B. Worthington, *Science in the Development of Africa* (London, 1958), p. 23.

[3] Douglas H. K. Lee, *Climate and Economic Development in the Tropics* (New York: Harper, 1957), p. 99.

tors that have kept the livestock industry in a low state of commercial development. An indirect effect of climate of considerable weight is the conduciveness of the constant warmth and, in the rainy tropics, of the constant humidity to promotion of insect pests and bacteriological life, to the detriment of man, plants, and animals.

VEGETATION

The striking fact about the vegetation of Africa from the standpoint of human use is the marginal utility of much of it. About 27 percent of Africa is classified as in forest, a lower percentage than for South America, but covering an area about equal in size. But much of the total is in savanna woodlands, whose trees, with few exceptions, are not suitable for lumbering. Nor are the trees of the tropical rainforest comparable in utility to those of the middle latitude forests, while they cannot be so easily exploited because of the necessity to maintain some canopy and some pioneer or nurse communities. Most of the grasses of the savannas and steppes are of relatively low nutritive value. And fire has degraded an enormous part of African vegetation.

On the more favorable side it should be noted that the rainforest remains as a very important resource, which may be far more effectively used than it is today; exotic species such as pines and eucalyptus grow with remarkable rapidity in extensive areas; it may prove possible to extend the distribution of several superior grasses to the advantage of soils, animal-support, and total productive capacity; and that some vegetation types have surprising recuperative capacities once they are protected from misuse. A recently appreciated advantage of considerable interest is use of the tree *acacia albida* in dry areas to increase yields of crops such as sorghum, millet, and peanuts grown beneath them.

Map 18, based upon a map prepared for the Commission for Technical Cooperation in Africa, depicts the vegetation distribution in Africa, but intensive surveys will be required before great certainty can be attached to vegetation zones. In West Africa, for example, there are differences on almost every map produced regarding the landward boundary of the rainforest regions. The accompanying photographs show some of the more important vegetation types of Africa.

SOILS

The picture with regard to soils is not quite so bleak, although the tropical latosols rate negatively, the soils of steppe areas do not compare in fertility and structure to the grassland soils of the middle latitudes, there are proportionately fewer young, rich, alluvial soils than on any other continent, and soil erosion has reached very serious proportions over broad areas. The soils of tropical Africa are almost universally deficient in nitrogen and have moderate to acute deficiencies in phosphate, while potash deficiencies are more localized. Latosolic soils provide the main reason for the dominance of bush fallow agriculture in tropical rainy and savanna areas, a system that presents severe limitations to raising rural standards.

There are, of course, some good to excellent soils, including the exceptionally rich alluvial soils of the Nile floodplain and delta, the soils of isolated volcanic peaks such as Mt. Kilimanjaro and Mt. Meru, and lowland soils of the Mediterranean regions. The soils of many highland areas are also relatively favorable, and the soils of swampy areas are worthy of greater study. Some soils, furthermore, display good recuperative powers once misuse is ended.

It is obvious that much more attention must be given to studying and protecting African soils. As Worthington states: "Soil can . . . be regarded as a kind of fulcrum on which the whole of land ecology is balanced,

Tropical rainforest or selva.

Savanna woodlands.

Four major vegetational types of tropical Africa

Savanna grasslands.

Steppe grasslands.

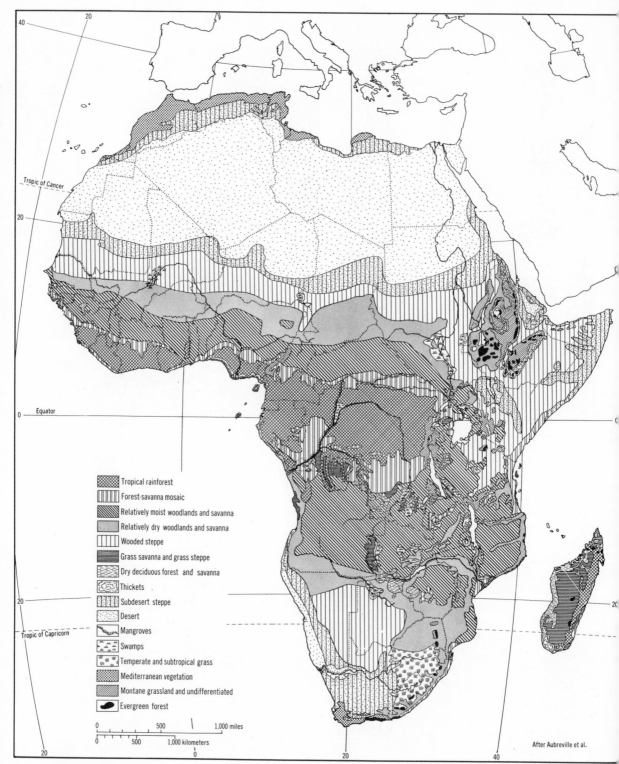

	Tropical rainforest
	Forest-savanna mosaic
	Relatively moist woodlands and savanna
	Relatively dry woodlands and savanna
	Wooded steppe
	Grass savanna and grass steppe
	Dry deciduous forest and savanna
	Thickets
	Subdesert steppe
	Desert
	Mangroves
	Swamps
	Temperate and subtropical grass
	Mediterranean vegetation
	Montane grassland and undifferentiated
	Evergreen forest

0   500   1,000 miles

0   500   1,000 kilometers
          0

After Aubreville et al.

Map 18. Vegetation zones of Africa

the physical environment on the one side and the biological environment on the other." [4] Finally, improved knowledge of proper soil practices should point the way to more effective use and care. Scientifically, the ideal way of maintaining and increasing soil fertility in the tropics is the combined use of organic manures, including composts, and chemical fertilizers. But while it is known that continuous cultivation is possible with efficient use of fertilizer, it is not always economically feasible under present conditions. Organic manures would be supplied by a mixed crop-livestock farming, but cattle cannot be kept in areas infested with the tsetse fly, which carries "nagana" or bovine trypanosomiasis. Thus, the solution to one physical problem requires solution of another, and this intimate interconnection is characteristic of many of the problems besetting Africa.

FAUNA

The cataloguing of African faunal species is not pertinent to this brief overview, which is designed to summarize in the broadest terms the most significant influences of the several physical elements on the human utility of the continent. The most important disadvantage regarding animal life is the existence of large numbers of deleterious animals, insects, and organisms. The anopheles mosquito carries malaria, which is still the number one disease problem in Africa. Although these pests can be completely eradicated with residual insecticides, many countries have not been able to mount the massive campaigns that would be required, and about 80 percent of the nearly 200 million persons living in malarious areas of Africa are still not benefiting from an eradication program. Nor is eradication without its drawbacks, including the deleterious ecological impact of insecticides and the developing resistance of some species to

insecticides and antimalarial drugs. There is hope, however, that sterilizing male mosquitoes with a chemical agent may reduce the mosquito population by as much as 99 percent in treated areas.

Schistosomes are parasitic worms that bring schistosomiasis, the greatest unconquered parasitic disease now afflicting men and animals,[5] to somewhere between 150 and 200 million Africans, including practically all peasants in Egypt and in other irrigated areas of tropical Africa. While it rarely kills it is increasingly debilitating with each succeeding infection and with the advancing age of the victim.

The tsetse fly carries both bovine and human trypanosomiasis; as a result cattle, with minor exceptions, cannot be raised over a vast savanna and rainy tropical area totaling at least four million square miles (10.4 million sq km). The fly *Simulium damnosum* carries onchocersiasis or "river blindness" to large areas in West Africa. A battery of UN agencies is working on a $120 million, 20-year program in the Volta Basin, where at least a million people have the disease; a biodegradable spray is being used. Trillions of locusts and billions of weaver birds or quelea-quelea birds consume large tonnages of crops—particularly sorghum, rice, and wheat—in semiarid areas. Other pests are termites, rodents, and baboons. Plant and animal diseases reduce yields markedly, often by 10 or 20 percent endemically and by higher percentages epidemically. Additional disadvantages include the fact that organized arriculture or the concentration upon one crop in a limited area almost assuredly leads to increasing plant disease, and the poor quality of most indigenous livestock.

On the favorable side there are, of course, possibilities for meeting some of these problems through such means as eradication or control of specific insects and organisms,

[4] Worthington, *Science in the Development of Africa*, p. 138.

[5] See John M. Weir, "The Unconquered Plague," *Rockefeller Foundation Quarterly*, no. 2, 1969, pp. 4–10.

perfection and dissemination of drugs and serums, and adoption of better livestock practices. In limited areas of Africa, the existence of big game provides an important attraction for the tourist industry as well as a potentially very great source of much-needed proteins. The coasts and inland waters provide opportunities for the fishing industry, which is, however, poorly developed except in South Africa, South-West Africa, Angola, and Morocco.

HYDROLOGY

The three main aspects of hydrology having an impact on the economic geography of Africa are the navigability of streams and other inland waters; the control of surface water and the tapping of underground water for potable water and irrigation or wet-land cultivation; and the harnessing of rivers for hydroelectric development. Aspects of less importance, though locally significant, relate to fish and aquatic animal resources (there is, for example, a regular offtake of hippo meat and crocodile skins in several areas), and the attraction of lakes, rivers, and waterfalls for recreation and tourism (Victoria Falls is a leading tourist site for Rhodesia and Zambia; Kabalega Falls [Murchison Falls] in Uganda, and Tisisat Falls on the Blue Nile in Ethiopia are other examples).

The large areas of arid climates and the small percentage of precipitation flowing to the sea have already been noted in the section on climate. These features are reflected in the very large areas without permanent streams and having interior drainage (covering about 52 percent of the continent), including the Sahara, Lake Chad Basin, Lake Rudolf and other parts of the Eastern Rift Valley, the Kalahari Steppe, and the Namib Desert. There are, however, a number of great rivers with large drainage basins and at least seasonally heavy discharges. The situa-

tion with respect to navigability has been briefly noted in chapter 1, while resources are discussed below under energy. With respect to irrigation potential, underground water resources have been best studied in the Maghreb and South Africa, though much remains to be done. Some aquifers have a considerable flow and are actual or potential resources of value. But even if they were all developed on a sustained yield basis they could add only a small total area to the irrigated lands of the dry regions, which are themselves a very small percentage of these areas. Some aquifers are too saline and other underground waters are of fossil origin and can be used but once. Some irrigation communities are already drawing water more rapidly than it can be replaced as is the case in parts of America and Australia. The Nile is by all odds the most important stream supporting irrigation communities; rivers of the Maghreb, the Senegal, the upper Niger, Shari, Logone, Tana, Rufiji, Zambezi, Orange, and other rivers supply smaller areas. There are very substantial opportunities for development of these resources, but no rapid extension is to be expected because of the high capital costs involved.

ENERGY RESOURCES

The dominant facts with regard to energy resources are the relative weakness in coal, the considerable wealth of petroleum and natural gas (limited, however, to the Sahara, the Niger Delta, and stretches of the coastal plains and offshore areas), and the great wealth of water power and fissionable raw materials.

Some good coal is found in southern Africa, particularly in the Republic, but most of the known coal is of poor quality in the south and in North Africa, Nigeria, Zaïre, Tanzania, Zambia, and Madagascar. Except for the Republic of South Africa, production remains low and may even decline in some countries with the provision of hydroelectric facilities. The meagerness of coal resources must be considered a major handicap to certain types of industrial development.

The Libyan, Algerian, and Niger Delta rate as major fields for petroleum and natural gas. Smaller fields are being tapped in Morocco, Egypt, Gabon, Congo, and Angola, and exploration is progressing in many new areas—particularly in the countries south of the Sahara and along numerous coastal stretches. But most of Africa is a basement rock area where petroleum and natural gas are not likely to be found. It must be expected, therefore, that most countries will be required to make greater expenditures of foreign exchange as their consumption of petroleum products continues to increase; the impact of markedly higher petroleum prices is a heavy burden on most of the underdeveloped nations of Africa.

The water power resources of Africa are tremendous and are estimated at about 40 percent of the world total. It is somewhat ironic that this continent, plagued with aridity over such great areas, should at the same time have such a magnificent hydroelectric potential. The high rainfall in the rainforest and savanna areas, their great expanse, and their high average elevation combine to explain this wealth. About 18 percent of the world and 45 percent of the total African potential is in the Zaïre Basin.[6] The 217-mile (348-km) stretch of the Lower Zaïre River from Kinshasa to Matadi has an estimated potential of 63 million kw.

Fissionable raw materials are known to occur in many countries. Zaïre's Shinkolobwe deposit held a leading position in production during World War II and the early postwar years. South Africa is now the leading producer, its output coming mainly as a byproduct of the gold ores. Gabon, the Cen-

tral African Republic, and Niger are other producers. The Republic of South Africa, Madagascar, and Senegal have large reserves of monazite sands. While the fissionable resources are not now significant in the field of energy production, they are a potential domestic raw material of no small dimensions.

Finally, based on reconnaissance surveys and infrared scanning, several regions are believed to have enormous potential in geothermal power, particularly the Rift Valley areas of Ethiopia and East Africa.

NONENERGY MINERALS

It is a commonplace that Africa is a mineral storehouse, and it has already been seen that the continent accounts for important shares of world production of a variety of minerals. The estimated position of African reserves in major minerals based on present knowledge is given in Table 7. Obviously, Africa rates favorably in the quantity and variety of its mineral resources. In the long run, possibly the most important minerals that would support manufacturing in Africa are coal and iron ore, needed for the iron and steel industry. The weak position of most countries in coal has already been noted; in iron ore, the continent is very well endowed.

Prospects for new discoveries may be considered good for several reasons: recent finds in areas thought previously to have been well prospected, the realistic expectation that large unmeasured reserves exist in many known bodies, and the large areal extent of promising geological conditions. The basement rock complex, comparable to the mineral-rich North American shield, forms the surface rock over much of the continent. But the full mineral wealth of Africa will not be known for scores of years because of the enormous size of the continent and the necessity for intensive study of small areas. Mineral investigation will require detailed geological mapping supplemented by aerial

[6] The Congo was unilaterally renamed the Zaïre when Congo (Kinshasa) changed its name to Zaïre. The renaming has been contested by Congo (Brazzaville).

Table 7. Estimated reserve position of Africa in selected minerals

Very large	Large	Average	Small
Bauxite	Asbestos	Antimony	Coal
Chromium	Beryllium	Barite	Gypsum
Cobalt	Columbium	Cadmium	Molybdenum
Copper	Graphite	Fluorspar	Nickel
Diamonds	Lead	Silver	Potash
Gold	Mica	Tin	Quartz crystal
Iron	Uranium	Titanium	Sulfur
Manganese	Vermiculite		Tungsten
Natural gas			Zinc
Petroleum			
Phosphates			
Platinum			
Vanadium			

photography; radiometry; magnetic, electromagnetic, seismic, and gravimetric surveys, and geochemical studies. These will require time, continuous application, and massive sums of risk capital.

THE UNKNOWN AND THE LITTLE KNOWN

In concluding this chapter it is well to emphasize that the physical background of Africa is still very inadequately mapped and catalogued and still very imperfectly understood. There is a need for scientific research and application. Answers to such questions as the following are required:

How can the limited fertility of the leached tropical latosols be preserved or how can it be improved?

How can the basic systems of bush-fallow rotation and shifting agriculture be altered to increase yields and productivity?

How can the menace of the tsetse fly be removed to encourage the introduction of a mixed agriculture and to help balance the protein-deficient diets?

In the widely represented steppe and dry savanna areas, what measures can be taken to prevent runoff, increase soak-in, utilize ground water, control the highly erratic and often silt-laden flow of wadis and mountain khors, prevent the more damaging effects of grass burning, improve the quality of the typically unnutritious and unpalatable grasses, and increase productivity by introduction of mechanical cultivation?

How can the livestock of the continent be improved through application of appropriate breeding, feeding, pasture management, and disease-control practices?

Many of the brief analyses given earlier in this chapter reveal that Africa rates unfavorably with respect to several important physical factors. This does not mean, however, that there are not enormous potentialities for improvement. The productive capacity of much of the area has neither been scientifically measured nor adequately tested. Many of the possible avenues for enlarging the production and realizing the productivity of the area will be examined in the ensuing chapters.

CHAPTER THREE

The Population and Peoples of Africa

The Population of Africa in mid-1975 was about 393 million. No accurate figure can be given because no censuses have ever been taken in some countries, many have had only 5 percent sample surveys, and the full censuses that have been taken are of varying validity.

It is difficult to obtain accurate demographic data in many African areas. Among the organizational problems are recruiting the desired number of qualified enumerators and central staff, the inadequacy of base maps, the scattered nature of many rural population nuclei, enumerating nomadic and seminomadic peoples, and achieving a sufficient degree of standardization to permit comparative studies. Another set of problems is associated with the lack of understanding by persons being enumerated, stemming largely from illiteracy and an average low level of educational achievement, but also reflecting suspicions that the census is being taken or will be used to increase taxes, recruit for labor or military service, or for other purposes detrimental to the respondent. In the 1962 and 1963 Nigerian censuses, however, there were serious overcounts, reflecting the knowledge that government expenditures and political representation would be related to the population totals of the enumerated areas. Indeed, these censuses became so embroiled in politics that the first was cancelled and "the very survival of the Federation was threatened." [1] The 1973 census, which contains several anomalies in state totals, has also been severely criticized; and it has not been accepted as official.

There are also serious problems in measuring fertility, mortality, and migration (which means that growth rate estimates are usually of questionable validity) and in obtaining information on such characteristics as age, ethnic classification, and religious affiliation. Despite the numerous problems, however, most of the more recent censuses have been reasonably and increasingly accurate and coverage of many African populations is now much improved.

DEMOGRAPHIC DATA FOR AFRICA [2]

Africa and its appurtenances had an estimated 374 million inhabitants in mid-1973, or about 9.7 percent of the world total (Table 8). Fertility and birth rates are among the highest in the world; birth rates average about 47 per M, or 38 percent higher than the world average; no other continent has such comparably high rates.

Explanations for the generally very high birth rates include the typically low marriage age, the high percentage of married women, the tolerance of illegitimacy, the characteristic desire for as many children as possible, wives' fears that husbands may leave them if they do not bear children, religious precepts, and the exceptionality of any interest in family limitation.

The existence of some relatively low fertility rates may be explained mainly by the

---

[1] R. K. Udo, "Population and politics in Nigeria," in John C. Caldwell and Chukuka Okonjo, eds., *The Population of Tropical Africa* (New York: Columbia University Press, 1968), pp. 97–105.

[2] Portions of this section are updated from William A. Hance, *Population, Migration, and Urbanization in Africa* (New York: Columbia University Press, 1970).

Table 8. Estimated population of Africa and the World for selected years, 1000–1973 and projections to 2000

Year	Population (Millions)		African percent of world total
	Africa	*World*	
1000	50	275	18.2
1200	61	384	17.5
1400	74	373	19.8
1600	90	486	18.6
1800	90–100	906–919	9.9–10.8
1900	120–150	1571–1608	7.6–9.5
1950	222	2517	8.8
1960	278	3005	9.3
1970	344	3632	9.5
1973	374	3680	9.7
1980 (est.)	475	4467	10.2
2000 (est.)	805		

SOURCE: William A. Hance, *Population, Migration, and Urbanization in Africa* (New York: Columbia University Press, 1970), p. 16; U.N. *Statistical Yearbook 1972* (New York, 1973), pp. 80–81.

high incidence of venereal disease in some regions. Other factors tending to reduce fertility include polygamy, certain sexual taboos—especially those connected with the prolonged breast-feeding of babies—the occasional use of contraceptives and abortifacients, and the separation of families owing to the preponderance of males in many migratory movements. Despite some evidence of a desire for fewer children as educational levels improve, urbanization increases, and incomes rise, fertility and birth rates are likely to remain high for at least several decades.

African mortality rates are also among the highest in the world, ranging from an estimated 17 per M for northern Africa to 24 per M for western and middle Africa, and averaging 21 per M for the continent as a whole. The last rate compares with 14 for Asia, 10 for Latin America, 9 for northern America, and 13 for the world. Estimates of African death rates are considerably less certain than those of birth rates and extrapolation is far more precarious, since significant

changes can occur in a very short period of years. The death rate on Mauritius, for example, fell 30.3 percent in one year, largely as a result of an intensive antimalaria campaign. The prevailing birth and death rates of Africa result in a very high percentage of young people and generally low life expectancies, averaging about 41 years south of the Sahara and 52 years in northern Africa.

The major explanation for the characteristically high death rates of most African countries is the high incidence of disease. But while disease-control campaigns may result in dramatic decreases in death rates, these are likely to remain relatively high for many years because of generally low standards of living, low educational levels, the scarcity of medical and health facilities, poor sanitation practices, inadequate clothing and housing, and malnutrition, and also because of the heavy expenditures required for suppression of even those diseases which are subject to control.

Only a small percentage of Africans are not handicapped by some endemic disease—more likely than not they have several diseases to which they have acquired some tolerance but which nonetheless sap their vitality and make them subject to the acquisition of other diseases, which may prove fatal. The pestilential diseases—plague, relapsing fever, smallpox, typhus, and yellow fever—have been attacked with considerable success, though cholera now appears to be on the increase. Malaria, despite some massive gains, remains the single most important disease problem in many African countries. While subject to complete eradication, the cost of the necessary programs remains beyond the reach of most peoples and over three-fourths of the population of tropical Africa have not yet benefited from an eradication program.

Communicable diseases such as leprosy, venereal disease, and yaws are now also subject to control, although most victims still go untreated and the incidence of venereal dis-

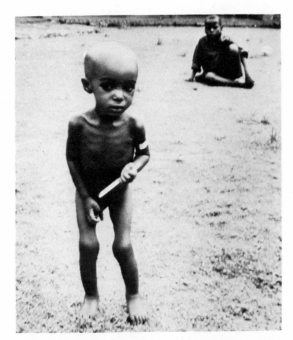

A child suffering from Kwashiorkor, an important diet-deficiency disease

ease appears to be rising in some areas. Other communicable diseases remain as serious problems: trachoma, bilharziasis, filariasis, trypanosomiasis, tuberculosis (ranking second to malaria as a killer), and measles. Nutritional problems contribute to ill health, particularly among children. Protein-calorie malnutrition (P.C.M.) of early childhood (leading to kwashiorkor, marasmus, and nutritional growth failure) and nutritional anaemia are the two general nutritional problems that affect all parts of Africa. There is increasing evidence that the contraction of P.C.M. in the first few years of life has an adverse effect on subsequent learning and behavior.

The third demographic dynamic, migration, is not significant as far as continental totals are concerned. But population movements have been taking place on a substantial scale within Africa since time immemorial. Map 19 shows some of the more important migratory patterns of more recent years. The predominant motive behind most

decisions to migrate is economic need and desire, and economic migrations include the movement of traders, fishermen, agricultural workers, mine workers, and very large and increasing numbers of those moving to urban centers.

Estimates of population growth rates for African countries must be accepted with caution, given the serious inadequacies of demographic data. The approximate rate of 2.5 percent in 1973, exceeded only by Latin America with an estimated rate of 2.8 percent, compares with 0.8 percent for northern America, 0.7 percent for Europe, and 2.0 percent for the world. Trends in growth rates are upward because of continuing high birth rates and lowering death rates. Expectations that the latter will decline further suggest that the average growth rate will rise for at least the next two decades.

POPULATION DENSITIES AND DISTRIBUTION

The average population density of Africa in mid-1975 may be estimated at about 33.6 per square mile (13.0 per sq km). This figure has very little significance other than to suggest a relatively low overall density, which in turn tends to contribute to the widely accepted myth that Africa does not have a population problem. The misleading character of crude density figures may be illustrated by the case of Algeria, where about 95 percent of the population lives on about 11.3 percent of the area and where over half the population resides at densities above 200 per square mile (77 per sq km). While the vast bulk of the area of Africa is indeed sparsely populated, of far greater significance is the fact that a very substantial percentage of the total population is actually experiencing high densities (Table 9). In fact, in 1967 over 80 percent of the total population lived at densities above the average and about a third resided at densities exceeding 300 per square mile (116 per sq km).

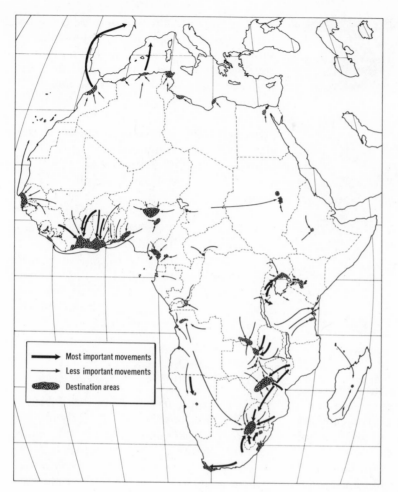

Map 19. Important population movements in Africa

Map 20 shows the distribution of population in Africa as of about 1960. Closer analyses will be made in regional chapters, but it may be noted here that there are numerous anomalies discernible in the overall pattern. While the effect of aridity is quite apparent in the low population of the Sahara and the Namib Desert and on the steppes of eastern Africa and the Kalahari, other areas with long dry seasons have unexpectedly high densities, such as parts of northern Nigeria, the Mossi country of Upper Volta, Senegal, and Gambia. Wet savanna lands, with longer rainy seasons, typically have lower densities than the drier savannas. Rainy tropical areas

show great variation; parts of the West African belt are densely populated, while the bulk of the rainforest belt of Cameroon, Gabon, Congo, and Zaïre has low densities. In some regions, as in the Maghreb, Egypt, or East Africa, correlations with physical factors are striking; elsewhere the relations are so complex as to defy accurate assessment, but it is clear that historical and social factors are of great significance in explaining the population patterns.

RURAL AND URBAN DISTRIBUTIONS

Africa is the least urbanized of all the continents; the rural population is about 80 per-

Table 9. Population density ranges for the continent of Africa, mid-1967

per sq. mile	per sq. km.	POP. %	AREA %	Pop. %	Area %	Pop. %	Area %
DENSITY [a]				**CUMULATIVE**		**REVERSE CUMULATIVE**	
−10	−3.9	5.86	59.21			100.01	99.99
10−	3.9−	6.72	13.75	12.58	72.96	94.15	40.78
20−	7.7−	5.20	6.55	17.78	79.51	87.43	27.03
30−	11.6−	6.62	5.41	24.40	84.92	82.23	20.48
40−	15.4−	5.23	3.33	29.63	88.25	75.61	15.07
50−	19.3−	4.32	2.22	33.95	90.47	70.38	11.74
60−	23.2−	4.00	1.73	37.95	92.20	66.06	9.52
70−	27.0−	1.75	0.68	39.70	92.88	62.06	7.79
80−	30.9−	2.73	0.89	42.43	93.77	60.31	7.11
90−	34.7−	4.56	1.30	46.99	95.07	57.58	6.22
100−	38.6−	14.59	3.00	61.58	98.07	53.02	4.92
200−	77.2−	7.54	0.86	69.12	98.93	38.43	1.92
300−	116.0−	3.21	0.27	72.33	99.20	30.89	1.06
400−	154.0−	3.51	0.23	75.84	99.43	27.68	0.79
500−	193.0−	2.50	0.13	78.34	99.56	24.17	0.56
600−	232.0−	1.47	0.06	79.81	99.62	21.67	0.43
700−	270.0−	1.29	0.05	81.10	99.67	20.20	0.37
800−	309.0−	0.58	0.02	81.68	99.69	18.91	0.32
900−	348.0−	1.44	0.04	83.12	99.73	18.33	0.30
1000+	386.0−	16.89	0.26	100.01	99.99		

SOURCE: William A. Hance, *Population, Migration, and Urbanization in Africa* (New York, Columbia University Press, 1970), p. 62.

[a] Crude density—28.4 per square mile (10.97 per sq. km.).

cent of the total. There is, however, considerable variation from region to region and by country. South Africa, Egypt, and Tunisia have over 40 percent of their populations urbanized, while Algeria, Morocco, Senegal, Congo, and many of the island appurtenances have urban percentages above 30. At the other extreme, most of the countries of the sudan-sahel belt, most of the eastern African nations, Swaziland, and Lesotho have urban populations below 10 percent of their total populations.

The main cities in each country are to an unusual degree the centers of modernization on the continent. They are the intellectual and social capitals, the seats of government, the main foci of political activity, and the economic capitals of their respective countries—the main assembly and break-of-bulk points, the great markets, and the main financial nodes; and they contain the vast bulk of the newer market-oriented manufacturing establishments as well as a considerable share of the raw-material-oriented plants. Indeed, one of the notable characteristics of many African countries is the rapid fading away of the signs of modernity as one leaves the urban centers.

Except for northern Africa, Ethiopia, the sudan belt running across the continent south of the Sahara, and traditional centers in the southern belt of West Africa, Mengo in Uganda, and Tananarive in Madagascar, the rise of urban communities is a recent phenomenon. European influence in sub-Saharan Africa first became marked by set-

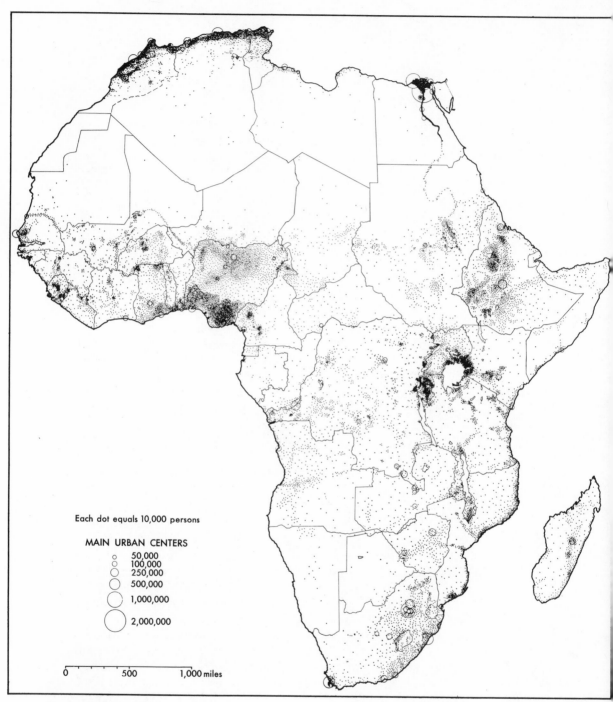

Map 20. *Population distribution of Africa*
Courtesy University of Stellenbosch

tlements in the sixteenth century, usually small coastal forts, trading posts, or way stations. Many of these early towns remained insignificant for centuries and some are still minor agglomerations. The eighteenth century saw relatively little town formation, while in the nineteenth century a good many towns in northern Africa, especially Algeria, were founded, transformed, or rejuvenated, new coastal points were settled, several major Yoruba towns were built, and a number of South Africa's major cities were established.

It was not until the last of the nineteenth century and the present century that many of sub-Saharan Africa's major urban centers were developed, and most of these remained quite small for several decades. The period between the two World Wars saw a steady if not spectacular increase in the large cities of northern Africa and South Africa. But since the last war the pace of urbanization has accelerated markedly and continues to do so in most countries. Particularly in tropical Africa this has been a period of mushrooming growth. At present, the rate of urban increase in tropical Africa is about 4 times the rate of rural increase and occasionally 6 to 8 times that rate. The first-ranking city of most countries is characteristically growing at a substantially more rapid rate than the rest of the urban communities.

A complex of advantages and disadvantages is associated with the rise of urbanism in Africa. On the one hand, urbanization tends to expedite the evolution to a modern economy, to loosen the hold of traditional beliefs and values, to permit a greater degree of specialization through the acquiring of new skills, to provide incentive for developing more diversified economies, and to develop concentrated markets for domestic produce. On the other hand, it may lead to the development of bidonvilles or shanty towns, subject the new urbanite in some cases to new forms of discrimination, stimulate inflationary forces, increase the prob-

Luanda, Angola, one of the many modern cities that have been growing at very rapid rates in recent years

lems of un- and under-employment, contribute to the rise of antisocial behavior, and lead to political unrest. Unfortunately, local economies have often been slow in rising to the new opportunities and needs created by urbanism, particularly in realizing new market possibilities.

The distribution of African cities is shown in Map 21. Several dominant patterns may be delineated. First is the large number of port cities. In the 33 countries of Africa that front on the sea, all but 8 have their largest city on the coast. The explanations for this dominant pattern, which exists despite the massiveness of the continent, are historical and economic. European interest in sub-Saharan Africa was for decades confined to coastal points; these points became foci of pathways and later roads and finally the coastal termini of the tentacle-like African railways. As commerce expanded they became the gathering points for domestic staple exports and the break-of-bulk centers for the greater variety of imported goods. In view of the relatively high percentages of the total money economies of Africa that are

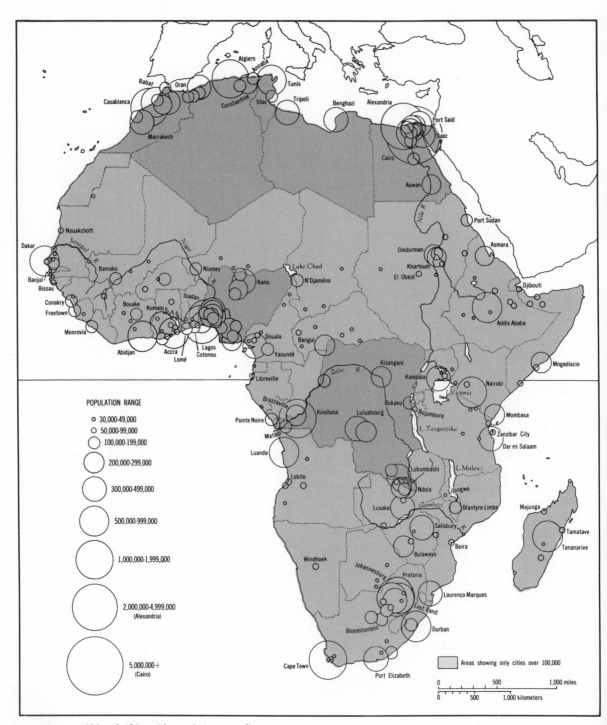

POPULATION RANGE

○ 30,000-49,000
○ 50,000-99,000
○ 100,000-199,000

200,000-299,000

300,000-499,000

500,000-999,000

1,000,000-1,999,000

2,000,000-4,999,000
(Alexandria)

5,000,000+
(Cairo)

Areas showing only cities over 100,000

0   500   1,000 miles
0   500   1,000 kilometers

Map 21. Cities of Africa with populations exceeding 30,000

represented by foreign trade, it is not surprising that ports have so frequently become leading cities.

It was logical that many of these cities be selected as colonial capitals. Both the commercial and the political functions resulted in disproportionate concentrations of the expatriates in these countries. This, in turn, gave many of the port-capitals considerable significance as consuming centers. Good examples are Casablanca, Dakar, Luanda, Lourenço Marques, and Dar es Salaam, which absorb some potential exports and a considerable percentage of total domestic imports within their urban boundaries. In post-independence years the importance of the capital cities was increased by the rapid burgeoning of national bureaucracies and the arrival of foreign diplomats and advisers, with some countries containing a larger number of expatriates than before.

All of these factors resulted in the provision of better services and amenities than are found in lesser communities—electricity, water, telephones, paved roads, superior housing, schools, and entertainment facilities. Once started, the development was self-generating.

A second group of important cities are inland ports—river and lake ports—which owe their location and their significance to the same assemblage and break-of-bulk functions that caused the coastal ports to grow. Seven of these—Bamako, Niamey, N'Djaména, Bangui, Brazzaville, Kinshasa, and the Three-towns conurbation of Khartoum–North Khartoum–Omdurman—are the prime economic cities and the capital cities of their countries.

A third group of African cities also owes its importance largely to the commercial function—the cities situated at important foci of land routes. This category includes some of the inland cities of the Maghreb, Ouagdougou, Sokoto, Kano, Lusaka, Salisbury, and Bulawayo. Some of these cities have grown up from indigenous market

towns. It is apparent from the number of leading African cities that fall into the first three categories—coastal ports, inland ports, and crossroad cities—that the most important explanation for their significance and their location is the commercial functions they fulfill.

There are only a few towns whose main function is manufacturing, and the classification of some of these is problematical. They include Mohammedia, Arzew, Melhalla-el-Kubra, Rufisque, Jinja, Gwelo, Que Que, Vanderbijlpark, Vereeniging, Germiston, and Sasolburg.

Surprisingly few mining towns with populations exceeding 20,000 are found. Most of them in tropical Africa are associated with the Shaba-Copperbelt complexes in Zaïre and Zambia. Petroleum operations have contributed strongly to the rise of certain towns, not always in the producing zones. South Africa, of course, has a considerable number of mining communities, though some have evolved to a much more complex pattern than existed when mining was their major raison d'être.

Only a few towns are classified as administrative cities, because other functions usually outweigh the political one. The large number of capital cities that have disappeared or declined over the years and the substantial number of countries that have switched their capitals—more often than not because of the desire to be located at the principal commercial node—suggest the greater significance of the transport-commercial function than the political function in the development of existing capital cities.

Several factors appear to be favoring the greatest urban growth in the already important cities: the tendency to suppress minor ports and concentrate traffic at fewer points, the increasing importance of the governmental function, and the desire of new consumer-oriented industries to locate in the main national markets. Inland cities may be expected to expand as internal exchange increases, new exploitations occur in mining, hydroelectric facilities are installed, and agricultural regions modernize. It is too early to tell what role governments will play in the attempted restriction of primate cities or the stimulation of lesser cities as growth poles. There is little doubt, however, that the process of urbanization will continue at a rapid if not an accelerating rate.

THE IMPACT OF POPULATION ON ECONOMIC DEVELOPMENT

The impact of the population factor on economic, social, and political development is pervasive, direct, and frequently complex. Population size is one important determinant of political power and prestige; it provides one measure of market size and thus strongly affects the economic potential of many of Africa's individual countries, in turn creating one of the greatest incentives to regional and subregional economic and political cooperation and consolidation. Population distribution affects political, social, and economic integration and the degree of land pressure that may develop within a country; it greatly influences the costs and difficulties of providing adequate social and economic infrastructures. This section will briefly examine several aspects of population having particular significance to economic development.

THE EFFECTS OF POPULATION GROWTH ON INCOMES

Population growth has a direct impact on economic development; it may increase the size of the market for certain items and the number of operatives available for productive output. However, too-rapid growth tends to increase the difficulty of development, because a substantial part of investment must go to provide the population increment with essentially the same services enjoyed by the existent population. When it is noted that it took until about 1650 for Africa's population to reach 100 million, the next 300 years to add the second 100 million, and only about 15 years to add the third 100 million, the significance of the growth factor may be better appreciated. Indeed, by 2010 the population of Africa may well exceed 1 billion, the world total only about 130 years ago.

Economic theory now holds that a high rate of population growth is itself an obstacle to advance. This is so because the capital needs of the increasing population are likely to absorb a substantial fraction of investment capital, thus reducing the amount available for "progressive" investment; it may be roughly estimated that the investment required to take care of the growth in population will absorb about 3 percent of the GNP for every 1 percent in the rate of population growth. Many African countries are not now, or are not capable of, investing enough to provide for both the increasing population and a satisfactory rate of economic growth.

Reduced birth rates affect the rate of growth of per-capita income in three obvious ways: the resulting lower population shares the national income, the lower number of children reduces the burden of dependency, and the long-term reduction in the labor force reduces the problems of unemployment and underemployment. Additional benefits may be better nutrition, health, and education, which would positively affect the quality of the labor force.

These considerations, it should be noted, apply to any country, but the higher rates of population growth in underdeveloped countries mean that a considerably higher proportion of their GNPs has to be invested in order to keep per-capita incomes at a constant level, while they have less capacity to invest and their need to increase incomes is very considerably more urgent. Dependency rates are unusually high in Africa, with about half the population of the continent in the dependent age group, compared to about 24 in Western Europe and 30 percent in the United States. The ratio of dependents to supporters may be expected to increase in the next decade or so.

IMPACT ON POPULATION PRESSURE

Rapid population growth will also have a deleterious influence by increasing pressure on the land. Contrary to the opinions of many, Africa is already experiencing population pressure. Certain patterns appear again and again in areas suffering from population pressure,[3] permitting the development of a typology of such pressure. Indicators that there *may* be excessive pressure in a given area include: (1) soil deterioration, degradation, or outright destruction; (2) use of excessively steep slopes and other marginal lands; (3) declining crop yields; (4) changing crop emphases, especially to soil-tolerant crops; (5) reduction in the fallow period without measures to retain soil fertility; (6) breakdown of indigenous farming systems; (7) food shortages; (8) land fragmentation, disputes over land, landlessness; (9) rural indebtedness; (10) unemployment and underemployment in rural and/or urban areas; and (11) certain types of out-migration.

Map 22 is an effort to depict those areas of Africa *now* affected by pressure of population. An estimated 47.1 percent of the area of Africa and 50.5 percent of its population was considered to be experiencing population pressure in mid-1967. Twenty countries were found to have more or less pervasive problems; twenty-five others had some portion of their areas and peoples living under pressure. When it is realized that the population on many of these areas will double within a generation, the seriousness of the situation becomes apparent.

OTHER IMPACTS OF POPULATION GROWTH

Increasing densities on the land tend to increase pressure on the land, particularly under existing systems and particularly when the rate of growth increases so rapidly

[3] For a summary of such evidence see Hance, *Population, Migration, and Urbanization in Africa*, pp. 393–413.

that adjustments through intensification are very difficult to achieve. These lead to increasing rural underemployment and unemployment, which in turn increase the necessity and propensity to migrate to urban centers. But the cities of Africa are not now capable of providing adequate employment to the stream of migrants coming to them and hence the unemployment is just transferred from rural to urban areas. Furthermore, even though urbanization is taking place at a very rapid pace it will be many years before the rural population reaches its peak level, given the high percentage of the population that is rural and the existing high rates of growth.

Pressure on national budgets is seriously affected by the increasing population totals. Several countries have, as a result, found it necessary to accept lower goals in the provision of education and health facilities. And there are almost no government plans that are capable of creating enough employment opportunities to absorb more than a portion of the new entrants to the job market. Nor would a decline in the fertility rate have a notable effect on the situation for several decades, because, those who will enter the market are already born.

THE PEOPLES OF AFRICA

About 98 percent of the total population of Africa is of African origin. The terms "European" and "Asian" are not entirely satisfactory because they include persons who have been born in Africa or whose forbears were born in Africa from one to many generations back. Other difficulties arise in differentiating among Asians of markedly differing racial origin, in the various ways of recording mixed ancestry, and in the inclusion of all whites under the term "European." A major problem exists for Madagascar, where it is not certain what portion of the population is descended from early Asian migrants and what portion from Africans.

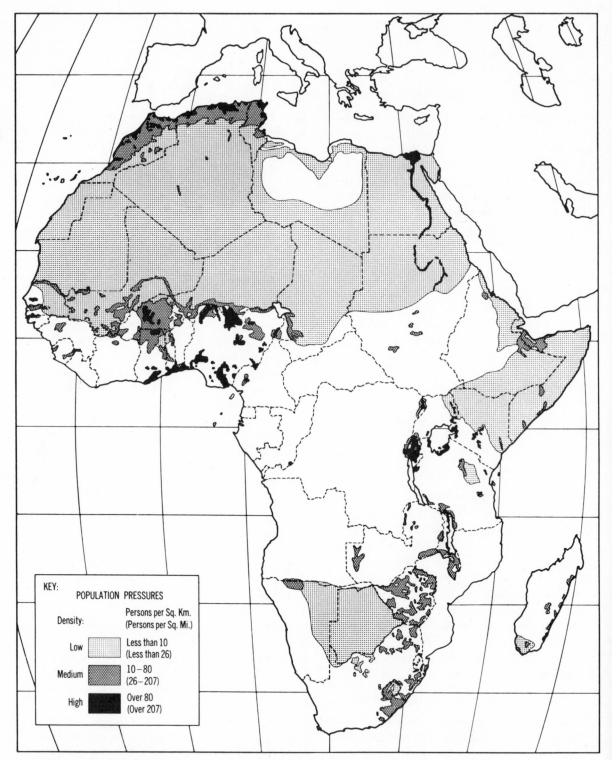

Map 22. Tentative depiction of areas experiencing population pressure at selected density ranges in Africa

KEY:

POPULATION PRESSURES

Density:

Persons per Sq. Km.
(Persons per Sq. Mi.)

Low — Less than 10
(Less than 26)

Medium — 10 – 80
(26 – 207)

High — Over 80
(Over 207)

AFRICANS

While it is possible to generalize that most of the indigenous population of African origin is rural, relatively unskilled, and still influenced largely by traditional ways of life, it must be noted emphatically that there is no such thing as *the* African. There is enormous ethnic variety. There are hundreds of tribes, about 800 languages (though many fewer language groups), numerous religions, and marked differentiation in occupation, educational achievement level, and land-use systems employed. Each group is advancing at a different pace; persons from within most groups have moved varying distances away from traditional patterns. In some cases the tribal pattern has remained dominant, elsewhere it appears to be crumbling. The resultant patterns are therefore complex in the extreme.

At the risk of overgeneralization, several broad contrasts may be made between "traditional" and "modernized" African societies and individuals. In most traditional societies the basic unit is the extended family, several of which combine into clans which in turn combine into more or less well demarcated tribes; there is often a degree of exclusivity in ethnic groups; the social role of individuals is determined by inherent status or custom; there are emphases on leisure and prestigious consumption; economies are more self-subsistent and less pecuniary. Characteristics of "modernized" societies include gains in significance of the nuclear family relative to the extended family; greater individualism, differentiation, stratification, and competitiveness; an increased acquaintance with and impact of markets, wages, laws of contract, and private as compared with communal ownership; a closer relationship to government; a notable emphasis on the importance of education and a growing gulf between the educated elite and the masses; and a more pecuniary economy. The dichotomy between traditional and modern is easily overdrawn; people are more likely to be in some part of a transitional spectrum than at either extreme, with persons in the urban communities likely to be nearer the modernized end, though most continue to recognize certain traditional values. The transition is sometimes difficult, even traumatic, and there are obvious gains and losses as far as economic development is concerned.

A number of factors affecting many Africans inhibit economic development—poor health and the high incidence of diseases, average low levels of educational achievement, the characteristic desire for large families. Language, religious, and ethnic differences also present impediments, sometimes quite severe, to economic advance. Serious tribal rivalries have caused difficulties in many African countries in the past decade or so, including Nigeria, Cameroon, Congo, Zaïre, Rwanda, Burundi, Sudan, Ethiopia, Uganda, Zambia, and Madagascar. It is no wonder that one of the main goals of modernist African leaders is the suppression of tribalism.

Two particular cultural features are restricting factors of serious proportions—the traditional systems of man-land relationships, affecting primarily the sedentary farmers, and the cattle culture, affecting most nomadic and seminomadic tribes south of the Sahara. The various and often complex systems of communal "land tenure" have important assets as long as a relatively stagnant demographic and economic situation prevails, but they have serious economic, agronomic, and psychological disadvantages as far as modernization is concerned. It does not follow, however, that change to a system of individual tenure can be easily made.

The common denominator of the various forms of cattle culture is the recognition of cattle as a store of wealth or prestige, not as an economic resource. This leads to the deterioration of livestock, to the prestige sym-

A Moslem sheik on pilgrimage in Cameroon

Inhabitants of Tamanrasset, Algeria

The diversity of African peoples is greater than on any continent, with the possible exception of Asia

A Karamojong pastoralist, Uganda

Dr. Liberty Mhlanga chairing a conference on development and the environment, Nairobi 1974

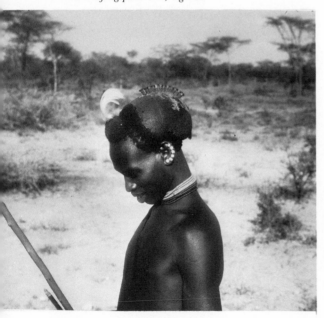

bols destroying their environment. Grazing land is almost universally in poor condition, and what could be one of the great assets of Africa becomes instead a destructive influence.

Other cultural features that are inhibiting to a greater or lesser degree include various taboos; certain aspects of the extended family system, particularly those which reduce the incentive to increase production and productivity; some features relating to the role of women, particulaarly their often excessive share of farm work; and the backwardness of most rural areas as far as practices and implements are concerned.

Education will be all-important in reducing the unfavorable impacts of these and other cultural factors, as is illustrated by the notable achievements of many of the educated elite in each of the African countries.

EUROPEANS

The population of Africa of European origin is very unevenly distributed, with the vast bulk in southern Africa. In the Republic, Europeans made up about 16.7 percent (4.16 million) of the total population in 1974. In tropical Africa in 1974, Angola had the highest percentage of whites, 6.3 (c. 400,000); Rhodesia had 4.5 percent (255,000) in 1972; and Mozambique had about 2.4 percent (c. 200,000) in 1974. Many white "settlers" left Mozambique prior to independence in that country; the outcome in Angola could go toward a multiracial nation on the pattern of Brazil or toward a massive exodus more like Algeria. Elsewhere there are relatively few European "settlers," most of them having left the Maghreb, Kenya, and other countries in the years immediately preceding and following independence. Most countries now have less than 0.2 percent of Europeans.

The composition of the white population in most independent African countries has changed markedly since independence. Administrative posts are now largely held by Africans; those Europeans employed by government are largely advisers, engineering and technical experts, medical personnel, and educators. Diplomatic staffs have, of course, increased, as have representatives of multilateral agencies and foundations. Mining and manufacturing ventures remain heavily dependent on expatriate managers, engineers, and technicians, though the percentage of foreigners to locals is decreasing. In those countries which have enjoyed more favorable rates of economic growth, the total number of whites is not infrequently equal to or greater than the total before independence.

Europeans tend to be dominantly urban based; a large percent are salaried or wage earners; and there is a considerable concentration in service occupations. The dominant position that Europeans held in almost every sphere until recent years engendered considerable resentment. Yet it is not possible at the present time to replace many of them with adequately trained Africans; indeed a decision to expedite localization is likely rather ironically to require more expatriate advice and training over the short run. And to be realistic the very large contribution made by expatriates in proportion to their numbers must be recognized.

ASIANS

Persons of Asian origin are concentrated in the east and the south of the continent, on Madagascar, and in the Mascarene Islands. Many Asians were first brought as plantation workers, particularly on the sugar plantations of Mauritius, Réunion, and South Africa, or to assist in the construction of ports and rail lines. Today they generally occupy a disproportionate number of intermediate positions as artisans, skilled and semiskilled workers, and clerks. They also have great importance in merchandising, including the operation of small shops in remote areas. Some have been extremely successful, being owners of hotels, movie houses, large stores, and manufacturing plants. They have, however, frequently been the subject of calumny on the part of Africans and Europeans or have been legally restricted from certain areas and occupations. In the early 1970s Uganda, in fact, expelled most of its Asian population, including many who had taken out citizenship in the country.

CHAPTER FOUR

The Political Scene in Africa

Concluding these introductory chapters is the following brief survey of certain features of the African political scene that have importance in economic and geographic terms. Rapid change, the rise of one-party and military regimes, and considerable instability have been keynotes of the political arena. Resentment of continued minority rule in southern Africa has been strongly stated in international forums. While it is possible to delineate some of the trends and many of the forces involved in the political field it is well to expect the unexpected in Africa.

THE COMING TO INDEPENDENCE

The dominant trend of the 1960s was the accession to independence of African countries (Maps 23 to 26). Before World War II, only Egypt, Liberia, and South Africa were independent, though Ethiopia had been one of the oldest nations in Africa until the Italian occupation of 1936. Ethiopia became nominally independent again during the war and was subsequently federated with Eritrea. Libya joined the small group of independent states in 1951, followed by Morocco, Tunisia, and Sudan in 1956. In the following year the Gold Coast became the first colony of Black Africa to achieve independence, joining with the trusteeship area of British Togoland to form Ghana. Guinea was the only territory to opt for independence in the French referendum of 1958 and was summarily abandoned by France. The early 1960s saw a rush to independence such that by 1964 about four-fifths of the area and nine-tenths of the people of Africa were independent. By 1974 only Spanish Sahara, the Por-

tuguese territories, the French Territory of Afars and Issas (formerly French Somaliland), the Comoros and Seychelles, and South-West Africa remained in a dependent status. The Portuguese areas witnessed revolts beginning in the early 1960s, which culminated in the overthrow of the metropolitan government in 1974, following which efforts were made to achieve a political rather than a military solution. Portuguese Guinea became independent as Guinea-Bissau in 1974, and independence was arranged for Fernando Po and Príncipe, Angola, and Mozambique later in 1975. The Comoros and the Seychelles were slated to become independent in 1975 or 1976. In November 1965, Rhodesia, which had been

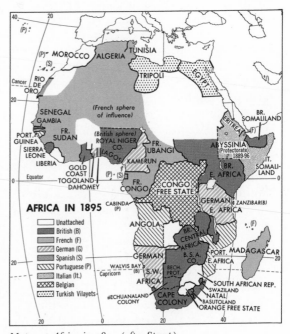

Map 23. Africa in 1895 (after Stamp)

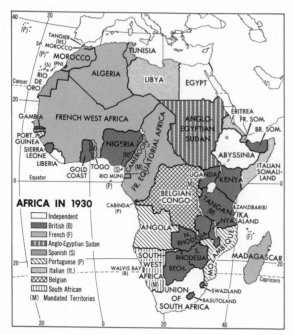

Map 24. Africa in 1930 (after Stamp)

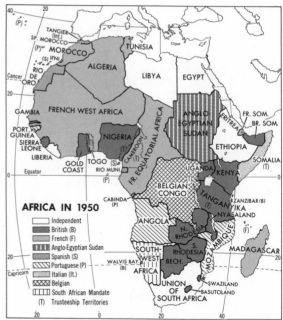

Map 25. Africa in 1950 (after Stamp)

a self-governing colony since 1923, declared its independence unilaterally; the action was immediately declared illegal by Britain; and Rhodesia has since been subjected to international sanctions (see chapter 19).

The most difficult position exists in the Republic of South Africa, which, although classified as independent, contains a majority group of black Africans who are denied representation except on a limited basis in a number of Bantustans and who are subject to economic and social restrictions that are a contradiction of the whole Western ethos. Nor would the formation of Bantustans in South Africa and South-West Africa, even if they were eventually to receive full independence, solve the problem.

THE GEOGRAPHIC PATTERN OF POLITICAL UNITS

The pattern of present national units in Africa presents several important characteristics. First, it tends to be peripheral; many

countries, particularly in West Africa, represent extensions from previously held coastal strips or nodes. This reflects the way-station character of early interests, the difficulty of exploration and penetration, the desire to control river mouths for more effective control of inland trade (e.g., Gambia and Nigeria), and the degree of competition among imperial powers. At the same time, Africa has by far the largest number of landlocked states of any continent—fourteen.

Second, political boundaries bear little relation to natural regions or ethnographic distributions. Almost every boundary has divided several tribes among two or more countries. The Ewe, Hausa, Zande, Bakongo, Masai, and Somali, to name but a few, were so divided. Third, it was characteristic for the imperial powers to have holdings in more than one part of the continent, and for each major region to be divided among several powers (Maps 23–25). This probably aided development to a degree in that each metropole strove to secure raw materials

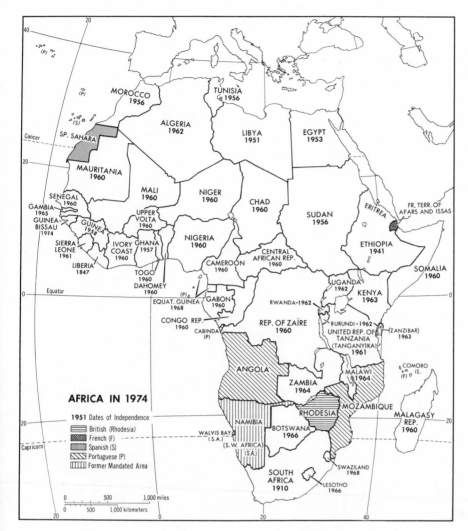

Map 26. Africa in 1974
Mozambique, Angola, the Comoros, and the Seychelles were expected to become independent in 1975.

On the map:

MOROCCO 1956
TUNISIA 1956
SP. SAHARA
ALGERIA 1962
LIBYA 1951
EGYPT 1953
MAURITANIA 1960
MALI 1960
NIGER 1960
CHAD 1960
SUDAN 1956
ERITREA
FR. TERR. OF AFARS AND ISSAS
SENEGAL 1960
GAMBIA 1965
GUINEA-BISSAU 1974
GUINEA 1958
SIERRA LEONE 1961
LIBERIA 1847
IVORY COAST 1960
GHANA 1957
UPPER VOLTA 1960
TOGO 1960
DAHOMEY 1960
NIGERIA 1960
CAMEROON 1960
CENTRAL AFRICAN REP. 1960
ETHIOPIA 1941
SOMALIA 1960
EQUAT. GUINEA 1968
GABON 1960
CONGO REP. 1960
CABINDA (P)
REP. OF ZAÏRE 1960
RWANDA-1962
UGANDA 1962
KENYA 1963
BURUNDI-1962
UNITED REP. OF TANZANIA (TANGANYIKA) 1961
ZANZIBAR 1963
ANGOLA
ZAMBIA 1964
MALAWI 1964
COMORO IS. (F)
MOZAMBIQUE
RHODESIA
MALAGASY REP. 1960
NAMIBIA
WALVIS BAY (S.A.)
(S.W. AFRICA) (S.A.)
BOTSWANA 1966
SWAZILAND 1968
SOUTH AFRICA 1910
LESOTHO 1966

AFRICA IN 1974

1951 Dates of Independence
▭ British (Rhodesia)
▨ French (F)
▨ Spanish (S)
▨ Portuguese (P)
▥ Former Mandated Area

0   500   1,000 miles
0   500   1,000 kilometers

---

from its own areas, and therefore a more intensive transport pattern was constructed; but it also resulted, in some cases, in unnecessary duplication and it contributed to the divisiveness that is related to the differing heritages from each of the metropoles.

Fourth, the continent is divided into an excessive number of political units, many of which are entirely too small to be economically viable or to have markets of sufficient size to sustain more than minimal industrial establishments (Table 10).

THE HERITAGE FROM THE COLONIAL PERIOD

The colonial period lasted but a brief moment in historical time, but its costs and contributions had impacts that are very apparent and that will remain so for many years. While Africa was conquered and occupied by European powers, only parts of it—South Africa, Swaziland, Algeria, and, to a lesser extent, Libya, Eritrea, Angola, and Kenya—were colonized in the sense that the

Table 10. Number of African political units by selected ranges of area and population, 1973

Population (millions)	Number of countries	Area		Number of countries
		1000 sq. km.	*1000 sq. mi.*	
<0.1	7	<1	<.39	5
0.1–0.4	5	1–	.39–	5
0.5–0.9	7	10–	3.86–	10
1–1.9	5	100–	38.6–	5
2–2.9	4	200–	77.2–	6
3–3.9	4	300–	116.0–	3
4–4.9	6	400–	154.0–	2
5–9.9	10	500–	193.0–	3
10–14.9	2	600–	232.0–	2
15–19.9	4	700–	270.0–	2
20–29.9	2	800–	309.0–	1
30–39.9	1	900–	348.0–	2
40–49.9		1000–	386.0–	9
50–59.9 [a]	1	2000–2500	772–965	3

SOURCE: United Nations *Statistical Yearbook* and *Monthly Bulletin of Statistics.*

[a] Preliminary figures released for the 1973 Nigerian Census gave a total population of 79 million in November.

Americas or Australia and New Zealand were. Algeria and Kenya no longer have significant settler communities, and in no case has the settler population been a majority of the total.

The attention accorded to the colonial territories varied not only between metropolitan powers but from colony to colony within each imperial holding. It was attenuated during the two world wars and the depression of the 1930s and reached its highest level in the decade or so preceding independence. The period brought a degree of modernization to many parts of the continent; infrastructures were developed, especially where activity was more intense; a market economy was widely diffused; modern consumer products and manufacturing techniques were introduced; Africans were involved in money economies, in part through the imposition of taxes and sometimes by labor corvées but increasingly on a voluntary basis; there was a notable emphasis on the export sector, which not only supplied desired raw materials to the metropoles but was also seen as the only way of providing adequate revenues for development and as the logical engine of growth; Western political institutions and social values were introduced; educational systems and mass media were instituted, providing perhaps the most important channels for diffusion of Western ideas and concepts.

Each of the metropolitan powers developed over the years a distinctive colonial policy, which has left very apparent legacies in the political, social, and economic structures of the independent African nations. Whatever the contrasts were, however, it is apparent that they made little difference in the desire of Africans for political independence.

The rapidity of the move to independence was unexpected as late as the mid-1950s and in no case was there sufficient preparation, despite more strenuous efforts after World War II and considerable devolution of power to African elites and legislative bodies.

It is true that Britain was long committed to the policy of evolution from colony to commonwealth or independent status; it is also true that postwar British governments genuinely sympathized with African political aspirations, but for some years it was thought that the colonies were being prepared for independence in the next century. Until a few years before independence France was committed to the goal of assimilation of African territories in a French Union in which all the citizens would become citizens of France. The goal was changed to one of association, but in 1960 an amendment to the 1958 Constitution permitted accession to complete independence without becoming dissociated from the Community (as Guinea had been in 1958), and shortly thereafter withdrawal from the Community became possible without impairment of relations with France. For Belgian territories, the rapidity of the move to independence was startling. Until about 1958 that country envisaged a very extended period before self-government was achieved, and its overall policy was predicated on a deliberate evolutionary pace; just two years later Belgium, quite surprisingly, offered complete independence in six months. The lack of adequate preparation contributed strongly to the chaotic conditions that ensued. Portugal continued to insist that all its overseas possessions were forever parts of the Portuguese empire until the military coup in that country in 1974. Spain has ceded Ifni and other holdings to Morocco, granted independence to Equatorial Guinea in 1968, and now holds only Spanish Sahara. Italy was divested of its holdings after the war but was made administrator of Somaliland until 1960.

Colonial governments exercised great authority; they dominated and directed not only political and social affairs but also many aspects of the economies. They were generally responsible for all the activities that national, state, county, and local administrations are in the United States and some that are frequently provided by private enterprise in developed countries. Productive sectors were, however, largely in private hands, including African producers of cash crops and petty entrepreneurs. The impact of expatriates was often reduced, however, by one or another form of indirect rule, which tried to make systematic use of the customary institutions of the people as agencies for local and regional rule and which permitted administration of considerable areas by a small number of colonial officials.

Contrasts among the policies of the major powers include:

(1) The decentralization and greater pragmatism of British colonial administration, which helps to explain the great diversity in anglophone countries in constitutions, franchise systems, and legislative bodies, as compared to the highly centralized control from Brussels of Belgian territories and more direct line of authority between Paris and the governors of French colonies.

(2) More open trade policies in British areas, where sales of African products were ruled primarily by world markets and prices, as compared to French policies, which always favored exchange with the metropole. In 1961, Commonwealth countries accounted for 58.6 percent of total exports from tropical Africa whereas French Community countries accounted for only 18.5 percent and former Belgian areas for 11.0 percent. About 70 percent of French overseas territories' trade was normally with the metropole, and while the territories benefited by guaranteed markets at high prices they were disadvantaged by mutual inflation, by the narrowness of the market, and by the reciprocal obligation to purchase their imports at high prices. This situation has been ameliorated by

devaluations and by the association of most francophone states with the European Economic Community (EEC), but the necessary adjustments were sometimes very difficult (especially for Senegal) despite grants designed to ease the transition and promote diversification in countries with heavy one-product emphases.

(3) British policy before the war was for each colony to pay its own way with only occasional subsidy. In postwar years local sources financed about two-thirds of government expenditures; the British government provided substantially increased grant and loan funds and the flow of private funds increased. The French, on the other hand, had long provided more direct support to colonial budgets. But investment in sub-Saharan French territories was very limited in prewar years, totaling an estimated $128 million in 1938 in French West and Equatorial Africa, and it tended to be concentrated in a few places, especially in Senegal and the Ivory Coast; after World War II, however, France greatly expanded its efforts. Since expenditures in British Africa were much greater over the long run, however, several of them inherited better infrastructures on independence. Belgian colonial policy gave first priority to the economic sphere. Their main forte was in the systematic and scientific development of the resources of the area, and there were few territories where a better effort had been made to use Africans in technical and industrial employment. Investments were considerable and substantial earnings within the Congo (now Zaïre) were reinvested to expand production. Large parastatal companies played an important role in mining, plantation agriculture, and transportation.

(4) In the important field of education, British policy favored the training of an elite, little technical education was provided, and more universities were founded in Africa. In French territories emphasis was placed in prewar years on the training of a small elite who were used in the government as auxiliaries, somewhat more successfully than by the British. In postwar years education was greatly expanded; several universities were opened, and larger numbers of Africans were accepted for higher education in French universities. Belgian educational policy fit logically into their overall plans. It called for a broad base and for some emphasis on technical education. Almost no effort was made to educate an elite; there were no universities in the country until 1954 and very few Africans attended universities in Belgium.

(5) Restrictive bars on Africans were more common in British and Belgian than in French areas. Particularly where white settlement had occurred, contrasts were sharp between the juxtaposed modern economies and the

Uhuru or Independence celebration of Kampala, Uganda, October 9, 1962

indigenous, largely subsistence economies. On the other hand, there was less interference with local cultures in British areas, while the French paid little attention to local cultures and languages, the goal being to make the African a Frenchman.

In each of the colonial territories political, social, and economic systems were likely to mirror or be extensions of comparable systems in the metropoles. The linguae francae were also borrowed from the ruling power and have been retained in post-independence years with some exceptions. The differing heritages in these institutions and languages have been divisive to a considerable degree, and certainly contribute significantly to the problem of achieving regional and continental cooperation.

POLITICAL CONDITIONS AFTER INDEPENDENCE—INTERNAL

Many African countries acceded to independence at the close of a decade of remarkable growth and at the beginning or during a decade of disappointing growth—the so-called "development decade." The educated and the ruling elites were in numerous cases pathetically thin. The problems of promoting viable nationhood, of suppressing tribalism, of meeting new international responsibilities, and of achieving a satisfactory rate of economic growth were often staggering. Optimism, which was high at the beginning, sometimes turned to disillusion. Rates of economic advance have in many cases been low, and nonexistent in a number of countries. The record has, however, been very uneven. Some countries have made advances comparing favorably with any countries in the world. Thus the years of independence have witnessed widening gaps, internally between the monetary and subsistence sectors, and externally between the richer and the poorer states. These trends are likely to continue if not intensify in the years ahead.

Among the casualties of the post-independence years was the demise of most of the democratic forms created on the handover. The Westminster model and its French and Belgian counterparts proved to be fragile, not surprising in view of the authoritarian heritage of colonial experience. By the end of the 1960s most countries had become one-party or military states (Map 27). While such regimes have certain assets and some can be reasonably responsive to public opinion, they are by nature authoritarian and rigid, so that inefficiencies are multiplied and corruption is more difficult to check. They can constitute vehicles for ambition and oppression, are all too likely to further alienate the elite from the masses, and do not provide ready solutions to the problem of achieving more democratic forms or of accession. Military regimes have also characteristically been weak in economic matters. The formation of one-party and no-party

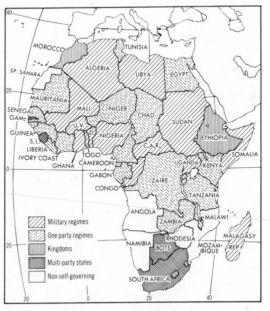

Map 27. *Types of government in Africa, 1974*
Ethiopia should now be classified as a military government.

states does, however, reveal the recognition that the maintenance of order is of prime importance.

A second notable trend has been the widespread acceptance of socialism as the ideal for political and economic development, though, as noted in chapter 1, a mixed economy is the characteristic form, the private sector has continued to dominate the productive economy, and practically all nations continue to invite further private investment. Disappointment with the rate of economic growth has, however, contributed to increasing dislike of neocolonialism: the concept that independence is being thwarted by the continuing presence of expatriates and foreign enterprise. The desire to consummate independence through the removal of neocolonialism takes several forms— required participation in various foreign companies, partial or full nationalization, prescriptions on foreign employment or investment in specific types of activity, and attempts to promote Africanism through au-

Ministerial buildings in the new capital of Lilongwe, Malawi Government staffs have increased markedly in post-independence years. Capital cities have seen much new construction for government offices and five countries have selected new sites for their capitals.

thentication of names and titles and removal of vestiges of the colonial past. A portion of the pronouncements in this matter are primarily for domestic consumption. There is a notable degree of realism among most leaders that expatriate know-how and capital can provide important contributions to political and economic viability. There may also be an increasing recognition that those countries which have followed reasonably orthodox economic policies have shown more satisfactory growth trends.

At the time of independence most countries of Africa were not true "nations," in the sense that they contained a body of people conscious of their unity. Governments faced the challenging and sometimes difficult task of promoting this unity, which has not yet been achieved in many. Part of the effort focused upon calls for constructive participation in nation building, part focused upon the evils of past colonialism and present imperialism and neocolonialism. While most states have shown commendable tolerance— religious, ethnic, and national—a few have become xenophobic, expelling foreign Africans, resident Asians, and Europeans.

Given the numerous difficulties faced, it is not surprising that there has been considerable internal instability in independent African countries, though anarchic situations have been rare. Some of the instability has been related to tribal conflict, which had been suppressed under colonial rule. These conflicts have sometimes been strictly nonpolitical; at other times they have been enmeshed in the governance of the new states. In some cases they have been short-lived or restricted areally; in other cases they have continued for years and can be classified as rebellions (Sudan, Eritrea, Chad) or civil wars (Nigeria, Zaïre, and the genocidal conflicts between Tutsi and Hutu in Burundi and Rwanda). The large number of countries that have experienced ethnic difficulties since independence is illustrated in Map 35.

Discord among various groups has also

Chief Pokkam, Chef Superieure de Bangangté, Cameroon
The conflict between modernist and traditionalist forces is
a common feature of the post-independence political
evolution of African countries.

have had a deleterious impact on economic
development, if for no other reason than
that they absorbed attention which might
have been more constructively placed. It is
often surprising, however, how little the productive sectors are affected by seemingly
rather traumatic political events.

POLITICAL CONDITIONS AFTER
INDEPENDENCE—EXTERNAL

Most of the independent countries of Africa
have adopted policies of nonalignment and
noninvolvement in disputes among the
major powers. All but a few maintained links
with the former metropoles, the anglophone
countries as members of the Commonwealth
and the former French colonies as members
or associates of the French Union. Relationships between the latter colonies and
France were more intimate and pervasive,
with important monetary, economic, cultural, and military links, and a heavy dependence on France for advisers and for
budgetary support, particularly for the
development budgets. There was also a
rather strong feeling of community among
the francophone African states. But as differing policies evolved in these countries this
feeling diminished, and the early 1970s saw
an increasing desire on the part of several of
them to gain greater control over their own
affairs. Some countries withdrew from the
franc bloc, there was a demand for greater
African participation and control of the
regional banks, increasing moves to restrict
the French presence in government, and the
ending of several military agreements, including the use of bases in Algeria and Madagascar.

contributed to instability. There have been
differences between traditionalists and mod-
ernists, between first-generation political of-
ficeholders and their would-be successors
(who are frequently better trained and not
very much younger), between military and
nonmilitary elites, and between the elite and
the masses. Contributing to the last are the
relatively very high government salaries, a
legacy from the colonial period, and a propensity toward conspicuous consumption
and ostentation, which was particularly apparent in the first years of accession to
power.

A measure of the political instability is the
number of coups and attempted coups that
have occurred since independence (Map 36).
A large number of plots have also been reported, but these are more difficult to assess
and quantify. In any case, it is obvious that
numerous internal political happenings must

Most anglophone states became members
of the Commonwealth on independence
(Map 32). The practical benefits of such
membership—tariff preferences, membership in the sterling bloc, and access to
special forms of development aid—have

tended to decline in significance both before and after the United Kingdom's entry into the EEC. Nor have the African nations found it possible to influence British policies with respect to southern Africa as strongly as had been hoped, and this has resulted in several defections. More and more the Commonwealth has had to be accepted solely as a free association of sovereign independent states without necessarily any very significant communality of interests.

Relationships between Zaïre and Belgium were largely severed in the early 1960s; they improved as Zaïre's leaders realized that Belgians were more readily available and more knowledgeable about their country than any other nationals, but almost all special connections were ended in 1974 as General Mobutu adopted more nationalistic and increasingly xenophobic policies. Nonetheless, there are now a considerable number of Belgian advisers, technicians, and educators in the country. Most Italians were expelled from Libya, Spaniards from Equatorial Guinea; a fair number of Italians remain in the Eritrean Province of Ethiopia and Italo-Somali relations have remained significant.

As the wave of independence spread across Africa, relations with nonmetropolitan powers naturally increased. The continent briefly became involved in the East-West conflict as the United States strove to counter Soviet involvement in such countries as Somalia, Ethiopia, Algeria, and, particularly, Zaïre. In more recent years there has been some rivalry between China and the USSR. It is perhaps regrettable that the interest of the United States and the USSR tended to diminish as the danger of conflict receded, but it is not regrettable that all of the major powers now maintain more "correct" diplomatic relations and have abandoned earlier simplistic expectations and occasional efforts at subversion.

Other nations that have increased their interests in Africa include the noncolonial countries of Western Europe, to a lesser extent the centrally controlled countries of Eastern Europe, Canada (whose French-speaking citizens have made special contributions in francophone African countries), Japan (which has greatly increased its purchase of raw materials and its investment in mining and industry, and which may be expected to become a more important source of aid in the years ahead), Israel (whose presence was greatly reduced after the 1973 Arab-Israeli War), and, more recently, several of the Middle East nations.

CONVENTION WITH THE EEC

Of particular importance in the external relations of many countries of Africa was their association with the EEC. These relations date from the founding of the EEC in 1957, when provision was made for preferential trade between the Six and their dependent overseas territories and for extension of aid to them through the European Development Fund (FED). After independence these arrangements were extended under the Yaoundé Convention of Association running from 1964 to 1969 and revised for the period 1969–75.

The main benefits to the African associates were the preferential access that they received to the rapidly growing market of the Community and the funds provided through FED, whose total allotments increased from $581 million in 1959–64 to $730 million in 1964–69 and $918 million in 1969–75, the vast bulk of which went to African countries. The Associates could levy tariffs on imports from the EEC on a nondiscriminatory basis to protect infant industries or relieve balance of payment difficulties.

While the EEC associates were without doubt greatly aided by the Yaoundé Convention there were accompanying problems for some countries. Former French territories had marketed a large part of their agricultural produce in France at guaranteed prices often well above the world market price. This did not conform with Community trade arrangements so, to cushion the adjustment to world prices, 31.5 percent of the FED allotment for 1964–69 was devoted to offsetting the losses sustained by specific countries, particularly by promoting diversification in them. Senegal had an especially difficult adjustment, because it had been selling its major export, peanuts, to France at as much as 75 percent above the world price.

Among the objections to the EEC association was the failure of the EEC to abolish all customs tariffs on shipments from the associates (competitive agricultural products are subject to tariffs), the extension of reverse preferences by African countries, the minor role played by the associates in the management of FED, and the divisive impact of the agreement within Africa.

When the EEC on January 1, 1973, was enlarged to include the United Kingdom, a new chapter with respect to relations with Africa was opened. It was now appropriate to include the former British territories, if they so desired, among the associates. After prolonged negotiations a new five-year treaty, the Lomé Convention, was signed in February 1975 with the African, Caribbean, and Pacific states (the ACPs). Included were the former associates, former British territories, and several countries which did not have colonial ties with the EEC nine.

The convention provided for elimination of reverse preferences for the EEC; duty-free entry of industrial products and 96 percent of agricultural goods from the ACPs; setting up a compensation fund to guarantee export prices for peanuts, cocoa, coffee, cotton, coconuts, palm products, hides and skins, timber products, bananas, tea, sisal and iron ore; and assistance totalling about $3.5 billion over the five year period, including $2.5 billion in grants and $450 million for the compensation fund.

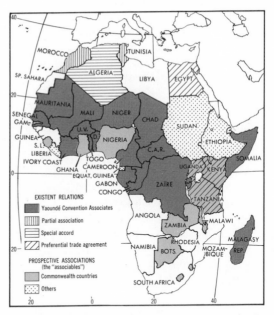

Map 28. *Relationship of African states to the European Economic Community, mid-1974*
The 1975 convention between the EEC and 46 African, Caribbean, and Pacific states, concluded after this map was drawn, included the associates and associables shown, although the word "associate" is no longer to be used. The Comoros and the Seychelles were expected to sign the convention on independence. Mauritius, not shown, was a signatory. Guinea-Bissau was a signatory and Angola and Mozambique were to be eligible on independence.

EFFORTS TO ACHIEVE COOPERATION AND INTEGRATION

Much attention has been given, particularly since independence, to the desirability of integrating numbers of African countries, politically and economically. There is a broad range of cooperative and integrative possibilities, ranging from various political joinings through economic and trade unions, common services organizations, and investment agreements, to projects involving specific sectors such as transportation, communication, agriculture, industry, education, health, and research.

The motives for emphasizing cooperation and unification are heavily political, associated positively with the ideological commitment to pan-Africanism and negatively with anticolonialism. There are also strong and often overlapping economic incentives:

(1) For locational reasons, most apparent for the landlocked states but present in many countries that have portions of their territories dependent on or best served by extranational routes (Map 29).

(2) For reasons of small size and the consequent desire to combine individual markets (see Table 10 and Map 1).

(3) For reasons of inadequate capital. This may be particularly important in highly capital-intensive activities such as airlines, certain regional industries, and water-control projects. Evidence has clearly shown that the larger countries, federations, and functioning economic communities are more attractive for foreign investors.

(4) For reasons of inadequate availability of indigenous skill and expertise, applicable to many of the avenues for regional cooperation and peculiarly important in certain research operations.

(5) To take advantage of such complementarities as exist. While interchanges among African countries are now relatively small and many economies are more competitive than complementary, there are opportunities to develop intra-African trade in animal and forest products, certain crops, and in manufactured produce, as is well illustrated by trade among members of the East African Community.

(6) Because development of shared resources requires cooperation if it is to be effective. Most important here is the multipurpose development of river and lake basins.

Turning to the accomplishments in the field of cooperation, it may be noted first that there have been few examples of political absorptions or joinings since or upon independence (Map 30). Indeed, unsuccessful efforts or actual breakups of formerly united territories are as numerous as the successes. In addition to those shown on Map 30, one might note the failure of the three East Afri-

Map 29. *Landlocked countries and regions of other states partly dependent on extranational routeways*

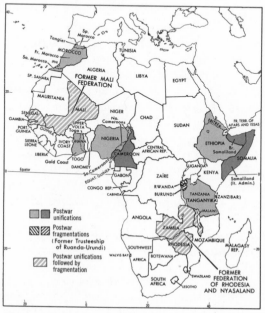

Map 30. *Modern boundary changes in Africa involving political consolidations or fragmentations*

can countries to achieve the announced goal of federating, various abortive efforts at federations among Libya, Egypt, Syria, Tunisia, and Sudan, and the creation of eleven countries from what had been French West and French Equatorial Africa.

The most important political association in Africa is the Organization of African Unity (OAU), founded in 1963 and now having 42 members (Map 31). The OAU has had some successes—mediating a number of disputes among its members, reconciling competing liberation movements, and coordinating the policies of African members of the UN. However, it has not yet played the role in African cooperation envisioned by its founders. Disputes and temporary defections have made it, at times, more a symbol of disunity; its Special Fund for Liberation is rarely paid up; and even allocations to sustain the secretariat in Addis Ababa have been ignored by some members. But it would be no more realistic to expect too much from an organization such as the OAU than to expect the UN miraculously to solve disputes among its members.

Many proposals have been made for

Africa House, Addis Ababa, Ethiopia
This is the headquarters of the Economic Commission for Africa of the United Nations. The Secretariat of the Organization of African Unity is also situated at Addis Ababa.

various political cooperative efforts, but none have gotten off the ground. The Afro-Malagasy Union (UAM) was succeeded in 1963 by the Afro-Malagasy Common Organization (OCAM) whose interests were more economic than political. Its cohesive elements were the important historical, cultural, economic, and linguistic links of former French and other francophone countries, and it grew to include Zaïre, Rwanda, and Mauritius as well as all of the former sub-Saharan French territories except Mali and Guinea. But as the member countries have evolved more divergent political and economic policies and the ties with France have weakened, the francophone emphasis of OCAM has been increasingly criticized as divisive in the African context, and six countries defected in the period 1972–74, including Madagascar, which suggests that the name is now properly the Afro-Mauritian Common Organization (Map 32). The Council of the Entente (Map 33) has increasingly forgotten its earlier political goals, and Nkrumah's United States of Africa is a dead letter. Thus, achievements in political unification have been less than spectacular. However, external aggression, which might have been one way of achieving integration, has been almost absent; there has been only one instance of conventional war between in-

Map 31. The Organization of African Unity (OAU)

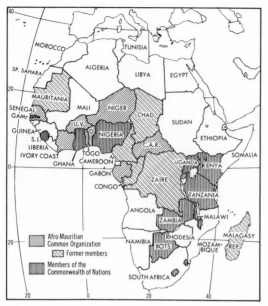

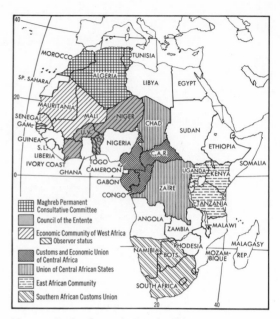

Map 32. The Afro-Mauritian Common Organization (OCAM); African members of the Commonwealth of Nations

Map 33. Regional organizations in Africa

dependent African states, the brief Moroccan-Algerian confrontation in 1963.

Most of the other organizations have been concerned solely with various forms of economic cooperation. Indeed, after disillusion set in over the difficulties of achieving political unification, this sphere has attracted the major attention. OCAM's more effective achievements, for example, have been in some of its specialized agencies, particularly Air Afrique, which has itself lost Cameroon as a partner and has been criticized by the Central African Republic and Chad, which withdrew temporarily in 1972.

A second trend coupled with the greater emphasis on economic goals has been toward regional rather than Africa-wide associations. These include a series of groupings with a variety of cooperative economic goals but with some emphasis on efforts to increase trade among the members of each group; the more important ones are depicted on Map 33. Most of them have had relatively little success in achieving effective customs unions or economic cooperation and

almost all have been plagued by disputes. As broader attainments have proved elusive, several of the groupings have tended to focus on more restricted goals, such as improving transport and communication links or training and research. The experience in francophone central Africa is illustrative of the vicissitudes facing economic cooperation. Before independence the area comprised French Equatorial Africa (AEF), the trustee territory of Cameroon, and Zaïre; AEF was succeeded by four independent states. In 1964 the five territories previously ruled by France formed UDEAC, which achieved some progress in the field of transport and even in the placement of a few regional industries. The landlocked states, however, concluded that they were being neglected to the advantage of the coastal states and in 1968 formed UEAC, joined by Zaïre; the Central African Republic withdrew from the new union and rejoined UDEAC in the same year, partly under pressure from France, leaving UEAC to consist of two separate states with very little in common. Chad, however, has since met with UDEAC and has indicated its willingness to cooperate with that organization.

The history of cooperative efforts in East Africa has similarly been plagued by internal difficulties. Kenya, Uganda, and Tanganyika already had, upon independence, a customs union, a common currency, and common services. The common currency rapidly disappeared and there were abrogations of the customs union. Indeed, what had been seen as a model for economic cooperation appeared to be facing total disintegration when, by an effort reflecting ideological dedication and requiring considerable skill, the East African Community (EAC) was formed in 1967. With the overthrow of the Obote government in Uganda in 1971 an extremely difficult period followed, marked by border closings, guerrilla incursions into Uganda from Tanzania, expulsion from and the disappearance of some Kenyan nationals in Uganda, a partial breakdown in trade, and the end to controvertibility of currencies. It was remarkable that the Community continued to exist, but it has, thanks to the good offices of President Kenyatta, the benefits perceived by all members, and the strong commitment to inter-African cooperation held by President Nyerere of Tanzania. Several other states have, in fact, applied for membership in EAC, with Zambia being the most likely candidate.

Other groupings involving cooperation among African states include a series of product-oriented bodies, a number of organizations concerned with combatting disease and pests, and several agencies devoted to the joint development of shared river and lake basins or specific ecological zones (Map 34). These last agencies must await intensive hydrologic, pedalogic, and other studies before the most appropriate installations can be determined. The history of one of them, the Organization for the Development of the Senegal Valley (OMVS), suggests the difficulties they may face and illustrates again

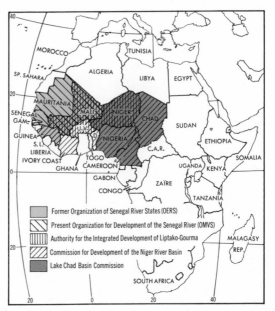

Map 34. Cooperative associations concerned with development of river and lake basins and other shared geographic regions in Africa

the trend to substitute narrower economic for broader political and economic goals. OMVS succeeded in 1972 the Organization of Senegal River States (OERS), which had been plagued by differences among its members; the new organization abandoned the more pervasive goals of its predecessor in favor of concentrating on the more technical aspects of developing the river basin itself, but withdrawal of Guinea, which contains the headwaters of that river, may preclude certain necessary features of an effective water-control program.

Mention must be made of the African Development Bank (BAD) to which 39 of the 42 OAU members belong. The bank, which was to be entirely owned and managed by Africans, was touted at its opening in 1966 as an outstanding example of pan-African economic cooperation, but many nations have not paid up their subscriptions and this has severely limited the loan capacity and made for extremely high administrative expenses. Pressure to accept capital subscriptions from the developed countries therefore arose, and in 1972 the African Development Fund was formed, with lending commencing in mid–1973. Hindsight revealed that the exclusivity of BAD, which was not followed by comparable banks serving other underdeveloped continents, had been disadvantageous to Africa; the Fund, which will make soft loans, has a relationship to the African Development Bank comparable to that of IDA to the IBRD.

Obviously, the success of most inter-African cooperative and integrative efforts has been moderate. This is not too surprising in some respects: customs unions, for example, could not be expected to have any great impact because of the generally competitive nature of African economies and their heavy dependence on extracontinental trade, and it is totally unrealistic to expect that African countries should achieve in a decade or two what the world has failed to accomplish in centuries. The question remains, why, given the strong ideological acceptance and the significant economic incentives of greater cooperation, have the achievements been so meager? Possible explanations include:

(1) Differences in ideology, not with respect to pan-Africanism per se, but with regard to how it should be achieved.

(2) The durability of existing states, which was not inherent, but rather required a considerable effort in the domestic arena and in foreign relations. As the legitimation of individual states strengthened, unity came "to be defined as unity of action and harmonization of policies among independent states, within limits imposed by state sovereignty, state ideology, and separate state leadership." [1]

(3) A series of minor and not so minor disputes between individual states,

[1] William Zartman, *International Relations in the New Africa* (Englewood Cliffs, N.J.: Prentice-Hall, 1966), p. 149.

Map 35. African countries experiencing internal dissension and cross-boundary disputes, 1960–1975

which exacerbated relations from time to time, thus reducing or delaying interest in cooperation. The widespread character of such disputes is illustrated in Map 35, which shows the boundaries across which they have occurred since independence. While many of the occurrences in southern Africa have been related to the widely accepted goal of supporting freedom movements in the white-dominated countries, the cross-boundary disputes that have adversely affected cooperative efforts have been between independent countries. Those mapped are ones that led to some rupture of relations and exclude incidents that caused only verbal expressions of animosity.

(4) Preoccupation with internal problems, which have often been so difficult as to preclude devoting adequate time and talent to external affairs. Evidence of this factor is seen in the countries that have experienced internal, mainly ethnic dissension (Map 35), and in the

number of coups, rebellions, and civil wars that have occurred since independence (Map 36).

(5) Certain considerations pertaining to cooperative and unifying efforts. These include the fears of smaller powers that they would be disadvantaged by joining with larger partners; unrealistic expectations of the benefits to be derived from economic cooperation, including the desire to see markedly more rapid economic growth as a predictable outcome and the notion that each partner must gain equally; and disputes regarding the placement of headquarters staffs or regional industries.

One must not exaggerate the situation. Many governments have been remarkably stable considering the magnitude of the problems they face and the multiple dilemmas defying short-run solution. International relations have been unstable, but external violence has been minimal. Furthermore, as negotiators become more experienced realism regarding economic potentialities will increase and the opportunities presented by cooperation will be better appreciated. Considerably more durable institutions of cooperation and greater progress toward the almost universally approved goal of cooperation may develop as a result.

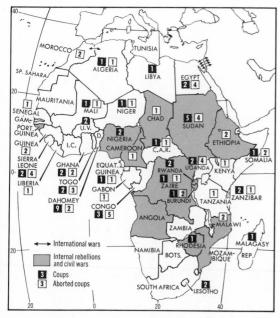

Map 36. *Coups, aborted coups, and international wars experienced by African countries from date of independence to mid-1974; rebellions and civil wars occurring in African countries, 1960–mid-1974*

CONCLUSION

These introductory chapters have been designed to sketch in the background for succeeding discussions of specific regions and countries. They contain a considerable number of generalizations, but it is well to repeat three pervasive points: Africa is a continent of enormous diversity; there is very much that is not known or adequately understood in all branches of African studies; the dynamic character of the African political and economic scene makes prediction very hazardous.

Part Two
NORTHERN AFRICA

Gathering lines at Zelten, Libya
Petroleum is now the leading single export by value of Africa.

CHAPTER FIVE

The Maghreb and Northwest Africa

The northwest segment of Africa has numerous physical and cultural characteristics that distinguish it from other parts of the continent. No criteria for defining the boundaries of the region are entirely satisfactory, but political boundaries permit a more effective use of statistical and other data (Map 37). The major political units are Morocco, which incorporated the French and Spanish protectorates and Tangier, and which became an independent kingdom in 1956; Algeria, which attained independence in 1962 after a 7½-year revolt; and Tunisia, which became a republic in 1957, the year after it had achieved independence. The main boundaries of these three countries, except for those of Algerian Sahara, date from the Turkish conquest over 400 years ago; Morocco, however, had existed as an independent state for a millennium before the French and Spanish protectorates were imposed. Minor political units of the region are Spanish Sahara and the "plazas" of Ceuta and Melilla and adjacent islands, considered as integral parts of Spain. With the important exception of oil and natural gas in the Sahara and in fields offshore from Tunisia and possibly Morocco, economic development is concentrated in the lands north and west of the high Atlas Mountains, in what is often referred to as Mediterranean Africa or the Maghreb.

SIMILARITIES WITH OTHER MEDITERRANEAN LANDS

The Maghreb is largely Mediterranean from the physical standpoint and it has many characteristics of other such lands. First, it faces the sea, which has for several millennia permitted contacts with other Mediterranean lands. Intermingling has also been facilitated by the closeness of Iberia and Sicily, and by movements along the northern coast of Africa from the Middle East.

Second, much of it has a climate with Mediterranean influences, either the true dry subtropical or the Mediterranean steppe. Exceptions include some of the high plateaus and mountains and the Sahara, which has a tropical desert climate (Map 38). The three main characteristics of the Mediterranean climate are moderate and unreliable precipitation concentrated in the winter with consequent droughts in the summer; hot summers and warm winters; and reception of a high percentage of the possible sunshine.

Third, there is a similarity in landforms between this area and the lands north of the Mediterranean, particularly in the existence of youthful mountain ranges. They are, in fact, sections of the western, or Alpide, portion of the Alpine system, which may be traced down the boot of Italy, across Sicily, and into North Africa (Map 39). The northern range bifurcates in northern Morocco. One branch curves around in the Rif Atlas and in the Sierra Nevada of Spain, eventually to reappear in the Balearic Islands; the other curves south in the Middle Atlas. The southern range, known as the Saharan Atlas in Algeria, becomes the High Atlas in Morocco. South of it, in Morocco, is the Anti-Atlas. As in the Mediterranean lands of Europe, the steepness of these mountains limits their utility, though there are surprisingly high population densities within them.

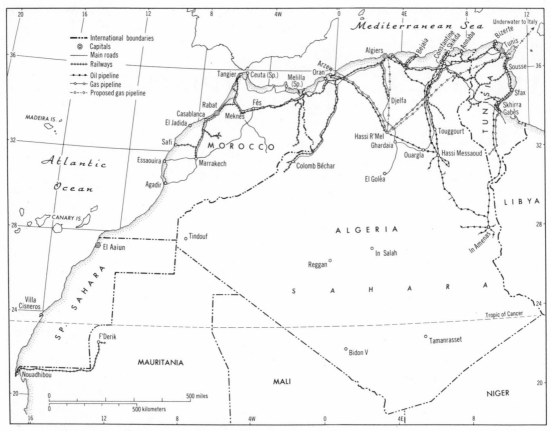

Map 37. Political units, railways, main roads, and pipelines of northwest Africa

Another topographic similarity is the existence of alluvial-filled piedmont and coastal plains, which are the sites of intensive irrigation agriculture. Fourth, the natural vegetation of the area is also Mediterranean.

The physical characteristics of the Maghreb result in a fifth similarity, that of land use. There are three distinct crop associations found in these and other Mediterranean lands, defined in accordance with the supply of moisture to the plants: (1) The rain-grown crops, which depend upon the winter precipitation. These crops, chiefly wheat and barley, cover by far the largest cultivated area. (2) The drought-tolerant crops, perennials which are adapted in one way or another to withstand the summer drought. These include the olive, fig, cork, and certain nut trees—some of the most distinctive Mediterranean plants. (3) The irrigated crops, which depend upon a supply of water beyond that from natural precipitation. They may be perennials (such as citrus and deciduous fruit, or the vine) or annuals (such as vegetables and flowers). While occupying only a small portion of the total cultivated area they account for a large percentage of the value of export crops. Other similarities with northern Mediterranea are seen in the trend toward Westernization, especially in Tunisia and Algeria, and the European character of modern industry and urbanization in such cities as Casablanca, Oran, Algiers, Annaba, and Tunis.

Population pressure is a final similarity. It threatens to be even more serious in North Africa than in other lands around the sea, however. In Algeria, for example, about 50 percent of the working-age men and 80 percent of the women are unemployed or underemployed, only about 30 percent of the working force is employed more than 50 days a year, at least six million people are victims of poverty, and the population is increasing by about 3.5 percent per annum, which, if maintained, would double the total in less than 21 years. About a third of the population in Tunisia and a quarter of that in Morocco may be considered unemployed, and poverty is the lot of large segments of the population. Population pressure in relation to resources and the present state of economic development is at the base of the severe economic problems that beset the area and, given the high rates of increase that characterize the Maghreb, it presents a frightening dilemma for the future. Furthermore, the extreme youth of the population (about 55 percent are under 20 years of age) means that the working age population must support a large number of dependents and that educational requirements are unusually high. Nor is the number of new jobs being created capable of absorbing more than a portion of new entrants in any of the countries.

CONTRASTS WITH OTHER MEDITERRANEAN LANDS

While there are significant similarities between northwest Africa and other Mediterranean countries, the contrasts are also quite striking. First, from the physical standpoint, Morocco was less influenced by sea contacts, partly because of the inhospitable character of its coasts. Climatic contrasts stem from the facts that the area is farther south and hence is both hotter and has less reliable precipitation, and that there are considerable areas of high plateaus which are colder in winter and more steppelike. Furthermore, the Mediterranean climate deteriorates into desert, while north of the Sea it borders more productive, humid climates. Hydrographically, therefore, the area is less favorably endowed. Only Morocco has permanent streams, and the opportunities for irrigation in each of the countries compare unfavorably with their counterparts in Europe.

There are also important human differences; the culture of the indigenous peoples of northwest Africa is in sharp contrast with those of northern Mediterranea. The Maghreb is the western wing of the Muslim world, if the Nile axis is the body and southwest Asia the eastern wing. While one often thinks of the lands east of Suez as the Muslim or Arab world, in fact the countries from Egypt and Sudan to Morocco have a Muslim population over twice that of Asia Minor.

The true "natives" of the Maghreb are the Berbers, whose origin is unclear. While there has been considerable admixture with the Arabs and the Islamic religion has been adopted everywhere, the Berbers remain dominant in some areas, particularly the highland redoubts of Morocco and the Kabylia and Aures Mountain areas of Algeria.

The majority of the area's people are Arabs, descended from groups that came in several waves from the eighth to the thirteenth century. About three-quarters of the population has Arabic as its mother tongue, a quarter Berber. Admixture with blacks has occurred in the oases and elsewhere. The number of Jews, many of whom were probably descended from early Berber converts, has declined greatly from the total of about 500,000 in early postwar years.

The European population of the Maghreb has also been greatly reduced from the years preceding independence, and has also undergone a notable transformation. In 1956 there were about 400,000 Europeans in Morocco, a million in Algeria, and 255,000 in Tunisia. Today the total foreign population is down to about 112,000 in Morocco, to

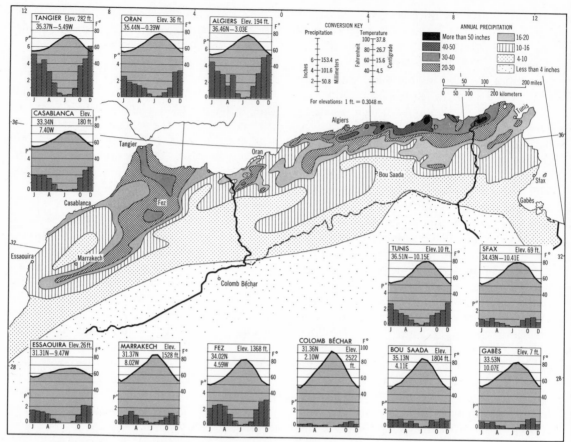

Map 38. Precipitation of the Maghreb and climographs for selected stations

about 186,000 in Algeria, and to perhaps 50,000 in Tunisia. The exodus of Europeans resulted in some gains for the indigenous people; it also entailed severe losses in skill, managerial ability, capital, and market capacity, and resulted in serious economic recession in each of the countries.

Another important sphere of contrasts with northern Mediterranea is in land use patterns. These reflect in part a response to the less favorable climate, but more importantly to the cultural heritage of the peoples of the Maghreb. The most significant contrasts include a greater concern with livestock grazing (especially the nomadic grazing of sheep and goats); an overweening interest in grains among the cultivated crops; the rel-

atively minor position of indigenous peoples in irrigation agriculture and production of high-value export crops prior to the exodus of the Europeans; and the more archaic methods they employ. There is relatively greater dependence on agriculture in all its forms than in European Mediterranea, but a higher proportion of it is at the subsistence level.

THE POSITION OF THE MAGHREB IN AFRICA AND THE WORLD

North Africa, like other parts of the continent, is underdeveloped, but it does compare favorably with other regions by several measures. The three main countries, with

9.9 percent of Africa's total area and 10.3 percent of its total population, accounted for about 14.5 percent of the total GNP of the continent in 1971. Per capita GNPs are well above the median, though still only about $270 to $400 per annum (1972). And the infrastructures of each are distinctly superior to those of tropical African countries.

Algeria ranks among the leading exporters of hydrocarbons and Morocco is the world's largest exporter of phosphates. The three major countries have also had great significance for France—in trade, colonization, and militarily. But history can be perverse, and developments in the Maghreb largely negated earlier expectations of the French. Rather than making a net contribution to France, it became an increasing burden on the French taxpayer; emigration was offset by a counter-migration to France and there are now about 750,000 Algerians, 240,000 Moroccans, and many Tunisians living abroad, predominantly in France; and in the military sphere, Maghrebians with experience in the French forces were among the leaders of nationalist revolts, while French bases in each of the countries have had to be relinquished.

After independence, each of the countries has wished to stress its own national identity, and each has followed quite different approaches in many respects. International relations have been broad-ranging, with foreign aid and technical assistance now coming from a large number of countries from both East and West. Relations with the former metropole have fluctuated wildly, but French influence remains very strong in all three. French continues as a diplomatic and commercial language of great importance; it will remain as the main language in secondary and higher educational institutes at least until a sufficient number of educators have been trained to permit a shift to Arabic; and there are still over 15,000 French teachers serving in the Maghreb. The area is, in fact, subject to conflicting pulls, which

divide its own people in their choice of goals. Some wish to emphasize more of the traditional aspects of Maghrebian culture, others favor strengthened ties to the Arab world, and still others, viewing these strategies as backward-looking, espouse the maintenance of close ties with France, and the use of whatever means are available to modernize the area. Internal problems, differing ideologies, and occasional disputes among the three—including the brief Moroccan-Algerian border war of 1963—have prevented any significant achievements in improving cooperation and integration among the three. In 1973 an announcement was made that Tunisia and Libya would be joined, but it soon appeared that this was no more likely than the previously announced joining of Egypt and Libya.

MOROCCO

Morocco, with an area of 172,413 square miles (446,550 sq km), had a population of 16.6 million in 1974. The estimated rate of population increase is about 3.0 percent, the population is extremely youthful, and there are over 480,000 births each year. Like the other Maghrebian countries it has a relatively high urban population by African standards, about 35.1 percent at the time of the 1971 Census.

Morocco probably has the greatest land use potential of the main countries, but it is the least advanced. This is explained by the greater physical isolation of the area, behind the mountain ramparts of the Rif and High Atlas ranges and with an inhospitable coast, and by the lateness of the date at which modern economic forces began to be introduced. The French were in Algeria 120 years, in Tunisia 80, but in Morocco only 44; all of it was not subjugated until 1934. At the inception of the Spanish and French protectorates in 1912, Morocco was still essentially a medieval, highly traditional Muslim theocracy.

Morocco is the most strongly individual of all the Maghreb states. It is part European, part Arab, part African in character. Fès is Arabic, Casablanca French, and Marrakech almost Sudanese. From the human standpoint Morocco has greater division between Arab and Berber, and between tribe and tribe, fewer educated *évolué,* and a greater retention of medieval culture and methods. While the king receives wide respect as a symbol of state and religious authority, his governments have been subject to strong criticism and an attempt on his life was made in an abortive military coup in 1971; the king has attempted to meet the unrest through several constitutional reforms, changes in government leadership, and repressive measures. Elements of disunity continue strong, however, including ethnic differences, political party rivalry, and conflicting positions of such groups as the labor unions, the modernized and urbanized population, the proletariat occupying the bidonvilles of Casablanca and other cities, the nomads, the mountain dwellers, and the inhabitants of the north who were under a Spanish rather than a French protectorate.

Morocco differs from the other Maghrebian countries in having considerably larger plains, the highest mountains, and in largely facing the Atlantic rather than the Mediterranean. Because the mountains more directly face the prevailing winter winds, there is heavier precipitation; their greater height brings more snow. As a result, the area has a better river system, including several permanent watercourses, and thus greater hydroelectric and irrigation potentialities.

THE REGIONS OF MOROCCO

THE COASTAL PLAIN. A 20- to 30-mile-wide (32–48 km) coastal plain (Maps 41–42) extends from Rabat to Essaouira, itself divided into two parallel belts: the sandy Sahel along the sea, with littoral dunes marking former shores of the ocean, and the "tirs," a black-soiled region suitable for grain production. The agricultural usefulness of the plain is somewhat restricted by the drying influence of the cool Canaries Current, though there are some irrigation communities and production of winter vegetables in favored zones. Behind Rabat and Kenitra are fairly extensive dry woodland areas contributing to the cork production of the country, and

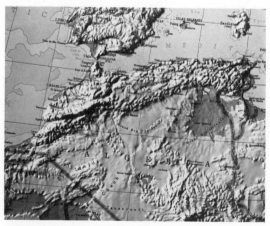

Map 39. Landforms of northwest Africa
Relief Map Copyright Aero Service Corporation

Production of winter vegetables on the coast south of Casablanca
Much of Morocco's coastal zone is made more arid by the effects of the cool Canaries Current, but in favored areas an intensive agriculture is possible.

Rabat has one of the largest cork factories in the world.

While the Canaries Current decreases the utility of the littoral zone, it is advantageous in supporting large schools of fish, which provide the base for one of the important fishing industries of Africa. Agadir accounts for half of total landings; Casablanca has the largest fish canning establishment; Safi and Essaouira are also important. The fishing industry has progressed well in recent years but requires modernization of the fleet and shore establishments; in 1972 the catch was 240,000 tons and 19,345 people were employed.

The coastal belt also contains the leading ports of Morocco. The construction of an artificial port at Casablanca greatly improved access to the country, which had been disadvantaged by the lack of harbors along the Atlantic coast; that city now accounts for about 75 percent of the seaborne trade of the country. It possesses over four miles (6.4 km) of quays with alongside depths of 30 feet (9 m) and the amount of cargo handled

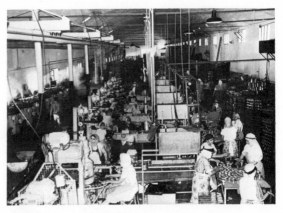

Sardine cannery, Agadir, Morocco
Agadir accounts for about half of the fish landings of Morocco, which has one of the important fishing industries of Africa.

grew from 200,000 tons in 1914 (the year after creation of an artificial port) to 9.5 million tons in 1961 and 13.8 million tons in 1971. Plans call for doubling the capacity of the port and for constructing facilities to accept 100,000-ton ore carriers. Casablanca is favorably situated with regard to the leading agricultural areas, the largest phosphate

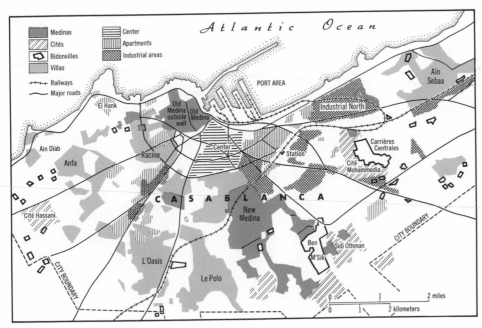

Map 40. Casablanca, Morocco

mine, the capital, and the most densely populated part of the country. The city (Map 40) plays a predominant role not only as a port, but in modern industry (it contains almost two-thirds of the total for the country), commerce and transportation, and services. The most modern of Moroccan cities, it also has the largest population living in bidonvilles and serious problems of unemployment, and it still suffers from recession resulting from the large-scale exodus of Europeans and Jews.

Safi, 120 miles (192 km) to the southwest, is Morocco's second port. It leads in shipments of sardines, but its exports are mainly phosphates from Youssoufia, some of which are processed in a large chemical complex near the port. Rabat has experienced a striking growth in recent decades, largely associated with its role as capital of the kingdom.

The main rail line and one of the major roads of the country follow the coast (Map 37). The main railway runs from Marrakech to Casablanca, Rabat, Fès, Oujda, and then connects with the Algerian line. Branch lines run from Tangier to Fès and from the phosphate deposits to their ports. The lines total 1,100 miles (1,770 km) and carry about 18 million tons per year, about two-thirds of which is phosphate rock.

THE MESETA. Inland from the coastal plain, a 60- to 80-mile-wide (100–130 km) raised platform of primitive rock overlaid with sedimentaries extends to the Atlas Mountains. It is dry and almost treeless except in the rugged northern part with its forests of holly and oak. Two major inland plains have agricultural importance—the Tadla plain on the Oum er Rbia and, in the southernmost part, the foundered plain of Haouz around Marrakech whose waters are derived from the Tensift.

The greater part of the meseta is inhabited by seminomadic pastoralists. Pastoralism, in fact, is *the* economy of the meseta and of a large part of the mountains as well.

Pastoralism in Morocco
The greater part of the Meseta and many of the mountain areas of the country are inhabited by seminomadic pastoralists. Physical conditions are very difficult and practices are poor in all regards.

Practices are poor in all respects: in breeding, in lack of shelter provided, in feeding, and in proper stocking. In areas with such variable precipitation stocking is, at best, extremely difficult. In arid years the grazier may lose or be forced to slaughter 40 percent or more of his flock. Sheep and goats are the most common livestock; camels, asses, and horses are kept as draft animals and beasts of burden.

Tillage agriculture on the meseta is found only on the better watered portions, in areas where orographic influences bring greater precipitation, in the incised valleys where water is available for irrigation, and around Marrakech where underground supplies are tapped. Although the Haouz Plain is in a region of low humidity, with inadequate and variable rainfall and subject to desiccating summer winds, the snows and the nearness of the Atlas provide surface and underground water. A system of open canals or "sequias" distributes the surface water; "khettaras" or subterranean channels, dug and maintained at an enormous expense of labor, control the underground supplies. Large date palm groves are supported in this area, which also produces olives, citrus fruit, and vegetables.

The chief crops of the meseta, and of Morocco, are raingrown barley and wheat, which occupy 85 percent of the total cultivated area, estimated at 12.5 million acres (5 million hectares—ha). About 96 percent of the total acreage in cereals is under dry farming. Over a five-year period, one very good harvest may be expected, three good to poor, and one bad to disastrous. Yields of only half the already low average are common; in the worst years only one-fifth to one-seventh of the average yields may be won. Despite efforts to achieve self sufficiency, Morocco is frequently a net importer of grains.

REGIONS OF NORTHERN MOROCCO. North of the Meseta is the basin of the Sebou River, richer and more populous than other parts of the country except the coastal belt. Stretching inland from the sea is the Rharb Plain, an ancient marine gulf filled with alluvium from the Sebou. This flat and sometimes marshy plain, a considerable part of which was developed by European farmers, has immense fields of wheat and barley and accounts for about 40 percent of citrus production, which has greatly increased in postwar years. The plain has been heavily dam-

Plowing a wheat field in Morocco
Barley and wheat are the chief rain-grown crops, occupying about 85 percent of the total cultivated area. Average yields are low and disastrous harvests are periodic.

aged by occasional severe floods. The central plain of the Sebou, or Saïs Plain, an ancient lake drained by the Sebou, is densely populated. Cereals and fruit trees are grown and there are many large, modern farms around Fès and Meknès. Fès, the religious and intellectual capital of Morocco, and Meknès, an important commercial crossroads point, are the chief cities of this subregion. The plain is enclosed by small massifs on the north and by a plateau attached to the Middle Atlas on the south. To the east, the Taza Corridor, which provided the routeway for early invasions, permits relatively easy access to Algeria.

In what was formerly Spanish Morocco, the Rif Atlas is the dominant physiographic feature. Not pacified until 1927, this region never received the attention accorded more favored areas, and the fiercely independent mountain tribes largely maintain their traditional systems. Some sections are very densely populated, especially considering the difficult landforms, and the region is constantly threatened with famine. Control of the Moulouya River has permitted irrigation of about 160,000 acres (64,000 ha) in a rich subregion.

The northern coast of Morocco is the site of several cities of some importance. Tangier, whose international status had attracted numerous phantom corporations and monetary manipulations, is now aided by its status as a free port and as a major entry point for tourists, and by the placement of several new industries in the area.

THE ATLAS MOUNTAINS. With the exception of the Rif, the mountains of Morocco consist of several roughly parallel Alpine ranges. The Middle Atlas, rising to 11,000 feet (3,350 m), is like the Jura in having high plateaus, limestone soils, subterranean drainage, and great forests. Efforts are being made to increase the output of cedar, establish new eucalyptus plantations, and develop chipboard, pulp, and box factories. The Middle and High Atlas are the reservoir of Morocco; their heavier precipitation feeds the most important streams. The High Atlas extends from the Saharan Atlas of Algeria to sea cliffs north of Agadir; over 9,800 feet high (3000 m) in the center and culminating in Mt. Toubkal (13,661 feet, 4,164 m), it is covered with vegetation on the north, rocky and bare in the south. The Anti-Atlas is formed of ancient rock, part of the Saharan shield, more recently uplifted and subjected to volcanic action.

The Atlas chains are an immense physical barrier, not so difficult to cross, but important in protecting the northwest of Morocco from desiccating Saharan winds. Some valuable timber is shipped from the north, but the mountains, which harbor some remark-

able population densities in the wetter areas, are locally self-sufficient.

SOUTHERN AND EASTERN MOROCCO. South and east of the Atlas ranges is pre-Saharan and Saharan Morocco, largely composed of "hammadas"—dry, bare, rocky plateaus. While the Atlas ranges protect seaward Morocco from the Sahara, they also shut off the inland areas from marine influences; hence most of the region is suitable only for nomadism and extensive stock breeding. Palm groves and intensive gardens are supported in wadi-type oases, which usually suffer from excessive population pressure. The main oases are along the Ziz and Draa valleys; new dams on both will help to even the flow in the oueds, permitting some rational-

Village of Imerghden in the Western Rif, Morocco
Many of the better watered highlands of Morocco have very dense populations. Pressure on the land is indicated by serious soil erosion and periodic threats of famine.

ization of agriculture and flood protection in the irrigation communities. The northeastern segments of the country prolong the regions of Algeria. High plateaus, with a semidesert climate, support some large herds of camels and sheep and yield a minor output of alfa grass.

AGRICULTURAL DEVELOPMENT IN MOROCCO

While agriculture is the occupation of about 60 percent of the population of the country, it accounts for less than 26 percent of the GNP. Farming is marked by the juxtaposition of two sectors: (1) Traditional farming, which has been described as a "world still enslaved by the conditions of a capricious climate," is largely self-subsistent, covers the majority of the farmed lands, is characterized by extreme subdivision of holdings (about 40 percent own less than 1.25 acres [.5 ha] and 54 percent have less than 10 acres [4 ha]), marked rural underemployment (a third of rural families have no land), a variety of tenurial systems, and poor methods. The wooden plow is used; there is little if any fertilization; dry-farming techniques are inadequately applied; and there is very little mechanization. The traditional sector occupies about 8.5 million acres (3.4 million ha). (2) The modern sector, with about 3.5 million acres (1.4 million ha), operates on large holdings, including those previously held by Europeans. This sector provides the vast bulk of commercial production and uses more or less modern techniques. The greatest concentrations are found in the Rharb and the region around Casablanca. The government was much more cautious in taking over European holdings than its neighbors, in part because of their importance in export earnings and as large employers of Moroccan labor, but almost all of them had reverted to Moroccan ownership by 1974. Some are now operated as cooperatives, some have been divided into 60-acre (24-ha) private farms,

and others have gone to wealthier Moroccans; the last disposition has been strongly resented by those favoring a more equitable redistribution.

Moroccan agriculture is, then, beset with serious problems. Despite substantially enlarged production of citrus fruit, fresh vegetables, sugar, and wine, output has not kept pace with the population increase. Nonetheless, Morocco must be credited with giving high priority to agriculture in its development plans. Efforts to improve the situation have included a variety of programs, sometimes quite massive, but results to date have been less than spectacular. "Operation Plow," designed to prepare mechanically a large acreage of smallholdings, was abandoned in 1962 after five years because of its high costs, inadequate provision for maintaining the equipment, difficulties of shifting peasants to a new type of farming, corruption, and pressure resulting from the large concealed and actual unemployment in the countryside. An associated tractor factory was also forced to close.

A second group of programs has attempted to make systematic use of Morocco's underutilized labor force. Most important was the *Promotion nationale* scheme created by royal decree in 1961, designed in part to slow migration to the cities by improving economic and social conditions in the countryside. Workers were paid a low wage plus wheat provided from the United States under PL 480. Its greatest successes have been in the building of schools and roads, though direct aid to agriculture has been made through work on soil erosion control, irrigation, some extension of cultivable lands, and afforestation.

Top priority in government agricultural investment has been given to the construction of dams and the development of irrigation perimeters, with some supplementary gains to hydroelectric capacity. In 1967 a plan was promulgated to bring irrigation to 2.75 million acres (1.1 million ha) in 25 to 30 years. By 1974 some 818,000 acres (327,000 ha) had been equipped, and by 1975 the number of dams in the country had been increased to 24. The Bin el Ouidane dam is by all odds the largest to date; constructed by the French in 1953 on the Oued el Abid tributary of the Rbia, it permits irrigation control of 370,500 acres (148,200 ha) and has an installed capacity of 135,000 kw. The Idriss I dam, 18 miles (30 km) northwest of Fès, is the first installation of thirteen to be constructed in a 27-year plan for the Sebou Basin. Project Sebou will add 200,000 acres (80,000 ha) to the irrigated lands of the Rharb, bringing the total to 625,000 acres (250,000 ha). In addition to the modern irrigation schemes, some 750,000 acres (300,000 ha) in the traditional sector are watered at least a portion of the year.

Irrigation projects are important to Morocco because of its problems with supplying food to a rapidly increasing population, but irrigation causes problems of its own. The use of deep plows on areas formerly in grass or only scratched by primitive plows can lead to serious erosion; there is the necessity to protect against the silting of reservoirs and canals; there are great difficulties in shifting untrained peasants to a more evolved agricultural system; and there are problems in securing adequate capital and developing schemes which justify the expenditures involved.

Other agricultural programs have been designed to increase the use of fertilizers, rationalize dry farming techniques, improve yields of olive trees and grains, enlarge the output of feedstuffs and promote dairy farming, set up agricultural machine centers, control erosion, and stimulate the production of industrial crops. Greatest success in the last has been achieved in the production of sugar from both beets and cane; cotton and oilseeds also receive attention. But progress comes very slowly in the areas where a largely illiterate peasantry is still confined by traditional habits and institutions.

SPANISH AREAS IN NORTHWEST AFRICA

Ceuta and Melilla are Spanish "plazas" and integral parts of Spain; they have been under Spanish rule since 1580 and 1497 respectively. Ceuta, with a mainly Spanish population of about 66,000 on only 7.6 square miles (20 sq km), serves as a transit point for tourists visiting Morocco; its port, with the best natural harbor on the Moroccan coast, has a significant bunkerage trade; aside from these activities and a handful of industries it, like Melilla, is heavily dependent on Spanish aid and the expenditures of the military personnel, which make up about a tenth of the population. Melilla, with about 65,000 people and 4.8 square miles (12 sq km), has an even more precarious economy, though some iron ore from the nationalized mines in the Rif continues to move through its port.

Spanish Sahara is a 102,700-square-mile (266,000-sq-km) chunk of the great desert. The figure of 60,000 for the indigenous, mostly nomadic, population is rejected by the resistance movement, which claims that there are numerous refugees living in adjacent countries. The territory has a tiny farming output, a fair number of animals of little commercial significance, a small export of fish, and the beginnings of a tourist industry. A search for oil in the 1960s was unsuccessful, although some useful water was found during the search. There is a 70-million-ton iron ore deposit in the middle, of no immediate value. Greatest interest attaches to a huge deposit of phosphate rock discovered in 1963 at Bou Craa; measured reserves total 1.7 billion tons, which ranks it second only to Morocco, but the total may prove to be much larger. An investment of about $300 million permitted the shipment of 3.3 million tons per year of 75–80 percent concentrates. Operations began in 1973, and plans call for increasing the level to 10 million tons. The rock is moved by a 60-mile (100-km) conveyor belt to the outport of El Aaiun, where loading facilities have been constructed for 100,000-ton carriers. Morocco is, quite naturally, concerned about the competition with its main export. Spain has also invested large sums in making the capital, El Aaiun, a colonial showplace; its main function appears to be that of a military base.

Morocco contests the continued presence of Spanish holdings in the north and is joined by Mauritania with respect to Spanish Sahara. While Spain has ceded Ifni to Morocco and has expressed willingness to give the population of Spanish Sahara the right of self-determination, it has never considered relinquishing the "plazas," despite their obvious comparability with Gibraltar, which it insists must be returned to Spain despite its having been under Britain since 1713. Solution to the dispute over Spanish Sahara has been hung up in recent years over differences as to how a referendum there would be administered, but Spain stated in August 1974 that it would proceed to hold a referendum in 1975.

ALGERIA

Algeria may be divided for examination into northern and Saharan Algeria. The north, with 95 percent of the total population (Map 41), covers an area of about 116,000 square miles (300,000 sq km) stretching for 620 miles (1000 km) along the Mediterranean and 120 to 210 miles (190 to 340 km) southward to the desert side of the Saharan Atlas, it is a land of mountains, interior plateaus, and limited coastal plains, subject to even greater climatic handicaps than Morocco.

Algeria long occupied the prime position among French overseas areas in investment, number of "colons," and general cultural attachment. Until 1957 France claimed that Algeria was an integral part of that country, but its special administrative regime, inequality of representation, and certain economic restrictions belied this assertion. Dissatisfied with the position of the Muslim majority, several leaders instituted a revolt in November 1954, which continued for over seven years, led to the collapse of the Fourth French Republic in 1958, and finally to independence for the country on July 3, 1962. The war is estimated to have cost France over $10 billion; over 200,000 persons were killed; over a million Muslims were dis-

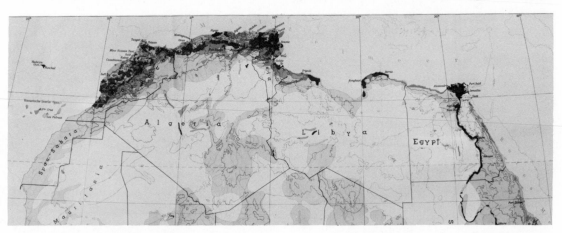

Map 41. Population of northern Africa

placed; the country suffered enormous material damage; and the heritage from the war continues to affect Franco-Algerian relations.

The end of the war created an entirely new situation in Algeria. Rivalry among various political and military leaders resulted in a chaotic period before effective control was achieved. Most of the Europeans fled, abandoning about three million acres (1.2 million ha) including a large part of the more productive farms, hundreds of small businesses and industries representing about 70 percent of such establishments, and a government apparatus bereft of most of its trained and administrative staff. Agricultural and industrial production fell alarmingly; construction declined so steeply that only 15,000 of the 120,000 normally employed were active by the end of 1962; total unemployment reached over 2 million, while the government stated that 4.5 million were without resources and dependent on food shipments from abroad, mainly from the United States.

After several years of continuing political difficulty and unsatisfactory economic order the army seized power in May 1965. Under Colonel Boumedienne it has since followed more pragmatic, less ideologically oriented, programs and has gradually suppressed all manifestations of opposition. Nationalizations of foreign holdings have been speeded and massive efforts have been made to train Algerians to assume the positions now held by expatriates.

Driven by the desire to modernize and diversify the economy as rapidly as possible and realizing that it must depend in the short run on foreign skills and capital, Algeria has amassed a small army of foreign technicians from a dozen countries. Despite repeated difficulties France still provides the largest number of expatriates working in Algeria; many are in education. France remains a major source of aid and investment and Algeria's number one trading partner.

Most French investments have, however, been nationalized, including the two major oil companies operating in the Sahara; trade is progressively being oriented elsewhere; and Algeria is no longer in the franc bloc. Relations with the eastern states, particularly the Soviet Union, have increased but may have reached a peak in 1969; that country has the second largest representation of technicians in Algeria. Although Algeria did not have diplomatic relations with the United States from the 1967 Israeli-Arab war to 1974, American influence is now more important than French or Russian in several sectors. This reflects the Algerians' desire to modernize in the most effective way, and their belief that American technology is the most advanced. Many United States firms are advising on industrial and other projects.

Algeria continues, however, to face enormous problems not amenable to a short-run solution; they cannot be financed with even the greatly increased revenues from the sale of hydrocarbons. As has been suggested, the demographic situation poses extremely difficult dilemmas, reflected in the high rates of unemployment and in the increasing pressure on the restricted agricultural resources of the country.

ALGERIAN AGRICULTURE

Agriculture is expected to support nearly two-thirds of the population, but it accounts for only about 15 percent of the GNP. An estimated 17.9 percent of Algeria's land is used in agriculture, 79.8 percent is nonagricultural, 1.3 percent is in alfa grass, and 1.0 percent in forest. Of the agricultural land (106 million acres, 42.5 million ha) only 14.7 percent is cropland and 43 percent of that is characteristically in fallow, while much is suitable only for dry farming and production of low-value and low-yield grains. Orchards and vineyards occupy about 1.3 percent of the useful land, the remaining 84 percent being in pasture and common grazings. Almost all of the cropland is in the north and over 90 percent is used for cereals, whose production is not adequate to meet domestic needs.

Contrasts between the traditional or private and the modern or socialist sectors of agriculture are striking; about 72 percent of the cultivable lands are in the private sector and 28 percent in the socialist sector. About 83 percent of the agricultural population is dependent on the 587,000 private holdings, which produce only about 40 percent of the gross revenues from agriculture. Theirs is a largely traditional world characterized by features similar to those prevailing in Morocco. The emphasis is on self-subsistence, but three-quarters of the farmers are estimated to be living at or below the subsistence level. Almost half of the rural population is unemployed or underemployed. In 1971 a program entitled *Revolution Agraire* was announced whose slogan is "give the land without peasants to the peasants without land." It is intended primarily to modernize the traditional sector by the introduction of new techniques, the eventual construction of 1,000 villages where markets, services, rural industries, social facilities, and administrative offices will be grouped, and by the redistribution of holdings. Thus it is intended that the rural exodus will be slowed and eventually stopped.

While it is too early to judge the effects of the agrarian revolution, it is difficult to be very sanguine because of the deep conservatism of the peasants, their fears that their lands will be nationalized, the inadequacy of the funds appropriated, and the impossibility of providing the number of viable holdings needed to absorb the nonlandholding residents. About 50 acres (20 ha) of unirrigated land are required to support a family, but 72.1 percent of the private sector holdings are under 25 acres (10 ha).

The 4.75 million acres (1.9 million ha) in the socialist sector largely represent lands formerly held by Europeans, who held many large individual or company estates but who also had over 7,000 farms of under 25 acres (10 ha). Their holdings tended to be concentrated in the more favored areas—in the coastal lowlands and the better-watered northern section of the plateau. Techniques were generally good on the larger holdings and yields of grain were over double those on Algerian holdings, only partly because of superior natural conditions. They also played a major role in the production of exports and of higher-value crops.

Today these lands have been allocated to about 2,200 "self-managed" farms, employing 135,000 permanent and 100,000 seasonal workers who support about a million people. The farms account for about 60 percent of the total value of agricultural products. The average value of production per worker is estimated to be 8.4 times that of the peasant in the traditional sector. The socialist farms have been plagued by numerous problems, including the lack of managers and bookkeepers, lack of acceptance by the workers, and high rates of absenteeism.

The government has also instituted a number of programs designed to improve agriculture, but progress tends to be slow— in part because of the relatively low priority given to agriculture, which receives only about 16 percent of total investment. In-

Turning the soil on a large socialist grain farm in Algeria
Most of the 4.75 million acres (1.9 million ha) in the socialist sector were formerly held by Europeans.

cluded are erosion-control programs, introduction of Mexican-type wheat seeds, conversion of many vineyards to cereals, and increases in the acreage under irrigation.

THE REGIONS OF ALGERIA

COASTAL LOWLANDS. The coastal lowlands of Algeria occur as a series of pockets or valleys between the encheloned prongs of the maritime Atlas (Map 42). From west to east the important lowlands are the plains around Oran and along the Sig River, the Chelif Valley, the plains of Algiers and the Metidja, and those near Bejaia and Annaba. Together, these make up one of the quality areas of Africa. On only 2 to 3 percent of the area of Algeria they contain over half its population and most of the 740,000 irrigated acres (300,000 ha) in the country, and account for a major share in the value of agricultural production and exports. Population densities average about 3,400 per square mile (1,300 per sq km) and there are over 2.5 million urban residents.

Advantages of these separated plains include the presence of excellent alluvial soils, a relatively favorable Mediterranean climate, and the availability of water for irrigation. The major crops of the coastal plains are the vine, citrus fruit, and vegetables; cotton, tobacco, and deciduous fruit are also grown.

To the French, the vine was the "soul of Algeria" and it was the most powerful attraction for many "colons." At the time of independence vineyards occupied almost a million acres (400,000 ha); three-quarters was in thirteen private and company estates but many smallholdings existed at the opposite extreme. The wine industry supported about 400,000 people and accounted for about 51 percent of the total value of exports excluding oil. Algeria ranked as the leading world exporter of wine; large quantities went by tanker to the ports of southern France, partly for mixing with French wines. Restrictions imposed by France, under pressure from domestic producers because of overproduction, have greatly reduced and sometimes closed this main market for Algerian wine. Efforts to secure alternate outlets have not been successful, though the USSR takes about three-fifths of production at low prices. It has been necessary, therefore, to plow up a large acreage. In 1973 the

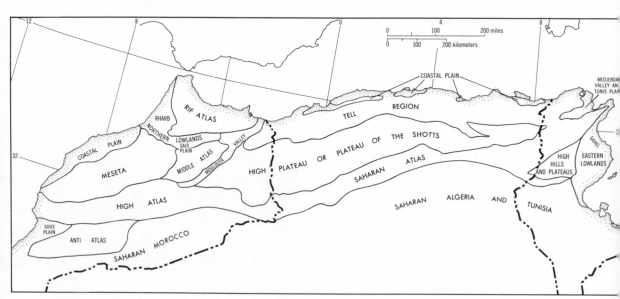

Map 42. Regions of the Maghreb

A small orange grove in Algeria
Citrus fruits are among the high-value irrigated crops of the country and are produced mainly in the coastal lowlands. Output has declined since independence.

acreage in vines had been reduced to about half the earlier figure and production of wine was down correspondingly. Replacement crops include wheat, potatoes, and tomatoes. There has doubtless been a significant loss in jobs because the vine requires over four times as many man days per acre as grains; and indication of the seriousness of the loss may be gained from the estimate that as late as 1967 no less than 53 percent of the work-days in the socialist sector were devoted to the vineyards.

The production of citrus fruit has also declined. The growing of market garden crops is markedly concentrated around Algiers because of the high dependence on speed of transport. Sometimes two or more crops are grown in a year, one on the winter rains, the remaining under irrigation. Irrigation in Algeria involves even more serious problems than in Morocco because so many factors combine to make for erosion: the extreme steepness of the Tell Atlas, the softness of its rocks, the sparse vegetation cover, and the characteristic deluges of the autumn and spring portions of the highly seasonal precipitation. Older reservoirs are already badly silted (the Oued Fodda reservoir lost over 47 percent of its capacity in 30 years) and multipronged prevention campaigns are required for all projects, among which are reforestation, terracing of tilled lands, control of grazing, and, in extreme cases, the movement of peoples from particularly vulnerable areas. Forced migration is difficult to impose and terracing is costly; protection of forests is hindered by indiscriminate cutting of trees (for firewood, tent poles, and tannin extract), destructive grazing, and fires.

The coastal lowlands are the sites of the major cities of Algeria, somewhat unlike Morocco, which has a number of large inland cities. Over two-fifths of the Algerian population is urban. Algiers (2.4 million in 1974) is the principal commercial, industrial, cultural, and political center. It does not handle the high percentage of national trade that Casablanca does for Morocco because of the scattered nature of producing areas along the coast and the greater importance of other outlets for mineral traffic, including hydrocarbons. Second-ranking Oran, with 328,000 people, has a better natural harbor than Algiers; in addition to serving its own productive hinterland it exports iron ore. The coastal cities are connected by a modern and well-equipped rail line; the north-south connections, however, which carry greater tonnages, are not so well developed as they should be. Despite the difficult terrain, the highway network is superior by African standards.

THE COASTAL TELL. This region may be divided into three zones: the littoral band where hills, known as the Sahels, plunge abruptly to the sea; the coastal plains and associated valleys already described; and the Tell Atlas. The Sahels are intensively used where possible for gardening, vineyards, and orchards; elsewhere they are tree-covered or in maquis, the characteristic scrub vegetation of Mediterranean areas.

The Tell Atlas consists of large tabular plateaus with some mountainous massifs often cut by deep valleys; it joins with the Saharan Atlas in Tunisia. Despite the inhospitable character of these mountains they are often densely populated. Kabylia, with about 4 percent of Algeria's area, has an estimated 15 percent of its population. It is in these mountainous areas, where land is being lost more rapidly than it is being reclaimed, that the worst problems of soil erosion exist.

There are notable contrasts between the western and eastern parts of the Tell. The west is split up into smaller regions and is less massive, drier, hotter, and more suited to pastoralism; the high plains are arid, very different from those of Constantine. The east is better watered, better suited for sedentary agriculture, more densely populated. It was also less affected by the European presence.

The main crops of the mountains are barley and wheat, which occupy 91 percent of the cultivated area of the entire country. Grains face many climatic handicaps in the Tell Atlas: cold winters, late frosts occurring even in May, damaging hailstorms, torrential rains, uncertain and widely fluctuating precipitation, and the desiccating sirocco. Also important in the mountains are viticulture in the west and olives and figs in the east. In Kabylia every fragment of usable land is employed. In the lower areas the land-use pattern is comparable to that in the littoral zone; at middle elevations figs and olives become dominant; the highest lands are in cork oak or grasslands. The production of olives and figs, carried on mostly by Berbers, who tend to be better farmers than the Arabs, still suffers from archaic methods. Failure to prune and spray the trees and to replant with younger stock, plus inadequate quality selection and packaging, reduces the market acceptance of these crops.

PLATEAU OF THE SHOTTS. Stretching across the country between the Tell Atlas and the Saharan Atlas is a region of semiarid high

Olives (above) and figs (below), both drought tolerant, are among the cash crops produced in the mountains of Algeria.

plateaus with an elevation of about 3,300 feet (1000 m). About 125 miles (200 km) wide on the west, it breaks up into a series of small valleys on the east and is only the width of a river valley at the Tunisian border. The elevation of the plateau makes for cool winters, while the mountains on the north condemn it to steppe-like conditions and overall low productivity. The average population density of the region is about 15 per square mile (6 per sq km).

Along the north of the plateau are large areas used for grain production. Here and there areas of more intensive use are found, usually associated with a stream bringing water off the back slopes of the Tell or in oases along the northern flanks of the Saharan Atlas. Nomadic and seminomadic herding is the most widespread activity of the plateau and becomes increasingly dominant toward the drier south. A large number of wells have been sunk to improve the domestic and stock water supply of the area. Considerable tonnages of alfa grass are cut and sold for the production of high-grade paper.

The Plateau of the Shotts derives its name from the basin of interior drainage, occupied by ephemeral lakes and salt flats. These shotts have little utility at the present time though it may prove possible to tap water beneath them in the future and even to develop hydropower by utilizing the drop from the higher to the lower plateaus.

THE SAHARAN ATLAS. The greatest significance of this southern range is the protection it gives to Algeria from the drying Saharan winds. It does support extensive grazing and some alfa grass production; its streams supply oases along both flanks and a few that stretch far into the desert. High-quality dates are exported from some of these oases, which have the advantage of relative nearness to coastal points and sometimes very good transport facilities. These factors normally stimulate the tourist trade as well.

Grazing of small stock on the Plateau of the Shotts in Algeria Nomadic and seminomadic herding is the main activity of this dry region between the coastal mountains and the Saharan Atlas.

THE ALGERIAN SAHARA. South of the Saharan Atlas is the enormous region of Algerian Sahara with an area of about 800,000 square miles (2.1 million sq km) and a population of 528,000. Following is a brief survey of the physical, human, and land-use characteristics of the area. They are similar to those in other parts of the Sahara, which covers an area astride the Tropic of Cancer of over 3 million square miles (7.8 million sq km) extending from the Atlantic to the Red Sea. (An account of oil and gas developments is given below, pp. 119–21.)

Physically, the most significant factor is, of course, the insufficiency of rainfall, which is under 4 inches (100 mm) a year over most of the area. Averages are meaningless, however, because extreme variability is characteristic; a period of years may be rainless, but when rain does fall it is likely to be torrential. While these rains may be destructive to clay houses and wadi farming, they are necessary to feed the subterranean reserves of water upon which many oases depend. Rain falling in gentle showers would be quickly lost to evaporation. Other climatic characteristics include a very low relative humidity (usually no more than 4 to 5 percent, very rarely above 20 percent); wide annual temperature ranges (frosts are not uncommon in winter, and night temperatures are consistently low; summer temperatures are very high); great diurnal temperature ranges, running as high as 60 to 70 degrees F (18 to 21 degrees C); and frequent winds of considerable velocity, often with great desiccating power and charged with sand and dust. These often brutal climatic conditions mean that man must be hardy, that he must be prepared in clothing and shelter for quick changes.

Topographically, the Sahara is mostly dry plains and plateaus with occasional mountains such as the Ahaggar, the Aïr and the Tibesti. Characteristic surfaces include sand dunes or ergs—such as the Eastern and Western Ergs of Algeria, each of which is about 20,000 square miles (50,000 sq km) in extent, or the Libyan Erg, which is larger than France—regs, or plains of stones and boulders; hammadas, or tables of denuded rock often with a polished surface; and basins of interior drainage, which may have salt pans within them. Gradational forces operate very differently in desert areas than in humid regions, giving special characteristics to both large and small features. Angularity is common, and may be seen in mesa-like forms and in wadi profiles.

The vegetation of the Sahara is characterized by its ability to withstand long periods without moisture. Among its adaptations are the sparsity of leaves, low habit, enormous root systems, wide spacing between plants, and the ability to go through the life processes with great speed once water is available.

The major impact of the harsh physical conditions of the Sahara is to limit severely the utilizable areas. In the whole Sahara, excluding the Nile Oasis, there are perhaps 500,000 acres (200,000 ha) under irrigation—less than 0.03 percent of the total area.

In the Algerian Sahara about 100,000 acres (40,000 ha) are cultivated—about 0.017 percent of the total. This means that there is less than a fifth of an acre (12.4 decares) of cultivated land per capita. Population pressure is a feature of most oases, some of which have densities exceeding 2,500 per square mile (965 per sq km). This in turn leads to standards of living "about as low as the average rainfall" and a continuing exodus to the coastal cities.

Inhabitants of the Algerian Sahara include Berbers, the oldest historically known white group, who may be either nomads or sedentary oasis dwellers; Arab-Berbers, who inhabit the northern regions; Negroid peoples, who are descendants of the earlier Saharan settlers and of slaves brought as workers to the oases; and Haratin, dark-skinned people who really represented a social class between the slaves and the upper classes. For centuries it was the nomads, mostly Berbers and Arabs, who controlled the oases through their control of the routes connecting them. The highly developed social stratification of desert life was severely disrupted by modern developments; French forces assumed the military function, while modern transportation methods broke down the caravan trade (one truck could replace hundreds of camels). As a result, much unemployment was created among the nomads. Many oases also declined: political changes cut some of them off from former orientations; some, which had only been way stations on caravan routes, found that their purpose for existence had disappeared; meanwhile the commodities oases could produce in surplus became of decreasing importance in a changing world. There were exceptions to these generalizations; some of the more accessible oases were greatly aided by modern transport, both for the export of high-quality dates and in the attraction of a fairly lucrative tourist trade. In the last decades, new employment opportunities have developed in the petroleum industry; the development of this industry has also led to significant improvements in transport routes.

The land-use patterns of the desert are noted by a marked contrast. On the one hand, there is very extensive nomadic grazing of goats, sheep, and camels, particularly on the margins. On the other hand, there is the highly intensive irrigation agriculture of the oases. It is estimated that about two-thirds of the inhabitants are oasis dwellers, one-third nomads. The oases of the Algerian Sahara support an estimated 4 million date palms and export considerable tonnages of high-quality Deglet-Nour dates. A notable variety of fruits, vegetables, and cereals are produced—mainly for local consumption—but the government is trying to stimulate the export of vegetables from some oases. Techniques are extremely backward; the tenurial system is highly complex; concealed unemployment is very high.

Considerable attention has been given to the potentialities for increasing the irrigated areas of the Sahara. Oases may receive their water supply in a variety of ways:

(1) Water is sometimes led from wadis to the gardens.
(2) Catch cultivation may be practiced in wadi floors after flooding. Sometimes small bunds or dikes are built to increase the soak-in. Usually only drought-tolerant crops, such as barley or wheat, can be grown.
(3) Water is often spring fed. Springs occur along faults, at the contact of permeable and impermeable beds, or where an aquifer comes to the surface.
(4) Wells dug to the water-bearing strata support numerous oases. Some are fortunate to have artesian or partly artesian wells, but in most the water must be laboriously drawn to the surface. This has frequently been done by manual power, sometimes by animal power, but until recently seldom by mechanical pumps. It is ironic that in

The oasis of Biskra, Algeria
High-quality dates are exported from a number of the Algerian oases.

areas underlain by millions of tons of petroleum, the high cost of fuel long precluded its use in obtaining water.
(5) Foggaras, or horizontal tunnels, are used to gather water from fans or from aquifers and direct it to surface gardens at lower elevations. Some foggaras are highly intricate and exceptionally well built, but the difficulty of maintaining them is leading to the decline of some oases.
(6) Under the bour system, which is found only where the water table is close to the surface, fields are excavated to sufficient depth to permit date palm roots to tap the table directly, though other crops may require supplementary watering. Here the main problem is keeping the sand out of the fields, which requires more or less constant attention.

The enormous effort expended to provide water and to maintain some oases is scarcely justified by the returns, a factor to remember when one assesses the potentialities for increasing irrigation in the desert. No accurate estimate of the possibilities can be made without very careful study of the underground supply. Numerous examples exist of oases being expanded through well drilling, only to be depleted later through overdrafts on the available supply. Elsewhere, drawing of water for a new oasis has resulted in the loss of water for existing oases, sometimes many miles away. Some concept of the difficulties of replacing water is gained when it is realized that it may require 100,000 years for water feeding the northern Egyptian oases to travel underground from its source in the sudan belt.

In the last 25 years a number of deep boreholes have been sunk in the Algerian Sahara, particularly in the Albienne Nappe, a thick aquifer extending southward from the Atlas Mountains. These deep wells are likely to be very costly and hence cannot always be justified. Furthermore, some water tapped is of fossil origin and once used is gone forever, while other water is excessively mineralized or too deep to permit economic pumping. Nonetheless, enough is known to suggest that the cultivated area of Saharan Algeria could probably be doubled, but this would still be only a tiny fraction of the total surface. Some water has been discovered and developed in connection with the development of the oil and gas resources of the desert. Some day it may prove possible to use desalinized sea water, but the cost will have to be markedly reduced before this can be economic in agriculture.

While the Sahara has undergone several revolutionary changes in the past (the introduction of the dromedary camel in Roman times, the subsequent replacement of cereals and animals by the date palm as the major concern of oases, the decline of caravan routes which was made certain by the development of sea trade, the abolition of slaving, the changing position of the previously dominant nomad, and the introduction of mechanical transport), the most dynamic change of all has come from the discovery (in 1956) and subsequent exploitation of hydrocarbons. Never before has so

much attention been focused upon the Sahara, and no period has seen equal changes in transport or in the socioeconomic position of many desert dwellers.

The Algerian Sahara, for example, now has three major highways to the south and about 2,000 miles (3,200 km) of paved roads. In 1971 the first segment of a modern trans-Saharan road was opened between El Goléa and In Salah. Numerous proposals for rail and road crossings of the desert have been made over the last century; by about 1980 a first-class road, 1,161 miles (1,858 km) of which will be in Algeria, will finally permit ready travel to Gao in Mali and Arlit in Niger. New vehicles have been constructed specifically for desert use. A number of commercial air fields and 135 private fields used by the petroleum companies now exist. Pipelines carry tonnages which each year exceed the total movements across the Sahara in all of history prior to the discovery of oil. Greatly increased attention has also been accorded housing, social work, and education, and to the provision of electricity and additional water to a considerable number of oases.

TUNISIA

Tunisia has much in common with the other countries of the Maghreb: its western Arab culture, its experience of French rule, its Mediterranean and Saharan climates, and its basic land-use patterns. But there are also profound differences. It is much smaller, less favored climatically because of its situation in the rain shadow of the Atlas ranges, more favored topographically because plains cover a very extensive part of the whole, giving a higher degree of penetrability. Culturally, it is more advanced, exhibiting markedly bourgeois and secular tendencies, and it is more solidly unified under its strongly centralized government, which has placed emphasis upon cultural as well as economic modernization. The women of Tunisia are freer from traditional customs and inhibitions. The country has placed greater emphasis on education and less on military forces than its neighbors.

The population of Tunisia was about 5.7 million in 1974; it is very youthful and increasing at a rate of about 2.2 percent. A relatively advanced family planning program, which has received support from traditional leaders, has contributed to a slight drop in the birthrate, a very rare phenomenon in Africa. Over 40 percent of Tunisians are urbanized; indeed the great age and importance of urban life and the significant commercial development of the eastern coastal communities are characteristics that stand in contrast with neighboring Algeria. Tunis and its suburbs have about a fifth of the total population. Underemployment affects over half of the active population; efforts to reduce its impact have centered on work projects (well and canal digging, terracing, reforestation, planting of fruit trees, road construction, and community development) and on encouraging labor migration to Europe (remittances provide about 8 or 9 percent of foreign exchange receipts).

Over 50 percent of the population is dependent for its living on the land, but agriculture provides under 20 percent of the GDP. Production varies greatly depending on climatic conditions, but Tunisia is increasingly dependent on imports of food and no longer has a surplus of grains for export. In bad years yields may be very low and as much as half of the livestock may be lost. For example, the average yearly production of wheat and barley in the ten years ending with 1972 was 625,000 tons, but in 1967 it was only 400,000 tons. Tunisia is affected not only by droughts but also by occasional disastrous floods, most recently in 1969 and 1973. As in the other Maghrebian countries, there is a dichotomy between the traditional and modern sectors.

The traditional sector comprises about 320,000 holdings (99.8 percent of the total) on 11.3 million acres (4.5 million ha), 86 percent of the area devoted to agriculture. 40.9 percent of peasant holdings are under 12.5 acres (5 ha), another 42.2 percent has between 12.5 and 50 acres (5–20 ha), generally too small to provide more than a meager subsistence, if that. The government, strongly dedicated to socializing the economy and to centralized planning, made a major effort starting in 1962 to combine peasant holdings into large cooperative or collective farms. In 1969 the program had brought the economy nearly to bankruptcy and a peasant revolt almost brought down the government. Efforts to organize the numerous merchants into cooperatives were also resented and contributed to the collapse. The peasant farms have since reverted to their former status, and Tunisia accepts that it is not practicable for a small group of government officials to plan, organize, and run the entire economy, a lesson from which it is suggested other countries might profit.

The modern sector comprises 590 cooperatives and state farms, most of which were previously held by Europeans. Some 400 are over 1,250 acres (500 ha), and their total acreage is 1.8 million (0.7 million ha). As in Algeria and Morocco, these farms occupy much of the best land, a substantial portion of the irrigated acreage, and account for a disproportionate share of the output of cash and high-value crops.

THE REGIONS OF TUNISIA

Tunisia may be divided most broadly into a relatively well-watered region north of the Atlas, where Mediterranean influences are dominant, and the semiarid to arid south, where Saharan influences are increasingly felt. A distinction may also be made, however, between maritime Tunisia, a series of plains which make up the most evolved and developed part, and the interior, which is larger, more diverse, and moderately mountainous. Over 70 percent of the population is concentrated in the northern third of the country, and there continues to be heavy migration to that area from the less favored south and the interior.

THE NORTH. The northwest or interior Tell prolongs the Tell Atlas and high plains of Algeria. It is largely forested, but subsistence crops are produced in the valleys; goats and sheep are grazed wherever possible, often to the detriment of the vegetation. Opportunities in this region include intensification and protection of forest output, planting of more deciduous fruit trees, installation of drainage works in the valleys, and development of summer crops in the valleys, including tobacco, corn, sorghum, and forage crops.

The northeastern Tell is the heart of Tunisia. It contains the only plains of the country enjoying a Mediterranean climate— the plains of Mateur and Bizerte, the lower valley and plain of the Medjerda, and the plains around Tunis and Cap Bon. Many of these lands are rich in underground water and possess fertile alluvial soils, though some are sufficiently heavy to require drainage. The region produces practically all of the wine, 95 percent of the citrus fruit, the majority of market-garden crops, and over a fifth of the olives of the country.

Aerial view of Tunis
The city has a notable degree of primacy as compared to the capital cities of other Maghrebian countries.

The Medjerda Valley is the focus of a major project extending over some 730,000 acres (296,000 ha). A series of storage and flood-control dams permits the irrigation of about 173,000 acres (70,000 ha). The scheme also involves the eventual reclamation of about 165,000 acres (67,000 ha) on the lower plains and delta, soil conservation measures in the headwater areas, and extensive flood-control works. Irrigation permits a transition from the former emphasis on extensive grain farming to a system of mixed farming based on vegetable, fruit, cotton, and fodder crops, and on dairying and poultry farming. Crop programs are determined by the authority not only on the state farms but also on the 12.5 to 25 acre (5 to 10 ha) individual farms, which are required to join cooperatives. Regulations prevent tenancy, subdivision, and absenteeism (common evils of traditional Tunisian farming); fines are imposed if standards are not maintained; and there are even restrictions against travel from the communities. The project is now about one-third completed. Significant gains have been made in increasing and diversifying production, in providing new farms and opportunities for employment of a large labor force, and in improving incomes of the residents. But the cost has been high and the project continues to depend on aid from a half dozen countries as well as on experts from abroad.

About four-fifths of the urban population of Tunisia is in the north, as are most of the modern and craft industries. Tunis and suburbs alone account for 54 percent of the value added in industry and has almost all of the head offices of enterprises operating in the country. Much of the country's individuality is, in fact, a product of Tunis City. The port of Tunis is well equipped but has inadequate depths alongside and in the canal connecting it to the sea; a major project has, however, developed the outport of La Goulette, which now handles about 70 percent of the total traffic for Tunis-Goulette; these cities together account for about 30 percent of the country's nonpetroleum overseas shipments.

SOUTHERN TUNISIA. This zone, which covers by far the largest part of Tunisia, may be divided into several subregions. The eastern plain stretches southward along the coast and is steppe-like in character. The Sahel area around Mahdia and Sousse has a quasi-monoculture of olive trees, as does the drier region around Sfax to the south. Tunis ranks second in the world in olive production, and olive oil is first by value among agricultural exports. Yields and exports fluctuate markedly, however, often in a three-year cycle in which there is one good crop followed by a bad one and a third with average or below-average yields. Olive groves have been extended by 40 percent since independence, but much of the new areas are in drier regions where high yields are less likely. Many of the olive groves in the south were introduced under the *m'gharça* system, whereby a Tunisian peasant would acquire half of an area in return for tending the trees for fifteen years until they ripened. Trees in the south are normally spaced about 75 feet (23 m) apart as compared with a 30-foot (9 m) spacing in the north, but yields per acre are superior because many trees in the north are older and less well tended.

There are some opportunities for small-scale irrigation of vegetables, forage crops, pomegranates, and dates in the maritime south. One relatively large-scale development based on the Oued Nebhana was undertaken without adequate study of stream flow or of the area to be served; its excessive costs in relation to the benefits reaped is a monument to national fervor, which does not always provide the wisest counsel.

West of the plains are steppe-like plateaus and hill lands, the domain of extensive livestock raising and dry farming. Stock breeding accounts for about a fifth of agricultural income in Tunisia and occupies about two-fifths of the utilizable lands. Alfa grass is also cut, some of it being pulped in a factory at Kasserine.

South of Gafsa only sporadic grazing is possible, mostly of camels. There are also a few oases around the immense Shott el Jerid and near Gabès. The Gulf of Gabès is the most important fisheries area for Tunisia; operations are usually small scale and concentrated on inshore waters. The Gabès region is the focus of a development plan whose features are modernization of the port, particularly for phosphate shipments, construction of a rail line from the phosphate mines near Gafsa, placement of cement and phosphate plants, and renewal of palm groves in the hinterland.

MINERAL PRODUCTION IN THE MAGHREB

North Africa has a significant production of petroleum, natural gas, phosphates, such metals as iron ore, manganese, lead, and zinc, and a minor production of various other minerals. In Morocco there are about 50 mines, many of which are distinctly marginal and about a quarter of which account for the vast bulk of production. Mining accounts for about 5 to 6 percent of the GDP; it has tended to stagnate, partly because of unfavorable world prices rather than reduced output. Employment has declined from about 40,000 in 1962 to 26,000 in 1972. The state owns at least 50 percent of all concessions and also all of the phosphate mines. Royalties, earnings, and taxes from mining account for about 10 percent of the ordinary government revenues; mineral carriage is 70 percent of rail tonnage, and 85 to 90 percent of port traffic at Casablanca and Safi; the industry consumes 20 percent of electric output and 40 percent of the fuel oil used; and mineral exports account for about 30 percent of the total value of exports.

The mineral position of Algeria is marked

by a spectacular wealth of hydrocarbons in the Sahara, but relatively meager resources in the north. Production of nonhydrocarbon minerals declined after independence, but several mines that were closed at that time have since been reactivated and several new mines have been opened. About 12,400 people are employed in the 22 mines in operation.

In Tunisia about 14,000 people are employed in mining; mineral traffic accounts for over 70 percent of total railway freight and 80 percent of port tonnages, exclusive of oil moving through La Skhirra; and mineral exports make up about 45 percent of the total value of exports.

Mining of phosphates is the only important economic activity in Spanish Sahara.

PHOSPHATES

Northwest Africa ranks as the world's leading exporter of phosphates, with shipments going primarily to Western Europe. The development in Spanish Sahara has already been noted. Morocco, with good-quality rock, is the main producer in the Maghreb, its output having grown from 1.3 million tons in 1937 to 8.3 million tons in 1962 and 19.5 million tons in 1973. Phosphate rock and concentrates have accounted for about a quarter of Moroccan exports by value, but a threefold price jump in 1974 will increase this share. Several deposits exist, but mining is centered in two areas: (1) Khourigba, with 74.5-percent tricalcium phosphate found in four layers of which the top two are being mined. This area accounts for three-fourths of the output, which is sent by rail 87 miles (140 km) to Casablanca. (2) Youssoufia, which produces a 70 percent product after screening and drying. Its output is sent 51 miles (82 km) to Safi for processing and export. A major chemical complex has been built at Safi, producing triple superphosphate, sulfuric acid, and phosphoric acid.

Algerian phosphates are of poor quality, but in order to sustain exports a $27 million investment was made to develop an enor-

mous 57-percent body at Djebel Onk, 205 miles (330 km) from Annaba, while the rock is treated at a new 550,000-ton fertilizer plant.

Phosphate rock covers about 380 square miles (980 sq km) in central Tunisia. The relatively low grade phosphates are mined mainly in the Gafsa area, but also in the north at Kalaa-Djerba. In the last half decade a heavy investment has been made to modernize the mining operations, process the rock to a higher content, and produce superphosphate fertilizers at large new plants situated at Sfax and Gabès. Output has risen from 2.3 million tons in 1967 to 3.9 million tons in 1972, and a further increase to 8 million tons is planned by 1980.

IRON ORE

Most of the iron ore of northwest Africa is hematite and limonite of good purity and with from 51 to 63 percent metal content. The bulk of the output is shipped to western Europe, but some is consumed in the steelworks at Annaba in Algeria. Production levels are about 3.3 million tons in Algeria, 1 million tons in Morocco (following the opening of a pelletizing mill at Nador in 1972) and 448,000 tons in Tunisia. Deposits are widely scattered, but production comes mainly from the Nador region in Morocco, the Ouenza deposit in Algeria near the Tunisian border (part of whose output is pelletized in a new plant), and at Djerissa and Douaria in Tunisia. A 3-billion-ton, 55-percent deposit exists at Gara Djebilet, 84 miles (135 km) south of Tindouf in the Sahara. Its remoteness and the past disputes between Algeria and Morocco militated against its exploitation, but there is now talk of a joint operation involving the mining of 700 million tons over a 60-year period; the ore is to be sent by rail about 250 miles (400 km) to Tarfaya in Morocco for shipment.

LESSER MINERALS

Morocco produces a considerable variety of other minerals of which manganese from

Imini and lead from numerous small workings are of some importance on a world scale; others are zinc, copper, cobalt, antimony, pyrites, baryte, fluorine, and salt. Anthracite coal of good quality but costly to mine because of the thinness of seams comes from Djerada; output ranged from 400,000 to 500,000 tons in the 1960s and was increased to 564,000 tons in 1973. About three-fifths of production is consumed at an on-site power station.

Algeria has a number of small lead, zinc, and copper mines, and produces minor amounts of baryte, kaolin, pyrites, mercury, and salt. A systematic survey is in progress over the whole country and several indications of potential interest have been reported both in the north and in the Hoggar. Low-grade coal was formerly mined at Kenadza in the Colomb-Béchar area, but the mine was totally uneconomic and was closed after development of the country's hydrocarbons.

In Tunisia, lead has been mined since Punic times; despite the inferior quality of the remaining ores output has been raised in recent years, as has that of zinc. Other minerals produced include mercury and salt, the last mainly from pans at Tunis and Monastir.

HYDROCARBONS

The enormous deposits of petroleum and natural gas in the Sahara have provided a most dynamic element in the economy of the Maghreb, particularly for Algeria.

ALGERIA. Serious efforts to find oil did not begin in Algeria until 1946, and were rewarded with a first strike in 1949, in the north. In the Sahara itself the first concessions date from 1952; four years later two important discoveries were made, at Edjeleh and Hassi Messaoud. These were followed by discovery after discovery—including enormous reserves of natural gas at Hassi R'Mel and Hassi el Gassi—and it soon became apparent that a major area of world reserves had been found.

Flaring gas at petroleum installation, Hassi Messaoud, Algeria
This center is situated on one of the major fields in the Algerian Sahara.

Geologists predicted many years ago that, because of the enormous size of the geological structures, if oil were found in the Sahara, it would be found in large quantities. The comparative richness of the area may be seen from the high average yield per well and the remarkably short period required before high production levels were achieved. Production of oil rose from about 450,000 tons in 1958 to 8.6 million tons in 1960 and 20.5 million tons in 1962. Output rose to 48.2 million tons in 1970 and has been about 50 million tons in recent years. Despite the great wealth in oil, production at a high rate probably cannot continue much beyond the year 2000; natural gas (see below) will last somewhat longer.

Numerous problems accompanied development. The Saharan environment is a difficult one: crews can work at full tempo only eight months of the year, they must be changed frequently, and heat and sand expose machinery to tremendous wear and tear. These factors result in high costs of drilling, which have, however, been more than compensated for by the results. The securing of an adequate water supply, recruitment of labor, and difficult transport to and from coastal points are other problems common to all sites.

The hydrocarbon domain in Algeria now

covers about 353,000 square miles (915,000 sq km) and exploration continues apace in scattered areas. There are 40 known oil fields of which five are particularly important (Map 37). Proved reserves were put at 1.7 billion tons in 1972. Four gas fields, of which Hassi R'Mel is most important, contain measured reserves of 4 trillion cubic meters, about an eighth of the world total. Large-capacity pipelines connect with Arzew, Bejaia, and Skikda, which are equipped to take giant tankers and large liquefied natural gas (LNG) ships, while one pipeline has its terminus at La Skhirra in Tunisia and smaller pipes serve the cities in northern Algeria.

The first developments in Algeria were made by the two French national oil companies, but capital requirements made it necessary to assign concessions to many other companies, both majors and minors. The Evian agreement, which ended the Franco-Algerian war, and the associated Saharan Petroleum Code were supposed to retain French and other foreign presence in the hydrocarbon industry; but succeeding years saw new arrangements increasing the profit ratio and the posted price on which profits were based, and then nationalization of the industry. Algeria, feeling that its national resources should belong to the state (particularly since oil and gas are basic to its economic development), and wishing to secure as large returns as possible from reserves that cannot last many decades, adopted what has been considered a radical policy. This has not, however, prevented the signing of numerous large contracts for supply of oil and LNG; indeed Algeria has acquired a reputation for very businesslike arrangements.

Algeria set up its national oil company, SONATRACH, in 1963 and began a massive program to train its citizens eventually to handle all of the technical and managerial aspects of the industry. Nationalization proceeded slowly at first, but picked up in 1967 when Esso Standard and Mobil were nationalized as a result of the Israeli-Arab war, while 14 other companies engaged in the distribution of oil and gas were taken over in 1968. The most bitter disputes concerned the nationalization of the French companies, which began in 1970 and was finally completed at the end of 1971. SONATRACH now holds sole rights on 91 percent of the hydrocarbon domain and rights in combination with foreign companies on the remainder. It produced 75 percent of the 1972 oil output and 100 percent of that year's gas output on its own, and participated with others in the remaining output of oil. SONATRACH's personnel increased from 3,000 in 1967 to 15,000 in 1972, 99 percent of whom were Algerians, but the company still faces difficulties in managing its enormous operations. Foreign companies remaining in Algeria normally hold a 49 percent interest, and Algeria expects to attract other partners on the same basis.

Investment in the industry has been enormous, and very large sums are required for new developments in exploration, transport, and processing. Hydrocarbons now receive over a third of Algerian industrial investment. Revenues to the country have greatly increased, moving from $80 million in 1966 to $1.1 billion in 1972; they may be expected to exceed $5 billion in the near future and surpass $10 billion with planned shipments of LNG and piped gas. Sales of oil have been increasingly oriented away from France, formerly the major market. West Germany has become particularly important among European countries, while large purchase contracts have been completed since late 1971 for delivery of crude to the United States.

Production of gas started in 1961 and rose slowly to only 15 billion cubic meters in 1972. But several large contracts with deliveries scheduled to begin after 1976 will greatly increase production and the consequent returns to Algeria. Shipment of gas requires construction of pipelines, liquefaction plants, LNG tankers, and facilities at landing points; hence gas shipment takes longer to arrange than export of crude oil. Contracts have thus far been signed with several West European countries and with two American companies for deliveries starting from 1975 to 1977 and totaling 77 billion cubic meters a year; the total investments, not all of which pertain to Algeria, probably exceed $10 billion. In addition, gas is to be delivered beginning in 1979 by a 1,550-mile (2,500-km) pipeline via Sicily to LaSpezia in Italy, and a second pipeline may go from Arzew across northern Morocco and under the Straits of Gibraltar to Spain.

While the hydrocarbon industry does relatively little directly to meet Algeria's employment problems, it is the major and a rapidly increasing source of expendable funds for the government and accounted for about three-fourths of total exports in 1973, before the notable price increases. It is also responsible for a considerable number of forward-linkage industries, including four refineries with a total capacity of 13 million tons; liquefaction plants at Arzew and Skikda with

Petrochemical complex at Arzew, Algeria
Algeria has three major petrochemical centers, each of which has large gas liquefaction plants.

others planned; and three petrochemical centers producing fertilizers, methanol, ethylene, plastics, nitric acid, ammonia, and urea. Some backward linkages have also been developed, including the production of pipe in the steelworks at Annaba.

While Algeria was the first nation in Africa to become a major producer of oil, it has since fallen to third place behind Libya and Nigeria, partly because of the problems associated with the nationalization program. The great value of its hydrocarbon resources, the country's desire to realize returns rapidly, and the non-Communist world's expected large-scale increase in consumption of oil in the 1970s all suggest notable growth in the immediate decades ahead. Whether the revenues will be adequate to modernize Algeria and create an economy that will be viable after the resources are exhausted, only time will tell.

TUNISIA. The hydrocarbon resources of Tunisia are considerably below those of Algeria, but reasonably significant finds were made in 1964 in the El Borma field, a field in the extreme south that straddles the border with Algeria, at Douleb in the west central part of the country, and around Sfax. Numerous foreign companies are exploring throughout the country and offshore, particularly in the Gulf of Gabès where some promising finds have been made.

Production reached 3.9 million tons in 1973 and is expected to increase to at least 9 million by 1976. Oil is brought from the south and from the Sfax field to La Skhirra. A recently constructed gas pipeline from El Borma supplies a new electric station at Gabès, while a small but declining gas deposit at Cap Bon has been used at Tunis since 1945. Petroleum now ranks as the main export of Tunisia; it accounted for about 33 percent of the total in 1974. That country has not followed Algeria in the nationalization of hydrocarbon resources, preferring a policy of association with foreign companies. It has, of course, benefited from price increases achieved by OPEC. Its refinery at Bizerte meets domestic needs, but plans call for increasing capacity either there or in the south to permit some export of products.

MOROCCO. In 1974 this country had only a tiny and declining production of oil, and a small output of gas from three fields. Exploration, however, was very active both on and off shore, with some favorable indications reported. Morocco's two refineries at Sidi Kacem and Mohammedia have recently been expanded; much of the oil consumed comes from the USSR.

INDUSTRIAL DEVELOPMENTS

Industry is in an early stage of development in the Maghreb, although it is ahead of that in most African countries. Unemployment, pressure on the land, sizable urban populations, and the desire to modernize their economies create strong incentives to expand and diversify industry, but many barriers impede a rapid buildup. Modern manufacturing accounts for about a sixth or seventh of the GDPs of each country, and employs about 125,000 in Morocco, 210,000 in Algeria, and possibly 100,000 in Tunisia.

Manufacturing was severely depressed following the large-scale exodus of Europeans, particularly in Algeria, where self-management committees were no substitutes for the former technicians and managers. While growth has since improved, in no case has industry actually fulfilled the goals of national planning. Industrialization policies and priorities differ from country to country. Morocco has followed a more ad hoc approach with less governmental participation; while progress has been relatively poor, the country does possess a fairly sizable and diversified industrial establishment.

Algeria has assigned high priority to industrialization, devoting 45 percent of plan investments to manufacturing and hydrocarbon development. Its goal has been rapid expansion on a centrally controlled socialist pattern; about 80 percent of manufacturing is owned by the government, though there are numerous mixed state/private enterprises, a few sectors such as textiles remain largely private, and many foreign firms are deeply involved in providing new plants, often on a turnkey basis. Particular emphases have been on technically advanced plants and on large basic industries. The theory is that these will make Algerian industry competitive and be the foundation of a more integrated manufacturing complex. Disadvantages of the policy are the high capital investment required, the relatively small number of jobs created, and the problems of finding export markets for some of the items produced.

Tunisia first opted for strong participation and control by the government, which owns the majority of plants, but encouragement is now extended to private investors, both Tunisian and foreign, particularly in light industry. Early placements tended to be too heavily oriented toward import substitution, with the result that their output is of high cost. More recently an effort has been made to promote export-oriented industries, for which special inducements were instituted in 1972. Both Tunisia and Morocco hope that their association with the EEC will stimulate the growth of low-wage-labor oriented industries catering to that market.

TYPES OF INDUSTRY

TRADITIONAL HANDICRAFTS. Industry may first be divided into the long-standing artisanal workshops and modern manufacturing. Despite competition from mass-produced items, the handicraft industries retain considerable importance, especially in Morocco and Tunisia. Both artistic and utility goods are produced. The latter have suffered most severely from rising competition; the former have actually increased in value of exports, including sales to tourists. Much of the work is carried on in the *souks* (markets) or in the homes of the craftsmen. Some fine leather goods, rugs, woolens, metal and wood articles, and pottery are produced, but much of the hand-crafted output is inferior in quality.

In both Morocco and Tunisia the number engaged in traditional crafts is larger than the number employed in modern manufacturing. But there is considerable underemployment in several guilds. Governments have attempted to aid this segment of industry by setting up special craft centers and instituting quality standards for products such as rugs.

MODERN MANUFACTURING. The largest segment of modern industry is still concerned with the processing of extractive output. Included are the already noted plants associated with phosphate and hydrocarbon production; metal concentrators and smelters; fish canneries; cork factories; pulp mills based on alfa grass and eucalyptus. And in the agricultural field there are flour, sugar, and oil mills; wineries; vegetable and fruit canning establishments; modern tobacco factories; and tanneries. A large number of raw material processing plants produce mainly for the export market.

The construction industry enjoyed a real boom in the postwar years that led to independence, suffered serious depression in the immediate years thereafter (85 percent of Algerian construction workers were unemployed in 1963), but has since improved to the point that new cement capacity has had to be built in all three countries in the 1970s.

Production of consumer goods in modern plants has grown modestly in the last decade. Industries in this sector include a variety of food and beverage plants; several spinning mills in each country and many plants producing clothing and other textile products; factories making containers of glass, plastic, wood, and metal; chemicals such as soap, insecticides, pharmaceuticals, paints and varnishes, matches, and the petrochemicals

noted earlier; miscellaneous industries producing furniture, glass, shoes, toys, office goods, and jewelry; and a fair representation in the field of metal fabrication. Most important in the last category are the vehicle assembly plants. Algeria hopes to produce automobiles the bulk of whose components will come from domestic supplies, but this goal may require some years before realization. Other metal products include household articles, hardware, furniture, wire, agricultural machinery, and radio and TV assemblies.

Like numerous other underdeveloped countries those of the Maghreb have wished to establish iron and steel mills as evidence of their maturity and as a basis for metal fabricating industries. Morocco has long considered a 240,000-ton steel mill at Nador, where a rolling mill provides about a quarter of the country's needs for steel bars and wire. Present thinking is that the mill would use pelletized ore from the Rif iron mine and natural gas from Algeria. Tunisia completed a small mill at Menzel-Bourguiba outside Bizerte using Djerissa ore, Algerian natural gas, and imported coke. It is too small to produce efficiently, which is likely to be the case for the Nador mill in Morocco. In late 1973 Tunisia signed letters of intent to construct a $55 million steel plant at Gabès, using a West German direct-reduction process to produce a million tons of sponge iron for export to Europe.

The largest industrial complex in Algeria is the El Hadjer iron and steel plant near Annaba, which uses ore from 94 miles (150 km) south, gas, and imported coke. The decision to build the plant was made before independence; progress was slowed for years thereafter, but an investment of over $300 million, financed by a large Soviet loan and supplier credits from four West European countries, has permitted installation of 450,000 tons of pig and 410,000 tons of steel capacity, hot and cold rolling mills, and a tube mill to supply the hydrocarbon sector. It is planned that capacities will be increased

Part of the iron and steel mill at El Hadjer, Algeria
With a capacity of 410,000 tons of steel, this is the largest industrial complex in northwest Africa.

to 1.2 million tons of pig and 2 million tons of steel by about 1978; these levels will require that export markets be secured for a portion of the output. It is not yet clear whether the complex will be able to produce at competitive prices.

The situation with respect to supply of energy to industry and for production of electricity has been greatly altered with the availability of low-cost natural gas in Algeria and Tunisia, but thus far only a few manufacturing plants specifically attracted by this resource have been installed. Hydropower provides for about 72 percent of electricity produced in Morocco, 17 percent of that in Algeria, and 6 percent in Tunisia. Morocco's linked plants on the Oued el Abid, Bin el Ouidane and Afourer, account for the largest part of that country's hydroelectric production. Indicative of the depressed condi-

tions in Algeria after independence was the drop in electric output from 1,387 million kwh in 1961 to 1,119 million kwh in 1966, but production has since increased to over 2,000 million kwh in 1973.

While considerable progress has been made in establishing modern industries in the Maghreb, several handicaps retard rapid expansion. Most important is the restricted size of the national markets not only in numbers but also in purchasing power. The range of industries that might be appropriate to the area could be increased by formation of a customs union, but at present each country is intent upon its own plans and there has been no joint planning or coordination. A second major problem is the shortage of skilled labor and managerial ability, but training programs and generally improved educational capacities are gradually meeting this problem.

LOCATION OF INDUSTRY

Modern manufacturing is heavily concentrated in a limited number of localities in Morocco and Tunisia and in a somewhat larger number in Algeria. About two-thirds of industrial employees in Morocco are in the coastal strip from Casablanca to Rabat and Tunis has about the same share of Tunisian industry. While Algeria had marked concentrations at Algiers and Oran, developments at Arzew, Bejaia, Skikda, Annaba, and elsewhere have spread investments more broadly. Some effort is being made in all countries to disperse new establishments, including financial incentives in Morocco to locate outside the Casablanca-Rabat corridor, the placement of large new complexes at Gabès and Sfax in Tunisia, and the scattering of new textile plants in Algeria among lesser centers.

TOURISM

One of the most rapidly advancing sectors of the economies of Morocco and Tunisia is the tourist industry, and it has been suggested for Tunisia that the three-star hotel and an oil derrick should replace olive oil and wine as the symbols of the country. Algeria has not made such strong efforts to stimulate tourism but its number of visitors is nonetheless increasing, and it has recently decided to devote more attention to the industry. The Maghreb has numerous significant attractions: the beauty and diversity of its landscapes, excellent beaches, a mild climate (though winters are not always so sunny as the advertisements would suggest), exotic oases in the Sahara, remarkably well preserved medieval Muslim cities, and historic ruins dating over several millennia.

The two major benefits to the economies are the employment created (roughly in the ratio of two workers per hotel room) and in the earning of foreign exchange. The number of tourists visiting Morocco has risen from 255,000 in 1962 to over a million in 1972 (the highest for any African country) and is expected to reach two million by about 1977. In Tunis the number of hotel beds increased from 4,000 in 1961 to 54,000 in 1972; it is planned that the total will reach 125,000 by 1980. Some 620,000 tourists visited Tunisia in 1971 and a rate of 1.5 million is expected by 1980; 1972 earnings from

Hotel at Timimoun in the Algerian Sahara
Morocco and Tunisia lead Africa in number of tourist visits. Algeria has recently decided to give higher priority to the industry.

tourism accounted for an estimated 18.3 percent of foreign exchange receipts, second only to the sale of petroleum.

There are, of course, problems associated with development of the industry, but the record of its growth has been so favorable as to suggest that continuing efforts should be made to attract tourists to these fascinating lands.

CHAPTER SIX

Libya

When Libya became independent by UN decree in 1951, it was incapable of supporting itself, had one of the lowest income levels in the world, a considerable annual budgetary deficit, a serious imbalance in trade, and no visible resources that could be developed readily to alter the situation. Over 90 percent of the population was illiterate, there were almost no indigenous professional or technical workers, and less than a score of Libyans were college graduates. As a former Italian colony, Libya had become the responsibility of the United Nations, where the only agreement among the majority was that the USSR should not become one of the trustees and this meant that independence was the only solution. One can only speculate what the disposition might have been if there had been knowledge of the wealth of oil resources beneath the miserably poor surface.

For centuries before the Italians began moving into Libya, in 1912, it had been ruled by the Turks, who were satisfied to hold the coastal cities and collect such taxes as were forthcoming. While the Italians improved the infrastructure in the coastal belt, little attention was given to improving the lot of the Libyans. Rather, the Italians promoted migration from their own overpopulated country to another overpopulated country, actually taking over a considerable acreage of land from Libyan peasants and graziers; by 1939 about 100,000 Italians were resident in Libya, a majority in the cities. During World War II, the Germans, Italians, and British marched back and forth across the country. Much destruction occurred and, since Libya was the territory of a defeated belligerent, damages could not be

collected after the war and reconstruction came very slowly despite the increasingly generous contributions of Western nations. After the war the country remained under British and French military rule until a UN commission was set up to organize its independence.

Libya had one valuable resource, which directly and indirectly supported its regular and development expenses in this period: a location that had strategic value for military and air bases. In the decade or so before oil revenues began to pour in, the country survived by subsidies received in return for the rights to these bases. Libya first became a kingdom with a federal and three provincial governments; in 1963 it was made a unitary state in an effort to reduce the centrifugal forces caused by differences between the more conservative Bedouins of Cyrenaica and the Berbers of Tripolitania. The separation of the two main provinces by nearly 500 miles of desert contributed to the lack of cohesion, while the remoteness of most of the oases meant that Fezzan and southern Cyrenaica played no significant role in the kingdom. While new administrative subdivisions cut across some former provincial boundaries, the old provinces still have reality as geographic regions (Map 43).

In September 1969, King Idriss was deposed by a military junta led by Colonel Qaddafi and the country became the Arab Republic of Libya. The present government is both radical and conservative. Believing that "the solution of all human problems is to be found in Islam," it espouses hard work, austerity, and spirituality. A second major orientation is to Arab nationalism and unification; both the capitalist West and the communist East are strongly criticized and policies have become increasingly xenophobic over the years.

Several efforts have been made to federate or unite Libya with one or more other Muslim states, but these have been rejected by each of the prospective partners, Libya has contributed to a wide variety of anti-Israeli and pro-Muslim causes, including the Egyptian military, Palestinian guerrillas, Muslim rebels in Chad, Eritrea, and the Philippines, and to African countries willing to withdraw recognition of Israel.

THE POPULATION OF LIBYA

The population of Libya was estimated to be 2.2 million in mid-1974. About 66 percent are in the old Tripolitania, 29 percent in Cyrenaica and 5 percent in the Fezzan. The population has been growing at an estimated rate of 3.7 percent per annum. The original inhabitants of northern Libya were Berbers; Arabs arrived in successive waves in the seventh, ninth, and eleventh centuries, displacing and assimilating their predecessors, while there was an admixture of Greeks in Cyrenaica in the fourteenth century and of blacks brought as slaves over many centuries from south of the Sahara. Today almost the entire population is Muslim and Arab-speaking. Italians, who numbered 110,000 in 1941 and 25,000 in 1964, have been further reduced as their lands have been expropriated; the Jewish population declined from about 30,000 in 1948 to only a few thousand.

It is difficult to estimate the percentage of the population that is urban because of the rapid influx to the major cities, but it may now be close to 50 percent of the total. Tripoli and Benghazi in particular have seen mushrooming development. Both have had their ports greatly improved, as has Tobruk in the east, and all coastal settlements have been connected by an upgraded highway extending 1,056 miles (1,700 km) across the country. At the time of the 1964 census 11.6 percent of the total population was nomadic, 9.2 percent seminomadic, and 52 percent were sedentary rural residents.

HYDROCARBONS IN LIBYA

The development of Libyan oil resources was even more recent and more rapid than Algeria's. It began in 1955 with United States companies obtaining the first concessions under a liberal law allowing for a 50–50 ratio plus depletion allowances. Two years later the first discovery was reported in the Fezzan, 560 miles (900 km) south of Tripoli. The first large find came in 1959 at Zelten in Cyrenaica only 95 miles (150 km) south of the Gulf of Sirte coast, following which a series of important fields were discovered. The dynamic nature of developments is revealed by the rise of oil exports from 0 in 1960 to 18.1 million tons in three years and to 83.5 million tons by 1967, and the increase in value of total Libyan exports from $8.7 million in 1960 to $137 million in 1962, $1.2 billion in 1967, and $3.6 billion in 1973.

In more recent years production has fluctuated around 105 million tons, but receipts have skyrocketed as higher prices have been imposed. Exports of LNG began in 1971. The per capita GNP increased by an annual average of 17.6 percent in the period 1960–1971, when it was estimated at $1,450 or about 6.5 times the average for the continent and about 36 times the level at the time of independence. The government budget, which was only $19.6 million in 1951, reached $3.5 billion in 1974.

Libyan oil production enjoys numerous comparative advantages: its closeness to the large West European markets, particularly as contrasted with shipments from the Middle East via the Cape; the shallowness of numerous pools; the relative closeness of producing fields to the coast and the absence of difficult barriers to cross as compared to

Algeria; the ability to load supertankers at offshore terminals, and the excellent quality of its oil, especially the low sulfur content. Oil is produced at exceptionally low real cost, about 10 to 20¢ per barrel. The country has not hesitated to use these assets to extract higher payments for its shipments, particularly in recent years.

Most production now comes from fields in the Sirte Basin, which are connected to five coastal points by oil and gas pipelines (Map 43). Libya has not seen processing developments comparable to those of Algeria, although the Marsa el Brega gas liquefaction plant is the largest in the world and there are separator plants, two small refineries, and a sizable petrochemical complex. This situation should change rapidly as various nations agree to provide industrial plants in return for contracted delivery of oil.

Relations between the government and the oil industry have undergone a dramatic change over the years. The first laws were very favorable to producers; to speed exploitation, they stipulated that quarters of original concession areas be relinquished after five-, eight-, and ten-year periods, and as a result each concessionaire tried to determine the most promising quarter as quickly as possible. Despite some difficulties in securing markets in the early years, numerous major and minor companies and consortia were attracted by the wealth and locational advantages of the Libyan fields; some offered extra inducements to secure concessions.

Libya joined OPEC in 1962, raised taxes on posted prices in 1965, and increased the posted price 80¢ per barrel after the Arab-Israeli War of 1967. The following year it set up what is now the Libyan National Petro-

Oil camp at Zelten
Oil was discovered in 1959 at Zelten, which is now one of the major producing fields of the country. Gathering lines from producing wells lead to gas-oil separation tanks in center.

leum Company (LNCO), which concluded joint agreements with four companies; it also managed domestic distribution.

Accelerated change began in 1969 with seizure of power by the military junta, nationalization of domestic distribution in 1970, and a reversal in the world demand-supply position in the same year. A sharp increase in demand in Western Europe and Japan, a break in a Syrian pipeline, and a temporary shortage of tankers combined to create a sellers' market, whereas previous decades had been characterized by surplus output and remarkably stable prices, which had made petroleum more and more attractive as compared to other sources of energy, particularly coal. The posted price was increased by 30¢ in 1970 and 90¢ in 1971, while the tax rate was raised to 55 percent with some retroactive payments required. Libya has since acquired the reputation of being the most difficult and demanding of all petroleum-exporting nations. Prices were increased from $4.90 a barrel in September 1973 to a peak of $16 in early 1974, with some oil going for even higher prices. By mid-1974, however, the world supply had recovered sufficiently for Libya to agree to sell oil at under $13 a barrel despite its prior statement that it would prefer to keep the oil in the ground rather than accept a lower price.

In 1971 Libya began to push for participation agreements, since which time companies that have refused to accept their suggested share (usually 49 percent) have been nationalized. The oil companies, although reluctant to accede to Libyan demands because the stipulations would have to be extended to other areas, have also been reluctant to abandon this favored area, so several have accepted the new terms and others have continued to manage their previous concessions without formal agreements. In the meantime the country opened several large blocks for exploration in 1974 with Libya to get 81–85 percent of whatever oil is found and the con-

tracting companies the remaining percentage free of any tax or royalty. The trend in Libya and elsewhere appears to be toward having the companies act as technical advisers, operators, and offtakers of oil and LNG under contract and not by concession. The desire to assure supplies of oil has led several countries and their oil companies to seek agreements that involve large-scale provision of industrial equipment, training of manpower, and supply of military equipment in exchange for promised tonnages of oil.

The impact of oil on the Libyan economy has been startling. It gave the country the chance to be self-sustaining for the first time in centuries. The national economy has been greatly stimulated by enormous expenditures on construction, road building, industrial plants, and other projects, while education and other aspects of the social infrastructure have received ever-increasing attention.

But hydrocarbon developments have brought new problems to the country. Only about 18,000 people are employed directly

Petroleum loading-terminal at Marsa Brega
The special bow mooring permits ships to swing a full 360 degrees while loading. Tug controls loading arm at right.

by the industry, and finding effective ways to diversify the economy and spend the large sums wisely is not easy. There is a strong tendency toward formation of a dual economy with a relatively small elite of persons associated with oil, the government, and tertiary activities contrasting sharply with the remaining traditional farmers and graziers and the burgeoning urban proletariat.

The exodus from the rural areas has been even more rapid than elsewhere in Africa, causing a decline in agricultural production, an increased dependence on imported food (now over two-thirds of total consumption), and an inability to provide adequate housing in the cities despite impressive construction programs. Inflation contributes to the difficulties.

The inability of Libya to staff the required bureaucracy and the inhibiting character of what had been an essentially static society are factors that cannot be overcome as rapidly as oil can be developed. There is a strong feeling against manual labor and a tendency to eschew investment in agriculture and industry in favor of speculating in construction and services. The speed with which the shortage of trained persons can be reduced is slowed by the circumscribed existence of women in this highly traditional society and by the desire to combine Arabic/Islamic religious instruction with modern science and teaching.

A final disadvantage, though it is not necessarily viewed as such by the junta, is the increased penchant for political adventurism. Increasing sums are being spent on military forces and, as noted earlier, on support of Arab and Muslim causes elsewhere.

LAND USE IN LIBYA

The area of Libya is 679,358 square miles (1,759,537 sq km) or about 2.5 times the size of Texas and 3.2 times the size of France. Ninety-five percent of the country is classified as desert and not much more than 5 percent can be put to economic use. Permanent cultivation is confined to less than .05 percent of the total area and probably cannot be more than doubled, though varying areas can be used for catch cultivation depending on precipitation. There are three regions of more intensive use: the coastal plain and Jefara of Tripolitania in the west, the Cyrenaican Plateau to the east, and the scattered areas in the south.

The former province of Tripolitania contains the best-developed regions of the country. With far more habitable space and a more socially advanced populace, it accounts for about two-thirds of the nation's cultivated acreage and the bulk of its agricultural production, including a greater variety of crops than the other regions. Its inhabited area consists of three parts. First is a broken line of oases extending for about 200 miles (320 km) along the coast between Zuara and Misurata (Map 43) and inland only six miles (10 km) or less. The resources of groundwater are fairly abundant, permitting the irrigation of about a quarter million acres (100,000 ha) and the most stable agriculture of Libya.

The second inhabited area is a triangular open plain behind the coastal oases called the Jefara and covering about 6,000 square miles (15,500 sq km). The main characteristic here, as in other areas of Libya outside the oases, is instability. Winds compound the difficulties occasioned from aridity; the hot, dessicating ghibli may raise temperatures well above 110°F (43°C) for days and cause additional damage by driving sand. Dry farming in the Jefara is concentrated on grains, particularly barley; the region is also the main olive producer. Last of the inhabited zones of Tripolitania is the Jebel, a plateau of 2,000–3,000-foot (600–900 m) elevation whose scarp bounds the Jefara on the interior. It is used mainly for grazing and the collection of alfa grass, but overexploitation of this resource has been a major cause of soil erosion in the Jebel.

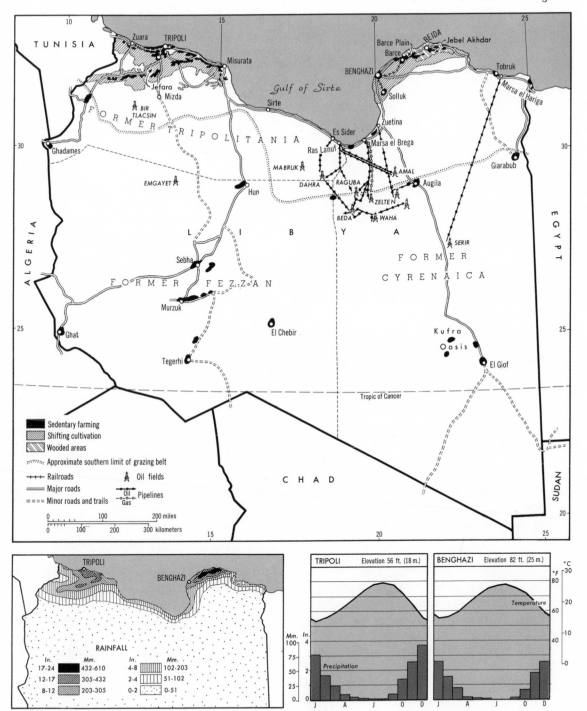

Map 43. *Economic map of Libya; precipitation and selected climographs*

Sheep grazing near the coast east of Tripoli
In 1964 a fifth of the population was nomadic or
seminomadic and half were sedentary rural residents.
The oil boom has, however, led to a massive exodus from
the country.

The main area of the Cyrenaican region is
the Barce Plain, a low limestone plateau running in a 150-mile (240 km) crescent along
the coast and having a maximum extension
inland of only about 30 miles (48 km). The
highest elevation is about 3,000 feet (900 m)
and the plateau falls abruptly to the sea in
the north. To the east and west it is fringed
by coastal steppes, which in turn merge into
the desert. Behind the Barce lies the Jebel
Akhdar, a Mediterranean steppe area which
supports the juniper bush, a few conifers,
and some production of cereals. But water
sources are deep and difficult to locate,
hence this former province has only an
eighth as much land suitable for sedentary
farming as Tripolitania. Grazing is therefore
almost universal.

The oases of the south, which are less numerous than in Algeria, have little economic
significance. Their most important crop is
dates, but yields and total production are low
and only a small export is possible. Paved
roads to some of the oases have increased

their accessibility and oil companies participated in an effort to modernize agriculture
in the Fezzan, but no major increase in commercialization may be realistically expected.

The main crops of Libya are cereals followed by olives, a variety of other tree crops
(figs, dates, almonds, citrus fruit), and peanuts. Many of the higher value crops were
concentrated on Italian farms, which
required subsidy even after Libya became independent. In the past decade production of
many crops has declined, farms and settlements have been abandoned, and exports of
agricultural products have decreased. The
government, disturbed by these changes in a
sector it views as significant in retaining desirable traditional values, has tried to reverse
the trends through subsidization, price supports, loans, land reclamation, sinking of
wells, rehabilitation of ancient water storage
facilities, renovation of abandoned farms,
and public works to improve the amenities
of rural life. The amounts allocated in relation to the agricultural population are almost
fantastic in comparison with those capable of
being made in most African countries, but
success is by no means assured. The sinking
of wells near Tripoli has already caused a
significant drop in the water table, while a
large scheme in the Fezzan is based on fossil
water.

SECONDARY ACTIVITIES

A small fishing fleet operates along the coast,
catching mainly tunny and sardines. Collection of sponges has been carried on for over
2,000 years, but is of little significance. Mining and quarrying are confined to the output
of natron in the Fezzan, salt from pans along
the Tripolitanian coast, and building materials. Large deposits of potash exist in the
Sirte Desert and a medium-grade iron ore
has been found in the Fezzan.

Manufacturing has only very limited significance. There are numerous small craft
shops; among the better products are Mi-

surata rugs and Tripoli leather and silver. Modern industry, other than the hydrocarbon installations, is concerned with the processing of agricultural produce (tanning, canning of fruits and vegetables, flour milling, olive oil pressing, etc.), production of relatively simple consumer products (food, beverages, cigarettes, textiles, clothing, shoes, furniture, etc.), output of construction materials (cement, bricks, tiles), and provision of various services (electricity, printing, repair facilities). About 56 percent of industry is in Tripoli and 14 percent is in Benghazi.

The small consumption of electricity is indicative of the low level of industrialization, but many new plants have been installed and a goal of the present plan is to electrify the entire country. The contribution of manufacturing to the GDP has declined from about 11.3 percent in 1958 to about 3 or 4 percent, but this reflects the enormous growth in hydrocarbons since the value of manufactured output has increased. The government has ambitious plans to diversify industry and has no real problems of capital availability, while the critical shortage of trained workers can be met by joint enterprises with foreign companies who provide the necessary skill until training of locals is accomplished. Serious problems facing industrial growth include the paucity of natural resources except for oil and gas, water shortages, and the small size of the national market. Another concern is the likelihood of surplus capacity developing in the world petrochemical industry.

While Libya has made very great expenditures in its efforts to diversify its economy and will have enormous funds to invest in the next 30 to 50 years, it will require great skill to develop activities that can maintain a high-level economy after the historically very brief period of the hydrocarbon bonanza.

CHAPTER SEVEN

Egypt

Almost all of the 37.6 million (1974) people of Egypt live in the Nile Valley or on its delta and are directly or indirectly dependent on the river. A fortuitous combination of physical features makes possible this greatest of all oases. First is the river, second longest in the world, with its headwaters in regions of abundant precipitation. One main branch has its source in equatorial lands 4,000 miles (6,436 km) from the Mediterranean, though the Nile proper measures 3,473 miles (5,588 km) from the outlet of Lake Victoria; the other branches, which contribute more water to Egypt, bring a tremendous seasonal flow from the Ethiopian Massif. Second is the presence of a topographic channel the Nile occupies in its lower course, which permits the river to flow all the way across the Sahara instead of spreading out and losing its waters by seepage and evaporation. Third is the existence of low-lying areas along the Nile and in its delta, suitable—indeed magnificent—for cultivation under irrigation. And last is the excellent temperature regime permitting double and even triple cropping.

The utilized area of Egypt is about 15,000 square miles (38,850 sq km), an area about as large as Switzerland and only 3.9 percent of the total area of the country. This gives an average actual density of 2,507 per square mile (970 per sq km), but rural densities over 3,500 per square mile (1,350 per sq km) are not uncommon. The farmed area of the Nile Valley is about 6.8 million acres (2.75 million ha), resulting in a man-to-arable-land ratio of one person to 0.18 acres (1 to 0.07 ha).

The raw figures of population density are impressive enough in their own right, but they reveal only part of the story. The evidence of extreme poverty, high disease rates, high infant mortality rates, low dietal standards, unsanitary living conditions in rural settlements, and heavy underemployment and unemployment reveal the consequences of such great densities, while a glance at the rates of population increase and an understanding of the improvements that are being made in certain health and disease conditions present a frightening prospect for the future.

The birthrate in Egypt is now about 37 per M, down somewhat in recent years but still high. The deathrate has declined from 29.4 per M in 1917 to about 16 in 1973. The rate of increase went up from 1.07 percent in 1917 to 2.54 percent in 1966 but declined slightly to an estimated 2.1 percent in 1973. While it is hoped that the birthrate will decrease with some rapidity, the family planning program does not appear to be adequate to the task and large families continue to be desired even in urban communities. Each year the population grows by over 790,000 and it may well reach 51.5 million by 1990.

There are perhaps five major possible solutions to the fantastic population problems of Egypt.

(1) Extend the land under cultivation.
(2) Intensify production on the already existing acreage.
(3) Absorb the excess population in non-agricultural pursuits—mining, industry, tourism, etc.
(4) Promote emigration.
(5) Promote the conscious limitation of population growth.

EXTENDING THE LAND UNDER CULTIVATION: CONTROL OF THE NILE

The major way of extending the cultivable land is by controlling the Nile, which has largely been achieved by construction of the Saad el Ali or High Dam. The construction of dams in the headwater areas would have been more effective, but these would have required a much longer period and political cooperation that might have been difficult to achieve. To understand the relative merits of the several possible approaches and the potentialities for further rationalization of water use, it is necessary to review briefly the hydrography of the Nile and the actual and prospective control measures.

THE WHITE NILE

The White Nile begins at Lake Victoria, which, unlike the other lakes of east-central Africa, occupies a relatively shallow uplifted basin rather than an elongated rift valley (Map 44). The lake, located in a region of high and fairly evenly distributed rainfall, has significance for Egypt in that it acts as a huge reservoir, evening the discharge into the Nile at Jinja. The big disadvantage, explained by its size and equatorial position, is the high evaporation loss, which exceeds 80 percent of the amount received by the lake in precipitation and from tributaries. Just below the outlet of the lake is the Owen Falls Dam, completed in 1954; it raises the level of the lake only three or four feet (0.9–1.2 m), but adds 200 billion cubic meters to its storage capacity. The Owen Falls Dam is a good example of international cooperation, all too often lacking in the 1,107,000-square-mile (2,867,000-sq-km) basin of the Nile. Egypt gained by more effective control, for which it contributed to the cost of the dam; Uganda and Kenya gained through the production of low-cost hydroelectricity.

The river flows through Lake Kioga, whose influence is imperfectly known, to Lake Mobutu (Albert), a distance of 254 miles (409 km) in which it drops 1,400 feet (427 m). Several dams may be placed along this stretch primarily to produce power, but planned construction of an underground plant at Kabalega Falls (Murchison Falls) has apparently been abandoned under pressure from conservationists, who objected to the impact on this famous tourist attraction and the game park that surrounds it. Lake Mobutu, a narrow, steep-sided, rift-valley lake, provides a suitable place for storage without excessive evaporation loss, but its ownership by two countries (Zaïre and Uganda) and the flooding of some utilized areas are impediments to such use. From this lake the Mountain Nile or Bahr el Jebel descends to the plains of southern Sudan, to the region known as the Bahr el Ghazal or "the Sudd." About a hundred miles (160 km) below the lake and just inside the Sudan border at Nimule is another site appropriate for a future dam.

In the Bahr el Ghazal region the river's gradient is so reduced that it forms tortuous and bifurcated passages, creating a vast swampy area for about 200 miles (320 km), full of papyrus reeds and floating vegetation called "sudd," which denotes blockage. This vegetation contributes to the water's spreading and interferes with navigation on the river. While the swamps act as a regulator, evening the flow of the river below them, of far greater significance are the enormous losses from evaporation that occur as the high waters spread over vast stretches of the region. One of the main functions of dams above this area would be to prevent such wide spreading by evening the flow into it. More effective would be the Jonglei Diversion Scheme or Equatorial Nile Project, which would involve cutting one or more 16-feet-deep (4.9 m), 390-foot-wide, (119 m), 175-mile-long (280 km) canals across the area. They would carry a portion of the flow while the main passages would continue to carry the rest, thus preventing the present dispersion with its consequent high losses to the sun. It is estimated that, with stabilization of flow into the area, the loss to evaporation in the region could be reduced to one-tenth of the present level.

This scheme, which has been studied off and on since 1904 and is now the subject of investigation, represents one of the few ways in which larger quantities of water could be passed down the river for later use in irrigation; most of the other control schemes are designed to even out the flow in Egypt or the Sudan. While they increase the so-called timely water—that is, water coming in the period when it can be used—they result in some actual loss in total flow owing to increased evaporation losses. This is a major disadvantage of Nasser Lake behind the High Dam, and it has been estimated that the Jonglei Scheme could just about make up for evaporation losses in that reservoir. Bringing the Jonglei Scheme to fruition will entail large expenses, political agreement between Egypt and Sudan, and the provision for some alternative method of livelihood for the indigenous peoples of the Bahr el Ghazal, whose way of life is closely coordinated with the seasonal ebb and flow of the Nile. The question should also be raised whether the waters saved should not be used in this area rather than further downstream. The remoteness of the area, its relatively sparse population, and the backward nature of the inhabitants argue against large-scale expenditures on intensive irrigation in this area, but it may also be held that control of these waters permits the development of an otherwise poor area that should not be condemned to poverty.

Where the Bahr el Ghazal and the Bahr el Jebel meet, the true White Nile begins. Map 44 presents in the water account insets a picture of the gains and losses at specific points along the Nile, and it may be seen that the river emerges from the Sudd region with only about 56 percent of its entering volume despite the precipitation that falls on the area and the existence of several tributaries. Soon the White Nile is joined by its major eastern affluent, the Sobat, which rises in southwest Ethiopia. The rains in that area are highly seasonal, concentrated in the months of April to October, and when the peak flow of the Sobat reaches the Nile in November and December its height acts as a dam to the main Nile, ponding its waters upstream from the confluence. The flow chart for Khartoum, showing the seasonal flow of the White and Blue Niles at their confluence, reveals the much more even regimen of the White Nile, explained by the more even precipitation in its headwaters, the reservoir effect of the lakes and swamps through which it passes, and the ponding effect of the Sobat. Before construction of the High Dam it was the White Nile that supplied the bulk of water to Egypt in spring and early summer, when it was most valuable.

Not far above the confluence of the White and Blue Niles is the Jebel Auliya Dam, completed in 1937. Its purpose was to conserve the White Nile waters during the high flow period of the Blue Nile. It served Egypt rather than the country in which it was built, though the Sudan was compensated for the lands that were flooded; Sudan uses the reservoir for support of pump schemes south of the dam and has installed a hydroelectric plant at the dam.

THE BLUE NILE

It is the Blue Nile that provides the spectacular element to the main Nile, contributing about four-sevenths of the total volume, or twice the amount from the White Nile. Its flow is highly seasonal, reaching a low in spring but rising from June to September to an average flow thirty times as great; the maximum flow can be 4,000 times the minimum. The Blue Nile caused the annual flood or Nili, which was recorded in Egypt from 3600 B.C.; it provided the chief supply of water to Egypt in August–September and brought down the tremendous quantities of rich alluvial silt which is the soil of Egypt. With the new High Dam this regime is now changed within Egypt, as will be seen below.

The Blue Nile rises in the Gojjam Highlands near Lake Tana, from which it flows through precipitous gorges onto the Sudan Plain near Roseires. Increased storage capacity in its headwaters would have been preferable to a reservoir in the middle of the desert, but the existence of shrines and religious monuments which would be flooded, the remoteness of the area, Ethiopia's lack of interest in major hydroelectric production, and political differences between Muslim Egypt and the Coptic Christian Empire of Ethiopia are factors that militated against a major scheme at Lake Tana.

Where the river descends to the Sudan, a new dam near Roseires creates a 3-billion-cubic-meter reservoir, permitting extension and intensification of the irrigated areas of Sudan, while at Sennar is the dam that stores most of the water for the Sudan Gezira. Under the 1929 Nile Waters Agreement the amount of water that could be utilized by Sudan was strictly limited. This agreement, which was completed when Britain was acting as the "protector" for Sudan and when little need was seen for supplying water to an underpopulated country, threatened in post-independence years to exacerbate Egyptian-Sudanese relations and to make "the unity of the Nile" an empty phrase. After several abortive negotiations, however, a new agreement was concluded in 1959, whose allocations with completion of the High Dam are as given below.

Distribution of Nile waters (billion cubic meters)

	1929 Agreement	1959 Agreement
Egypt	48.0	55.5
Sudan	4.0	18.5
Flow to sea	32.0	
Loss to evaporation		10.0
Total average flow	84.0	84.0

The 1959 Agreement also provided for a permanent commission to supervise the distribution and for payment of $43 million by Egypt for the evacuation and resettlement of about 70,000 Sudanese residing in the area of Sudan flooded by the High Dam reservoir.

THE MAIN NILE

At Khartoum, which means "elephant trunk," the two branches of the Nile converge. Two hundred miles (320 km) downstream the last of the great east-bank tributaries, the Atbara, joins the main stream. Its regime shows the same marked variations as the Blue Nile; its contribution to the river provides about an eighth of its average flow at the confluence. Water is extracted at the new Khashm el Girba Dam to irrigate 518,000 acres (210,000 ha) in the scheme of the same name. From that point the Nile flows 1,600 miles (2,570 km) to the Mediterranean without a single tributary other than wadis, with only very irregular if any flow.

The Nile traverses the Nubian Desert in a great S curve. In places there are high and barren cliffs and the desert comes right to the river's edge; elsewhere there are narrow bands of irrigated lands, especially in the Dongolo region between the third and fourth cataracts. The sixth to the third cataracts occur before the S bend ends, three of which may eventually be the sites of storage and power dams. The Aswan dam was at the first cataract, 216 miles (348 km) downstream from the Sudan border. First of the major dams to be constructed, it ponded back the waters over 100 miles (160 km) when full; built in 1902, the dam was later raised to a maximum head of 76 feet (23 m) to provide greater storage capacity. The flow at Aswan had both great seasonal range and great variation from year to year. In a 50-year period, the flow varied from 42 to 130 billion cubic meters, and since the storage capacity of the dam was only 5.3 billion cubic

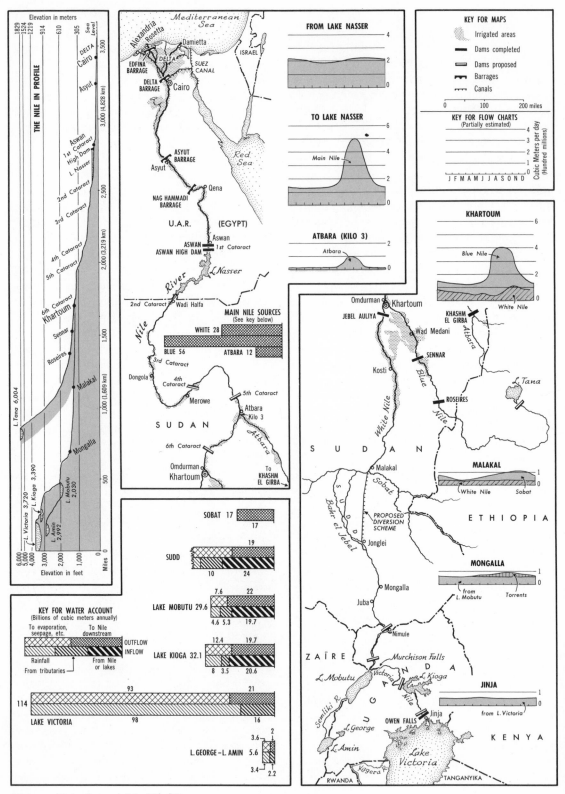

Map 44. Major features of the Nile River

meters there was an average wastage to the sea of 32 billion cubic meters.

Most of the present and prospective storage dams thus far mentioned were parts of the so-called Century Storage Plan, worked out by the Egyptian Water Department. It probably represented the most effective control system for the Nile, but faced with the need to provide more water to extend irrigation rapidly, Egypt opted for the grandiose High Dam, which could increase the irrigated acreage by about 1.3 million acres (526,000 ha) after ten years. This does not mean that the other projects are all passé, though certainly the need for them has been postponed. Storage above the Sudd region and the Jonglei Scheme, however, have no alternatives and, as noted earlier, would add to the volume of water available for Sudan and Egypt.

THE HIGH DAM. This dam, which was inaugurated in 1971, is situated five miles (8 km) south of the Aswan Dam. It is 2.6 miles wide (4.2 km), 364 feet high (111 m), 3,900 feet thick (1,189 m) at the base, and 126 feet thick (38 m) at the top, thus being the world's largest earthfilled structure. When full it can store 130 billion cubic meters of water in the 1,544-square-mile (3,994-sq-km) Lake Nasser, which extends southward some 310 miles (499 km). Unlike the Aswan Dam it does not have sluices, which would permit the flood waters with their valuable load of silt to pass downstream; now the silt is deposited at the upper end of the lake.

The following advantages are claimed for the High Dam: (1) the extension of the cultivated area by 1.3 million acres (526,000 ha); (2) the conversion from annual to perennial irrigation on 700,000 acres (283,000 ha) (see below, p. 142); (3) the securing of all Egypt from floods; (4) the improvement of drainage on existent land; (5) the provision of a considerable amount of hydroelectric capacity (see below, p. 148); (6) the improvement of navigation below the dam; and (7) the provision of a new source for fishery development. The dam and lake have also be-

The Saad el Ali or High Dam near Aswan
Inaugurated in 1971, this dam is the largest earth-filled structure in the world.

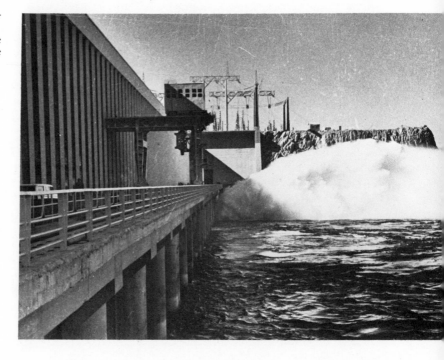

come a new tourist attraction and will probably attract a larger number of visitors to the
famous sites at Luxor and elsewhere in the
south.

The dam does have certain disadvantages
and has indeed been subjected to strong criticisms by some observers. It is not absolutely
certain that the dam would prevent the highest floods from wreaking havoc, despite the
provision for storage of 30 billion cubic
meters for flood protection; and it presents a
strategic target whose destruction could
wash away the country. A major disadvantage is the high loss to evaporation in this
mid-desert area, estimated at about 10 billion cubic meters a year or almost double the
capacity of the Aswan Dam. It has also been
claimed that seepage has been greater than
anticipated; the validity of this assertion cannot as yet be determined but it may be noted
that some water entering underground
aquifers could later be tapped in the so-
called New Valley west of the reservoir or
elsewhere.

Another major disadvantage is that the
huge reservoir acts as a desilting basin, de-
priving the country of its yearly supply of
enriching silt and gradually filling the reservoir. Allowance was made for the accumulation of silt at an annual rate of 60 million
cubic meters a year for over 500 years, but
increased applications of organic and inorganic fertilizers will be required to offset
the loss of the enriching silt.

Among other disadvantages are an increased erosion of the Nile banks below the
dam, which is being offset by the construction
of additional barrages; increased incidence of
bilharzia in the lands converted to permanent irrigation (claims that the dam caused
an explosion in the incidence are exaggerated because a high percentage of fellaheen
were already affected); increased incursion
of salt water along the edges of the Delta; a
decline in the sardine fisheries in the eastern
Mediterranean because of the reduced discharge of silt and organic matter supporting
the plankton population; and burial of
numbers of archeological sites below the
waters of the lake, though the most valuable
temple at Abu Simbel was raised in a notable
internationally supported effort.

The High Dam has figured importantly in
Egyptian and international politics. Angered
by withdrawal of offers to extend loans for
construction of the dam by the United States
and Britain, who felt that Egyptian commit-
ments to purchase arms from the East precluded its covering the loan, Egypt nationalized the Suez Canal. This, in turn, led to the
aborted Anglo-French-Israeli invasion of
Egypt in late 1956. The Soviet Union then
extended loans of about $240 million and
provided the necessary technical assistance
to help complete the dam on an accelerated
basis. The remainder of the $782 million
cost was paid by Egypt, which has already
paid back about half of the Soviet loan.

BELOW THE HIGH DAM. Between Aswan and
the Delta the river flows 605 miles (973 km)
with an average gradient of 1 in 13,000.
High cliffs rise on each side, confining the irrigable area to from only a few yards to 10
miles (16 km) in width; the long band of
green in Upper Egypt is divided with razor
sharpness from the desert. There are no
more storage dams below Aswan, only a
series of barrages that raise the level of the
river high enough to feed the canals on either bank and, later, in the interdistributary
areas of the Delta. Egypt has about 9,000
miles (14,500 km) of large canals and 48,000
miles (77,000 km) of small canals and
ditches.

About 60 miles (96 km) south of Cairo, a
canal takes off from the left-bank canal and
cuts through the narrow gravel ridge separating the Nile Valley from the Faiyum
Depression (Map 45), one of a number of
structural depressions whose surface had
probably been further lowered by the wind.
About 3,600 years ago one of the Pharaohs
conceived of using the depression as a kind
of safety valve in the event of excessively

high floods and had the canal connection dug. Some thousand years later it was decided to make the Faiyum an integral part of the Nile and the necessary canals were constructed. Meanwhile, many feet of rich Nile mud had been deposited. Today, over a million people live on the 700-square-mile (1,800-sq-km) irrigated area of the Faiyum, whose distinct advantage is that most irrigation can be done by gravity as the basin slopes gradually toward the lake, Birket el Qarun, which is 140 feet (43 m) below sea level.

Just below Cairo the Sweetwater Canal leads northeastward and across the desert to the Suez Canal area. Utilizing in part the bed of an early canal connecting the Red Sea via the Nile to the Mediterranean, this canal provides water for support of the communities along the Suez and for a shelter belt against windblown sands.

Twelve miles (19 km), below Cairo are the Delta Barrages, built in 1835 and 1884; the older is no longer in use. Canals from above the later one serve each side of the Rosetta and Damietta distributaries. These barrages are of particular interest because of their age and because the Delta contains about 60 percent of the cultivated land of the Egyptian Nile. The barrage at Edfina, near the mouth of the Rosetta Branch, has somewhat different functions from the run-of-river barrages; in addition to raising the level to ease irrigation it holds sea waters out and increases the deposition of silt to help build up the Delta.

One cannot respond to the original question—how much land can be added to the irrigable area by improved control of the Nile?—with precise figures because the investigations have not been sufficiently precise and because some crops, particularly rice, require considerably more water than others. The High Dam will add about 1.3 million acres (526,000 ha) to the present area of 6.0 million acres (2.4 million ha), an increase of 21.7 percent. A good part of this

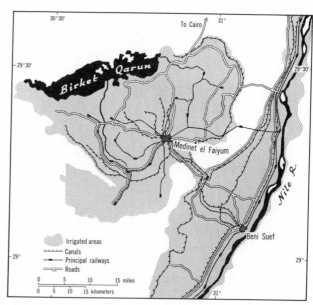

Map 45. The Faiyum Oasis

will be in the sandy area west of the Delta, part will be in reclaimed lands along the seaward margins of the Delta, and part may be taken to the New Valley in the western desert. Reclamation costs about $1,000 per acre ($2,500 per ha) and it requires about six to nine years to achieve full productivity on the new land. As of 1974 about a million acres (405,000 ha) had been added to the previous total. The High Dam will also permit intensification on a considerable acreage, a point covered in a later section of this chapter. Completion of the Jonglei Scheme might add 7 percent to the water available to Egypt, assuming a division of waters on the present ratio. The possibilities are, then, substantial; but when projected against the population growth it becomes apparent that they are nothing but stopgap measures. Indeed, the population increased during the construction period of the High Dam by a percentage higher than that of the additional area irrigated. The Jonglei and other schemes would obviously only give additional short-period respite to the inexorable population pressure.

EXTENDING THE LAND UNDER CULTIVATION: OTHER METHODS

There are other possibilities for extending the cultivated area of Egypt. One is the so-called New Valley Project, which involves tapping the groundwater supplies in the desert between the Kharga and Dakhla oases. Early estimates that as much as 1 to 3 million acres (400,000–1,200,000 ha) could be watered appear to have been grossly overoptimistic; a figure of 100,000 acres (40,000 ha) might be more realistic. The disappointing results of this project may have contributed to the proposal to bring Nile waters to the area.

A second possibility would be to tap the underground water of the Nile itself, which is probably substantial. Again, however, no firm estimate is possible. In the meantime, this subterranean water is being tapped to provide potable water. In view of the present practice of taking drinking water from the Nile and also using it for sewage disposal and bathing, these wells are making an important contribution to the improvement of health conditions.

Third, there is some possibility of increasing the utility of the narrow steppe-like bank along the Mediterranean coast west of the Delta, formerly a granary of considerable significance. Lastly, there is the longer-term potentiality of desalinizing sea water and pumping it on the land. Just when this may become practicable is a matter for speculation, but it quite obviously provides the only really substantial solution to extending the productive area of the country.

INTENSIFYING PRODUCTION ON THE EXISTING AREAS

Increasing the output on the areas that are now subject to irrigation may be accomplished in several ways: by extending perennial irrigation, by avoiding wasteful application of water, and by intensifying methods on these areas.

PERENNIAL AND BASIN IRRIGATION

With the exception of a minor production of catch crops sown on areas along the river as the high waters recede, there have been two main systems of irrigation employed in Egypt. Before the barrages and the Aswan Dam were constructed the basin system was dominant. Each year the Nile flooded carefully prepared basins; when control was possible, the water was drained off, usually after two to three weeks, and as soon as the soil could be tilled the crop was planted and grew solely on the basis of the moisture the soil was capable of retaining. The system permitted only one crop a year and only drought-tolerant crops such as wheat or barley. Modern engineering developments, occurring mainly after 1900, brought an agricultural revolution to Egypt by permitting the widespread application of the perennial system. By storing the waters and releasing them as required, lands could be irrigated all year, permitting multiple cropping. Such crops as corn and cotton could be grown in the summer, while grains and bersim (Egyptian clover used as fodder) could be grown

Fields irrigated under the basin system
With increased control from the High Dam such areas are being converted to perennial irrigation, permitting multiple cropping and production of higher value crops.

in the winter. This system permitted a striking increase in production. Perennial irrigation brought other problems: associated control works increased evaporation losses and
decreased the silt supply; the incidence of
disease, particularly bilharzia, increased.

Nonetheless, one of the important contributions of the High Dam is the virtual
elimination of the basin system, permitting
more intensive use and the planting of
higher yielding crops on some 700,000 acres
(320,000 ha). The more even year-round
supply of water should also lead to its less
wasteful application. The tendency of the
fellah has been to take more than is required
when it is available; this has led to waterlog-
ging, reduced yields per acre, and higher incidence of plant diseases. A program to
deepen the field drains is also improving this
situation.

MORE INTENSIVE PRACTICES

A second way to increase production is
through more intensive agricultural practices: increased fertilization, improved control of insects and plant diseases, more careful seed selection including use of new
high-yielding and hybrid seeds, etc. Egypt
consumed large quantities of fertilizers even
before the High Dam greatly decreased the
annual addition of enriching silt—particularly nitrogenous fertilizers coming from
plants at Suez using refinery gas, at Helwan,
and at Aswan, where a new large nitrogen-
fixation facility uses hydroelectricity. In addition, local fertilizing agents are used as
available: silt cleaned from canals, mixed
with grass; stable manure; rubbish mounds;
material taken from the neighboring cliffs in
Upper Egypt; and pigeon droppings from
the huge rookeries that are often the most
noble edifices in the rural landscape. These
rookeries may yield as much as one to six
tons of extremely rich manure each year.
The treatment of sewage from Cairo and
other cities could provide an additional
source.

A pigeon rookery in the Faiyum oasis
Such large rookeries may yield from one to six tons of
guano yearly and are a locally important source of
fertilizer.

Plant insect and disease control presents
some significant possibilities. Cotton is often
protected by meticulous removal of individual leaf worms or infected leaves. Even this
extremely labor-intensive system proved inadequate in severe attacks of leaf worm,
which can destroy as much as a third of the
crop, so the fellaheen are now provided with
subsidized pesticide spray and equipment.
Disease control is made more difficult by the
practice of storing cotton twigs and corn
stalks on house roofs, but the absence or cost
of other sources of fuel has made it impracti-
cable to destroy these breeding sources. The
Nile sparrow, related to the quelea-quelea
bird, is a serious pest; they travel in flocks of
hundreds of thousands and consume enormous quantities of grain. None of the
methods employed to reduce their devasta-
tions—clubbing, spraying, noise-making—
have been adequate, while destruction of
their natural enemies through DDT inges-
tion has apparently increased their numbers
in recent years.

Use of high-yielding seeds is estimated to

have made a major contribution to increasing yields of wheat by 50 percent, corn by 80 percent, and rice by about 75 percent in the period 1952–1970. Other ways of increasing yields are to switch the crop emphasis to more intensive crops and to intensify the rotation. Conversion of basin to perennial irrigation has permitted switching to rice and corn, which give higher yields per unit area than wheat, barley, or sorghum. Increased production of vegetables is another form of intensification.

There are numerous rotational systems used in Egypt, but a common one would be: April to October, cotton; November to April, wheat; May to November, corn; and December to March, bersim. Rotational intensification does result in greater loss from insects and requires heavier fertilization.

It is sometimes suggested that cotton and other cash crops be replaced by food crops in view of the growing population and the present necessity to import foodstuffs. On a large scale, this would tend toward a low-level subsistence economy and reduce foreign exchange earnings, which can purchase more than the value of the replacement food crops. Cotton now provides about 45 percent of exports, while cotton manufactures account for an additional 14 percent. Sugar, potatoes, peanuts, onions, fruits, and other vegetables are also exported. The unusual quality of Egyptian cotton, with 1.5–2 inch (37–51 mm) staples, gives it high value; that country accounts for about two-thirds of the world's long-staple cotton and one-third of its medium/long cotton. While the heavy dependence on cotton has the usual weaknesses of concentration, it is not easy to find substitutes of equal value.

REPLACEMENT OF ANIMAL POWER

Still another way of increasing the agricultural production of Egypt would be to substitute mechanical power for animal power, thus releasing the area now devoted to feedstuff production, particularly bersim, to other crops. There are an enormous number of animals in Egypt: the water buffalo or *gamoose* is used in plowing and for drawing water; sheep and goats supply meat, milk, wool, and hides; the camel is a beast of burden and a source of meat; and the ass is used for transport.

The application of mechanical pumps would have a considerable effect on the need for animal power, but would also markedly increase the amount of rural unemployment. Despite the modern development of large-scale engineering works, the methods of lifting water to the fields are still very primitive. The most rudimentary system employs the *nattala,* a shallow vessel about 18 inches (46 cm) in diameter suspended on ropes; two men dipping it into the water, raising it, and emptying it into the irrigation runnel can irrigate about one acre (0.4 ha) in 12 hours of strenuous labor. The *shaduf,* of which there are some 50,000, is a crude balanced-weight lift, which permits two fellaheen working in two- to three-hour shifts to water about an eighth of an acre (0.05 ha) in a 12-hour day. In upper Egypt, two to three shadufs may be used in tandem to get the water to the necessary height. Where water need be raised only about 30–40 inches (75–100 cm) the Archimedes screw is commonly used (there are about 250,000 in the country). This is a wooden cylinder about six to nine feet long hooped with iron; inside is a spiral or helix in which the water rises as the cylinder is turned. While the Archimedes screw costs more and is heavier to work, it can irrigate two or three times as much as the shaduf. The most complex of the ancient water-lifting systems is the *saquia (sagia, sakieh),* used since the time of the Ptolemies, consisting of a horizontal cogged wheel turned by an ox, buffalo, donkey, or camel, which engages a vertical wheel with wooden or galvanized iron pots or pockets which in turn pick up water at the bottom and tip it out at the top into channels running to the field. The saquia, being expensive, is often

A saquia in use in the Nile Delta
About 16,000 of these devices are used to raise water to the fields in Egypt. Each saquia can supply water to about five acres.

owned by several fellaheen who bring their animals in turn; it can water about 5 acres (2 ha). There are said to be about 16,000 saquias in Egypt.

Pumps are surprisingly few in number. The average fellah can neither afford to purchase one nor to buy the fuel to keep it running. However, it is said that about a quarter of the power from the High Dam will be used for lifting water, reducing the cost of irrigating an area by an estimated 65 percent.

There is also much excess use of man and animal power in plowing the fields. Many of the plows are no better than those used thousands of years ago, except for the metal edge on the share. While replacement of water-lifting devices and plows with mechanical pumps and tractors would undoubtedly release land now used to support the animal population of Egypt, numerous considerations inhibit this course of action: the lack of capital, the low wages of rural labor or use of unwaged family labor, the cost of fuel, the often tiny holdings, the impediments of numerous ditches and canals, the desirability of maintaining animals for provision of meat, milk, skins, hides, wool, and manure, the im-

portance of bersim as a nitrogen-fixing agent in the rotation, and the desire not to increase rural unemployment.

MORE EQUITABLE LAND DISTRIBUTION

It has often been suggested that subdivision of larger holdings would result in more intensive use, through the greater care that would be taken by individual owners. Certainly the distribution of agricultural holdings appeared to be highly inequitable before the land reform began in 1952, when less than 0.1 percent of owners held more land than the 72 percent with the smallest holdings, and the 1.2 percent with over 20.8 acres (20 feddans, 8.3 ha) held about 45 percent of the land. At present an individual can own no more than 50 acres (20.2 ha), a family is limited to 100 acres (40.5 ha). By 1972 about a million acres (405,000 ha) had been distributed to smallholders, but little more than 10 percent of the rural population was benefited and some evidence suggests that both production and employment were adversely affected.

Actually, it would be impossible to divide the land so that each family had a viable holding. With a farm population of 20–21 million there is only about a third of an acre (0.13 ha) per capita. Nonetheless the land reform program, the first undertaken in modern times, had great political value, as did accompanying programs designed to prevent subletting at usurious rates, rural indebtedness, and absentee landlordism. Some confiscation of land for purely political reasons was, however, deeply resented and caused owners to reduce desirable investments on the land; these sequestrations begun under Nasser have been discontinued under the present regime.

To summarize the possibilities of increasing Egyptian agricultural production, it is apparent that the greatest gains have already been made by extending the cultivated area, conversion to perennial irrigation, and use of higher yielding seeds. The full area that can be irrigated is not yet under control, and additional gains can be made by more effective and complete use of improved seed, better pest and disease control, more double and triple cropping, and increased application of fertilizers. But such measures can scarcely keep pace with the population increase for more than a decade or two. When it is recalled that there is already great poverty, high rural unemployment, and even greater concealed unemployment, it may be concluded that the agricultural field can provide only a temporary palliative to the population problem.

A geophysical crew exploring in the western desert
Numerous finds have been made in this and other areas of Egypt in recent years.

PROMOTION OF NON-AGRICULTURAL EMPLOYMENT

What, then, are the opportunities for meeting the Egyptian population problem by the development of other economic sectors? Nonagricultural employment is already relatively high by African standards. An estimated 990,000 persons are employed in industry and the population of urban areas has increased from 19 percent of the total in 1882 and 1907 to 25 percent in 1937, 33 percent in 1957, and about 41 percent today. This increases the pressure to develop secondary activities and, since a continued exodus from the land may be expected, it is a trend likely to accelerate in the years ahead.

MINERAL DEVELOPMENT

As in other North African and Middle East countries, oil is the most important mineral produced in Egypt. It was first discovered in the Suez area, then in the Sinai Peninsula, which was the major producing area when the June 1967 war occurred, after which those fields were taken over by Israel. The El Morgan field in the Gulf of Suez quickly replaced this loss and total production increased from 5.6 million tons in 1967 to 9.6 million tons in 1973. Output in the Suez fields has begun to decline, but substantial new discoveries and intensified exploration

make it realistic to expect increases in production in the coming years.

Exploration is proceeding in the Suez area, in the Nile Delta, in offshore areas, and in the Western Desert by American, Russian, Japanese, Italian, and other companies. Very interesting finds have been made at several fields in the Western Desert, all of which are readily tied to the sea terminals at El Alamein and Al Hamra. Gas has been discovered at several points in the desert and the Delta, and a 168-mile (270 km) pipe is being laid from the Abu Gharadik field in the desert to Helwan. Plans also call for substituting gas for oil where possible in domestic consumption to release more oil for export. Oil in the Suez area has a high sulfur content requiring special refining processes; that in the west is comparable in quality to Libyan and Algerian petroleum.

Relations with foreign oil companies have altered considerably in the last two decades, fluctuating from excessively rigid regulations and threatened nationalization to additional incentives in 1972 to stimulate more rapid exploration and increased production. At present, companies pay exploration costs but are reimbursed for half of exploration and development expenditures by up to 40 percent of oil production. The Egyptian General Corporation for Petroleum enters at the production stage, usually with a 50–50 sharing agreement, a tax of 55 percent on net profits, and a 12.5 percent royalty on output. Concession areas are restricted to a maximum of 7,700 square miles (20,000 sq km), minimum outlays are specified, and holders give up 25 percent of the area every three years if no discovery is made, while if oil is found the contract is extended for three years.

Egyptian consumption of petroleum is relatively high, in part because of military requirements, and imports have normally exceeded exports by value, in part because of the difficult quality of Suez crudes which were shipped to Italy for special desulfurizing treatment. It is expected that net exports will be increasing after 1975. In the meantime refinery capacity is being increased to 11.5 million tons, including facilities to treat part of the Suez crude.

Other minerals produced are iron ore, mined since 1955 about 30 miles (48 km) east of Aswan and more recently in the Bahariya oasis, which has been tied to the Helwan steel complex by a 542-mile (337 km) rail line; phosphates, which come from the Red Sea coast, the 1972 output having been 1.2 million tons. Discoveries of phosphates have also been made in the eastern desert, at Isna in the Nile Valley, and between the Kharga and Dakhla oases, where a 1 billion ton deposit is to be tied by a 320-mile (515-km) railway to the Red Sea port of Quseir. About 5 to 10 million tons a year of 72–74 percent concentrate will be moved on this line. Other mineral finds have been reported, including a possibly important deposit of rare earths near the Red Sea. While petroleum may be contributing significantly to export revenues by 1980, the mineral sector does not appear to promise much additional employment.

INDUSTRY

Egypt places major emphasis on expansion of manufacturing, which now accounts for almost an equal share of GDP with agriculture. The value of industrial production in 1972–73 was estimated at over $4 billion, but this apparently included mineral output as well. In 1963, there was a total of 3,280 factories employing ten or more people; employment totaled 457,600, and capital investment was estimated to be $1,254 million.[1]

THE INDUSTRIAL PATTERN. Egypt has an unusual range of manufacturing establishments. The food processing and textile industries were developed earliest; more recent emphasis has been on heavy industry and on military production. Food, beverages, and tobacco accounted for about 29.4 percent of industrial production in 1970. The textile industry ranks a close second, with 28.7 percent of the value of industrial production in 1970. Most plants in these sectors are small, and a considerable part of output is still on a handicraft basis, but there are large textile centers at Kafr el Dawar and at Melhalla el Kubra in the Delta. Cotton yarn and woven cotton are now exported, accounting for a seventh of the total value of exports.

Metal and metal fabricating industries, which are better represented than in most African countries, accounted for an estimated 12.1 percent of the value of industrial production in 1970. Pride of place goes to the Egyptian Iron and Steel Company works at Helwan, where production of pig began in 1958. Output of steel in 1970 was 250,000 tons, but additional furnaces are expected to bring the capacity to 2 million tons by 1980. Iron ore comes from Aswan and Bahariya, gas is to be piped from Ghadarik, and coking coal is imported. Production costs have been high but should be reduced with the new supplies of ore and gas and the larger capacity plant, but export markets will have to be found for about a third to half of the output. Metal fabrication includes vehi-

[1] K. M. Barbour, "The Distribution of Industry in Egypt: A New Source Considered," *The Institute of British Geographers Transactions and Papers*, no. 50, 1970, p. 159.

Part of the Egyptian Iron and Steel Company plant at Helwan
Production costs have been high, but expansion of capacity to 2 million tons by 1980 and use of natural gas for part of the energy requirements should improve the viability of this plant.

maceuticals. This sector accounted for 11 percent of the value of industrial production in 1970. Miscellaneous industries include printing and publishing, leatherworking, production of pressed wood, porcelain, and glass.

Indicative of the rapid increase of industrial output is the increase in production of electricity from 1.2 billion kwh in 1953 to 7.6 billion kwh in 1971, at which time 62 percent was hydroelectric. The High Dam has brought substantial improvement in the energy field, as will the newer petroleum and natural gas finds. A serious disadvantage of the High Dam installation is the substantial seasonal drop as stored water is released in the period of low reception. The full capacity of the High Dam is expected to be utilized by 1980 when new hydrocarbon or nuclear electric plants will be required. It has been suggested that water be brought from the Mediterranean to the Quattara Depression in the west over a 735-foot-high (230 m) ridge to develop hydropower from the drop into the depression, which is 427 feet (130 m) below sea level.

Greater Cairo and Alexandria, with estimated populations of 5.4 and 2.3 million in 1972, are credited with about 70 percent of all factories and factory workers in the country, but for only about half of the capital invested in industry. Suez is an important chemical center while Aswan is advantaged by the use of hydroelectricity in new large plants. There is, however, a wide scattering of industrial establishments, particularly in the towns of the Delta. Free trade zones have recently been established in certain cities on the Suez Canal where especially favorable import, exchange, and tax concessions are expected to attract many new industries. A number of investors, including some from Japan, see Egypt (and to a lesser extent Algeria) as having an economically very strategic location for catering to the European, African, and flourishing Middle East markets.

cle assembly; construction of rail cars; assembly of household equipment; production of metal furniture, boilers, sanitary cast iron, wires, containers, hardware, and razor blades; shipbuilding; and output of military equipment.

The construction industry is fairly well represented, but because most Egyptians live in houses of sun-baked clay, markets are confined mainly to the large cities and new industrial projects. Output of cement reached about 3.7 million tons in 1972–73, second only to South Africa on the continent; output is expected to reach 7.4 million tons by 1980.

The chemical industry has seen a considerable expansion in recent years. In addition to the petroleum refineries and fertilizer plants, Egypt produces soap, perfume, caustic soda and chlorine, petrochemicals, pulp and cellulose, rubber, pesticides, and phar-

THE ASSETS AND HANDICAPS OF INDUS-
TRIALIZATION. Egypt faces serious problems
in its drive to industrialize. Until recently,
the high cost and deficit position of the energy sector was one, but this has been greatly
ameliorated by recent developments in hydroelectric and hydrocarbon production.
The necessity to import coke is a continuing
problem and some of the imports from Eastern Europe have been of poor quality. Location of the iron and steel complex at Alexandria rather than near Cairo might have
proved more economic from the standpoints
of lower delivered costs of coke and natural
gas.

The lack of a diversified raw material base
is a major handicap to industry. The list of
minerals is not impressive as far as basing
large-scale industry is concerned, though the
existence of iron ore is helpful, at least in
saving valuable foreign exchange. Some agricultural commodities do support important
processing industries—sugar mills, cotton
spinning and weaving, etc.—but much grain
is ground into meal and flour by primitive
domestic techniques. The growth of processed cotton exports is a favorable feature.
The country is also attempting to push the
sale of fruits and vegetables in the West European market, and a large canning industry
may eventually develop.

While the position regarding scientific and
engineering talent, semiskilled and skilled
labor, and managerial talent is much better
than for most African countries, there remains a serious shortfall, as is evidenced by
the dependence on foreign engineers in construction of the High Dam and in numerous
technically advanced industries. This situation is improving with the greater attention
that has been given to technical and on-the-
job training and as more Egyptians graduate
from scientific and engineering departments
of Egyptian universities. Restrictions on the
hiring of expatriates—policies that led to the
emigration of experienced minority groups
(particularly Greeks, Italians, and Jews)—

and discrimination against the approximately 2.4 million Coptic Christians who are
native Egyptians adversely affected the industrial and commercial life of the country
up to 1974. But in that year a rather remarkable switch occurred in Egyptian policies.
Previous restrictions on foreign investment
and on the employment of foreigners in industrial development were relaxed.

The situation with respect to investment
capital also changed dramatically in 1974.
While the government had succeeded before
that in channeling rather large sums into
manufacturing and had received considerable aid from the USSR and some of the
Arab states, in 1974 a veritable flood of prospective investors were negotiating for a
wide variety of manufacturing and other
ventures involving total investments of over
a billion dollars. The emphasis on socialization of industry was relaxed, and private investment from Egyptian sources has been
encouraged.

Perhaps the most severe handicap to industrial development has been the restrictions of the domestic market, despite the relatively large population. Most fellaheen and
city dwellers have such low incomes that they
can afford little beyond the bare necessities.
Nevertheless, with the second largest population in Africa and a per capita GNP of
$220 (1971), Egypt has a substantial market
for basic consumer goods. It has, for example, the largest national textile market in
Africa and its industrial consumption of cotton is about 4.4 times that of South Africa
and greater than that of the United Kingdom. Future growth must, however, be increasingly oriented to manufacturing for
foreign markets.

An advantage in this respect is the very
low wages for labor, only slightly above those
of India. Indeed, Egypt and Mauritius appear to be closer to a Hong Kong type of
economy than any other African countries,
though they do not yet possess comparable
skills. There is also much child labor, on the

An Egyptian textile mill
Egypt is one of the very few African countries that has a significant export of manufactured goods based on use of low-cost labor, and cotton textiles are the leading such export.

farms and in the factories. The considerable amount of underemployment and the continuing influx to the main manufacturing centers promises to sustain labor docility and pressure to depress wages. The reverse of this coin is, of course, the low purchasing power of the average worker.

Certain industries have gained advantage from government protection, particularly in earlier postwar years when the emphasis was on import substitution. Such protection is not always beneficial to the economy as a whole. If tariffs were used to stimulate infant industries that became progressively more competitive they would surely be justified, but all too frequently, as in many other countries, inappropriate industries are protected and the infants require bottle feeding to senility.

SUMMARY. Despite the rather dynamic growth of the manufacturing sector it cannot be expected to expand with sufficient rapidity to meet the population problems Egypt faces. The index of employment in industry has risen somewhat more rapidly than population since 1960, but the annual population increase is now about 80 percent as large as the total number employed in industry. Some modern manufacturing has a depressing effect on employment by substituting more productive techniques for manpower in agriculture and in the large handicraft section.

OTHER ACTIVITIES

The possibilities of increasing employment substantially in other occupations do not appear to be very great. There is already an excessive number of government employees and low-paid domestic servants. Fishing and collection of sponges and shells are minor activities and estimates of the size of catch to be expected from Lake Nasser appear to have been grossly exaggerated. The unique attractions of Egypt for tourism need not be belabored. The major cities and such spots as Luxor are well equipped and, under propitious conditions, thousands of Egyptians cater, sometimes excessively, to the wishes of the visitor. But political uncertainties have prevented the growth of this sector as rapidly as in other countries of North Africa.

PROMOTION OF EMIGRATION

Since 1967 the government has supported emigration as part of its demographic policy. But this means of easing population pressure does not now appear to offer much hope. Adjacent countries have their own population problems or, as in the case of Sudan, would not be likely to welcome any large-scale influx of Egyptians. Nor do Egyptians show great propensities for migrating from their homeland. The only significant emigration in recent years is that of minority groups, which could be welcomed or accepted in other countries.

CONSCIOUS LIMITATION OF GROWTH

It is difficult to see how any solution other than deliberate limitation of the birth rate can do anything but temporarily alleviate the population pressure in Egypt. But it will take decades before such a program can bring significant surcease, because those who will enter the job market in the next 15 to 20 years are already born and because the necessity and desirability of limiting families are not yet widely accepted. The average fellah continues to consider a large family a must, his only hope of assistance in the arduous work of raising water, controlling cotton insects, tilling the soil, harvesting the crop, and providing support in old age.

Progress is being made, however, in promotion of family planning. Birth control clinics were opened in 1955; a Supreme Council for Family Planning was established in 1965, and over 2,700 centers are now in being. Egypt has a large network of health facilities that can be used in family planning services and an unusual number of trained medical personnel for an "underdeveloped" country. But the program has been hurt by shortage of funds, dependence on volunteer and overtime work, and failure to make provision for enlisting the support of husbands for smaller families.

Perhaps the greatest hope lies in the rapid extension of educational facilities, not only in this sphere but for many aspects of Egyptian life. In the late 1950s, 76 percent of the population and 95 percent of the fellaheen were illiterate. There were 5.9 million students enrolled in educational establishments in 1974, primary education is now nearly universal, and Egyptian universities graduate a proportionately larger number of students than do the United Kingdom or the Scandinavian states.

TIMELESSNESS AND CHANGE

One tends to think of Egypt as a place of timelessness. Certainly the fellaheen remain much as they probably were in the years when the Great Pyramids were built. They use the same implements: the sickle, the shaduf, the straw basket. They still level the fields and dig the runnels as their ancestors did; they still keep the same jealous, even selfish, watch on the water supply. An Egyptian woman striding gracefully along the banks of a canal carrying a water jug on her head brings to mind scenes depicted in bas-relief on the walls of the great temples.

Yet many important changes have occurred. The masters of the fellaheen have changed from Persians to Greeks, Romans, Byzantines, Arabs, Turks, French and British, to Egyptians. The main religion has changed from Totemism to Christianity to Islam, the language from Egyptian to Arabic. Corn was introduced in the nineteenth century; cotton assumed its great importance even later. The buffalo became a major beast of burden in recent times and the perennial system of irrigation is largely a product of the present century. Evidence suggests that the fellah is quite prepared to change once the reason for change becomes apparent, but it has not been for very long that any interest has been taken in his education or betterment. Here may hopefully be the greatest potential contribution of the post-monarchical governments: they have imbued the country with a national pride, adopted programs to move away from the inequitable status quo, and fostered a massive educational program, which cannot fail to have profound effects on the entire Egyptian milieu.

THE SUEZ CANAL AND SUMED

An additional aspect of the Egyptian economy requires brief attention—the Suez Canal, which was entirely within the country but not "of" it, and its proposed partial replacement, the Sumed pipeline. The importance of the canal stemmed from the savings in time and distance it permitted (Map 46). When the Suez Canal was completed in 1869 it was not the first time that the Red Sea had been linked with the Mediterranean. An inscription at Karnak suggests that a canal connected the Nile with what was then the head of the Red Sea via Wadi Tumilat in 1380 B.C. This and two other canals dug in the fifth century B.C. and the seventh century A.D. went via the Nile rather than directly across the Isthmus; part of the last one was used for the Sweetwater Canal, which provides water to the canal zone.

The 103-mile (165-km) Suez Canal could accommodate vessels of 24.6-foot (7.5-m) draught when first completed. Eight improvement programs under the Suez Canal Company and continued dredging after nationalization in 1956 gradually increased the depth and width to permit ships with 38-foot (11.6 m) draught and 65,000 deadweight tons to negotiate the canal. While use of the canal increased more or less continuously from its opening, the sensational Middle East oil developments in postwar years brought canal traffic to record levels almost every year to its closure in 1967 as a result of the Arab-Israeli war. In 1966, the last full

year of operation, the canal transited 194 million tons of cargo from south to north and 48 million tons in the opposite direction, with 72 percent of the total being petroleum. The gross revenue to Egypt from canal tolls in 1966 was $220 million. This source of revenue had particular importance because it was primarily in freely convertible currencies. Saudi Arabia, Libya, and Kuwait agreed in 1967 to compensate Egypt for loss of these revenues, but greatly enlarged military expenditures more than absorbed the amounts received and the USSR required payment for military equipment that was not donated in hard currency.

The impasse between Israel and Egypt over Israeli occupation of the Sinai precluded reopening of the canal until the Israeli withdrawal from the zone in early 1974. And after crash clearance the canal was reopened in 1975. In the meantime advances in the design and size of tankers and bulk carriers have made it increasingly obsolete. Supertankers were beginning to carry large tonnages around Africa before 1967 but now less than 40 percent of tankers could use the canal at its last depth. Plans have been made, however, for a massive 5- to 6-year program which would widen, deepen, and straighten the canal to permit the passage of 250,000-ton vessels. The cost has been estimated at about $1 billion.

A partial alternative to the canal was also talked of before 1967 but has been actively pushed in recent years—a pipeline running from the Gulf of Suez to the Mediterranean. Approval of the Sumed project faced numerous difficulties, which led to repeated delays in its acceptance. Efforts to secure the necessary financing were first assumed by a French-led consortium, which had to raise $380 million (up from the original estimate of $280 million). In October 1973 an American concern was reported to have received a contract for its construction; but after the October war it appeared that Egypt would secure sufficient funds from other Arab

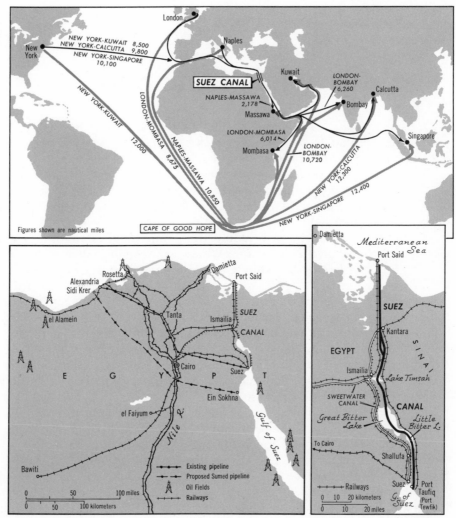

Map 46. The Suez Canal and proposed Sumed pipeline

states to finance the pipeline. A second factor that delayed the lines (twin pipes with an initial capacity of 80 million tons a year were planned) was the necessity to secure firm contracts for oil movements to ensure the ability to repay the loans; this required several years but was achieved in 1973; however, whether reopening of the canal may cause second thoughts is not known. A third concern was with the poor credit position of Egypt, met by stipulations that the bulk of receipts would go to repayment for the fi-

nancing. Finally, the fear of renewed hostilities, which did in fact occur, probably caused hesitation among some financing agencies and potential users. Even the reopening of the canal, it should be noted, does not rule out use of the pipelines, because tankers incapable of transiting the canal could be unloaded and loaded at deepwater terminals near Suez and west of Alexandria. Egypt could be expected to earn approximately $70 million a year once repayment of the financing had been completed.

CHAPTER EIGHT

The Republic of the Sudan

The Sudan, with an area of 956,218 square miles (2,505,813 sq km), is the largest of African countries. Its population was estimated at 17.5 million in 1974, with a growth rate of about 3.0 percent per annum. There is very uneven distribution, the highest densities being found along the Nile, including the Gezira area between the Blue and White Niles, and in an east-west belt from Gedaref to El Fasher.

Like other countries sharing the same latitudes, the Sudan contains climatic, vegetation, and land use regions that range from desert and steppe to savanna and rainforest; it is like them also in having an Islamized north and a pagan or Christian south. But it differs from the lands to the west in that it has a great through-flowing river, in its transport orientation to the east, in the marked isolation of its south, and in the relative position of its northerners and southerners. Here the former are the more evolved, the more politically conscious, and they have had longer and greater contact with the outside world. Westward from the Sudan—including the West African states, with the exception of Senegal—the situation is reversed.

The more permanent heritages from rule by the British under the Anglo-Egyptian Condominium, which lasted from 1899 to independence on January 1, 1956, were probably the installation of important irrigation works and a reasonably efficient rail system. Sudan has, since independence, experienced serious political difficulties including early disputes with Egypt, a series of coups and attempted coups, and a seventeen-year revolt in the south, which absorbed substan-

tial energies and funds and required the stationing of about half the army in that area. The two most constructive political achievements have been the conclusion of the 1959 Nile Waters Agreement with Egypt, which permitted the country to move forward with important extensions to its irrigation program, and settlement of the north-south dispute in 1972.

While the positions of the several governments have fluctuated considerably with respect to relations with Western and Eastern nations, they have rarely strayed too far from a pragmatic stance whose main emphasis was the economic development of the Sudan. When the "Revolutionary Government" of President Gaafar al-Nimeiry seized power in 1969 it instituted a series of nationalizations of foreign banks, commercial houses, and industries and of domestic pump schemes, and received a large influx of Russian aides. After an unsuccessful coup in 1971 in which Sudanese Communists were involved, the government adopted a more detached ideological position, arranged to compensate firms that had been nationalized, and introduced a new code designed to attract both Sudanese and foreign private investment.

One of the great concerns of Sudan has been its heavy dependence on cotton. When prices were high in the early 1960s the economy boomed; as prices declined in the 1960s and marketing difficulties arose, barter agreements were made with seven Communist nations, reducing flexibility in imports and resulting in the receipt of some machinery and capital equipment of inferior quality. Long-staple cotton, which was the main export for years and in which Sudan accounted for about a third of world supply, has had a fairly static market in recent years, in considerable part because of the use of synthetic substitutes. As a consequence, Sudan has tended to emphasize medium staple cotton in newer developments. But agro-economic studies have revealed that cotton is by far the most valuable crop in irrigation areas, and dependence on it has continued to increase. In recent years it has accounted for about 60 percent of total exports by value and a fifth of government revenue.

REGIONS OF THE SUDAN

Because the vast bulk of the Sudan is an immense plain land use is primarily affected by climatic factors and the major regions are vegetation-climatic belts running roughly east-west across the country. Availability of water from the Nile offsets this pattern to an important degree where it is used to support irrigation (Map 47).

THE DESERT AND THE NILE
NORTH OF KHARTOUM

The northern 30 percent of the Sudan is desert or semidesert, including almost all of the Northern Province, northern Darfur, and the northern half of Kassala Province. This section of the country is largely empty except for a few oases, a discontinuous strip along the Nile, and the southern part, which is used by nomadic pastoralists in the rainy season.

Cultivation along the 1,000-mile (1,600-km) stretch of the Nile from Khartoum to the border is restricted to a narrow and disconnected belt varying in width from a few feet to a maximum of 2.5 miles (4 km). A total of only about 550 square miles (1,425 sq km) supports over 750,000 people. Part of this is *seluka* land—that is, land along the banks and on islands which is flooded in the high water period and subsequently used for a catch crop, usually millet or wheat. Some is

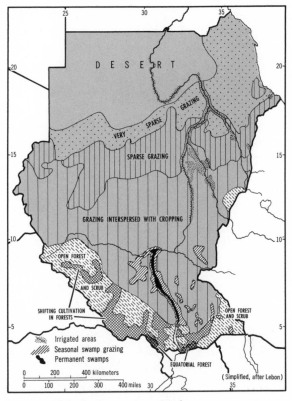

Map 47. Major land use regions of Sudan

saquia or shaduf land, subject in area to the vagaries of the river. Basin irrigation is utilized on an average of about 80,000 acres (32,400 ha) but the area flooded may vary enormously from year to year, and over half of the land is now irrigated by diesel pumps.

Despite having superb date and citrus soil (it is described as "the dream of perfection") most of the inhabitants of this belt live on the verge of poverty. Land fractionation, jealously maintained, "has a stranglehold on agricultural advancement" and rural indebtedness is almost universal. Failure to replace old date palms, which provide food, drink, construction material, and cash income for over half of the people, has reduced total production and commercialization. Toward the south of the river belt the date palm disappears and is partially replaced by fruits and vegetables, which are marketed in the

Three Towns (Khartoum, North Khartoum, Omdurman). A canning factory at Kereima, 200 miles (322 km) north of Khartoum, produces about 3 million tins of fruit, vegetables, and tomato paste yearly.

Lake Nasser extends southward into Sudan, covering an area formerly occupied by about 41,000 persons. A new Wadi Halfa port was constructed a few miles south of the old city; many of the residents were resettled at Khashm el Girba, and others chose to accept compensation for the loss of their properties. The resettlement project involved construction of the Khashm el Girba Dam on the Atbara east of Khartoum. The dam is capable of storing 500 million cubic meters to control about 500,000 acres (200,000 ha), and about 75 percent of this area is now developed. Completed in 1966, the dam cost about $20 million and the 26 settlement villages $36 million, bringing the total cost well above the $42 million compensation made by Egypt. A $28 million sugar factory at the site processes cane grown on 45,000 acres (18,000 ha) of the scheme; production costs for sugar have proved to be very high, over double the world price, but it is well below the cost at the earlier sugar area at Guneid (see below). The Khashm el Girba cane is produced with direct labor as opposed to a tenancy arrangement at Guneid; less mechanization has also been used, and lower costs have resulted. The main part of Khashm el Girba follows a more intensive rotation than earlier schemes, which has given higher returns. Cotton, wheat, and peanuts (usually called groundnuts in Africa) are the major crops, and the scheme accounts for about 10 to 12 percent of Sudanese cotton production; horticultural crops occupy about 7 percent of the cropped area.

ACACIA DESERT SCRUB BELT

South of the desert proper is a semidesert belt with erratic precipitation averaging between about 4 and 14 inches (102 to 356 mm) yearly. Covering the northern part of Kordofan, central Darfur, and much of southern Kassala provinces, the bulk of this area is inhabited by seminomads who settle in the winter near a permanent water source around which a wide belt of grass has been carefully left ungrazed during the summer. The graziers are seldom able to save enough fodder, however, and characteristically bring their animals through the dry winter on a semistarvation basis. When the rains begin to move up from the south, the tribal units split up; they first move southward and then swing around and proceed as far northward into the desert as the rains permit, sometimes to 18° N latitude. Only when there is no more forage available do they return to the *dammars* (settlement areas). Although this is a largely self-sufficient economy, there is some sale of sheep, wool, skins, and camels, the last sometimes being driven overland to markets in southern Egypt.

ACACIA GRASS STEPPE AND SAVANNA

Paralleling the previous belt is a broad region of steppe and savanna with an average of from 14 to 30 inches (356 to 762 mm) of precipitation. Including southern Darfur, most of Kordofan and Blue Nile provinces, and part of southern Kassala to the east of the Blue Nile, this belt contains the most developed parts of the country, including major irrigation schemes based on the Nile which have, of course, an atypical land use pattern within the whole region. It also includes several mountain massifs rising above the generally monotonously flat terrain. While most tribes in this area are still seminomadic because of the precarious precipitation, rain cultivation becomes important and accounts for a growing value of Sudanese commercial and export production.

Dura (great millet), millet, corn, sesame, and peanuts are grown on the *qoz* lands, former sand dunes that have been fixed by vegetation, while the areas in between are used for grazing livestock. Remoteness of the western portion of the belt restricts the possibilities for sale of crops and livestock, though some tobacco is shipped to Omdurman while cattle are driven to El Obeid. Commercial production becomes more significant in Kordofan, where better transport is available both to the largest domestic market and to Port Sudan for export. In addition to sale of the major staples, gum arabic and melon seeds are sold at auction in the markets of El Obeid, the latter being exported in considerable quantities to Egypt. Gum arabic, of which the Sudan accounts for 75 to 85 percent of total world exports, is a major source of income. Most of the good quality gum is collected from trees in enclosed tracts called *geneinas*. It is sold at auction under government supervision, cleaned, and shipped in bags to Port Sudan. Used mainly in candy, but also for adhesives, textile sizings, ink, polishes, insecticides, and beverages, gum arabic normally ranks second among Sudan's exports, but production fluctuates rather widely in relation to rain in

Gum arabic
The Sudan accounts for over three-fourths of total world exports of gum arabic, which is obtained from a species of *Acacia* tree.

the previous season and to prices offered. Efforts are now being made to replace destructive use of the axe by a special tool in collecting the gum, and to develop gum acacia plantations in several parts of the country to permit more effective collection. On the Kordofan Plateau, American upland rain-grown cotton is produced on an area varying from about 150,000 to 250,000 acres (60,000–100,000 ha); most is used in the two large textile mills in North Khartoum. The separated islands of higher land, the Nuba Mountains south of El Obeid and Jebel Marra in Darfur, were places of refuge for the Nuba and Fur tribes who were driven there in the Arab invasions eight or nine centuries ago. Some intensive practices were developed, such as terracing on the higher slopes, but overuse led to soil erosion and eventual deterioration of the terraces. Many people have moved to lower areas where water is available and now ship peppers, small tomatoes, cotton, and tobacco to Sudanese markets. The Baggara, formerly entirely nomadic, graze their livestock on the clay plains among the hills and grow cotton and grains where permanent water supplies are available.

Nomadic and seminomadic tribes also utilize the very favorable grazing lands along the White Nile and the less desirable lands along the Blue Nile, but there is increasing settlement and tillage agriculture in much of this whole area. Commercialization of livestock is poorly developed in relation to the large numbers of animals in the country. Improved transport has stimulated additional offtake, but opportunities in the steppe and savanna areas have only begun to be realized.

The special *harig* system of farming is practiced in several parts of this belt, though it now appears to be declining. Under this system, old grass is saved for several years until a dense mat is formed; after new grass has sprouted with the first summer rains the old grass is fired; this controlled firing kills off the new shoots and provides an ash into which grain or cotton seeds are immediately planted. Specialists feel that most of the disadvantages of grass firing are avoided by burning at this time. A grain crop can then come up without further attention until harvesting, while the indigenous tribes can move off with their flocks until that time approaches.

In 1944 the so-called Mechanized Agricultural Scheme was started in an area northwest of Gedaref, where the natural grasses could be easily cleared. Until 1953 the government participated by mechanically disking and sowing the lands, cultivators handled the remaining operations and shared equally with the government in the proceeds. This arrangement, patterned in part on the partnership prevailing in the Gezira, proved too expensive for low value crops such as dura and sesame and was abandoned in favor of the current private schemes. At present, about 2 million acres (810,000 ha) are under mechanical cultivation in the region around Gedaref, or some 70 percent of the mechanically farmed areas of the country. Dura, sesame, peanuts, and cotton are the major crops and, while returns have compared favorably with those on irrigation schemes, there are opportunities for substantial improvement through use of higher yielding seeds and better timing of field operations.

THE PROBLEM OF WATER SUPPLY. As in other steppe and dry savanna areas of Africa, the availability of water is of prime importance in most of the region under consideration. Around El Obeid some man-dug wells are 250 feet (76 m) deep. To the west, where there were fewer wells, baobab trees whose hollowed trunks may hold up to 1,000 gallons (4,500 liters) of water, were sometimes used as storage tanks. The Dar Hamar are said to be so saving of water that they make children whose heads are to be shaved run around until they are perspiring heavily.

In recent decades much attention has been given to increasing and conserving water in many of the semiarid areas of Africa, by techniques it is appropriate to summarize at this point because the Sudan has applied most of them in a variety of areas and programs. Thousands of wells and boreholes have been dug in many African countries in recent years, some much deeper than earlier sources. Africans must often walk long distances to secure water; frequently it is impure, which contributes to the high incidence of internal diseases. One of the most valuable material improvements that can be provided is a certain and clean supply of domestic water, although Africans sometimes complain that clear water lacks the taste of their previous supply. The multifaceted aspect of many African problems is well illustrated by these simple wells. On the one hand, they bring safer water, reduce disease, often permit grazing of livestock where it

A baobab or tebeldi *tree*
The hollowed trunk of the baobab, which is found in dry savanna and steppe areas throughout tropical Africa and Madagascar, is sometimes used to store water for the dry season.

Securing subsurface water from the stream bed of the Bubu River, Tanganyika
This illustrates the difficulty of finding adequate water for domestic and stock use in the arid and semiarid areas of Africa. In some cases such pits are so deep that two or three men working at different levels are needed to lift the water to the surface.

would otherwise not be possible because of an inadequate stock or human water supply, and sometimes permit the growth of fodder so badly needed in the dry season and required for a real upgrading of the livestock. On the other hand, they may contribute to greater soil erosion, degradation of the vegetation, and to possible overpopulation. In all too many cases microdeserts have been created by failure to control the use of wells. The point is that African problems need to be attacked on a broad front; an effort must be made to envisage the results of even the simplest of development projects so that their undesirable effects may be minimized.

Important techniques of conserving surface water include bunding, digging of surface reservoirs, construction of small dams both above and below the beds of streams and wadis, and larger scale irrigation projects utilizing modern control works. In the Sudan, *terūs* or earth banks are used extensively on the clay plains to decrease runoff and increase soak-in. A large number of surface reservoirs, called *hafirs,* have been provided; these large rectangular basins, about 15 to 20 feet (4.6–6.1 m) deep, may be hand dug, but are now usually machine excavated. Used to collect surface water in the rainy season, a hafir of 15,000-cubic-meter capacity can meet the minimum water requirements for about 2,500 people. Problems connected with hafirs are the prevention of silting, protection of the banks from destruction by man and animals, and avoidance of water pollution. The first can be met by providing desilting basins, the others by fencing, piping of water to a nearby tank, and control of numbers using the hafir, which is often very difficult. A somewhat specialized type of water collection is the use of inselbergs as a sort of roof supply, in which water is captured at the base to serve local communities.

The Sudan has several very interesting examples of mountain streams whose waters have been more effectively utilized with minor engineering works, though in some cases no control is considered feasible. The Tokar and Gash Deltas are important examples of flush irrigation. The former, situated near the Red Sea south of Suakin, is fed by the Baraka River, which drains a large part of the Eritrean Highlands (Map 48). It is dry most of the year, but from mid-July to mid-September violent spates arrive, sometimes lasting only a few hours, sometimes a few weeks. The silt carried by the Baraka may reach one part in 10, or 48 times as much as is carried by the Blue Nile. Irrigation on the Tokar Delta is an example of wild or uncontrolled flooding, there being only a slight control at the apex of the delta to give initial direction to the flow. The area watered varies from about 30 to 150 thousand acres (12,000–60,000 ha), but the soils are so rich and so remarkable in water-hold-

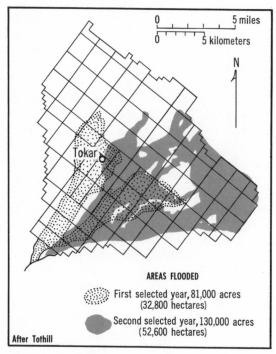

Map 48. The Tokar Delta

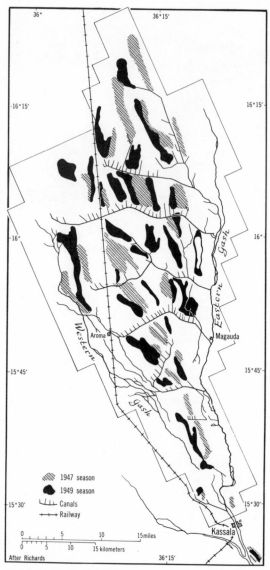

Map 49. The Gash Delta

ing capacity that cotton can be grown after only ten days of flooding. Land is leased from the government on an annual basis by registered holders. Cotton, castor seed, and dura are the main crops.

A similar but larger area is that of the Gash Delta, north of Kassala and due east of Khartoum (Map 49). An average of about 50,000 acres (20,000 ha) are flooded yearly, though the gross area is about 700,000 acres (280,000 ha). Unlike the Tokar Delta, no water ever reaches the sea from the Gash. Although there is greater control of the flow of the Gash, this is still irrigation in its simplest and roughest form. The flow, which is highly variable and lasts from 68 to 111 days, is carried down the eastern side of the delta in a natural watercourse; five main canals leading across the delta take off the relatively silt-free top water, and channels with masonry headwork lead from there to the fields. The soils are exceptionally rich and the cotton produced is very high grade,

but yields under this type of irrigation are only about half those of the Gezira. Land in the Gash Delta is distributed by the government on a tribal basis after the area that has been flooded has been delineated. Castor seed, cotton, and dura are the major crops.

A third flush irrigation scheme has been developed in postwar years on the Khor Abu Habl, which drains the northern portion of

A newly constructed well in northern Nigeria
Thousands of such wells have been dug in postwar years, greatly improving the supply of sanitary drinking and stock water.

A wadi near Omdurman, Sudan
Conservation of water from intermittent streams is one method of improving water supplies. Small surface and subsurface dams may be used, but silting often presents a serious problem.

the Nuba Mountains. Two small dams and canals permit the flush irrigation of 7,000 acres (2,800 ha), with an additional 3,000 acres (1,200 ha) soon to be added. But the soils did not permit cropping comparable to that of the Gash and Tokar Deltas so it was decided to provide a series of basins not exceeding 30 acres (12 ha) where, after thorough flooding, cotton would be produced. The relatively low incidence of cotton pests has led to use of the area to provide seed for the Nuba Mountains.

Another method of conserving water is the rejuvenation of underground storage both natural and artificial, the second being particularly applicable to North Africa. Some of the techniques noted above contribute to underground supply.

Large- and small-scale irrigation projects based upon gravity flow from permanent streams are receiving attention in many parts of Africa. While irrigation agriculture permits high intensity, it is also likely to require substantial capital outlays in relation both to the areas served and to the farmers engaged, so most countries must proceed to develop their irrigation potential over a period of years or decades, particularly when returns from rain growing can be just as high. The Sudan has greater dependence on irrigation than any African country except Egypt and a

large potential, which can be brought under control relatively easily. It has, in the Gezira Scheme and its extensions, by far the most important irrigation project in sub-Saharan Africa, as well as several lesser but still important developments such as the Khashm el Girba, Guneid, and large areas in pump schemes along the Blue and White Niles.

THE GEZIRA SCHEME. The most important economic "island" of the Sudan is the area of the Gezira Scheme, which took its name from "the gezira" or "island" lying between the Blue and White Niles south of Khartoum. With its Managil extension it accounts for about 60 percent of the cotton grown in the Sudan and a higher percentage of the value because of the higher-value, long staple cotton produced (Map 50).

Before the scheme was begun the semipastoral population of the area lived a traditional life, still enslaved by the requirements of a capricious climate. The rainfall permitted the planting of drought-tolerant grains, but at least two years in five produced poor crops. Today the gross cultivable area of the Gezira-Managil is about 1.96 million acres (790,000 ha) and the annual area in cotton is over 520,000 acres (210,000 ha). Cotton is

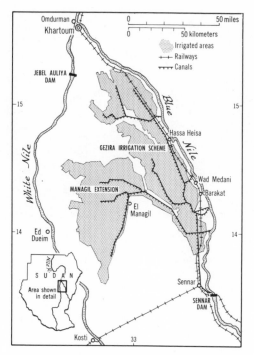

Map 50. The Gezira-Managil Scheme

the economic mainstay of the operation, but there is also an assured production of food and fodder. A highly organized, highly productive irrigation has replaced a low-productive, precarious semipastoralism. The ability to support people has been enormously increased, and per capita incomes are about three times the average for the country.

Despite certain problems associated with environment, the Gezira-Managil has unquestioned physical advantages. Topographically, the gentle downward slope from south to north and from the Blue Nile to the White greatly reduced expenses of leveling and permitted gravity irrigation, while the banks of the Blue Nile were sufficiently high above Sennar to make practicable the control and storage dam there. More recently, construction of the Roseires Dam at the only suitable site above the Sennar and within Sudan has permitted the storage of a much greater quantity of water for the Gezira-Managil and for other gravity and pump

schemes. Climatically, the prolonged winter aridity permitted a far more effective control of plant pests and diseases; this involves strenuous efforts during the dry period to uproot and destroy all cotton plants. The dry period also causes the soil to crack, so the greater penetration of water and air is possible. The soil, deposited many years ago in great depth when the Blue Nile flooded annually across the Gezira plain is high in minerals, and its high impermeability permits negligible loss by seepage, making the lining of canals unnecessary. Difficulties associated with the physical environment have included the effect of summer rain on weed growth and incidence of plant diseases, the constant battle against pests and diseases, the danger of waterlogged soils, and the necessity to provide drainage, especially from the lower-lying areas.

The two principal engineering features of the Gezira-Managil are the Sennar and Roseires Dams. The Sennar Dam, constructed in 1925 and later raised, has a storage capacity of 930 million cubic meters. The Roseires Dam, completed in 1966 at a cost of over $100 million, has a capacity of 2.3 billion cubic meters which can be raised to 4.4 by heightening the dam; it permitted the irrigation of an additional 300,000 acres (120,000 ha) in the Gezira and about three times that much elsewhere.

In the main Gezira the flow of water is continuous during the irrigation season in the canals and major distributaries, which command almost the entire area by gravity flow. A unique system of "night storage" is employed whereby water is fed to the minor distributary canals only by day, which assures easy flow, simplifies supervision, and prevents wastage. Continuous watering has been tried on half of the 100,000-acre (40,000-ha) northwestern extension. Water is applied to the individual fields on a carefully controlled and scheduled basis.

The rotation employed results in a crop division about as follows: ⅓ cotton; ¹⁄₆ wheat;

¹/₆ peanuts, vegetables, and fodder; and ¹/₃ fallow. Production of wheat is subsidized; its introduction is fairly recent and reflected the desire to reduce dependence on imports. All of the cotton grown is derivative of Egyptian cotton; the ecological suitability of the area to extra-long and long-staple cottons is an important element in the economic success of the scheme. Yields vary rather sharply depending upon differences in precipitation and the incidence of pests and diseases. The fight against the chief diseases, black arm and leaf curl, has been one of the most interesting and successful phases of the operation.

Some of the farming processes in the Gezira are mechanized, including most of the cultivation, crop spraying, and the digging of field irrigation channels. Most of the work, however, is done by hand with the use of relatively primitive equipment. The cotton is brought to the ginneries by a light railway run by the Sudan Gezira Board, which has the largest such enterprise under single management in the world.

Picking cotton in the Gezira-Managil Scheme
Fields are mechanically cultivated and sprayed, but most other operations are done by hand. Many migrant workers are employed.

Aerial view of fields in the Gezira-Managil Scheme
This scheme produces about 60 percent of the cotton grown in the Sudan; it is the largest irrigation project south of the Sahara.

Two of the most important features of the Gezira Scheme are the tenurial arrangements and the tripartite partnership under which it has operated. Use of the land was first nationalized, and the owners received a rent equivalent to the highest market rate before the scheme started (most has since been purchased); tenancies were then allotted, with priorities going to the former owners and their relatives. The tenants' real return comes from their share of the work, not through ground rent. These arrangements permitted direction of land use and prevented land speculation. Under the partnership arrangements, tenants receive a 44 percent share in the cotton crop plus full right to the other crops, the government receives 42 percent, 10 percent goes to the managing Sudan Gezira Board (in the original concession which expired in 1950 the private managing board received 20 percent of the gross profits), and the remaining 4 percent is shared between local councils and a social development fund.

The advantages of the original partnership arrangement were that it protected the interests of the indigenous people, permitted the long-term financing of capital expenditures, and secured the services, for a specific period, of an outside, independent, commercial organization with managerial skill and experience. The scheme, however, cannot be considered an unmitigated success. It has been likened to a huge cotton factory where there is little chance to develop initiative. There has also been criticism for failure to give adequate concern to social development. But the most serious difficulty is the danger of excessive population resulting from the practice by tenants of employing workers to do agricultural work, which is considered as socially demeaning. About 80,000 tenancies exist on the whole scheme, but about half of the 40-acre (16 ha) tenancies on the main Gezira are operated as half-tenancies, so the effective number is nearer 96,000; the Managil tenants have 15-acre (6-ha) holdings. In addition, the wage labor

force numbers perhaps 400,000. While the aversion to work may have the effect of leveling incomes, it also results in greater production costs, failure to realize the possible benefits of the operation, the development of an absentee landlord pattern, and the persistence of a rigidly class-conscious social system.

The general successes of the Gezira Scheme have led to its being copied in numerous schemes in Africa, most of which have failed, except in areas reasonably comparable to the Sudan itself. The features that appear to be necessary for success of a partnership scheme are the ability to markedly increase productivity, to focus on a high-value crop, and to control the cash crop to prevent illegal marketing and to ensure that receipts will be properly divided among the partners; in addition there is a need for specialist technical or managerial skill, for capital investment the farmers cannot themselves provide, and for sufficient control of farming practices to protect the soil, reduce damage by plant pests and diseases, and improve quality of the product. The Gezira met all of these prescriptions, but it is clear that its partnership and tenurial arrangments provide no panacea for agricultural development in tropical Africa.

THE RAHAD SCHEME. The Sudan was delayed by capital shortages in taking advantage of the capacity of the Roseires Dam. In 1973 funds totaling $158 million were finally secured from the IMF, the United States, and Kuwait to develop and provide equipment for the Rahad Scheme. Water is to be brought by canal and the River Rahad to an area between the Rahad and Blue Nile; 330,000 acres (134,000 ha) will be controlled in the first phase to grow medium staple cotton, peanuts, and vegetables. About 70,000 people will be settled in the area and seasonal employment will be available for about 90,000 workers.

PUMP SCHEMES. The first pump schemes of significance in the Sudan were developed be-

A large ginnery in the Gezira
The Sudan Gezira Board operates the largest ginning establishment under one management in the world.

tween 1917 and 1928 in the north and on the White Nile. Development on the Blue Nile was slower because of its relatively deep valley and a concern to protect the Gezira from infestation. By 1944 some 372 pump schemes containing a gross cultivable area of 181,000 acres (73,250 ha) had been installed. Postwar years have seen a remarkable extension, largely under private investment, and about 1.5 million acres (610,000 ha) are now under pump schemes. The acreage devoted to cotton is about 225,000 (91,000 ha) and the estates have accounted for about 20 to 30 percent of long staple production in recent years. Dura, fodder crops, and a variety of food crops for the Three Towns market are also grown. The Guneid Scheme, which is government-owned, was selected as the site for the first major sugar production in the country. The 15-acre (6-ha) tenancies proved to be unsatisfactory for cane growing and the cost of sugar produced by the Guneid factory has been about 3.5 times the

world price. Peanuts, beans, and fodder crops are also grown, but the scheme has revealed the difficulties of trying to diversify crop emphases. It is also disadvantaged by the relatively high cost of supplying water, which must be lifted as much as 65 feet (20 m).

There are substantial possibilities for increasing both pump and gravity irrigation in the Sudan. At present about 4 million acres (1.6 million ha) are controllable, 62.5 percent by gravity and the remainder in pump schemes. With about 2.5 million acres (1 million ha) actually irrigated, about 44 percent of the waters allocated to the Sudan is being used, while about 5 percent is lost to evaporation in the Roseires, Sennar, and Khashm el Girba reservoirs. Assuming a somewhat lower application per unit area in much of the lands to be irrigated in the future (e.g., in the Keneina area south of the Gezira) because of their heavier precipitation, Wynn has calculated that an additional 4.4 million acres (1.78 million ha) might be brought under control, of which about 82 percent would be cropped.[1] As elsewhere, the controlled area must exceed the irrigated area to allow time for the clay soils to dry and crack between crops, to provide a fallow at least one year in six to maintain soil structure and control perennial grasses, and to provide a "dead season" to reduce pest and disease losses.

SOUTHERN SUDAN

The bulk of the remaining area of the country is savanna, sometimes predominantly grassy, elsewhere mainly savanna woodlands, plus the enormous area of the Bahr el Ghazal subject to seasonal flooding and bits of rainforest in the extreme south. Most of it is remote and very undeveloped, contributing very little to the economy of the Sudan and less to the exports of the country.

Nilotic tribes such as the Dinka, Nuer, and Shilluk occupy the lands along the Nile and the great swamp of the Bahr el Ghazal. All of them have a "cattle culture," which is characteristic of pastoral tribes in most semi-arid areas of sub-Saharan Africa. The common feature of the cattle culture is the consideration of cattle as a sign of wealth and prestige, only rarely as a source of meat or an item for commercial sale. Cattle have particular significance in payment of the "bride price" and in ceremonial functions. The milk, which often has a high butterfat content, is drunk, though yields are typically low, averaging perhaps about five pints (or liters) a day. Some tribes also regularly draw blood from their cattle as an item of their diet.

The Dinka and Nuer live on the huge, flat, grassy plains between the watercourses of the Bahr el Ghazal, building their huts and grazing their livestock on the higher

parts and using the lower lands when the water recedes. Some of their practices reveal an excellent adaptation to the environment, such as knowledgeable selection of soil and the tethering of cattle so that land is evenly fertilized for the planting of dura. But the remoteness of this area and the character of the inhabitants suggest that the region will remain one of the more backward parts of Africa for some years, though there are possibilities for considerable intensification and particularly for the production of rice and sugar.

Equatoria Province, largely low, undulating plateau country rising in the southwest to the watershed between the Nile and Congo systems, is the wettest part of the Sudan, having an annual precipitation ranging from about 24 to 80 inches (610–2,030 mm). A great melange of tribes inhabits the province, of which the largest is the Zande, also found in adjacent parts of Zaïre and the Central African Republic. This is largely a region of bush fallow agriculture, with the major subsistence crops being those typical of rainforest and wet savanna regions elsewhere in Africa. Its extreme remoteness

A typical hamlet in southern Sudan
This remote area has very little participation in the money economy of the country.

[1] See R. F. Wynn, "Prospects for the Further Development of the Sudan's Share of the Water Resources of the Nile Basin," *East African Journal of Rural Development* (1971), pp. 37–66.

meant that for years it received minimal attention; expatriates were limited to a few missionaries and district officers. Juba is over 900 miles (1,450 km) and 12 days by river from Kosti; it requires two or three days by road to reach Nzara from Juba, while the region to the east of the river is very poorly served by any transport. The area was severely affected by the long revolt, which started before independence and erupted sporadically until a settlement was reached in 1972. International aid is helping in the reconstruction and the government is attempting to reinstitute the cash production of tobacco, coffee, and chillies and to develop a forest industry.

The rather interesting Zande Scheme, situated in the remotest part of Equatoria, represented the only major effort to upgrade the area. Begun in 1946 as "an experiment for the social emergence of indigenous races in remote areas," its goals were to improve the standards of local self-sufficiency and to provide for the purchase of those few necessities that cannot be produced domestically through a limited exchange with the north. Agriculturally, a system of organized bush fallow was applied under which the area was subdivided into rectangular blocks and each householder was allotted from 25 to 40 acres (10–16 ha) laid out in long, narrow plots running at right angles to the roads or cycle tracks on which the huts were erected. Each holding was further subdivided by distinctive hedge strips into a series of fields, which were opened up in succession, used for a few years, and then permitted to revert to bush fallow. Farmers in each block opened up the same plots each year, which at least made it possible to contemplate the use of machinery.

Each tenant was required to grow cotton on a half acre (1/5 ha) for sale to a mill built at Nzara, which produced cloth for local use and sale to the north. Subsistence crops were similar to those already grown, but special efforts were made to introduce higher yield-ing varieties, while scattered community oil palm plantations were set out. The scheme was well received at first when cotton prices were high, but there was retrogression after independence, particularly after large-scale textile mills at Khartoum eliminated that market, and it collapsed during the revolt. An effort is now being made to rehabilitate the scheme, but just which of its original concepts will be retained is unclear.

MINERAL PRODUCTION

There is at present very little production or export of minerals from the Sudan. Claims of important finds are issued periodically, but there is insufficient evidence to assess their significance. A small amount of iron ore comes from the Red Sea hills; the search for oil by Italian and American companies has thus far been unsuccessful. Exploration is being aided by the UNDP and the USSR and occurrences of mica, asbestos, gold and copper have been reported.

INDUSTRIAL DEVELOPMENT

Manufacturing development in the Sudan is less well developed than in at least ten other tropical African countries, and progress since independence has not been particularly favorable. There were some 300 plants in 1970, but over half were more traditional than modern. The gross value of production was estimated to be $146 million, of which 80 percent was in consumer goods, 14 percent from the processing of raw materials, and 4 percent from the manufacture of equipment. Industry was estimated to account for 12 percent of the GDP, but handicraft production accounted for the bulk of this contribution.

As in other parts of Africa, most modern plants are owned and managed by outsiders, despite the nationalization of a number of firms and initiation of about a dozen industries by the government. The present policy

A crude sesame seed oil press ot Khartoum
Handicraft production contributes more than modern manufacturing to the GNP of Sudan.

but builds passenger and freight cars for the Sudan Railways.

Production of electricity is still at a low level, but it did rise from 16.7 million kwh in 1948 and 103 million kwh in 1961 to over 400 million kwh in 1972. Only the larger towns have distribution of electricity and many factories must provide their own power. The hydroelectric potential was not tapped until 1962, when a 15,000-kw station was installed at the Sennar Dam. A station at Khashm el Girba added 17,000 kw, but neither of these plants has firm power. The Roseires Dam can support a capacity of 210,000 kw, but only a small part has thus far been developed. The total hydroelectric potential of the Nile Basin in the Sudan is only about 2 million kw, which reflects the general flatness of the country; a major segment of this occurs between the sixth cataract and Lake Nasser.

The Three Towns have the lion's share of consumer-oriented manufacturing, with the industrial estate at Khartoum North being particularly significant. This conurbation at the confluence of the Blue and White Niles has the advantages of central location, exis-

is to attract private investment, except in specified industries, or to have government participation in new ventures. Many of the state-initiated establishments have not been successful, in considerable part because the selection was ill-advised.

Industries based upon the processing of domestic raw materials include cotton ginning, flour milling, sugar refining, meat packing, tanning, canning of fish, fruit, and vegetables, and production of vegetable oils. Most of the plants that are essentially market-oriented have been constructed since independence. Their products include beer, soft drinks, cigarettes, shoes (two major plants including a Bata shoe factory, which has been nationalized), assembled air conditioners and water coolers, tires, batteries, paints, enamelware, aluminumware, pharmaceuticals, matches, containers, hardware, bricks, tiles, cement, refined petroleum products, and textiles (two relatively large spinning and weaving mills). The most important plant in the service industries is the railway workshop at Atbara, which not only maintains the locomotives and rolling stock

A modern cotton textile mill
Several large textile mills are situated at Khartoum North, which has the most important industrial estate in the country.

tence of the country's largest market in the cities themselves and in the nearby Gezira and pump schemes, relatively good transport links with the rest of the country, and proximity to the national administration. This triumvirate of cities had an estimated population of 691,000 in 1971. Pt. Sudan ranks a poor second in manufacturing and Atbara has special significance as the site of the railway workshops and a cement mill. The raw material processing plants are much more widely dispersed; the Gezira has the most important segments of the cotton ginning industry, which ranks first among this type of manufacturing.

TRANSPORTATION

The Sudan is relatively well off by sub-Saharan standards in the provision of transport routes and particularly in its considerable mileage of railways, totaling 2,955 miles (4,756 km). By all odds the most important means of physical communication, the railways are run by the largely autonomous Sudan Railways, which also operates Port Sudan and services on the Nile. Its staff of about 30,000 is a significant fraction of the total wage employees in the country. Moving almost all of the foreign traffic of the country, the railways usually operate at a profit if not always on time. Among the reasons for their overwhelming importance, especially as contrasted with road transport, are the great distances of producing areas from each other and from the port, the bulk character of the main exports, the relative ease of constructing lines across the vast plains of the country, and deliberate suppression of roads that might compete with the railways.

The railways were begun as a military line from the Egyptian border to Atbara in 1898 and Khartoum was connected with Port Sudan at a relatively early date. More recent additions to the railway go to Nyala in the west and to Wau in the southwest. Plans call for continuing the latter line to Juba and for

connecting the Sudan Railways to the East African railways at Gulu. There has been talk for some years of linking the western line across Chad to the Nigerian railways at Maiduguri, though it is difficult to see how the level of potential traffic would justify the investment.

The Sudan has a total of about 2,325 miles (3,740 km) of inland navigable waterways, two-thirds of which are open year round. Services operate year round on the White Nile between Kosti and Juba and on the main Nile in the north between Merowe and Kerma, and seasonally on the Sobat at Gambeila in Ethiopia and via the Bahr el Ghazal and Jur Rivers to Wau. Water transportation has always provided the chief link between central and southern Sudan, but the services are run at a loss because of the small traffic generated (c. 150,000–190,000 tons). Operation in the Bahr el Ghazal region is sometimes rendered difficult by floating vegetation, worsened by the appearance of water hyacinth in the late 1950s.

Few countries are so lacking in roads as the Sudan. Only around and within the main cities and in the south are there all-weather roads. In the dry season, however, vehicles can travel over large parts of the country. A long-delayed road between Khartoum and Port Sudan is now under construction, and construction of a more direct link to the south is planned. Camels and other animals are still significant in the movement of goods to railheads and in the more remote areas of the country, but because of their high cost, they are gradually being replaced by motor vehicles.

Port Sudan handles almost all of the foreign shipments of the country. Constructed to replace Suakin, which was inadequate except for dhows and small ships, it can accommodate 14 ships at its two quays. It has been handling about 2.5–3 million tons per annum in recent years, but has frequently been plagued by congestion, in part because of inadequate berthage, in part

Aerial view of Port Sudan
The only major ocean terminal of Sudan, this port has experienced severe congestion in many years, leading to a plan to construct a modern port at New Suakin.

because the rail line to Khartoum has been overloaded. In late 1972 arrangements were made to construct a pipeline from Port Sudan to Khartoum. Construction of a new port has been under consideration for some years; Suakin was rejected because of its difficult entry, but a recent decision calls for developing a nearby site, called New Suakin. In the first phase this port will have a capacity of 1 million tons and will handle bulk grains and minerals since it will be able to take vessels of up to 100,000 tons compared with the 30,000-ton limit at Port Sudan. Like many other countries in Africa, the Sudan is attempting to build up its own merchant marine and has purchased six small vessels to begin the fleet.

The Sudan airways, run as an independent corporation, handles internal and some international air services. It employs about 1,050 persons of whom only 38 are expatriates.

Part Three
WEST AFRICA

Traditional fishing at Kpémé, Togo, with mechanical phosphate loading pier in the background

CHAPTER NINE

West Africa: Introduction

With some 16 countries covering 20.3 percent of the continent, having about 31.2 percent of the population of Africa in mid-1973 and 23.5 percent of its exports by value in 1972, West Africa is one of the most interesting and most important parts of the continent. The tremendous diversity, especially in the political sphere, creates difficulties in writing about the area. A region-by-region or country-by-country approach would be excessively lengthy and repetitive; the functional approach is not entirely satisfactory. In this and the next three chapters a compromise is attempted: certain features, such as transportation, land use, mining, and industry, are treated in separate sections for the whole of West Africa, but analysis is given of the relative importance of individual nations and subregions in various economic activities.

West Africa, parts of which have long had contact with the Western world, has important contrasts with other areas of the continent, including its generally lower elevations, its greater political subdivision, an earlier political evolution, the predominance of smallholder production in commercial agriculture, and, until recent years, its relatively unimportant position in mining.

THE AREA AND ITS POLITICAL UNITS

West Africa comprises a total area of about 2.37 million square miles (6.14 million sq km). Considerable parts of the three largest countries—Mauritania, Mali, and Niger—are desert, however, so the utilizable area is roughly three-eighths the size of the United States. The area measures about 2,100 miles (3,380 km) from Dakar to Lake Chad, equivalent to the distances from Madrid to Moscow or Boston to Salt Lake City; the usable north-south distance is about 600–700 miles (965–1,125 km). Politically, the region is made up of fifteen countries (Table 11). Nine were under French rule prior to independence, eight of these being parts of French West Africa and one, Togo, a United Nations trusteeship. Four were under Britain, Guinea Bissau was Portuguese until 1974, and Liberia has been an independent country for over 125 years.

MAURITANIA

About two-thirds of the Islamic Republic of Mauritania is part of the Sahara. Some 80 percent of its population is Moorish (Arab-Berber and mixed racial origin) and 20 percent consists of distinct tribes of Negraic origin. The difference between the two major groupings is a major divisive factor, but intertribal and intratribal differences also exist. About two-thirds of the population is engaged in livestock herding, and most of the balance is sedentary farmers. Four-fifths of the population is concentrated in the south, especially along the banks of the Senegal.

The modern sector consists of enclave-type mining developments, which provide the main portion of government revenues and foreign exchange earnings, but which have almost no significant links with the predominantly subsistent and traditional economy. France has provided the bulk of aid since independence, but monetary and other ties were loosened beginning about 1972.

Table 11. West African political units: area, population, per capita GNP, trade

	Area		Population	GNP per cap	Foreign trade			
					Exports		Imports	
Country	Sq. mi.	Sq. km.	Mid-1974	1971	1961	1972	1961	1972
	(thousand)		(thousand)	$ U.S.	(million $ U.S.)			
Mauritania	398	1.031	1,280	170	2	92 [a]	31	71 [a]
Senegal	76	196	4,319	250	124	195	155	364
Mali	479	1,240	5,483	70	10	54	36	115
Upper Volta	106	274	5,844	70	3	24	28	63
Niger	489	1.267	4,468	100	...	56	19	68
Gambia	4	11	506	140	9	25	13	31
Cape Verde Is.	2	4	285	180	11	10 [b]	11	15 [b]
Guinea (Bissau)	14	36	580	250	4	3 [b]	11	27 [b]
Guinea	95	246	4,289	90	61		50	
Sierra Leone	28	72	2,759	200	82	132	91	158
Liberia	43	111	1,496	210	66	289	91	192
Ivory Coast	125	323	4,746	330	177	857	150	710
Ghana	92	239	9,679	250	316	594	394	449
Togo	22	56	2,178	150	19	61	26	101
Dahomey	44	113	2,975	100	13	47 [a]	25	94 [a]
Nigeria	357	924	61,040 [c]	140	482	3,385	623	1.877

SOURCE: U.N., *Statistical Yearbook, 1973* (New York, 1974);—*Monthly Bulletin of Statistics,* 1974; I.M.F., *International Financial Statistics,* XXVII, No. 4 (April, 1974); I.B.R.D., *World Bank Atlas* (Washington, 1973).

[a] 1972

[b] 1970

[c] Preliminary figures from the November 1973 census give the population at that time of 79,758,969

Relations with the Maghreb were difficult until a detente was reached with Morocco in 1970 and Algeria offered support in the Franco-Mauritanian differences in 1973. The meager resources of the country as far as land use is concerned pose very difficult problems for development.

SENEGAL

One of the two most important economic units of former French West Africa, Senegal accounts for a very high percentage of peanut-product exports from this area, and these account for over half of Senegalese exports by value. Indeed, the very heavy dependence on one crop, which has been only moderately reduced by mining and fishing developments and by efforts to diversify ag-

riculture, is a major economic problem of the country.

A notable factor in the importance of Senegal is the presence of Dakar, which has the bulk of the country's 40,000 to 50,000 non-Africans, including perhaps 29,000 French, a disproportionate share of manufacturing industry in the former French areas, and a commercially very strategic location in ocean and air transport. Dakar's position as the capital of the huge French West Africa, together with its unusually attractive climate, accounts in considerable part for its having so large an expatriate populace and such a relatively well-developed interest in manufacturing. But despite continuing heavy subsidization, the economy has been somewhat stagnant in postindependence years, and

Dakar possesses a degree of artificiality that is causing uncertainty about its economic future. Greater Dakar, with a population of 497,000 in 1970 (or 14.8 percent of the total), continues to receive a large influx of migrants despite the lack of accommodations and job opportunities, a phenomenon present in most of the prime cities in Africa.

MALI

Largely desert and steppe, Mali is one of the poorest areas of Africa, though it has interesting potentials in mining and in further irrigation of the inland Niger River delta. Mali early set itself resolutely upon a course designed to create a planned, highly centralized, socialist economy. Despite considerable aid from Communist countries the effort was unsuccessful, and following a coup in 1968 the government has followed more moderate lines. While Mali depends heavily upon peanuts as its dominant export, the bulk of the population is engaged in self-subsistent grazing and farming.

UPPER VOLTA

The Republic of Upper Volta, like Mali, participates to only a minor degree in the commercial economy of West Africa. Much of the farmed area has poor soils, which have been seriously degraded by overuse, while use of the better soils in the river valleys is precluded by the prevalence of onchocersiasis (river blindness) and other diseases. Over 98 percent of the agricultural produce is self-consumed and imports of food greatly exceed agricultural exports by value. Much of the country is suitable only for grazing, and animals and animal products provide the major part of its small exports.

The Mossi, whose kingdom was developed over 900 years, compose almost half of the total population. They are concentrated in the central portion of the country at unusually high densities for the latitude. A very important measure of relief and significant exchange earnings derive from the large

A boy guiding two blind victims of onchocerciasis or river blindness in Upper Volta
Some 20 million people in West Africa are said to suffer from the disease, whose incidence keeps some otherwise favorable areas of Upper Volta empty of people.

number of Voltans who migrate to adjacent countries to seek employment. Estimates of the numbers involved vary greatly from about 450,000 to a million; about 150,000 migrate on a seasonal basis, 200,000 for periods ranging from one to five years, and an indeterminate number have located definitively in other countries. About three-fourths of the emigrants characteristically go to the Ivory Coast and the bulk of the remaining quarter go to Ghana. Studies have revealed that agricultural diversification faces severe difficulties in Upper Volta, but there are opportunities for replacing much of the imported foodstuffs by domestic production.

NIGER

The Republic of Niger, like Mauritania and Mali, is mainly desert. Its economically important area is limited to the lands along the Niger River and bordering Nigeria, where peanuts and cotton, the main exports, and grains are produced. About 52 percent of the population is concentrated on 8 percent of the area having over 21.5 inches (550

Concentration plant at uranium mine, Arlit, Niger
This and other uranium mines, situated in the desert, promise to bring new wealth to this poor country.

mm) of precipitation annually; 83 percent of it is found in the 24 percent having over 13.7 inches (350 mm) of rain. Only about 2 percent of the total surface is cultivated; livestock grazing covers the bulk of the utilized area, which comprises something over half of the country. Nonetheless about three quarters of the population is sedentary and follows activities well differentiated from the predominantly nomadic ethnic groups. Niger has attracted considerable interest in recent years as a potential source of minerals and several deposits of uranium ores have been developed.

GAMBIA

A riverain enclave in Senegal, Gambia has been called the peanut colony, a term apt for its size as well as for its overwhelming dependence on that crop. While there was some expectation that it might join with Senegal in a Senegambian Union, it now appears determined to retain its full independence. It has followed conservative economic policies, but

continues to rely on financial aid from Britain to a significant degree.

THE CAPE VERDE ISLANDS

This archipelago has a total area of 1,557 square miles (4,033 sq km). Long a Portuguese possession, after the 1974 coup in Portugal it negotiated for independence, which is expected to be granted. Ten of the islands are inhabited, some at unusually high densities, by a mixed population of Portuguese-Guinean origin; many Africans were originally brought as slaves to what had been unpopulated islands when they were discovered. The islands are mountainous and of volcanic origin; Mt. Fogo, at 9,279 feet (2,828 m) the highest point, was active in the seventeenth century and again in the 1950s.

The leeward group of islands, arid and sparsely populated, is used mainly for goat grazing. The windward group, also severely disadvantaged by aridity, produces a variety of crops including coconuts, sugar cane, bananas, oranges, and coffee, which is the main export. Fishing and bunkering are of some importance on São Vicente. The keynote, however, is poverty, related both to the meager resources and the excessive population pressure.

GUINEA-BISSAU

This territory is a relic of early contacts made on the West African coast by the Portuguese, who claimed it from 1462 and who developed Bissau as a slave port from 1690. A revolt against the Portuguese began in 1961 and liberation forces gradually occupied about half of the country, tying down some 20,000 to 30,000 Portuguese troops. In 1973 the government-in-exile was recognized by many African governments and admitted to the OAU; like its close relation the Cape Verde Islands it negotiated for full independence in 1974. The economy of the country is dependent mainly on the export of peanut, coconut, and oil palm products.

GUINEA

This country was the only French African area to opt for independence in the plebiscite of 1958, whereupon it was summarily abandoned by France. Declaring itself a revolutionary socialist state, the government nationalized much of industry and services and even attempted to bring retail trade and agriculture under state control. Much aid and many technicians were provided by Communist states, but relations deteriorated when Soviet aides were accused of political machinations. Despite aims to the contrary, the economy has become increasingly neocolonial in character. Agriculture has stagnated, industry operates at only about 20 to 40 percent of capacity, corruption and clandestine marketing are widespread, and some 300,000 Guineans are believed to have emigrated to seek employment, mainly to the Ivory Coast.

Guinea has enormous bauxite reserves, and is heavily dependent on returns from the mining and processing of this main source of aluminum by expatriate firms. Despite this marked dependence on foreign enterprise the government maintains a consistently militant socialist stance and the party attempts to supervise practically every aspect of life. That this militance is partly verbal is suggested by the recent conclusion of new mineral contracts and an investment code designed to attract foreign companies to specific activities.

Guinea has often severely criticized the neocolonialism of its neighbor Ivory Coast, and it has followed policies of Africanization, nationalization, and socialization far more vigorously. Ironically, its functioning economy is more dependent on foreign enterprise, its physical and social infrastructure is less developed, and its rural residents are less involved in commercial production. In the long run Guinea has excellent potential, with great wealth in bauxite, iron ore, and diamonds, a varied physical milieu capable of producing numerous crops and timber, and substantial hydroelectric resources.

SIERRA LEONE

Freetown was established in 1788 by British philanthropists as a settlement for liberated African slaves, and the "Creoles" who live there are descended from those settlers. More advanced than the tribal peoples of the hinterland, they faced a difficult transition period before and after independence as increasing democratization weakened their position. This has contributed to the considerable instability that has characterized the political scene to the present. Relatively poor economically despite a broad range of exports, Sierra Leone has become increasingly dependent on shipment of minerals, particularly diamonds, which make up about three-fifths of total domestic exports. Illicit diamond operations and smuggling to adjacent countries have adversely affected the economy and perhaps even the psychology of the country; the expectation that output can last at high levels for only a decade or so raises serious questions for the future. The agricultural economy has remained rather stagnant despite efforts to increase production of cash crops and rice, the main subsistence crop.

LIBERIA

This country is one of the oldest independent states in Africa; Monrovia and other coastal points were settled with American aid by freed slaves whose descendants are called Americo-Liberians. Until after World War II there was very little economic and social development, with the exception of the large Firestone rubber plantations. In postwar years, greatly increased American aid and investment brought it rapidly into contact with the economic world and, for the first time in its history, efforts were made to bring the tribal interior into the national economy. The "open-door" policy adopted by the government also attracted investment from Western Europe and Japan.

The most sensational development has been the opening of iron mines, which have made Liberia one of the leading exporters of iron ore and pellets in the world. Efforts to expand agricultural output have been less successful. Forestry, fishing, and industrial developments have been started where almost none existed before. The country has also seen great improvements in its infrastructure and in its social services. The size of the labor force employed in the money economy increased from only about 30,000 in 1950 to about 165,000 in 1972. Government revenues increased from $4 million in 1950 to $32.4 million in 1960 and $96 million in 1974, and the country now has the fourth highest GNP per capita of West African states. Much work remains to be accomplished, of course, especially in bringing the people of the hinterland into the picture, but Liberia has made remarkable strides and is no longer the butt for criticism of other African countries.

IVORY COAST

This republic, which became independent in 1960, is the richest of all the former French areas in sub-Saharan Africa, producing large quantities of coffee, cocoa, and tropical woods and increasing tonnages of palm products, bananas, pineapples, and cotton. The government has been pragmatic, supportive of free enterprise, and receptive to foreign investors and technical assistance, though it has participated strongly in the economy, particularly through price supports in agriculture. There has been a massive influx of capital to the Ivory Coast, about 80 percent of which has come from France. The importance of the foreign presence is suggested by the size of the French population, about 40,000 in 1973 or double the level at independence, and by the estimate that Europeans receive about three-eighths of the national wage bill.

The Ivory Coast policies have subjected it to criticism by the more socialist-oriented states and by some Ivoirians, who favor a more rapid localization of personnel and a reduced foreign presence. President Houphouet-Boigny long resisted such moves but his government is now giving increased attention to Ivoirization through the reservation of certain new activities in agriculture, forestry, and services to citizens, to government participation in mixed corporations operating in all sectors, and to the encouragement of private local investment.

Criticism also arose over what was considered to be an excessive concentration upon Abidjan, whose modernity and dynamism faded out rapidly in the hinterland. In 1963, for example, the south received 90 percent of total investment and Abidjan alone got 66 percent. Since then major efforts have been made to introduce cash crop production in the north and two massive regional development schemes are in being, one centered on the new port of San Pedro in the little-developed southwest and the other based on the Kossou Dam on the Bandama River.

Whatever the faults of the Ivory Coast economy may have been, it cannot be denied that remarkable progress has been made in practically every sector. During the period 1960–71 the GNP per capita increased by an estimated 4.6 percent per annum as compared to 0.1 percent for Guinea and 0.0 percent for Ghana, both of which countries followed very different policies and both of which were highly critical of the Ivory Coast. The strength of the economy is suggested by the attraction of about 1.5 million foreign African workers including about 600,000 Voltans, 300,000 Guineans, 200,000 Malians, and 200,000–300,000 Ghanaians. Indeed, about 65 percent of unskilled workers in the country are non-Ivoirians. While the Ivory Coast is certainly not yet ready for take-off, as has sometimes been claimed, its productive capacity and efficiency of production have been greatly increased in the past score of years.

GHANA

The first British colony in tropical Africa to achieve independence (March 6, 1957), Ghana was formed from the Gold Coast and the Trustee Territory of British Togoland. The world's largest producer of cocoa, West Africa's main mineral exporter at the time, and having reserves of $700 million at independence, Ghana was one of the richest of black African countries. President Kwame Nkrumah, however, in an effort designed to create state socialism as rapidly as possible, expended vast sums on state industries, collective farms, work programs, and the military so that by early 1966 the country had used up its reserves and accrued a debt estimated at about $520 million, three-fourths of which was in suppliers' credits. While the government posture and emanations from the press and radio were Marxist, anti-Western, anti-European, and anticolonialist, and

there was a strong presence from Communist countries beginning in 1961, economic actions were seldom so radical as the verbiage. There was, however, an increasing penchant for repressive legislation and for deification of the leader.

By 1966 Nkrumah's policies had been repudiated by most Ghanaians and early that year he was ousted in a military coup. The military leaders followed a more conservative, pro-Western policy, and in 1969—after adoption of a new, revised constitution —relinquished their authority to a civilian government. Both regimes were plagued by the difficult financial position they inherited, which was only partially relieved by a two-year moratorium and successive reschedulings of the nation's debts. Worsening the situation were lower world prices for cocoa and stagnation of mineral sales. The civilian government was overthrown in a bloodless military coup in January 1972, when power was assumed by the National Redemption Council. A massive devaluation followed, some debts were abrogated, minimum wages were raised, and the status of the military was reestablished. Financial problems, including inflation and a weak trading position, remain serious, but efforts have been taken to adopt programs that will achieve more effective growth. In the meantime, the image of Nkrumah, who died in exile in Guinea, has been partially restored, particularly for his early promotion of Panafricanism.

Despite its political and economic vicissitudes Ghana continues to be one of the richer and better developed of African countries. It has important assets, not the least of which is a considerably larger proportion of educated persons than most African countries.

TOGO

A small sliver of territory between Ghana and Dahomey, this former French Trusteeship relies mainly on sales of coffee, cocoa, palm products, and phosphates for its international exchange. With about 78 percent of its population engaged in agriculture, the country's production of food has not increased so rapidly as its population, and food imports have increased as a result. Greatest importance attaches to the need to rationalize agriculture; only an estimated 11 percent of the area is cultivated, while an additional 26 percent is estimated to be cultivable.

DAHOMEY

The smallest of the former French West African countries, Dahomey is a minor producer of palm products and miscellaneous other tropical commodities. It remains very heavily dependent on agriculture, but only about a ninth of the cultivable area is under use. Dahomey has experienced one of the most unstable political histories of African countries since independence; there have been nine changes of government, including five military coups, from 1960 to 1975.

NIGERIA

With the largest population of any African country and over half the population of all of West Africa, Nigeria has considerable physical diversity and a multitude of tribal groups. Political rivalry among the major ethnic groups accounted in a major way for two military coups in 1966 and the succeeding civil war, which lasted from May 30, 1967, to January 15, 1970, when the Republic of Biafra surrendered. Estimates of the casualties vary from about 1 to 2 million; most of the deaths were caused by starvation. The military government under General Yakubu Gowon, which has ruled since the second coup in 1966, has stated that it plans to return the government to civilian rule in 1976, but it now appears this goal will be postponed. Meanwhile the previous four regions have been replaced by twelve states

in an effort to reduce friction among the major ethnic groups and to satisfy some of the lesser groups who resented the domination of others.

While the per capita GNP of Nigeria is well below those of Ghana or the Ivory Coast, its much greater population alone makes it one of the leading economic units of tropical Africa. Its main agricultural exports normally comprise three crops—palm products, cocoa, and peanut products; it also has significant exports of rubber, tropical woods, sesameseed, cotton, and hides and skins. The agricultural sector has, however, witnessed certain stagnation in recent years, which causes concern because of the continuing high percentage of the population engaged in agriculture (c. 70 percent in 1972) and the high rate of population growth; in 1972 production of agricultural export commodities was only 36.4 percent of the 1960 level and total agricultural output was estimated to be expanding by 2 percent per annum as compared to a 2.5 percent yearly increase in population.

The most dynamic economic development has been in petroleum, for which Nigeria rivals Libya among African producers. Petroleum accounted for only 17.9 percent of exports in 1968 but for about 80 percent in 1973, at which time it provided over three-quarters of government revenues. The increase in the posted price of Nigerian oil from $4.29 a barrel in August 1973 to $14.69 on January 1, 1974, is expected to increase oil revenues to about $7 billion in 1974–75 or seven times the level in 1971–72. The natural gas resources have had only minimal use to date, but large contracts for export of LNG may be expected in the near future, and work has begun on a 960,000-kw gas turbine electric plant at Ogorode in Midwest State. Other minerals play only a minor role in the economy.

Nigeria experienced a rather dynamic industrial growth before and after the civil war and its large potential market continues to

Drilling for oil in southern Nigeria
Petroleum now accounts for over 80 percent of Nigerian exports by value.'

attract investment. The country faces continuing serious problems, however, including the necessity to control inflation, rapid urbanization, unemployment, the shortage of technical specialists, the relatively poor showing in agriculture, and the disposition of an idle army of about 240,000. While its large population has been seen as a major asset, it also presents severe problems, particularly since an estimated 46.5 percent of the population is now affected by pressure on the land.

THE PATTERN OF POLITICAL UNITS
IN WEST AFRICA

West Africa has the greatest political fragmentation of any African area, explained in part by the length of its contact with Europe. European countries were early interested primarily in coastal points where trade might be effected; if there had been an interest in

colonization some of the larger powers might well have absorbed the holdings of lesser nations. Division of the interior came partly in response to demands to quell tribal wars and to prevent what were considered to be barbaric practices, but largely to satisfy imperialistic appetites or to offset those of others. During the period of division France displayed the greatest initiative, moving behind the coastal extensions of other powers to consolidate the large, contiguous French West Africa.

The superimposed political pattern has had significant effects on the economy of the area, aside from its failures to divide the area into logical physical units or to recognize the various tribal areas, which were often split by artificial boundaries. One of the unfortunate effects has been the projection of economic nationalism to West Africa, seen best in trade and transport policies. Differing administrative, legal, educational, and economic systems and differing linguae francae also tend to divide the area. All countries have rejected possible changes in their artificially superimposed boundaries, and there has been relatively limited progress in achieving economic cooperation. The boundaries are sometimes rather porous, however, and significant quantities of goods move clandestinely across them.

Since independence the southern tier of countries has tended to outpace development in the northern tier, particularly in the three landlocked countries, and all signs point to the widening of gaps among the several countries. Despite the contrasts stemming from political differences in West Africa, the basic land use and economic pattern is much more affected by climate. Broadly speaking, the same products and the same types of agriculture are found in the similar climatic belts of the various countries. These climatic belts run predominantly east-west, more or less counter to the direction of most political boundaries, especially along the Guinea Gulf.

PHYSICAL OVERVIEW
OF THE AREA

LANDFORMS

While West Africa may be divided into hundreds of fairly well-defined topographic units, the overall pattern is fairly simple; most of it is plains and plateaus under 1,500 feet (460 m). A typical traverse from south to north would cross, first, a relatively narrow, low-lying zone, consisting of sandbars often backed by lagoons and mangrove swamps; second, a coastal plain varying in width and rising gradually inland to a more or less well-marked escarpment; third, the West African Plateau, composed primarily of an erosion plain of pre-Cambrian rock bordered by plateaus whose sandstone layers form scarps falling abruptly to areas of less resistant rock; and fourth, a lower basin of immense alluvial plains occupied in part by the Senegal and Upper Niger rivers and Lake Chad and often marked by volcanic inselberg. Only in a few areas do higher lands occur: the cuestal Fouta Djalon Plateaus and Guinea Highlands on the west, rising to over 6,000 feet (1,830 m); the Togo and Atakora ranges; the Jos Plateau in Nigeria; a series of mountains and high plateaus running along the eastern border; and Saharan massifs north of the utilized zone.

From the landform standpoint alone, West Africa is one of the more favored of African areas. There are relatively few landform barriers; scarps are not so formidable as in the remaining portion of the west coast and most of southern and eastern Africa. The average low elevation also suggests the easier overland movement that characterizes the area, while it further explains in considerable part the paucity of white settlements, which led to difficult political problems in other parts of the continent. Perhaps the greatest topographic handicap of the area was the absence of good natural harbors or of protected bays along much of the coast, particularly along the Gulf of Guinea; the

construction of new ports has greatly reduced the impact of this handicap.

CLIMATE

The most important physical factor in explaining the general pattern of commercial agriculture of West Africa is climate—particularly precipitation. Some climatic regions are relatively well defined because of topographic conditions, but there is a gradual transition from zone to zone in most of the area (see Maps 15–17). A certain arbitrariness therefore attaches to a delineation of specific regions, particularly because precipitation may vary greatly from year to year or in cycles of several years.

The major characteristics of precipitation are as follows:

(1) A progressive decrease as one goes from south to north. Rainfall is especially heavy where the land rises to create an orographic influence. The average annual precipitation drops off relatively rapidly toward the interior—for example, from 110 inches (2,800 mm) on the Niger Delta coast to 80 inches (2,030 mm) at Benin 65 miles (105 km) inland, 49 inches (1,245 mm) at Bida, and 33 inches (838 mm) at Kano. Rainfall on the coast of Liberia is 120 inches (3,050 mm), while at the inland boundary it is down to 60 to 80 inches (1,500 to 2,030 mm). An exception to this general pattern is seen in the corridor of drier climate extending from Cape Three Points in Ghana eastward to Dahomey along the Gulf of Guinea coast. It is not fully understood why this anomaly exists, but the facts that winds at this section of the coast are more parallel to the coast and that cool waters upwell along it are probably important.

(2) A progressive decrease in the length of the rainy season toward the interior. Along most of the southern coast

it lasts from eleven to twelve months; about 150 miles (240 km) inland it is only about nine months, while at Kano it is down to five or six months' duration. In some sections, double maxima occur as the sun's direct rays move north and then south in the summer. The length of the rainy season is a critical factor for vegetation and crops. For example, crops associated with the tropical rainy climate may be grown when precipitation is no more than 40 to 60 inches (1,000 to 1,500 mm) if the dry season is of moderate duration and not entirely dry. Lengthier periods or double maxima also may permit double cropping, whereas more northern areas are limited to one crop per year unless irrigation can be applied.

(3) A decrease in dependability of precipitation toward the north. The generalization that the drier the climate the greater the variability of precipitation applies to this area as it does to most parts of the world, but there is somewhat greater dependability owing to the monsoonal influence on the area. It is this influence which also explains the short dry period in the southern portions of the area. Recent years have presented a tragically dramatic example of the great variability of rainfall in the drier areas. In the seven-year period 1967–73, a wide belt bordering the Sahara experienced drought conditions in five years, and the 1972–73 crop year was the most disastrous since 1913. Over 25 million people were affected; it is believed several hundred thousand died of famine, while millions of livestock also perished.

Temperature conditions are of minor importance in influencing land use in West Africa. Temperatures tend to be high throughout, but there are contrasts between coastal and interior regions. In tropical rainy areas temperatures are high year round, averaging about 85°F (30°C). Coupled with the characteristic high humidity, they produce high sensible temperatures and hence considerable discomfort. During the short period when the winter monsoon, the *harmattan*, is blowing, humidity is somewhat reduced.

In the interior, a slight seasonal temperature curve appears, but the summer rains have a cooling effect, which flattens the curve in that season. The diminution of vegetation, clear skies, and low humidity make for high diurnal temperature ranges. Nights can be uncomfortably cold; days may have temperatures rising to 100°–120°F (43°–49°C).

VEGETATION

The vegetation zones of West Africa show the same transition as the climate (see Map 18). The neat transitions suggested by any of the systems of classification used often do not exist in fact. Use by man and the effects of fire, usually set by man, often distort the pattern, resulting particularly in a sharp division between grasslands and the tropical rainforests. Several interesting problems pertaining to the vegetational pattern of West Africa exist: How rapidly is the rainforest disappearing and with what results? How serious is grass burning? How can the vegetation and soils of the area best be preserved? What human practices are tending to cause deterioration in the natural landscape? Is West Africa suffering from progressive desiccation with a consequent invasion of Saharan conditions into the steppes and steppe conditions into the savanna? Most of these questions will be examined in conjunction with discussion of the land-use regions to which they apply. The last question may be briefly examined at this point.

Until a few decades ago it was quite generally accepted that climatic desiccation was taking place here and in other parts of the continent, and some students continue to maintain that West Africa is suffering from progressive desiccation. In support of the contention it is claimed that the long period of tribal southward migration has been required by desiccation, that sand is actively drifting in some areas, that Saharan drift is gradually filling the Lake Chad basin, and that alluvial soils underlie sands in parts of Senegal. It is further pointed out that ancient rock engravings in the Tibesti Mountains and other parts of the Sahara depict elephants, hippopotamuses, and other animals that could no longer be supported there, and that some stream profiles in the desert are characteristic of such profiles in more humid areas.

Much of this evidence is not questioned by those who believe that climatic desiccation is not now taking place in the area. There is agreement, for example, that the desert was wetter in Quaternary times, 10,000–15,000 years ago, but many question whether it is a continuing phenomenon. They note that rain-eroded dunes may be seen along the desert border, that a blanket of sand found from the Sahara to about 10°30′N, and presumably blown there, is everywhere fixed by vegetation, and that the extreme youth and infantile character of stream lines in the Fouta Djalon, one of the wettest areas in Africa, suggests that this region was drier in the past than it is now.

Whatever the fact may be—and more evidence is needed for certainty—it is agreed that bad human practices have led to degradation of vegetation, which would give the impression that there was desiccation. These malpractices, which will be covered later, include deforestation, overcropping, overgrazing, and grass burning. There is little question that man is slowly, and in some cases not so slowly, downgrading a vast part of the West African environment.

SOILS

The generalities regarding tropical soils given in chapter 2 pertain to West Africa and need not be repeated here. It should be stressed, however, that increasing evidence suggests the necessity for more detailed soil studies, including analysis of the association

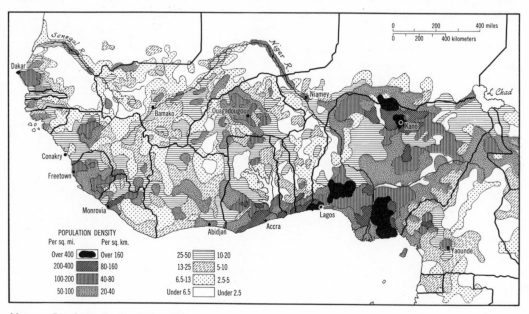

Map 51. Population density of West Africa

of important cash crop areas with distinctive soil regions and examination of some fairly intricate soil selection systems.

POPULATION PATTERNS

The distributional pattern of West African population is not easily discernible or readily explained (Map 51). The broadest-scale features are the large empty area in the north, the heavier densities toward the southeast of the region (Ghana, Togo, Dahomey, and Nigeria, with 21.7 percent of the area, have about 70 percent of its total population), and the relatively high rural-to-urban ratio (c. 5 to 1 if urban centers above 5,000 are used). Senegal, Ivory Coast, Ghana, and southwestern Nigeria have higher urban percentages, though the last includes many large Yoruba towns, which still contain a high percentage of agriculturalists.

Observers have sometimes attempted to delineate parallel belts associated with the east-west climatic/vegetation zones. Further analysis suggests that the patterns are much more complex, but a four-zone division not rigidly confined to physical zones is useful as a generalizing device. Very briefly these belts and the major explanations for their population characteristics are as follows: [1]

(1) The northern desert/sahel belt, roughly north of the 15°N parallel, which is empty or very sparsely populated except for bands along the Senegal and Niger Rivers and lesser nodes associated with the Aïr and Adrar des Iforas Massifs. The general sparseness or lack of population is obviously climatically controlled. The major exceptions are explained by the presence of through-flowing streams, availability of ground water, or higher-than-average precipitation due to the oro-

graphic influence. Minor exceptions are explained by the occurrence of mineral exploitations, and by fishing activities along the Mauritanian coast.

(2) A northern belt, containing a series of high-density nodes, including those in Senegal and Gambia, the Fouta Djalon Mountains of Guinea, around Bamako in Mali, the Mossi country of Upper Volta, around Korhogo in northern Ivory Coast and Navrongo in northern Ghana, the Kabrai country of northern Togo, the Atakora region of northern Dahomey, and regions around the emirate cities of northern Nigeria. The areas of relatively high

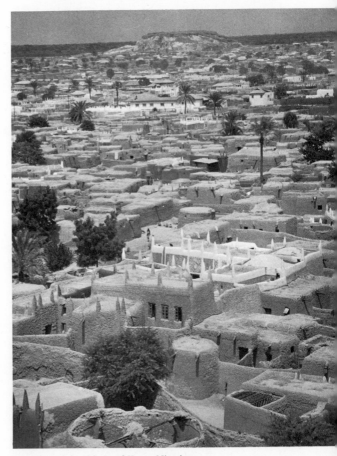

A portion of the old city of Kano, Nigeria
The close settled zone around this ancient market center is one of the important nodes of West Africa.

[1] See William A. Hance, "Population and Resources" in John C. Caldwell, ed., *Demographic Growth and Socioeconomic Evolution in West Africa* (New York: Columbia University Press, 1975).

density around Kaduna and on the Jos Plateau may also be assigned to this belt. Explanations for the series of high-density nodes characterizing this belt include the following: ability to grow such subsistence crops as sorghum, millet, peanuts, and beans and such cash crops as peanuts and cotton; the presence of some through-flowing streams and of easily tapped underground water in several areas; the availability of some relatively good soils; the use of some relatively advanced agricultural techniques, including heavy application of animal and human waste permitting annual cropping without fallows in some areas; the relative ease of movement across the belt; the early development of commercial nodes; the buildup of artisan activity in the urban centers; the more recent provision of improved transport to specific centers; the atypical advantage of the westernmost nodes of closeness to the coast; the existence over several centuries of powerful states permitting and sustaining the economic development required to support dense populations; and the

An aerial view of Ibadan, Nigeria
This is the largest of the many Yoruba towns of southwest Nigeria.

concentration of some groups in refuge areas in several parts of the belt.

(3) The so-called "middle belt," which occurs south of the northern high-density belt and which is generally more sparsely populated than the belts to either side. Overall densities are higher in the east of the belt than in the west, particularly in Nigeria, and regions of higher density occur in Sierra Leone, upper Guinea (the inland southeast), around Bouaké in Ivory Coast, and in Tivland, Nigeria. The population pattern of this belt is considerably more difficult to explain than any of the others. Significant factors include presence of the tsetse fly, precluding movement into the area by cattle herders and leading to a higher incidence of human trypanosomiasis; the widespread absence of permanent water supplies; the restricted suitability of the area for many valuable tree crops; the existence of powerful kingdoms on both margins and their general absence in this belt; the impact of slave raids and intertribal conflict on the area; the less developed transport pattern; and the meager production of cash crops in the belt. In toto, these do not provide a fully satisfactory explanation, however, because there are a number of exceptions and because the belt is ecologically suitable for a rather broad range of crops, which presumably could have supported considerably higher densities.

(4) A southern belt, which is sparsely populated to the west and densely populated to the east. It contains zones of particularly high density as follows: Ashanti, Ghana; a coastal belt running from southern Ghana through Togo, Dahomey, and western Nigeria, and coalescing with Yorubaland in the Western State and its extension into Kwara State in Nigeria; Iboland in the

East-Central State; and Ibibioland in the South-Eastern State. West of Ghana there are lesser nodes in southeastern Ivory Coast and around Monrovia, Liberia. Factors explaining the high-density portions of this belt and many of the lesser nodes include the ability to produce crops on a year-round basis; the comparative ease of producing such subsistence crops as manioc, corn, yams, plantains, and palm oil; the ecological suitability of the belt for many important export crops; the particular attractiveness of the rainforest/savanna border zone for ease of clearing; the presence of some relatively favorable soils; the occurrence of most of the mineral-producing areas of West Africa; the historical existence of a number of kingdoms whose political stability permitted the rise of more intensive and diversified economies; the longer and more intensive contacts with the outside world during the modern period when technological advances have been most dramatic and levels of trade far higher than when the sudan belt was trading by trans-Saharan caravan; the concentration of modern political, commercial, financial, manufacturing, and cultural activities at coastal points for all of the states fronting on the sea; and the greater impact of in-migration in this belt.

There are relatively few non-Africans in West Africa; the only large groups are found in Senegal and the Ivory Coast. Europeans are concerned with technical, educational, and advisory assignments in government, with foreign missions, with the larger commercial houses, with mining except for small diamond workings, and with modern industry. Lebanese and other Mediterraneans and some Indians are important in some large and many small commercial operations.

TRANSPORTATION

Transportation is of such fundamental importance to economic development in West Africa that a survey of its major components will reveal many features of significance to land use in the area. Before we turn to the specific ports and land transport routes, a few generalizations may be made:

(1) The area was not favored as far as access by sea is concerned owing to the prevalence of longshore bars, silted river channels, and the heavy movement of sand and the general absence of indentations along the Guinea Coast. Several solutions to this problem will be detailed below. It is interesting that Britain secured most of the superior coastal points, including the Gambia, the estuaries of Sierra Leone, and the distributaries of the Niger.

(2) The inland waterways are also navigable only with difficulty. Most streams flowing south have only short negotiable stretches. Even the better waterways are usually troubled with great seasonal fluctuations, shifting channels, and interrupting rapids.

(3) Overland movement is not generally difficult so far as landforms are concerned, except for the high expense of bridging the numerous streams. Great vegetational and climatic handicaps do exist.

(4) The growth of traffic on roads and railways and at ports has been extremely rapid, sometimes causing severe congestion. Road traffic has been gaining relatively over rail, especially on short hauls; this has created difficulties for some of the nonmineral railways, many of which have always run at a loss. An example of the shift to road use is seen at the Apapa wharves at Lagos, Nigeria, where only 44 percent of export traffic and 32 percent of import traffic moved by rail

in 1967–68 as compared to 94 and 90 percent respectively in 1948–49. Heavier use of roads has also required an accelerated program of upgrading and paving, as laterite or gravel surfaces become inadequate with intensive use.

(5) Improvements to existing transport means and extensions of routes, particularly roads, have also been great. Nigeria's road mileage, for example, has increased over 70 percent since World War II, and bitumenized roads have increased from 500 miles (800 km) in 1946 to about 9,750 miles (15,700 km) in 1972, though a substantial part of these roads is in poor condition. Bridge construction in many countries has eliminated traffic bottlenecks. There has been increasing attention in most territories to the construction of access roads, which have made a principal contribution to expanded production and exchange of export crops and domestic foodstuffs. While there have been relatively few extensions to rail lines, there has been much replacement with heavier track, some realignments, heavy accessions of rolling stock, improved signalization, and dieselization.

One of the striking features of postwar years has been the greatly reduced significance of lighterage ports of any type. Such ports in Africa vary from completely open roadsteads to protected bays and inlets where vessels can anchor in safety. Ships may be loaded and unloaded by surfboats paddled by a crew that doubles as stevedores, or by modern motor lighters. Installations at the shore may be almost nonexistent, in which case the surfboats land on the beach through the surf, or there may be a metal or wooden jetty along which lighters may tie up to take on or discharge their cargoes. Sometimes small cranes are available to assist with heavier goods. Lighterage ports are an anachronism in modern times. They are slow, dangerous, and expensive as far as cost of handling is concerned. Lack of protection for piers is another disadvantage; numerous examples exist of partial or severe damage from heavy storms to such installations. By 1962 the last of the major surfports had been replaced in West Africa and by 1966 all of the countries fronting on the sea had at least one modern port with deepwater facilities.

Expenditures on the transport infrastructures of West African countries have perforce absorbed substantial percentages of development funds. In the first postwar decade transport and communications received as much as 36 to 64 percent of development funds. While there is considerable variation from year to year because of the lumpiness of major projects the portions of budgets allocated to transport and communications have been ranging from about 4 to 12 percent of recurrent budgets and 3 to 33⅓ percent of development budgets, with total budgetary expenditures usually running from 17 to 30 percent of the totals. International agencies have contributed heavily with loans and grants to transport developments.

(6) There has been a notable increase in air transportation in postwar years. By 1950, arrivals and departures by air had exceeded those by ship. Domestic routes have been markedly expanded in many countries, but many of the national lines appear to be overequipped and to require excessive subsidization.

(7) Despite the expenditure of large sums of money, the needs in the field of transport are still very great. Some areas remain remote from any modern means of communication; many

roads are impossible for varying periods in the rainy season; and many of the trunk lines must be reconstructed to higher standards. The long-agreed construction of first-class road linkages between neighboring countries has yet to be achieved.

NATIONAL TRANSPORT SYSTEMS

In the following presentation of specific features of the transport picture of each country, the focus (moving from west to east) will be upon the ports and their connecting routes. Maps 9 and 10 may be helpful throughout the following section.

MAURITANIA. Most of Mauritania's small international trade formerly moved through Dakar, except for fish landed at Nouadhibou (Port Etienne). Two occurrences have altered this position: the development of Nouadhibou as the outlet for large tonnages

Loading iron ore at F'Dérik, Mauritania
Trains of 180 wagons and 1.25 miles (2 km) long deliver the ore to a mechanical loading wharf at Nouadhibou.

of iron ore from F'Dérik (Fort Gouraud) and as a modern fishing port, and the construction of facilities at the new capital at Nouakchott. In 1966 the latter was equipped with a pier now capable of handling 150,000 tons a year, including copper concentrates from Akjoujt; construction of a deepwater port, which would be difficult, is under study; there is some expectation that the Chinese would provide the required capital. Mauritania is attempting to redirect as much of national overseas traffic as possible via Nouakchott, but the relatively small scale of traffic has meant that a considerable tonnage still moves via Dakar. In addition, the government tries to control and also subsidizes internal transport, which a World Bank team suggested was economically unjustified.

Nouadhibou possesses the only sizable bay between Casablanca and Dakar with good depths near to shore, shelter from the strong Atlantic swell, and a location close to rich fishing grounds. A well-equipped commercial wharf and special fishing facilities handle about 100,000 tons of cargo; about 6 miles (10 km) south of the town a mechanized mineral wharf permits 100,000-ton ore carriers to be loaded at a rate of 6,000 tons per hour. A 419-mile (675-km) rail line connects with the iron mines from which two 180-car trains deliver about 30,000 tons of ore daily to the storage area, where each car is unloaded in 66 seconds. This line has proved to be about twice as costly to operate as had been estimated because of the high wear and tear of the desert environment.

SENEGAL. Dakar ranks as one of the great ports of Africa. It is largely an artificial port enclosed by breakwaters, though the Cape Verde Peninsula does provide protection from the west and north (Map 52). There is quay space for 46 vessels; many of the quays are very modern, as is the mechanical equipment of the port. Bunkerage facilities are well developed; there is special gear for handling phosphate exports, a well-developed fishing port, a large drydock for repairs

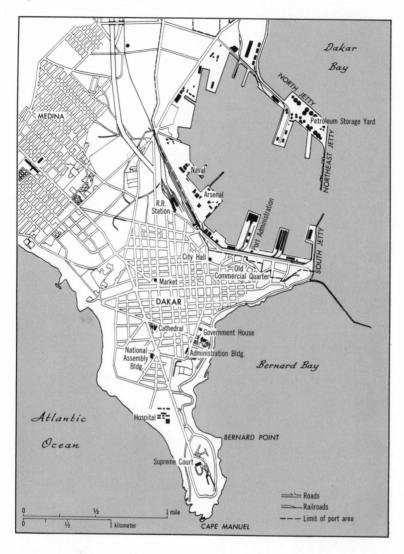

Map 52. *The port of Dakar*

(plans call for building new drydocks to repair large ships and tankers of up to 500,000 tons), and a floating terminal capable of taking supertankers.

The high tonnages handled at Dakar (5.3 million tons in 1972) tend to exaggerate its importance in the West African economy, because about three-eighths is bunkering traffic, which yields low returns and concerns only the port, and additional large movements are in low-value phosphate rock. The hinterland is a big producer of peanuts and phosphate, but of not much else, though

Dakar does ship some manufactured goods. It was aided by closure of the Suez Canal in 1966, but has been hurt by diversion of some traffic for Mali to Abidjan despite the designation of a customs-free zone for that country, by Mauritania's efforts to handle its own traffic, and by the switch from decorticated nuts to peanut oil.

The Dakar–Niger Railway, which operates at a deficit despite regulations requiring its use for movement of peanuts, extends to Bamako and Koulikoro on the Niger River in Mali, while branches run to Saint-Louis and

Linguéré. The French originally planned that the rail lines would extend from points on the West African coast to navigable stretches on the Niger and Senegal Rivers, which would serve as the backbone of the whole transport system. In fact, neither river is very adequate, nor are the interior areas along them generators of important traffic. Saint-Louis, near the mouth of the Senegal River—the former capital of both Senegal and Mauritania and an ocean port of some significance in earlier years—has handled no overseas traffic since 1965. Steamers handle about 21,000 tons a year on the Senegal, some 250 large pirogues carry another 10,000 tons, and smaller boats move an indeterminate amount of river traffic. The river is navigable all year from Saint-Louis to Podor (170 mi, 274 km) and from early August to late October to Kayes (580 mi, 933 km).

Minor ports on the Saloum and in Casamance are all seriously encumbered by shifting sandbars at channel entrances, difficult navigation, and shallow waters. The road system of Senegal is relatively well developed, and a comparatively high percentage of the main roads is paved.

MALI. The major outlets of this huge landlocked country have been the Dakar–Niger Railway and the Senegal River, though some traffic from the south has been moving via Abidjan. Mali has two navigable stretches of the Niger River, connected by the Sotuba Canal around the rapids between Bamako and Koulikoro. The stretches from Kouroussa or from Kankan on the Milo tributary in Guinea are little used. Eastward the Niger is navigable on a very uneven basis to Gao and sometimes to Ansongo, a total distance of 869 miles (1,400 km). Despite the difficulties of operating on the river, the main river ports have been improved and the total river traffic in Mali runs around 45,000–60,000 tons per annum.

GAMBIA. Banjul (Bathurst) is the only significant ocean port for this small country; its

deepwater facilities are being extended to accommodate several vessels. The Gambia River, navigable to the eastern border, handles most of the country's peanut traffic plus clandestine movements from Senegal and distributes imports not consumed in the capital city itself. The Gambia might have been the main outlet for most of the peanut zone of Senegal and Mali had it been incorporated in Senegal, but desire to develop and support national transport systems led to French construction of the rail line, which essentially parallels the river.

GUINEA (BISSAU) AND THE CAPE VERDE ISLANDS. The main port of the former is Bissau, with deepwater facilities accessible to vessels drawing up to 36 feet (11 m). Navigability for about 100 miles (160 km) of its three main rivers facilitates assemblage and delivery of goods. As has been noted, a fueling station exists on São Vicente in the Cape Verde Islands, but it ranks well below competing ports at Dakar and on the Canary Islands, in part because of poor water supplies.

GUINEA. Conakry is the dominant port of this country. Situated on one of the few segments of hard rock providing a firm base along this coastal stretch, it is a fairly modern port which has been improved to handle larger tonnages of iron ore and alumina. Its protection is not so good on the north as would be desirable, and its problem of shallowing depths requires continuous dredging. A new port has been constructed at Kamsar on the north coast to handle large tonnages of bauxite from Boké, to which it is connected by a new railway.

Guinea also has a 375-mile (603-km) narrow-gauge line from Conakry to Kankan on the Milo tributary of the Niger with an 87-mile (140-km) branch to the bauxite-alumina operation at Fria. The main line runs through very difficult, precipitous country, much of which is not highly productive. Except for the stretch from Fria the line has not been adequately maintained, a considerable portion of equipment is very old, and operations have suffered greatly in efficiency.

The arcuate shape of Guinea means that a large part of the interior is far removed from a national port. The natural outlets for much of southeastern Guinea would be Freetown, Monrovia, or Buchanan. But strenuous efforts are made to route traffic from interior Guinea by truck to Kankan and Dabola and then by rail to Conakry. This does contribute to support of domestic transport and conserves foreign exchange, but it places a considerable burden on the remote region, which must to some degree reduce its ability to trade. The prospect of developing iron ore across the border from the Mt. Nimba workings in Liberia raises the same issue; the most practical outlet would be via the Liberian line to Buchanan, but studies are being made of a 540-mile (870-km) line to Conakry, which would cost an estimated $450 million.

Guinean roads are very inadequate for a country of its size; the shortage and poor quality of overland routes is one of the main limiting factors in development of the hinterland. Yet road construction has been given low priority and existing roads are more often than not poorly maintained.

SIERRA LEONE. Freetown, capital of Sierra Leone, has one of the finest natural harbors in Africa and the third largest in the world, created by submergence along this segment of the coast. It is further advantaged by deep water along the shore maintained naturally by tidal scour. Nonetheless, it was only a lighterage port until as late as 1954, when the Queen Elizabeth Quay was opened; it required extension in 1970 to relieve increasing congestion and can now berth six ships. The port also has a tanker jetty and bunkering facilities.

Freetown is the terminus of the Sierra Leone Railway, a narrow-gauge line which has always operated at a loss. Trucks are gradually taking over its traffic and the line

should soon be discontinued. Across the bay from Freetown is Pepel, which handles iron concentrates brought on a 52-mile (83-km) railway from Marampa. The channel to Pepel has been deepened several times and can now take 90,000-ton ships, which are mechanically loaded at a rate of 2,750 tons per hour.

Sierra Leone's drowned coastline provides opportunities for coastal navigation; there are at least 680 miles (1,100 km) of navigable waterways along the coast and on the lower courses of the rivers, though some are utilizable only part of the year. As elsewhere in Africa, however, most of the smaller ports have been abandoned, handle only limited tonnages, or, as in the case of Point Sam, are used to lighter minerals to vessels standing in deeper water.

Roads were given minor attention in Sierra Leone before the war, but under a greatly accelerated program the road mileage was doubled between 1948 and 1960 and has since been further extended. A problem of increasing proportions is how to maintain the typical laterite roads under the increasing traffic they are called upon to bear. Sierra Leone roads are among the most poorly marked in Africa and connections with adjacent countries are inadequately developed.

LIBERIA. Before the war, Liberia's port facilities were practically nonexistent; since then it has had one of the more dynamic construction programs in Africa. Its first modern port, Monrovia, was constructed during the war by the United States; an artificial port protected by two breakwaters, it has a general cargo quay, three finger piers for mechanical loading of iron ore, three jetties used by fishing boats, an oil jetty, bulk facilities for loading liquid latex, and a 150-foot (46-m) drydock. Monrovia is connected by private lines to three iron mines. Farrell Lines operates a coastal feeder service, which includes the movement of latex from Harper and from the Farmington River. The main

Loading rubber at the Firestone Plantations dock at Harbel, Liberia
The rubber is taken by this small coastal vessel to Monrovia for shipment overseas.

roads of the country focus on Monrovia and have brought additional traffic to the port.

Three other ports have been built in Liberia in recent years and a fourth may be constructed in the not too distant future. The port at Buchanan was built for the iron mines at Mt. Nimba in the extreme east of the country, to which it is connected by a 175-mile (282-km) railway. Buchanan loads almost as much iron ore and pellets as Monrovia because the Mt. Nimba operation is the largest in the country, but its general cargo is very much less. The facilities at Greenville were built to serve a German banana plantation, which has since converted to rubber, while those at Harper are now serving relatively small shipments of rubber and cocoa. Both have rather minimal facilities and will require alterations if large-scale exploitation of the rich forests in their hinterlands takes place. Robertsport, to the west of Monrovia, may be selected as the outlet for the potentially very large iron mine at Wologisi in the northeast of the country.

In 1926 Liberia had one 27-mile (43-km) road and five cars; in 1945 it still had no paved highway and only 206 miles (330 km) of unimproved roads. A major effort to correct this situation began in 1952 and by 1974 there were about 2,500 miles (4,000 km) of public roads plus 1,300 miles (2,100 km) of concession roads and approximately 28,000 vehicles, up from 7,900 in 1961. Large parts of the country remain essentially unserved, particularly in the sparsely populated southeast. Construction of a road linking the coastal points is planned.

An interesting feature of the transport industry of Liberia is its growth in postwar years as a nation of registry for foreign-owned oceangoing vessels. This concentration upon Liberia as a "flag of convenience" was stimulated by very attractive laws and rates (an initial fee of $1.20 per net ton plus an annual tax of 10¢ per ton, guaranteed not to change for 20 years after registration), said to have been suggested by the Farrell Lines. In 1948 Liberia had a registered fleet of 1,000 gross registered tons (GRT); by 1962 it ranked fourth in the world with 10.6 million GRT, and in 1973 it was in first place with almost 59 million GRT. Receipts from registration were $5.3 million in 1971.

The use of "flags of convenience" has met strong opposition from such countries as Norway and the United Kingdom, whose shipping earnings are an important source of revenue, and who view such registration as a subterfuge to escape taxation and more rigid safety regulations. Seamen's unions have also attacked the device as a threat to wage levels and employment. Owners of ships registered in Liberia, most of whom are Americans, reply that their wage rates and safety regulations maintain accepted standards (Liberia has recently adopted a safety inspection system based at various world ports), that subsidization of many national fleets makes competition impossible unless some way is found to reduce tax burdens, and that, rather than reducing the fleet in the event of war, they permit maintenance of a much larger tonnage than would otherwise be possible. They see the "flag of convenience" as really a "flag of necessity."

IVORY COAST. This country had no modern port facilities until the 1950s, when Abidjan was opened to oceangoing vessels. Its coast is the epitome for West Africa: straight, sand-bar ridden, and subject to enormous movements of sand. The western stretch was served only by lighter ports, of which Sassandra was foremost until the modern port of San Pedro was opened in 1972. The first major development occurred at Abidjan, which was finally opened to ocean traffic when the Vridi Canal was cut through the coastal bar (Map 53), thus giving access to the large, well-protected, and relatively deep Ebrié Lagoon. The problem of silting by longshore currents, the curse of the whole Guinea Coast, was met by extending the canal's western dike further than the eastern dike and by narrowing the seaward lip of the canal, thus creating a current to direct any sand from the mouth to a deep offshore fosse known as the *trou sans fond*. Some con-

cept of the nature of the problem may be gained from earlier experiences in cutting through the bar at Abidjan. The first attempt, made in 1904–7, was badly sited and the canal silted up in a few months. In 1933 a yard-wide ditch was cut along the abandoned canal to release unusually high flood waters from the lagoon. In eight days the ditch had widened to 375 yards (340 m), but six months later the outlet had already been blocked by the action of the sea. The port of Abidjan has been equipped with 6,560 feet (2,000 m) of general cargo quays, two banana piers, a fishery quay, a wood depot, and, on the south of the lagoon, a mineral loading post and an oil berth for smaller tankers. Larger tankers are served by pipeline from offshore. Construction is proceeding on additional quayage, which is being built on an infilled section of the Ebrié Lagoon between the main port and the Vridi Canal.

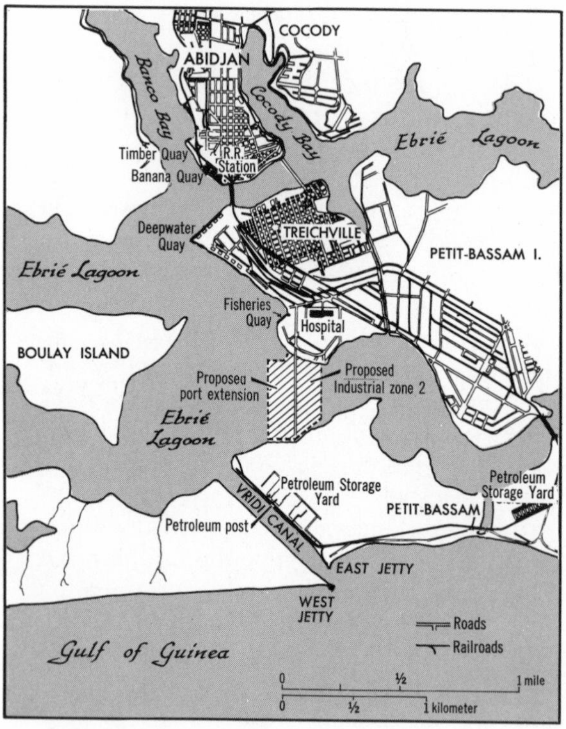

Provision of modern port facilities at Abidjan illustrates well the stimulus that such facilities can bring. Between 1948 and 1952, the capital had a threefold increase in population; by 1961 it was, at 190,000, twelve times its prewar size; and its population was estimated to be 650,000 in 1971. Port facilities were originally planned to handle 850,000 tons a year; the million-ton level was reached in 1954 and 5.4 million tons were moved in 1971. Particularly notable has been the increased movement of logs and lumber, much more difficult with more primitive facilities, the attraction of previously nonexistent modern fishing operations, and the stimulation of cash crop production in an enlarging hinterland. In the meantime the city itself has been largely restructured and enormously expanded; it stands today as one of the most striking examples of modernization in tropical Africa.

Abidjan's ability to serve its hinterland is aided by a series of lagoons linked by canals extending about 180 miles (290 km) along the coast, a road system that has been greatly extended and improved and is now one of

Map 53. The port of Abidjan

the best in Africa, and the Abidjan-Niger Railway, which traverses the country and extends into Upper Volta to Ouagadougou. The rail line, however, operates at a deficit mainly because of increasing competition from truck traffic.

San Pedro is considered to be a growth point for the large, sparsely populated, forest-rich southwest, which is under the administration of the Authority for Development of the Southwest Region. Its site features were favorable, requiring only protection and dragging of the entrance to the lagoon. Two quays totaling 1,102 feet (336 m) and six posts for loading logs have been provided, as have shore facilities and a new town. Investment had totaled about $83 million when the port was opened in 1972, and even larger sums were projected in the next few years. While logs will probably make up the main cargo in early years, plans call for construction of paper and cement mills and for a link to a large iron-ore body to the north. The capacity of the present site can be taken to 2 million tons; further expansion would require construction of an artificial basin along the sea.

UPPER VOLTA. This landlocked territory is served by the railway to Abidjan, which was extended from Bobo Dioulasso to Ouagadougou after the war, plus road connections to the four countries lying between it and the sea. Traffic on the railway, which is the major routeway, totals only about 400,000 tons with exports being only about a quarter of that.

GHANA. Ghana's first modern port was Takoradi, an artificial port constructed in 1928 and extended in 1955. It now has nine berths for large vessels, five for small, mechanical loading facilities for manganese and bauxite, ten moorings for vessels loading logs (which are frequently floated alongside vessels to be deck-loaded by the ship's gear), a jetty for unloading clinker used in a nearby cement mill, a lighter wharf for moving cocoa and sawn timber, an oil berth, and a small shipyard and drydock. Takoradi's traffic has been about 2 million tons a year in recent years, consisting to a great extent of minerals and timber which are mainly produced in its hinterland and a considerable share of the country's cocoa exports.

Until the opening of Tema in 1962, seven surf ports handled what was not moved by Takoradi. Accra, the most important, handled the surprising total of 1.03 million tons in 1961. Tema, situated 17 miles (27 km) east of Accra, gave Ghana a second modern port with deepwater facilities. While it was originally conceived as a port for the Volta Scheme, it became increasingly clear that eastern Ghana needed improved facilities regardless of that scheme. Tema—entirely artificial and protected from the sea by 7,200-foot (2,200 m) and 4,800-foot (1,460 m) breakwaters—was designed to permit an eventual capacity of 5 million tons, with 20 deepwater berths. At present there are 12 general cargo berths, a tanker berth and a special berth for importing alumina, a drydock, and a fishing harbor in the lee of the eastern breakwater. Tema ships over half

A "mammy wagon" near Kumasi, Ghana
Such busses and comparable trucks play a very important role in West African transport.

Surf boats working at Accra, Ghana, in 1962
Completion of the new port at Tema resulted in elimination of this slow, dangerous, and costly mode of handling cargo in Ghana.

the country's cocoa, but is primarily used for general cargo and alumina imports, which run well above the unloaded tonnages at Takoradi. Except for fishing, all of the Ghanaian surf ports discontinued operations after Tema was opened.

Ghana's 617 miles (993 km) of rail lines are concentrated south of Kumasi; the main lines form a triangle with Takoradi, Tema, and Kumasi at the angles. Branch lines serve the mining towns in the southwest. While there has been increasing competition from road traffic, the minerals and timber moved by rail (roads are sometimes closed after rains to prevent their destruction by heavy trucks) help to make the railways profitable. There has long been talk of extending the rail line to Tamale in the north and on to Ouagadougou in Upper Volta, but the likelihood of generating sufficient traffic to justify the investment appears small. Ghana has an unusually dense network of roads in the south, a considerable mileage of which is paved.

Navigable waterways are of very minor importance. Lake Volta has a steamer service

operating between Akosombo at the dam and Kete Krachi, about halfway up the lake, and plans call for extending this service; but it is not likely that much tonnage will move via the lake between the north and the south.

Ghana was among the first African countries to establish a national airline and merchant marine. The airline has proved to be a costly venture; distances between the main southern cities are too short, traffic with the north is low, and some of the planes have not been suitable for the flight lengths or level of passenger and freight traffic. The Black Star Line, established in 1957, had 16 vessels in 1971. It is probably too capital intensive for a developing nation, which can ill afford to subsidize a fleet.

TOGO. Lomé is the capital and major gateway of this small territory. It was a lighter port with a single pier until a small artificial port was completed in 1968. It has a total capacity of 500,000 tons, which is expected to be reached by 1990; extension of the eastern jetty and construction of a new mole could then raise the capacity to 750,000 tons. A 3,280-foot (1,000-m) metal wharf at Kpémé

Mechanical phosphate loading wharf at Kpémé, Togo
This 3,280-foot (1,000-m) wharf can load rock at a rate of 2,500 tons per hour.

loads phosphates on a fully mechanized basis and takes oil ashore.

Togo's railways total only 288 miles (464 km) and extend inland only 172 miles (277 km); except for the short line serving the phosphate workings, the lines are obsolete and uneconomic and may well be abandoned as trucks and busses take over an increasing share of cargo and passenger movement.

DAHOMEY. Cotonou, the major port of Dahomey, was long dependent on a wharf and lighters that were totally inadequate to serve the growing level of traffic. This condition led to the imposition of surcharges of as much as 100 percent on goods moving through the port. An intensive study began in 1952 to determine where a new port to serve Dahomey and Togo might be placed and how the problems of shallowing water and coastal sand movement might be met. Marked by sometimes bitter rivalry among various localities and interests, the debate ended with a decision to construct a new artificial port at Cotonou. Three types of installation were studied: a manmade island connected by bridges to permit the free movement of sand along the coast; continuous piping of sand around a jetty port; and a port with projecting jetties west of which sand would simply be permitted to accumulate. The first two solutions were rejected as being too expensive, and the first had the added disadvantage of congested access to the port area.

The third alternative was selected, but dredging will probably be required every five years, as all of the sand movement has not been controlled. The new port was completed in 1965 at a cost of about $32 million. With four general cargo quays, a fishing wharf east of the main jetty, and special posts for the bulk loading of vegetable oil and the unloading of petroleum, the port has a total capacity of 1 million tons. Traffic increased from 351,000 tons in 1965 to 683,000 tons in 1972. This was a more rapid increase than expected, and it resulted partly

Aerial view of Cotonou, Dahomey
The new port, protected by two jetties, is typical of several such facilities constructed in postwar years in West Africa. The old unprotected wharf, which could only take lighters, is in the foreground.

from mineral developments in Niger, which required large imports. Plans call for the addition of two general cargo berths, one mineral post, and two posts for containers, at a cost of about $22 million, and for construction of a new basin west of the main jetty and a groin to give better protection from the sand.

Cotonou is the starting point of the 273-mile (439-km) Benin-Niger Railway, which ends at Parakou (approximately in the middle of the country), and of shorter lines which serve the south. The railway operates at a heavy loss and needs upgrading, but it may eventually be extended to Niamey in Niger, which will finally justify the railroad's name.

NIGER. Most of Niger is best tied to the sea by routes through Nigeria, the natural outlet for the east and center but superior even for Maradi, because 80 percent of the 895-mile (1,440-km) distance is by rail compared to 30 percent over almost the same distance to Cotonou. Niger and Dahomey subsidize the route via Parakou and Cotonou under "Operation Hirondelle" (Operation Swallow), the first to conserve foreign exchange, the second to stimulate use of its rail line and port. Niger accounts for about a quarter of the traffic at Cotonou, but about two-thirds of its peanuts go out via Nigeria, as does an indeterminate but considerable amount of clandestine traffic. Southwestern Niger has been benefited by the Niger River, which is navigable from Niamey to Gaya. While there is much local movement on the river, it lost most of its attraction when a bridge at Gaya cut the delivery time to Niamey to a half day as compared to two days by water.

NIGERIA. Lagos is the leading port of the Federation (Map 54), although loadings of petroleum from onshore and offshore facilities have a considerably greater tonnage. Despite improvements and extensions, Lagos

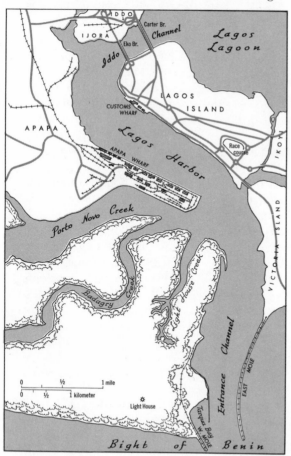

Map 54. The port of Lagos

was still plagued by congestion both before and after the civil war until a particularly efficient military officer assigned to manage the port succeeded in making order out of chaos. The main berths are along the 5,000-foot (1,524-m) Apapa Quay, which can accommodate 10 vessels; the old Customs Quay on Lagos Island has three berths used principally for incoming generaly cargo. Additional facilities include two petroleum posts, fishery facilities, mooring buoys, and a floating dock for vessels up to 4,000 tons, while a second terminal to handle 70,000 containers a year is planned. There is ample room for expansion of the port in Badagry Creek. Lagos, as the name implies, has a lagoon harbor; its access to the sea is protected by several jetties. Physical problems associated with the port include erosion of Victoria Island to the east and a relatively shallow depth requiring periodic vigorous dredging.

Port Harcourt, 41 miles (66 km) up the Bonny River, ranks second to Lagos as a general cargo port. It has a 3,500-foot (1,067-m) quay for 7 ships plus a coal-conveyor berth, a bulk palm oil berth, and mooring buoys. The port was damaged during the war, but was rapidly restored thereafter. At Bonny, near the mouth of the river, a petroleum depot—dredged to a 37-foot (11-m) depth at a cost of $7.8 million—dispatches crude gathered from the Delta fields (Map 60). Offshore terminals are increasingly important, however, as they can serve larger tankers.

A series of lesser ports are situated on various watercourses of the Niger Delta, though many points used in earlier years have been abandoned. Those reached by the Escravos entrance are Sapele, site of a large United Africa Company lumber mill and plywood plant; Burutu, terminus of the Niger River fleet; Warri, selected as the main port for this area and used for imports needed by the petroleum industry; and Koko, little used despite construction of a new quay. A 5-mile (8-km) mole was constructed at a cost of $22.4 million on the south side of the Escravos Estuary in 1964 in an effort to maintain depths over the bar of 20 feet (6 m) at all tides, which would have permitted the loading of up to 5,000 tons instead of the previous 1,400-ton limit. Calabar, on the Cross River to the east of the Delta, has seen little use in recent decades, but it has been selected for substantial improvements as part of a plan to develop the Cross River Basin.

Nigeria is something of an exception to the rule as far as inland waterways are concerned. A network of navigable creeks in the south permits movement from inside Dahomey (this section having been used for movement of petroleum products from Lagos until Cotonou was improved) to Eket, 60 miles (100 km) from the Cameroon border, while the Cross River is used to bring exports to Calabar. But it is the Niger and its main eastern affluent, the Benue, which are the leading interior waterways. The Niger is navigable from June to March to the confluence of these streams at Lokoja, 360 miles (580 km) from the sea, but depths vary greatly above Onitsha. Above the confluence, the Niger is navigable from August to March a further 206 miles (331 km) to Jebba, but boats seldom go farther than Baro, 74 miles (119 km) from Lokoja, where a rail branch permits more direct contact with the north.

The Benue is navigable 150 miles (240 km) to Makurdi from June to November, 520 miles (837 km) to Yola, near the border, from July to October, and for six to ten weeks between July and October to Garoua in Cameroon, 612 miles (985 km) from Lokoja and 927 miles (1,564 km) from the sea. In addition to an intensive canoe and barge traffic, a regular river fleet handles traffic on the Niger and the Benue, but tonnages moved have declined from about 270,000 in the early 1960s to about 46,000 tons in recent years. Difficulties of navigation, its limi-

tation to daytime hours, and problems of integrating the traffic flow result in freight rates not far below those on the railways, whereas waterway rates in more developed countries are normally only a quarter to a half of rail tariffs. As additional dams are built on the Niger and Benue beyond the first dam at Kainji (see page 266) their navigability will be improved; but because the two rivers traverse the relatively poorly developed middle belt, traffic levels will remain low for some years.

Nigeria has 2,178 route miles (3,504 km) of rail lines. The western line runs 700 miles (1,126 km) from Apapa to Kano with branches from Zaria to Kaura Namoda, from Kano to Nguru, and from Minna to the river port of Baro. The eastern line runs 569 miles (916 km) from Port Harcourt to Kaduna; its Bornu extension, completed in 1964, terminates at Maiduguri in the northeast. A feature of the traffic pattern in recent decades has been a marked loss of short-haul and high-value freight to road haulers, which has contributed to the difficulties of operating the railway without serious losses. Nonetheless plans have been drawn to construct a 600-mile (965-km) link from the west to Calabar and a 130-mile (209-km) extension from Gusau to Sokoto in the northwest.

Nigeria's road net doubled from 1951 to 1968, when it totaled 55,300 miles (88,980 km) of which about a sixth was paved. The civil war interrupted road construction and led to the destruction of some roads and key bridges, but in 1971 a ten-year program involving an estimated $700 million was adopted; it is to provide a grid of four main south-north roads—starting at Lagos, Warri, Port Harcourt, and Calabar—and four east-west roads for a total of 6,100 miles (9,800 km) of primary roads, while about 7,000 miles (11,260 km) of roads would connect important towns not served by the trunk routes. The southernmost road of the grid will serve as the western end of the Trans-

A busy street scene in Ibadan, Nigeria
A very large number of petty traders are found in most West African countries.

African highway connecting with Mombasa in Kenya. Costs have greatly risen since inception of the plan and expenditures exceeded the original estimate in the period 1970–74. Efforts are being made to increase the capability of Nigerian contractors to participate in the road building operations. The north of Nigeria is advantaged in the use of vehicles by the decision to sell gasoline at the same price throughout the country.

Nigeria, like Ghana, has its own merchant fleet—with some 13 ships totaling 87,600 registered tons—and its own airways. The size of the country justifies a fairly well-developed internal network.

One feature deserving of note in concluding this section is the markedly national orientation of almost every country's transport system. There is no road connecting the major coastal cities; there is no easy route from Dakar to Lagos. And despite numerous agreements to construct connecting links plus desires to develop the tourist industry, the red tape involved in moving across boundaries or entering various countries by air is often appalling. Nonetheless, the progress that has been made in improving transport, particularly in port and road programs, has been striking.

TRADE AND COMMERCE

Space prohibits a detailed examination of the foreign trade and domestic commercial patterns of countries in West Africa, though many specific features will appear in the following discussions. Country imports and exports for 1961 and 1972 are given in Table 11. West Africa is like other parts of the continent in the high dependence of its money economies on foreign trade, in the limited number of commodities appearing on the list of exports, and in exports resulting almost entirely from extractive activities. The orientation of trade has shown a relative decline in the importance of the former metropoles. This is particularly true for francophone countries, whose trade with France has declined from about 70 percent of the total in 1961 to about half that level, though France remains the number one trading partner for most of them. The U.K.'s share of trade with Commonwealth West Africa has long been lower than that of France with its territories, but it too has tended to decline though at a slower rate. The continental members of the EEC, excluding France, and the U.S. have increased their shares of francophone West African trade; the U.S. has also taken a larger share of Nigerian exports. Japan remains a minor partner, though the rate of increase of sales and purchases has been high. Most West African countries have had unfavorable trade balances in recent years (Table 11). As noted earlier, there is a substantial amount of unrecorded illegal trade operating across West African borders; it has been recently estimated that clandestine movements between Dahomey and Nigeria, for example, involved $10 million worth of cocoa moving to Nigeria and $14.4 million worth of tobacco, alcoholic beverages, textiles, and other manufactured goods going to Dahomey.

Features of the domestic economy are the very large number of petty traders found in most countries and the increasing replacement of large expatriate firms in wholesale trade, inporting, and exporting. Government marketing boards or agencies have taken over much of the handling of staple exports formerly done by such companies as the United Africa Company, Compagnie Française de l'Afrique Occidentale (CFAO), Société Commercial de l'Ouest Afrique (SCOA), or by Syrian and Lebanese companies. Legislation is increasingly restricting the role of foreigners in commerce, and there have been several large-scale expulsions of foreign Africans engaged in trading and other activities in various countries.

TOURISM

When compared to the Maghreb or East Africa, West Africa has not been favored by visitors; but substantial efforts are now being made in a number of countries to develop the tourist industry, which can make a significant contribution to foreign exchange receipts. The Ivory Coast hopes to increase the number of visitors (it is not generally possible to distinguish tourists from business and other travelers) from about 53,000 in 1971 to 300,000 in 1980. It is developing beach resorts on the coast near Assinie and a huge residential-recreation complex known as La Riviera Africaine on the Ebrié Lagoon east of Abidjan, while that city's Hotel Ivoire—with three towers, five restaurants, a casino, two pools, an international conference center, and a permanent ice-skating rink—is unique in tropical Africa. Senegal has plans to upgrade facilities at Saint-Louis, Zuiguinchor, Tambacounda (in the east near a national park), and on the historic Ile de Goree, as well as at Dakar. Even tiny Gambia has been discovered by Swedish tourists, who come in charter flights to beach hotels; about 1,100 persons were employed in the industry in 1972–73 and plans call for 8,000 hotel beds by 1978, which would give employment to about 5,000 people. Many West African countries have also striven to set aside or develop game parks, while regular tourist circuits are established to such attractions as Mopti, the Dogon country, and Timbuktu in Mali despite rather meager hotel accommodations.

CHAPTER TEN

West Africa: Land Use

With the background provided by the previous chapter, it is now appropriate to examine the major land-use patterns of West Africa. Large climatic-vegetation regions will be used as the basic units for this discussion. Emphasis is upon commercial development and potentialities, but it should not be forgotten that the primary concern of most farmers and graziers is still subsistence production, and very large regions produce no surplus for sale outside their immediate settlements.

COASTAL AREAS

The coastal regions have certain economic uses with which they may be associated, including, in addition to the port functions already discussed, ocean and lagoon fisheries, the growth of swamp rice for domestic use, and some production of coconut products.

FISHING

There is a considerable and growing interest in fishing along many parts of the West African coast; some ethnic groups are almost solely devoted to this activity. More modern commercial operations (often dating only from the 1960s) now land larger tonnages than traditional fisheries in many countries. The importance of fishing varies considerably from country to country. In Mauritania fish exports rank second but well below iron ore; the fishing industry, like mining, is essentially an enclave activity. Senegal has seen a remarkable increase in fishing in the last decade, and ocean products now rank second to peanut products among its exports. About 70 percent of the catch is exported,

including shrimp, of which it is Africa's leading producer. Domestic consumption is also high; the country ranks fifth in the world in per capita use of fish. Gambia's catch is also increasing steadily and a new company is processing crabs, oysters, and lobsters.

Other West African countries tend to be net importers, though many export tuna and shrimp. Liberia, Ghana, Togo, and Dahomey have many communities engaged in fishing. In Ghana fish consumption exceeds that of meat and the industry is credited with about 1.2 percent of the GNP; nonetheless, it imports sizable tonnages. Nigeria, the population giant of Africa, consumes the continent's largest amount of fish, but about three-quarters is imported. Foreign vessels play a significant role in sea fisheries, particularly in more distant waters and for the tuna catch.

Reliable data are not always available on the numbers engaged or dependent upon fisheries or on the size of catches, but the following estimates may be cited. Mauritania's fish landings increased fivefold from 1965 to 1972, when they were 33,800 tons valued at about $10.2 million; most of this small catch is landed at Nouadhibou by the Canarian fleet, while the vast bulk of the estimated 300,000 tons caught off Mauritania goes directly to metropolitan countries. Some 28,000 fishermen operate in Senegal, while 400,000 permanent or temporary workers find employment in the industry; record catches were made in 1972, 248,000 tons by the modern fleet and 197,000 by local fishermen for a total value of about $100 million. About 60,000 full- and part-time fishermen work the sea and inland waters of the Ivory

Coast; their catch is about 100,000 tons a year and increasing rapidly. In Ghana about 200,000 persons in 200 fishing villages and towns earn their livelihood from fishing; approximately 200,000 tons are landed or about three times the 1960 level. Togo claims about 12,000 fishermen and Dahomey 23,000, of whom only 3,000 work at sea. Nigeria lands about 130,000 tons, 2.5 times the 1964 level.

Methods in the traditional fisheries are largely primitive; operations are usually carried on from pirogues launched through the surf and paddled by hand. The catch per man and per boat is very low, about one ton per fisherman per year. Wastage on shore is often very high owing to inadequate preservation, storage, and marketing. Traditional techniques of conservation include sun drying, charring, smoking, and salting; fish to be sold for local consumption are smoked for about eight hours, while those destined for the interior may be smoked for several days. The high demand for fish and the desirability of promoting their consumption to increase protein intake have stimulated efforts to rationalize the industry, though modernization sometimes adversely affects employment in the traditional sector.

As far as fishing itself is concerned, emphasis has been placed on the use of bigger boats, the introduction of powered vessels or outboard motors on pirogues, and the adoption of better and bigger nets, particularly shark, set, and encircling nets. A notable recent development has been the presence of tuna vessels from France, Spain, Japan, and the United States, which operate off practically the entire coast; some of the West African countries are beginning to operate their own tuna boats. Some conception of the progress being made may be derived from the following: 57 percent of the 5,617 pirogues operating in Senegal had been motorized by 1972, while that country operated 3 of the 5 sardine catchers, 14 of the 86 trawlers, and 14 of the 41 tuna boats using

A fishing village near Tema
At present most fishing is primitive and characterized by low production per man and per boat.

Senegalese bases; over half of the 10,000 pirogues in Ghana, two-thirds of those in Togo, and a tenth of those in Dahomey are motorized; distant fishing began only in the 1960s in several of the countries, including Liberia, Ghana, and Dahomey; tuna boats now operate out of at least seven ports in as many countries; and Ghana now has the best equipped trawler fleet in the region, though overinvestment in the state-owned segment has required some retrenchment and contracting with Norway to manage the fleet and train its workers.

Provision of better port facilities for fishing vessels has received increased attention and catches have gone up with sufficient rapidity that extensions have already had to be planned for some of the newer installations. Modern fishing ports have been constructed at Nouadhibou; Dakar; Freetown; Monrovia; Abidjan (one is scheduled for San Pedro); Takoradi, Elmina, Tema, and Ada in Ghana; Lomé; Cotonou; and Lagos. Trawler and tuna fishing require large-scale and modern shore facilities, but any real improvement in any kind of ocean fishing requires better and safer port installations.

With regard to shore establishments, attention is being given to the erection of canning, processing, and fish meal facilities, plus provision of cold stores which usually also produce ice for the boats. Nouadhibou's shore facilities are grossly overbuilt, with a capacity ten times the present catch; Senegal, the Ivory Coast, Liberia, and Ghana have had to extend their shore installations. Gambia got two modern plants in 1971–72. Despite the advantages of reduced losses and an increased widening of the market permitted by canning and proper preservation, some plants have had difficulties because of erratic supplies, insufficiently advanced marketing facilities, and consumer preference for smoked rather than fresh fish. The installation of cold store circuits does permit, however, not only consumption of fresh fish in the hinterland but also delivery of such fish to areas where they can be prepared to local tastes with less wastage. Such circuits have been established in four cities outside Dakar in Senegal; by refrigerated trucks in Liberia, Ivory Coast, Dahomey, and Nigeria; in an unusually large number of centers in Ghana, which has among the best developed systems in tropical Africa; and in a three-town circuit in Togo.

Relatively little is known regarding the potential of the ocean waters off West Africa, particularly those beyond the immediate coastal waters to which the smaller boats and canoes are perforce restricted. It is apparent, however, that tropical waters are far richer than previously believed and that resources off West Africa are very substantial. Not all signs are favorable, however. Concern has been expressed about overfishing of the inshore waters at several points and off the coast from Rio de Oro to Liberia. Several countries have complained bitterly of the large catches by foreign vessels, including one Norwegian factory ship servicing 17 catchers reported to be taking a daily catch equal to the total Liberian consumption and converting it to fish meal. Mauritania, Senegal, and Ghana have unilaterally claimed

national control over fishing as far as 30 to 110 miles (48 to 177 km) offshore, but whether they have the capacity to enforce these claims or to conduct a fish war is questionable.

While sea fishing contributes the greatest share of the total fish catch in West Africa, fishing is also widespread in the lagoons, lakes, rivers, and seasonally inundated lands throughout the area. In addition to fishing from canoes, hand lining, seining, and draining of ponds, fish are trapped in crude raffia nets. Togo and Dahomey have an unusual number of inland fishermen, though the catch is declining in the latter because of overfishing and the increased salinity in the lagoon behind Cotonou associated with construction of the new port. Nigeria's inland catch is thought to be about equal to that of its marine fisheries. The reservoirs of Lake Volta and behind the Kainji Dam may be expected to yield 10,000–20,000 tons. Some 500 fishermen migrate seasonally to the Nigerian side of Lake Chad, catching an estimated 10,000 tons, while those from Niger land about 3,500 tons. The potential of that lake-cum-swamp is estimated at about 150,000–200,000 tons; Lake Chad is very

Ganvié, Dahomey, a lagoon fishing village

shallow (usually 3.3 feet, 1 m, deep and a maximum depth of 23 feet, 7 m) and varies in size from 4,250 to 8,500 square miles (11,000–22,000 sq km). It is hoped that construction of a new access road and improved smoking kilns will reduce the characteristically high handling losses. Ironically, the three landlocked countries of West Africa export fish; Niger and Upper Volta have small catches, but Mali takes about 100,000 tons, shipping a considerable quantity of dried, smoked, and salted fish to the south.

In postwar years many countries of tropical Africa have built and stocked large numbers of fish ponds. The favorite species is *tilapia,* which multiplies rapidly, provides good eating, reduces the mosquito population, and keeps the pond free of weeds. These ponds serve multiple purposes, particularly in the drier savanna and steppe areas; in addition to providing an important source of protein, they can improve the domestic and stock water supply, help retain a high water table, and reduce erosion.

THE USE OF COASTAL MANGROVE SWAMPS

The mangrove forested areas of West Africa are usually very sparsely populated, and the trees are used only to a minor degree for fuel, for construction poles, and the peeling of bark for tannin extract. Sierra Leone has done the most to reclaim littoral swamps, but mangrove swamps account for only about 4 percent of its total rice production. Mangrove trees are difficult to clear and, so far, most reclamation has been by hand. Attention should perhaps be given to adapting swamp buggies such as are used in petroleum exploration in the bayous of Louisiana to the large-scale clearing of swamp areas. A second problem is the necessity to construct bunds to keep out salt water and protect against tidal scour. There must then be an adequate flow of fresh water to flush out the salt and prevent a toxic accumulation on the empoldered land. Finally, production of wet rice tends to result in a greater incidence of bilharzia and malaria. There are enormous

Fish traps used in seasonally inundated land in northern Nigeria

potentialities for the extension of rice cultivation in coastal swamps and in the Niger Delta. An expert studying the mangrove swamp area of that delta, however, concluded that large-scale reclamation was not now practicable because of the low resident population, difficult working conditions, problems of weed growth, poor soils, and the presence of deleterious insects and birds.

THE COCONUT

The coconut is best adapted to coastal lands and in West Africa is concentrated in various places along the margin of the sea. The Portuguese introduced the coconut to Dahomey in the sixteenth century and the Germans and French promoted its production in Togo and Dahomey in the present century. Togo's coconut groves were seriously afflicted by disease in the mid-1960s and now cover only about 10,000 acres (4,000 ha), though efforts are being made to replant with resistant stock. Dahomey has about 30,000 acres (12,000 ha) of coconut plantings but yields are low and exports are declining as domestic consumption increases. Nigeria has coconut groves near Lagos and elsewhere along the coast; it exports some copra. The Ivory Coast is establishing industrial coconut plantations surrounded by village plantings, and hopes to increase the area under coconut palms from 69,000 acres (28,000 ha) in 1971 (only 28.6 percent of which were in selected species) to about 126,000 acres (51,000 ha) by 1980.

HUMID TROPICAL REGIONS

Extending inland from the coastal zone, which in part has a humid tropical climate, is a belt of varying width including the remaining portions of the tropical savanna climate region of West Africa (see Map 16c). This belt accounts for a very large portion of the vegetable product exports of the area, particularly of the important tree crops such as cocoa, coffee, palm products, and rubber.

The rainforests, which once covered most of the area, are included in this zone.

EXPLOITATION OF THE TROPICAL RAINFORESTS

The extent of the rainforest in West Africa is inadequately mapped. Furthermore, much of the area depicted as forested is no longer so covered and from one-third to two-thirds of what is forested is in inferior secondary growth. Some experts maintain, in fact, that the forest never completely recovers after it has been cropped. The following factors work toward a reduction of the total area under forest: the increasing population (forcing the use of a greater area for farming), the extension of commercial crop production, overexploitation in some regions, and damage caused by fire. Clearing of forests for subsistence and commercial crops does yield a temporary supply of "salvage" timber, which may convey a false picture of the productivity of an area. Removal of the forest for farming has gone particularly far in Ibo, Ibibio, and Yoruba country in Nigeria, the eastern and Ashanti regions of Ghana, the original forested area of Sierra Leone (of which only about 5 percent still remains), and in the Ivory Coast where the extent of dense forest declined by an estimated 44 percent from 1956 to 1973.

Despite the diminishing extent of tropical

Stripping logs prior to shipment near Kumasi, Ghana
Removing the bark reduces the loss from insect attack.

rainforests, very large areas are still available for exploitation in West Africa. Guinea has a sizable unused area; about a third of Liberia is in high forest; southwest Ivory Coast has 7,700 square miles (20,000 sq km) of exploitable forest; about 25 percent of Ghanaian high forests remain intact; and Nigeria has extensive reserves in the Midwest, Rivers, and Southeast States.

Exploitation of West African forests has been characterized in recent years by the following:

(1) A great increase in production both for export and for local use. Very large but unmeasured quantities are used for fuel, and many areas have an urgent need for more fuelwood plantings. But the focus here is upon commercial production, where the rate of increase has been higher than any other major vegetable product from the area. The Ivory Coast, now the primary exporter of forest products in Africa, saw a tenfold increase in shipments from 1945 to 1959 and then a 5.5-fold increase to 1972, when they totaled 2.6 million tons; investment in the forest industry is now over $27 million and more than 100,000 people are employed. Ghana's exports of timber doubled in value in the period from 1966 to 1972, when 858,000 tons were shipped.

   Only the Ivory Coast, Ghana, Nigeria, and Liberia are large-scale shippers. Timber products usually rank second to coffee in export value in the Ivory Coast but surpassed that product in 1969 and 1972 (their share is about 25 percent of total exports). They account for about 9 percent of exports in Ghana and rank second to cocoa by value; Nigerian timber accounts for only about 3 percent of nonoil exports, but unlike the other vegetal products, their value has in-creased in recent years; Liberian exports of logs and lumber are about 3 percent of the total. Liberia has just entered a boom in forest output, and log exports have risen eightfold from 1967 to 1972, when they totaled 406,000 tons valued at $20 million. Guinea should join the list of important shippers, as it has contracted with Bulgaria to supply about 100,000 cubic meters of hardwoods yearly.

(2) The acceptance of a wider variety of species. Rainforests are characterized by a large number of species; it is not uncommon to find 100 species in a relatively small area. In the Ivory Coast 61 families of trees (divided into 276 genera, with 596 species) have been described. In prewar years mahogany accounted for over 90 percent of timber exports from West Africa. Today about 30 species are regularly marketed and mahogany is only one among the ten leaders; most of these species were practically unknown before the war. Not only are fine hardwoods exported, but there is an increasing sale of general-purpose timbers. Ghana is trying to encourage the use of secondary species by lowering the tax, export duty, and rail transport on them; the Ivory Coast has raised the tax on the more valuable species while a French research organization is pushing the testing and certification of previously unused species.

(3) An increasing use of mechanization in cutting, hauling, and shipping. Logging is not easy in the tropical rainforest environment. Dense undergrowth (often matted together by creepers), swampy terrain, the absence of waterways, the high cost of providing and maintaining roads and railways, and the highly selective cutting that market acceptance requires make the costs of tropical forestry high, de-

spite the low wages of labor and the great total volume available per acre. The buttresses many species have—they may rise to 10 or 15 feet (3–4.6m) on the trunk—make cutting more difficult. Most trees are still cut by axe from stages that are nailed to the tree to permit cutting above the buttresses. Sawed into manageable lengths, the logs are hauled by Caterpillar tractor and "logging arch" to the nearest road, rail, or river. (This part of the operation was formerly carried on by large gangs of men who dragged the logs to loading or floating points). Except in Nigeria, most logs are moved overland to sawmills or to port; in that country most are rafted and towed by small tugs to the mills or shipping points, a process much faster than floating. Very dense logs or "sinkers," which were formerly attached to two "floaters," are now often lightered. Transport savings are of great importance for this bulkiest of exports; even with recent improvements in handling, transport takes 50 to 100 percent of the f.o.b. price depending on the species.

(4) An increased processing of timber products. A large part of exports is still in logs; the best logs are usually rafted or lightered alongside ships, though the giant hardwood may first be squared to remove most of the sapwood, thus reducing the danger of insect attack and permitting better stowage aboard ship. But an increasing percentage is exported as sawn lumber, plywood, and veneer (Map 63); processing increases the value per cubic meter felled by about 2.5 times. The share of sawn wood in total timber exports from Ghana increased from 7 percent in 1946 to 23 percent in the early 1970s. In Nigeria no sawn wood was exported in prewar years,

A sawmill in Ghana
The timber industry in Africa has been characterized in postwar years by greater mechanization of operations from cutting to milling.

while lumber now represents about 21 percent of total timber shipments. Less than 10 percent of Ivory Coast exports of forest products in 1972 were not in logs. Governments, wishing to benefit from the higher value of processed exports and to stimulate local industry, are beginning to institute requirements to this end. Ghana banned the export of some species of log beginning in 1972, specifying that they must be processed locally to be eligible for sale; its long-run goal is to ship only processed timber. The Ivory Coast has made the granting of new concessions contingent upon the processing of a certain portion.

The number of sawmills has greatly increased in the last decades, though some pit sawing still takes place. Most of the sawmills are small, some catering primarily to the domestic market. Liberia has 19 sawmills, 5 of which produce three-fourths of the lumber. The Ivory Coast has 57 sawmills, 4 of which handle about half of the sawn wood; that country's forest product industry accounts for about 9 percent of the total value of industrial output. There are dozens of sawmills in Ghana and Nigeria, the largest mill in the latter being the United Africa Company plant at Sapele, which also produces plywood; a similar plant is situated at Samreboi in Ghana, while the Ivory Coast has three plywood mills, one of which produces particle board, and one is planned for the Harper area in Liberia, which would also produce hardboard.

Production of plywood, veneer, and pressed wood represents relatively new phases of processing. While there is a strong incentive to produce such commodities close to the forests so that the large waste need not be shipped, there are difficulties associated with their production in West Africa. The high temperatures and humidities require a more expensive glue, which makes the plywood more costly, while shipping of veneers is more difficult than the basic log as far as protection from damage and insect attack is concerned.

There are interesting potentialities for continued growth and evolution of the forest industries. Demand may be expected to increase despite substitution of other materials for wood. Exhaustion of valuable species can be offset by setting out plantations; Ivory Coast, Ghana, and Nigeria all have such programs to produce mahogany, teak, and other species. Plantations of selected woods can be sited at preferred locations with respect to transportation, and they eliminate the high cost of selective cutting; but the danger of insect attack is increased when one species is dominant in an area. A more intensive attack on insect pests would also save timber now wasted. The loss in West Africa attributed to the ambrosia beetle alone is estimated at 15 percent of the value of forest products. Methods of attack include spraying from the air, dipping, and incorporating chemicals in the glue used in plywood manufacture.

A great advance would be achieved if it

Unloading logs at Abidjan, Ivory Coast
The logs are assembled in rafts and floated alongside vessels lying in the harbor for loading by the ships' gear.

became practicable to exploit forests for the production of pulp. The characteristically shorter fibers and the hardness of tropical woods militate against such use, but pilot operations have shown that it is possible to produce pulp and paper from mixed tropical woods. Perfection of the technology and the erection of large-scale plants where cheap power is available might produce pulp economically and would permit great savings through the rationalization of forest operations. An area of appropriate size to permit natural regeneration or to justify replanting could be cut over, the better logs being used for lumber, plywood, and veneer, the other logs for pulping. The Ivory Coast has plans for a special plantation in the southwest to support a pulp and paper mill at San Pedro. The forest industry will also benefit from improved transport, including the laying of roads and rail lines not intended primarily for that industry, such as the mineral lines in Liberia. The existence of programs to train forest workers and specialists in several West African countries augurs well for the accelerated adoption of improved practices, and should also contribute to the more rapid Africanization of the industry. In the Ivory Coast, for example, about 60 small timber companies are now owned by Ivoirians, but the large foreign companies account for about 70 percent of total production.

AGRICULTURE OF THE HUMID TROPICAL REGIONS

Agriculture, including the production of tree crops, is the main activity of the bulk of the inhabitants of West Africa and especially of humid tropical areas. But it does not occupy a high percentage of the area except in the more densely populated regions.

BUSH FALLOW AGRICULTURE. The basic agricultural system was originally a self-sufficient bush fallow on small farms with a great variety of crops often grown on the same patch and with a considerable interest in tree crops. Planted crops included corn, yams, manioc, sweet potatoes, rice, vegetables, and pineapples. Tree crops used for food were palm oil, coconuts, bananas, mangoes, guava, and citrus fruit. For the annual crops, openings were laboriously cleared in the forests and planted. Care had to be taken to assure a continuous supply of food because storage was extremely difficult.

Clearing in a rainforest area, first stage in the practice of bush fallow agriculture
This low-productive agriculture is necessitated by the poor quality of the leached latosolic soils.

After from two to five years the soil would be exhausted, forcing the clearing of new patches and permitting the abandoned plot to revert to bush.

The explanation for the bush fallow system is the poverty of the regional latosolic soil. It supports the world's most vigorous vegetation growth (the tropical rainforest), which misleads the uninformed about its fertility—it is actually one of the world's poorest soil types. This seeming contradiction is readily explained by the different character of forest and cultivated vegetation. Under forest, the cycle of decay maintains an adequate supply of humus to support rapid growth, while tree roots extend sufficiently far below the surface to secure the necessary minerals. Cut the forest down and replace it with annual plants, however, and this cycle is destroyed. Shallow roots cannot supply minerals from below, the available humus is rapidly used up, exposure to the sun and rain speeds up decomposition and increases leaching. Yields drop off rapidly, usually making it desirable to abandon the plot in less than five years. This process is one of the most fundamental factors limiting African agricultural production.

Under present conditions, a number of changes have been made in the basic pattern of bush-fallow agriculture, resulting from the superimposing of a money economy on a primarily subsistence agriculture and from an expanding population. Production of some commercial crops has developed monocultural tendencies resulting in the soil's depletion if not in its actual erosion. Overemphasis on chemical crops has sometimes led to a narrower range of food production, again leading to accelerated depletion, and incidentally to a lowered dietal standard. Expanding populations have forced a lengthening of the growing period and a reduction of the bush fallow period. In Sierra Leone, for example, the average fallow period has been reduced from seven years to no more than four.

It is generally accepted that the bush fallow system is inadequate, even with a full fallow and despite the fact that it displays a reasonably good adaptation to the physical environment. In addition to its physical disadvantages, it also requires so much labor in relation to what is produced that it almost precludes the high productivity that would be required to raise living standards beyond a certain level. Yet no really effective alternative has been developed, so the proper utilization of tropical latasols continues to be a prime problem for Africa.

Research institutions, particularly the U.S.-financed International Institute of Tropical Agriculture outside Ibadan in Nigeria, and experimental farms are devoting increasing attention to this problem, but to date only palliatives have been developed. These practices include:

(1) Green manuring, which involves the turning under of a leguminous crop. This is beneficial, but not adequate, and it is often unpopular because the indigenous farmer cannot see the value of planting a nonproductive crop.

(2) Composting, which is often better, but which is seldom practiced.

(3) The use of legumes, which is not so beneficial as in the middle latitudes since local legumes do not have equal nitrogen-fixing power.

(4) Litter farming or planting of cover crops to reduce leaching and exposure to the sun. The common practices of close-planting of crops and growing a mixture of crops in the same field have the same advantages.

(5) Application of inorganic fertilizers, which is increasing, particularly on commercial operations, but which is usually too expensive, even with subsidization, to permit adoption by peasant cultivators. Relatively little is known about the effect of various chemical fertilizers, but the rapid leaching in tropical rainy areas does require heavier doses than would be necessary in the middle latitudes. While fertilizers may provide a scientific solution to maintenance of soil fertility, numerous other changes are required before they can meet the practical requirements.

(6) Application of organic fertilizers. Rapid oxidation and leaching reduce the value of organic applications, but it is the inability to keep cattle, other than dwarf species such as the N'Dama and West African Shorthorn, in the humid tropics that prevents the development of a mixed agriculture comparable to that of the middle latitudes. The use of household wastes and droppings of small animals does permit continuous cultivation on tiny kitchen gardens near the dwellings.

(7) Copying the natural pattern as closely as possible. This involves a concentration on tree crops, among which the oil palm is least demanding on the soil. Mixed cropping (typically at least three crops but sometimes ten to fifteen crops are planted in a field) is also a good adaptation to the natural environment but can only slightly lengthen the cropping years.

(8) Coppicing, which is an adaptation of bush fallowing. A rosaceous bush is cut back when planting takes place and permitted to grow again in the fallow period. Its deep roots secure minerals, but it is easier to cut than large trees and has the additional advantage of checking gullying in badly eroded country. Coppicing is practiced only in some densely populated parts of Ibo and Ibibio areas of Nigeria.

In addition to the basically physical problem of farming tropical soils, a host of human problems beset West African agriculture including ignorance and apathy, the scattered nature and small size of holdings farmed, problems associated with tenurial arrangements and communal man-land relations, excessive reliance on female labor, and market difficulties.

COMMERCIAL CROP PRODUCTION

The vast bulk of crops produced for sale in West Africa comes from African smallholdings; probably less than 5 percent are produced on plantations and estates. The area ranks as one of the most important in the world in commercial tree crop production, particularly in cocoa and palm products. The three outstanding crops of the humid regions are these two, plus coffee.

OIL PALM. The oil palm, native to West Africa, is found in a belt varying from a few miles to 150 miles (240 km) in width from eastern Nigeria to Ghana and in a less concentrated belt from Ghana to southern Senegal. The oil palm finds its ideal habitat in a belt 5 degrees to each side of the equator, with heavy precipitation, no marked dry season, and average high temperatures and humidity. Selected trees grown on plantations yield in the fourth year, reach a productive peak at about the tenth year, and are normally replaced at age 30 to 35. The selected

The fruit of the oil palm
Growing wild over large parts of the wetter regions of West Africa, the oil palm makes a major contribution to food supply and exports.

varieties are much shorter than the wild trees and hence easier to harvest. It is 10 to 12 years before a wild palm bears fruit, and then it yields much less than selected trees. Oil is secured both from the outer fruit and from the palm kernel. Palm oil is used chiefly in soap and in edible oils and fats. The sap of the tree is also tapped, sometimes destructively, for palm wine, while the fronds are commonly used for hut walls and roofs.

Nigeria has long been the leading exporter of both palm oil and palm kernels, but its output was severely cut by the civil war and then adversely affected by low producer prices, while exports have been further reduced by increased domestic consumption. Local and FAO experts have both warned that Nigeria will become an importer of palm oil by the 1980s unless vigorous steps are taken to stimulate this valuable export. Most of palm oil shipments have characteristically come from the so-called Oil Rivers District in the eastern states, and most of the kernel oil and kernels from the Western State (Map 55).

Dahomey is the leading producer among the francophone countries; indeed, palm products are the fundamental richness of the country and account for over a third of exports by value. An estimated 100,000–124,000 acres (40,000–50,000 ha) are under oil palms in the south, but less than half are productive. To arrest a general decline in shipments the National Society for the Rural Development of Dahomey (SONADER) was created in 1961; by 1974 about 77,500 acres (31,400 ha) had been planted in blocks of some 990 to 1,730 acres (400–700 ha) associated with equal areas of food crops, and palm product exports appeared to be increasing.

The Ivory Coast has had a major program under a state organ, SODEPALM, created in 1963 to increase oil palm production; indeed this effort has been the most important single part of its agricultural diversification program. About 30,000 acres (12,000 ha) of industrial plantations had been set out by 1972, and the goal was to treble this by 1980. Village outgrowings are associated with most of the plantations. SODEPALM's program, which also involves 50,000 acres (20,000 ha) of coconut plantations, will comprise 50 villages housing some 200,000 people by 1985, when about $230 million will have been expended. Since less than half of the plantings had reached yielding age in 1972, a large increase in production may be expected in the years ahead. Exports of kernels have already increased from about 8,000 tons in the first postwar decade to 20,000 tons in 1972; palm oil exports grew from about 1,000 tons to 47,000 tons in the same period.

Palm products were the greatest resource of Sierra Leone from the mid-1800s to the 1930s and kernels ranked as the leading agricultural export until 1972, when they were replaced by coffee; but production, which is richest in the far southeast, has tended to stagnate or decline. Several plantations have been started to attempt to reverse this trend.

Other West African countries that have minor commercial production of palm prod-

ucts are Gambia, Guinea, Liberia, and Ghana. Local factors other than the physical environment that influence the production pattern and sales of palm produce include:

(1) The use of palm oil for food. Palm oil is an important subsistence crop; as much as 100 pounds (45 kg) per person per year is consumed in some areas. The kernel, being too hard to crack without a machine, is almost always exported. Local demand for oil sometimes confines exports to kernels.

(2) Competition from other crops, particularly coffee and cocoa, which are preferred crops in several areas that formerly exported palm products.

(3) The failure of smallholders in some countries to take advantage of free seedlings or to properly tend those that have been planted. Many farmers consider the wild palm as a gift of God, which need not be tended or cultivated.

(4) The bulkiness of the crop, which tends to concentrate production where transport is readily available. Exports

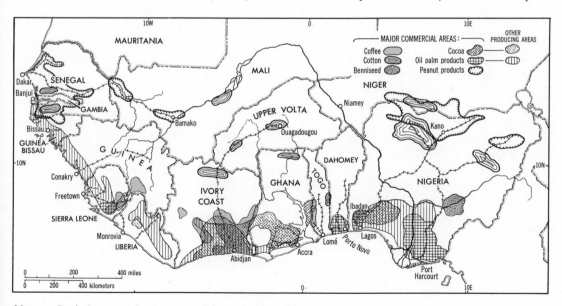

Map 55. Producing areas of major commercial crops in West Africa

are more important from the coastal side of the palm belt and come from interior points only when a river, road, or railway provides superior conditions. There have been considerable improvements in postwar years in the handling of palm products, including the use of bulking stations, tank trucks, and pipelines for loading palm oil in the tank holds of vessels equipped for this trade.

(5) Quality improvement has aided some areas more than others. The introduction of Pioneer Mills, hand presses, and the use of incentive payments resulted in sensational increases in the percentage of top-grade oil in Nigeria. Pioneer Mills can extract 85 percent of the oil content of the fruit, hand presses 60–65 percent, and traditional techniques only 45–60 percent. Speed in delivery of the fruit is important in retaining a low free-fatty-acid content. In very recent years several West African countries have erected palm kernel mills, increasing the value and reducing the bulk of exports; only Zaïre among African countries had long extracted oil from kernels.

(6) Some governments have had much more active programs than others. The Ivory Coast now appears to be making the greatest progress. The efforts in Liberia and Sierra Leone have been far less successful, while a very intensive program in Nigeria was seriously set back by the war.

In the early 1920s almost all of the palm products entering world trade came from their native West Africa. The oil palm was introduced to Malaya and Indonesia on a plantation basis and by 1939 Asia was producing about 40 percent of the world's supply. Plantation methods resulted in higher yields, a higher extraction rate, and a better grade oil. Yields on southeast Asian planta-

tions were as much as twenty times those from West African smallholdings. Zaïre, then Belgian Congo, also became a strong competitor, with most of its oil coming from large plantations. Despite strong protests from commercial operators, the British refused to permit the introduction of plantations in West Africa, a decision that explains the concentration of United Africa Company oil palm plantations in Zaïre, operated by a subsidiary. Because of World War II and continued postwar difficulties in Asia, Africa partially regained some of its losses. In the meantime, improvements in the quality of peasant production made it more competitive, thus helping to sustain the importance of Nigeria in total production. The advantages of the plantation system (see pp. 322 to 324 for discussion) have led many West African countries to experiment with cooperative, partnership, or commercial plantations, and with outgrowing schemes which are now altering the traditional smallholder pattern to an important degree.

Palm oil is only one of many sources of edible and industrial oil. Competing for one or another use of palm and kernel oil are butter, lard, and whale oil among animal products; peanuts, coconuts, sesameseed, cotton seed, rapeseed, soya beans, shea nuts, sunflower seed, and corn oil among vegetable products; and synthetic detergents made from petroleum. This makes for an immensely complex competitive position and means that market limitations are sometimes the main factor restricting expansion. But the oil palm gives the highest yields per acre of any oil-bearing plant and the consumption of edible oils and oilseeds has been increasing rapidly enough to justify the expansion efforts being made in West Africa.

COCOA. By common usage the plant cacao and its product the cacao bean have become cocoa in African terminology. Cocoa is produced in the tropical rainforests and the transition zone to the savanna. Its needs for well-drained soil keep it away from low-lying

Cutting pods from the cocoa tree
Because cocoa trees are usually grown under the shade of partially cleared forests and in untilled land, the appearance of a "cocoa farm" bears little resemblance to the usual concept of a farm.

coastal lands. A short dry period is desirable to permit harvesting, and drying in the sun is aided by the winter monsoon or harmattan. Other requirements are protection from winds, considerable shade, and high temperature. Production is greatly affected from year to year by weather conditions.

Today, Africa accounts for over 70 percent of total world cocoa production. West Africa accounts for about 85 percent of African output and the three leading producers—Ghana, Nigeria, and Ivory Coast—for 96 percent of that (Map 55). But cocoa was not introduced to the mainland until 1879, when it was brought from Fer-

nando Po to what is now Ghana. The first exports of that country were in 1896, when 100 tons were shipped; tonnages increased thereafter to 40,000 in 1911, 118,200 in 1921, and to 311,200 tons in 1936, a figure not surpassed until 1959. In recent years Ghanaian output has generally been between 400,000 and 460,000 tons. Cocoa is the premier cash crop and the chief source of the comparative wealth of Ghana, which now accounts for 25 to 35 percent of total world production. It furnishes a livelihood to over a quarter of the population, 55–65 percent of the country's exports, 6–23 percent of government revenue, and about 10 percent of the GNP.

The center of production in Ghana has been shifting from the Eastern Region to the Ashanti and Western Regions owing to losses from disease in the older areas and the opening of virgin lands to the west. Over 4 million acres (1.6 million ha) are now in cocoa, but a "cocoa farm" bears little resemblance to the concept of a farm in the middle latitudes. Trees are often more or less haphazardly placed under natural shade trees, and underbrush may contribute to the general appearance of a woodland landscape. There are many types and sizes of cocoa farms: in the oldest areas a farmer may have one or more holdings from about 0.5 to 3 acres (0.2–1.2 ha) in size while his wives may have smaller plots of their own; in the formerly uninhabited forest of southern Ghana farms were opened by strangers; elsewhere, groups or companies of farmers developed cocoa farms, some of very considerable extent. Laborers from the north of Ghana and from Upper Volta are nearly equal in number to the farmers themselves; they may operate on a sharecrop basis, or as annual, contract, or daily workers. While they are perhaps fortunate to share in the wealth of this crop, they also reflect the desire of owners to avoid work themselves and often to reside away from their farm holdings.

Cocoa is also the leading export crop of

Nigeria, accounting for 35 to 42 percent of nonpetroleum exports in recent years; its production has been fluctuating around 60 percent of the Ghanaian total, which ranks the country second among world producers. About 1.25 million acres (500,000 ha) are estimated to be under cocoa in some 350,000 smallholdings, about 95 percent of which are in the Western State.

Cocoa was for years the leading export of Ivory Coast but has ranked after coffee and timber in recent years, accounting for about 20 to 25 percent of total exports. About 1.38 million acres (560,000 ha) are planted to cocoa. Cocoa and coffee, which are both grown on many farms, are estimated to involve about half of the active population in that country plus an additional 275,000 foreign workers. Production and exports have increased more rapidly than in other West African countries in postwar years, exports having gone from a yearly average of 62,000 tons in the 1950s to 190,000 tons in the early 1970s, and production is expected to increase to about 300,000 tons by 1980, when exports may again surpass the value of coffee. Large-scale plantings are planned for the southwest.

Togo has about 55,800 acres (22,600 ha) in cocoa, almost all in the Palimé-Atakpamé-Badou triangle, on the plateau. Recorded exports have ranged from 18,000 to 30,000 tons in recent years, but as much as 10,000 to 20,000 tons are believed to come from Ghana illegally. Dahomey, Liberia, Sierra Leone, and Guinea are minor producers of cocoa.

Cocoa in West Africa presents an excellent example of the ever-present danger of serious losses from pests and diseases, which characterizes the production of crops in tropical areas much more than in areas where winter cold results in at least a seasonal halt to their attack. The more concentrated production becomes, the more likely it is that disease will strike.

Swollen shoot has been the most serious of cocoa diseases; it threatened for a time to wipe out the industry in Ghana. This virus disease, carried by the mealy bug, gradually kills the tree and spreads rapidly from tree to tree. Despite much research, no real solution has been found other than to cut out and destroy the diseased trees and to replant with resistant species. As the tree continues to yield for some years following attack, cocoa farmers have sometimes bitterly opposed compulsory cutting programs, despite compensation and provision of superior stock for replanting. Before independence Nkrumah had gained support by promising to eliminate compulsory cutting, but was forced to adopt it again after a voluntary program had failed. Over a hundred million trees were cut out in Ghana by 1961. Swollen shoot remains a serious threat in some areas, particularly in Togo and the western end of the cocoa belt in Yorubaland, Nigeria.

The Cocoa Research Institute at Tafo, Ghana, while unsuccessful in finding an antidote to swollen shoot disease, did show the way to increased production of cocoa. It was found that seedlings brought from the Amazon were resistant to swollen shoot; these were propagated and improved so that it is now possible to replant with stock that yields

Spraying against the capsid bug on a cocoa farm in Ghana
Losses from insects and diseases present an ever-present danger and have been exceptionally large in some years.

in four instead of seven years, is disease resistant, and gives a higher yield per tree. Governments have also promoted the adoption of better methods, which farmers are gradually accepting, of caring for the cocoa tree. It is believed that farmers could treble yields with proper methods, instead of practicing what sometimes amounts to little more than a gather economy.

A second serious disease is caused by the capsid bug, which has caused serious losses, in some years running to as much as 40,000 tons a year in Ghana. Spraying with Gamma BHC has proved effective if carried on by teams at the proper time and if a good canopy has been formed. There is evidence of resistance developing if these practices are not followed. Sale of the insecticides and hand-spraying equipment at subsidized prices encourages farmers to spray their trees; but they tend to slack off when cocoa prices are low.

Fungicides have been developed to cope with a third enemy of the cocoa tree, black-pod disease. Replanting and disease protection require continuous attention; in Nigeria, for example, it is estimated that half of the cocoa area needs replanting and that a sixth is disease infected, while Ghana has large areas in the Eastern Region in need of rehabilitation despite large-scale programs begun in 1969.

The high dependence of Ghana and western Nigeria on cocoa is a good example of the "one crop" character of many African economies, already seen in the concentration of Egypt and the Sudan on cotton. Such dependence contains both physical and economic dangers. From the physical standpoint, it is likely to deplete the soil and lead to soil erosion, and plant disease is more likely to occur and to be more damaging when it does. From the economic standpoint, the country's economy is subject to erratic world markets over which it has very little, if any, control. Cocoa prices fluctuate more than those of many tropical products; in the last decade the average change in price from year to year was 33.7 percent while the highest annual price was more than 4.5 times the lowest.

The supply of cocoa is relatively inelastic, as new plantings cannot affect the market for from four to seven years, while supply from any given area may be seriously affected by weather and disease. Demand for cocoa is more elastic, as high prices will cause confectioners to turn to substitutes, reduce the amount of cocoa in a candy bar, or reduce the size of the bar. Low prices will not only reduce the producer's income, unless there is some cushioning arrangement, but may also greatly affect government revenues, which are likely to be highly dependent on export taxes and on duties on imports. Low cocoa prices played an indirect role in the overthrow of both the Nkrumah and Busia governments in Ghana. Evils of a one crop economy are not confined entirely to periods of low sales and prices; high prices may result in an overemphasis on the cash crop to the detriment of the local food supply and may introduce inflationary conditions, which curtail development. And conspicuous consumption is favored by the lack of tradition for saving that is likely in an underdeveloped community.

All African countries are cognizant of the dangers of a one crop economy and most have adopted measures to alleviate them. Five main avenues have been followed: (1) diversification of the economy, not only through the introduction of new crops but also by development of mining, industry, and other sectors; (2) promoting a better balance between domestic and export production; (3) attempting to emphasize quality production so that the individual country will be less affected than competitors in the event of reduced markets or prices; (4) the establishment of systems to even out receipts of the producers; and (5) efforts to secure international agreements designed to stabilize markets and prices.

Many if not most African countries have adopted some kind of marketing arrangement designed to sustain the income level of peasant farmers, such as statutory marketing boards or stabilization funds. Fixed prices are often announced before the buying season, with the government selling the produce and, after covering the cost of transport and handling, placing any surplus in a reserve fund. Producer prices were not always raised when prices were high, and some of the marketing boards accumulated such large reserves that governments began to use them as sources of revenue. The original purpose of the boards thus became obscured and the system became one of taxation rather than stabilization, although some funds were used to combat disease, promote quality standards, and otherwise benefit producers of the specific crop. Low producer prices, meanwhile, had the disadvantages of reducing rural incomes, decreasing the incentive to expand and improve production, and causing resentment that might lead to political unrest. The Federal Government in Nigeria has recently taken control of the marketing boards and substituted a 10 percent ad valorem export tax in an effort to stimulate farmers to increase production and also to eliminate friction among the states, which had had their own boards.

Repeated unsuccessful efforts have been made since 1960 to create some kind of international control agreement for cocoa. To date, the world's two largest consuming nations—the United States and West Germany—have been reluctant to endorse the proposed agreements, while higher prices and disappearing stocks in 1973–74 greatly reduced the interest of producing nations in the agreement.

COFFEE. Coffee grows wild in the forested area of West Africa up to about 8°N, but exported coffee comes from small farms and a few plantations. Most coffee produced is robusta, which is more suitable to lower tropical areas than the higher value arabica, which is the main type grown at higher elevations in Africa. Africa has assumed much greater significance as a coffee producer in postwar years. In 1934–38 it accounted for about 5.1 percent of world production; by 1961–62 the percentage had increased to 17.3 and it has been about 31 percent in recent years. Two factors were of particular importance in permitting this notable increase: the lower price of African robustas than of Latin American coffees, and the increased use of solubles, for which robustas are as good as or better than higher priced coffees.

The African share of world production would probably have been higher than it is if 62 producers and consumers had not joined the International Coffee Organization (ICO) in 1962, assigning quotas to producing nations in an effort to bring stability to an increasingly chaotic market. The ICO arrangements were long held up as the great example of a successful commodity agreement, but when the price began to rise persistently the consuming nations—led by the U.S., which alone absorbs 40 percent of world coffee—refused to renew their part of the restrictions in 1973 (see Chart 2). An agreement is still in being, but quotas are maintained solely by the producing nations, which means that a single medium-scale producer could disrupt world markets. It is not unlikely that a new agreement will be effected, but it will probably require some softening in world prices.

Coffee is now the leading export by value in eleven African countries and a major export from four others; it provided a major engine of growth in a number of these countries until market restrictions imposed limits to expansion. Ivory Coast is by far the leading coffee producer in West Africa and has consistently outproduced its African rival, Angola, in recent years to achieve third place in world output after Brazil and Colombia.

Coffee was introduced to the Ivory Coast in 1880 but did not see its first important expansion until 1926. Exports remained below 20,000 tons until 1941 and below 100,000 tons until 1956, however. In recent years they have ranged from about 200,000 to 300,000 tons depending mainly on climatic conditions. Plans now call for maintaining a level of about 250,000 tons, for rehabilitating some older areas, and for shifting some production to the southwest. Over 1.48 million acres (600,000 ha) are planted to coffee, the vast bulk of which is in African smallholdings.

Other West African producers, all of which are minor, are Guinea, Liberia, Sierra Leone (where coffee has recently replaced palm products as the leading agricultural export), Ghana, Togo (which has about 100,000 acres [40,000 ha] in coffee, 90 percent of which is in the southwest plateau area, where coffee accounts for about 31 percent of the country's total exports), Dahomey, and Nigeria (where some robusta is grown in the Western State and arabica trials are underway in the Mambilla Plateau near Yola in the Northeast State).

SECONDARY COMMERCIAL CROPS

Palm products, cocoa, and coffee are the big three among commercial vegetable crops coming from the humid tropical areas of West Africa. Bananas and rubber have a secondary role, and a long list of lesser crops have tertiary importance.

BANANAS. The plantain or cooking banana is grown over a large part of West Africa for subsistence purposes though it nowhere assumes the importance that it does in parts of East Africa. Bananas produced for export include the dwarf or finger banana and the Gros Michel. The needs for strict quality control, for a well-coordinated and reasonably continuous delivery to port, and for special handling facilities have tended to restrict the areas of production and to give plantations a major share in output. Excepting British Cameroon, the southern part of which is now part of Cameroon, the main

Bananas being grown under partial irrigation near Kindia, Guinea
Most bananas exported from West Africa are grown on plantations and estates rather than on indigenous smallholdings.

West African producer in prewar years was French Guinea, which had almost doubled exports by 1958. Guinea bananas come mostly from small estates about 60 miles (100 km) inland from Conakry in the Kindia district and around Mamou, and are grown under a system of partial irrigation.

The Ivory Coast has inherited much of the interest focused upon Guinea and now ranks first among African exporters, excluding the Canary Islands. Exports grew from about 29,000 tons per annum in the 1950s to the present level of about 160,000 tons. Bananas are the fourth ranking export from that country and provide employment for about 10,000 people. Production is strictly regulated; bananas cannot be grown more than 124 miles (200 km) from the port and growers must show that they can provide a certain level of high-quality fruit. The National Packing Agency maintains a factory to produce cartons and a score of packing centers where the bananas are cut in hands, chilled, packed, and dispatched in refrigerated trucks to Abidjan, which has modern chilling facilities and a conveyor system for loading refrigerated vessels. Liberian efforts to grow bananas for export were unsuccessful because of an attack of Panama disease. Ghana has increased its exports considerably but is still only a minor shipper.

RUBBER. Owing to uncertainties about supplies from the Far East and a continuing increase in the world consumption of rubber, various African countries have been promoting the production of rubber with some intensity in postwar years. The African share of natural rubber supplies increased from 1.3 percent in 1938 to 7.3 percent in 1970, but fell to 5.9 percent in 1973. Nigeria, Cameroon, Liberia, and Zaïre account for the bulk of African output, but several other countries have also begun to produce in recent years.

The Firestone Plantations Company has played a notable part in the history of Liberia, having been the largest employer, exporter, and trainer of skills in the country. The company's decision to produce natural rubber in Liberia resulted largely from a desire to escape the Stevenson Plan of 1922–28, which regulated rubber exports from British producing areas with the aim of sustaining world prices. After some difficulties, Firestone secured a 99-year concession on one million acres in late 1926, and by 1940 some 72,500 acres (29,340 ha) had been planted, 54 percent of which were yielding. The importance of Firestone Plantations increased enormously during the war, and production—with some emergency overtapping—reached 20,000 tons in 1945. A major planting program was again undertaken in 1950 and regular replanting with higher-yielding stock began in 1954. Today, about 90,000 acres (36,400 ha) are under rubber in the two Firestone plantations, about 86 percent at Harbel and the remaining 14 percent at Cavalla.

The Harbel plantation is said to be the finest and most modern rubber plantation in

Tappers on the Firestone rubber plantation at Harbel, Liberia
About 16,000 Liberians are employed on this plantation, one of the most modern in the world.

the world; its yields are the highest in the world. About 16,000 Liberians are employed on the plantation and in its associated factories, including the world's largest latex processing plant, a small hydroelectric installation, a sawmill, a brick and tile factory, and a workshop making latex cups, soap, and rubber sandals. Before the development of iron mining Firestone was responsible for as much as 85 percent of total Liberian exports and 39 percent of government revenues.

Five other rubber plantations have been started in Liberia since 1960, bringing the total plantation area to 138,000 acres (56,000 ha) by 1972; of this about a quarter was not yet yielding. In addition some 4,200 Liberians own rubber holdings ranging in size from 1 to 2,000 acres (0.4–800 ha) and totaling about 150,000 acres (60,700 ha); free seedlings are provided to these farmers. There is, however, a need to increase the effectiveness of the smallholdings, where yields are about half those of the plantations and on which many trees go untapped because of the lack of labor; about a third of their area also needs replanting with high-yielding budded stumps. Nonetheless, the record of devolving production to private farmers has shown commendable progress: the number of independent farms grew from 150 in 1941 to 777 in 1951 and to the present 4,200; their share of production has increased from 4 percent in 1955 to 14.5 percent in 1960 and 25–30 percent in the early 1970s. Tappings are delivered to plants at the big concessions, two-thirds going to Harbel for processing. While rubber now accounts for only 14 to 17 percent of total Liberian exports (iron ore takes the dominant part) the industry is of considerably greater importance to the economy from the standpoint of labor employed. A total of 42,000 Liberians—or about two-fifths of the paid labor force—are engaged, five-eighths on the concessions and three-eighths on the private farms.

In Nigeria there are several plantations plus smallholdings in the Sapele-Warri-Benin area of Midwest State, which produces about 80 percent of Nigerian rubber, and additional plantations near Calabar. About 450,000 acres (182,000 ha) are in rubber, the vast bulk in some 100,000 smallholdings ranging in size from 1 to 10 acres (0.4–4 ha). Yields on the African farms are low, about a quarter of what they could be with selected seedlings, and the quality is inferior, but conditions should improve with the refurbishing of war-damaged plantations and facilities, the distribution of seedlings, and the establishment of modern processing plants. Partnership plantations in Western State have not, however, proved very effective.

Other West African countries have lesser interests in rubber. About 32,600 acres (13,200 ha) had been planted by 1973 in Ivory Coast, most by two large companies of which one is 50 percent government owned. Plans call for trebling the area by 1980 with a large acreage to be planted in the southwest, partly on a big new Michelin holding. Exports, which were nonexistent in the 1950s, reached 13,700 tons in 1972 and are expected to be about 23,000 tons by 1980.

Ghana had 24,000 acres (9,700 ha) planted to rubber in 1972 and is establishing a 25,000 acre (16,000 ha) estate in partnership with Firestone north of Sekondi-Takoradi as the nucleus for smallholder production. Sierra Leone has a small output and one latex plant.

The fears of intensifying competition from synthetic rubbers and of lower prices have been a deterrent to increasing rubber plantings, and prices prevailing in the 1970s caused losses on some plantations, reduced tapping by smallholders, and lowered export earnings. The share of natural rubbers in total elastomer fell from 45 percent of the total in 1961 to one-third in 1970, but total consumption of both increased and the FAO forecasts the ability to sell about 50 percent more rubber in 1980 than world production in 1970.

COMMERCIAL CROPS OF LESS
IMPORTANCE

A number of lesser commercial crops are produced in the humid tropical regions of West Africa. They will receive only brief attention.

KOLA NUTS. These nuts are one of the few agricultural products shipped in quantity from the humid south of West Africa to the drier north. The Ivory Coast, for example, has shipped an estimated $7.2 million worth to Mali and Upper Volta in recent years, though only about half this value was in recorded exports. Chewing kola nuts releases caffeine and theobromine—mild stimulants that reduce the feeling of hunger. Also used in extract for soft drinks, kola nuts are shipped overseas in minor quantities from several West African countries.

GINGER. Sierra Leone exports a small amount of this crop, which is also produced around Zaria and on the Benue Plateau in northern Nigeria.

PIASSAVA FIBER. This fiber, taken from the leaves of the raffia palm, is used for making brooms. The Sherbro district of Sierra Leone has practically a world monopoly in the sale of best-quality piassava fiber, whose marketing is handled by a successful African cooperative. Exports were valued at $800,000 in 1972.

FRUIT. There is a considerable local trade in citrus fruit and pineapples in West Africa and a small export of fruit, preserves, and extract from a half dozen countries. The quality and appearance of much of the fruit requires improvement if substantial growth in sales is to be achieved. The Ivory Coast has had considerable success with pineapples, grown on the plantations and by outgrowers and canned in three factories. But canning establishments elsewhere have had difficulty in securing sufficiently continuous supplies to assure their viability.

CASHEW NUTS. Produced in considerable quantities in eastern Africa, the cashew nut

has been introduced to West Africa as one of several crops that would help to diversify production. About 4,000 acres (1,600 ha) of plantations were set out in southern Nigeria in the 1960s, and Ghana has also experimented with this crop. In Dahomey a FED-financed program aims to plant 42,000 acres (17,000 ha) in the center and north; production reached 800 tons in 1972 but is expected to increase fivefold by 1975. Two factories to process the nuts, which have previously been shipped to India for decortication, have been set up at Parakou and Bohicon.

TOBACCO. With the large increase in the consumption of cigarettes, many nations of West Africa have wished to increase domestic production in order to save foreign exchange and to provide a new cash crop for peasant farmers. Tobacco manufacturers have had well-managed programs to assist in promoting production. Much of West African tobacco is grown in the drier north, but several humid tropical regions now produce and intend to expand output of tobacco. Dahomey, the Ivory Coast, Ghana, and the Western State of Nigeria all have plans to expand tobacco production in southern regions. Only Dahomey has a small export; most countries still import up to three-fourths of the amount used in local factories.

RICE. This is the favored staple in the countries from western Ivory Coast to Senegal, which, despite long-standing programs to stimulate domestic production, have been importing increasing quantities. About two-thirds of production is from upland rice, a quarter from swamp rice, and the rest from paddies. In Sierra Leone rice is the most widely grown crop (produced by about 86 percent of all farmers) and it accounts for an estimated 45 percent of the total value of agricultural production. The government's rice policy has been to encourage the expansion of swamp rice. Advantages of wetland cultivation include the presence of some good to excellent soils which, being under water most of the time, are not subject to erosion, and the ability to obtain a considerably higher yield per acre from rice than from other major food crops. Irrigated rice is superior to swampland rice because it gives the highest yields and matures in 3½ months instead of 6. Rice is also comparatively easy to store and nutritionally superior to yams or manioc.

Incentives to expand production of rice in the several countries of the western portion have included price supports, control of imports, subsidization of fertilizers, land clearing, subsidization of mechanical plowing, and demonstration schemes run by Taiwanese or mainland Chinese. The cost of the rice produced is higher than that of imports and the cost to the consumer is high because of protection. Evidence also suggests that mechanization increases costs. Most of the countries feel that they must strive for self sufficiency, however, to conserve foreign exchange and Sierra Leone even hopes to again become an exporter, which it was in the early 1950s. Ghana and Nigeria, whose rice is not the major staple, are also expanding production, the latter partly by setting up a 20,000 acre (8,100 ha) plantation in Lagos State.

There is, of course, considerable commercialization of other crops for local sale, particularly in southern Ghana and southern Nigeria. Concerned with the increased imports of foodstuffs, many countries are turning to the promotion of import substitution, which has been a major theme in manufacturing for several decades. In addition to tobacco and rice, noted above, sugar and cereals have received attention.

ANIMAL INDUSTRIES

The humid tropical regions of West Africa are deficient in animal products, chiefly because the tsetse fly precludes keeping most species of cattle. Most small livestock (mainly chickens and goats) are kept on a haphazard basis with little thought for improvement or

commercialization. There is often a shortage of meat, dairy produce, and eggs in urban centers. Efforts to correct this situation have thus far been inadequate. They include the setting up of poultry and pig farms, and the upgrading of dwarf cattle, which are resistant to trypanosomiasis. The dwarf N'Dama and West African Shorthorn are very poor milkers and have very low carcass weights, but with proper selective breeding they might be bred up to more satisfactory weights and milking yields.

ORGANIZATIONAL APPROACHES TO
AGRICULTURAL DEVELOPMENT

The recognized inefficiencies and limitations of traditional farming practices have led to the adoption of a variety of new approaches to farming in West Africa. Efforts vary from simple extension work to state farms comparable to the sovkhoz of the Soviet Union. Many of the programs were not clearly thought out or preceded by pilot operations and appear to have represented more a desire to start something new than to face up to the real problems of African agriculture.

Most of the countries have promoted plantations for such crops as rubber, sugar, cashew nuts, pineapples, bananas, and oil palm, though efforts are commonly made to reduce the significance of foreign presence through government participation, partnership arrangements, or provision for outgrowing by smallholders. When based upon adequate previous preparation, the plantations have generally been reasonably successful.

Cooperative farming has been tried in Sierra Leone, Ghana, and Nigeria, but cooperatives have only been successful in the marketing of crops, not at the production level. Cooperative schemes for tractor services, which represented a major investment in Ghana in the early 1960s, were almost a total failure; most of the machines were out of service within a few months. The former Eastern and Western Regions of Nigeria placed great emphasis on resettlement schemes patterned on the Israeli Moshav multipurpose cooperative farm. These, too, were unsuccessful, the cost per settler (c. $10,000–$14,000) being totally out of line with the local economy or with the results. Ghana is now starting a large settlement scheme in the Keta-Avu area of the Volta Region.

Ghana under Nkrumah attempted to form numerous state farms, while Guinea at one point tried to nationalize all farm operations. Totally unacceptable to indigenous farmers, this approach has been abandoned, although some of the resettlement schemes in Nigeria have been converted essentially to state plantations.

Francophone countries have frequently contracted with private and mixed French corporations to assist in rural development. These and national organizations may be focussed on specific crops, or on regions, while French research and technical organizations have also been important. Some of these approaches have had considerable success, particularly when care has been taken to work with groups of farmers with some previously existing cohesion.

Modernizing indigenous agriculture in Africa is a formidable task, and no easy solutions to the problems involved have been found. Collective and state farms have not worked any better than they have in the Soviet Union. The high capital intensity of settlement schemes almost condemns them to failure in areas where land and labor are relatively cheap, at the same time reducing the attention that can be given to general farming. It should not be forgotten that peasants have been the main dynamic force in production of export crops; they have invested savings in small-scale agriculture and agricultural expansion, and have demonstrated a willingness to adopt new crops or new techniques when these are seen to be beneficial. The need would appear to be to avoid "projectitis" and to make a greater effort to de-velop the initiative and knowledge of the individual farmer.

TRANSITION ZONE TO THE DRIER SAVANNA

The middle belt of West Africa, lying between the humid tropics and the drier savannas and steppes, is of little commercial significance. As already noted, it is less densely populated than the zones to the north or south. It cannot easily participate in livestock production because of the tsetse fly. It is often further disadvantaged by inadequate transportation. But even in Nigeria, where two great rivers plus roads and rail lines linking the north and south are available, there is relatively little production for sale outside the area. Nonetheless, it must be expected that these areas will gradually become more important if for no other reason than that they provide room for an expanding population. The middle belt in Nigeria, for example, contains about two-fifths of the country's area but less than a fifth of the population.

Crops that are produced in surplus in the transition zone include grains, yams, soya beans, tobacco, cotton, and sugar. The area inhabited by the Tiv along the Benue in Benue-Plateau State is an important producer of sesameseed, locally called benniseed. Nigeria's main sugar plantation is at Bacita on the Niger in Kwara State, while a 33,000-acre (13,400-ha) sugar estate began production at Numan on the Benue in 1974. The latter is expected to support a 100,000-ton refinery, the total cost of the installation to be about $120 million.

Dahomey has an important program for the integrated development of the Zou region in the center of the country. Allen cotton is the cash crop, plantings having gone from almost nothing in 1965 to 52,000 acres (21,000 ha) in 1971 when production reached 18,000 tons. Ghana is planning a large ranching scheme plus production of soya beans, cotton, and tobacco in its part of the middle belt.

THE DRIER SAVANNA AND STEPPE BELT

The regional occupation of this area is grazing, and this pastoralism covers a far wider area than sedentary agriculture. Its contribution to the money economies of West Africa, however, is much less than that of the cultivators. The large number of tillage farmers, far more concentrated than the graziers, produce the bulk of the food consumed in the area and of the goods exported from the region. Even in a dry country such as Niger, with only 2 percent of its area under cultivation, tillage agriculture provides the main occupation of about two-thirds of the population.

Most of the inhabitants of this belt are Muslims, though clusters of pagan peoples reside in the higher sections. With the exception of Senegal and the capital cities, the area has been less influenced by contact with modern forces than southern zones; traditional life gives it a conservatism that contrasts with the dynamism of the south.

PASTORALISM

Pastoral tribes of West Africa, of which the Fulani or Peuhl is the most important, range widely over the dry savanna-steppe belt and into the desert; they keep cattle, sheep, and goats in large numbers. Some sedentary tribes, including the settled Fulani, also keep cattle and smaller livestock. Asses are commonly used in the north and camels are still used to carry produce to and from some of the market towns. Graziers can be broadly divided into three groups:

(1) The true nomads, such as the Fulani Bororo or Bush Fulani, who are the most tradition-bound. On the move with their large herds almost continuously, they consider sedentary graziers and tillage agriculturalists to be practicing inferior occupations. Theirs is a largely self-sufficient existence and cattle are sold only with great reluctance.

(2) The transhumants, including certain Fulani, who congregate around rivers, lakes, and marshes during the dry months and disperse during the rains. Many transhumants are involved in a symbiotic relation with sedentary farmers. In northern Nigeria, for example, many Fulani spend part or all of the dry period grazing their herds on the stubble or the fallow fields of the Hausa, being paid a small amount for the enriching manures provided.

(3) Sedentary graziers, who are largely settled Fulani or mixed Fulani groups who have converted to tillage agriculture but who continue to keep as many animals as can be supported in their area. Some of the farmers belonging to Peuhl society are descendants of slaves originally taken from various sedentary tribes.[1]

The problems of these pastoralists are legend and may be considred as typical of those besetting graziers in most of the drier areas of tropical Africa.

PHYSICAL PROBLEMS. Climatic difficulties are among the most severe handicaps to grazing in steppe areas. Water supply is highly irregular, because violent showers are often followed by a long period with no rain at all. Consequently, stock water is often in short supply in the dry period, and vegetation dries up, becoming less palatable and less nutritious. Furthermore, precipitation is unreliable, and in dry years there may be a wholesale loss of animals. Not only does rainfall vary from year to year, but it is also

[1] See deRouville C. Queant Thierry, *Agriculteurs et Eleveurs de la Région du Gondo-Sourou* (Ouagadougou: Centre Voltaïque de Recherche Scientifique, June 1969).

likely to fall very unevenly within the rainy period, whose start and ending are not readily predictable. Relatively little can be done about these natural conditions; the major palliative is to provide better stock water supplies by one or more of the techniques noted in the chapter on the Sudan. In addition, temperatures are generally too hot for quality animals from the middle latitudes. They cannot sweat adequately, develop an artificial fever, lose their appetites, and become more susceptible to disease.

The vegetation of steppe areas also leaves much to be desired. Some grasses have a high protein content when they sprout after the early rains, but as they grow older the fiber content increases and the protein and water content decreases. They tend to be high in silica and low in phosphorus and calcium. They are coarse and hard, unpalatable to quality animals, particularly in the dry season. Shrub vegetation, particularly thorn bush, also covers large expanses and, although eaten by sheep and goats and even by cattle when grass is not available, it too is poor animal fodder. It would obviously be very difficult to improve indigenous grasses, as the low value of lands does not justify the expenditure required. Relatively little study has been made of tropical grasses, though some superior species have been delineated. Under controlled conditions it might be possible to sow superior grasses under a system comparable to that of *harig* farming in the Sudan, while eventually it might prove practical to carry on a large-scale program of replanting if a sufficiently cheap method of tearing out brush vegetation becomes available. The absence of goats might also preserve the vegetation, but their elimination is not likely to win the acceptance of graziers, who use the goat milk and see the animals as a kind of insurance during extremely dry cycles. One of the most effective improvements that could be made with regard to cattle feeding would be provision of larger quantities of fodder in the dry season. This would require development of a kind of mixed agriculture whereby sorghum and millet could be ensilaged for winter consumption while livestock grazed on the natural vegetation in the summer. Senegal proposes to plant many miles of shelter belts to protect the vegetation and soils in the grazing areas.

A great variety of diseases, parasites, and insects afflict tropical livestock, presenting one of the greatest physical barriers to improvement. In the past, epidemics of catastrophic proportions have hit the livestock of Africa, influencing the whole history of certain areas, such as parts of the East African highlands. Much progress has been made in combating animal diseases. Efficient vaccines now exist for rinderpest, bovine pleuropneumonia, and other diseases; tick-borne diseases are being controlled by regular dipping and spraying, which is compulsory in many areas. But it is an enormous task to treat every animal in the huge grazing areas and success has been by no means complete.

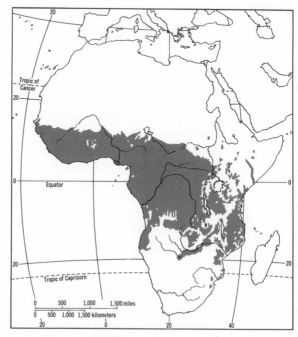

Map 56. Areas infested by tsetse fly in Africa

Furthermore, effective treatments have not yet been developed for certain diseases. The most limiting of such diseases, not necessarily in the steppe areas but in the continent as a whole, is bovine trypanosomiasis, otherwise known as nagana or sleeping sickness.

Carried by the tsetse fly, trypanosomiasis makes vast stretches of Africa untenable for man and beast, affecting in one way or another the livelihood of perhaps 45 million people in West Africa alone. As may be seen from Map 56, various species of the tsetse fly are found in a broad belt of Africa from about 12°N to 15°S, with certain areas within, especially at higher elevations, being tsetse free. Generally speaking, cattle cannot move into tsetse-ridden areas except on a seasonal basis, hence in West Africa grazing is rather effectively restricted to steppe and drier areas and to higher lands such as the Jos Plateau. As has been noted, this restricts the possibility of developing a soil-conserving mixed agriculture in tropical rainy and savanna areas.

Among the numerous techniques that have been tried in the battle against the tsetse fly are biological control by using the fly's parasites and predators, interbreeding to promote sterile offspring, and gamma ray bombardment to render male flies sterile; destruction of wild game carriers or fencing to protect cattle from close contact, which is practicable only in limited areas; painting cattle with DDT, since spraying of large areas is too expensive; development of a serum to give immunity. None of these methods has proved effective, though atabrine serum gives temporary protection and reduces losses of cattle trekked overland to markets within the tsetse belt. Thus far the only effective answer appears to be the clearing of bush where breeding occurs and the immediate use of the area to prevent reversion. This applies more to human trypanosomiasis than to nagana, for only the intensity of use involved in tillage agriculture is sufficient to stem renewed invasion by the fly.

Clearing of bush is expensive, however, and cannot be considered a practical method of opening the savanna grasslands to cattle raising.

That no real solution to the tsetse fly has been found is not entirely to be regretted, because opening of the huge areas now infested would subject these lands to the same destructive practices that characterize the steppe regions. To prevent this, it will be necessary, when a remedy is found, to mount a multifaceted campaign to ensure that proper land use is practiced in the newly opened areas.

Returning to the physical difficulties affecting grazing in the steppe lands, we find that poor quality of the indigenous livestock is also a major deterrent to advancement. This applies to all types of livestock; about the only quality products coming from West Africa are goatskins from the Red Sokoto or Maridi goats from Nigeria and Niger. Indigenous breeds of cattle, sheep, and goats do have the advantages of adaptability to local conditions, an instinctive avoidance of toxic and other undesirable vegetation, a high-butterfat milk, and resistance to some diseases. But they also have slow growth and breeding rates, low carcass weights and milk yields, and high mortality rates among young animals. The poor quality of livestock in West Africa and other parts of the tropics stems from both physical deficiencies of the area and poor human practices. Considering the lack of attention livestock receive, the poor quality is not surprising; evidence shows that there is a favorable response to simple improvements in husbandry.

For years it was thought that crossbreeding with quality animals from other areas would be the best method of upgrading tropical breeds. But it is now accepted that adaptability to the tropical environment is as important as production per se. The best chance for improving tropical livestock would appear to be in the upgrading of beef cattle. What they consume is relatively abundant as compared to fodder for dairy cows; they occupy the cheaper lands; dietal deficiencies can be corrected with minor additives; and the heritability of meat production is high. Improvement of dairy cows is much more difficult, and yields comparable to those of the middle latitudes, say 10,000 pounds (4,500 kg) per year per cow, can probably never be attained except under strictly controlled artificial conditions. Such levels require full feeding—three to four times more than maintenance feeding—but under tropical conditions such feeding increases the heat burden on the animal, increases its discomfort, and results in a loss of appetite.

HUMAN PROBLEMS. Human problems connected with livestock raising are equally impressive. Lack of knowledge and certain cultural practices lead either to serious malpractices or to lack of interest in improving livestock or in commercializing their production. As has already been noted, most pastoral tribes of tropical Africa practice a "cattle culture." While the nomadic Fulani, for example, may ostensibly pay a great deal of attention to their cattle (calling each by

Fulani cattle in northern Nigeria
Some Fulani are true nomads, others are transhumants who graze their cattle in the dry period on the stubble and fallow fields of the sedentary farmers, and still others have settled permanently, continuing to keep animals on their farms.

name) and may even practice a kind of selective breeding, their practices are anything but modern, and selection is sometimes guided by desire for a particular color or for large horns as much as for meat and milk yield. However, many practices associated with "cattle culture" are not irrational, as has sometimes been suggested. Cattle play a vital role in the lives of the graziers, providing milk, meat on ceremonial occasions, and hides—plus a store of wealth, and the bride price in marriages. Even the desire for large numbers may be rationally explained as reflecting the desire to ensure against losses from disease and drought, though smaller, healthier herds might in fact be more practical.

Grass burning is one of the most widespread African malpractices. It is believed that the resultant ash is beneficial to the soil and vegetation, and burning is also practiced to reduce the insect population. But burning also consumes humus in the upper soil (making it more subject to erosion), favors the spread of fire-resistant bush and poorer grasses, and destroys helpful worms and organisms. No adequate alternative for grass burning is yet known, and despite its illegality in many countries it is still widely practiced as the only way to control the insect population and bush growth. Carefully timed late burning may reduce its harmful effects, and burning once in four years is probably adequate to control bush, but this would require no grazing in the previous season to provide sufficient combustible material for a fierce fire. Here is another good example of the close interplay of physical factors in Africa, the need for attacking a problem from many angles, and the need for scientific research to find a solution to physical problems before human problems can be satisfactorily met.

Overgrazing is another malpractice prevailing over most grazing lands in Africa. It has contributed greatly to degradation of vegetation in the steppes and dry savannas

Stock water tanks at Gajiram, northern Nigeria
Overuse of the tanks had led to destruction and degradation of the vegetation over a considerable area years before the long drought of recent years.

and to its destruction along the desert margins. Livestock are selective eaters, choosing the more palatable and nutritious grasses, which, unless proper stocking is practiced, are gradually replaced by less nourishing plants. The accelerating expansion of woody, and often thorny, growth is a curse in many grazing areas of Africa. Overstocking is explained by lack of knowledge, failure to allow for periodic and certain droughts, presence of the "cattle culture"—with its emphasis on numbers as much as quality—and by the practice of bringing cattle through the dry period on a near starvation basis. Any stocking practice based on wet-season carrying capacity is bound to impair growth and productivity in the wet season and be destructive in the dry period. Overgrazing has become so severe in certain parts of Africa that governments have had to introduce voluntary or compulsory culling to prevent outright destruction. This is often very difficult because it runs counter to the traditions of the pastoralists. Somewhat ironically, control of cattle diseases has contributed importantly to overstocking, revealing again the necessity to operate on several fronts when attempting to ameliorate problems in Africa. The greatest check to overgrazing is, of course, the periodic severe droughts, which greatly reduce animal populations. In West Africa before the recent catastrophic drought cycle, there were at least twice as many animals as there should have been.

Skinning and drying of hides and skins is also characterized by inadequate techniques, though veterinary officers have succeeded in introducing improved methods in some areas, including the stretching of hides on frames rather than on the ground and provision of shade drying.

Despite the many problems that exist, grazing remains the most widespread activity of the western sudan belt, and there is some export of hides and skins from all countries within it. There has also been an increasing sale of live animals and meat as the "cattle culture" has broken down in West Africa, though annual offtake remains at only about 7 to 10 percent compared to about 25 to 30 percent in Europe and Anglo-America. Much of the sale goes unrecorded and international movements are difficult to estimate because of clandestine crossings. Efforts to increase commercialization include the installation of modern abattoirs, provision for shipping frozen carcasses by air or rail to southern cities, and the setting up of a number of ranching schemes. While the total exports of livestock products from West Africa have never been large on a world scale, livestock can become a resource of potentially great importance once practices have been improved.

SEDENTARY AGRICULTURE

Tillage agriculture in the drier savannas and steppes contrasts strongly with that in the humid tropical regions. The tilled fields are more comparable to those of the middle latitudes than to the cleared patches in the south. The staple crops are sorghum (guinea corn), millet, and peanuts, plus sweet potatoes, cassava, sesame, beans, and cowpeas. In limited areas, where water is available, rice is a leading crop, while tomatoes, onions, and sugarcane are also produced. The calabash vine is grown, often on the huts, to provide containers and utensils of varied size and shape; henna produces a favored dye; peppers and tobacco are commonly grown for local consumption. Much more labor is required than in the humid areas, where manioc can be grown by sticking shoots in the ground and many cash and food items are tree crops.

Many improvements are possible in agriculture through the association of farming and cattle keeping, use of bullocks for plowing and of ridging plows, application of manures and chemical fertilizers with the aid of carts, some hand spraying, use of improved seeds, better rotational practices, and more systematic distribution of the *acacia albida* or *Faidherbia* tree, which enriches the soil beneath it. It has been shown that such practices can raise yields per unit area and per man hour by two to five times with modest capital inputs. But they have not been widely adopted as yet. In northern Nigeria, however, the number of farmers using bullock-drawn plows increased from about 2,000 in early postwar years to 20,000 in 1960 and some 75,000 in 1973. Animal traction is also being pushed in Gambia, Mauritania, Senegal, and Upper Volta.

Conditions for tillage agriculture are often very difficult in the region. Aridity and uncertainty of precipitation increase toward the

Aerial view of a small village and surrounding fields near Katsina, northern Nigeria
Peanuts are the main cash crop in this region; sorghum and millet are the main subsistence crops. The regular field pattern contrasts strongly with that of the rainforest and wetter savanna areas.

desert. Many of the soils are poor and some are severely eroded. In the middle latitudes soils that have formed under natural grassland have a higher humus content than forest soils and are usually more fertile; "in tropical Africa, the reverse is true. The savanna soils have less humus and are often less productive than the forest soils." [2]

Only about 2 percent of Mauritania, 29 percent of Senegal, 41 percent of Upper Volta, and perhaps 16 percent of Niger are cultivable, though often less than half of available lands are cropped and much less than that effectively cultivated. While the vast bulk of the farmed area produces rain-grown crops, catch crops are grown on the floodplains or *fadama* lands after the seasonal inundations, and seasonal or perennial irrigation is practiced by traditional methods or in some modern schemes. Use of *fadama* lands has the advantages of high yields, use of good soils, and production of crops in the off season.

While certain common denominators can be recognized over the area, practices vary greatly among ethnic groups and among farmers in the same groups, as has been shown by several investigators.[3] Considerable skill and intensive practices are shown by such groups as the Serer in Senegal, who are the only people to practice a real mixed agriculture in West Africa, the Songhay and Dogon in Mali, and the Hausa in Nigeria, who farm much of their lands in the close-settled zones around the emirate cities on a permanent basis, made possible by heavy applications of wastes available from these urban communities.

COMMERCIAL CROPS. While there is a considerable amount of local sale and some

[2] Peter M. Ahn, "Some Soil and Related Problems Affecting Development," East African Seminar on Development and the Environment, April 15–24, 1974.

[3] See Polly Hill, "The Myth of the Amorphous Peasantry: A Northern Nigerian Case Study," *The Nigerian Journal of Economic and Social Studies*, 10, no. 2 (July 1968): 239–60.

shipment of country produce to the humid tropical areas to the south, the two major crops shipped to world markets from the drier north are peanuts, by all odds the more important, and cotton.

Peanuts. West Africa ranks as the largest exporter of peanuts in the world, normally accounting for about 70 percent of the total, though it produces only about a fifth of world output. Two major production zones may be delineated: that of Senegal and Gambia extending inland to Mali, and the nodes of production in northern Nigeria and adjacent Niger (Map 55). In the western area peanut production is sometimes monocultural, which has resulted in such deterioration as to require a progressive move to the interior.

Peanut production started in Senegal in the mid-1800s and that country was producing about 125,000 tons in the early 1900s. The output had increased to 550,000 tons before World War II and has averaged about 912,000 tons in the last decade, with considerable fluctuations from year to year depending mainly on climatic conditions. About 2.5 to 3.0 million acres (1.0–1.2 million ha) are planted with peanuts each year.

While the total value of sales has been quite constant in the decade to 1972, their share of exports was almost halved in the period to 35.6 percent of the total. Characteristically low yields, small holdings, and the necessity to adjust to lower world prices have meant that the average cultivator gets only a small return, despite efforts to aid peanut production through the distribution of better-yielding varieties, subsidization of fertilizer sales, and reorganization of marketing arrangements to reduce the grossly excessive number of middlemen.

Peanut products provide about 93 percent of the value of Gambian exports. Peanuts are grown there on light sandy soils away from the riverside swamps. Mali produces about 100,000 to 130,000 tons yearly of which 40 to 60 percent is commercialized. Until recently a large share of peanuts from West Africa were decorticated and shipped in bags; an increasing share from all producing countries is now converted to oil and cake. Beginning in the early 1970s Senegal discontinued all shipment of nuts; most of its peanuts are processed in large mills in Dakar, and this industry is the largest in Senegal.

Peanuts being collected in southern Senegal for shipment to an oil factory
Peanuts are the major crop export of Senegal, Gambia, Mali, Niger, and northern Nigeria.

In the eastern producing zone, the area around Kano and Katsina is most important, but large areas in the north are devoted to the crop. Shipments have declined drastically in recent years because of increased domestic consumption, low producer prices, and poor weather conditions. Peanuts are the most important cash crop in Niger and account for about a third of total exports.

Other West African countries are promoting peanut production, including the Ivory Coast, Togo, Dahomey, and Upper Volta. The last three have small exports, but most of the output is consumed domestically.

Cotton. A major part of the cotton produced in West Africa is consumed by traditional spinners or in the rapidly expanding modern mills. Production and export levels were severely reduced from 1970 to 1974 by the prolonged drought affecting the sudan-sahel belt south of the Sahara. The most important producing region is centered in the western part of Kano State in Nigeria, but extends westward to Sokoto and south to Zaria. Cotton fits in well with peanuts as it is planted and harvested later. Nigeria's goal is to meet the growing demands of the textile industry and to maintain exports by extending the area under cotton and by improving

Charcoal brought to Kano for sale
The need for cooking fuel has put great pressure on the woodlands of the dry savanna and steppe areas of Africa.

yields, which can be markedly increased with small applications of fertilizer and use of pesticides.

Numerous other West African countries grow cotton and several have programs to increase its production. Mali has been the chief cotton producer in francophone West Africa, with output running about 60,000 to 70,000 tons a year; the producing zone extends from south of the Niger and Bani Rivers and east from Bamako to the Ivory Coast and Upper Volta borders. Senegal has promoted production as part of its diversification program, the commercialized output having increased from nothing in 1960 to 21,200 tons in 1972; over 300,000 persons are involved in its production on about 69,000 acres (28,000 ha) and further increases are expected. Upper Volta increased production from 1,000 tons in 1960 to around 25,000 to 35,000 tons in recent years, permitting cotton to become a significant export crop. Niger's cotton is produced mainly in dried-up river beds in the south; yields are variable and costs are high, but output has increased to about 10,000 tons. The Ivory Coast sees cotton as becoming a major cash crop in the relatively undeveloped north; production increased about ninefold in the decade to 1973 when it was 52,800 tons, almost all of which was consumed in local textile mills. Ghana appears to be copying the Ivory Coast's pattern by encouraging smallholder production in the north. Cotton accounts for about 22.5 percent of Dahomeyan exports. That country's drive to raise output started in 1951 and was accelerated in 1966; it succeeded in increasing marketed output by ten times in the decade to 1972, when it was 47,300 tons derived from about 132,000 acres (53,600 ha). Togo's small output has remained at about the same level, averaging about 7,000 tons in the last decade.

Production of cotton in most of the francophone countries has been aided by French technical organizations, Compagnie Française pour le Développement des Fibres Textiles (CFDT) and Institut de Recherches du Coton et des Textiles Exotiques (IRCT), which have introduced new high-yielding, disease-resistant species, promoted the use of fertilizers and insecticides, and advised on agronomic practices including especially the timing of planting. Their success may be illustrated by the case of Mali, where cotton yields per unit area increased by 312 percent in the six years to 1972.

Other cash crops. Other commercial crops of the northern zone are of minor importance among exports but some are of considerable actual or potential significance in national markets. They include gum arabic, kapok, sisal, shea products, tobacco, sesameseed, market garden produce, sugar, and rice. Mauritania, Senegal, and Nigeria export small quantities of gum arabic; Mali is the main producer of kapok; and Mali and Upper Volta have small sisal plantations. Shea products come from the shea butter tree, which is native to the savanna and steppe areas and is another of the many sources of vegetable oil. The existence of many trees across the area periodically leads to suggestions that the tree be more intensively commercialized. But its oil is not so rich as palm oil, its extraction requires much labor, and long distances to shipping points increase the costs of marketing. Senegal is attempting to promote the output of market produce, most of which comes from the Cape Verde Peninsula, and hoped to export about 40,000 tons a year to Western Europe in 1975; Upper Volta has been airlifting stringbeans to France in the off season.

As was true for the humid areas, efforts are being made in the savanna-steppe belt to promote self-sufficiency in such crops as tobacco, sugar, and rice. Upper Volta is developing a 5,560-acre (2,250-ha), $22 million irrigated sugar plantation at Banfora, expected to save about $2 million a year in foreign exchange. Mali has developed sugar estates on the perimeter of the Office du

Niger (see below) and in the Kayes region. Niger is studying the placement of an 800-acre (2,000-ha) sugar plantation in the Tillabéri region on the Niger River, expected to produce about 15,000 tons yearly, which would more than replace the present imports of about 9,000 tons. The Ivory Coast, which now imports all of its sugar, has a major scheme under development at Ferkessédougou on the rail line in the extreme north; expected to produce 60,000 tons beginning in 1976, the project involves the construction of two dams, laying out a 4,700-acre (1,900-ha) plantation, which would later be more than doubled, and provision of processing and refining plants at a total estimated cost of $66 million.

Senegal has a number of schemes run by different societies aiming to increase the domestic production of rice, but despite some success imports have continued to rise. Mali produces rice in the Office du Niger and from the live delta of the Niger as well as along the valleys of the Niger and smaller rivers, and now has a project in controlled submersion. The Ivory Coast produces about 300,000 to 350,000 tons, mostly in the north, but still imports about 100,000 tons a year; efforts are being made to increase irrigation, which now accounts for only about 5 percent of the area in rice, because yields are about four times those of raingrown rice, and it is possible to produce two crops a year. Production of rice in Niger increased fourfold in the 1960s; almost all of it comes from the areas around Tillabéri and Niamey. It is not always easy or sensible to increase the area of irrigated rice because of the high costs involved. In Niger, for example, it costs an estimated $860 per acre ($2,128 per ha) to develop an irrigated area whereas treating a regular crop area costs only 5 percent as much.

A rising menace to rice and other grains in the zone is the weaver or quelea-quelea bird; swarms numbering in the millions devour these crops. In the Senegal Valley alone hundreds of millions of weaver birds consume an estimated 100,000–200,000 tons of grain a year. All manner of weapons have been used to reduce their depredations: poison bait, smoke, noise, biological control, explosive charges in the nesting and roosting sites, flamethrowers, and air spraying with chemical poisons. None has proved adequate, but the last is probably the most effective method to date.

SPECIAL DEVELOPMENT SCHEMES

The northern belt is the scene of a number of special projects of considerable interest, most of them associated with development of river basins in the area.

OFFICE DU NIGER. The Niger Project was before independence the largest single agricultural development scheme in all of the French areas south of the Sahara; over $100 million was devoted to it up to 1959 (Map 57). The inland delta of the Niger, the site of the project, was created by the Upper Niger before its capture by the Lower Niger

created an outlet to the sea for part of its waters. It is divided into two parts, the dead delta on the right bank and the live delta, subject to annual flooding by the river. The purpose of the Office du Niger, created in 1932, was to revivify the dead delta and to use the live delta more effectively. The major engineering work is the 8,580-foot long (2,615 m), 16-foot high (5 m) Sansanding Dam, started before World War II and completed in 1947 and situated at the apex of the delta. Its reservoir feeds the canals, which use in part the channels of former distributaries and lead to the irrigated areas, while a 43-mile (69-km) dike protects the irrigated areas from seasonal flooding. About 124,000 acres (50,000 ha) are controlled but only two-thirds of that are actually irrigated in the scheme.

The original plan was to concentrate on cotton under a partnership arrangement somewhat comparable to that of the Gezira in Sudan—indeed, the French hoped that the Office du Niger would become a second Gezira—but the scheme has never approached its original goals. Cotton proved too difficult to produce and was finally abandoned in 1970; rice has been the major replacement crop, and now occupies about 95 percent of the irrigated area; the remaining area is in sugar, a recent introduction.

The major explanation for the disappointing results of the Office du Niger is probably the poor quality of much of the soil, which is far from matching the fabulous Nile mud. Other physical problems have included the incidence of cotton parasites and attacks by locusts and quelea-quelea birds. Human problems have included the difficulty of training tenants to a new system of agriculture and a certain disinterest in producing the major cash crops. Economically, the scheme has been excessively costly both in relation to the number of persons involved and to the output levels achieved. Mechanical operations were very expensive and have generally been replaced by manual labor.

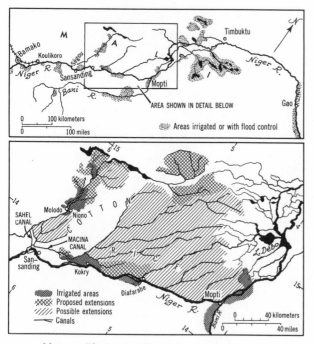

Map 57. The interior Niger Delta area in Mali

Despite the disappointing achievements of the Office du Niger, the government plans additional features, including the Selingue Dam on the Sankarani tributary, which would improve the security of supply and support an additional 100,000 acres (40,000 ha) under irrigation.

A large number of small irrigation enterprises started in the live delta and other parts of the river valley have proved cheaper and more productive, and now cover more than three times as much area as the Office. A good deal of floating rice is produced on the inner delta. To grow floating rice the field is hoed before the first rains and sown dry, and the rice germinates on the first rains; the rice must grow fast enough to keep pace with the rising waters when the flood arrives; four months of inundation give good yields; harvesting often is done from pirogues.

SENEGAL RIVER SCHEMES. There are a number of development projects in being and others under construction or proposed along the Senegal River in Senegal, Mauritania, and Mali. The largest to date and the most important project ever undertaken in Senegal is the Richard-Toll Scheme, which involves use of Lac de Guiers (which fills up during the high-water period), as a reservoir from which water is pumped to irrigate an area along the river (Map 58). The project was begun in 1947 to provide domestic rice; about 20,000 acres (8,100 ha) of saline, infertile soil were developed and mechanically cultivated, but costs have been excessive and some soils have had to be abandoned because of salinity and wind erosion while rice production suffered severely from losses to the quelea-quelea bird. In late 1972 it was decided to convert much of the area to sugar; production of about 100,000 tons on 18,500 acres (7,500 ha) is planned for 1976. The total investment for plant and equipment will be about $50 million.

Another scheme on the lower Senegal has involved bunding a large area along the

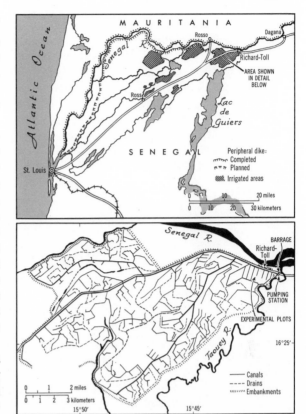

Map 58. Irrigation projects on the Senegal River in Senegal

river north of Saint-Louis and pumping water from the river to permit controlled irrigation instead of the previous highly uncertain planting of catch crops on a widely varying area. By 1977 it is expected that about 25,700 acres (10,400 ha) will be producing rice in this area. A small area has been successfully used to produce tomatoes under irrigation in the dry season. Mauritania also is developing several small hydro-agricultural schemes along the Senegal and its tributaries, partly to produce sugar and rice.

There are substantial additional possibilities for increasing irrigation along the Senegal, but this must await construction of a large regulatory dam at Manantali on the Bafing tributary in Mali, three submersion dams in the section between Bakel and

Dagana, and a barrage above Saint-Louis to prevent encroachment of sea water.

SOKOTO RICE SCHEME. Another specialized scheme for the production of rice was situated north of Sokoto in northwest Nigeria. Each year large sections of floodplain of the Sokoto River and some of its tributaries, especially the Rima, are inundated in the rainy season. After the water recedes, crops are grown on the saturated soils, but primitive techniques limited the amount of these fadama lands that could be cultivated. The scheme simply involved the mechanical plowing of as much riverain land as possible, after which the local farmers performed the rest of the work. The scheme was quite successful in its early years, but then declined in importance apparently because of the traditional conservatism of the farmers, interfer-

ence with cattle herding, and objections to the charge for plowing. In 1973 Nigeria created the Sokoto-Rima Valley Development Authority to oversee the integrated development of this densely populated zone.

THE CHAD BASIN AUTHORITY. Nigeria is promoting the production of crops on bunded fields along its portion of the Lake Chad basin and plans to have outputs of 60,000 tons of wheat, 75,000 tons of rice, and 7,500 tons of cotton by 1980.

RESETTLEMENT SCHEMES. Numerous such schemes have been instituted in the savanna-steppe zone, particularly in northern Nigeria. Designed for one or more of the following purposes—to reduce pressure on the land in densely populated areas, to keep the tsetse fly under control by concentrating settlement, to demonstrate the possibilities of

A Tuareg encampment near Timbuktu, Mali
This nomadic group has suffered severely from the prolonged drought of recent years.

mechanical cultivation, to provide models for settlement organization and agricultural practices—most of the schemes failed and have been abandoned. Sometimes the physical conditions proved to be unsatisfactory, particularly the soils; capital costs were often too high; mechanical equipment could not be properly maintained and cost too much to operate; occasionally there was difficulty marketing the products; more frequently the settlers did not appreciate or understand the purpose of the schemes and were quick to abandon them or to revert to previous methods. The most successful schemes have characteristically been more heavily dependent on the initiative and work of the settlers themselves, not only for the farm work but also for construction of dwellings and other community buildings.

COLLECTIVE FIELDS. Mali adopted after independence a system of village cooperatives and of collective fields to which each resident was supposed to devote one day a week. These *"champs collectifs"* were widely misunderstood; to many they seemed little different from work requirements that had been imposed by the colonial government.

Hence work was only grudgingly provided. It is possible that such fields might have been effective in the introduction of better practices if they had been organized on a lineage basis.

WEST AFRICAN DESERT AREAS

Most of Mauritania, Mali, and Niger are parts of the Sahara; all of their northern boundaries extend to or beyond the Tropic of Cancer. The southern portions are invaded each year by nomads; scattered oases support a sedentary agriculture whose chief resource is the date palm; and a few higher massifs, such as the Aïr or the Adrar des Iforas, give meager sustenance to permanent but possibly declining populations. The effect of increasing aridity toward the heart of the desert is strikingly illustrated by the case of Niger, where only 3.33 percent of the total population resides on that half of the country which receives less than 4 inches (100 mm) of precipitation yearly. Minerals would appear to hold the greatest potential for these areas.

CHAPTER ELEVEN

West Africa: Minerals and Mining

In the past two decades, mining has been increasing rapidly in significance in many West African countries. Minerals provide over 70 percent of total exports in five of sixteen countries (Mauritania, Guinea, Sierra Leone, Liberia, and Nigeria), for over half in Togo, and for from 10 to 25 percent in Senegal, Ghana, and Niger. Nigeria leads all other countries because of its large and growing exports of hydrocarbons. As in most other parts of Africa, the two most important contributions made by the mineral industry are to government revenues and to earnings of foreign exchange, both of which may be very important. It rarely provides employment for more than several thousand people.

Minerals sometimes play a significant role in transport; they sustain rail lines in seven countries and contribute to the profitability of at least fourteen West African ports. Most of the region's minerals are exported crude or with only simple washing and grading, but there has been a trend toward further processing, such as beneficiation and pelletization of iron ore. There are very few examples of backward-linked activities.

Geological prospecting has been and is being pursued with increasing vigor throughout the area; many new deposits have been brought into production, others are nearing this stage, and still others await further investigation and infrastructural planning.

Mining enterprises are usually foreign owned and managed, though governments are increasingly becoming joint owners. Greater attention is being given to the Africanization of all aspects, from taking higher profit shares, joint participation, and outright nationalization to progressive replacement of expatriates at many levels. Heavy dependence continues on foreign capital and on technical, engineering, and managerial expertise in all phases from exploration to development and operations.

The most important minerals now produced (in order of value of exports) are petroleum, iron ore, diamonds, bauxite and alumina, phosphates, gold, tin, uranium oxide, and manganese. Almost all mining output is exported except for coal and a small fraction of hydrocarbons in Nigeria, common building materials, and salt.

A country-by-country analysis will provide a better basis for estimating the present importance and some of the future potentialities of the mining industry in West Africa.

MAURITANIA

IRON ORE

The story of opening the iron ore deposits around F'Dérik illustrates well the time lag that often occurs between discovery and development of an ore deposit and the difficulties associated with such development, especially in the Sahara. The existence of a mountain of iron was noted in 1912 and confirmed in 1937. In 1949 Bethlehem made an appraisal of the body. A development company, Miferma, was formed in 1952, but it was not until 1960 that sufficient funds, including a $66 million loan from the World Bank, became available to permit installation of the necessary facilities and infrastructure. By this time Miferma was owned by a West European consortium in which the French held a major interest. In late 1974

Mauritania nationalized Miferma, promising adequate compensation to its former owners.

Iron ore occurs in several bodies atop a range of mountains. Reserves of 63 to 64 percent iron content are now about 150 million tons; they will be exhausted in a dozen years, but there are also at least 4.5 billion tons of lower grade ore in the vicinity and around Zouérate, where a new mine is being opened. Bench mining is used in the four operating quarries. Among the problems encountered in opening up the bodies (in addition to that of amassing the $190 million investment required) were securing water, very high summer temperatures, blowing sand, and availability of transport and of market. Sources of water were found near the mine and inland from Nouadhibou, from which a rail line was constructed, while substantial markets were assured by inclusion in the consortium of steel companies, which contracted to take specific shares of output. The first shipment was made in 1963 and output has been about 8–9 million tons a year from 1968 to 1974; plans call for increasing the level to 11–12 million tons, which will require an additional investment of $20 million.

The iron mines have not proved so profitable as had been hoped: desert conditions required about twice the estimated expenditures for maintenance and repair of mechanical equipment and the railway; iron ore prices declined on the world market; and because of their enclave character the operations had a lesser effect on the Mauritanian economy than had, perhaps mistakenly, been believed. Nonetheless, they greatly raised government revenues, increased exports from $16 million in 1963 to $90 million in 1973, and now provide direct employment for about 4,000 Mauritanians plus about 375 Europeans.

COPPER

Surface deposits of copper near Akjoujt, about 168 miles (270 km) northeast of

Blasting in the iron mine at Tazadit, Mauritania
The nationalized Miferma mines produce about 8 to 9 million tons a year of high-grade iron ore.

Nouakchott, have been known since 1931 and were studied over recent decades by several companies. But they were not opened until 1970 because of the problems of concentrating the complex ore. An international consortium, Somima (in which Anglo-American's Charter Consolidated holds the largest share), concentrates the difficult ores with the new Torco process; it plans a yearly output of about 50,000 tons of 80 percent concentrates containing a valuable content of by-product gold. Output from 1971 to 1973 was disappointingly low, however, because of strikes and environmental problems. About 28 million tons of 1.79 to 2.7 percent ore have been proved. The concentrates are taken by road to Nouakchott for shipment abroad.

OTHER MINERALS

A small output of gypsum has come from a deposit north of Nouakchott and has been used for plaster during construction of the capital; it is hoped that about 15,000 tons a year can be trucked to Senegal. Yttrium, used in color television tubes, was mined briefly in 1967 and 1968, and a few hundred tons of salt are produced. There are promising indications of other iron and copper deposits and of phosphate and rutile. Prospecting for oil, thus far confined to geophysical work, has been conducted off and on since 1961 in the Tindouf Basin along the Algerian border, in the valley of the Senegal, and along and off the coast.

SENEGAL

PHOSPHATES

Senegal has two phosphate deposits under exploitation (Map 59). Aluminum phosphate is quarried near Thiès by a Pechiney subsidiary at a rate of about 220,000 tons a year. At Taiba, some 70 miles (113 km) from Dakar, a vast phosphate deposit has been worked since 1960, the level of output having increased to about 1.3–1.5 million tons with plans to raise this to 1.8 million tons. After removal of the overburden and mining of the rock by draglines, it is concentrated on site. The product of both operations is sent by rail to Dakar, where it is mechanically loaded aboard bulk carriers.

OTHER MINERALS

Oil exploration began in 1952 and several small finds of oil and gas have been reported. Some concessions have been relinquished, but prospecting by French and American companies is continuing. An iron ore deposit of 60 percent content and containing an estimated 1.25 billion tons is under study at Falémé in the southeast. Tentative plans suggest construction of a 435-mile (700-km) rail line, which might be extended to similar deposits inside Mali; use of electricity from a hydroelectric plant at Manantali in Mali, 75 miles (120 km) away; and pelletization of the ore. A copper deposit of potential interest has been found at Gabou in the east, where a Russian team is prospecting for gold. Senegal produces about 125,000 tons a year of salt from salines, some of which is exported to neighboring countries. Rutile has been mined at the mouth of the Saloum, but the restrictions of the shallow bar have discouraged production.

MALI, UPPER VOLTA, AND NIGER

The remoteness of these landlocked countries had a depressing effect even on the exploration for minerals, which was not undertaken with any intensity until the late 1950s. Deposits of all but the higher-value minerals would have to be more than average in quality to justify the heavy cost of transport to seaboard.

Deposits of possible future interest in Mali include iron ore near the Senegal border, noted above; an 800 million-ton 40–45 percent bauxite reserve in the southwest; and a manganese body near Ansongo on the Niger. There has been some prospecting for oil. Very small amounts of salt and gold are produced.

In Upper Volta a manganese deposit at Tambao in the extreme north has received considerable attention since its discovery in 1960. Containing at least 10 million tons of good-grade ore (52.45 percent Mn) the proposal is to ship about 400,000–500,000 tons per year. This would, however, require a lengthy rail extension costing at least $30 million, rather high given the size of the body even though it might carry some other traffic (including limestone for a domestic cement mill). Upper Volta is strongly interested in exploitation of the deposit because it would raise the GDP substantially and exports by about 60 percent. Other minerals of possible interest are: vanadium, found at 11 sites near Oursi, west of Tambao; copper, under study at Gaoua by an international team; nickel; antimony; bauxite; and gold, which was produced in small amounts to 1966.

The most interesting development in the three countries has been that of uranium

The mining community of Arlit, Niger
Supply for the uranium mine and community at Arlit, situated 155 miles (250 km) north-northwest of Agadès, presents severe problems.

ores at Arlit in Niger, situated in the desert 155 miles (250 km) north-northwest of Agadès. Research by the French Atomic Energy Commissariat (CEA) revealed reserves greatly superior to those in Gabon, where production was taking place. A consortium with French leadership, Somair, was formed in 1968 and in 1972 production reached the capacity-level of 867 tons of concentrate, valued at $6.3 million. At that time the French cut the price unilaterally and delayed plans to double capacity, which created some strain in its relations with Niger despite increased aid extended by France. Prices for uranium concentrates are expected to go up after 1980 and Niger may expect to benefit greatly thereafter, as its total reserves are now estimated to be one-fourth those of the United States.

The deposits at Arlit are sedimentary, like those of the Colorado Plateau. After removal of the 115–131-foot (35–40-m) overburden, mining takes place by open pit; the ore is concentrated *in situ,* which required construction of a sulfuric acid plant and a 9,000-kw carboelectric station, and the discovery of enough water to provide 2 cubic meters per ton treated. Water now comes from pits and from an aquifer uncovered in the mining. Unlike the situation for most

minerals, inward shipments greatly exceed the tonnages exported. To produce the planned 1,500 tons requires imports of about 40,000 tons, mainly of sulfur and oil, but also of equipment. In addition, practically all the supplies of the supporting town, which houses about 2,000 persons, must be imported. Such shipments add to the already high costs of operating in the desert since they must travel about 250 miles (400 km) by rail from Cotonou and 994 miles (1,600 km) by truck. The total investment of Arlit has been about $53 million.

Uranium bodies have also been found at Akokan, 12 miles (20 km) to the southwest; these are under study by a CEA-Japanese-Nigerien group. Another discovery is at Imouraren, near . Agadès. Underground mining will be required at Akokan, whose opening would require an investment of about $100 million.

Niger also produced 136 tons of cassiterite in 1972, double the 1961 level; mined in the Aïr 155 miles (250 km) north of Agadès, it is taken to Nigeria for smelting. Niger has attracted much prospecting in recent years and numerous discoveries of a considerable variety of minerals have been reported. Of possible interest are three 100–200-million-ton bodies of medium-grade iron ore in the Niger Valley, a manganese occurrence northwest of Niamey, and a small coal field. Four companies hold permits for petroleum exploration.

GUINEA

Guinea appears to be one of the more favored mineral areas in West Africa and is becoming one of the leading bauxite producers in the world. The break in relations with France at the time of independence, plus nationalization of many segments of its economy (including one bauxite producer), undoubtedly reduced the interest of potential investors. But the consistent honoring of contracts with two mining companies and signing of a huge contract with Halco have altered these attitudes, and it is not unlikely that additional large investments will be forthcoming. Mining operations are extremely important to Guinea for the revenues and convertible currencies they provide and for their dynamic growth in what has otherwise been a stagnant if not regressive economy.

BAUXITE

Guinea has the largest known reserves of this most important ore of aluminum. Bauxite was first mined in 1952 on an island off Conakry by an Alcan subsidiary; the operation was nationalized in 1962 and shipments from this small but well-situated deposit fell off thereafter.

A 300 million ton, 42–44-percent-alumina-content bauxite deposit is under exploitation at Kimbo in the Fria region. The operation has been the main backstop of the Guinean economy since 1960, and provided 60–70 percent of foreign exchange earnings from 1964 to 1972. An international consortium, FRIA (headed by Olin Corporation but including aluminum companies from four West European countries), developed the mine, the associated alumina plant (the first in Africa), and the required transport infra-

A portion of the alumina plant, Fria, Guinea
Guinea is one of the richest countries in the world in bauxite. A portion of the output is converted to alumina and it is likely that an aluminum smelter will be constructed in the near future.

structure beginning in 1957; in 1973, responsive to the desire of Guinea for participation, a 49 percent share was assigned to the government in return for concessions on additional bauxite reserves and the new joint company became FRIGUIA. A limited overburden permits economical stripping of the ore, most of which is converted to alumina in a large nearby plant whose original capacity of 480,000 tons has been increased to 700,000 tons. Output is now running around 2.1 million tons of bauxite and 660,000 tons of alumina.

One of the world's largest deposits is found in the north at Sangaredi near Boké. Reserves are placed at over 700 million tons of 45 percent or better ore, with a low silica content. After a false start by the Alcan subsidiary, Bauxites du Midi, and lengthy negotiations the Compagnie des Bauxites de Guinée was formed with the government holding a 49 percent share and Halco, a consortium including most of the world's leading aluminum companies and managed by Harvey Aluminum of Pittsburgh, holding the remaining 51 percent. The $187 million complex at Boké was inaugurated in mid-1973; after mining by open pit the ore is railed to the new port of Kamsar; output is expected to be about 2 million tons a year.

The USSR has developed another bauxite body near Kindia, about 61 miles (98 km) from Conakry; shipments began in late 1973 at an undisclosed rate. About $138 million has been invested, including funds to improve the rail to Conakry and port facilities there.

Guinea has stated that within a decade it expects to be shipping 26 million tons of bauxite and 2 million tons of aluminum annually. Converting alumina to aluminum requires large amounts of electricity, which could be produced at the Konkouré Falls and elsewhere in Guinea. It is likely that Guinea will continue to push for processing of bauxite to alumina and aluminum and for as rapid localization of staff as possible, but it apparently decided after meeting with other major world producers in March 1974 that an international cartel could not be formed to raise prices as had been done for oil by OPEC. Two factors that were probably important in this conclusion were the widespread occurrence of bauxite and the expectation that it may soon be practical to extract aluminum from clays and shales containing only 20 percent alumina.

IRON ORE

Laterite ores of 48–55 percent metal content are mined on the Kaloum Peninsula only seven miles from Conakry. Hematite underlies the surface ores and probable reserves are put at 2 billion tons. Additional iron ore bodies exist in lower Guinea.

Considerable interest has been focused on large deposits of high grade iron ores in the Nimba Mountain and Simandou Range in extreme southern Guinea, across the border from existing operations in Liberia. Negotiations with several African countries and foreign firms, including a major Japanese company, have been carried on sporadically since 1961. Ore could most readily be shipped via Buchanan in Liberia, but Guinea apparently decided to push for construction of a 540-mile (869-km) rail line to Conakry, which would be equipped with a second deepwater port. It is planned that the rail line will have the capacity to handle 50 million tons, including 15 million tons of iron ore and 30 million tons of bauxite from deposits at Toughe and Dabola. There has also been talk of an alumina plant and aluminum refinery, and of a steel mill. Just how much of these extremely ambitious plans will be realized in the next decade remains to be seen; the capital requirements would be enormous and it would require about eight years to construct the railway, which would itself cost about $450 million.

DIAMONDS

Diamonds come from the interior southeast of Guinea where exploitation of surface deposits was first undertaken by two companies. In 1956, when thousands of Guineans who had been digging for diamonds in Sierra Leone were expelled from that country, many of them began potholing in the Kerouane area. Clandestine movements of diamonds are so complex among the countries in this part of Africa that published figures of production are very questionable, but Guinea has been credited with an output of about 70,000 carats in recent years.

GOLD

A large number of African panners work in the Siguiri area near the northeast corner of the country, but the level of output is not known.

SIERRA LEONE

The growth of mining in Sierra Leone in postwar years was spectacular. In 1929 the value of mineral exports was about $1,000 for 26 ounces of coarse platinum; the next year it increased to $30,000 with the addition of gold exports. Diamonds were first produced in 1932, iron ore in 1933, and chrome ore in 1937, but total exports were only about $9 million in 1940 and $9.2 million in 1950. By 1959, however, exports had increased to $38.4 million and continued sharply upward to $72.6 million in 1960; in the period 1969–1973 they averaged about $85 million and accounted for four-fifths of total domestic exports by value. Mining is also the main source of government revenues. Unfortunately, the impact of mining has not been entirely favorable, and diamond production cannot be expected to remain at high levels for many more years.

DIAMONDS

Diamonds rank first among Sierra Leone's minerals and account for about three-fifths of domestic exports. Very widespread alluvial diamond deposits occur on or near the surface of Sierra Leone. Some fields are unusually rich, containing up to 250 carats per cubic yard (300 carats per cu m), and some exceptionally large stones, from 50 to 100 carats, have been secured from them. In 1945 the 770-carat Victory was found and in 1972 the 968.9-carat Star of Sierra Leone, the third largest ever discovered in the world, was uncovered. Diamonds were first discovered at Kono and it was thought that they were confined to that area. Mining began in 1932 after the Sierra Leone Selection Trust (SLST) had been given exclusive rights for diamond exploitation over the entire colony. Production grew to an average of 748,000 carats in 1936–38 and went on normally until 1952 when potholing, or surface digging, led to a veritable diamond rush and to a great increase in illicit sales, which in a few years involved about 75,000 Africans.

Various methods of dealing with this mining, which abrogated the agreement with SLST, were attempted. But powerless to stop it, the government finally concluded a new agreement reducing the company's rights to a 450-square-mile (1,166-sq-km) area in compensation, for which it was paid $4.4 million and promised control of illegal mining in that area. Part of the remaining diamond area, covering at least 9,000 square miles (23,310 sq km) was opened to licensed African diggers, who were supposed to deal exclusively with the official purchasing organization.

Illegal mining and sales continued apace, however, the latter often exceeding the legal sales. From 1957 there were varying periods during which the situation became chaotic; pitched battles took place with the police, numerous potholers were expelled from the company area, and several dealers were deported. De Beers' Central Selling Organization was later invited to manage the Government Diamond Office, not to combat illicit mining but to reduce smuggling. Even after taking a 51 percent interest in SLST in 1970, with compensation being paid over eight years in the amount of $6.6 million, the government was unable to control illicit mining on the areas of the new National Diamond Mining Company (NDMC).

While NDMC operations are highly mechanized and efficient and many of the licensed diggers also win a high percentage of the stones in the areas worked, illicit miners have often picked the eyes out of deposits, leaving a substantial content that could have been won by modern techniques. In most cases the deposits so worked are rendered uneconomic for further operations. The impact on the landscape of these operations is deplorable. Valleys and terraces have been broadly scarred, leaving a pockmarked, battlefield-like surface. Ponds are left which become insect infested and disease ridden. River drainage is impeded, which has caused flooding and undermining of homes and settlements, while the increased sediment load of streams affects potable water supplies adversely.

In addition to the physical damage occasioned by potholing, the social and economic costs are severe. The large number of people attracted to the mining areas live under primitive and sometimes lawless conditions. Migration of young men from their home areas has contributed to a decline in farming output and standards. Losses to the government have been very large, including a loss in prestige resulting from the inability

Potholing for diamonds in Sierra Leone
Illicit miners often pick the eyes out of deposits, leaving a substantial content that could have been won by modern techniques.

to enforce laws or even to maintain peace. Observers have suggested that contempt for law engendered by the illicit diamond operations may have contributed to increasing graft, violence, and malaise throughout the country.

The future of diamond mining in Sierra Leone is difficult to assess. P. K. Hall of the Geological Survey of Sierra Leone has predicted a progressive decline in the 1970s to a low level of output for another decade. The production did drop from 1,031,090 carats in 1971 to 569,808 carats in 1973, though the impact was cushioned by a 220 percent increase in the average value per carat. A large part of the NDMC remains to be worked and it may prove possible to mine deposits under a heavy overburden, but this would require a substantial new investment.

IRON ORE

Iron ore has supplied a less spectacular but much steadier support to the Sierra Leone economy. It is mined by the Sierra Leone Development Company (DELCO) from two deposits located 52 miles (84 km) from Pepel in the Marampa area. A 47-percent ore readily concentrated by washing to 64.5 percent metal content is mined at a level of 2–2.5 million tons and railed to Pepel; iron ore has accounted for about an eighth of Sierra Leone's exports in recent years. A deposit at Tonkolili has attracted several studies in the past dozen years, but potential investors have been reluctant to invest in this larger but lower-grade body.

OTHER MINERALS

Bauxite mining began in the 1960s in the Mokanji Hills in the south, where other deposits exist. After bench mining the ore is trucked to newly constructed facilities at Port Sam for lighterage to ships lying off Sherbro Island. Shipments have increased steadily and accounted for 3.5 percent of domestic exports in 1972, when they reached 660,000

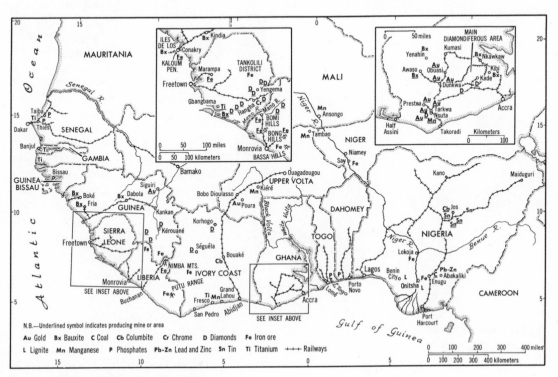

Map 59. Mining and minerals in West Africa

tons. It is hoped that a deposit at Port Loko will soon be developed.

The world's largest known deposits of rutile, an ore of titanium which is used in the production of paint, occur in the south, and Sherbro Minerals had an operation at Gbangbama from 1966 to 1971 when heavy losses because of inadequate extraction methods resulted in its sale to Sierra Rutile Ltd. The revised operation is expected to produce 75,000–125,000 tons a year beginning in 1974, and to yield returns about as large as those from iron ore. A German group is also prospecting in the Bonthe District for rutile. Gold and chrome mining has been discontinued. Other minerals of possible interest include corundum, columbium, and molybdenum. Two companies are exploring for oil.

LIBERIA

Liberia presents an excellent illustration of the great dynamism mining can introduce to an economy. In a country with no significant mineral output before 1950, iron ore accounted for 47.4 percent of the value of exports in 1961 and about 70 percent in recent years.

IRON ORE

Liberia is now Africa's leading producer of iron ore and fourth among world producers; it exported a record 25 million tons in 1973. There are now four operations with a fifth under active consideration. The Liberia Mining Company (LMC), the first to enter iron mining in the country, began operations in 66 percent ore at Bomi Hills, 42 miles (68 km) north of Monrovia in 1951, where it erected a plant to beneficiate its lower grade 42 percent ore in 1958. The second development was by the National Iron Ore Company (NIOC) on the Mano River, 40 miles (64 km) from Bomi Hills. Its 54 percent ore is improved to 57–58 percent by scrubbing and washing.

The iron mine of LAMCO at Mt. Nimba, Liberia
This is the largest of the four iron mining operations in Liberia, which is Africa's leading producer of iron ore.

The third and now the largest venture is that of the Liberian-American-Swedish Minerals Company (LAMCO), managed by the Granges Company of Sweden; Bethlehem Steel holds a 25 percent share of LAMCO. Its mines are in the Nimba Mountains, 168 miles (270 km) across the country from its port at Buchanan, both at the site opened in 1963 and since 1973 at Tokadeh, 25 miles (40 km) away. The ores (65 percent metal content at the main body and 40–55 percent at Tokadeh) are bench mined, crushed, and conveyed to a railhead at the base of the mountain. Part of the ore is pelletized in a 2.2 million-ton plant at Buchanan. Lastly, the German-Liberian Mining Company (DE-

LIMCO) operates a mine opened in 1971 in the Bong Range, 50 miles (80 km) east of Monrovia. The ore is relatively low-grade (37 percent) but is upgraded and pelletized before shipment.

These four mines employ about 10,000 Liberians. Investments have been very large, about $126 million for DELIMCO and $285 million up to 1972 for LAMCO. Liberia owns 50 percent shares in NIOC and LAMCO and receives 50 percent of net profits of the other two, making iron ore by all odds the most important source of government revenues.

Efforts have been made in recent years to secure sufficient funds (c. $550 million) for the Liberian Iron and Steel Company (LISCO) Wologisi Range concession in the northeast about 135 miles (216 km) from the coast. This would involve processing magnetite to 66–68 percent pellets, constructing a double track railway, and installing loading facilities at Williamsport to take 150,000-ton carriers.

OTHER MINERALS

Like Sierra Leone, Liberia has a large area where alluvial diamonds may be dug close to the surface, principally along the Lofa River in the north. When, in 1957, the attraction of diggers to the area threatened to replicate conditions in Sierra Leone and to disrupt both the rubber and iron mining concessions the government closed the area. The fields have since been reopened on a controlled basis with an estimated 10,000 diggers engaging in mining or potholing on a seasonal basis. But just how much of the $2–$3 million in recorded diamond exports come from domestic sources is impossible to tell, while actual exports are thought to be as much as five times the recorded figure.

Small quantities of alluvial gold are also won in various areas. A Dutch company was reported to be extracting rutile from sands in eastern Liberia in 1973. Three companies have been searching for oil in offshore waters and several shows have thus far been claimed.

IVORY COAST

The Ivory Coast was a relative latecomer among African mineral producers and has not yet achieved any really significant output. Diamonds are mined by two companies,

The pelletizing plant of LAMCO at Buchanan, Liberia
An increasing amount of iron ore is being beneficiated before export from Africa.

one near Korhogo, the other at Seguéla, and potholing takes place in the latter district. Output has been about 330,000 carats in recent years, but the known deposits are quite small and may last for only another decade. A manganese deposit mined at Grand-Lahou along the southern coast was closed in 1970.

Greatest hope attaches to the possibility of opening a one-billion-ton, medium-grade iron ore body at Klahoyo. Tentative plans, which would require an investment of at least $170 million, are to produce 5–10 million tons of enriched ore and to pipe it or rail it to San Pedro for pelletizing and export. Intensive prospecting is under way in several parts of the country but no important new finds have been reported.

GHANA

Ghana was for long the leading mineral producer in West Africa, though the great value of its cocoa production tended to obscure the significance of mining output. The industry has been stagnant in more recent years, but minerals have still accounted for about 10 to 14 percent of total domestic exports. About 25,000 persons are employed in mining companies in Ghana and possibly 20,000 individuals dig or pan for alluvial gold and diamonds. Gold, diamonds, and manganese have been the big three among Ghanaian minerals; bauxite has been less significant but should soon rank with the leaders. In 1972 Ghana stated that it would take a 55 percent share in private mining operations, with the owners being partially compensated and retained as managers.

GOLD

Gold has been worked in Ghana since the tenth century and the Portuguese traded for it as early as 1471. In the early 1700s, the British, Dutch, and Danes are estimated to have secured about £250,000 a year from the area, giving an obvious justification for the name Gold Coast. Modern mining began in 1880, and today Ghana is the fifth-ranking gold producer in the noncommunist world, with an output of about 700,000–730,000 fine ounces (21,800–22,700 kg) a year. Placing second to cocoa, gold exports were valued at about $25 million per annum before the notable increase in the price of gold in 1973. Until then production had been declining and several mines required government subsidies.

In 1961 Ghana nationalized most of the producing mines, which had been unable to find the necessary capital to continue development after a 30 percent wage increase had made some ores unpayable. Compensation to the previous owners was considered eminently fair in the London market. Five mines are now producing, all within a 60-mile (100-km) radius of Dunkwa, and 90 percent of output comes from underground. Alluvial workings are centered along the Ankobra River in the west and the Ofin River in the central south.

DIAMONDS

Ghana ranks fourth among world diamond producers with about 5 percent of total output. Most of the stones, which are predominantly industrial, are produced by four companies; Consolidated African Selection Trust has accounted for about 45 percent of output from its modern workings at Akwatia, where about 2,500 workers are employed. The companies complained of the impact of higher taxes on operating costs in the early 1970s but rising prices have since redressed the balance at least partially. Enormous tonnages are moved in recovering diamonds, which are roughly one part in six million of the original gravel. Possibly 12,000 African potholers work mainly along the Birim River and the lower part of the Bonsa River.

MANGANESE

One of the world's largest manganese mines is situated at Nsuta, within sight of the gold mines at Tarkwa. The 50–57 percent ore occurs in 100-foot (30-meter) beds along the tops of five ridges, permitting low-cost bench mining. Output began in 1916 and reached a peak in 1953; it has been about 190,000 tons a year in recent years, which ranks Ghana seventh among world producers with roughly 2.5 percent of total output. As the ore at Nsuta is nearing exhaustion operations are expected to move to a new site. The operating company, a subsidiary of Union Carbide Corporation, was nationalized in 1973. A manganese body of very high grade has been located at Kibi in the Eastern Region.

BAUXITE

At present there is only one mine producing bauxite, at Kanayerebo, 50 miles from Dunkwa in the Western Region. The ore is bench mined from the top of a hill, moved by aerial ropeway to a washing plant at Awaso, and then railed to Takoradi. Output has been about 330,000 tons per annum in recent years. But bauxite deposits containing very large reserves are known to occur in several areas, and Ghana wishes to develop further production and convert a portion to alumina for the Valco smelter at Tema, which now uses alumina from Jamaica.

In 1972 a license was issued to the Bauxite Aluminium Study Company Ltd. (BASCOL) whose major shareholders are Kaiser, Reynolds, and a Japanese consortium, and which has begun preliminary development of deposits near Kibi containing 150 million tons of 45 percent bauxite. Plans call for an investment of about $200 million, including $140 million for a 600,000-ton alumina plant, which would meet the full needs of Tema and have a surplus of about 200,000 tons for export. Another international consortium is prospecting a deposit at Nyinahin, west of Kumasi, which has a potential of about 450 million tons of 50 percent bauxite.

OTHER MINERALS

Some nine companies have oil prospecting licenses and one offshore discovery of potential significance was made in 1970. Two fairly large iron ore deposits are known, and plans call for quarrying limestone at Nauli in

A diamond dredge at work on the River Ankobra in Ghana
Recovery rate of diamonds is about one part per six million of the gravel worked.

the extreme southwest. This would be converted in a $24 million plant to clinker to be fed to cement factories at Tema and Takoradi now dependent on imports.

TOGO AND DAHOMEY

PHOSPHATES

The only important mineral operation in these two countries is the exploitation of phosphates behind Lake Togo where a 100 million-ton reserve has been proved; potential reserves are very much larger. Operations began in 1961 and exports of enriched rock increased steadily to 2.44 million tons in 1973 when they accounted for 52 percent of total Togolese exports by value. About 1,100 people are employed in the operation, which represented a total investment of $44.5 million when it was nationalized in 1974.

OTHER MINERALS

A project in Togo calls for quarrying limestone from below the phosphate to supply a 100,000-ton clinker factory, which would export a portion of its output to other West African countries. A variety of other mineral occurrences are listed for Togo but little is known regarding them. Three companies have been prospecting for oil in southern

Wheel shovel opening the phosphate rock deposit at the Hahotoé quarry, Togo
This large machine can move 1,000 cubic meters per hour.

Dahomey and in offshore waters, where a find was reported in 1968.

NIGERIA

Minerals contributed only about 11 percent of the value of exports and 1 percent of net income in Nigeria in 1960. But the rapid development of hydrocarbons in succeeding years has made that country the leading mineral producer in tropical Africa and second or third producer on the continent.

OIL AND NATURAL GAS

The search for oil began in Nigeria in 1937 and was renewed after the war, but it was not until 1957 that the first finds were made, near Port Harcourt (Map 60). A Shell-BP subsidiary played the dominant role in early exploration and now accounts for about 60 percent of production; it has since been joined by over a dozen other concessionaires, of which Gulf and Mobil have been most successful. Development of oil resources in Nigeria has been a long, difficult, and costly affair because of the complex subsurface geology, the relatively low productivity per well, and the inhospitable environment in the swampy delta area, not to mention the interruption of the civil war. Actual costs of production are estimated at about 55¢ per barrel, well above those in Libya.

But efforts have been well rewarded and Nigeria now ranks eighth among world petroleum producers. Its output has increased from 2.3 million tons in 1961 and 16.8 million in 1967 to 101.8 million tons in 1973. The location of the fields and pipelines as of 1973 is shown in Map 60. In 1972 about 28.4 percent of the oil came from offshore wells. Nigeria has three advantages over Middle East producers: its closeness to North Atlantic markets, the very low sulfur content of the crude, and its noninvolvement in Arab-Israeli affairs.

Very little has thus far been done with the natural gas, considerable quantities of which

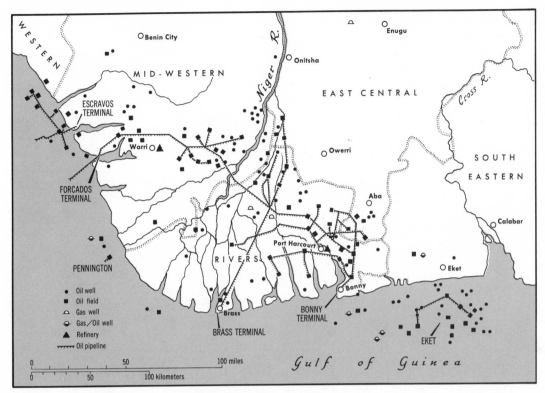

Map 60. Oil in Nigeria

Drilling for oil in southern Nigeria
The environment, particularly in the swampy Niger Delta, presents many difficulties in exploration and transport.

are flared off. Small amounts are used by the operators and in local industry, but large-scale sale awaits arrangements for LNG plants. Negotiations have been undertaken with several prospective investors and it is likely that several contracts will be completed involving several billion dollars if the costs of tankers and processing facilities at both ends are included.

Nigeria's first refinery, built near Port Harcourt in 1964, was damaged during the civil war but renewed production in 1970. After much delay occasioned by regional rivalries it was decided that a second refinery would be situated at Warri and a third somewhat later at Kaduna in the north.

Nigeria is very conscious of the wasting nature of oil and gas resources and is determined to derive as much benefit as it can during the production period. In 1974 the government took a 55 percent participation in all producing companies and indicated that it would probably take full control when it is capable. Much attention is being given to training Nigerians to assume the various functions required in the industry.

Returns to Nigeria have risen dynamically, mainly because of increases in the posted price from $4.29 a barrel in 1973 to $14.69 a barrel in 1974, but also because of higher profit shares and increased production levels. Earnings were about $1.5 billion at the lower posted price, but are expected to be $7–7.5 billion in 1974–75. Returns from petroleum will thus account for at least 80 percent of government revenues and over 90 percent of exports.

TIN AND COLUMBITE

Tin was long the backbone of mining in Nigeria; it is produced and smelted on the Jos Plateau mainly by large-scale hydraulic mining but also from small pit operations. While production levels and employment have tended to decline in recent years, the industry has been aided by higher world prices and new royalty arrangements. Tin ranks fourth among Nigerian exports but accounted for only 1.3 percent of the total in 1972 and 7.1 percent of the non-oil total. Exploitable reserves are not extensive and the industry may phase out in another decade. Nigeria ranks third in the free world in the production of columbite, used in the making of stainless steel; it is mined as a by-product of tin.

COAL AND LIGNITE

Nigeria has indicated reserves of subbituminous coals totaling 242 million tons in three areas. Coal has been mined for over 55 years near Enugu, but operations have been subsidized since 1950 and the availability of oil and natural gas now threaten to further reduce production, which fell from 924,000 tons in 1958 and 597,000 tons in 1961 to 324,000 tons in 1973. Coal is still needed in the production of cement, while use of a special process for coking the coal could again stimulate interest in an iron and steel works. Nigeria also possesses substantial deposits of lignite but these are not likely to be exploited for many years.

OTHER MINERALS

A lead and zinc operation at Abakaliki started in 1966 was disrupted by the civil war and production has not been resumed. Iron ore bodies exist in several areas; most of the ores are low to medium grade and some also contain an unfavorably high level of phosphorus and sulfur. Limestone is quarried to support the large cement industry.

CHAPTER TWELVE

Industrialization in West Africa

While the industry that now exists in West Africa is rather insignificant from the world standpoint, the strong desire of most nations to expand the manufacturing sector of their economies, the considerable number of industries that have been set up in the area, and the problems that impinge upon further expansion make it important to analyze the present and potential industrialization of the area. In doing so, West Africa will be used as a case study for manufacturing development in all of tropical Africa.

INCENTIVES TO INDUSTRIALIZE

There are many explanations for the keen desire of African nations to expand industry:

(1) Such development presents a major avenue for strengthening the economy by broadening its base and reducing the overwhelming dependence upon extractive activities, most especially agriculture.

(2) The desire to provide employment for an expanding population and one in which unemployment and underemployment on a considerable scale are characteristic. While this is an entirely valid objective, the ability of industrialization to meet it can easily be exaggerated because of the relatively small numbers required to operate a modern factory and losses in employment that may result in the handicraft sector. A hypothetical but not unrealistic case illuminates this point. If a country had 80 percent of

its labor force in agriculture in 1970, if its population increases at 3 percent per annum, and if the rate of growth in nonagricultural employment is 5 percent per year, the agricultural population would still be 64 percent of the total in the year 2000. Furthermore, the maximum absolute size of the agricultural force would not be reached until 2028, when it would be 2.71 times the 1970 level.

(3) The wish to earn more foreign exchange by the further processing of raw materials and export of higher valued products. There are distinct opportunities for such processing, but tropical Africa is not yet at a stage of industrialization where export of manufactured consumer goods to world markets can be expected.

(4) The wish to save foreign exchange through import substitution. While the import list is valuable for indicating possible industries and many consumer-oriented plants have successfully replaced imports, evidence suggests that import substitution has already been pushed too far in some countries. And it must be remembered that industries with such a basis usually require other types of imports so that the net gain is sometimes rather minimal.

(5) The realization that it is usually easier to increase productivity in manufacturing than in agriculture.

(6) The desire to have industry as evidence of modernity. This last reason, which stems from a simple comparison

of developed and underdeveloped countries, may be more emotional than logical. If the desire to emulate advanced manufacturing nations leads either to neglect of agriculture or to the establishment of inappropriate industries it is likely to be detrimental rather than helpful to the economy.

Agriculture is the primary activity in most African countries and the first manufacturing industries have often been concerned with the processing of agricultural exports. Agriculture can continue to stimulate growth by providing the main market for consumer-oriented industries, labor, exchange for the capital equipment and imported raw materials needed in new industries, and funds for investment in industry itself or in the infrastructure that manufacturing requires.

What is required for a rational development is the selection and stimulation of industries that are most suitable at a given time in a given country, and, conversely, the eschewing of investment in, subsidization, and protection of industries that are likely to become a perpetual charge on the national economy. A great deal can be learned from studying industries that are already in place in underdeveloped countries, from analyzing the factors that justify their presence or that have caused their lack of success. By examining the list of industries, it is often possible to discover prospects that the nation seeking to broaden its manufactural base might profitably study.

It is also necessary, of course, to know the economy of the country as thoroughly as possible, including its resources, its transportation and power facilities, its position with respect to labor and management, the availability of capital, the size, wealth, and stage of development of the market, the governmental attitude toward industry, and other elements. Each of these factors changes in time, hence continuing or periodic surveys are required; an industry that is inappro-

priate today may later become justified by the discovery of new resources, by the provision of additional or lower cost energy, by a developing or diversifying agriculture, by improved standards of living, and by many other dynamic conditions.

STATUS OF INDUSTRY IN WEST AFRICA

Manufacturing in West Africa may be broadly divided into craft industry and modern industry. There is a great variety of traditional craft industries, including spinning, weaving, dyeing, pottery work, cord making, wood cutting and carving, production of furniture and household utensils, gold- and silversmithing, ironworking, tanning, leatherworking, preparation of calabash containers, raffia weaving, etc. Some handicraft work goes on in practically every village; other specialities vary from group to group, within communities on a guild, clan, or caste basis, and from area to area. Certain of the handicraft work displays a high level of skill and artistic taste, but much is relatively crude. A series of newer handicraft activities are also sometimes included under manufacturing, although many are really services—vehicle, radio, and watch repair, photographic shops, tailoring, plumbing, electrical work, etc.

Generally only hand tools are employed in the making of handicraft items. There are few "transition workshops," where some mechanical power equipment replaces traditional hand operations. A major exception is the ubiquitous Singer sewing machine, found in the markets and bazaars all over Africa and used to sew up simple garments on demand.

Handicraft production has survived remarkably well in the face of low-cost manufactured imports and output from new domestic plants. There is, however, a tendency for the traditional crafts to decline except for the artistic output, but the new crafts are offsetting the losses to a degree. In Senegal an estimated 50,000 persons are involved in handicrafts in 8,000 enterprises (3,850 production, 2,550 service, and 1,600 artistic), and the value of production was estimated at between $10.8 and $15.5 million in 1970–71. In Niger there are over 22,000 artisans. In Ghana there were 95,167 small artisan establishments with no paid employees in 1962. In Nigeria a 1965 survey revealed that 900,000 rural households were engaged in manufacturing activity, while 100,000 small-scale enterprises of an artisanal nature were found in the cities; craft production accounted for 35.2 percent of the value of industrial output in 1962–63 and 19.3 percent in 1967–68.

Modern industry, with which this chapter is mainly concerned, is in an early stage of development in West Africa, though each year sees the introduction of new manufacturing establishments. There is not yet sufficient data to permit a statistical comparison of all of the countries; the following fragmentary data will, however, help to suggest the status of manufacturing in West Africa.

In Senegal the industrial sector excluding construction accounted for 10 percent of the GDP in 1959 and 18 percent in 1967, when some 400 enterprises employed about 30,000 workers including seasonal laborers. About 133 of these establishments, with 17,700 wage employees, accounted for 97 percent of the total turnover in that year of $243 million. The volume index of manufacturing output was 209 in 1972 with 1959 the base year. The largest components are the peanut oil mills, production of construction materials, and textiles. Senegal had the leading manufacturing establishment in francophone West Africa when Dakar was the capital of the Federation; loss of markets in the independent states caused a degree of stagnation in some industries and less rapid overall growth, and in 1968 the Ivory Coast surpassed Senegal in industrial output.

Industrial establishments in Mauritania,

Dying vats at Katsina, northern Nigeria
Handicraft industries continue to have considerable significance in many parts of West Africa and employ a larger number of people than modern industry.

Gambia, Mali, Upper Volta and Niger are all very small. Industry in Gambia accounted for 3.1 percent of the GDP in 1970–71. Mali's 35 modern plants accounted for about 5 percent of its GDP in 1972. Before independence most plants were concerned with the processing of agricultural produce; during the 1960s several consumer-goods industries were started with aid from the East, including a $10 million textile mill and a $26 million cement mill, the two largest industrial investments in the country. But most plants are small, depend on aging equipment, and operate below capacity. In Upper Volta transformation industries accounted for about 4 percent of GDP and employed only 1,288 people in 1971. Many of the 40 or so plants operate well below capacity. Niger had only about six factories in 1960 of which three were oil mills; in 1971 there were 40 enterprises with a value of production of about $21.6 million, and manufacturing's contribution to the GDP was estimated at 5 percent. The number of employees increased from 550 in 1963 to 1,880 in 1967. As is the case in Mauritania, the shortage of water is a deterrent to industrial expansion.

Guinea's industrial sector is relatively poorly developed, although the general pattern by types is comparable to that of other West African states. It reflects the somewhat disorganized and generally isolated economy of the country. Sierra Leone's manufacturing complex is also small, with perhaps three dozen modern plants, as is that of Liberia, where industry counted for 4.8 percent of GDP in 1968.

The Ivory Coast has seen more dynamic growth, with manufacturing accounting for 6 percent of GDP in 1961 and 14 percent in 1971; the number of plants increased from 180 in 1960 to 355 in 1971, of which the 26 largest accounted for 61 percent of total output; gross production increased sevenfold in the same period; and the number of wage employees rose from 22,000 in 1965 to 39,000 in 1971. Over four-fifths of industry is foreign owned, the largest holdings being French; about 4 to 5 percent of the employees are non-Africans, but they receive about 35 percent of total wages.

Ghana's drive to industrialize has been much more erratic. In the period 1961–1966 public policy concentrated on government investment in manufacturing, which resulted in the establishment of a good many poorly selected, inefficiently operated, and unproductive plants, and in 1966 less than half of the existing capacity in total industry was being used. Most of the state-run plants have since been sold to private operators, but the generally tight financial situation in the country has curbed investment in recent years.

Togo had some 28 modern industrial establishments in 1974, when industry accounted for about 11 percent of GDP. The recency of the manufacturing buildup is striking: only 1 enterprise dates from the 1940s and 4 from the 1950s, while 15 were started in the 1960s and 18 in the four years to 1974. Dahomey has about 40 modern installations employing some 2,100 wage earners; the value of output was $24 million in 1970 and the contribution to the GDP in 1971 was 12 percent.

Among tropical African countries, Nigeria has been particularly attractive for industrial investment because of its large potential market. The contribution of modern industry to the GDP was 0.6 percent in 1950, 4.7 percent in 1965, and about 12.7 percent in 1973–74. Manufacturing and crafts accounted for an estimated 9.6 percent of the gainfully employed population in 1967. At that time there were about 464 modern establishments employing over 10 people. Industrial growth was slowed by the civil war, which saw the partial destruction of a good many plants in the east. Since 1970, however, boom conditions have prevailed in many sectors and a sizable portion of the greatly enlarged revenues from oil are likely to support continuing rapid growth.

Most of the modern manufacturing establishments in West Africa are owned by European companies. The largest are usually associated with the processing of agricultural, forestry, or mineral output, and, for the two latter fields, are owned by the concessionaires. Increasingly important, however, are plants owned or partially owned by governments, while private investment by national citizens is encouraged in several ways. Regulations restricting certain enterprises to ownership and management by nationals are now common. In Nigeria, for example, a 1972 decree established two schedules of enterprises with respect to localization: the first lists 22 enterprises, of which 10 are manufactual and are reserved exclusively for Nigerians; the second lists 33, of which 20 are in manufacturing and are barred to aliens if the paid-up share capital does not exceed $608,000 or turnover does not exceed $1.5 million. Much room remains in Nigeria and elsewhere, however, for foreign investment in more sophisticated industries and such investment continues to be solicited by almost all governments.

Most plants are still managed by non-

Africans, although the number of African staff workers is increasing. But a rapid replacement of European personnel cannot be expected, primarily because it takes time to develop technical and managerial talent, but also because relatively few Africans are as yet able or willing to invest in their own establishments.

INDUSTRIAL REPRESENTATION

There are numerous ways by which the industries of West Africa may be classified: by product (shoes, beer, lumber, etc.); by end use of product (food, clothing, industrial raw material, etc.); by industrial group (forestry, textile, chemical, metal fabrication, etc.); by source of raw material (domestic or foreign); or by major locational orientation (raw material, labor, market, power). There are lessons to be learned from any of these systems, but the last is revealing because it helps to *explain* the presence of specific industries. While many industries have a complex of factors that influence their placement, it is usually possible to distinguish among primary, secondary, and tertiary attracting forces.

RAW-MATERIAL–ORIENTED INDUSTRIES

The earliest and still a very important group of industries in West Africa is that concerned with the primary processing of raw materials. Indeed, this group remains predominant in the smaller and less industrialized countries such as the landlocked states, plus Guinea, Sierra Leone, Liberia, Togo, and Dahomey. Since the money economies of all the countries have a strong involvement in the export of staples, and since the production of raw materials usually requires at least some preliminary processing, this is not surprising. Because it is often claimed that manufacturing should be carried forward to the finished product stage, however, it is important to seek the specific explanations for raw material orientation.

Raw materials tend to be processed *in situ*

for several reasons: (1) to make the product less perishable (examples would include the canning of fish, fruit, and vegetables, or the extraction of oil from palm fruit, where speed is essential in avoiding a too-high free-fatty-acid content); (2) to reduce the bulk of the product and thus save on transport costs (examples are the processing of meat and the squaring of logs); (3) to reduce the weight of the material shipped for the same reason (as, for example, in the extraction of vegetable oils, ginning of cotton, concentration of minerals, and production of sawn timber, plywood, and veneers). In some cases, manufacturing reduces bulk, weight, and perishability, as in the case of canning pineapples or tuna.

PROCESSING OF AGRICULTURAL RAW MATERIALS. Most of the plants in West Africa devoted to the processing of agricultural products confine their operations to the early stages of processing. Exceptions include the production of certain food items, such as canned fish or processed meat, where there is essentially only one step between raw material and finished product, or the manufacture of soap and margarine, which may be produced very simply in adjuncts to an extraction plant.

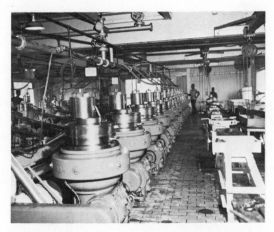

Centrifuges used to process concentrated latex at Harbel, Liberia
An example of agricultural processing industries, widely scattered in West Africa.

The most important industries in West Africa concerned with the processing of agricultural raw materials are the oil extraction mills, particularly for peanuts and palm oil (Map 61). In Senegal the peanut oil mills are the single most important industry; they also export oil cake used as cattle feed in Western Europe. Processing of oil palm products is usually confined to small plants such as the pioneer oil mills, of which there are over 100 in eastern Nigeria; palm kernel crushers are a recent addition and are much larger. A few small mills in scattered localities also process shea nuts, cashew nuts, coconuts, and cottonseed. Soap and margarine are occasionally produced in conjunction with oil milling, but the newer factory-scale soap plants tend to locate in market areas.

Other relatively important processing industries in this category include rubber processing; cotton ginning (Map 61); rice mills (Map 62); sugar mills; meat packing and processing; and tanning. Modern abattoirs and tanning establishments are not so important as the traditional, more primitive methods in output of meat, hides, and skins.

Less important industries processing agricultural produce include vegetable and fruit canning (mainly pineapples and fruit juices) (Map 62), kapok cleaning, sisal decortication, processing of manioc, production of cocoa butter and cake, and output of dairy products. There are also a large number of very small and a few large grain mills in a number of countries producing flour from locally produced millet, sorghum, corn, and peanuts.

PROCESSING OF FISH AND FOREST PRODUCTS. As has been noted, a number of tuna fish canneries and fish and shrimp processing factories have been set up at various places along the coast (Map 62). Opportunities in this area should increase as the industry is modernized and marketing facilities improve.

The growth and broadening of forest industries has already been described. The savings in bulk and weight are particularly notable in this group, which is concerned with the bulkiest of West African exports. Usually at least half of the weight is lost, sometimes more. While most of the sawmills of West Africa are small scale, a fair number of them catering to the domestic market, this

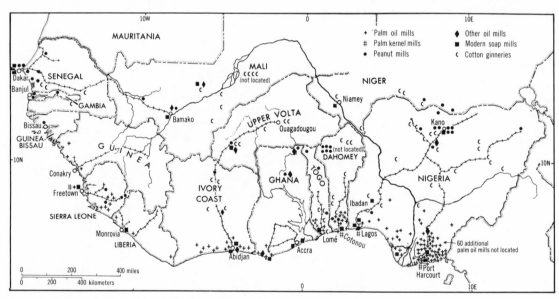

Map 61. Industry in West Africa, I: Vegetable oil processing plants, cotton ginneries

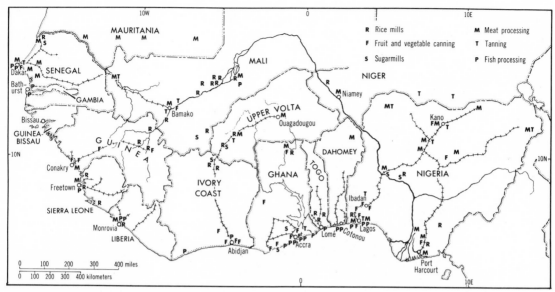

Map 62. Industry in West Africa, II: Rice mills, sugar mills, canning of fruits and vegetables, meat and fish processing

The United Africa Company lumber and plywood mill at Sapele, Nigeria
This ranks as one of the largest industrial establishments in West Africa.

category includes several of the largest plants in the area, including several of the plywood plants (Map 63).

PROCESSING OF MINERALS. Most ores contain a relatively low mineral content; the remainder is waste. There may be, for example, only from .25 to 1 ounce of gold per ton of ore, while diamonds occur in even lower ratios to the surrounding rock or alluvium. Therefore, the original concentration

usually takes place as close to the mineral deposit as possible. In the case of alluvial gold and diamonds, the process is sufficiently simple to permit separation by African diggers, but the separation of gold, uranium oxide, or copper from ores or winning a higher percentage of diamonds from alluvial material requires more elaborate equipment.

Significant plants for concentration, washing, and sorting of minerals include (Map 59) those for phosphates in Senegal and Togo; the important alumina plant in Guinea; iron ore beneficiation and/or pelletization plants in Mauritania, Sierra Leone, and Liberia; plants for extraction of gold in Ghana; the collieries at Enugu; tin smelting on the Jos Plateau; and the uranium oxide plant at Arlit, Niger.

As it happens, however, several of the more important minerals in West Africa are shipped raw, or only after simple crushing, washing, and sorting, including much of the iron ore, bauxite, and manganese. These minerals have a sufficiently high metal content to permit direct shipping, though there is a distinct trend toward beneficiation of iron ores and production of alumina within the

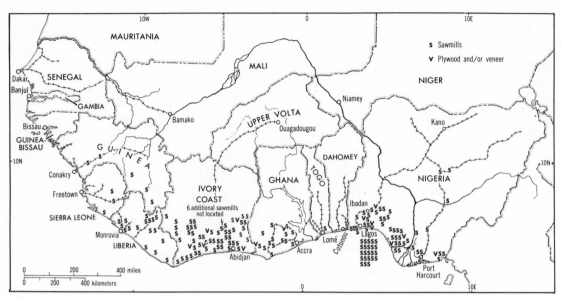

Map 63. Industry in West Africa, III: Sawmills; plywood, veneer, and chipboard plants

producing countries, which increases their returns and sometimes improves the marketability of the product.

LABOR ORIENTED INDUSTRIES

Industry is attracted to labor for two rather different reasons. First, if labor costs represent a high percentage of the total cost of production and if relatively unskilled labor can be utilized, the industry will tend to be attracted to areas of low-cost labor. Examples are attraction of the textile industry to India, Japan, and Hong Kong, or from New England to the American South. Second, if an unusual skill is required, the industry will tend to locate where that skill, which is usually rather immobile, is available. Examples are the production of optical goods in Rochester or New York City, the production of watches in Switzerland, or the manufacture of high-grade steel products in Sheffield and Solingen.

One would not expect to find the latter type of industry represented, but it may be surprising that examples of low-cost labor orientation are absent. Low wages are a significant consideration, but there are no in-

stances in which plants have been attracted to the area specifically to take advantage of low-cost labor. If this were an asset of the region one would expect to find industries producing consumer goods for sale in foreign markets. Instead, protection is required for many industries even to sell their goods in the domestic market. One should not, of course, confuse low-wage labor with low-cost labor. And by the time West African labor has gained experience, it may be hoped that wages will have improved sufficiently so that the area will not have to rely on low-cost labor to attract further industry.

MARKET-ORIENTED INDUSTRIES

Almost all of the remaining manufacturing establishments in West Africa may be classified as market-oriented industries. And it has been in this category that the greatest advances have been made in the past decade. Manufacturing tends to be drawn to the market under a variety of circumstances: (1) if the product is made more perishable by manufacturing; (2) if the product is more bulky than the raw materials from which it is made; (3) if the product is more fragile than its raw materials; (4) if the industry is a service activity; (5) if it is cheaper to import the raw materials than the finished product because of lower transport or lower tariff rates on the raw materials; and (6) if the industry is protected by tariffs or other means. The last two factors, be it noted, may be largely artificial, whereas the first four are based upon physical factors.

The strength of the pull toward market, raw material, power, or other orientation varies from product to product, and an examination of the industrial representation in any tropical African area will reveal that those industries which are most strongly affected by a specific locational factor are the ones which appear in the industrial roster and are also the ones which are likely to be more widely spread within individual countries. Good examples of this point would be cotton gins for the raw material category and carbonated beverages for the market-oriented group of industries.

The large number of market-oriented establishments makes it desirable to subdivide this category.

FOOD AND BEVERAGE MANUFACTURE. Examples of industries whose products become more perishable after manufacture include baking and ice cream making. The first is often carried on in small shops, but larger mechanical bread and biscuit factories have been introduced in recent years in many major cities.

Beer and carbonated beverages are examples of products that, because of their large water content, offer substantial transport savings if good water is available in the market area. It is not surprising, therefore, that many of the largest cities of Africa have big breweries (Map 64). The production of carbon dioxide as a by-product makes it logical to have associated facilities to produce carbonated beverages, but the much lower capital and skill required to produce soft drinks and their lower value per bottle as

A modern bakery near Lagos, Nigeria
An example of an industry whose product is more perishable than its raw material and that must therefore be situated close to its market.

contrasted with beer have permitted and led to a much wider distribution of plants producing colas, lemonades, and other such bottled drinks. Breweries are frequently among the larger, more prosperous, and more rapidly growing industries in West Africa.

Other market-oriented food and beverage industries include confectionaries, milk reconstitution (usually based on imported pow-

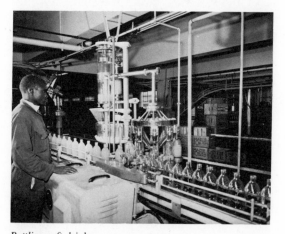

Bottling soft drinks
The high content of water in such products gives them a very strong market orientation because large savings in transport costs are possible.

dered milk), and a few examples of tea conditioning, production of soluble coffee, margarine making, and distilling.

TOBACCO PRODUCTS. The production of cigarettes and some pipe tobacco is favored by transport savings resulting from use of domestic leaf or imported leaf in bulk and more importantly by protection accorded this industry, whose significance as a source of government revenue contributes to the desire to reap the full benefit of its manufacture. Cigarette factories are among the larger plants in several West African countries and, as has been noted, manufacturers often have programs designed to increase the share of local tobacco consumed. Those cities with modern cigarette factories are shown in Map 64, a and b.

TEXTILES AND CLOTHING. At the present stage of economic development in West Africa cloth and simple clothing are two of the most important items of expenditure by the average citizen, and they frequently loom large on the import list. There is a natural desire, therefore, to produce them domestically. Factors helping to make this possible include the relatively small capital investment required, the fact that labor can be relatively easily trained to the necessary tasks, the low wage levels prevailing, and, in some cases, the existence of domestic supplies of cotton and the remoteness of some market areas from textile producers in Europe and Asia. An advantage of the textile industry over many others is the relatively large number of persons required to produce a given value of output, a factor of some significance given the unemployment rates in many countries. Despite these considerations, the textile industries of West Africa have required tariff protection it has not yet been possible to reduce, despite some fairly early entries in this sector.

The textile industry has shown notable growth in recent years in many countries. In francophone West Africa there were only about six textile mills in 1960; by 1972 the number had been increased to about 25 and output was estimated to cover about 60 percent of local needs; Dakar and Rufisque in Senegal and Bouaké in Ivory Coast are the leading textile centers in this area, and they export a portion of their output to other West African countries. The mill at Bouaké, started in 1922, is one of the earliest such plants in tropical Africa. The Ivory Coast plans one of the world's largest spinning and weaving mills at Dimbroko and another large mill at Agboville costing about $35 million. Textiles and clothing combined are the second ranking industry in Senegal; in the Ivory Coast in 1972 the 27 plants in this sector accounted for about one-ninth of investment and one-tenth of value of putput in industry.

Ghana made a somewhat belated entry into textiles but has greatly strengthened and diversified this sector in the last decade. The government participates in a number of the larger plants. Nigeria has the most important textile development in West Africa and that industry is the largest employer of labor among manufacturing sectors in the country, with over 50,000 workers in 1973. The industry was temporarily hurt in 1971 by reduced protection intended to combat inflation, but the government has apparently decided that the importance of the textile and clothing sector as a growth industry justifies higher protection to absorb unutilized capacity and stimulate further expansion. Nigeria's textile and clothing industry has a broader range of processes and products than most other tropical African countries and handles a greater variety of raw materials, including cotton, wool, jute, coir fiber, and synthetic fibers. Total investment in the industry was estimated to be $290 million in 1972, when there were 20 knitting mills, 50 apparel makers, and 20 factories producing made-up textiles like tarpaulins and towels as well as the basic spinning and weaving mills.

SHOES. Many shoes are produced by artisans in West Africa. They use leather, plastic, and old tires usually to make sandals, but several modern plants are found in the area. The Bata Shoe Company has had particular success in introducing the relatively complex processes of shoe manufacture into quite a few African countries, primarily by breaking them down into individual operations and by careful attention to adequate training of personnel; the company also operates surprisingly extensive chains of very modern retail outlets (Map 64).

HOUSEHOLD FURNISHINGS. It is primarily the great increase in bulk after manufacture that makes it appropriate to produce many household furnishings in the market area. Certain items, such as wooden furniture, can also be produced on a semi-handicraft basis. Wooden furniture has been produced for many years in West Africa and almost every city has some shops engaged in this industry; a more recent introduction is the manufacture of metal furniture. Steel utensils, enamelware, and aluminum hollow ware and utensils are other household furnishings produced.

Metal doors and window frames—desirable because of their freedom from warping and from insect attack—and metal burglar guards have been among the first metal products manufactured in tropical Africa. They are easily assembled and their bulk is enormous in relation to the space occupied by the metal strips, sheets, and bars of which they are made. Plastic combs, cups, buckets, garden hose, toys, suitcases, etc. are produced in many plants. Nigeria alone has over 60 companies making plastic articles.

METAL AND MECHANICAL INDUSTRIES. In addition to the steel and aluminum furnishings noted above, there are a number of metal fabricating industries and a growing number of plants assembling vehicles, agricultural implements, household appliances, radios,

A small furniture factory at Accra, Ghana
The great increase in bulk after manufacture makes it appropriate to produce many household articles in the market area.

An automobile and truck assembly plant at Tema, Ghana
While the increase in bulk of assembled products favors a market location, many assembly plants in Africa require tariff protection because their small size makes for high operating costs.

and television sets. These industries are desirable, because many more components can be packed in the space that a fully assembled product would occupy. Only a few components are manufactured locally, though new contracts call for a gradual increase in their number in the larger countries. Despite the transport savings involved, most assembly plants require protection, and several are too small to operate efficiently.

Metal drums, cans, and other containers were among the earliest metal items produced in West Africa, since it is particularly costly to ship empty containers. In more recent years, pipes, storage tanks, metal boxes, wires, rods, bottle tops, and simple hardware items such as nails, nuts, bolts, and screws, tools, and locks have been added. Corrugated iron and aluminum sheets, used widely for roofing, and metal gutters are also produced in a few plants.

Small steel-rolling mills are situated at Tema and Emene, near Enugu, using scrap as the raw material, while each country has at least one all-purpose metalworking shop, often quite small.

CHEMICAL PRODUCTS. In addition to plastic products already noted, a variety of chemical products are produced in West Africa. Most common are plants producing paints and varnishes, matches, cosmetics, perfume, soap and detergents, and industrial gases—mainly oxygen and acetylene. Other products prepared for the deomestic market are pesticides, pharmaceuticals, and tires. Liberia has an explosives factory supplying the mining industry in that country and elsewhere in West Africa. Only a few fertilizer plants have thus far been constructed in the area. Senegal plans to set up a sizable petrochemical plant to produce several major plastic raw materials, with the hope of supplying the plastic product industries of the region.

In recent years there has been increased interest in petroleum refineries, though only the Nigerian plants use domestic feedstock. Some of the refineries are more status symbols than viable plants, because their small size makes them uneconomic without protection. Companies have often vied to place refineries in African countries, however, to secure privileged positions in the national markets.

CONSTRUCTION MATERIALS. Because of their great weight in relation to their value, there is a strong incentive to produce such materials as bricks, tiles, cement, and cement products as close to their markets as possible. Bricks and tiles can usually be readily produced near their markets, since clay is common and only a small-scale operation is required. Cement cannot be so readily decentralized: a large investment is required if the plant is to be economic, suitable limestone is not so widespread as clay, coal should be available at a reasonable cost, and there must be good transport facilities to the market. These sometimes conflicting requirements have meant that the number of cement plants in tropical Africa is usually limited to one or two mills in a country, but that a cement plant is likely to be established as soon as the size of the national or regional market justifies the investment. Where suitable limestone is not available, clinker-grinding plants using imported material are likely to be erected. Cement shapes are produced in a much larger number of facilities, while specialized products such as asbestos cement sheets and pipes come from a few plants.

MISCELLANEOUS PRODUCTS. A few products manufactured in West Africa do not fit conveniently into any of the above categories. These include paper products and stationery items; glass bottles, tumblers, and panes; and neon signs, phonograph records, batteries, and ceramics.

SERVICE INDUSTRIES. Industries in this category, if they are to exist at all, must be in the market area, for a variety of obvious reasons. The largest such establishments in West Africa are the railway workshops. The shop at Ebute Metta in Greater Lagos, headquarters of the Nigerian Railways, is one of the largest industrial plants in Nigeria. The shops of the Public Works Departments, equipped to repair many types of mechanical gear but particularly vehicles, are a similar type of service industry. Vessel and aircraft repair facilities are also in this category.

Other service establishments include refrigeration and ice plants, printing works, carpentry shops, and tire retreading plants, as well as the industrial gas plants mentioned above.

Electricity. Perhaps the most important of all service industries is the production of electric power, an essential part of the infrastructure of any nation that wishes to industrialize. The availability of power for general industry is not to be confused with large-scale production of low-cost hydroelectricity to attract power-oriented industries (see below). Power consumption may represent only a small fraction of the inputs in most industries and need not necessarily be especially low in cost; but it must be available either from public lines or by private installation.

Power in West Africa is usually confined to the larger cities and towns and the mining communities. It is often very expensive at inland points, but is relatively less costly where larger and more efficient plants have been installed. Diesel oil is the most common fuel; gas turbine plants are situated in Nigeria (at Afam and Ughelli); some sawmills and peanut mills use plant waste for a portion or all of their fuel requirements; only Nigeria still has a coal-fired station.

Though the capacity and production of all but a few electric plants are now small in all countries, both are expanding rapidly in the main countries. For example, the Ivory Coast got its first important electric plant in 1950 at Abidjan, but by 1962 twelve internal communities had received plants; total production of electricity increased from 1.8 million kwh in 1952 to 120.0 million kwh in 1962 and 661 million kwh in 1973.

A number of small hydroelectric plants are now in being, while three large ones have been installed. Developments in the northern tier of countries are restricted because of their aridity and the small domestic markets. Nonetheless, Mali has 1,000-kw and 5,000-kw plants on the Niger and a rather large plant may eventually be placed near Gouina on the Senegal. Upper Volta has been studying the possibilities of locating a plant on a site on the Black Volta or importing electricity from Ghana, but its total consumption is too low to justify large investments. Niger could install about 100,000 kw in stages on the Niger south of Niamey and is considering a 15,000-kw installation at Dyodonga on the Mékrou, 132 miles (212 km) from Niamey; that country could also import electricity from Kainji in Nigeria, while a German study is being made of an 80,000 kw nuclear-electric plant at the capital.

Guinea has a 10,000-kw plant at Grand Chutes on the Samou River. Sierra Leone produces some power from its Guma Valley Dam (its primary purpose is to supply water to Freetown) and is constructing a $41 million, 50,000 kw station at Bambuna Falls in the Tonkolili District to supply the mining areas and the capital. Liberia has a 4,200-kw station serving Harbel and a 34,000-kw plant at Mt. Coffee, 25 miles (40 km) northwest of Monrovia; the latter was constructed in 1966 at a cost of $29 million and its capacity can be increased to 102,000 kw. In the Ivory Coast two installations with a combined capacity of about 50,000 kw are situated at Ayamé on the Bia River, 68 miles (109 km) east of Abidjan. In 1972 a 174,000-kw station at Kossou on the Bandama River 180 miles (290 km) northwest of Abidjan was completed at a cost of about $100 million; it more than doubled the installed electric capacity of the country. The mile-long (1.6 km), 187-foot-high (57 m) dam created a lake three times the size of Lake Geneva, which will hopefully support 4,000 fishermen and a yearly catch of 15,000 tons; about

100,000 persons were displaced by the reservoir.

Togo has a 1,600-kw plant at Kpimé, near Palimé. It has examined the possibility of a multipurpose dam at Ngambeto on the Mono River, which would have a 31,000-kw capacity; three smaller dams could increase this to 100,000 kw. But Togo and Dahomey decided instead to import power from Ghana, and the transmission lines were completed from Akosombo to Lomé and Cotonou in 1972.

Nigeria has one of the largest general-purpose hydroelectric stations in tropical Africa at Kainji on the Niger River in Kwara State. The keystone of the 1962–68 plan, this project cost about $235 million and was completed in 1968; the installed power is 480,000 kw but it can be doubled, while additional dams at Jebba and the Shiroro Gorge on the Kaduna River would add another 960,000 kw. The latter would permit firmer power at Kainji by further evening the flow. Kainji electricity is transmitted south to Lagos and north to Kano and to the intermediate cities. The installation has sub-

sidiary advantages: partial flood control on the plain below Jebba, improvement of navigation as far upstream as Niamey, some regularization of navigability below the dam (though real improvement would require additional dams on the Niger and the Benue), provision of a new road-crossing over the Niger, greater potential for fishing, and possible attraction of a small tourist industry. Ironically, the decision to construct Kainji was made before the great wealth of oil and natural gas of Nigeria was appreciated. It is doubtful that the same decision would have been made five or ten years later, though its existence now permits larger sale of petroleum at prevailing high prices.

LOW-COST-ENERGY–ORIENTED INDUSTRIES

Industries are attracted to sources of low-cost energy if the cost of energy represents a considerable fraction of the total cost of the finished product. Examples are such electrometallurgical industries as those refining aluminum (one ton requires 14,000–16,000 kwh) and certain other metals and such electrochemical industries as those producing calcium carbide, nitrogen by fixation, heavy water, and enriched uranium.

The Akosombo Dam of the Volta River Project, Ghana
The main purpose of this project was to attract low-cost-energy-oriented industry to Ghana.

The reduction plant of the Volta Aluminum Company at Tema, Ghana
Producing aluminum ingots, this plant consumes about 74 percent of the electricity produced at the Akosombo power plant.

At present there are very few examples of strictly low-cost power-oriented industries in Africa and only one in West Africa, the $154-million smelter of the Volta Aluminum Company (VALCO) at Tema, Ghana. This plant, owned jointly by Kaiser Aluminum and Chemical Corporation and Reynolds Metal Company, was attracted by the possibility of securing a large bloc of low-cost electricity from the installation on the Volta River at Akosombo, while construction of the dam and power plant was in turn justified by a contract from VALCO to "take or buy" about 85 percent of the output. The plant's capacity, now 145,000 tons, is slated to be raised to 220,000 tons; it permits Ghana to rank first in Africa in the production of aluminum.

The Akosombo Dam, completed in 1965 at a cost of $182 million, is the key engineering component of the Volta River Project, whose main purpose was the production of electricity to attract industry. Subsidiary benefits included irrigation on part of the Accra Plains, an improved water supply to greater Accra, and navigation, fishing, and recreation on Lake Volta, which covers about 3.3 percent of the country's area and extends inland about 250 miles (400 km). The hydroelectric installation now has a capacity of 882,000 kw; about 74 percent of production goes to the Tema smelter, 50,000 kw of capacity is allotted to Togo and Dahomey, and the rest is transmitted to the main centers of southern Ghana.

The availability of excess power did not attract other low-cost-power–oriented industries as had been hoped, which helps to explain the desire to export electricity to neighboring countries. As has been noted, Ghana is also pushing for the development of an alumina industry using domestic bauxite, which was part of the original proposal made as long ago as 1915. While the entire project has yielded a low rate of return to Ghana, it did attract large-scale investment for the construction of a major power installation that has provided low-cost electricity and a major consumer, the aluminum industry, which assured repayment of the debt. It will return a higher yield after 1976.

LOCATION OF INDUSTRY IN WEST AFRICA

A clear distinction between the locational pattern of raw material and of market-oriented establishments is apparent in West Africa. The processing of agricultural, forestry, and mineral raw materials tends to be in or very close to the producing area or mine. The locational pattern for these industries, therefore, shows considerable dispersion, as can be seen in Maps 61 to 63. With the major exceptions of Dakar, with its large peanut mills, and Lagos, with many small lumber mills that cater mainly to the domestic market, most such industries are found outside the major urban communities.

The market-oriented industries, on the other hand, are concentrated in the major cities, which are also often ports and capitals of their countries (Maps 64 a and b). Dakar at one time had over half of industry in French West Africa, and Abidjan ranked second; these two cities continue as the leaders in francophone West Africa, though Abidjan has gained many more new industries than its older rival. In 1970 Abidjan accounted for an estimated 58 percent of the value of manufacturing production of the Ivory Coast, but an effort is now being made to disperse new plants to reduce its share to about 42 percent by 1980; two major growth poles are to be developed centered on the new port of San Pedro and on Bouaké, the second largest city, with lesser growth nodes at Dimbroko (a huge spinning-weaving mill), Kossou (hydroelectric facilities), Ferkessédougou (sugar), Agboville, and other towns.

Accra-Tema have attracted the lion's share of consumer industries in Ghana, while Tema also has several heavy industries such

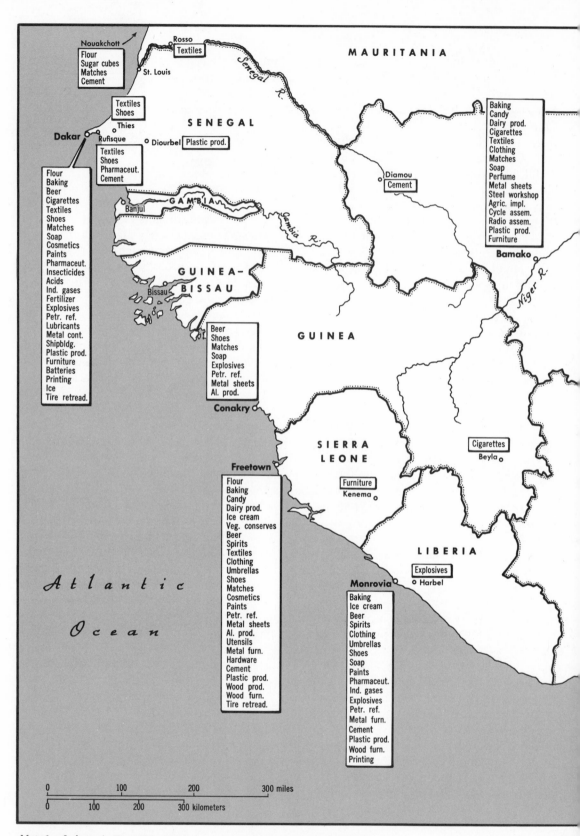

MAURITANIA

Nouakchott
Flour
Sugar cubes
Matches
Cement

Rosso
Textiles

St. Louis

Textiles
Shoes
Thies

SENEGAL

Dakar
Rufisque
Diourbel Plastic prod.

Textiles
Shoes
Pharmaceut.
Cement

Flour
Baking
Beer
Cigarettes
Textiles
Shoes
Matches
Soap
Cosmetics
Paints
Pharmaceut.
Insecticides
Acids
Ind. gases
Fertilizer
Explosives
Petr. ref.
Lubricants
Metal cont.
Shipbldg.
Plastic prod.
Furniture
Batteries
Printing
Ice
Tire retread.

GAMBIA

Banjul

Gambia R.

GUINEA-
BISSAU

Bissau

Diamou
Cement

Baking
Candy
Dairy prod.
Cigarettes
Textiles
Clothing
Matches
Soap
Perfume
Metal sheets
Steel workshop
Agric. impl.
Cycle assem.
Radio assem.
Plastic prod.
Furniture

Bamako

Niger R.

GUINEA

Beer
Shoes
Matches
Soap
Explosives
Petr. ref.
Metal sheets
Al. prod.

Conakry

SIERRA
LEONE

Cigarettes
Beyla

Freetown

Flour
Baking
Candy
Dairy prod.
Ice cream
Veg. conserves
Beer
Spirits
Textiles
Clothing
Umbrellas
Shoes
Matches
Cosmetics
Paints
Petr. ref.
Metal sheets
Al. prod.
Utensils
Metal furn.
Hardware
Cement
Plastic prod.
Wood prod.
Wood furn.
Tire retread.

Furniture
Kenema

LIBERIA

Monrovia

Explosives
Harbel

Baking
Ice cream
Beer
Spirits
Clothing
Umbrellas
Shoes
Soap
Paints
Pharmaceut.
Ind. gases
Explosives
Petr. ref.
Metal furn.
Cement
Plastic prod.
Wood furn.
Printing

Atlantic

Ocean

0   100   200   300 miles
0   100   200   300 kilometers

Map 64. Industry in West Africa IV: Industries catering to the domestic markets
a. Senegal to Dahomey and western Niger

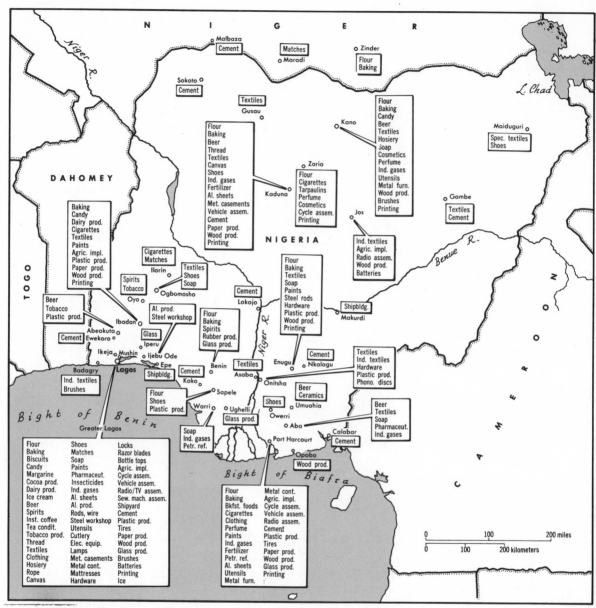

b. *Nigeria and southeastern Niger*

as the steel rolling mill and the VALCO smelter. Takoradi is a secondary center, but neither it nor Kumasi have kept up with greater Accra.

Lagos and its adjoining towns, Apapa, Ebute Metta, Mushin, and Ikeja, comprise the largest industrial nucleus in Nigeria. In 1965 Greater Lagos accounted for 28 percent of the 776 plants employing ten or more people, 32 percent of the employees in these plants, and 37 percent of their gross output. But the very substantial authority of the former four Regional and now the twelve State Governments has resulted and is continuing to result in a greater dispersion of industry than would otherwise have occurred. There is, in fact, a degree of state "me-too-ism" in the growth of certain industries, particularly textiles, and each state is striving to attract new factories to its centers. In every case the capital city is the major industrial center, but a number of states have secondary centers of greater or lesser importance (Table 12 and Map 64b and 65).

In the smaller countries, the capital city usually has the vast bulk of market-oriented manufacturing establishments. Only Upper Volta presents something of an exception, which can be explained by the early rise of Bobo Dioulasso as the commercial center before the rail line was extended to Ouagadougou. Industries producing consumer goods are attracted to the capital cities because they have the largest markets in their countries, the best transport facilities both for import and for distribution to the hinterland, the best-developed power facilities, the greatest number of semiskilled workers, and the advantage of proximity to government departments.

FACTORS INFLUENCING DEVELOPMENT OF INDUSTRY

The preceding analysis of present industries has pointed to several important considerations pertaining to the suitability of specific industries for the stage of development now existent in West Africa. Now, an examination of the major factors influencing manufacturing will permit a further assessment of the potentialities and limitations for continued industrial development in the area.

RAW MATERIALS

Developments already noted suggest that there are further opportunities for the exploitation and processing of raw materials. Particularly important will be the new mineral bodies coming into use, the liquefaction of natural gas in Nigeria, the expansion of rubber production, increased processing of forest output, extension of plants expressing

oil (from peanuts, palm fruit, and palm kernels), sugar mills, and ginning of cotton where that crop is being extended. But suggestions that the key to industrialization in underdeveloped areas is the placement of agricultural processing plants in rural areas do not appear to be valid for West Africa, because in many areas such plants already have more capacity than is being utilized.

It is often assumed that, because the raw material is available, the producing country should not only perform the preliminary processing but should also manufacture the finished product and export it to the world market. Several examples will illustrate the necessity of examining the appropriateness of each successive stage of manufacture, for the locational factors may be entirely different from one stage to the next. It is sometimes suggested, for instance, that Ghana, the world's largest producer of cocoa, should make chocolate and other finished products in which cocoa is consumed. There are numerous reasons why this is not practicable, except insofar as the chocolate and candy would be marketed within Ghana itself: the storage of raw cocoa is difficult, that of the finished product even more so, especially under tropical conditions; cocoa is only one raw material in chocolate, sugar and milk being more important from the standpoint of volume; there is only a very small internal demand; the finished product requires greater space in shipment, is more perishable, and is subject to greater damage in handling; and the skill required in manufacture is not yet available in Ghana.

A second example is the iron and steel industry, whose appropriateness for Nigeria has been the subject of several investigations. There is no question that it is physically possible to produce steel from the raw materials in Nigeria. But the iron ore is only of moderate grade and the coal is low quality and non-coking. Natural gas could be used for a substantial part of the fuel requirements but would not satisfy the needs of a blast fur-

Table 12. Selected data on the twelve Nigerian states

State	Percent of total area	Percent of total population, (1963) 1973	Percent of total of state budgets, 1972–73	Estimated share of export production, 1971	Percent of 71 types of industry present, 1971	Percent, gross output of industries employing over 10 workers, 1965 [a]	Capital	Other important cities	Main commercial products and interests
Lagos	1.5	(2.5) 3.2	5.0	0.2	76.1	37.4 [a]	Lagos	Mushin, Ebute Metta, Yaba	Federal capital, #1 port, #1 industrial center
Western	7.6	(17.1) 11.5	12.5	12.5	47.9	9.9 [a]	Ibadan	Abeokuta, Oshogbo, Ondo, Akure, Ife	Cocoa (95%), palm products, cola, timber, #2 industry
Midwest	4.0	(4.5) 4.2	10.0	43.7	15.5	4.9	Benin	Sapele, Warri, Burutu	#1 petroleum, rubber (c. 80%), #1 timber
East-central	2.3	(13.0) 10.4	8.7	1.2	22.5	9.7	Enugu	Onitsha, Aba	Palm products, coal, some petroleum
Rivers	1.8	(2.7) 2.9	11.1	32.3	38.0	6.3	Pt. Harcourt		#2 petroleum, #2 port, palm products
Southeast	3.6	(6.5) 4.9	4.4	5.0	5.6	0.4	Calabar		Palm products, petroleum, rubber, timber
Kwara	7.5	(4.3) 6.0	6.3	0.1	8.5	0.5	Ilorin		Sugar, yams
Benue-Plateau	10.9	(7.2) 4.1	7.7	1.2	8.5	6.8	Jos	Makurdi	Tin, columbite, cotton, sesameseed
Northwest	17.0	(10.3) 10.9	8.5	0.4	2.8		Sokoto	Gusau	Peanuts, cotton, livestock
North-central	7.8	(7.4) 8.7	8.9	1.2	14.1	8.1	Kaduna	Zaria, Katsina	#1 cotton, #2 peanuts, livestock, ginger
Kano	4.4	(10.5) 14.0	7.6	1.8	32.4	13.1	Kano		#1 peanuts, #2 cotton, livestock
Northeast	31.6	(14.1) 19.8	9.2	0.3	5.6		Maiduguri	Bauchi	Peanuts, livestock, fish, cotton
Totals	'000 357 sq. mi. 924 sq. km	million 79.8 [b] (55.7)	million $U.S. $691	million $U.S. $1,811	types 71	million $U.S. $622 [c]			

SOURCE: National statistical sources

[a] Ikeja included in Greater Lagos

[b] Provisional: total of provisional state figures does not jibe with provisional country total

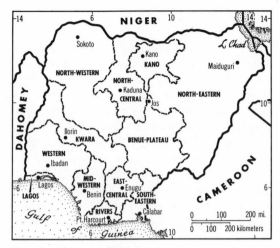

Map 65. The states of Nigeria

nace. Here the alternatives would appear to be using a special process to cokefy the coal or using one that does not require coke, but such processes have not proved economic elsewhere, even in Norway where extremely low-cost electric power is used. An even more serious limitation is the size of the Nigerian market, which is too small to justify a mill large enough to permit savings in scale; and the record of small mills in Africa, South America, and elsewhere indicates that their output is not competitive with the imported product. Unlike cotton textile mills, which can be constructed at relatively low cost, utilize unskilled labor, and have a market for the finished product, an integrated iron and steel mill involves a very heavy investment, requires additional investments in iron, coal, and limestone mining, needs a substantial number of skilled workers, and has a ready market only for simple products such as wires, bars, and sheets. A fully integrated mill presupposes a substantial buildup in metal fabricating industries. It might well be more advantageous at Nigeria's present stage to stimulate the growth of metal fabricating industries by using the lowest-cost steel available until such time as the market justified construction of domestic facilities. In the meantime,

small rolling mills (such as those at Tema and Emene) based upon domestic scrap might satisfy the desire for possession of some steel-making capacity. Another possibility for West Africa would be the construction of a large mill based on the availability of high-grade iron ore, backhaul of coke, and export of the ingots to world markets.

ENERGY

Little need be added to what has already been noted regarding the energy position and potentials of West Africa. Provision of power for small industries is proceeding satisfactorily. Hydroelectric capacity to support an aluminum smelter comparable to that at Tema is being developed in Guinea. Nigerian resources of natural gas might attract a number of industries seeking low-cost energy, but its export as LNG would probably yield a greater return. Discovery and development of hydrocarbons in other West African countries may greatly improve their energy positions, particularly those in the interior where power is very expensive.

MARKET

In numbers, only Nigeria has a sizable market, though various customs unions would create reasonably large market populations. Unfortunately, little progress has been made either in forming such unions or in harmonizing industrial development programs to permit the introduction of regional industries serving the West African market. The wide dispersion of peoples, particularly in francophone West Africa, is disadvantageous as far as distributional costs are concerned, and transport facilities are often far from optimal as far as marketing is concerned. The marked concentrations of people in some parts of the area, on the other hand, favor the development there of consumer-oriented establishments; most favored from this standpoint are the three main clusters in

Nigeria, the south of Ghana, and the area dependent upon Dakar.

From the standpoint of purchasing power the average per capita incomes are very low, which means, as far as market-oriented industries are concerned, that emphasis should be on relatively simple items, which enjoy a broad market—clothing, shoes, household goods, etc. The Ivory Coast, Senegal, and Ghana have the highest average incomes, which helps to explain their relatively good positions among West African countries in manufacturing. The considerable number of Europeans with high incomes in Dakar and Abidjan gives these two cities special positions of considerable actual value to manufacturers.

Limitations of size of the national markets is probably the most significant handicap to industrial development. This is particularly true for the smaller and poorer countries where even the present plants are often plagued by overcapacity.

Restricted size of domestic markets has led several nations to consider the possibilities of producing consumer goods for export to world markets. Some success has been achieved by a few plants selling cloth, but the prospects for large-scale expansion of such exports is not now too promising. It has been estimated in the case of textile mills that costs of construction, power, maintenance, and transport are about 13 to 15 percent higher for African textile mills than for those of the EEC countries; wages of workers are a fourth as high (as compared to a seventh or an eighth in such countries as Korea and Formosa) but those of expatriate supervision are three times those in Europe; the older plants have only half the productivity of the newer, but labor fights efforts to increase productivity. If West African mills are to be competitive in European markets, it was concluded, they would have to receive continued preferences, achieve greater efficiency, replace expatriates as training progresses, and mount well-organized sales campaigns.

LABOR

There is no shortage of labor willing to work in factories in West Africa; there is a shortage of skilled workers and a severe shortage of engineering and managerial talent, which will require many years to supply from nationals of the area. Earlier problems associated with migratory labor have largely dissipated in manufacturing, where turnover rates are often remarkably low. Absenteeism does, however, remain as a significant problem, poor health being the greatest single cause. African productivity in simple repetitive operations is usually very good, sometimes higher than that in Europe, but productivity may fall off by a half or two-thirds in more skilled work.

Efforts to improve the output and skill of labor have received much greater attention in recent years. Increased emphasis has been placed on the selection of workers, on technical training in government facilities and within industry, and on provision of various incentives to stimulate productivity and steady attendance.

The subject of wage rates in Africa must be studied in the context of African conditions. Direct comparison of wages in Africa with those in Europe or North America is misleading. Purchasing power should be calculated in terms of local food and housing, and industrial wages should bear some relation to earning capacity in other activities, including agriculture. Low wages also, of course, reflect low skill and productivity. If wages are artificially forced to too high a level the economy is likely to be distorted and the chances for introducing new activities in many segments will be reduced.

Government should, of course, protect against exploitative wages and practices. But concentration of wage earners in the large cities—in government, services, and market-oriented manufacturing—means that periodic mandated raises for wage earners add to the dichotomy between rural and urban areas, tend to create a government and urban elite increasingly out of touch with the bulk of the citizenry, and stimulate further migration to the cities with all of its attendant problems.

PROVISION OF CAPITAL

The securing of adequate capital is another problem for African countries wishing to industrialize. Capital available from domestic sources is usually quite inadequate, though increasing amounts are being allocated from governmental revenues for investment in industry. In francophone countries, mixed companies have long been a common feature. Ghana made direct investments in a considerable number of plants, many of which were ill-chosen. Nigerian Regional Governments also encouraged investment by providing a portion of the capital required for new industries in partnership with foreign and domestic companies.

Practically all African governments have provided various incentives to attract additional foreign capital, including temporary exemptions from taxation, allowances for depreciation, remission of customs duties on imported equipment and raw materials, increased tariff protection for the product manufactured, and free transfer of dividends and repatriation of capital. In addition, attention is given to improved financial and advisory services, to the creation of a financial infrastructure more responsive to the needs of modern industry, and to the establishment of industrial estates, which obviate the necessity for prospective industrialists to worry about otherwise difficult tenurial situations and provide space where electricity, water, and other services are available. Finally, efforts have been made to provide

economic resource surveys and other data of importance to potential investors. On the other hand, countries have imposed regulations on hiring and training of locals, have restricted investment in some specific sectors, and have sometimes abrogated agreements regarding repatriation of earnings and capital investments. It has been much less common to nationalize industries than mining enterprises, though participation of governments in private manufacturing establishments has increased.

The climate for investment, of course, differs widely from one country to another, and may change abruptly in any country. One would not have expected private capital to move to Guinea or Mali when these countries were nationalizing enterprises in a variety of fields and were set upon a course of state socialism. The antipathy toward Western institutions expressed by the Ghanaian government and media under Nkrumah also undoubtedly discouraged interest in that country. But these countries and others have come to realize that the task of developing industry requires the utilization of many resources, that private enterprises can often operate more efficiently than state-run establishments, and that their qualitative contribution in technology is often of prime importance.

In conclusion, it is apparent that many difficulties exist to inhibit the rapid expansion of West African industry, but that constructive steps are being made to reduce their dimensions. The opportunities for growth are undoubtedly very substantial, in at least a number of the larger countries, and there is no permanent impediment to the achievement of fully integrated industrial establishments in the area.

Preparing to fell a large okoumé tree in Gabon
The largest area of rainforest in Africa is situated in the equatorial belt facing the Atlantic.

CHAPTER THIRTEEN

Middle Africa from Chad to Congo

There is no convenient term for the group of states covered in this chapter. Included are four republics derived from the former French Equatorial Africa; the United Republic of Cameroon, consisting of the former French Trusteeship and part of the former British Trusteeship; Equatorial Guinea, previously Spanish and consisting of the mainland Rio Muni and the island of Macias Nguema (Fernando Po); and the Portuguese islands of São Tomé and Príncipe. The French heritage of five of these countries and the actual or near juxtaposition of the rest make it appropriate to include them in one grouping. The islands are treated in brief subsections and do not necessarily share in the generalizations made for the continental areas.

The six independent countries have had widely varied political experiences since independence and their rates of economic development have differed considerably. Gabon and French Cameroon have been politically stable, though the French intervened in Gabon in 1964 in support of the incumbent president and a revolt in the Bamiléké country of Cameroon flared sporadically for some years. Congo has witnessed a series of coups and attempted coups. Since 1965 it has been strongly oriented to Communist countries with either Cuban, Russian, or mainland Chinese influence being dominant at different periods; France nonetheless continued to provide the largest financial and technical support in most years, though relations were strained in 1972 by a series of anti-French measures. The Central African Republic (CAR) has had an unfortunate history of exploitation over several centuries: before colonization there was slaving in the

south by northern tribes; then there were depradations by the 27 concessionaires that were given rights to "develop" the area; under colonial rule there was forced planting of cash crops and labor corvées, particularly to construct the Congo-Ocean railway; and five and a third years after independence in 1960 a coup brought to power an autocratic and repressive government, which had no more concern for the common man than its predecessors.

Chad has suffered from long-standing rivalries between the population of the north—nomadic and seminomadic Muslim groups and—that of the south—Bantu, Christian or animist, sedentarists. Some northerners, who have a warrior tradition (they were not definitively subjugated by the French until 1937), engaged after independence in banditry, which progressed to guerrilla warfare over fairly broad areas of the north and east. Attempted suppression took as much as half of the national budget, and several thousand French troops came to the aid of the government in 1968. In 1971, after the warfare had diminished in intensity, an agreement was reached which brought the northerners half the government ministries; and the government has since received aid from Libya, which had previously been supplying arms to the dissidents. Equatorial Guinea, which became independent in 1968, experienced a crisis shortly thereafter that resulted in the exodus of almost all of the Spaniards, severe harassment of many Nigerian plantation workers, and a collapse of the economy and services. Its president is one of the most authoritarian on the continent.

The impact of these political events has affected the economies differentially. Gabon, with great mineral and forest wealth and a small population, enjoyed a 5.2 percent annual increase in per capita GNP from 1960 to 1971, at which time, at $700, it ranked fourth in Africa. Cameroon has made commendable progress in expanding and diversifying its largely agricultural economy and has built up its industry more rapidly than any of the others. It has been successful in gradually bringing together the economic, social, and political forms of East and West Cameroon with their very different colonial heritages and, in 1972, the former Federation became the United Republic of Cameroon. Both Cameroon and Gabon have retained a strong French presence; in 1972 there were about 13,000 French citizens in Cameroon and some 14,300 in Gabon. In the latter country Gabonization has progressed in government, which had in 1972 only 300 Europeans as compared to 12,613 Gabonese of which 1,189 were in upper level positions; in the mines, oil industry, forestry, public works, and services, however, there were only 200 locals in a managerial and engineering cadre totaling about 1,520.

Congo has made reasonably good progress in some economic sectors, partially through financial and technical aid from Communist countries, but investment has not come forward for the massive Kouilou hydroelectric/industrial scheme and other enterprises. A series of nationalizations plus restrictions on repatriation of profits in the last few years are likely to have an adverse effect on the economy, which remains excessively dependent on forestry.

The CAR and Chad, always severely disadvantaged by their remoteness, have made only modest progress, their per capita GNPs having increased by average rates of 0.4 and 0.6 percent respectively from 1960 to 1971. Both have highly unfavorable trade balances, serious budget deficits, heavy reliance on low-value cotton, and dependence on France for support, particularly for develop-

ment budgets. Chad has suffered severely in recent years from the prolonged sub-Saharan drought.

GENERAL CHARACTERISTICS OF THE AREA

Before beginning an examination of the regional economies, a few keynotes and generalizations may be noted: the enormous heterogeneity of the area; the marked isolation characterizing much of the region; the generally low densities of population; the relatively low stage of economic development over large parts, even by African standards; and the marked concentration of commercial activities in limited sections.

HETEROGENEITY OF THE AREA

The countries of this region have exceptional physical heterogeneity, not surprising since the area is almost a third the size of the United States and extends from about 5° South to 20°20′ North and from the Atlantic about half way across the continent. Geologically, it has portions of the Zaïre and Chad Basins, of the pre-Cambrian basement complex, of recent volcanics, and of the Atlantic coastal plain. As regards climate and vegetation, it stretches from the equatorial rainforest through savanna and steppes to the heart of the Sahara. Precipitation varies from 400 inches (10,160 mm) to practically nothing. Some areas have a surplus of water and wood; in others the scarcity of both presents major problems. Even the individual nations have relatively little homogeneity.

The human aspects of the area show similar diversity. There are, for example, an estimated 140 ethnic groups in Cameroon, 40 in Congo, and 40 in Gabon, though often only a few groups are of major importance. Ethnic groups range from Negrillos in the rainforests to more or less fractionated peoples of Negro, Bantu, and Hamitic stock. Contrasts in stage of development and degree of contact with modern forces are as sharp as anywhere in Africa. These contrasts tend to accentuate tribal differences and help to explain some of the political problems of the area.

IMPORTANCE OF ISOLATION

Chad and CAR lie close to the center of Africa, and it is not surprising that they suffer from isolation. N'Djaména is almost 700 miles (1,126 km) from the sea and still possesses no easy routeway to it. Delivery of gasoline from Nigerian ports to that capital city raises its price eightfold; cement from Douala costs four times as much at N'Djaména. About half of the population lives over 500 miles (800 km) from the sea. Isolation is further increased by the general sparseness of the road net, landform barriers, and, for some areas, an extremely dispersed population. Despite heavy expenditures on transportation both before and after independence, the element of isolation remains as a prime factor inhibiting development over much of the area.

LOW DENSITY OF POPULATION

Most of the region stands in strong contrast with many of the countries thus far examined from the standpoint of population density, although the islands are exceptions to this generalization.

The market at Abéché, Chad
Much of the remote parts of middle Africa have a locally self-subsistent economy.

Congo, Gabon, the CAR and Chad have among the lowest densities on the continent, but crude density figures conceal marked disparities within each nation. The French were long concerned with the low populations of these countries, seeing them as inhibiting development. Without doubt, there are many problems in providing adequate physical and social infrastructures in sparsely populated regions, but the nature of the resource base and the economies of Gabon and Congo in particular permit quite satisfactory development with the existing numbers; it is highly likely that much larger numbers would merely have increased the percent of

Table 13. Area and population of middle African countries

Country	Area Sq. mi. '000	Area Sq km '000	Estimated Population, 1974 (million)	Population Density per sq. mi.	Population Density per sq km
Chad	496	1,284	3.96	8.0	3.1
Central African Republic	241	623	1.75	7.3	2.8
Gabon	103	268	0.52	5.0	1.9
Congo	132	342	1.30	9.8	3.8
Cameroon	183	475	6.21	33.9	13.1
Equatorial Guinea	11	28	.30	27.3	10.7
Rio Muni	10	26	.22	22.0	8.5
Fernando Po	0.8	2	.08	108.5	41.9
São Tomé and Príncipe	0.4	1	.07	187.0	72.0

subsistence farmers and lowered the average incomes.

In Congo about 70 percent of the population resides in the southern portion. The Brazzaville district, with 302,000 in the city itself (or 23.2 percent of the 1974 population of the country) and with a relatively dense peripheral zone, is the most heavily populated part of the country. The coastal zone is the second relatively populous area, while the forest zone and the Batéké plateaus are almost empty. Congo has an unusually high urbanized population, about 40 percent of the total; but unemployment has existed in Brazzaville since at least 1952 and has led to more than one serious riot and periodic political unrest.

Gabon's population is markedly aligned along the waterways and the relatively few roads, usually in small dispersed hamlets. Much of the center and east is virtually devoid of people, while about 23 percent of the population is urbanized.

The Central African Republic has no densely populated areas, with the exception of a few regions along the Ubangi River and the border with Chad. The eastern three-eighths of the country has only one-sixteenth of the population. CAR has an urban population of about 30 percent. Bangui, with about 240,000, is by far the largest agglomeration; its growth has been excessive in relation both to the opportunities for employment and the needs for development in rural areas.

Cameroon has a higher crude density than the other main countries, but two-thirds of the population is concentrated in three areas: the north from the Benue to Lake Chad; the western highlands; and the southwest, including the two major cities of Douala (350,000) and Yaoundé (190,000). The Adamoua Plateau and the southeast have much lower densities. About 24 percent of Cameroon's population is urbanized.

Chad's population is concentrated in the wetter south. About 46 percent of the people live in the wettest 10 percent of the area; there are also concentrations along the Chari and Logone Rivers. Only about 9.0 percent of this country's population lives in urban communities.

The population of Equatorial Guinea is very unevenly distributed between the more developed island of Macias Nguema (Fernando Po) and the mainland, Rio Muni. Ile Macias Nquema, like many of the island appurtenances of Africa, has a relatively high population density (108.5 per square mile, 41.9 per sq km); Rio Muni's is only one-fifth as high. Sâo Tomé and Príncipe had a crude density of about 187 per square mile (72 per sq km) in 1974.

It should be emphasized that population figures for the region are based mainly on sample surveys, some of which were taken in the early 1960s. Populations are characteristically very youthful and believed to be increasing rapidly, except in Gabon, where there is a high rate of infertility, explained largely by the incidence of venereal disease; Gabon's estimated annual growth rate is 0.8 percent.

RELATIVE UNDERDEVELOPMENT

By most measurements, the bulk of this middle African area rates low on the scale of economic development. French Equatorial Africa and Cameroon long remained the Cinderellas of the French Empire. This is explained only in part by the disappointing failure of many early concession companies, by the low population, difficulty of transport, and generally unhealthy conditions of the coastal stations. Other French overseas areas were considerably more attractive for the fairly limited prewar investments—particularly Indochina, but also Senegal, Ivory Coast, and Madagascar. Except in a few instances, no real effort was made to develop the economies of these areas until after

World War II. As late as 1936 there were only 4,749 Europeans in all of French Equatorial Africa. The Congo-Ocean Railway was not completed until 1934; the modern port of Pointe Noire began operating in 1939.

And despite a huge effort in postwar years, the economic positions of the several countries rated poorly at the time of independence. French Equatorial Africa and Cameroon accounted for only 3.3 percent of imports and 3.1 percent of African exports in 1961; these figures were less than those for Ghana alone. The remote areas had very moderate levels of exchange; the bulk of the populace remained in the subsistence sector.

While considerable progress has been made since 1960, much of the area remains little touched by modern forces. Chad and CAR account for only 15 percent of the area's exports, and there has been little diversification in their economies. Congo has seen no significant increase in the value of exports, which have been artificially inflated by inclusion of diamonds smuggled from CAR and Zaïre; in the period 1963–1968 these accounted for about 40 percent of total exports, but with better controls in the producing countries they fell to under 8 percent in the early 1970s.

Cameroon has made substantial progress in regional development with a notable increase in cotton production in the north and a sizable relative gain in cocoa production in Africa. Gabon has gained substantially from mineral developments, but its agriculture has remained static and its manufacturing sector has only recently begun to expand with vigor.

MARKED CONCENTRATION OF COMMERCIAL PRODUCTION

This keynote will have begun to be apparent by inference from the statements regarding varying population densities, isolation, and the relatively minor importance of the landlocked countries in export trade. It will be illustrated more fully in connection with the regional discussions.

LAND USE REGIONS

THE PREDOMINANTLY RAINFOREST SOUTH

Much of Congo, almost all of Gabon and Rio Muni, a narrow strip in southwest CAR, and a large swath across the south of Cameroon are in tropical rainforest (Maps 66–68). The only large savanna areas within this section occur in Congo, where the coastal plain is largely untimbered, as are the Niari Valley, the hilly region around Malebo Pool (Stanley Pool), and the Batéké Plateaus toward the north.

TOPOGRAPHY. This whole area may be fairly easily divided into a number of roughly parallel topographic regions. First is a generally narrow low coastal plain, extending about 40 miles (64 km) inland in Congo, from 18 to 60 miles (29–97 km) inland in Gabon—except in the lower Ogooué where it is about 125 miles (200 km) wide—and from a few miles in the extreme south of Cameroon to about 60 miles (100 km) in width in the Douala area. From Congo to Cape Lopez the shoreline is almost straight; here sand from the Zaïre and other rivers carried by the Benguela Current has enclosed a series of lagoons, which once were bays. North of Cape Lopez there are many capes, bays, and estuaries, but access to the latter is often reduced by bars and shallow water.

Inland from the coastal plain the land rises more or less abruptly to a series of hills, mountains, and plateaus, running parallel to the coast. In southern Congo this is the Mayombé Massif, consisting of a succession of sharp ridges with elevations from 1,000 to 2,600 feet (490–790 m) cut deeply by the gorges of the Kouilou. In Gabon, almost all of which lies in the Ogooué Basin, mountains rise in various parts above the plateau extending over a portion of the south and all of the northern and eastern sections of the country. North of the Mayombé in Congo and extending into Gabon are the Chaillu Mountains, forming Gabon's main watershed and giving it its highest elevation, 5,168 feet (1,575 m). In the north of Gabon the Crystal Massif lies inland from the coastal plain. The plateau of Cameroon has an average elevation of about 1,600 feet (490 m) but is quite hilly over large stretches. Much of this zone is deeply carved by the numerous rivers of the area.

On its interior side, this region of mountains and plateaus forms the western margin of the Zaïre Basin. The terrain along the Zaïre River is quite hilly where it cuts through the system at Malebo Pool. This gives way northward to the monotonous Batéké Plateaus, separated from each other by the deep valleys of the northern tributaries of the Zaïre, while northeast of the Sangha River in Congo is a part of the true Zaïre Basin, where the rivers form a jumble of branches linked with one another and often flowing through a dense forest, which is seasonally inundated.

TRANSPORTATION. This southern region, enjoying the best contacts with the sea, contains, with the exception of the western highlands of Cameroon, the most important actual and potential economic regions, accounting for a very large portion of the exports of the whole area.

Starting in the south, the first port is Pointe Noire, the best-equipped of the ports. Entirely artificial and protected by dikes, Pointe Noire was not started until 1934, and not opened until 1939 (before it was fully completed). The occurrence of a submarine rock spur along an otherwise difficult coastal stretch facilitated jetty and pier construction. The port now has seven berths, mechanical loading facilities for manganese ore from Gabon, a 5,300-foot (1,620-m) pier extending into deep water for the mechanized loading of potash from Congo, petroleum facili-

Wharf for mechanical loading of potash near Pointe Noire, Congo
Pointe Noire is the best equipped port of this region. Most of its cargoes are timber and minerals, including manganese ore from Gabon.

ties, and a small fishing port. Almost all of Congo's forest products are brought by rail and road to Pointe Noire, where the logs are often floated alongside vessels at mooring posts.

Traffic at Pointe Noire has grown from 225,000 tons in 1948 to 793,000 tons in 1961 and 3.4 million tons in 1972. Exports greatly exceed imports because of the heavy tonnage of minerals and logs, a reversal of the position before 1953, when imports were higher than loadings. A major investment program in the early 1970s increased the capacity of the port and its connecting railway by 40 percent.

The 321-mile (517-km) Congo-Ocean Railway (CFCO), a single-track meter-gauge line, was started in 1921 and finished, at an immense cost in lives and money, in 1934. It provides, like the Matadi-Kinshasa line in Zaïre, an outlet from the areas tributary to the Zaïre River, whose extensive inland navigable stretches begin at Malebo Pool. A small amount of produce from eastern Gabon, the vast bulk of the trade of CAR, and a significant portion of Chad's overseas trade employs the former "Federal Route" via the Zaïre and the CFCO, while a 178-mile (286-

km) branch to M'Binda carries manganese from Gabon. From time to time the line has been used to carry traffic for Zaïre, but the desire of that country to handle as much of its traffic as is practicable and recurrent disputes between the two countries since independence have prevented any large-scale movement, even when the Kinshasa-Matadi axis has been congested. The CFCO follows a difficult route through the Mayombé Massif, necessitating 12 tunnels—one about a mile (1.6 km) long—92 bridges, steep gradients, and excessively narrow curves. Shortage of funds accounts for the somewhat inadequate alignment, which required slow operating speeds; aided by international agencies the line is being converted to standard gauge, modernized, and partially realigned.

The city of Brazzaville has grown with great rapidity in postwar years, first as a government capital, second as an industrial center, and third as a river port. Its port has been improved by construction of quays and provision of mechanical gear, but its cargo tonnage is far below its counterpart on the other side of the pool, Kinshasa.

The Zaïre and Ubangi Rivers form the border of Congo and CAR with Zaïre. They are navigable to Bangui, a distance of 740 miles (1190 km) from Brazzaville. A shelf at Zinga, 60 miles (100 km) below Bangui, which inhibited navigability in low flow periods, has been removed to make Bangui accessible all year. There the Ubangi is interrupted by rapids, upstream from which the river is not well suited for regular traffic. Many of the right-bank tributaries of the Zaïre in Congo are navigable for some distances at least seasonally. The Sangha, the most important, is open to smaller boats as far as Ouesso. With forest operations beginning to expand in the regions served, including southwest CAR, traffic should increase on these streams, which have heretofore been used mainly by dugout canoes for local interchange.

The rail-river route from Pointe Noire to Brazzaville to Bangui is, then, one of the main routeways of the whole region. The relatively low levels of river traffic, about 350,000–450,000 tons at Brazzaville and 200,000 tons at Bangui, reflect one of the keynotes mentioned—the relatively low stage of economic development in the areas served. They also suggest, when compared with the figure for Pointe Noire, the significance of Brazzaville as a consuming center and the importance of the seaward areas as producers.

Port Gentil and Libreville are the main ports of Gabon, which has one of the least adequate transport systems in Africa. Port Gentil is a shallow-water lighterage port equipped with three moles, one for general cargo, one used by the large plywood-veneer mill, and one serving the oil companies. An oil terminal at Cape Lopez, linked by pipes to various fields, can handle 80,000-ton tankers, but plans call for providing a loading buoy for supertankers. Except for petroleum, most of the traffic at Port Gentil consists of okoumé logs rafted down the Ogooué system or products from the mill.

Libreville is also a lighter port, situated along the northern shore of the Gabon Estuary, actually a deeply indented bay. Its facilities were very modest until 1954, when a new mole was completed providing berths for six small vessels drawing less than 6.6 feet (2 m). Okoumé roundwood, which accounts for a high proportion of outgoing traffic, is handled at Owendo, 10 miles (16 km) upstream; rafts of 40 to 60 logs are sorted and reassembled into larger rafts called *dromes*, which are pulled by tugs alongside ships lying at anchor. New facilities have just been constructed at Owendo, where a 1,493-foot (455-m) quay can accommodate three 12,500-ton vessels on one side and two smaller ones on the other side. A timber port, two quays, a fishing pier, and a shipyard are to be built, while a pier extending to deepwater will be constructed at Santa

Clara west of Libreville to handle large tonnages of iron ore from northeastern Gabon.

Detailed studies have been made for a 362-mile (582-km) rail line from Santa Clara and Owendo to Bélinga. It will be designed to move as much as 25 million tons a year of iron ore plus logs and possibly pulp. Since the iron ore may not be developed for some years, Gabon has been seeking funds to build the first 210 miles (338 km) of the railway, to Booué, to open up Forest Zone III. Three-fourths of the line's hinterland is forested and an estimated 1.2 million tons of logs would move downline. The first phase, which includes 250 miles (400 km) of forest feeder roads, would require five or six years and an expenditure of about $210 million. Later it is suggested that a branch line would go 200 miles (320 km) to the manganese mine at Moanda to handle ores above the capacity of the present route to Pointe Noire.

Gabon benefits considerably from the presence of navigable waterways, whose total length is about 1,060 miles (1,700 km). The Ogooué and its main affluent, the Ngounié, are particularly important in the movement of logs. The roads of Congo and Gabon have provided only a rudimentary system. Construction and maintenance costs are very high, especially in the rainforest regions. Most of Gabon's roads date from after World War II (there were only about 60 miles open in 1936). But, with much foreign aid, several very substantial road programs have been started in recent years in Congo and Gabon. Because of the difficulties of overland movement, and the large number of small, privately owned craft, Gabon has an unusually large number of towns served by air. Rio Muni's main gateway is Bata, a lighterage port using a small concrete jetty. Some of the country's rivers are used for floating logs. Roads are reasonably good, in part because the villagers along them are responsible for their maintenance.

Cameroon has one major and three minor ports. Kribi is a small lighterage port at the mouth of the Kienké with a 460-foot (140-m) pier used primarily for shipment of logs and of cocoa from the Boulou district. The main gateway is Douala on the estuary of the Wouri River. It is accessible by a channel that requires periodic deepening but that is now capable of taking the largest cargo vessels used in western African trade. Douala has eleven quay posts; one is specially equipped to handle forest products and another is used for unloading petroleum. The port also has a transporter for alumina and fishing facilities. Bananas are loaded at Bonabéri across the estuary, linked by bridge with Douala. Douala-Bonabéri accounts for the vast bulk of Cameroonian imports and exports, its traffic having increased from 937,000 tons in 1961 to 1.9 million tons in 1972. Some transit traffic for CAR and Chad is handled at the port.

Douala is the terminus of two rail lines, both of which are high-cost operations requiring high rates to cover expenses. Largely constructed by the Germans before World War I, the railroads have been modernized and dieselized, and the longer Eastern Line is now being extended. The Northern Line runs 107 miles (172 km) to N'Kongsamba, serving the most important banana and coffee producing areas of Eastern Cameroon. A new 19-mile (30-km)

Part of the port of Douala, Cameroon
The port handles the vast bulk of traffic for Cameroon plus a portion of the overseas trade of Chad and the Central African Republic.

branch to Kumba taps part of the traffic from West Cameroon.

The Central Line runs 191 miles (308 km) to Yaoundé and has a 23-mile (37-km) branch to Mbalmayo. It serves the aluminum smelter at Edéa, the capital at Yaoundé, and the main cocoa producing area of the southern plateaus. In 1962 a decision was made to continue the line from Yaoundé to N'Gaoundéré on the Adamoua Plateau. Constructed under loans and grants from France, the United States, and international agencies, this Transcameroon Railway reached N'Gaoundéré in 1975, 398 miles (641 km) from Yaoundé. Whether the line becomes profitable will depend on how much traffic it draws from CAR and Chad, how much it will stimulate development on the plateau, and whether a bauxite deposit is opened up in the Martap region, the only known resource whose development could yield really large-scale tonnage traffic. There has been discussion over several score of years regarding the desirability of tying the Cameroon line to Bangui in CAR and to Moundou or Sarh (Fort Archambault) in Chad. Extension of the line has renewed interest in such connections, but France has decided that financing the line to Bangui would not be justified at present, a decision that was accepted with poor grace by President Bokassa. It is more likely that the railway will be extended to Moundou, but no definitive steps have thus far been taken.

The roads of southern Cameroon are sparsely distributed and often in poor condition. For decades the main road from Douala to Yaoundé was poorly maintained in order to reduce competition with the railroad. The importance of roads in permitting rural areas to commercialize a part of their production has been better appreciated in more recent years, and Cameroon has undertaken a major road building program, doubling the mileage in a decade. Cameroon should gain as much or more than other countries from completion of the Trans-African Highway in 1978; this 4,400-mile (7,080-km), $26 million, route is being financed by a number of foreign government and international agencies as well as the countries to be served. Starting at Mombasa in Kenya, it will traverse Uganda, Zaïre, CAR, and Cameroon, and end at Lagos in Nigeria. The main road will extend 819 miles (1,318 km) in CAR and 665 miles (1,070 km) in Cameroon, while feeder roads may be constructed from Chad, Gabon, and Congo.

West Cameroon has two very inadequate ports, Tiko and Victoria-Bota. Recent construction of a direct road to Douala has linked the region to the hinterland of that port, though logs and plantation products continue to move via the local harbors, and new deepwater facilities are planned for the area. The roads of West Cameroon are reasonably well developed in the south though large trucks hauling logs have often destroyed the surface. Some 91 miles (147 km) of narrow gauge lines serve the coastal plantations.

THE SUBSISTENCE ECONOMY. Although the rainforest areas of the middle African region have the greatest commercial development of any parts, and although a considerable number of people have been attracted to the forest industries and to the urban centers, a high percentage of the population is still engaged primarily in subsistence production. The small Negrillo population follows a semi-gather economy, but the more widespread agricultural system is the traditional bush fallow which, in some of the more remote areas, is probably as little changed as anywhere in Africa. Production from the laboriously cleared forest plots is supplemented, where possible, by products gathered directly from the forests, by fishing, and by hunting. Some areas have the advantage of having a forest-savanna mosaic, in which case tree crops are likely to be grown in the forests and field crops to be grown in the grasslands.

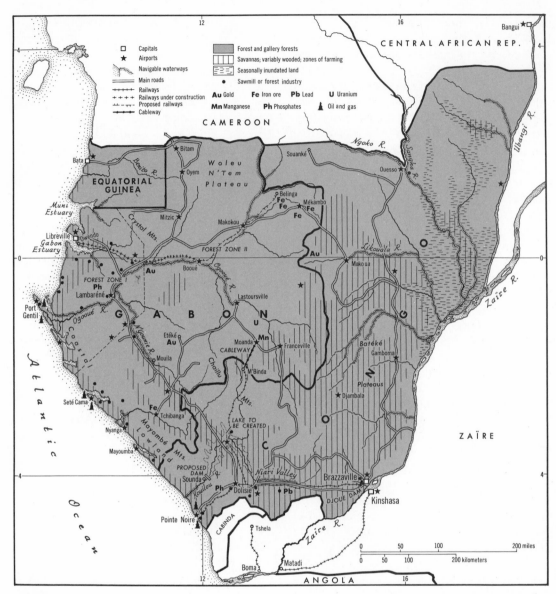

Map 66. Economic map of Gabon and Congo

FORESTRY. High forests cover half of Congo and about 85 percent of Gabon and Rio Muni. They provide the major exports of Congo (c. 56 percent in 1971) and the main nonmineral exports of Gabon; in the latter country they accounted for 61 percent of total exports in 1961 but their share dropped to about 23 percent in 1971, not because of reduced production but because of the rising importance of minerals, which is continuing. About 5 percent of CAR is in rainforest and, despite the handicap of remoteness and limited capacities of the CFCO, exports increased from 22,000 cubic meters in 1967 to 157,000 cubic meters in 1972. Forests cover about a third of Cameroon; while the importance of its forest industries has been more modest than in

Gabon and Congo, production and exports have grown rapidly in recent years and there are now 60 enterprises and 53 sawmills in the country.

In general, exploitation of the rainforests in this region has the same limitations and potentialities as were detailed for West Africa. However, the area does have certain distinguishing characteristics.

First, the forest area is greater and less degenerated. This is explained in part by a much lower density of population in the forest zone. Nonetheless, there has been such intensive exploitation in the so-called Forest Zone I in Gabon, particularly of the favored okoumé, that operations have had to shift to Zone II, which is further from the coast, in more rugged terrain, and not accessible by waterways. The coastal areas are now increasingly being reserved for small Gabonese operators who account for about 30 percent of total output. The expense of constructing roads and the much greater investment in heavy mechanical equipment required in Zone II, and soon in Zone III, make it difficult for locals to operate in these areas.

Second, some species occur in somewhat greater proportions than is typical of most tropical forests. Particularly favorable is the extent of okoumé, which is not found in

Plywood factory at Port Gentil, Gabon
While much of the timber exported from the region goes in the form of logs, there is increased processing and this plant is one of the world's largest plywood factories.

other tropical forests and which is particularly suitable for peeling to produce plywood. Gabon has most of the okoumé trees, and operators there have tended to exploit it excessively. Gabon has instituted regulations regarding the size of trees and the number per unit area which may be cut and all timbers are bought and sold by a government agency. It is also planting okoumé in solid stands for future use and it is estimated that a given area of plantations can produce about 50 times as much as the natural forest. Some 56,800 acres (c. 23,000 ha) had been planted by 1972 and they will reach exploitable age in about 50 to 60 years.

Third, some areas of this region are particularly favored by the existence of an intricate system of waterways, particularly the northern embayments and the Ogooué system in Gabon. And fourth, the labor supply has, at least in the past, been more of a limiting factor than in West Africa.

Other special features of forestry in this area include the existence of one of the world's largest plywood plants at Port Gentil and a pulp mill near Pointe Noire, which produces cellulose. The mill is small but it hopefully represents the beginnings of much larger scale operations, which would also be based on low-cost power. Gabon, for example, hopes to establish a 100,000-ton ca-

Tractor used for constructing rough forest roads and hauling logs, Gabon
Gabon has been favored in transport of logs by many navigable waterways, but overland movement becomes more important as exploitation moves farther inland.

pacity mill at Kango using power from a hydroelectric installation. Other somewhat specialized plants include a mill in south-central Cameroon producing knocked-down wooden fruit and vegetable crates for sale in France, a large mill at Eséka in Cameroon mass-producing pre-fabricated houses, and a mill at Foulenzem in Gabon cutting and impregnating rail ties for the French Railways and prefabricated items for buildings. Despite the existence of these and other plants, most of the forest products of the region are exported undressed or only semi-finished.

COMMERCIAL CROP PRODUCTION. Small tonnages of various tropical vegetable products are commercialized in the rainforest region under consideration, but only cocoa, palm products, coffee, and bananas are of any great importance.

The major producing area for cocoa is the region around Yaoundé in southern Cameroon, though plantings are found in a much wider belt across the south. An estimated 220,000 to 250,000 planters produce about 125,000 tons yearly on small holdings, often intermixed with bananas and cassava. Yields are low and quality often unsatisfactory, but the distribution of selected seedlings, premium payments for higher quality beans, and programs to reduce losses to black pod disease and capsid bugs have raised quality and yields in recent years. Returns to producers have been set by stabilization funds since 1935, but they are well below the f.o.b. Douala price because of taxes, transport costs, and middleman fees. Cocoa was long the prime export of Cameroon, which ranks fifth among world producers, but coffee now frequently exceeds cocoa by value of exports.

Gabon produces about 6,000 tons of cocoa yearly from the Woleu-N'Tem and Ogooué-Ivindo regions in the north. Production, which began under a compulsory planting program, gradually caught on, and now these regions are about the only areas of

Cocoa growing in a partially cleared rainforest area near Yaoundé, Cameroon
Cameroon ranks fifth among world producers of cocoa.

Gabon with a stable and relatively prosperous indigenous agriculture. A very small output of cocoa comes from northern Congo.

Robusta coffee comes from the same general areas as cocoa in each country, but most of Cameroonian coffee is produced outside the region under discussion. Banana production is concentrated mainly on plantations and smallholdings in a narrow band along the Northern Line of the Cameroon railway. Production has been hurt by inadequate demand and incidence of Panama disease, which started in 1956. Palm products account for a small percent of exports from Cameroon and Gabon. Cameroon hopes, under "Plan Palmier," to increase the area under oil palm to 100,000 acres (40,470 ha) by 1977, mostly in the west. Gabon has one

state-owned and one private palm plantation, together covering an area of about
3,000 acres (1,200 ha), on the banks of the
Ogooué above Lambaréné. Palm products
rank first among agricultural exports from
Congo, but account for only about 7 percent
of the total value of exports.

The Niari Valley in Congo, inland from
Pointe Noire, has a special position in that
country because of a favorable climatic
regime, which permits two planting seasons
with intermediate dry seasons for harvesting,
a substantial flattish area with good soils,
ready access to the CFCO, and a better than
average supply of labor. Numerous crops
are grown in the valley, some on large,
mechanized farms. Most important is cane
sugar, whose output increased from 15,000
tons in 1960, when about 6,000 acres (2,400
ha) were cut, to 95,000 tons in 1969, but it
fell in succeeding years to 75,000 tons.

There is a small commercial production of
numerous other agricultural products, which
may be briefly summarized as follows:(1)
about 4,000 tons of rubber come from a
plantation at Dizangué on the banks of the
Sanaga near Douala; plans call for renewing
about 15,000 acres (6000 ha) by 1990 at
which time production would be about
16,000 tons; (2) tobacco is produced in the
Djambala and Batéké Plateau regions in
Congo; (3) livestock, including several thousand head of the dwarf N'Dama cattle, are
raised on ranches in the Niari Valley and in
Gabon; and (4) peanuts are exported from
the Niari Valley in small tonnages.

Gabon and Congo have such poorly developed agricultural bases that they are quite
heavily dependent on food imports. Production for the domestic market is restrained by
the difficulty of farming in rainforest areas,
the very small total areas in food crops (only
235,000 acres [95,000 ha] in Gabon), by the
shortage of labor and dispersion of the rural
population, by the poor transport facilities,
by the lack of tradition of agricultural markets, and by the adult males' aversion to

farm work. Efforts to stabilize rural communities and to expand production of food
crops have frequently been frustrated.
Congo has promoted "mutuels," incipient
cooperatives, with only moderate success.
Gabon has pushed the introduction of irrigated rice in three centers with the aid of
Formosan experts and has also tried to stimulate market garden operations in several
places. Given the importance of nonagricultural activities and the relatively high urban
populations, particularly in Congo, there are
obviously substantial opportunities to produce cash crops for the domestic markets,
but the response to these demands has thus
far been inadequate.

FISHING. Relatively little is known about
the fish population of the offshore waters
and commercial landings are quite small at
the modern facilities that have fairly recently
been installed at Pointe Noire, Port Gentil,
Owendo, and Douala. Inland fisheries have
long been locally important and are particularly intensive on the Zaïre near Brazzaville;
Congo has a modern smoked-fish factory on
the Zaïre at Mossaka.

EQUATORIAL GUINEA

As both the mainland and islandic portions
of Equatorial Guinea and the Portuguese islands have tropical rainy climates, it is appropriate to examine them briefly at this
point.

Rio Muni, the mainland portion of Equatorial Guinea, long remained largely untouched because of the more attractive character of Ile Macias Nguema (Fernando Po).
It did, however, witness a considerable investment in postwar years. Its main interest
is focused on timber products and coffee.
Forest operations are concentrated near the
coast, to which the Benito River and Muni
Estuary are used to float logs; production
was running around 300,000 tons before independence, but dropped to a third of that
level after the flight of foreigners in
1968–69. Exports of coffee also dropped

from about 6,500 to 4,800 tons after independence; the crop was produced on smallholdings and Spanish-owned estates, mainly in Rio Muni. Palm products and bananas are also grown on large plantations. Per capita incomes of the Rio Muni population, which is predominantly part of the Fang ethnic group, the largest single tribe in Gabon, are still only about a seventh of those on Ile Macias Nguema.

That island, about 44 miles (70 km) long and 22 miles (35 km) wide and with an area of 785 square miles (2,034 sq km), is the more important part of Equatorial Guinea. The smaller islands, some of which were infamous slaving depots, are now of little economic importance. The islands, of volcanic origin, are related to the Cameroon Mountain system. On Macias Nguema a narrow coastal plain rises gradually behind steep shores to mountain slopes reaching the 9,350-foot (2,850-m) level.

Ile Macias Nguema had a population of about 85,200 in 1974. About a quarter are descended from the original Bantu-speaking Bubi, 72 percent are Nigerians who migrate to the island to work on the estates, and the remainder are "Fernandinos," descendants of English-speaking Creoles who arrived during British hegemony in the nineteenth century and who form a kind of African bourgeoisie. The Spanish population, which was about 6,000 at independence in 1968, is now below 2,000. In 1973, Nigeria suspended the recruitment of laborers for Macias Nguema pending an investigation into the reported mistreatment of workers on the island.

Cocoa is the main crop of Macias Nguema, which was one of the first areas in Africa to become interested in its production. Despite its small size, the island accounts for about 2.9 percent of the world's output. Most of the cocoa comes from large farms (fincos) ranging from 100 to 2,000 acres (40–800 ha) situated all along the western, northern, and eastern coasts up to 2,000 feet (600 m) eleva-

tion. Quality of the beans is high and the island has remained free from swollen shoot disease.

Ile Macias Nguema also produces about a third of the coffee of Equatorial Guinea: it is grown up to about 5,000 feet (1,500 m) elevation and almost entirely on estates. Beginning at about 3,000 feet (1,000 m), but largely between 4,000 and 5,250 feet (1,200–1,600 m), is a region of pasturelands, frequently enveloped in mist, which have for the most part been carved from the forests. Their output permits Macias Nguema to be self-sufficient in dairy products, but the country must still import meat.

On the northern tip of the island lies Malabo (Santa Isabel), the capital; the city is the commercial and financial center, and main port. It has docking facilities for two freighters, which is considered adequate for the foreseeable future. The island is well served by hard-surfaced roads and there is an airport at Malabo. Most of Equatorial Guinean trade has long been with Spain, which paid above world prices for its produce. The per capita GNP, estimated at $210 in 1971, is above average for tropical Africa.

SÃO TOMÉ AND PRÍNCIPE

Situated about 180 miles (290 km) off the coast, these islands were discovered by the Portuguese in 1470. They are expected to become independent in 1975. Both islands are volcanic, which explains both the richness of their soils and the high percentage of lands in steep slopes. Precipitation varies from 100 inches (2,500 mm) to 12 inches (300 mm) depending on aspect of slope and elevation. The lower and most intensively used sections, which are on the northeastern parts of the islands, have densities reaching about 780 per square mile (300 per sq km). Cocoa was introduced in 1822 and the two islands ranked as the second world exporter in 1907; annual production is now about a third that of Equatorial Guinea. Coffee is the second main crop, and there is a small output of coconuts, palm oil, cinchona, vanilla, rubber, cinnamon, and kola nuts.

Despite a population density of 187 per square mile (72 per sq km) on the islands, there is a shortage of farm workers, and about 25,000 contract laborers are recruited from other Portuguese African holdings to work on the cocoa and coffee plantations.

THE REGIONS OF SOUTHWESTERN CAMEROON

The southwestern regions of Cameroon are distinguished by their elevation, their volcanic heritage, and by population densities considerably higher than in most of the middle African area. Starting with Mt. Cameroon (13,370 feet, 4,075 m) there is a whole line of volcanoes running northeastward, sometimes rising from and bounded by high crystalline and volcanic plateaus. Many of the volcanic soils are superior in quality, while the area is ecologically suitable for a great variety of tropical crops, including those of higher value usually associated with tropical highlands. Development in the former French and former British parts of the area differs greatly, making it desirable to divide the area into two regions (Map 67).

FORMER FRENCH SECTION. Comprising less than 5 percent of the area of East Cameroon, this region contains close to 22 percent of its total population. The major ethnic group is the Bamiléké, who occupy a series of plateaus with an average elevation of 4,000 to 5,000 feet (1,220–1,520 m). About 690,000 members of this group live in an area of 2,390 square miles (6,200 sq km), giving an average density of 288 per square miles (111 per sq km), while some parts have densities up to 900 per square mile (350 per sq km). It is no wonder that there is heavy out-migration from the region, there being many Bamiléké in Douala and in the Mungo area. North of the Bamiléké country is the Bamoun Region, which represents a prolongation of the plateaus to the north. It

is less humid and less densely populated and is occupied by the Bamoun, an Islamized tribe of Sudanic origin.

The Bamiléké have organized the totality of the surface they occupy, with each topographic zone receiving a particular use: the convex middle and upper slopes of the many rounded hills are in pasture; the lower concave slopes, which have the better soils, are the farming area par excellence; and the often marshy valley bottoms, which were formerly in gallery forests, are now planted to raffia palms, which provide material for buildings and enclosures, wine, and textiles. This group has remolded the environment more than most African tribes. They practice agriculture and grazing of sheep and goats in association; elaborate systems of pathways within live hedges doubled by raffia palm mattings separate each exploitation, permit the livestock to go to pasture

A market in Bamiléké country, Cameroon
This intensively cultivated area has exceptionally high population densities. Palm oil and palm wine are frequently sold in gourds such as those shown.

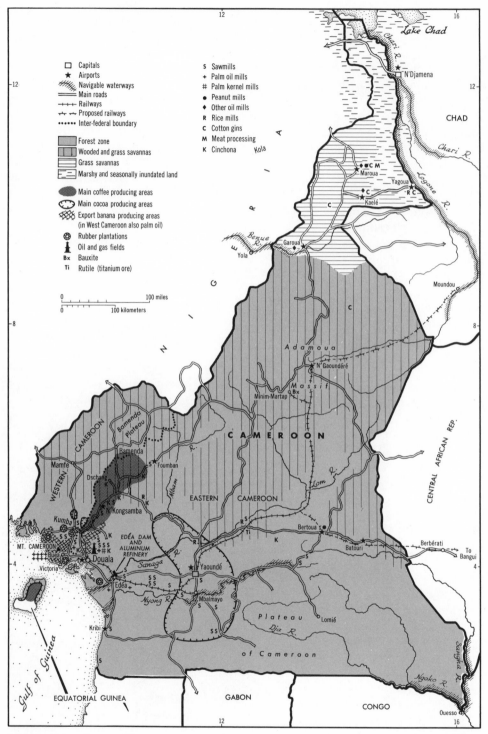

Map 67. *Economic map of Cameroon*

Robusta coffee growing under bananas in the Bamiléké country near N'Kongsamba
Coffee is the main cash crop of the area but its extension has contributed to degradation of the previously well adapted system of agriculture.

unobserved, serve as anti-erosion barriers, and provide firewood. Unfortunately, what was a well adapted system has become progressively degraded by two major forces—increasing population pressure and the allocation of too much land and attention to coffee—both of which have reduced the pasture area and the care given to the livestock.

Commercial products from the region include coffee, palm products, cinchona, kola nuts, bananas, and tea. The area around N'Kongsamba is excellent for coffee, both from the standpoint of climate and because of the presence of some very fertile soils. Robusta and arabica coffees are produced, and the region accounts for about 75 percent of Cameroon's coffee.

WEST CAMEROON. That section of the British Cameroons which is now West Cameroon has an area of about 16,791 square miles (43,489 sq km). Its topography is very accidented. North of Mt. Cameroon is a lower region extending to Kumba and corresponding to the valley of the Mungo across the border; the land then rises to the high grasslands of the Bamenda Plateau, while the frontier with East Cameroon, starting

about 60 miles (100 km) north of Douala, is the chain of extinct volcanoes already noted.

The population of West Cameroon is about 1.3 million; it contains a complex of tribes speaking over 100 vernaculars, and the demographic scene is further complicated by several migratory patterns including Fulani from northern Nigeria moving to the high plateaus, plantation workers coming from the densely populated parts of eastern Nigeria, and movements of indigenous peoples to the more developed south. While some sections are densely populated, the overall density is 77 per square mile (30 per sq km) and, in contrast to the Bamiléké country, there are large areas of unused land of good quality.

The backbone of the West Cameroonian economy has been the plantations of the Cameroons Development Corporation (CDC). Plantations were introduced to the area before World War I by the Germans. The lands were taken over by the British during that war, and were sold back to their previous owners by 1924. Again taken over in World War II, they were put thereafter under a statutory corporation, the CDC, which received aid from and was managed by the Colonial Development Corporation. After independence the CDC was further

A young palm plantation in West Cameroon
Plantations of the Cameroons Development Corporation produce large tonnages of palm products, bananas, and rubber.

Workers' housing on one of the Cameroons Development Corporation plantations
This state-owned corporation provides houses, schools, and medical facilities for its approximately 15,000 employees.

aided by the FED and the World Bank and now has over a score of plantations covering about 81,500 acres (33,000 ha), with the numbers employed (c. 15,000) ranking second only to those working directly for the Cameroon government. The CDC provides roads, houses, schools, and medical facilities in its zone of operation and remits profits and taxes to the government, though it appears that a greater share of profits should have been retained for investment.

The CDC produces large tonnages of bananas, rubber, and palm products and lesser quantities of tea, pepper, and cocoa. Bananas, which were also produced on private plantations and by smallholders working under the Bakweni Cooperative Union, were the leading export from West Cameroon, but output was seriously affected during the 1960s by several disease attacks, losses to high winds, and marketing difficulties; production in recent years has been less than half the 1962 level. There has, therefore, been a switch to other crops, particularly to the oil palm and to rubber, which is also produced on private plantations and by an increasing number of outgrowers. Tea plantings have been made on the slopes of Mt. Cameroon and in the Bamenda area, with production reaching 646 tons in 1972. Experience with coffee and cocoa has been less successful. In toto, the CDC has made a sig-

nificant contribution to the economy of West Cameroon, but more recent management has been criticized as wasteful and a large retrenchment was necessary in 1972 to improve the financial position of the corporation and permit repayment of the loans.

A very substantial increase in timber output has taken place in the last two decades. Several thousand people are now employed in logging, which has a good future so long as access roads are made available. It is questionable whether forest concessionaires are now paying an adequate share of road building and maintenance costs in view of the heavy damage caused by log haulers.

The most intensive developments in West Cameroon are found within 50 miles (80 km) of the coast. The Bamenda Plateau, which could be an important surplus producer of corn, peanuts, and cattle, is inhibited by poor roads, as was the area between Kumba and Mamfe until connection was improved in recent years. The area is well suited to cattle-keeping, but the local peoples have shown little interest in it and even express repugnance toward it. Greatly improved connections with East Cameroon after about 1969 have opened up large new markets for garden crops, markets that were never available in Nigeria for this remote region.

THE ADAMOUA MASSIF

This plateau, a granitic massif with basaltic outflows in the highest part near N'Gaoundéré, ranges in elevation from 2,500 to 4,500 feet (760–1,370 m) and extends more or less east-west across Cameroon and into the Central African Republic, presenting a formidable barrier between the north and south of the country. It is a savanna region with rather heavy precipitation owing to its elevation, but with a more clearly marked dry season than in the south.

Population densities on the Adamoua Massif are low, only about 8 to 13 per square mile (3–5 per sq km). It is like northern Nigeria in having two distinct population groups: Mboums, agriculturalists of Sudanic origin related to the Hausas, and Peuhs and Bororos, related to the Fulani pastoralists. The main richness of the massif is its 600,000 animals, which are only poorly commercialized, though some are moved south for sale in the protein-deficient rainforest areas. A local company, La Pastorelle, runs a 45,000-acre (18,200-ha) ranch and has an abattoir at N'Gaoundéré, from which refrigerated carcasses are flown to Douala, Yaoundé, and Brazzaville. There are also a number of private ranches, only a few of which are well managed. Unlike most livestock areas of Africa, the plateau is apparently not overstocked. It is also favored, more so than the north, in its freedom from rinderpest and foot-and-mouth disease, probably owing to its elevation. It is also tsetse free, but there is a high incidence of parasitic infestation. Efforts have been made to upgrade the stock with French Montbeliard and Texas Brahman cattle, to introduce better pasture plants, and to control grass firing. With continued improvement in practices and particularly in pastures, and with better transport facilities, this region could become a favored area.

THE LOWER SAVANNAS AND STEPPES

Included within this region are all of the Central African Republic (with the exceptions of the small tropical rainforest zone in the southwest, the extension of the Adamoua Massif in the western Yadé Massif, and the Bongo and Dar Challa chains in the northeast which prolong the Darfur Massif of Sudan); the southern half of Chad; and part of Cameroon north of the Adamoua Plateau (Map 68).

THE CENTRAL AFRICAN REPUBLIC. The dominant topographic feature of the CAR is a gently undulating plateau, with elevations ranging from 2,000 to 3,300 feet (600–1,000 m), which is a sill between the two great river systems of the Chad and Zaïre basins. Vegetation varies from gallery forests along the southern rivers to savanna woodlands and grasslands, which gradually decline in height and quality toward the drier north. Larger than France and the Low Countries combined, its population was only about 2.1 million in 1974. The rate of population growth is estimated to be 2.2–2.5 percent per annum, and 42 percent of the people are under 15 years of age. The GNP per capita in 1971 was thought to be about $150 and it increased by only 0.4 percent per year from 1960 to 1971. There is an enormous disparity in incomes between government employees and peasants, which contributes to a generally poor contact between the administration and the people. The country is heavily dependent on foreign aid, including subventions from France. Exports, much of which have required direct or indirect support, have been about $37 million yearly, while imports have been about $44 million.

The CAR is essentially an agricultural country, though diamonds rank as the main export. Cotton and coffee are the major commercial crops, accounting for about 45 percent of total exports. Cotton production began about 1926 on a compulsory basis, which, together with the laborious work involved and the low returns, made it a chronically unpopular crop. Cotton production was organized here and in Chad by four concessionaires who employed African supervisors, or "boys coton," in local areas to enforce planting requirements. Africans who failed to obey instructions were fined and on occasion physically punished, a procedure that contributed to a sizable migration from the countryside. The companies were charged with collecting, processing, transporting, and exporting cotton from their respective zones; purchasing prices and profit percentages were set each year by the government. No profits were, in fact, earned until 1936, when over 100,000 planters produced about 16,000 tons.

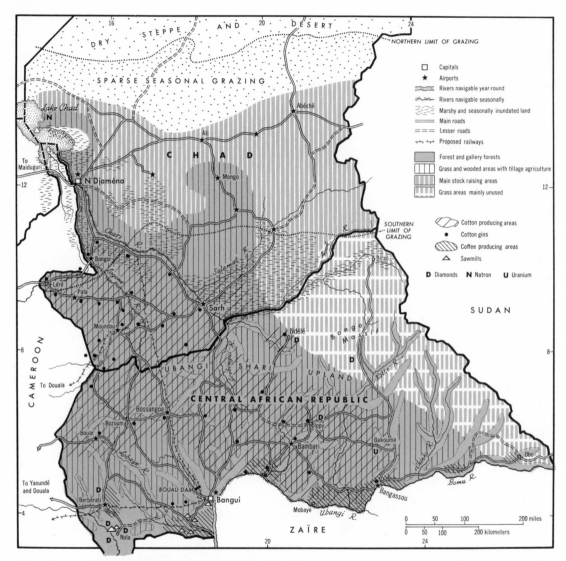

Map 68. Economic map of Chad and the Central African Republic

Despite the objectionable features associated with cotton production, it gradually caught on and its output was sustained and increased after abolition of forced labor in 1946, though production declined as independence approached and in the immediate years thereafter when a near anarchic situation prevailed. Today cotton is indispensable to the economies of CAR and Chad, whose governments have attempted to promote a continued increase in production.

The crop illustrates well the precarious nature of the economies of these two countries. On the one hand, technical progress has trebled yields and improved quality, and at least 1.5 million people in the CAR and Chad are wholly or partly dependent on cotton for whatever cash income they enjoy. On the other hand, returns to the farmers are low and have scarcely increased since independence, the cost of transportation to world markets keeps the export price low, and the loss of direct price support from France and a gradual adjustment to world prices in accordance with the trading regulations of the EEC permitted only a very small increase in the producer price, one that did not nearly match the rate of inflation. The production of cotton (and also coffee) declined significantly after a decision was made in 1970 to dispense with foreign technical assistants even though the decision was later annulled.

Cotton is now cultivated by about 300,000 planters, or about half of the total number of farmers in the CAR. Production has ranged between 45,000 and 60,000 tons in recent years. The producing zone of this country and Chad is indicated on Map 68, which also shows the location of cotton ginneries. Some cotton is absorbed by local textile mills in both countries.

Coffee production was begun about 1925 on European plantations and by 1938 about 1,500 tons were produced on 11,140 acres (4,510 ha) of European estates and 27,200 acres (11,000 ha) of smallholdings. A serious blight then largely eliminated African production, but a second boom occurred in postwar years, with robusta replacing the former excelsa coffees. Output totaled about 10,500–13,900 tons in the late 1960s, but declined to 6,500 tons in 1972; it comes from some 34,600 acres (14,000 ha) of plantations and 81,500 acres (33,000 ha) of family holdings in scattered areas along the southern margins of the country.

A small amount of peanuts, rubber, and cocoa are also exported. The livestock population is inadequate to meet local demands, and the deficit is normally filled from Chad. Some N'Dama have been imported to try to build up a cattle herd in this tsetse-ridden country.

CHAD. Southern Chad is a vast peneplain cut by the valleys of the Chari and Logone Rivers, with a low sedimentary basin around Lake Chad. Around the lake and between the two rivers enormous swamplands modify the scene and are inundated for about half the year. Several massifs rise above the plain. The richest and most densely populated part of the country is the southern tenth, which has about 47 percent of the total population and contains the most important agricultural zone. North of this is a broad sudano-sahelian belt occupying about three-tenths of the country and containing about 47.5 percent of the population. Stock keeping is the main activity, but millet, cotton, and peanuts are grown along the wadis. Sedentarists and semi-nomads along the south give way to grand nomadism on the vast denuded plateaus of the sahelian portion.

The modest level of the Chadian economy is revealed by the low per capita GNP (c. $80 in 1971), which grew at a rate of only 0.6 percent per annum from 1960 to 1971; a government budget totaling only about $ 98 million; very heavy dependence on foreign aid, particularly from France and more recently from Libya; and a heavy imbalance in external trade, with exports running between $29 and $39 million and imports ranging from $60 to $82 million in recent years. The country was severely hurt from 1970 to 1974 by recurrent drought.

Chad has an almost exclusively agricultural economy, with a high percentage still concerned with local subsistence. This area and northern Cameroon are like the Adamoua Plateau or northern Nigeria in their strongly Islamic populace, their ethnic groups ranging from purely cultivators to purely pastoralists, and their scattered pagan elements. Because of its remoteness and the very late date at which any real concern was given to its development, Chad remains much less changed by modern forces than most African countries.

Cotton is even more important to Chad than to the CAR, for it accounts for about 65 to 70 percent of Chadian exports and one-third of its budget resources. The crop was introduced in 1929 on the same basis as in the CAR. Some 740,000 acres (300,000 ha), farmed by about 450,000 families, are now planted to cotton, which is usually grown in rotation with peanuts. New varieties and the use of pesticides have brought about significant improvements in yields—which fluctuate rather widely, however, in accordance with weather conditions. Some cotton is grown under irrigation in the Chari-Logone region—where, with more effective control, a rather intensive agricultural region could be developed—but most is grown in the southwest on a basis entirely comparable to that of the CAR. A record production of 150,000 tons of seed cotton was reached in 1968–69, but aridity reduced this to 95,000 tons in 1970–71. Despite the difficulties and low returns associated with this crop, the government, aided by foreign technical and funding agencies, plans to increase the area devoted to cotton by 20 to 33.3 percent.

Livestock products rank second among exports from Chad, which had an estimated 4.6 million cattle, 4.2 million sheep and goats, and 370,000 camels in 1972; heavy losses from drought and starvation have occurred in recent years. Over 340,000 people are engaged in stock raising. They are faced with all the problems noted in the discussion of livestock in West Africa, plus the additional deterrent of an unusually remote location.

Postwar efforts to improve the livestock economy have included an intensive and generally successful antidisease program, cross-breeding with imported stock, digging of numerous wells, provision of modern processing and refrigeration facilities, and promotion of air transport to equatorial markets. About 150,000 head of cattle, including many on a clandestine basis, are moved out on hoof to Nigeria and the CAR in a normal year, while about 12,000 tons of carcass and dressed meat are also exported.

A small amount of peanuts and of gum arabic are the only other crops exported, but distance handicaps the ability to compete with other producing areas. Efforts have

been made in recent years to increase production of rice for local markets by polderizing fields along the margin of Lake Chad; about 148,000 acres (60,000 ha) are considered to be available, of which some 14,800 to 19,800 acres (6,000–8,000 ha) are now producing about 6,900 tons of rice, most of which is locally consumed. About 30,000 families produce rice in the Logone area, but primitive methods and inadequate control result in poor yields. An improved dike at Bongor provides flood control to about 145,000 acres (58,700 ha), part of which is used for production of cotton and rice and part for millet. Eventually, lengthy dikes might bring approximately 2.6 million acres along the Logone (1.05 million ha) under control.

Fish are a resource of some value in Chad and adjacent parts of northern Cameroon, the rivers and lakes having an abundant population. In Chad, about 5,000 regular plus many seasonal fishermen take an estimated 110,000 tons from the rivers and Lake Chad, of which about 16,000 tons are exported.

NORTHERN CAMEROON. The northern part of Cameroon consists of the Benue Valley, a gentle rise between that and the basin of the Logone River and Lake Chad, and the Massif of Mandara along the Nigerian frontier. It contains about 27 percent of Cameroon's total population.

The lower parts of the area are quite comparable to adjacent areas of Chad and CAR in ethnic, physical, and economic conditions. Cotton, introduced in the early 1950s, has caught on well and is now produced by about 100,000 families; output was about 72,000 tons in 1971 but only 43,000 tons in 1972 because of unfavorable weather. Peanut production has also increased, with the commercialized share fluctuating widely; IDA is assisting in a 7,400-acre (3,000-ha) rice irrigation project, and there is a fairly sizable shipment of animal products and fish; but the potentially valuable livestock population is only meagerly commercialized.

The granitic Mandara Mountains extend from the Benue northeastward to the Mora Plain. South of Mokolo, plateaus dominate; they are inhabited by cultivators who sought refuge from the Foulbé in the mountains but who descended to the plateaus when peace was secured during the colonial period. North of Mokolo mountains dominate and one finds a melange of pure montagnards of which the main group is the Magoumaz. Living at densities ranging from 180 to 635 per square mile (70–245 per sq km), they practice a very elaborate crop/livestock agriculture involving the construction of spectacular terraces and small bunds in the individual fields. They also carry soil to replenish what has washed downslope, and they engage in permanent cultivation, careful weeding, fertilizing and composting, the stabling of cows, and the keeping of sheep, goats, and chickens. Millet is the main crop and peanuts the cash crop; sorghum is also grown, as are garden crops including tobacco and condiments, while rice is grown in the valleys. This unusually intensive agriculture can probably not be further monetarized without threatening the subsistence output, and there is no room for extension in the area. There is some protection against the rising population, now believed to be growing by about 2.5 percent per annum, because the Magoumaz have evolved a kind of individual tenure and do not permit subdivision of inherited holdings below about 5 acres (2 ha); nonrecipients are forced to migrate from the area.

TRANSPORT IN THE NORTH. The element of isolation is such a marked keynote of this entire region that it is deserving of reemphasis. There are four major outlets available to the region, none of them particularly favorable or adequately developed at the present time:

(1) The old "Federal Route," involving road transport to Bangui, a distance of 740 miles (1,190 km) from N'Dja-

A cotton market in northern Cameroon
Cotton is the main export crop of the Central African Republic, Chad, and northern Cameroon, but remoteness makes the returns to producers very low.

ména, the same distance by river to Brazzaville, and 317 miles (510 km) on the Congo–Ocean Railway. While about two-thirds of Chad's overseas traffic and the vast bulk of CAR's shipments move over this route, it is obviously inadequate from the standpoints of distance and the multiple transshipments required.

The river port of Bangui, Central African Republic
Situated at the head of navigation on the Ubangi tributary of the Zaïre River, Bangui is a major transit point for C.A.R. and Chad traffic.

(2) West and south to Douala via road to the Transcameroon railway or entirely by road. Douala is the closest port to most of the interior north, but it is poorly connected by roads. From N'Djaména the distance by road is 973 miles (1,566 km) or by road and rail about 1,175 miles (1,891 km). Its share of Chadian traffic has been very small but may increase, especially if the railway is taken to Moundou.

(3) The western route by road to Garoua and then by river to Nigerian ports. From Garoua to Burutu is a distance of 930 miles (1,496 km). This is the cheapest route for much of the area but it has severe disadvantages: the upper Benue is navigable only 7 to 10 weeks a year, thus requiring expensive storage facilities; it has sometimes proved difficult for the Niger River fleet to provide adequate services to Garoua and too expensive to provide Cameroonian barges, which would only be used on a seasonal basis; the route is an extremely slow one, requiring as much as 11 months for shipments of cotton from Chad to be delivered to French ports. Despite its shortcomings this route handles about a seventh of Chad's overseas trade plus some shipments for CAR and Cameroon. In the future, control of the Benue might permit its use for perhaps 18 or 19 weeks.

(4) The western route via Maiduguri in northeastern Nigeria. About a quarter of Chad's imports and a tenth of exports move via this route.

Other routes are road to the western railhead in Sudan, used for a very small tonnage of Chad's traffic; trans-Saharan roads, of no measurable value at the present time; and air freighting which, although it does move some meat, is obviously no answer to the problem. Studies suggest that the best of the several alternatives would be by road to Moundou when and if the railhead from Douala reaches that point.

The internal transport net of the northern regions is only reasonably satisfactory, though the long dry season permits use of readily constructed routes. The CAR has the best network, but its claim of having "seven excellent main roads" is grossly exaggerated. Roads in the bulk of the area are poorly maintained, bridges are hazardous, and some ferry service is rather rudimentary. Waterways are of local and seasonal significance. The Chari may be used from August to December from N'Djaména to Sarh. The Logone is too winding, though it is navigable for small craft to Bongor and Moundou for a few weeks. The rivers of CAR are not particularly valuable except, of course, for the Ubangi downstream from Bangui.

THE DESERT OF NORTHERN CHAD

About three-fifths of Chad is desertic, with very meager or no vegetation, except at the base of the Ennedi and Tibesti Massifs. The approximately 50,000 people, mostly nomads, who inhabit the area, make a very minute contribution to the economy of the

The Benue River near Garoua, Cameroon
Open to navigation only 7 to 10 weeks a year, the Benue nonetheless handles about a seventh of Chad's overseas traffic plus some shipments for C.A.R. and Cameroon.

country. The Tibesti Mountains reach 11,204 feet (3,415 m) in Emi Koussi and have sufficient orographic precipitation to support a small permanent populace, a mere remnant of what the area apparently supported several millennia ago.

MINING IN MIDDLE AFRICA

The countries treated in this chapter were latecomers in the production of minerals. But one of them, Gabon, has become a leading world producer of manganese, has a significant output of petroleum, and may become a major iron ore shipper within a decade. Congo has an important output of potash; diamonds are the first-ranking export of the CAR. The other countries have still to develop production of world significance.

GABON

PETROLEUM AND NATURAL GAS. Gabon has been experiencing an oil boom in the last decade, and petroleum has become the country's most important export in recent years. Development of hydrocarbons began in 1957, and output increased from 800,000 tons in 1960 to 7.6 million tons in 1973. New finds are constantly being reported, especially offshore, where in 1973 the largest

The petroleum refinery at Port Gentil, Gabon
Petroleum has recently become the leading export of Gabon.

field thus far discovered was found 50 miles (80 km) southwest of Port Gentil; this deposit alone is expected to produce 4 million tons by 1975, so that, together with the higher prices prevailing, oil may be expected to provide considerably more than the 40 to 50 percent of total exports it contributed in the early 1970s. The substantial gas resources are thus far used only by the refinery at Port Gentil, whose capacity is being expanded, and the electric plant for that city. Exploration for oil is also proceeding in Rio Muni.

MANGANESE. One of the largest bodies of manganese ore in the world is situated at Moanda near Franceville in southeastern Gabon. The deposits cover four plateaus and are very favorably disposed for open pit operations. Operated by La Compagnie Minière de l'Ogooué (Comilog), in which United States Steel, the French Government, and private French interests respectively hold 49, 22, and 29 percent of the shares, output of the 48–52 percent ores has increased from nothing in 1960 to 1.94 million tons in 1972, making Gabon the world's leading shipper and third-ranking world producer. Plans call for increasing production to 2.5 million tons after 1975. About 3,000 workers and a staff of 300, most of whom are expatriates, are employed.

The area around Moanda is very sparsely populated, covered with dense forest, and very rugged. This helps to explain two features of the operation, the highly mechanized mining and the use of a 47-mile (76-km) cableway from the mine to M'Binda. From there the ore moves by rail to Pointe Noire in Congo. The transport infrastructure absorbed about two-thirds of the original $99 million investment. As noted earlier, a branch of the proposed rail line from Libreville may be constructed to Moanda to permit an increase in the level of shipments.

URANIUM. Ores of uranium were discovered, after ten years of exploration by the

The uranium workings at Mounana, Gabon
The quarry, older pits, the concentration plant and part of the employees' housing are shown. A spontaneous atomic reaction is thought to have occurred in the Precambrian era in the deposit.

French CEA, in 1956 at Mounana in the Franceville region. Mining was first open pit but went underground in 1966; this mine was closed in 1971 because of the discovery of a nearby deposit at Oklo which could be worked at the surface, while ores have also been discovered about 60 miles (100 km) away and exploration continues elsewhere. High costs, about three times those in South Africa, low market prices, and discovery of superior deposits in Niger have depressed production levels in recent years, but output and returns will probably increase when higher prices prevail toward the end of the decade. In 1970 uranium concentrates accounted for about 3.6 percent of Gabonese exports. Some 900 locals and 100 French specialists are employed.

IRON ORE. The iron ores of Bélinga in the Mékambo region of northeast Gabon have been known since 1895. Reserves are estimated at 860 million tons of 64 percent ores; they occur to a depth of 100 to 200 feet (30–60 m) on the crests of elongated hills rising over 1,640 feet (500m) above the surrounding peneplain, and are underlain by 45 percent ores. The company slated to work the Bélinga ores is the Société des Mines de Fer de Mékambo (Somifer); it is co-owned by Bethlehem Steel (50 percent), the French government (12 percent), French banks (13 percent), and a group of steel producers in Western Europe (25 percent). Production would be 10 million tons in the first phase but might rise rapidly to as much as 25 million tons per annum; the heavy infrastructural costs require a high level of output to justify the operation. About $285 million will be needed to open the mine, construct a 348-mile (560-km) rail line from Santa Clara near Libreville, and provide concentrating and loading facilities there. Production is not likely to start before the early 1980s.

OTHER MINERALS. There is a small production of gold in Gabon, and in 1972 a marble quarry near Libreville and a factory there began producing about 30,000 to 40,000 tons a year of slabs and tiles. Prospecting for a variety of minerals is continuing apace.

CONGO

PETROLEUM. The first discovery of hydrocarbons was made in 1957 and production began in 1960, with most of the small output coming from Pointe Indienne near Pointe Noire. These older fields reached a peak output of 123,000 tons in 1962, after which there was a decline to only 19,000 tons in 1971. In the meantime, however, finds were made offshore and output from the country reached 336,000 tons in 1972 and 2.1 million tons in 1973. In early 1974 foreign oil companies working in Congo were nationalized.

POTASH. One of the world's largest deposits of potash was opened in 1969 at Holle, about 25 miles (40 km) northeast of Pointe Noire with production planned at a level of 500,000 tons a year. Preliminary studies turned out to have been misleading and much of the original equipment had to be replaced. In 1972 output reached 473,000 tons, valued at about $10 million. Total investment, much of which was made by French government bureaus, was over $123 million by 1970.

Potash plant at Holle, Congo
One of the world's largest deposits of potash is found at this site just 25 miles (40 km) from Pointe Noire.

OTHER MINERALS. A small lead-zinc-copper deposit has been operated at M'Passa in the Niari Valley. It was closed in 1969 and reopened in 1972; it now works lower grade ores at a reported rate of 87,000 tons. The body will be exhausted in a few years. Diamonds figure importantly in Congo's exports, but they are all received on a clandestine basis. Bauxite and uranium ores have been found in several localities.

CAMEROON

This country produces very small quantities of gold and cassiterite. Of interest for the future are scattered and uneven bauxite deposits at Minim-Martap about 60 miles (100 km) southwest of N'Gaoundéré. Containing perhaps a billion tons averaging about 42 percent aluminum oxide and occurring under a shallow overburden, the ores have been the subject of several studies. They cannot be exploited unless a branch from the Transcameroon railway is constructed. Cameroon would like to produce alumina from domestic sources to replace the imports now smelted at Edéa, but a larger tonnage output might be necessary to make the operation viable and this would require additional facilities at Douala. Another much smaller bauxite deposit occurs near Dschang.

Indications of other minerals have been reported, including iron ore near Kribi, rutile, and copper. Oil prospection is underway along the coast, with high hopes that significant finds will be made—not unrealistic hopes given the successes in Nigeria and Gabon. In 1973 a find was reported in an offshore drilling near Victoria.

CENTRAL AFRICAN REPUBLIC

DIAMONDS. First found in 1913, diamonds have been mined in the Mouka region north of Bria—and after 1937 in Haute-Sanga—at fluctuating levels in succeeding years. In 1962 a virtual diamond craze occurred when some rivers opened to African prospectors proved to be richer than had been expected. Four mining companies are operating in the country, but the bulk of output is won by an estimated 50,000 African potholers. As in Sierra Leone, the attraction of diamond digging has led to social disintegration in the producing zones, to depressed agricultural output, and to graft at every level. Smuggling has been high both into CAR from Zaïre and out of the country. There have been repeated efforts by the government to organize the collection and marketing of diamonds and to impose higher and back taxes on the companies, while in 1972 a decree was issued restricting diamond collection to citizens. Recorded production rose from 265,400 carats in 1962 to 609,400 in 1968; since then they have dropped, apparently because of increased smuggling, and recorded output was down to 523,000 carats in 1972. Output from the companies was stopped in 1969 and all foreign workers were expelled, but an agreement to reopen the industrial operations was concluded in 1972 with the concurrence of the then-unemployed former workers. Evidence suggests that digging in the small rivers cannot continue at high levels and that a substantial investment in modern equipment would be desirable, but the numerous uncertainties, fluctuating governmental policies, and the likelihood that potholers can no better be controlled than they were in Sierra Leone militate against the rational exploitation of diamond resources.

URANIUM. In 1957 a CEA mission discovered uranium ores in CAR, the most interesting of which is 19 miles (30 km) north of Boukama in the east. Difficulty in finding a suitable process for concentrating the refractory ores plus poor economic conditions delayed proceeding to development, but an agreement was reached in 1969 with a French society to start production in 1972. This fell through and a contract was signed with an American concern looking toward commencement of shipments by 1977. Development will require a more direct and better road from Bangui, mining lignite in the area to power an 8,000-kw electric station, installing a concentrator with a capacity of 500–700 tons of uranium oxide, and mining by dragline because of the problems of inundation; all of these improvements will require an investment of about $44 million.

CHAD

The only mineral produced in Chad is natron (sodium carbonate), used for cattle licks; output is about 10,000 tons a year. Indications of tin and tungsten have been noted in the Tibesti Mountains.

INDUSTRY

Manufacturing is less well developed in this region than in the four leading West African nations, but considerable progress has been made since independence in diversifying the establishments in all of the countries. Cameroon has the largest development and now ranks after the Ivory Coast and Senegal among former French tropical African countries (Map 69). About 20,000 workers are employed in manufacturing, whose value of output was about $150 million in 1971, when industry was credited with contributing 15.5 percent of the GDP; this probably exaggerates the role of manufacturing, since it in-

cludes other activities. About a hundred modern enterprises are in being, with a third having been started since 1964. Of particular importance are brewing, the aluminum smelter at Edéa, and a large textile mill. Douala, with its suburbs of Bassa and Bonabéri, is the leading center, but Yaoundé also has a fairly diversified range of industries.

Congo ranks second in manufacturing in the region with an output valued at about $88 million in 1970—four times greater than the 1962 level. About 85 establishments existed in 1970, 25 of which had been started since 1964. Agricultural processing and food industries are the most important segment, with the sugar mill in the Niari Valley the largest single factory. Congo has received aid in establishing new industries from a variety of countries including, in recent years, several Communist countries. Several companies have been nationalized in the last few years; other plants are owned by mixed companies in which the state holds varying shares. Brazzaville and Pointe Noire are the major centers, but the Niari Valley towns are also of some significance.

Gabon, which had only forest product industries for many years, has made a major effort since about 1967 to diversify its manufacturing representation. Port Gentil has two of the more important establishments—the plywood mill mentioned earlier and the petroleum refinery, plus a number of other plants—but Libreville-Owendo has gained from the drive to introduce market-oriented establishments. The small size of the total population even with its relatively high average income, however, greatly restricts the possibilities for achieving a fully integrated industrial establishment.

The other countries of the region have lower development, as would be expected by their remoteness and small markets. In 1971 the value of production in CAR was perhaps $16 to 18 million. The textile industry, started in 1953 at Boali because of the installation there of a hydroelectric plant and

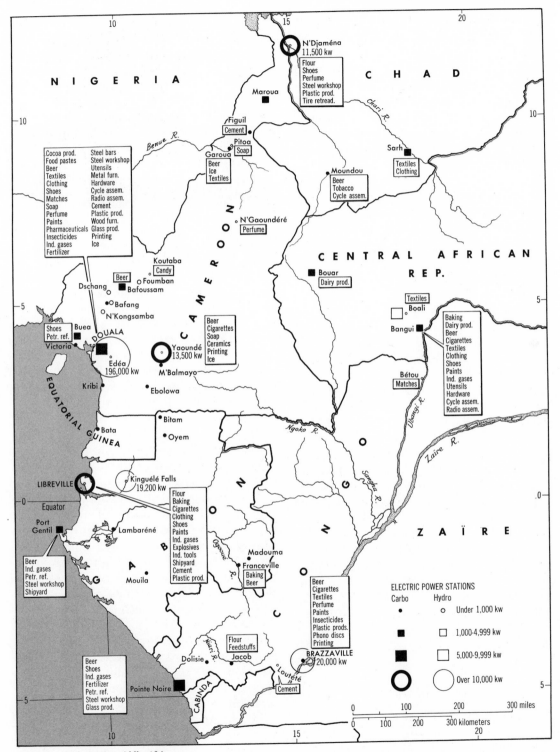

N I G E R I A

C H A D

Maroua

N'Djaména
11,500 kw
Flour
Shoes
Perfume
Steel workshop
Plastic prod.
Tire retread.

Chari R.

Figuil
Cement

Pitoa
Soap

Garoua
Beer
Ice
Textiles

Sarh
Textiles
Clothing

Moundou
Beer
Tobacco
Cycle assem.

Cocoa prod.   Steel bars
Food pastes   Steel workshop
Beer   Utensils
Textiles   Metal furn.
Clothing   Hardware
Shoes   Cycle assem.
Matches   Radio assem.
Soap   Cement
Perfume   Plastic prod.
Paints   Wood furn.
Pharmaceuticals Glass prod.
Insecticides   Printing
Ind. gases   Ice
Fertilizer

Benue R.

C A M E R O O N

N'Gaoundéré
Perfume

C E N T R A L   A F R I C A N
R E P.

Bouar
Dairy prod.

Koutaba
Candy

Dschang   Beer
        o Foumban
        Bafoussam
    Bafang
    N'Kongsamba

Textiles
Boali

Bangui

Baking
Dairy prod.
Beer
Cigarettes
Textiles
Clothing
Shoes
Paints
Ind. gases
Utensils
Hardware
Cycle assem.
Radio assem.

Shoes
Buea
Petr. ref.
DOUALA
Victoria

Edéa
196,000 kw

Kribi

Yaoundé
13,500 kw

Beer
Cigarettes
Soap
Ceramics
Printing
Ice

M'Balmayo

Bétou
Matches

Ubangi R.

Ebolowa

E Q U A T O R I A L   G U I N E A

Bata

Bitam
Oyem

Ngoko R.

C O N G O

Zaïre R.

Sangha R.

Z A Ï R E

LIBREVILLE

Equator

Port
Gentil

Kinguélé Falls
19,200 kw

Flour
Baking
Cigarettes
Clothing
Shoes
Paints
Ind. gases
Explosives
Ind. tools
Shipyard
Cement
Plastic prod.

Lambaréné

G A B O N

Ogooué R.

Madouma

Franceville

Baking
Beer

Beer
Ind. gases
Petr. ref.
Steel workshop
Shipyard

Mouila

Beer
Cigarettes
Textiles
Perfume
Paints
Insecticides
Plastic prods.
Phono discs
Printing

ELECTRIC POWER STATIONS
Carbo   Hydro

Niari R.

Flour
Feedstuffs
Jacob

Dolisie

BRAZZAVILLE
20,000 kw

Loutété
Cement

Under 1,000 kw

1,000-4,999 kw

5,000-9,999 kw

Over 10,000 kw

Beer
Shoes
Ind. gases
Fertilizer
Petr. ref.
Steel workshop
Glass prod.

Pointe Noire

CABINDA

0   100   200   300 miles

0   100   200   300 kilometers

Map 69. Industry in middle Africa

expanded by a large plant at Bangui in 1967, is the leading sector. Bangui has the vast bulk of the newer plants. Industry in Chad was estimated to account for 8.5 percent of GDP in 1970, when about 70 plants were in being. Two textile mills accounted for 60 percent of total industrial output, estimated at about $7 million in 1970. N'Djaména followed by Sarh are the leading centers, with Moundou being of secondary importance.

The types of industry represented in the region are very comparable to those in West Africa except that there is not so broad a range. The locational pattern is also similar, with primary processing being dispersed and the newer, market-oriented factories being heavily concentrated in the ports and capital cities.

Aluminum sheet mill at Edéa, Cameroon
The reduction plant at Edéa, which consumes over four-fifths of the electricity from the nearby plant on the Sanaga, is the largest industry in Cameroon.

PRIMARY PROCESSING

A brief examination of Maps 66 to 69 will reveal the close relationship between industries in this category and the major products exported from each of the countries. Most of the processing plants are small, the major exceptions being the plywood mill at Port Gentil, a few sawmills, and the sugar refinery in the Niari Valley. Several large plants will probably be added to this list in the years ahead, including a plant to pelletize iron ore when Bélinga is opened up, an alumina plant in Cameroon, a cellulose mill in Gabon, and a gas liquefaction plant in Gabon. The cellulose mill would be based not only upon the forest resources of that country but on installation of a hydroelectric station at Fiengene in the Kamlo region; the plant, to be located at Kango, would have a capacity of 250,000 to 500,000 tons of bleached pulp requiring an investment of at least $120 million.

POWER-ORIENTED INDUSTRY

The aluminum smelter at Edéa, Cameroon, was the first important low-cost-power–oriented plant in sub-Saharan Africa. The

simple run-of-stream dam on the Sanaga River was built in the early 1950s, partly on faith, with an original capacity of 22,000 kw; success of the venture was not assured until the smelter was erected. In succeeding years, capacity of the electric plant has been increased to 196,000 kw and the flow of the river regularized by a dam at Mbakou completed in 1975 to provide firm power at Edéa. Output of the smelter has been about 50,000 tons in recent years, but will probably be increased; it absorbs about 83 percent of the power produced at Edéa.

A white hope for future industrialization in Congo is development of the Kouilou River. Studies began in 1954 and plans drawn up before independence called for construction of a 410-foot-high (125-m) dam where the river leaves the Crystal Mountains in the Sounda Gorge, installation of an 820,000-kw station (which would be one of the largest in Africa), and a variety of electro-process industries. Repeated delays in moving forward with the scheme were occasioned by the difficulty of attracting enough large-scale consumers to assure absorption of the power and, after independence, by uncertainties regarding security of the very large investment that would be required—at least $400 million.

MARKET-ORIENTED INDUSTRY

Almost all of the consumer-product industries of the region have been installed in the last two decades and are concentrated in a limited number of centers (Map 69). Most of them are small scale and require aid or protection to permit profitable operation. The more important plants include the breweries, textile mills, three Bata shoe factories, the cement mills, and the petroleum refineries.

A rather unusual feature of manufacturing in the region has been the construction of a few regional plants catering to the combined markets of the five main countries. These include the refinery at Port Gentil, but when its capacity was fully utilized there was a bitter dispute over the location of a second plant; and it now appears that larger refineries will be sited both at Pointe Noire and at Victoria in Cameroon. Textile mills in CAR and Chad were also supposed to serve regional markets for certain products; the plant at Sarh, constructed in 1967 and extended in 1972, is a remarkable example of international cooperation and of the association of public and private capital, involving as it does shareholdings by the

Production of suitcases at Douala, Cameroon
Containers of many types are strongly oriented to the market because of the enormous increase in bulk of the finished product.

Chad and Cameroon governments and by German and French groups. Whether additional regional plants may be constructed in the future will depend on the adherence of the five states to UDEAC, which has had some stormy periods in the last decade.

POWER FOR INDUSTRY

The rapid growth of industry in recent years has required the provision of additional power, and the wealth of hydroelectric resources in the southern parts has focused attention on several installations. Brazzaville receives power from a 20,000-kw station on the Djoué River, which flows into the Zaïre just west of the city, and there is discussion with China regarding construction of a 70,000–80,000 kw plant at Bouenza, about 250 miles (400 km) west of the capital. Gabon has recently built a 19,200-kw plant at Kinguélé Falls on the M'Bei River about 60 miles (100 km) east of Libreville which may later have a capacity of about 76,800 kw, and, as noted above, the country is planning a sizable installation at Fiengene.

The CAR has an 8,700 kw hydro plant at Boali, about 65 miles (105 km) northwest of Bangui, and is considering one new plant above Boali and another on the Kotto River in the east. Cameroon uses the power not consumed in the aluminum mill from the Edéa installation and has five very small hydroelectric stations in the west. It is examining several additional plants for future development.

CONCLUSION

The very large, heterogenous, generally underdeveloped area of middle Africa is one of the most fascinating on the continent because of the sharp contrasts from region to region in ethnic makeup and in stages of economic development. There are sufficient known and undeveloped resources to justify optimism regarding its future, particularly that of mineral-rich Gabon. The greatest possibilities appear to be in opening up the mineral reserves, developing the immense hydroelectric resources, and further exploiting the forest wealth, but none of the countries can afford to neglect the agricultural sector. The northern and interior portions will probably fall further behind the coastal regions unless unforeseen resources are discovered. An improved transport net is a requirement for the whole area, but particularly for the landlocked Chad and Central African Republic.

CHAPTER FOURTEEN

Zaïre

The Democratic Republic of Zaïre (formerly Belgian Congo and then the Democratic Republic of Congo, or popularly Congo-Kinshasa) experienced one of the most traumatic series of events in its early years of independence of any African country. The decision to grant independence to Congo on July 1, 1960, came abruptly and represented a drastic change in Belgian policy. To some extent it represented an abrogation of responsibility, because the Congolese were very inadequately trained to assume management of such a large, economically complex, and culturally heterogeneous area. Two grave errors were made by the Belgians: first, the failure to change their approach in consonance with the rapidly evolving political scene in Africa, and second, having decided in 1959 that such an error had been made, the failure to institute a crash program to prepare Congolese leaders for independence.

Until 1959 the Belgians' primary emphasis was on developing the Congolese economy, which they contended must precede extension of political rights. And credit is due the Belgians for the achievements made in the economic sphere, as well as for the attention given to primary and technical education, health and medical advances, and other social betterments. It is true that the eventual objective was self-government—hopefully with some more or less formal relation with Belgium—but even in the mid-1950s officials spoke of reaching this goal in 50 to 100 years. But after two universities were opened in 1954 and 1956, labor unions were permitted in 1957, and elected councils were introduced in the three main cities in 1958, the Belgians decided the pace of devolution must be accelerated; in January 1959 they announced a program under which local, provincial, and central legislative bodies were to be successively introduced. Riots occurred just before and after this pronouncement. While it appeared that Congo had moved almost overnight from a rather tranquil state to one of political turbulence and dynamism, it became obvious that the Congolese had not remained politically unconscious of the rapid strides being made in other African areas.

In December 1959 the Belgians promised to grant independence by July 1960. They were doubtless influenced by the relatively uneventful evolutions that had been occurring in British and French territories. They believed that this massive gesture would win the confidence of the Congolese, who would continue to rely heavily on Belgian administrative, technical, and financial support following independence. But the Congolese lacked political experience and the number of potential leaders was pitifully small. In 1959 there were only 19 college graduates among the African population, and there were only 21 Africans enrolled in the universities; in 1960 only 161 Congolese graduated from secondary schools.

POLITICO-ECONOMIC EVENTS AFTER INDEPENDENCE

Within a week of independence there was a collapse of authority and a mutiny by the Force Publique. There followed a five-year period of violence and chaos marked by the exodus of most Europeans, intercession by the United Nations to maintain order and assist in administration, the secession of Katanga Province (now Shaba) from 1960 to 1963 ended by UN military action, the establishment of separate administrations by rival factions in various parts of the country, conflicting support of outside powers threatening to involve Zaïre in the cold war, renewed outbreak of rebellion in the northeast and east in 1964, and a succession of governments, none of which was successful in ending the troubles or correcting a distorted economy.

By the mid-1960s an economy that had been one of the strongest in Africa was severely disrupted. Agricultural exports were down by 50 percent, copper sales did not return to their 1960 level until 1966, consumer prices increased sevenfold from 1960 to mid-1967, a virtual breakdown in controls on expenditure had occurred, many plantations, mines and other productive facilities had been damaged, the transport had been destroyed, and the per capita GDP had declined from $95 in 1959 to an estimated $80 in 1966. Only manufacturing, particularly the heavy concentrations in Kinshasa (Léopoldville) and Lubumbashi (Elisabethville) increased its output, under the stimulus of inflation and protection.

On November 25, 1965, General Mobutu seized power in a bloodless coup, beginning a firm, often repressive, but dynamic administration, which substantially changed the nation's structure and policies. While a revolt by mercenaries occurred in 1967 and a series of plots have been reported in the years since, Mobutu has succeeded in bringing the police and military forces into effective roles, has suppressed regional and ethnic rivalries (in part by rotating government officials and making frequent changes in government portfolios), and has moved to instill a feeling of loyalty to the nation throughout the country. The latter has involved a "return to authenticity," in which many names have been changed from the days of colonialism, in-

cluding not only the names of the country, the river for which it was named, and many regions and cities, but also personal names (the General's own became Mobutu Sese Seko Kuku Ngbendu Wa Za Banga) and titles. In 1970 Mobutu was unanimously elected to a new seven-year term.

In the economic sphere a stabilization program, adopted in 1967 with support from the International Monetary Fund and from the United States, reestablished monetary stability (a new currency unit, the zaïre, was introduced at a substantially devalued conversion rate), imposed stringent wage, fiscal, and credit policies, ventilated the economy by removing certain tariff restrictions, and brought the budget into equilibrium. Also of great importance in the succeeding boom were the enunciation of a liberal investment code and conclusion of agreements in 1967 and 1969 with Union Minière du Haut Katanga (UMHK) regarding compensation for its takeover by the government in 1966. The latter removed an element of concern to foreign investors and freed the situation to permit development loans from international agencies. A marked increase in the price of copper in this crucial period was important to the achievement of financial stability and generally favorable balance of payments and reserve positions.

The years after 1967 have seen rapid growth in several sectors, particularly mining and manufacturing. Foreign investment increased dramatically, development budgets rose substantially, and the per capita GNP grew at an average annual rate of 3.6 percent from 1965 to 1971. The record is by no means entirely favorable, however, and it will take some years to remove the heritage of the early chaotic post-independence period.

The physical and social infrastructures require very heavy investment even to regain the 1960 levels; the monetized agricultural sector has made very uneven progress, particularly that portion previously provided by smallholders; the still very large population dependent upon subsistence agriculture has essentially been ignored; the problem of servicing the debt is one of serious proportions; Mobutu and his colleagues have adopted an ostentatious lifestyle and corruption is rampant; and the per capita GNP in 1971 was only about $90, below that of most African countries.

One of the stated keynotes of government policy is Zaïrization of the economy. This was reflected in the nationalization of UMHK and other enterprises in the 1960s, but major new measures were announced in late 1973, which may reduce the interest of foreign investors and are likely to depress agricultural exports significantly. These measures included 100 percent nationalization of the major diamond company, a required offering of 50 percent equity in all other mining companies still privately owned, the nationalization of foreign-owned plantations and estates, restrictions on foreigners' trading in a wide range of basic commodities, and prohibitions against Portuguese, Greeks, and Asians' living in five of the nine provinces. Whether these measures will be rigidly applied is not yet apparent.

Zaïre has also taken steps since 1965 to reduce the heavy dependence on Belgium by encouraging investment from and contacts with other western countries, particularly the United States, and in 1974 all official traces of a special relationship between Belgium and Zaïre were ended. At that time between 30,000 and 40,000 Belgian citizens were still resident in the country, comprising the largest number of expatriates in the educational, technical, scientific, and missionary fields.

THE ECONOMIC-GEOGRAPHIC KEYNOTES OF ZAÏRE

Before examining the sectors of Zaïre's economy, certain keynotes of its economic-geography will be presented; some of these special features will provide additional back-ground information required to understand the country.

EQUATORIAL POSITION

Zaïre's equatorial location (about a third of the country lies north of the equator and two-thirds south) is the most important factor in setting the climatic pattern; it is a strong indirect determinant in the gross distribution of vegetation and soils and, hence, greatly influences the land-use pattern of the country.

Climatically, a large tropical rainy area straddles the equator, skewed somewhat to the Northern Hemisphere because of the greater land mass. A small region of similar climate is found in Lower Zaïre (see Map 73 for Regional boundaries). Precipitation in these regions ranges from 50 to 60 inches (1,270–1,524 mm) a year up to about 80 inches (2,032 mm) and 10 to 12 months are humid. Most of the remainder of the country is tropical savanna except for zones of tropical highland climate along the east-

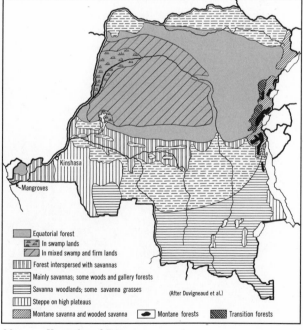

Map 70. Vegetation of Zaïre

ern margins. Only very small areas of tropical steppe climate are found, in the deep rift valleys along part of the eastern border.

The vegetation pattern of Zaïre is shown in Map 70, adapted from several sources. Again there is a marked symmetry to the north and south of the equator. The southern savanna zone is predominantly grassy in a belt south of the rainforest, but mainly savanna woodlands in the southern Bandundu, Kasai East, and Shaba Regions.

Latosols are dominant among the soils of Zaïre, which means that the country is poorly endowed in this resource. Only limited areas have good soils: alluvials in the central basin and along some of the main watercourses; soils regenerated by erosion in Lower Zaïre; and especially the volcanic lands of Kivu, some of which are periodically enriched by volcanic ash carried by the wind.

HYDROGRAPHIC UNITY

A second keynote is the hydrographic unity of the country, which gives Zaïre greater geographic unity than most of the other areas of the continent. The vast bulk of Zaïre is within the Zaïre River Basin and the bulk of that basin is within Zaïre (Map 71). Much of the area is a huge fan-shaped, shallow

Map 71. Landforms of Zaïre
Relief Map Copyright Aero Service Corporation

saucer, covering about 400,000 square miles (1,036,000 sq km), whose base is at the point where the Zaïre breaches the Crystal Mountains just below Malebo Pool (Stanley Pool). The fact that this basin also straddles the equator contributes further and importantly to the symmetrical pattern already noted for climate, vegetation, soil, and land-use patterns.

From the geological standpoint, Zaïre rests on an ancient foundation, revealed in the mountainous rim circling the vast depression of the central Zaïre basin. Most of the mineral wealth of the country is found in the peripheral rim. The basin is covered by more recent sediments of varying thickness deposited in the last 200 million years. The effects of erosion during this period have left great accumulations in the depressions and sometimes terraced plateaus, especially in the south.

The geological history of Zaïre explains its relief features and the hydrographic basin. The central depression has an average elevation of about 1,300 feet (400 m). It is lowest in the Lake Tumba and Lake Mayi Ndombe (Lake Leopold II) region, both lakes being the remnants of a once vast inland sea. Plains and terraced plateaus extend to the rim of the basin. The rim consists of more or less dislocated plateaus, lowest on the west and north but rising progressively as they approach Shaba to heights of 5,500 to 6,500 feet (1,680–2,000 m) in several masses. On the east, the great longitudinal fractures of the Central Rift, 875 miles (1,400 km) long and 25 to 30 miles (40–48 km) wide, distinguish that area from the remainder of the country. The rift valley is bordered on both sides by uplifted mountains, while extinct and semiactive volcanic formations exist, particularly near Lake Kivu. The magnitude of the fractures is suggested from the height of the block Ruwenzori Mountains—whose elevations generally range between 6,500 and 13,000 feet (2,000–4,000 m) with a maximum elevation of 16,800 feet (5,120 m) in

Mt. Marguerite—and from Lake Tanganyika's maximum depth of 4,800 feet (1,460 m).

The country may, then, be broadly divided into three landform regions: (1) the concave central plain, containing some higher parts such as the hills of Zongo; (2) the rather circumscribed coastal zone consisting of an elevated plateau west of the Crystal Mountains, which terminates in an escarpment, sometimes quite high, along the Atlantic beaches; and (3) the peripheral higher hills, mountains, and plateaus. This pattern provides generally well-marked boundaries, accentuating the geographic unity created particularly by the cuvette of the Zaïre.

THE EXISTENCE OF A GREAT SYSTEM OF INLAND WATERWAYS

Closely related to the previous keynote is the presence, within the Zaïre Basin, of the most important system of inland waterways in Africa (Map 73). Although by no means perfect, the Zaïre has permitted greater ease of access than in most interior African areas and, just as important, rather cheap means of communication. The series of thirty rapids and falls in the 217-mile (349-km) stretch between Matadi and Kinshasa where the Zaïre cuts through the Crystal Mountains did, however, long discourage penetration,

A landing stage on a small tributary of the Zaïre
Zaïre is unusual among African countries for the vast extent of its inland waterway system.

and the potential value of the middle Zaïre was not appreciated until Stanley, on his second trip, came down it from the east. Navigable stretches on the 3,000-mile-long (4,830 km) Zaïre total about 1,700 miles (2,735 km), the two most important being the 80-mile (129 km) maritime section to Matadi and the 1,085-mile (1,746 km) section between Kinshasa and Kisangani (Stanleyville). The total distance of navigable waterways within the Zaïre Basin is about 7,565 miles (12,174 km). Contributing to the suitability of the Zaïre for navigation is the unusual evenness of its flow, explained by the presence of tributaries on both sides of the equator. The average flow at Kinshasa is 40,000 cubic meters per second; the ratio of low to high flow is only 1:3 as compared to 1:20 on the Mississippi or 1:48 on the Nile. The development of commerce on the Congo and its position within the whole transport complex of the republic is treated in a later section.

GREAT MINERAL AND HYDROPOWER WEALTH

The final physical keynote of Zaïre is its great richness in minerals, including the "white gold" of falling water. Zaïre ranks fifth among mineral producers in Africa: it is fifth in the world in output of copper, its most important mineral, and seventh in tin; it leads the world in production of industrial diamonds and cobalt; and it has an important output of a variety of other minerals. Exports of minerals now account for four-fifths of total exports by value, and the mining industry makes the most important contribution to the revenues of the country.

In hydropower, the country has the greatest potential of any nation in the world and some 40 percent of the estimated total for Africa. While developed facilities have harnessed only a fraction of the enormous potential, Zaïre has witnessed one of the more dynamic hydroelectric programs in Africa. The details of mining and hydroelectric development are given below.

LOW AVERAGE DENSITY OF POPULATION

The final keynote is the generally low density of population, averaging about 26.7 per square mile (10.3 per sq km) in 1974. The low population presented certain problems for plantations in the equatorial rainforest zone and for mining companies in Shaba, and the literature on Belgian Congo gives frequent reference to what was long considered a population crisis. At the end of the nineteenth century Stanley made a purely speculative estimate of the population, placing it at 28 million. This and equally unscientific estimates were frequently cited as a fact and contributed to a rather widely accepted belief that the population was actually declining, threatening the whole base for economic development. It is probably true that the death rate was extremely high, especially during the Leopold period when there were few restrictions on the use of labor, when men were frequently removed from their communities for long periods, and when the incidence of disease was heightened by migrations and the introduction of previously absent maladies. Experts now calculate that the population was about 9 million in 1900 and about the same in 1925.

Belgian concern regarding the supposed population problem led to measures intended to correct it, including a relatively advanced health program and particular efforts to reduce the originally high infant mortality rate, limitations on the numbers of workers that could be recruited from an area, and the common practice of recruiting on a family basis with employers being required to provide housing and other amenities in nontribal areas. The rate of population growth gradually increased from an estimated 0.6 percent per annum from 1925 to 1940 to the present level, which is variously estimated at 2.1, 2.8, and 4.2 percent. The population in mid-1974 was estimated at about 24.2 million. It now appears that the population problem, and particularly the problem of worker shortage, was

exaggerated. There were some real advantages to a low population density: it encouraged the adoption of labor-saving equipment, which permitted training of more technicians; it stimulated the desire to upgrade agricultural methods; and, for the future, there was a greater possibility of achieving a good standard of living rather than simply that of a glorified peasantry. Even with its present low average density, Zaïre may prove to have a problem in its very rapid rate of increase—far greater than it faced in adjusting to real and imagined shortages of manpower. Over 51 percent of the population is under 18 years of age, and there has been unusually rapid migration to the cities since independence, partly in search of security during the 1960s. These facts suggest that problems of absorption will grow apace in the years ahead.

Use of the average density figure has the usual disadvantages. As elsewhere, there is great unevenness; over half the country has only about 3 per square mile (1.2 per sq km), while districts in Lower Zaïre and the eastern highlands have average densities running to about 78–111 per square mile (30–43 per sq km). About 15 percent of the total population lives in the eleven largest cities. Kinshasa has seen a particularly striking increase in recent years, from 403,000 in 1959 to 1.3 million in 1970 and possibly 1.6 million in 1974.

A great mosaic of people exists in Zaïre, but the vast bulk of the population is classified as Bantu-speaking. Of the 69 more important ethnic groups 49 are Bantu, 16 located along the northern border are Sudanese, one in the east is Nilotic and another Hamitic in origin, and there are four small groups of pygmies and pygmoids. While there are several hundreds of dialects, one or more of four African languages are spoken or understood by most people: Kikongo, Kiluba, Lingala, and Kiswahili. French is the official language.

About 93,000 non-Africans were estimated to be residing in Zaïre in mid-1971, which compared with the preindependence level of 118,000 at the 1958 census. There have been several large-scale migrations of refugee African groups both out of and into Zaïre since 1960; in 1970 it was estimated that about 440,000 foreign refugees were residing in that country.

THE TRANSPORT COMPLEX

The main features of the transport network of Zaïre may be revealed by tracing the major routes into and within the area (Map 73).

THE MATADI-KINSHASA AXIS

Matadi is the only port of real importance within the national territory. Zaïre has a very small coastline on the Atlantic, the use of which is further restricted by the freak boundary situation in the area in relation to the topography. Most of the southern shore of the Zaïre River in its lower navigable stretch belongs to Angola (Map 72). Zaïre has the northern bank in this stretch but to use it for a major port would have required either a bridge at Kinshasa and a rail line running through Congo (Brazzaville) or a bridge at Matadi and a right-bank railway within Zaïre; both schemes would be very costly. A third alternative would be to use the river port of Brazzaville and the Congo–Ocean Railway to Pointe Noire.

Located 80 miles (129 km) up the maritime section of the river, Matadi occupies the very small area belonging to Zaïre on the left bank and on the lower navigable stretch. This site has three main disadvantages: (1) the impediments to navigation on the lower Zaïre including powerful currents at the entrance, shifting sandbanks, sharp turnings in the constricted rocky passage below the port, and the whirlpools of the Devil's Cauldron between Ango Ango and Matadi, which are difficult for smaller vessels to negotiate, the maritime stretch was, however, considerably improved over the years by deepening the

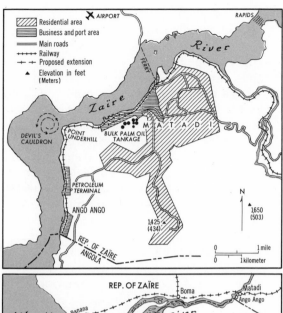

Map 72. *The port of Matadi and the approaches to it*

channel, continuous dredging, and improved channel markings, including the installation of luminous buoys which permitted night navigation under pilotage after 1955; (2) the difficult topography at the port, where blasting of the steep hills rising abruptly from the river was necessary to extend the very limited flat land, and where the narrow width and velocity of the river precluded anything but marginal quays; and (3) the restricted area available for expansion, since the proximity of the Devil's Cauldron prevented continuation of the quays to Ango Ango, whose site features are comparable to those at the main port.

Matadi now has about 6,560 feet (2,000 m) of deepwater quays, bulk palm oil facilities, a petroleum port at Ango Ango connected by pipelines to Kinshasa, and a fishing installation, also at Ango Ango. Lighters are extensively used to expedite loading and unloading when worked on the outboard side of berthed vessels.

Traffic at Matadi/Ango Ango increased

from 531,000 tons in 1938 to 1.75 million tons in 1959, but declined to 1.06 million tons in 1962; it was 1.25 million tons in 1972. Congestion was severe between 1946 and 1955, forcing an increased use of foreign outlets. It has again been congested in recent years, and the government has therefore decided to bridge the Zaïre at Matadi, construct a rail line over difficult terrain, and install a major new deepwater port at Banana, near the mouth of the river. At present, Boma handles most of the traffic for this region; it is the terminus of an 87-mile (140 km) rail line, augmented by a highway, running to Tshela.

The immediate hinterland of Matadi is extremely restricted; its traffic largely comes from or is destined for the interior Zaïre Basin. It may, in fact, almost be considered as an outport of Kinshasa, the inland terminus of the Matadi-Kinshasa line (CFMK), and the base of navigation for the whole middle Zaïre River system. Of the cargo handled at Matadi, 95 percent normally moves over the CFMK. This 227-mile (366-km) line cuts laboriously through the Crystal Mountains on a route that involved much cutting, filling, and bridging. Despite marked improvements, train speeds and axle loads on the CFMK are still restricted. Traffic on the CFMK was 2.13 million tons in 1959, fell to 1.12 million tons in 1962, and was 1.62 million tons in 1972.

Kinshasa, the interior terminus of the axis, is situated on Malebo Pool, where the vast fanlike interior Zaïre River system converges. It is Africa's most important river port, normally handling more than most ocean terminals in Africa. Its port facilities include about 3,210 feet (980 m) of quays, bulk storage for palm and palm kernel oil and also for petroleum products, loading beaches, and shipbuilding and ship repair yards. The port is well equipped with mechanical gear, but does suffer from inadequate space behind the river front. Expansion of the port will require construction of

new facilities at Kimpoko, 21 miles (34 km) upstream. Kinshasa, with the largest population of any tropical African city, is an important consuming center both for goods moving inland from Matadi and for those coming down the Zaïre. Almost all of its transit traffic moves on the river.

There has been much discussion over the years of connecting the CFMK to the line serving Shaba, which would eliminate the necessity of transshipments at Ilebo (Port Francqui) and Kinshasa. Two routes have been studied, one paralleling but not along the Kasai, the other going southward via Inkisi to Kananga (Luluabourg). Both connections would be very expensive, but the northern route has the advantages of being shorter (533 miles as compared to 845 miles, 857 km compared to 1,360 km), less tortuous, and of requiring fewer terraces and bridges. Construction of the line, which will cost an estimated $160 million, would permit use of national routeways for the vast bulk of Zaïre's traffic, a long-standing goal. But the distance from Lubumbashi to Matadi would be 1,751 miles (2,817 km) as compared to 1,308 miles (2,104 km) to Lobito in Angola, and the increased traffic would require use of the new port at Banana.

THE ZAÏRE WATERWAY SYSTEM

Almost the whole transport system of Zaïre was originally designed to take advantage of the extensive navigable waterways of the great Zaïre Basin. The particular features of the road and rail system were, generally speaking, determined by the specific features of the Zaïre and its tributaries; since water transport normally costs less than overland movement, the roads and railways were frequently built to supplement the waterways. Rail lines were placed at breaks in the navigable river or were directed to it from regions that were not served by navigable streams.

The Belgians were also interested in focus-

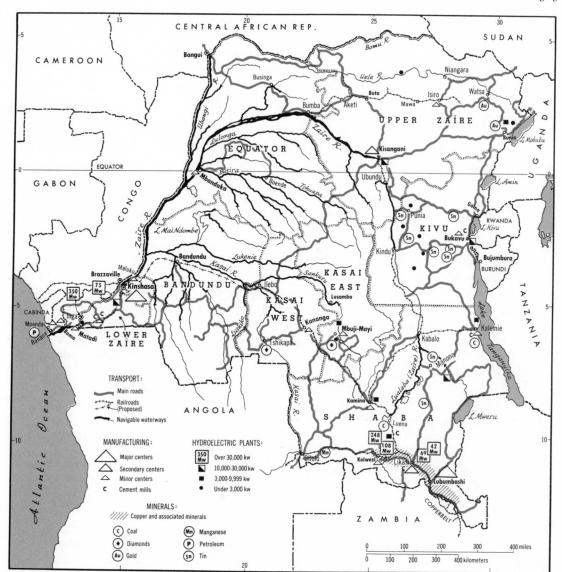

Map 73. Economic map of Zaïre

ing traffic on the Zaïre because use of that route involved support for the colony's port. Special rates were instituted on combined routeways to induce trade to use the "voie nationale" instead of shorter and less costly foreign routes. Waterborne traffic maintained its 40 percent share of total traffic from prewar to postwar years, but must be expected to suffer a relative decline in the decades ahead, when the road system has been improved and the CFMK and southern line have been connected.

The two most important sections of the middle Zaïre are the 1,085-mile (1,746-km) stretch from Kinshasa to Kisangani and the 378 miles (608 km) of the Kasai River from its confluence to Ilebo. Map 73 shows the navigable streams of Zaïre. In the early years little was done to improve navigability. Flat-bottomed sternwheelers towing strings of low-draft barges operated on the main streams; small tugs drawing less then 2 feet (0.6 m) towed those barges on the shallower streams. Important changes in this general pattern were not made until after World War II, when luminous beacons and mirrors were installed and vessels were equipped with radar; other improvements included dieselization, use of the push-tow system, and removal of shoals. Unfortunately, the whole system deteriorated in the 1960s and there were losses to the fleet; as yet, adequate funds have not been provided to rehabilitate the waterways and the waterway agency created in 1971 has been able to maintain dredging on only half of the distance required.

Many shipping points, some well equipped and others only landing stages, occur along the waterways. Apart from Kinshasa, the two most important are Kisangani and Ilebo, the latter being mainly a transshipment point between the rail system of the south and the river. Kisangani, an important regional capital, is located where a series of seven rapids interrupts navigation. They may eventually be canalized but are now bypassed via a 78-mile (125-km) rail line to Ubundu. Above the rapids the Upper Zaïre, also called the Lualaba, is navigable to Kindu and again from Kongolo to Bukama. The river flows across a vast plain in this region and spreads out each year into numerous lakes, some of which are permanent. It has been necessary to cut and maintain a passageway through the floating vegetation and papyrus on these stretches. A portion of the Chemin de Fer des Grands Lacs (CFL), running from Kindu to Kabalo, connects the two navigable sections of the Upper Zaïre, and other rail sections continue from the latter town to Kalemie on Lake Tanganyika and connect with Kamina on the Shaba line (Map 73). When the Kabalo-Kamina link was completed in 1956, most of the CFL system was converted

from meter gauge to 3-foot, 6-inch track. The CFL river-rail system was seriously interrupted from 1961 to 1965 and has not yet been fully rehabilitated; a British firm, however, is now rebuilding the stretch from Kabalo to Kalemie and is providing new rolling stock and a 500-ton river craft.

A 637-mile (1,025-km) rail system (the CVZ) in the northeast of the country was also built largely as a feeder to the Zaïre waterway, to which a direct connection was completed in 1973. It serves the Uele cotton producing region. Elsewhere in the basin, roads link interfluvial areas to available river shipping points.

It may be seen that the Zaïre waterway system and its supplementary road and rail connections cover a very large portion of the country, even servicing portions of the south via the Kasai and Lualaba routes. But use of the Zaïre is not entirely appropriate for some sections of eastern Zaïre, which are nearer the Indian Ocean than they are to the Atlantic.

ROUTEWAYS TO THE EAST

Several routes leading to ports on the eastern seaboard are open to use for Zaïre traffic but are normally of only minor significance. The eastern edge of Zaïre is the most remote area of the country; 1½ to 2 years may elapse before goods ordered in Europe reach their destinations in that region. The Kivu area may be taken as an example of the inadequacy of the national routes, which are, however, frequently used. Goods moving from Bukavu would go by (1) truck to Usumbura or Uvira on Lake Tanganyika, (2) lake vessel to Kalemie, (3) rail to Kindu, (4) river boat to Ubundu, (5) rail to Kisangani, (6) boat to Kinshasa, (7) rail to Matadi, and (8) ship to destination. An alternative national route, reducing the number of transshipments, became available after 1956—namely rail from Kalemie to Ilebo. An improved road is now under contruction from Goma, north of Bukavu, to

Kisangani which would further reduce transshipments.

Possible extranational routes available to various parts of eastern Zaïre include (1) road to the Nile in southern Sudan and then via boat and Sudan Railways to Port Sudan (very seldom used); (2) road and ferry services to rail heads in Uganda and then to Mombasa; and (3) road or rail to Lake Tanganyika, steamer to Kigoma in Tanganyika, and Central Line to Dar es Salaam. The last of these routes is the most important.

TRANSPORT IN SOUTHERN ZAÏRE

As has been noted, much traffic of southern Zaïre uses the Zaïre-Kasai waterway. But this area, and particularly Shaba, is least well served by water routes, not only because of their distance from the producing areas but also because transshipments are more difficult for bulk mineral traffic than for other types of cargo. The transport backbone of Shaba is the Kinshasa-Dilolo-Lubumbashi line (KDL—formerly the BCK), with 1,642 miles (2,642 km) of 3-foot, 6-inch gauge track. The KDL leads all Zaïre railways in freight transported, mainly because most of its traffic is in minerals and metals. The KDL greatly raised its capacity in postwar years by re-equipping and relaying with heavier track and by electrification of a considerable section.

The KDL and the Benguela Railway to Lobito is the most direct outlet for Shaba, and as a result that Angolan port normally ranks second among African ports in tonnage of traffic for Zaïre. After closure of the Rhodesian-Zambian border in 1973 the KDL transited 724,000 tons for Zambia to and from Lobito. Efforts to release capacity on the Benguela Railway by transferring Zaïrean traffic to the national route were somewhat frustrated in 1973 by low water on the Kasai-Zaïre river section. The status of Angola as a European-controlled territory provided additional incentive to Zaïre to link its national rail lines; the Benguela Railway

A train operating on the electrified section of the rail line in Shaba This line handles almost all of the mineral traffic of Shaba. It has connections with the Benguela Railway, the Zambia Railway, and the line in southeastern Zaïre.

was cut on several occasions by guerrillas but the traffic was never stopped for more than a brief period.

SUMMARY

It is probable that the center and north of Zaïre will continue to rely on the Zaïre and the south on the KDL with the proposed connection to the Matadi-Kinshasa line. But it may be expected that road transport will become of increasing significance as it has elsewhere in Africa. A particular feature of Zaïre transport has been the very limited importance of trucking, measured by percent of total ton-mileage of freight carried. A number of factors disfavor the use of roads more in Zaïre than in most African countries: its interior location, which means that trucks have less chance of competing with railways and waterways for the long-distance movements required; the considerable bulk of much of the traffic; the fact that a substantial part of the highest value agricultural traffic is generated in the more remote eastern reaches; the unusual extent of navigable waterways; and the usual physical problems which beset road building in all wet tropical areas. During the 1960s the roads deteriorated very badly and only now are funds being allocated in adequate amounts to rebuild some sections. There are few lengths of paved highways outside the main centers, though construction is underway on a road from Matadi to Kinshasa and eastward to Kikwit, a total distance of about 680 miles (1,094 km). A trans-Zaïre highway in the south is rendered particularly costly because it runs counter to the general north-south grain of the country and involves a multitude of expensive bridges. Zaïre will have about 966 miles (1,554 km) of the Trans-Africa Highway running across the northeast.

The great distances between centers and the slowness of surface connections have meant that airways have played an important role in movement of persons and very high value freight. There are now three international-class airports plus about 175 other fields. Air Zaïre serves about 44 centers on a regularly scheduled basis.

AGRICULTURE IN ZAÏRE

A high degree of local self-sufficiency is still characterisic of Zaïre agriculture, though a higher percentage of farmers had been brought into the money economy at the time of independence than in most of tropical Africa. About 70 percent of the population is engaged in agriculture, but less then 5 percent of the land is under cultivation. The most important subsistence crops are manioc, by far the leader though several regions have been hit by manioc blight in recent years and resistant stock has not yet been widely disseminated; bananas; corn, which is the basic crop in the southwest; and millet. Livestock is relatively unimportant to the Af-

ricans except in the higher, tsetse-free Kibali-Ituri area. Plantations are more important in Zaïre than in any of the countries thus far studied except Liberia, while European estates also figured significantly in commercial production. Before 1960 agricultural exports accounted for about 40 percent of total exports by value, a not unimpressive share given the great importance of the mining industry. Agriculture, and in particular commercial production by smallholders, was severely set back by the events after independence. Whereas marketed agricultural produce accounted for 21 percent of the GDP of the money sector in 1959, it accounted for only 10.8 percent in 1972, by which time its share of exports had fallen to 17.5 percent of the total.

DEVELOPMENT OF PLANTATIONS

European plantations and estates exploited over 1,500 square miles (3,885 sq km) of crop lands and 450 square miles (1,166 sq km) of forests and pasturelands in 1959, employing a labor force of about 300,000 Africans. Many of the plantations and estates were abandoned in the early 1960s, but most were again in operation and others were being rehabilitated when the government nationalized all plantations and estates in late 1973. It is too early to assess the impact of this potentially very damaging measure. Large companies were especially important in the production of palm products, rubber, robusta coffee, and cocoa; colons were concerned with specialized crops and cattle.

Plantations had their debut in 1890 when the governor ordered his district commissioners to establish cocoa plantations at each post. Although over 4,500 acres were planted to cocoa by 1898, the results of this program were disappointing, owing to the lack of knowledge and of qualified personnel and to the fact that some areas were not ecologically suitable to the cocoa tree. In following years, efforts were made to introduce plantation production of rubber, sisal, ramie, and cotton with only limited success, despite the issuance of various decrees from 1891 to 1903 requiring the local chiefs to supply labor to produce plantation crops on lands belonging to the state.

Early efforts to stimulate African output of commercial crops were also disappointing, owing to indifference, lack of skill, and inadequate knowledge regarding suitable crops. Convinced that it could not move forward with sufficient rapidity by this method, the government adopted a policy favoring the introduction of company operations. In 1911 a convention was signed with Lever Brothers involving rights of exploitation of natural and planted oil palms in large blocks of land. The spectacular results of this entente, which increased exports of palm oil and palm kernels with great rapidity, were a dominant factor in the continuing emphasis given to plantation production. The period from 1925 to 1930 witnessed a particularly rapid increase in extension of plantations; banana plantations were started in Mayumbe, sugar plantations in Lower Zaïre, and acreages were considerably increased in rubber, coffee, cocoa, and other crops. Many of these new plantations were closed during the great depression, despite government aid through reduced taxes, lowered freight rates, and assistance funds. In the late 1930s and particularly after World War II, there was renewed interest in plantations, and an especially notable increase took place in coffee between 1950 and 1957. In 1959, production on European plantations and estates accounted for about 75 percent of the value of agricultural exports.

The question whether plantations should be promoted in developing areas has been a subject of discussion for many years and is worthy of parenthetical attention at this point. The physical and economic advantages weigh heavily in favor of plantation methods. It is much easier to apply scientific techniques on plantations, including careful seed selection, proper rotational and fertil-

An oil palm plantation at Yaligimba when three years old
Foreign-owned plantations and estates, which accounted for about three-fourths of Zaïre's agricultural exports before independence, were nationalized in 1973.

ization practices, and scientific disease control. As a consequence of using more advanced methods, plantations typically have yields ranging from two to ten times those of indigenous producers. They also often bear fruit earlier and have a longer productive life.

In addition, the use of large-scale techniques often permits lower cost production. Orderliness facilitates control of competing growth; harvesting can be systematic, swift, and cheap; special transport can be provided, such as narrow-gauge lines or overhead conveyors; assurance of a regular supply may justify provision of processing mills, which, in the case of palm oil, for example, permit cleaner, richer oil and extraction of about twice as much oil as is normally obtained by cruder indigenous methods; processing plants may also be able to utilize by-products that would otherwise be wasted; finally, there is usually a marked saving in manpower. The United Africa Company, for example, has reported that only one-third as many laborers are required on its plantations to produce a given amount of oil as would be required in indigenous production. Another advantage of plantations is that they permit a rapid increase in production of cash crops.

Plantation operators claim that there are also social benefits of considerable value associated with the plantation system. Governments usually stipulate that certain standards of schooling, medical facilities, housing, and food supply be maintained and many plantation companies make such provisions on a level superior to the legal requirements.

The arguments are by no means all in favor of plantations. From the physical standpoint, there is a somewhat greater possibility of disease devastating a plantation than the scattered, mixed holdings of indigenous farmers. The United Africa Company states rather strongly that "new diseases, not found in natural circumstances, are bound to attack." Also, concentration on one crop may be harder on the soil than mixed agriculture, though the prevalence of tree crops in plantations makes for a closer comparison to the climax vegetation.

From the economic standpoint, there are the disadvantages of too great dependence on one crop and overhead expenses not present in peasant farming. And there is the danger that too great a portion of earnings will leave the country.

From the social standpoint, there is disruption of the existing way of life, possible conflict with traditional man-land relations, and sometimes the separation of families when labor is recruited from a distance and without provision for family living.

From the political standpoint, enterprises run by outside companies are always open to charges of exploitation—whether these charges are justified or not. In many areas where plantations were important, they have been the subject of increasing resentment as nationalism increases. This is perhaps the most potent argument aginst them. Yet to eliminate plantations because of political objections to them is possibly to deny a country the advantages of what can be a highly efficient and competitive way of increasing production and productivity. Possible solutions to this dilemma include the setting up of partnership or cooperative plantations, participation of governments in ownership, requirements regarding training of locals for eventual management, and agreement regarding level of profits permitted. It is also possible to promote outgrowing by smallholders around a nucleus plantation.

Plantations appear to be most appropriate when a new crop is being introduced to an area and when the processing facilities require both substantial capital and skill. Cocoa and coffee are perhaps least appropriate; tea, sisal, rubber, and bananas are most appropriate. The first three need fairly sizable plants, and bananas have to be supplied in high quality on a regular basis to meet transport and marketing schedules. All of these crops have, however, been produced successfully on an outgrowing basis and some regions in Africa are now sufficiently familiar with tea growing to justify the setting up of factories supplied entirely by local farmers.

ESTATE AGRICULTURE

The term "estate agriculture" is used here to refer to farms managed by their owners, as contrasted to the usually much larger plantation operations, which are typically company-owned and managed by a paid staff. The first effort to promote estate farming in Zaïre was made in Shaba. Mining brought a big influx of Europeans and Africans to the area after 1908; sparsely populated, the region did not have agricultural surpluses available for provisioning the mining communities, while its remoteness made importing of food very costly. To stimulate local food production, and also to neutralize what was then considered the serious danger that Shaba might be attracted to Britain by penetration from the south, it was decided in 1910 that Belgian colonists would be encouraged to settle in the area. Despite subsidization only 18 farms were occupied and they met with very disappointing results. Shaba soils are generally poor, precipitation is less regular than in other areas of Zaïre, labor was hard to find, and it proved difficult for the colons to produce foodstuffs that could compete either with those of local Africans or with imports from South Africa. By 1926, however, some 115 colons had become established in Shaba, which remained one of the two regions for European estate farming.

The second area for European colons was in the higher eastern part of the country, in Kivu and Ituri (Map 74). In 1959, most of the 1,899 colons in agriculture in the country were in the east. A variety of assistance programs inaugurated in 1945 had helped to increase their number from 657 immediately after the war, but estate agriculture in Zaïre was usually carried on by settlers who had substantial financial resources; it bore little relation to pioneer farming in the American West.

Colonization in the east was in strong contrast to that in Shaba. In the latter the farmers raised cattle and hogs and produced food crops for sale to the large mining com-

A coffee estate in the eastern highlands
Estates in this area produce high-value crops such as coffee, tea, pyrethrum, medicinals, and perfume essences.

munities. Settlers in the east concentrated on high-value, low bulk crops, most of which were destined for foreign markets. The remoteness of the area required such an emphasis, but the region is also ecologically well suited to specialized tropical crops. Possessing some excellent soils and having a tropical highland climate with moderate temperatures, it was far more attractive for European settlement than other parts of Zaïre. Labor was also much less of a problem, the east frequently having a high population density. Coffee was the prime crop on European estates in the east; tea, medicinals, pyrethrum, perfume plants, and tobacco were also of importance, while some farms specialized in the raising of cattle. What percentage of these estates are functioning in 1975 is not known, but apparently a sizable number had been rehabilitated when nationalization occurred in late 1973.

DEVELOPMENT OF INDIGENOUS AGRICULTURE

The first effort to stimulate African peasant farming in Zaïre, both for production of commercial crops and for improved local standards, was the imposition of obligatory duties. This system led to abuses which, after

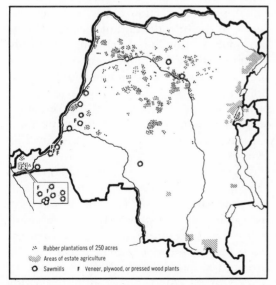

Map 74. Estate areas; rubber plantations (c. 1960); sawmills and other forest industries (1973) in Zaïre

.·: Rubber plantations of 250 acres
Areas of estate agriculture
○ Sawmills   F Veneer, plywood, or pressed wood plants

severe criticism, were gradually removed. The reform, however, led to what the Belgians considered economic disaster: many Africans simply stopped working. Obligatory duties were reinstituted during World War I, with the provisions that they were to have an educational goal and that the work was to be for the exclusive profit of the indigenous producer. From 1927 to 1930 each African was required to plant 10 palm trees a year around his village; after 1935, increased efforts were made to stimulate African production of coffee and systematically to establish indigenous palm plantations. While required work did lead to a big increase in production of cotton and other crops, achievement of the educational goal was only moderately successful.

In 1936 an experiment was undertaken near the Gandajika station of INEAC, the agricultural research organization of Zaïre. In the experimental area each volunteer was allotted a plot divided into yearly pieces to be cultivated according to a rotational system determined by the station. This modest effort was important in that it presaged the large-scale introduction of paysannats after the war.

After World War II the Zaïre peasants

Terraced agriculture in Kivu Region
Care must be taken in this densely populated highland area to prevent soil erosion.

gradually abandoned their attitude of passivity and took increasing initiative in their own advancement. Imposition of duties was necessarily replaced by persuasion and demonstration; the duration of *travaux educatifs* was reduced to 45 days a year in 1955 and they were largely suppressed after 1957. But indigenous methods evolved relatively little and increased production was offset by losses to erosion, deforestation, and degradation of soils. It was with these considerations in mind that the Belgians decided to introduce a system of paysannats. Paysannats involved all manner of technical and social improvements in the indigenous agricultural system, but the distinctive feature was the institution of organized bush fallow agriculture, sometimes called the "corridor" system. While there were several variants, the system basically involved organizing land into corridors whose fields were systematically opened and permitted successively to revert to bush on an overall 20-year cycle. The advantages of the corridor system, which represents a compromise between technical agronomy and the natural regenerative capacity of the soil, are that it tends to stabilize the rural population, permits an orderly rotation, makes plant protection easier, can be conducive to the introduction of mechanization, and may be instituted without drastically changing the indigenous approach to land and agriculture. Resettlement of farmers along roads or tracks may also facilitate improvements in transport and the provision of social amenities. Some of the paysannats also involved the introduction of individual tenure.

By the end of 1958, some 194,000 peasants were associated with the entire paysannat program, and the paysannats were responsible for a disproportionate share of indigenous agricultural production. In 1957, when they included less than a tenth of the agricultural population, they produced 15 percent of the corn, 22 percent of the paddy rice, and 27 percent of the cotton. Incomes of paysannat members were typically well above those of nonmembers in the same areas. In addition, thousands of miles of road had been constructed to serve the paysannats, hundreds of wells had been sunk, and thousands of fishponds had also been prepared and stocked. The second ten-year plan called for the accelerated application of the paysannat system.

Following the collapse of authority in Zaïre, all of the monitors who were responsible for scheduling work on the paysannats were dismissed, very few Belgians who had been associated with them remained after the first few months, and machinery quickly fell into disrepair. The need for supervisors and skilled technical advisers militates against the possible reintroduction of the system, which also was resented by some members because of the extra effort required. Just what traces of the paysannat system remain is not known, but it is likely that there have been residual benefits from the well-digging, road-building, and erosion-control measures.

Indigenous agriculture has received low priority in the programs designed to rebuild the Zaïre economy. Despite the growing shortfall of food moving to the urban areas, which has necessitated large-scale imports, the agricultural sector has been allotted only a very small portion of government expenditures—2.5 percent in 1973. Repeated statements that agriculture is to be given top priority appear to have been ignored. Stimulation of this sector must involve rehabilitation of the transport infrastructure, rebuilding the marketing system to assure the availability of goods desired in the rural areas (without which there is little incentive to produce crops for sale), and a multipronged program devoted to agriculture per se. A variety of international agencies are now beginning to assist in the major task of rehabilitating agriculture in Zaïre, but it may require many years before satisfactory advances are achieved.

THE COMMERCIAL CROP AND LIVESTOCK PATTERN

All of the major tropical crops are produced in Zaïre, giving that country a somewhat broader range of exports than is enjoyed by most African countries. But four products—coffee, palm oil and kernels, rubber, and cotton—have been of outstanding importance among vegetable-product exports (Table 14).

COFFEE. Now the single most important agricultural export, coffee is produced in a more dispersed pattern than that of any other export commodity; many areas are situated great distances from the national seaboard. Robusta coffee is native to Zaïre and can be cultivated up to 5,000 feet (1,500 m). It is grown both on smallholdings and plantations in the wetter savanna areas of the Zaïre Basin (Map 75), the Uele region having witnessed a particularly rapid extension from 1945 to 1958. Most of the robusta coffee is processed in Kinshasa.

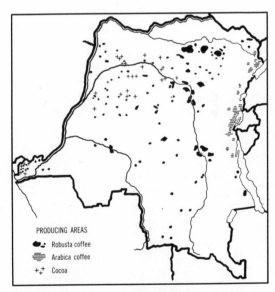

Map 75. Coffee and cocoa producing areas in Zaïre
While this map is based on data for the immediate pre-independence period the pattern is unlikely to have changed significantly, since there have been few new plantings since 1960.

Table 14. Volume and value of agricultural and forest product exports of Zaïre, selected years 1950–1972

	Volume—thousand tons				Value—mil $
	1950	*1959*	*1968*	*1972*	*1972*
Robusta coffee	33.2	93.4	44.1	61.5	47.6
Arabica coffee			9.0	12.5	8.5
Palm oil	132.0	185.5	141.2	87.0	18.6
Palm kernel oil	14.6	60.7	48.5	37.3	6.6
Palm kernel cake	20.8	100.2	44.9	44.6	1.3
Rubber	8.2	40.2	40.1	37.7	11.2
Logs	⋯	162.5	49.6	30.8	5.8
Sawn wood	⋯		36.1	37.5	
Quinine	⋯	1.8	0.3	0.5	5.2
Tea	0.6	3.6	5.1	7.1	4.7
Cotton	51.0	52.8	—	4.6	3.1
Cocoa	1.7	3.9	5.1	6.0	2.2
Urena and punga	⋯	4.2	—	1.2	a
Copal	12.5	4.1	0.5	0.1	a
Pyrethrum	1.7	0.2	—	—	—
Bananas	—	31.1	3.1	—	—
Other					10.7
Total					125.5

⋯ = Not available.
— = None or negligible.

SOURCES: Banque Centrale du Congo et du Ruanda-Urundi; République du Zaïre, Department de l'Economie Nationale, *Conjoncture Économique*, No. 13, November 1973; Banque du Zaïre, *Bulletin Trimestrial*, 3rd trimester, 1973.

a Included under other.

Arabica coffee, cultivated between 3,300 and 8,200 feet (1,000–2,500 m), is mainly grown on estates in Kivu and Ituri, where it is the main cash crop. Arabica coffees are produced and graded at Goma, though significant amounts have been smuggled to East Africa in recent years. About 235,000 acres (95,000 ha) are devoted to robusta and arabica coffee and plantations and estates account for at least two-thirds of output.

While Zaïre had long had an export of coffee, the crop only became important after World War II and particularly during the 1950s boom in African coffee sales. Exports trebled in tonnage from 1945 to 1959 when they were 93,400 tons; they fell off drastically in the following years but had recuperated to 74,100 tons in 1972.

PALM PRODUCTS. Palm oil, palm kernel oil, and palm kernels were the leading agricultural export by tonnage and value from at least 1887 to the mid 1960s. In addition, the oil palm is the chief source of edible fats within Zaïre and the raw material for domestic production of soap and margarine. It is grown wild in the equatorial basin and has been planted there and in parts of the savanna zones; the total area in plantations and individual holdings is estimated at about 1,112,000 acres (450,000 ha). Plantations have accounted for about three-quarters of output.

Before independence Zaïre was the best-equipped country in Africa for the treatment of palm produce. Most kernels were processed there rather than in Europe; bulk-

Robusta coffee nursery in the Uele district of Upper Zaïre Region
Coffee now ranks as the leading agricultural export of Zaïre.

ing and loading facilities were well developed. Palm products rely more heavily on the Zaïre waterways than any other crop, and the low cost of shipment over this system was significant in encouraging the expansion of such exports.

During the 1960s several large plantations were unable to maintain operations and there was little replanting with high-yielding stock. In more recent years expansion and rehabilitation were avoided because of low world prices and government requirements that half of plantation output must be sold locally at prices below the cost of production. The tonnage of palm oil exports, which was 185,500 in 1959, fell to 141,200 in 1968 and 86,971 in 1972, and the fear has been expressed that Zaïre may become a net importer within a decade unless steps are taken to rejuvenate production.

COTTON. Cotton was normally the most important commercial crop grown primarily by Africans. Its production, which the Belgians considered to be of prime importance in maintaining village life and in introducing the African to the modern economy, was stimulated by provision of selected seed, establishment of ginneries, elimination of

middlemen, and creation of a marketing board with reserve funds for poor years.

In the period to 1960 cotton became an important crop over a large part of the non-equatorial regions (Map 76). It does best north of 2°5′N, where dry seasons permit easier harvesting. About 9,000 miles (14,480

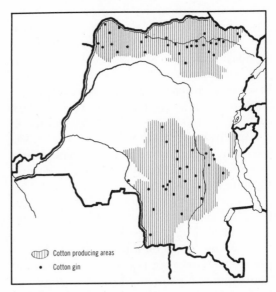

Cotton producing areas
• Cotton gin

Map 76. Generalized cotton producing areas and new cotton gins in Zaïre

km) of access roads were built by cotton societies, as was the CVZ rail line. By 1960,
some 700,000 planters were growing cotton
on an estimated 825,000 acres (334,000 ha).
Yields ranged from 66 to 110 pounds of
fiber per acre (75–125 kg per ha). About
four-fifths of production was exported, one-
fifth being absorbed by domestic spinning
factories.

Since independence there has been a dis-
tressing drop in production and exports,
and cotton has had to be imported in some
recent years. At present an estimated
280,000 acres (114,000 ha) are planted to
cotton; 21,500 tons were produced in 1972,
only one-third the 1959 level. The FAO and
World Bank are aiding a 75,000-acre
(30,000-ha) program in Equator Region, on
which it is hoped that 50,000 farmers will
produce about 35,000 tons of cotton. The
government has taken over the purchasing
and marketing of cotton but its agency,
ONAFITEX, has had financial problems and
its long delays in paying producers have depressed interest in the crop. Sixty of the former 125 ginneries are being modernized.

OTHER CROPS OF THE RAINFOREST AND LOW-
LYING SAVANNAS. Rubber exports increased
from 1,000 tons in 1938 to 40,200 tons in
1959, giving Zaïre third place among African producers and a 1.8 percent share of
world production. Shipments were reasonably well maintained during the 1960s and
stood at 37,700 tons in 1972. About 150,000
acres (60,000 ha) are planted to rubber, most
of which comes from ten plantations in
the northwest where considerable replanting
has been undertaken in recent years. About
a tenth of production is used by two domestic shoe factories.

Cocoa proved a very difficult crop for
Zaïre because of its susceptibility to disease.
But its production was increased from 1,300
tons in 1938 to 4,960 tons in 1957, and it is
one of the few crops that now has an output
above pre-independence levels—about 6,000
tons in 1971 and 1972. It is produced almost

entirely on estates covering about 15,000
acres (37,000 ha), chiefly in Mayumbe and in
the equatorial zone from Lukolela to
Kisangani.

Bananas have been grown mainly on plantations but also on African collectives in
Mayumbe and Lower Zaïre. Exports of Gros
Michel bananas increased from 2,400 tons in
1939 to 31,000 tons in 1959, when they met
about a quarter of Belgian needs, but they
fell to only 3,000 tons in 1968 and none
were exported after 1970.

Sugar production, from two plantations
covering 17,800 acres (7,200 ha)—one near
Tumba on the Matadi-Kinshasa line and the
other in Kivu—covers about two-thirds of
domestic needs. An additional plantation is
planned. Urena and punga fibers are grown
entirely by Africans in parts of Bandundu
Region, and a bag factory using these fibers
is situated at Kinshasa; earlier hopes that
urena would become an important crop have
not been met. Exports of peanut products,
never very large, ended in 1971. Tobacco
production has increased from 148 tons in
1968 to 1,000 tons in 1973; smallholder
growing is being promoted by the British-
American Tobacco Company in southern
Shaba.

COMMERCIAL CROPS OF THE HIGHER LANDS.
Arabica coffee, as has been noted, is the
chief commercial crop of the higher eastern
lands. Efforts were also made to extend the
acreage under tea with considerable success,
production having increased from 300 tons
in 1948–52 to 3,600 tons in 1959. Despite
damage during the early 1960s, exports
reached 7,065 tons in 1972. Some 21,000
acres (8,500 ha) are planted to tea in Kivu,
where FED has been assisting in the restora-
tion of plantations and a processing factory
and in the promotion of smallholder production.

Other specialized crops of the eastern
highlands have tended to decline, particularly where synthetic substitutes have increased competition. This is true for cin-

chona, pyrethrum, derris root (another natural insecticide), and perfume essences. Tobacco has fared somewhat better and has a growing internal market. With the exception of tobacco, most of the specialized crops have been produced primarily on European estates.

LIVESTOCK. Cattle are precluded from most of Zaïre by presence of the tsetse fly but efforts to increase the numbers of cattle and other livestock met with some success and permitted reductions in meat imports before independence. Most of the European-owned herd, totalling 469,000 in 1958, were raised on ranches in Lower Zaïre, Kivu, Ituri, and the high plateaus of Shaba and Kasai, but standards on many ranches left much to be desired. Cattle are mostly of African breeds, though some farms specialized in European and Asian breeds.

African-owned cattle are concentrated primarily in the northeast; their number increased from 194,000 in 1935 to 1,006,000 in 1958. The pig, sheep, and goat population also increased rapidly in postwar years. The livestock population is now estimated to be about 10 percent below the pre-independence level. Zaïre remains heavily dependent on imports of meat and greater attention must be given to raising livestock before it can provide a much larger share of its needs for animal products.

FOREST AND FISHERY RESOURCES

Forests, ranging from tropical rainforest to sparsely wooded savannas and montane forests, cover about 45 percent of Zaïre. Much of the total area is in woodlands whose trees are suitable only for firewood, but straddling the equator is one of the largest zones of selva in the world, while a much smaller region of rainforest exists in Lower Zaïre (Map 70). Only the latter has a location competitively comparable to those of the leading African producing nations. That

area has long been the dominant source of exports, but reserves of limba, which had made up 80 percent of exports, were so depleted that it was necessary to impose restrictions on output in 1955.

Production from the main forest area of the Zaïre Basin has been primarily for domestic needs, the high costs of transshipping having made export difficult until recent years. Exports of logs and sawn timber have fluctuated considerably in the last decade—owing in part to restrictions imposed to assure domestic supplies—but have increased markedly in 1972 though they remain well below the 1959 level.

Forest exploitation has undergone most of the trends noted for West Africa. Semiprocessed products provide a larger share of exports; new sawmills and peeling, veneering, plywood, and particle-board plants have been constructed (Map 74). Plans have been made to establish tree plantations in Lower Zaïre, where resources have been seriously depleted.

The fishery resources of Zaïre are considerable, and have been expanded through the introduction of fishponds. Sea fisheries, conducted largely from Banana, account for about a ninth of the total catch, lake fisheries for about two-ninths, and river fisheries for the remaining two-thirds, but only about half of the total catch is commercialized. The number of fishponds increased from 47,000 in 1952 to 122,400 in 1958 covering over 10,000 acres (4,000 ha). From 880 to 1,320 pounds can be obtained from an acre (c. 1,000–1,500 kg per ha) if the pond is regularly fed with household waste, banana leaves, fresh manioc, etc. Systematic feeding can greatly increase these yields.

MINING AND METALLURGY IN ZAÏRE

Mining has been the most dynamic and most important element in the modern economy of Zaïre. Its output increased tenfold from

1920 to 1950 and doubled again in the next six years. There are several hundred mines, of which a small number account for the bulk of output. About 100 processing plants treat the mineral output of the country. Mineral output was remarkably well sustained in the main producing area, Shaba, through the 1960s, though the central government was deprived of revenues from that area during the period of secession. Mining and metallurgy now account for about 23 percent of the monetary GDP, over 80 percent of exports by value, and a very important share of government revenues, though these fluctuate with changes in the world prices of metals, particularly copper. Total value of mineral exports was about $559 million in 1972 but was over $700 million in 1973 as prices rose substantially.

Mining employs a relatively small number of people considering its importance. Before independence about 100,000 Africans and 3,500 Europeans were engaged in the industry; in 1973 the total was 66,621, down partly because some mines were still not functioning but more importantly because of higher productivity. Mining and metallurgy have greatly stimulated transport and hydroelectric developments. While adding to the problems of the industry itself, the fact that the more important producing regions are far in the interior has meant that Zaïre transport was much better developed than it otherwise would have been. It also stimulated the desire to process ores in the area, thus contributing to the expansion of industry and increasing the value of exports.

Almost all mining is concentrated in the foundation rock formations girdling the Zaïre Basin. Minerals are scant in the ancient rock on the edge of the central depression, especially in the north and northeast, but the ranges of strongly folded sediments peripheral to the basin have great mineral wealth, especially in Shaba and the Kibara Mountains. Tin, tungsten, niobium, rare earths, gold, and bismuth occur in these zones. A

third element of the Congo's foundation mass is composed of other layers encircling the depression—layers which, like the rest of the base, were folded and dislocated time and again and then peneplained over an extended period. The most important portion of this last element is found in Shaba, including systems that are abundantly mineralized with copper and its allied minerals. The younger sediments of Zaïre contain the coal beds of Shaba and oil along and off the coast.

The chief minerals of Zaïre by percent of the total value of mineral exports in 1972, when minerals accounted for 81.5 percent of exports, are shown below:

Copper	71.8%	Gold	1.2%
Cobalt	9.4%	Manganese	0.9%
Diamonds	7.5%	Silver	0.3%
Zinc	4.2%	Germanium	0.1%
Tin and cassiterite	3.2%	Others	1.4%

By grouping those minerals produced in the Shaba copper mines (cobalt, zinc, cadmium, germanium, lead, silver, and palladium) we find that they accounted for over 86 percent of the total value of mineral exports in 1972.

THE SHABA COPPER COMPLEX

The Shaba copper mines are distinctly the most important segment of the Zaïre economy and developments are underway that likely will increase their status. Mineralization occurs discontinuously in a belt about 50 miles (80 km) wide extending about 340 miles (550 km) across the south of Shaba. With the continuation in Zambia it forms one of the most remarkable metallogenetic zones in the world, second only to the Rand–Orange Free State gold complex among metallic-mineral producing regions in Africa. The Shaba-Zambian Copperbelt areas are, indeed, the most important economic "island" north of the Republic and south of the Sahara.

Mining and metallurgy in Shaba were

The open-pit copper mine at Kamoto
Gécamines has three underground and nine open pit mines in Shaba.

developed by Union Minère du Haut-Katanga, founded in 1906. Zaïre nationalized UMHK in January 1967 but signed an agreement with affiliates to market the output and, after difficult negotiations, to manage the operations under contract. In 1969 a definitive agreement was concluded whereby the affiliate would receive 6 percent of the value of minerals produced for fifteen years to compensate for the takeover and cover the management fee, and 1 percent thereafter to cover technical cooperation. The Zaïre company, Gécamines, now owns the former UMHK installations and was responsible for all copper output up to 1972; in that year its mineral production accounted for two-thirds of all the exports from Zaïre. The company employed 28,250 people in 1973, or 42.4 percent of the total number employed in mining.

The Gécamines operations are divided into three groups centered on Lubumbashi, Likasi, and Kolwezi (Map 73). There are three underground mines and nine open pits, including new mines in both categories not yet in full production. While the ores are very rich by comparison with most copper regions, the occurrence of the surface ores in irregular veins means that much in-

terlarded waste must also be mined. Reserve figures are not available, but probably at least 750 million tons of ore have been proved.

The copper ores of the area are divided into sulfides and oxides. Sulfides come from the Kipushi mine, where an unusually rich vein extends almost vertically downward; a new shaft has recently permitted mining to a depth of 3,770 feet (1,150 m) below the surface. Associated with the copper of this mine are zinc, cadmium, and germanium, plus small amounts of gold, silver, platinum, and palladium. The complex ore is concentrated at the mine, with the zinc concentrates being roasted and the copper concentrates being smelted at Lubumbashi to produce 99 percent blister copper. The lesser by-products are recovered from flue dust, refuse from electrolyzing zinc, and in the final refining, which takes place both in Shaba and in Hoboken, Belgium, while sulfurous gases produced in treating the zinc concentrate are used to make sulfuric acid.

The oxide ores, of sedimentary origin and containing 4 to 6 percent copper with which cobalt is associated, come from surface mines, chiefly in the western group. Operations are highly mechanized, permitting high productivity per worker. The oxide ores of the western mines are concentrated at Kolwezi or Kamoto and refined at Luilu, whose capacity has recently been enlarged. In the central mines, sulfureted ores are concentrated at Kambove and treated in the metallurgical complexes at Likasi-Shituru and Likasi-Panda. These installations, making up the most important industrial establishment in Zaïre, produce 99 percent pure cobalt, 99.95 percent copper, and cobalt-copper alloy—which is forwarded to Belgium or the United States for refining. Slag is used to produce metallurgical cement at Likasi. In addition to the metallurgical plants, Gécamines has workshops, an acetylene plant, a factory producing explosives, and important hydroelectric facilities.

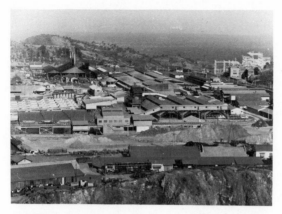

The Shituru metallurgical complex
This installation, which produces copper and cobalt, is the most important metallurgical plant in Zaïre.

The story of the establishment of the whole mining-industrial-urban complex of Shaba in a remote, sparsely populated part of Africa is a fascinating one. Numerous problems had to be overcome over the years, and all of them were, with considerable success, except for the relations with the independent government. The real or suspected role of UMHK in the Katanga

The electrolysis hall at the Luilu plant
The copper-cobalt refinery serves the western group of mines in Shaba.

secession, resentment of its paternalistic attitudes, and the desire to have national ownership of a wasting resource all contributed to the takeover of UMHK, but there is no question that its technical achievements bequeathed a major resource to Zaïre.

One of the major problems faced in developing Shaba was that of securing an adequate labor supply. That region was sparsely populated and could not support the numbers required. A recruitment program was instituted, which involved moving workers from distances up to 500 miles (800 km) away. The shortage was also responsible for the decision made as early as 1927 to recruit on a family basis, and doubtless contributed to the efforts to make living and working in the area as attractive as possible by the provision of housing and other amenities. Certainly the record of labor turnover compared very favorably with that of comparable enterprises. The labor situation also helps to explain the efforts to improve productivity, which made the mining and metallurgical operations as modern as can be found. The very high wages required to induce European employees to this otherwise unattractive part of the world probably also contributed to the desire to upgrade African employees, who at early dates performed many tasks often assigned to Europeans in other African mining communities.

A second great problem involved the provision of transportation for this area located 1,200 miles (1,930 km) from the coast. At the start of operations, the nearest railhead was 425 miles (684 km) south, and in 1910 only that rail line extended to Lubumbashi. With construction of the KDL and then the Benguela Railway, available transport was much improved, but the problem was not really solved until postwar years, because Rhodesia Railways was periodically incapable of delivering adequate amounts of coal. The Belgian policy of favoring the "national route" militated against the Shaba mines, because it required costly transshipments at

Ilebo and Kinshasa. This mineral output was permitted to use the direct rail route to Lobito.

A third problem was largely technical and has been solved—namely, the development of new methods required to treat some of the more difficult ores and concentrates.

The most difficult supply problem was that of providing adequate power and coking coal (the latter is required for smelting operations). This continued to be a severe problem until about 1956, occasioning periodic slowdowns and restricting overall activity for considerable periods. The mines relied first on coal or coke from Europe, then mainly on the Wankie collieries in Rhodesia, but in the first postwar decade the latter could not supply sufficient quantities, sometimes because there was inadequate output, sometimes because of congestion on the railways. It was necessary, therefore, to use large tonnages of wood, which is expensive, and to import coking coal from South Africa or the United States via the Benguela Railway, a very costly haul for this bulk commodity. After the unilateral declaration of independence in Rhodesia in 1965 there was again periodic difficulty with the receipt of coal from that country, but even after the Rhodesia-Zambia border was closed in 1972 the Zambians permitted coal to move across its territory to Shaba.

To reduce the pressure on overall energy demands, output of coal from the Luena coal field to the north was increased from about 100,000 tons to about 460,000 tons in 1955, but this coal is of very low grade and could not be used in smelting operations. Its major role was to produce electricity and to provide fuel for the railways and a cement plant at Lubudi, and thus to reduce the need for imports. With the opening of additional hydroelectric plants, partial dieselization of the railways, and more steady delivery of Rhodesian coal, output of coal in Zaïre declined to 247,000 tons in 1959, and by 1972 it was down to 113,000 tons.

Construction of hydroelectric facilities has been the most important solution for energy problems and permitted the extension of electrolytic refining in the region. Four plants in Shaba with a total capacity of 467,000 kw now serve the copper industry (see Map 73). A portion of the output of the largest of these, Seke, was transmitted to the Copperbelt in Zambia with the agreement that it might later be repaid by reverse transmission from Kariba. It is somewhat ironic that until recently three-fourths of Zaïre electric power should be produced in an area not particularly well suited for hydroelectric installations because of the highly seasonal precipitation and the great year-to-year fluctuations in amount of flow, while the magnificent hydroelectric potential of the Zaïre River below Kinshasa was almost untouched. This is explained by the limitations on economical transmission distances and the great consumption requirements of Shaba, where 4,000 to 5,000 kwh are needed to produce one ton of copper or zinc. This

Penstocks and generating hall of the Seke hydroelectric plant
This is the largest of the stations constructed in Shaba to serve the copper industry. Power for increased needs will soon be transmitted across the country from Inga on the lower Zaïre.

situation is about to change, however, as power from Inga (see below) near Matadi will be transmitted to Shaba to support increased needs of Gécamines and other producers by about 1977.

Two potentially large new producers of copper have entered the scene in Shaba in recent years. The Société Internationale des Mines du Zaïre (SIMZ), established in 1969 with a 15 percent holding by Zaïre and 85 percent by a Japanese consortium, has a 13,900-square-mile (36,000-sq-km) concession southeast of Lubumbashi. Two mines are being developed—at Musoshi, where a concentrator was completed in 1972, and at Kinsenda. Reserves are very large, varying in grade from 2 or 3 to as high as 8.4 percent.

The Société Minière de Tenke-Fungurume (SMTF) has even more exciting possibilities, having proved 45.7 million tons of 5.5 percent copper, 0.44 percent cobalt ore, and other very rich ores. SMTF is headed by Charter Consolidated but has strong American participation. An investment of about $350 million is being made in the Tenke-Fungurume area north of the rail line between Likasi and Kolwezi to develop open pit mines, the full range of processing facilities, and the necessary infrastructure.

Output of copper in Shaba was about 495,000 tons in 1973 and is expected to reach 800,000 tons by 1977, when Gécamines plans to produce 500,000 tons, while SIMZ will reach an estimated 200,000 tons, and SMTF 100,000 tons. Zaïre has indicated that all minerals must be refined domestically when Inga power reaches the area. Copper and its associated minerals will then play an even more important role in the economy of Zaïre than they now do.

OTHER MINERALS IN SHABA

Uranium was mined until 1961 at Shinkolobwe, 15 miles (24 km) west of Likasi. This mine is of great historic importance because it supplied a very important share of uranium ores utilized by the United States in World War II. Its unusually rich pitchblende ores were mined at the surface until 1950, after which it was necessary to turn to underground operations, whose reserves were exhausted after about a decade. An Italian company has recently been studying unexploited uranium reserves in the Gécamines area.

Lead is mined at Kengere, 34 miles (55 km) from the Manika station of the Shaba railway. As has already been noted, coal of mediocre quality is mined at Luena in an opencast operation. Small quantities (29,000 tons in 1972) of low-grade coal also come from an underground mine at Lakunga near Kalemie on Lake Tanganyika. Salt and iron ore deposits are known but not exploited.

Manganese is mined at Kisenge in southwest Shaba. After washing, a product containing 51 percent Mn is exported via the Benguela Railway. Output has increased from 6,400 tons in 1948 to 207,000 tons in 1960, and reached 369,000 tons in 1972, when 569 workers were employed.

THE EASTERN "TIN ZONE"

A second mineral region is found in an area extending more than 600 miles (965 km) north from northern Shaba through Maniema, the leading producing area, to northern Kivu and about 200 miles (320 km) east and west. It has scattered but sometimes rich deposits of tin, tantalum, columbium, tungsten, beryllium, bismuth, gold, and rare earths. Tin is extracted from cassiterite, sometimes in primary mineralized formations, sometimes in alluvial and eluvial sediments. Often associated with it are wolframite (containing tungsten) and tantalocolumbite (containing tantalum and columbium), each of which is extracted and exported as concentrates. For the future, the deposits at Manono and elsewhere may become an important source of lithium. The pegmatites there contain 15 percent spodumene, containing lithium, making the re-

Tin mine at Manono
The tin mining zone extends from northern Shaba to central Kivu and contains very large reserves.

serve one of the more important in the world.

Extraction from small detrital deposits has characteristically been by hand; working of the large detrital and the primary deposits is achieved either by hydraulic mining, or by the use of shovels and draglines. Most of the output is sold as "marketable" cassiterite, containing 12 to 76 percent tin, though electric foundries produce 99.9 percent ingots at Manono and Lubudi. In 1973 about 23,500 people were employed in about 20 operations mainly in a belt stretching across the center of Kivu but also in two districts of northern Shaba. Output totaled 5,850 tons of Sn in 1972, including the tin content in the cassiterite exported.

Operations in this area were severely disrupted and installations suffered considerable damage in the 1960s. Investors are showing renewed interest in the region, and the very large reserves suggest that production will grow in the years ahead. A Union Carbide–Belgian–Zaïre consortium is examining a columbium body at Bingo in Kivu, said to be one of the richest in the world.

THE NORTHEASTERN GOLD AREA

The gold zone of Zaïre runs in a broad belt along the east, overlapping the tin zone in Kivu, but extending across Upper Zaïre Region to the Sudan border. About two-thirds of production now comes from a nationalized company mining high grade primary deposits underground at Kilo. A small output comes from Kivu. Alluvial deposits are largely exhausted. Gold mining suffered heavy damage in the 1964–65 rebellion in the east and output has only regained about half of the pre-independence level. About 6,625 people were employed in gold mining in 1973.

THE KASAI DIAMOND AREA

Alluvial diamonds are mined in two zones, between Tshikapa and Luebo in Kasai West and in the Lubilash River Basin in Kasai East, while mining of kimberlitic pipe diamonds takes place at Mbuji-Mayi in Kasai East. The vast bulk of output (12.2 of 13.4 million carats in 1972) comes from the Lubilash-Mbuji-Mayi region, where the Société Minière de Bakwanga (MIBA) has highly

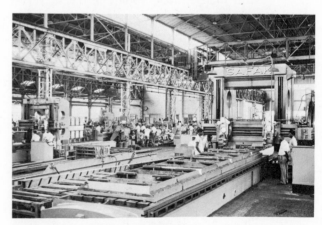

Diamond washing installation in Kasai Region
Zaïre is the world's leading producer of industrial diamonds.

mechanized operations. In 1972 it exploited 1.6 miles (2.5 km) of the Bushimayi River by drying the bed and excavating the diamond-bearing gravel; in 1973 it worked a slightly longer stretch and opened a new mine on the Sankuru River. MIBA, which is government-owned, employed 4,258 people in 1973. Its operations are supported by two hydroelectric stations near the Lubilash Falls with a combined capacity of 8,400 kw.

Most of Zaïre's small output of gem diamonds comes from artisanal alluvial workings in the Tshikapa area; production fluctuates wildly from year to year, having ranged from 551,000 to 2.5 million carats in the five-year period to 1973. Discovery of two potentially important deposits was announced in 1969 in Bandundu Province, 300 miles (480 km) east of Kinshasa.

After independence, operations at Tshikapa became more and more chaotic and those in Kasai East were interrupted in 1960. Illicit sales were heavy during the 1960s when Congo (Brazzaville) suddenly became a significant exporter of diamonds. Despite improved controls and marketing arrangements, substantial smuggling continues, though at a greatly reduced rate. Zaïre produces about half of the world's industrial diamonds; while the value of diamond sales has fluctuated rather widely in recent years, they ranked fourth among Zaïre's exports in 1972, with a total value of $42.8 million.

MINERALS IN THE LOWER ZAÏRE REGION

Oil in commercial quantities has been found along and off the shore of the small ocean frontage of Zaïre; production is expected to begin in mid-1975 and should reach about 1.2 million tons a year within a few years. Sands containing 15 percent bitumen have been mined along the Atlantic coast to provide surfacing, but the operation is small scale.

Plants near Ngungu in Lower Zaïre have treated lead and vanadium on a small scale from local deposits. Two bauxite bodies are known and might be opened to feed an aluminum mill based on power from Inga. A Japanese company has been studying an iron-silicon deposit in the area for its suitability for use in a glass factory.

INDUSTRIALIZATION IN ZAÏRE

Manufacturing increased rapidly in Zaïre in postwar years; its total development compares favorably with other tropical African countries except Rhodesia. Output increased somewhat erratically in the 1960s, but has been booming since 1969. Existing plants have expanded and a wide variety of new factories have been built and are planned. With 1970 the base year, the production index stood at 132.6 in mid-1973. Excluding the very large metallurgical industry, manufacturing accounted for an estimated 8.1 percent of the money-sector GDP in 1972. About 100,000 people were employed in manufacturing in 1970.

Outstanding features of Zaïre's industrial pattern and contrasts between it and those of other tropical African countries include the following:

(1) Mineral processing has particular importance. Unlike other countries, it is usually excluded from data on manufacturing. It is pertinent to note that the mining-metallurgical complex of Shaba has given rise to several linked industries including industrial gases and acids, explosives, and mining tools.

(2) Oil mills have represented the largest processors of vegetable raw materials but they have been hurt by the decline in output of palm products. There is the usual range of plants, usually quite small, processing coffee, grains, cotton, rubber, fish, timber, and animal products plus some large sugar, tea, and forest-product installations.

(3) Production of beverages ranks first among manufacturing groups. Some

12 plants brew beer, which has become the national drink, and many plants make soft drinks.

(4) The textile and clothing industry ranks second. Integrated establishments are found at Kinshasa, Lubumbashi, Kalemie, and Bukavu. Kinshasa has the largest plants and the greatest diversity of output, with five textile mills and numerous clothing plants.

(5) The shoe industry is relatively well developed, with three factories in Kinshasa and two in Lubumbashi including a Bata plant at each center.

(6) The metal fabricating and mechanical industries are reasonably well represented. In addition to the usual products such as containers, metal furniture, doors and windows, there are naval works at Kinshasa, copper and bronze foundries and an output of prefabricated metal buildings at Lubumbashi, and a burgeoning vehicle assembly industry, mainly located at the capital. British Leyland and Fiat opened assembly plants at Kinshasa in 1972; Renault-Peugeot, General Motors, Ford, and Nissan will follow. That each of these must be small is sug-

gested by the fact that the yearly market for vehicles is estimated to be only about 60,000, though it may be expected to increase rapidly. These companies have all been motivated by a desire to get within tariff barriers; Leyland, for example, can import knocked-down Land Rovers at a 7 percent tariff as compared to 70 percent on a fully assembled vehicle. The largest single manufacturing investment thus far made in Zaïre is the $120 million, 125,000-ton steel mill built by Italian-German interests at Maluku, upstream from Kinshasa.

(7) The chemical industry is represented by plants producing industrial chemicals and gases, explosives, tires, and the usual range of consumer products. A refinery at Moanda now has a capacity of 600,000 tons.

(8) The construction industry is notable for the large number of cement mills, seven (see Map 73).

(9) There is an unusual development of hydroelectric facilities. In the period 1950–59 twenty-four stations were either constructed or extended and at independence the installed hydroelectric capacity was greater than that of any other African country.

HYDROELECTRIC DEVELOPMENTS

Zaïre's total hydroelectric potential is estimated to be about 16 percent of the world total and 40 percent of the African total. In the 217-mile (349-km) stretch of the Zaïre River between Malebo Pool and Matadi, in which the river drops 886 feet (270 m) in thirty falls and rapids, a capacity of no less than 103 million kw could be developed at average flow. This figure is equal to the total installed capacity of all electric plants in France and the United Kingdom in 1970. There were no developments on this stretch, however, until the 1970s.

The installation of hydroelectric stations

A shoe factory at Kinshasa
Kinshasa has the best developed market oriented industrial mix in Zaïre.

up to independence was marked by: (1) an effort to provide large supplies for the electrolytic refining of copper in Shaba, for which four plants were built with a total capacity of 467,000 kw (Map 73); (2) construction of about nine stations, of which the largest had a capacity of 29,000 kw, to support other mining operations; and (3) installation of a number of plants to serve the larger urban centers of the country. Most of the last were very small, with even the Zongo station on the Inkisi, serving Kinshasa, having a capacity of only 75,000 kw.

Production of electricity totaled 3.6 billion kwh in 1972, when Inga I entered service. The four Shaba plants accounted for 71.5 percent of the total with mining and metallurgy taking 91 percent of that. Kinshasa absorbed 12.5 percent of the total.

Development of the main Zaïre below Kinshasa did not begin until recent years for several reasons: it was too far away from the

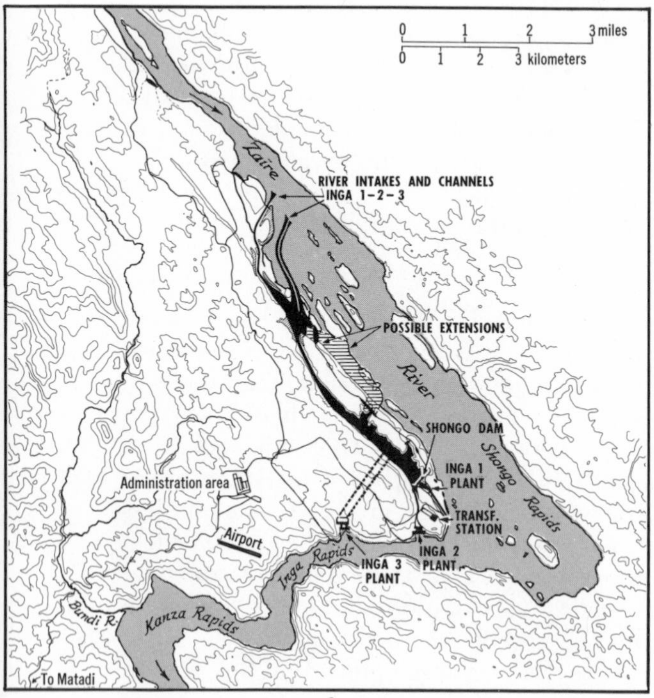

Map 77. The Inga hydroelectric development on the Zaïre River

major mining areas, there were no large-scale consumers in the area, and it was technically easier to install small stations on minor tributaries. Studies of the river date as far back as 1887, but a decision to proceed with a first stage of development, called Inga I, was not made until shortly before independence, construction did not begin until the late 1960s, and it was not completed until 1972. Inga has a number of features that give it particular interest: there is a drop of 335 feet (102 m) in a 9-mile (15-km) stretch; the average flow of 42,000 cubic meters per second, a volume exceeded only by the Amazon, permits a production potential of 370 billion kwh; and the site is unusually well suited to gradual hydroelectric development involving a relatively simple diversion scheme using an early channel of the river (Map 77). Inga I had an original capacity of 175,000 kw, and was increased to 350,000 kw by 1975; Inga II, slated to open in 1976, will eventually have a capacity of 1,272,000 kw; Inga III, not yet committed, would add another 1,200,000 kw; while later, at the Great Inga Stage, a dam would be constructed across the main stream.

While most of the output from Inga I could be absorbed in Kinshasa, Matadi, and towns in their immediate hinterlands, only a decision to transmit electricity 1,131 miles (1,820 km) to Shaba justified construction of Inga II. This unusually long line will cost about $250 million; it is being constructed by an American firm and is scheduled for completion in 1976. Other large-scale users that have been considered would be situated in Lower Zaïre and would be attracted by the availability of low-cost power; they include an aluminum refinery (at first using bauxite from Guinea) and a uranium enrichment plant. It is also possible that the Kinshasa-Matadi railway will be electrified.

LOCATION OF INDUSTRY

Most of the market-oriented plants in Zaïre have been located at Kinshasa and at Lu-

New construction along the main street of Kinshasa
The capital of Zaïre has experienced an unusually rapid growth in the years since independence.

bumbashi, the latter being the major city of the Shaba area and the former headquarters of UMHK. While Lubumbashi has not grown so rapidly as Kinshasa, its population rose from 98,000 in 1950 to an estimated 318,000 in 1970; it serves the reasonably concentrated, relatively high income mining-metallurgical region of southern Shaba. Kisangani and Likasi are secondary centers (Map 73).

Concerned that too many industries were siting at the two main cities, Zaïre has selected Kisangani and Bukavu as major growth poles for the northeastern and eastern parts of the country. Matadi has been a minor manufacturing center and it is likely that the new port at Banana plus power from Inga will attract new plants to these Lower Zaïre cities. Kananga in Kasai West and Kalemie on Lake Tanganyika are tertiary centers of some significance.

TOURISM

Prior to independence Zaïre had a budding tourist industry with the highly attractive and diverse areas north of Bukavu having a reasonably well developed circuit. The country has now decided to make significant investments to stimulate tourism not only in the major centers, whose increasing number of visitors include more officials than tourists, but also in the more scenic portions of the country, including its national parks. There is every reason to expect that Zaïre will develop this sector, which has the potential to become a very important contributor to foreign exchange earnings and to employment of locals.

SUMMARY

Zaïre has usually had favorable trade balances, but not always favorable payments balances. While agricultural exports declined seriously after independence and have still not regained their previous position, mineral exports have sustained the economy remarkably well and should increase dramatically in the present decade. Zaïre was never heavily dependent on the Belgian market and, unlike French territories, its commodities were normally sold at prevailing world market prices.

Zaïre probably experienced greater disruption of its economy in post-independence years than any African country. Yet there was surprising resilience in the more highly organized segments of the economy, especially large-scale mining and manufacturing but also in many of the plantations and estates. While conditions now appear reasonably favorable for rapid growth, one must remember to expect the unexpected in Africa. The highly authoritarian nature of the Mobutu government and its favoring of a small elite may lead to political difficulties. There is a particular need to devote more attention to agriculture, upon which the bulk of the population is dependent, and greater attention must be given to the redevelopment and extension of the transport and social infrastructures.

Part Five
EASTERN AFRICA

Part of the central business district of Nairobi, Kenya

CHAPTER FIFTEEN

Ethiopia

With the exception of a brief period under Italian rule from 1936 to 1941, Ethiopia has existed as an independent entity for over 750 years, by far the longest modern period for any African country. It cannot be said, however, that the very old traditions of Ethiopia have brought it to a more advanced stage than younger African countries; rather it remains as one of the least developed nations on the continent, a country whose complex of customs and institutions are encrustations that have inhibited economic, social, and political advance. In 1952 Ethiopia was federated with Eritrea, which had been an Italian colony from 1890 to 1941 and under British administration thereafter. In 1962 the federal system was abolished, and Eritrea became a province of Ethiopia. In February 1974 a government takeover began that led some months later to the arrest of Emperor Haile Selassie and assumption of power by the military.

The cornerstone of Ethiopian society had been the emperor, the church, and a feudal land system. The emperor, who was considered to have descended from a divine line, long dominated the political and social scene. He acceded in postwar years to some progress toward a constitutional form of government, though the Parliament was only partly elected and its acts were subject to his veto. Haile Selassie, in fact, made greater strides toward opening the country to modern influence than all of his predecessors. But the balance to be achieved between the impatient demands of the younger, educated, modernist elite and the older, deeply entrenched traditionalists was a delicate one and any step forward was too little for the former and too great for the latter. Arrest of the emperor (he had not been officially deposed in early 1975) will not automatically lead to rapid modernization nor suppression of the more traditional elements of society, because there is an intimate web of connections between the nobility, the church, and parts of the military leadership. The change of government has, however, weakened control over the country, which has in turn stimulated efforts by the Eritrean Liberation Front to gain independence for that province and led to numerous strikes and much unrest in many areas and in many sectors of the economy.

Coptic Christianity is the officially established religion of the country and the ruling Amhara have been Christians since the fourth century. The church has been a major source of national culture and a unifying force for perhaps half of the population, but it is also a source of conservatism and a preserver of the archaic land tenurial arrangements, having rights to revenue on at least a quarter of agricultural lands in the highlands.

The feudal system is highly complex and is intertwined with other components of the culture. It is represented by an unusually conservative aristocracy, a long heritage of serfdom, and a caste-like social stratification which is dependent upon a wide variety of physical, religious, and political criteria.

The land tenure systems, which vary greatly from one part of the country to another and which are still very poorly understood, have an unusual degree of complexity and the inequitable distribution associated with feudalism. Large parts of the farmlands of Ethiopia were owned or controlled by the royal family, powerful and often absentee landlords, or the church. Most Ethiopians had their holdings or their rights of use under tenancy systems that kept them at least partially subservient to the owner.

In the last decades the government attempted in a series of moves to modernize the land tax system as a first step to major land reform, but each step was militantly resisted. Hoben maintains that official policies have been based on inaccurate assumptions and that the necessity for land reform, widely accepted as a prerequisite to development, has been grossly exaggerated.[1] He sees the *rist* system of land use rights as flexible, not creating a landless class, and strongly preferred by the peasants who see it as guaranteeing their liberties and the opportunities they cherish, including pretensions to use of more land than is actually held.

Despite the unifying influences of church and empire there were and are great divisive

St. George's Coptic Christian Church at Addis Ababa
Ethiopia's highly developed religious, imperial, and feudal institutions have inhibited the introduction of modern economic, social, and political advances.

[1] Allen Hoben, "Social Anthropology and Development Planning—a Case Study in Ethiopian Land Reform Policy," *Journal of Modern African Studies*, 10, no. 4 (December 1972): 561–82.

forces in Ethiopia. The majority of the population still lives in tribal societies, noted for feuds within and between groups. Amharas, who make up about a third of the population, are the dominant group; they and the Tigreans speak Semitic languages, are mainly Christians, and have been relatively well united in facing threats from other ethnic groups. When the government speaks of unifying the country, it views it as a process of Amharization.

The greatest threats to the Amhara, other than revolts from within, are from the Galla, a mainly Muslim (although some are Christian), Cushitic-speaking people somewhat more numerous than Amharas; the Somalis, living in the eastern Haud and Ogaden, suspect because of their relatives in Somalia who dispute the south-eastern boundary; and from a portion of the Muslim minority in Eritrea, where guerrilla activity has flared off and on since 1962, tying down a sizable part of the military forces. Most of the remaining ethnic groups are Cushitic speaking, though some Negro peoples live along the low western borderlands, remote from the control or ken of the central government. In all, there are perhaps 100 tribes in Ethiopia, 95 languages, and two important religions plus numerous local cults. Levine labels the common claim that the large number of tribes of Ethiopia are a fractious and disunifying element, however, as a myth;[2] he maintains that a relatively comparable historical experience has contributed to unification though a complex net of intertribal relations, shared elements of a common culture, and some long-established recognition of a national center.

Ethiopia is still in an early stage of educational development, with a high percentage of illiteracy. But the number of students in government schools has increased from 20,000 in 1944 to over 700,000, though only about 17 percent of the children attend pri-

[2] Donald Levine, "The Roots of Ethiopia's Nationhood." *Africa Report,* May 1971, 12–15.

mary schools and 2 percent attend secondary schools. A large part of the population, in fact, remains very little touched by outside forces, unaware of the modernizing influences in parts of the country, entirely dependent upon their own efforts for subsistence.

Many groups and nations have been trying to help this complex and stultified society move more rapidly into the modern world. The United States, the World Bank, and the USSR have led in providing assistance, but Israel, Sweden, West Germany, Yugoslavia, Czechoslovakia, China, and others have contributed in one way or another.

Working in Ethiopia has proved frustrating for some foreign technicians and advisers. The government has been described as a tangled combination of modern bureaucracy, feudalism, and primitive society where red tape is a major activity and procrastination a password. But progress is clearly discernible; a uniform legal system has been adopted, provincial tax collection has been reformed, a modern money and banking system has been organized, a rudimentary frame of good roads has been constructed, the country has an excellent airline, exports have increased and diversified, and most important, education is being sharply extended. Nonetheless, Ethiopians remain among the poorest of Africans, the per capita GNP having been only about $80 in 1971.

Given the rigidities of Ethiopian society and policy, the resentment of property division, privilege, and corruption, and the growing desire of people, particularly the young and educated elite, for human freedoms, it was no wonder that widespread unrest—marked by a brief army mutiny, numerous strikes and one general strike, and demonstrations—began in February 1974. While the sparks which ignited these actions may have come from the inept response to a severe drought, which killed at least 100,000 people and caused serious inflation in food prices, there is little question that the multi-tude of long-standing inequalities fanned the flames.

Responses to the events of 1974 have included, in addition to the arrest of the emperor, the naming of new premiers and cabinet officials, the execution of 59 former government and military officers in late 1974, pay raises for the army and others, the preparation of a new constitution, announcement of plans for a one-party socialist state that would directly control all economically useful property, and promises of many reforms including the distribution of "excess holdings" to tenant farmers. It will not be possible, however, to accede to all of the demands for reform and for higher pay, though increased coffee prices have eased the economic situation of the country.

THE PHYSICAL BACKGROUND

The conservatism and relative backwardness of Ethiopia may be attributed in part to the isolation of the country, explained in turn by the topographic pattern of its core areas—highlands and plateaus of high average elevation with difficult access and sharply distinct from adjoining regions.

LANDFORMS OF EASTERN AFRICA

Before looking more closely at Ethiopian topography, it may be noted that it shares certain characteristics of the area extending from the Ethiopian Highlands to South Africa. This is high Africa, the bulk of which is a great uplifted land mass. Steep scarps along the coasts and/or at irregular intervals inland are typical throughout. The landform pattern shows signs of extreme geologic youth: freshness of scarps, steepness of canyons, disturbed drainage patterns including developing areas of inland drainage, and lack of extensive alluvial depositional forms.

Great rift valleys run from north of this region all the way to the Zambezi River (Frontispiece map), giving another similarity in structural character to the area. Occupied in places by long, narrow, deep lakes, the rift valleys are a series of grabens or down-faulted trenches cut in the high plateaus, often bounded by upfaulted blocks or horsts. The Western or Central Rift System runs from the Upper Nile to the Zambezi, joining with the Eastern Rift System just north of Lake Malawi. Lake Tanganyika in the Central Rift, draining west to the Zaïre, illustrates an extreme of displacement: the floor of the lake is 2,172 feet (662 m) below sea level, the lake itself about 2,540 feet (774 m) above sea level, and mountains on both sides attain elevations of 8,000–9,000 feet (2,440–2,740 m). The lake is also a biological enigma, with a unique fauna that has developed in part along independent lines.

The Eastern Rift System begins in the Dead Sea, is occupied by the Red Sea, splits Ethiopia into two massifs of unequal size, and is well marked across Kenya but less easily recognized in central Tanganyika. The system bifurcates south of Mbeya, one branch being along the Luangwa trough, the other extending along Lake Malawi and the Shire River to the coast of Mozambique. Many of the lakes in the Eastern Rift do not drain to the sea and are becoming increasingly saline.

Another characteristic common to many eastern African areas is the significance of volcanism. However, there is a gradual transition from the north, where recent volcanism has molded the land intensively, as in Ethiopia and much of Kenya, to the center, where mighty volcanic masses lie sometimes far apart, and to the south, where there are no recent magmatic outpourings, as in Rhodesia, Zambia, and South Africa.

The structure of eastern Africa has important effects on use of the area. It often makes road and rail construction difficult; it results in an almost total absence of navigable streams, though many of the lakes function to a greater or lesser degree as transport arteries; and it has a strong influence on climatic and hence on land-use patterns.

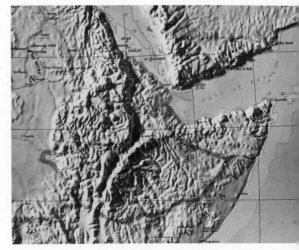

Map 78. Landforms of Ethiopia and the Horn of Africa
Relief Map Copyright Aero Service Corporation

LANDFORMS OF ETHIOPIA

The highlands of Ethiopia form the most extensive upland of the entire African continent (Map 78). Most of the main western massif is formed of "plateau basalts," which reach thicknesses of several thousand yards (meters) in some places; these are also found in the more thoroughly eroded remnants of the eastern or Harar Massif stretching from Harar to Sidamo. Elsewhere, the highlands are composed of crystalline rocks. Elevations in the north are commonly 8,000 to 9,000 feet (2,440–2,740 m) but rise to 14,000 and 15,000 feet (4,270–4,570 m) in the higher peaks; they are several thousand feet (c.600–900 m) lower in the south. While the massifs of Ethiopia are frequently called plateaus, this conveys an erroneous impression of the present topographic situation because their surfaces are rarely flat. Both massifs are broken by hills, mountains, peaks, and cliffs, and are rent by canyons sometimes of great depth. The main highland block is sharply divided from adjacent regions: on the east a 2,000- to 4,000-foot (600–1,200 m) escarpment drops to the Red Sea plains and the Rift Valley; on the west there are a series of terraces and much broken land leading to

Highlands in the Blue Nile region
The Ethiopian massifs compose the most extensive upland of Africa. Scarps separating them from adjacent regions and a highly accidented topography within them help to explain the isolation that characterized the country for many centuries.

the Sudan plains. The Harar Massif has similar sharp topographic boundaries on the north and west but slopes fairly gradually toward the southeast, where it merges into the Somali Plateau—broken and eroded country mainly covered with bush, thornbush, and coarse grasses.

The topographic pattern of the Ethiopian massifs makes them among the most isolated parts of Africa and still among the most poorly known. This isolation is compounded by zones of aridity around the base, the 100-mile (160-km) wide belt on the south being one of the sharpest ethnologic divides on the entire continent.

The Rift Valley is a distinct topographic region, dividing the massifs but also providing, in the Awash Valley, a kind of gateway to the western highland. At the southern border the floor of the valley is about 2,000 feet (600 m) above sea level, while in the section near Addis Ababa it is about 6,000 feet

Rocky fields near Asmara
This region of the massif is arid and subject to periodic severe droughts.

(1,800 m) above sea level. Not much farther north it opens on to the Danakil Plains, which form a sizable region along the northeast of the country. Elevations drop to 381 feet (116 m) below sea level in the Danakil Depression, east of which the volcanic Red Sea mountains parallel the Red Sea coast, rising to an elevation of 6,700 feet (2,040 m) in the south. Along the west of Ethiopia are several small segments of the Sudan plains, vastly different in every way from the other parts of the country.

CLIMATE

The climate zones of Ethiopia are based primarily on elevation, though the use of the several zonal terms is not uniform throughout the country. Lands above about 11,500 feet (3,500 m) are called *wirch;* they are of little use and the problem of atmospheric rarefaction is a permanently limiting one. From about 7,900 to 11,500 feet (2,400–3,500 m) is the *dega;* from about 5,000 to 7,900 feet (1,500–2,400 m) is the *woina dega,* which would be comparable to the *tierra templada* tropical highland zone. This zone has a very low temperature range but a high diurnal range and contains much of the more favorable lands of Ethiopia from the climatic standpoint. From sea level to about 5,000

feet (1,500 m) is the *quolla* zone, which includes the *bereha* or desert areas.

Precipitation varies greatly, but is usually over 40 inches (1,000 mm) on the highlands (Map 79). There is a marked maximum in the summer of the Northern Hemisphere, but in the highlands the Ethiopian names for the seasons are almost reversed owing to higher temperatures in the clear dry period and lower temperatures, especially sensible temperatures, under the rain and cloud cover of June to September. Much of the moisture reaches Ethiopia from the Atlantic Ocean, which means that the heaviest precipitation is on the southwest facing slopes.

SOILS

Soils formed from the "plateau basalts" are generally young, rich, and of excellent structure, and their sizable representation gives Ethiopia soil conditions superior to most African countries. The soils of the crystalline rock areas are much less favorable. The finest soils of the country are probably the alluvials of the Nile, Awash, and other rivers, but they cover a relatively small total area.

The Ethiopian massifs are, then, generally favored by climate and soils, but much less favored by landforms, which are a major cause of isolation with respect to both the

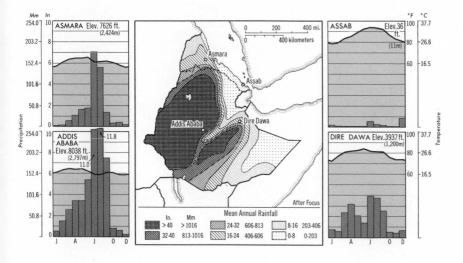

Map 79. Precipitation in Ethiopia; climographs for selected stations

outside world and internal regions, and which reduce the utility of otherwise high-quality areas. The physical attributes of Ethiopia have led many people to conclude that it could one day be one of the great granaries of the world. Much closer study would be required, however, before this assessment can be accepted.

THE POPULATION OF ETHIOPIA

No one knows how many people live in Ethiopia, which had not had a national census until 1970. This was a sample census covering a high percent of both urban and rural people but excluding Bale and Eritrea. Based upon these incomplete figures the population may be estimated at 27.3 million in mid-1974, and to be increasing by about 2.5 percent per annum. About 72 percent are under 30 years of age. The urban population was estimated to be 9.4 percent of the total in 1970, with Addis Ababa having 796,000 people, almost double the number at the 1961 census of the capital.

Owing to poor health conditions, mortality rates are high. Malaria is endemic up to 5,000 feet (1,500 m) and affects 8 to 10 million people. Venereal diseases affect a third to a half of the adult population; 30 to 40 percent of the total population is thought to have tuberculosis; and intestinal parasites are widespread, especially in urban communities.

The distributional pattern of the population is shown in Map 80.

Map 80. Population density map for Ethiopia and Somalia

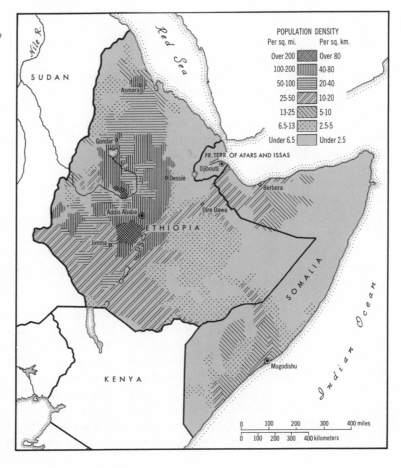

AGRICULTURE

About 85 percent of Ethiopians are engaged in agriculture, which accounts for over 90 percent of exports and about 53 percent of GDP. But most farmers cultivate for subsistence only; much of the main export "crop," coffee, is actually more of a gather product than a cultivated crop. Poor transport, isolation, obscure and inequitable land rights, and lack of knowledge reduce the motivation to produce above subsistence levels.

An estimated 9.5 percent of the total area is in field crops or permanent cultures, 54.6 percent in pasturelands, 6.5 percent in forests, and 29.3 percent is largely wasteland. Individual farms have about 3 to 5 acres (1.2–2.0 ha) in crops, usually in separated fields. Agriculture is characterized by primitive techniques, including the use of crude implements, poor seed selection, lack of integration of livestock and tillage agriculture, very low levels of fertilization, and failure to protect against soil erosion, which is frequently of serious proportions. Ox-drawn plows are widely used in the highlands, but plowing is typically so shallow that fields must be plowed three times to prepare them for planting. The government had given relatively little attention to agriculture until the last plans. One of the problems which is shared with other underdeveloped areas is the antipathy toward working with one's hands; few graduates even of the Agricultural Technical School at Jimma actually go back to farming.

Special interest attaches to the "agricultural minimum package project," initiated by the Swedish International Development Authority in 1971 and now receiving assistance from IDA. It is hoped that about 40 project areas will be in operation by 1980, with each having about 10,000 farm families on blocks of 185,000 acres (75,000 ha) situated along all-weather roads. The program has three phases: in Phase 1, lasting two years, field tests are made and observers look for potential model farmers; in Phase 2, allotted one year, demonstrations are made through the model farmers, and if these are successful the area is designated a minimum package area and Phase 3 begins. It is then assigned a supervisor, five extension agents, and five marketing agents. At first the area is provided only with fertilizer and seeds and with marketing assistance; later advice is given on crop storage, erosion control, livestock development, and home economics. These arrangements are designed to recognize and reward those farmers who are prepared to innovate. They appear to avoid many of the pitfalls of capital-intensive schemes introduced in various parts of Africa, and it will be most interesting to see how well they succeed in the years ahead.

MAIN CROPS

About three-fifths of the cultivated area of Ethiopia is planted to grains, and half of that is devoted to teff, a very fine-grain cereal which is particularly important on the higher lands. Barley, sorghum, corn, and wheat are other grains of significance. About a quarter of the cereal output is lost to insects and rodents, while birds and locusts sometimes cause huge losses. Grains paid as rent to landowners account for the bulk of the grain that reaches the market. Other subsistence crops include a variety of dry legumes, oil seeds, vegetables, and ensete, the false banana, from whose leaves a flour is derived. Considerable quantities of honey are produced, particularly in Gojjam, much of it being used to make *tej*, a mildly intoxicating beverage. In 1973 a $4 million tea project was launched, intended to produce for the Ethiopian market and thus save an estimated $7 million in foreign exchange. The physical conditions of the *woina dega* zone permit a very wide variety of middle-latitude, subtropical, and tropical crops.

COFFEE. Ethiopia is believed to be the original home of arabica coffee, and it continues to grow wild in the forests of the main high-

Map 81. Economic map of Ethiopia

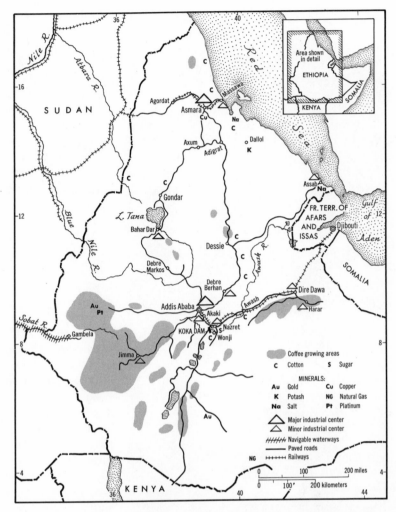

lands, particularly in the southwest. Much coffee is still harvested from wild plants. Formerly, these were stripped bare in one picking, which resulted in a mixture of ripe and immature beans and in greatly reduced yields the following year. Efforts to improve picking techniques have met with some success, but little care is accorded the bushes, which would yield much more if the tall trees were pruned and competing growth were occasionally removed. Better still would be the replacement of wild coffee with regularly planted fields. The government promotes this replacement by distributing selected seedlings. In the Harar Highlands, coffee is produced on large and small plantations, and the government has also favored the introduction of some large foreign estates.

Exports of coffee account for about 55 percent of total export value and some 5 million people are connected with its production and marketing. Output grew during the 1950s and 1960s, but restrictions under the ICO and low prices in some periods tended to depress both the local and national economies. Yields per tree are generally low and quality is inferior, but Ethiopian coffees are favored in the American market because they are the cheapest arabicas and are ex-

A coffee-producing forest near Jimma
Much of the Ethiopian coffee still comes from plants that grow wild in such forests.

Asses bringing fodder into Addis Ababa
More work animals are used in Ethiopia than in most African countries, in part because the road system is still only moderately developed.

Ethiopian exports, accounting for about a fifth of the total in recent years. A great variety of other crop and gather products are exported—including fruits and vegetables, sugar, chat (a mildly narcotic plant whose leaves are shipped from Harar to Arabia), and cereals—but their total contribution is only about 5 percent of exports.

IRRIGATED CROPS. While precipitation is adequate on the highlands to produce rain-grown crops, it is quite inadequate in the lowland areas. Since streams flow from the more humid uplands into these areas and there are some excellent soils available, increasing attention has been paid to the installation of large-scale, usually foreign-owned or managed, plantations. At present about 135,000 acres (54,600 ha), half in the Awash Valley, are under irrigation, but at least eight times that area could be developed. Production of irrigated cotton and sugar is primarily for the domestic market. About a fifth of Ethiopia's cotton is rain-grown and comes from smallholders who use the lint themselves; estates producing cotton are most important in the lowlands of Eritrea and in the Awash River Valley at Tendaho. While output has increased rapidly in the last decade it is not adequate to meet the needs of the well-developed textile industry.

A coffee plantation in the Harar area
Efforts are being made to replace wild coffee with higher-yielding, better-quality plants.

cellent for blending. IDA has recently extended a $10.3 million credit to provide 100 new coffee washing plants.

OTHER EXPORT CROPS. Pulses and oilseeds have been of increasing importance among

The Wonji Plain, downstream from the Koka Dam on the Awash River, is the site of the largest investment in Ethiopia—sugar estates and factories owned jointly by Ethiopian shareholders and a Dutch company. The first sugar was produced in 1954 and output now provides most of the country's needs and permits a small export. About 12,000 people are employed on the Dutch estates.

A variety of vegetables and fruits are grown on irrigated estates near Lake Awasa and in the lowlands of Eritrea. Some produce is marketed in the cities, some is processed and canned for export.

LIVESTOCK INTERESTS

Ethiopia has the largest livestock population in Africa, including about 26 million cattle, 23 million sheep and goats, 5 million donkeys and mules, 1.4 million horses, and 1 million camels; the equines and camels provide the main means of transport in most of the country. On the highlands, most farmers keep some animals, but they are not well integrated with tillage farming. At lower elevations grazing is the dominant interest, though carrying capacities are very much lower in the steppe and desert areas. Here nomads and seminomads depend on sheep, goats, and camels, frequently fighting with their neighbors over pasture and water rights. Overgrazing is usual in these areas.

Practices are poor in all aspects: feeding, breeding, disease control, use of manure, preparation of hides and skins, etc. Animals are characteristically scrawny and diseased, and the cows are poor milkers; most wool is so short it cannot readily be marketed. Animal products provide only about 12 to 14 percent of exports though they could equal coffee in value. Hides and skins account for three-fifths of the total and are followed by preserved meat, live animals, beeswax and honey, civet, a perfume base taken from the civet cat, fish, and eggs. Efforts are being made to achieve greater commercialization

The Elaboret irrigation estate west of Asmara
Much of the modernized agriculture of Ethiopia is concentrated on foreign-owned plantations and estates.

of livestock products, including the provision of a government-owned meat extract factory said to be the largest in Africa, a big tannery capable of treating 25 million skins, and the setting up of a number of large ranches. But it will take some decades before the average farmer is using his livestock effectively.

MINING, FORESTRY, AND FISHING

Mining is thus far of only minor significance. Output is confined to common building materials, salt, small quantities of gold and platinum, and copper. The last has been developed at Berbarwa, 18 miles (30 km) south of Asmara, by a Japanese firm; the first shipment was made in 1974 and it is hoped that an output of 40,000 tons a year can be achieved. After many years of unsuccessful searching by several companies, a promising gas find was made in the Ogaden by Tenneco in 1973. This find has stimulated interest not only in adjacent areas of Ethiopia but across the borders in Somalia and Kenya. Concessions to mine large deposits of potash in the Dallol area of the Danakil Depression were relinquished in the late 1960s. A government geology survey office was not set up until 1969 and its studies may stimulate further mineral prospecting in the years ahead.

Most of Ethiopia's forests, which cover perhaps a seventh of their former extent, are too remote to be of more than local interest. Eucalyptus trees grow rapidly in the highlands and plantings, particularly in and around Addis Ababa, and provide important quantities of firewood. About 7,000 registered fishermen land an annual catch of about 8,000 tons from coastal waters; about a quarter of the catch is exported.

INDUSTRIAL DEVELOPMENT

Modern manufacturing accounted for 4.6 percent of the GDP in 1969 and 1 percent of wage employees. Traditional craft industries accounted for 5.2 percent of GDP in the same year and employed many more people. The figure for employment in modern industry is inflated by inclusion of sugar plantation workers. A 1967–68 survey revealed that there were 395 manufacturing plants of which the 84 employing over 100 people accounted for 83.4 percent of the total employment. There has been a rapid increase in numbers of industries in recent years and value of production grew by an average of 16.4 percent per year in the 1960s. Most plants are foreign owned and managed, though the government has a part or whole share of a considerable number of establishments. Italians founded many of the industries in Eritrea, which continues to have a disproportionate share of the country's manufacturing plants. Greeks are also represented, while A. Besse and Co.—a French concern, which is the largest import-export firm—has important investments in food processing and other industries. The Wonji sugar factories and the petroleum refinery at Assab are the largest individual establishments.

The corridor running along the rail line from Addis Ababa through Akaki to Nazret has an estimated 47 percent of total industrial output in the country (Map 81 and Table 15). Asmara accounts for about 25 percent. Bahar Dar is a minor textile center; development there is based in part on a hydroelectric plant at the Tisisat Falls on the Blue Nile. The government has favored development there to offset overconcentration at Addis and to intensify growth in the Blue Nile Province, but the remoteness of the town has made investment unattractive.

As may be seen from Table 15, Ethiopia has a range of industries characteristic of a relatively important tropical African country, but the scale of many plants is smaller than in other countries and the chemical and metal fabricating industries are not so well represented. The food and textile industries account for about 70 percent of total employment in manufacturing, with the latter having undergone notable expansion in the last score of years. Many branches of manufacturing are characterized by excess capacity, explained by a shortage of managerial and technical staff, inefficient operations, inadequate supplies of raw materials especially in the agricultural processing sector, and by the still small size of the national market despite the country's having the third largest population on the continent.

Ethiopia has ranked very low in per capita consumption of energy. While there is a very large hydroelectric potential, only a fraction of the total has been developed, and the remoteness and extremely difficult access of many sites militates against their use. The three most important hydroelectric installations are (1) a new $81 million, 100,000 kw plant, which utilizes a 600 foot (180 m) fall on the Finchaa tributary of the Blue Nile about 180 miles (290 km) west of Addis Ababa; the Koka Dam and other supplementary works on the Awash River about 50 miles (80 km) southeast of Addis, constructed in part with Italian reparations and now having a total capacity of 126,000 kw; and the already noted 14,000 kw station at the Tisisat Falls.

The discovery of natural gas raises the

Table 15. Industrial representation in Ethiopia, c.1974

Industry or product	Present in		
	Addis Ababa	Asmara	Other
Raw material processing			
Vegetable products			
Coffee washing	Many		Scattered
Freeze-dried coffee	P		
Grain milling	P		Akaki, etc.
Cotton ginning			Upper Awash Valley, etc.
Oils and seeds	P		Nazret, Dessie, Harar, Debre Markos, etc.
Veg./fruit canning			Awasa, Elabaret
Sugar			Wonji, Shoa, Metahara
Dom palm buttons		P	
Pulp and paper			Nazret (bagasse)
Sawmills	P		Jimma, etc.
Animal products			
Abattoirs		P	Keren, etc.
Meat processing		P	Massawa, Dire Dawa, Mogo Wando
Tanning	P	P	Holletta (reptile skins)
Shell buttons			Massawa
Fish meal			Massawa
Dried eggs		P	
Market-oriented			
Food, beverage, tobacco			
Macaroni, biscuits	P	P	
Confectionery			Wonji
Modern dairies	P	P	Elabaret
Brewing	2	P	
Spirits	P	P	
Soft drinks	P	P	Nazret, etc.
Cigarettes	P	P	
Textiles, clothing, etc.			
Textiles, cotton	P	P	Akaki, Dire Dawa, Bahar Dar
Textiles, woolen			Debre Berhan
Textiles, silk, rayon	P		?
Clothing	P	P	
Blanket, rugs	P		Debre Berhan
Cords and sacks	P	P	Akaki
Umbrellas	P		
Shoes	3	2	
Household articles			
Furniture, wood, metal	4	P	etc.
Plastic	2	P	
Aluminum	P	P	
Ceramic		P	
Chemicals			
Matches		P	
Soap, cosmetics	P	P	etc.
Paints & varnish	P		
Pharmaceuticals	P		

Table 15. Industrial representation in Ethiopia, c.1974 (Continued)

Industry or product	Present in		
	Addis Ababa	Asmara	Other
Chemicals			
Industrial gases	2	P	
Ammunition	P		
Petroleum refinery			Assab
Metal products			
Metal sheets	P	P	
Metal containers		P	
Hardware	P	P	Akaki
Vehicle assembly	P		
Construction			
Cement and products	P	P	Massawa, Dire Dawa
Bricks, tiles	P	P	
Plywood, particle board	P	P	
Miscellaneous			
Tires	P		
Paper products		P	
Batteries		P	
Service			
Printing	7+	4+	etc.
Tire retreading	P	P	
Ship repair			Massawa

P = present

Hydroelectric station at the Tisisat Falls on the Blue Nile
The small plant near the outlet of Lake Tana provides power for the minor industrial center of Bahar Dar.

possibility of using this for power production. Crude oil might also be piped to Addis Ababa, thus avoiding the difficult route from Assab. The UNDP has been studying the geothermal resources of the Rift Valley area, which appear to be extremely large.

TRANSPORT

The characteristically precipitous edges of the Ethiopian highlands and the highly accidented interior topography make for difficult and expensive transport. Considering the ability to practice local self-sufficiency and the long period of isolation from outside forces, it is little wonder that the transport net is grievously inadequate.

ROADS

Italy did construct a fairly good road system in Eritrea—no mean accomplishment given the mountainous terrain—and built 4,000 miles (6,400 km) of roads in Ethiopia, half of them paved, in its brief period of colonial rule. These roads opened the country for the first time, but most of them were permitted to deteriorate in the 1940s.

A really major program of road building was begun about 1950 with aid from the World Bank and the United States, and by 1971 the Imperial Highway Authority had constructed about 5,000 miles (8,000 km) of all-season roads, of which about half were paved. In addition there are about 11,200 miles (18,000 km) of second-class roads and tracks of highly variable negotiability. There are plans to double the number of first class roads by 1990, with a particular emphasis upon feeder roads, the greatest need.

Road construction is probably the single most important contribution that has been made to Ethiopian economic development. It has provided relatively good contact between most of the major cities, opened up rich agricultural regions, and improved contact among the various ethnic groups. Transport costs on maintained all-weather roads have been calculated to be only 5 percent of those on dry-weather roads, and such roads have reduced trucking times between Addis and the following cities as shown: Assab, 50 percent; Asmara, 60 percent; Debra Marcos, 85 percent. Most of the country, however, still remains dependent upon animal transport, which has been estimated to carry to market about three-fourths of the produce.

RAILROADS

Ethiopia has two of the most spectacular railroads in Africa from the standpoint of topographic profile. The narrow-gauge (95 cm) Eritrean line was constructed from 1899 to 1911 between Massawa and Asmara with later extensions bringing the total length to 190 miles (306 km) by 1932. Its most difficult section, between the coast and Asmara, at 7,694 feet (2,345 m), has extremely sharp curves, gradients of 3.3 percent, and 30 tunnels. Its freight and passenger traffic is quite low and the line is experiencing increasing competition from truck traffic, which can use a roughly parallel paved road.

The other line serving Ethiopia is that from Djibouti in the French Territory of Afars and Issas (F.T.A.I.) to Addis Ababa. Begun in 1896, the line reached that city in 1917, but the French company's inadequate capital resources had dictated construction at as low a cost as possible and the railway still suffers from excessively sharp curves and light rails. After Ethiopia and Eritrea were federated in 1952 and Ethiopia ceased being a landlocked country, it favored use of its national ports to the detriment of the line to Djibouti. An agreement was accordingly made with Ethiopia which provided that the line be jointly owned, that Ethiopian customs inspectors could be stationed at Djibouti, and that transit privileges through F.T.A.I. were free. These provisions made use of the line, now called the Franco-Ethiopian Railway, more attractive and resulted in profitable operating years after 1958.

The Franco-Ethiopian Railway is a meter-gauge line totaling 487 miles (781 km) in length (Map 81). Since World War II it has been modernized and dieselized, but it remains one of the most difficult and most costly lines in the world, and transport charges are a heavy burden on both import and export commodities. It has lost revenue to road traffic between Assab and Addis Ababa after completion of an improved road via Dessie, while a new, less precipitous road following the rift valley from Assab to Nazret will further disfavor the railway. Completion of a good road between the capital and Dire Dawa also provided increased competition, while political uncertainties regarding F.T.A.I. have undoubtedly strengthened Ethiopian desires to use its national routes and ports, though its attitudes toward Somalia have meant that it has approved the continued French presence at Djibouti.

PORTS

Massawa is the best developed of Ethiopian ports and has a natural harbor superior to Assab. Providing the major outlet for the north of Ethiopia, Massawa has accommodations for six large vessels and five smaller ones plus bulk facilities for petroleum unloading and salt loading. Assab was reconstructed from 1958 to 1961 and is now capable of handling six large vessels and a number of coastal craft. A Japanese team has been surveying both ports with regard to future developments, including provision of ship repair facilities. The special privileges at Djibouti make that port to all intents and purposes a third port of Ethiopia. There is one river port in the country, Gambela in the extreme west, which is open to navigation to the White Nile when the Sobat is at the high water stage.

The nationally owned Ethiopian Shipping Lines has a fleet of six or seven small vessels and one 33,000-ton tanker.

AIRWAYS

Ethiopian Air Lines (EAL) was founded in 1945 and was first managed and partly staffed by TWA, which continues to serve in an advisory capacity. It runs scheduled services to over 40 domestic towns and has an unusual length of international connections. Its unduplicated route miles increased from 2,494 miles (4,013 km) in 1946 to 34,665 (55,776km) in 1972, while operating revenues increased over 60-fold in the same period. In 1972 the staff of 2,424 was 90 percent Ethiopian and 58 percent of the pilots were nationals. The poor surface transport helps to explain the importance of airways, which play a prime role in maintaining contact with otherwise very remote points. EAL is very important in plans to develop the tourist industry, which has been expanding rapidly in recent years. A World Bank study recommended recently that $50 million be allotted in the next 10 years to provide improved facilities for this industry.

CHAPTER SIXTEEN

Somalia

On July 1, 1960, two of the lands inhabited predominantly by Somalis became the independent Somalia, a country with an area (246,199 sq mi, 637,657 sq km) as great as California and Oregon, with a population estimated to be about 3.1 million in mid-1974. Despite the low density, 12.6 per square mile (4.9 per sq km), it appears to have too many people in relation to its meager natural resources and its population is thought to be increasing by about 2.4 percent per annum. After an army coup in late 1969 the country became the Democratic Republic of Somalia, under a military-socialist government which has nationalized most foreign enterprises and placed many aspects of the economy under its direct control.

One of the poorest countries in Africa, having had an estimated per capita GNP of $70 in 1970 (lower than in 1960), it is dependent on foreign aid for almost all development expenditures and has received since independence more aid per capita and per unit area than any other African country. At present, the USSR and other Communist countries provide the largest share of grants and loans, but the government has declared a policy of nonalignment and continues to receive aid from many western nations, including Italy, and from FED, the World Bank, and other international agencies.

THE SOMALI PEOPLES

The Somalis are an eastern Hamitic people, relatively homogeneous in religion, language, and culture, but often split by primary loyalty to the clan, by intratribal friction and territorial disputes associated with claims to water holes and pasture rights. They do have a strong ethnic consciousness in relation to other groups, evident in the drive to independence and support for a Greater Somalia. The Somali flag contains a five-pointed star, each point representing the Somali residing in particular areas: one each for former Italian and British Somalilands now united in the Republic, and one each for those groups in northern Kenya, eastern Ethiopia, and F.T.A.I., formerly French Somaliland. The unification of all the Somali clans is the one goal that unites all Somali, though it has not prevented their participation in pan-African movements or ratification of the Convention of African Unity. The desire for unification led in the early 1960s to excessively heavy expenditures on military and paramilitary forces and support for guerrilla activities in Kenya and Ethiopia; after the country had reached near bankruptcy, less aggressive policies were adopted and detentes were achieved with its neighbors, but relations with Ethiopia have deteriorated in recent years.

Somalia has not yet had a census and estimates of its population vary widely. Almost no other African country has so high a percentage of its population in one tribal group, for in addition to the Somalis and a few indigenous Negro peoples there are only about 30,000 Arabs plus a few thousand Indo-Pakistanis and Europeans, whose numbers have declined since the takeover of foreign enterprises and commerce beginning in 1970. The minority groups had been responsible for most of the economic and commercial life of the country, particularly its international trade and transactions.

Somali nomads at a new well
About three-quarters of Somalia's population is nomadic and faces very difficult physical conditions in this dry country.

Occupationally, the population may be roughly divided into (1) nomads, making up about three-fourths of the total, and roughly divided in a 60:40 ratio between nomads and seminomads; (2) settled Somali and other agriculturalists, possibly 12 percent of the total; (3) government employees, about 6 percent; (4) traders, about 4 percent (among whom the Arabs handle petty trade, while Pakistanis and Goans participate in much of the more important commercial operations); (5) fishermen, about 1 percent; (6) handcraftsmen, about 1 percent (this group includes most of the small Negro population, who are hunters, iron and leather workers, and the chief collectors of gum and resin, and who have generally been considered as inferiors, even outcasts, by the dominant Somalis); and (7) Europeans, who, unlike the other groups, cannot be considered as permanent residents. Italians have played a significant role in the larger commercial and transportation services, a dominant role in irrigation agriculture, and an important one in government advisory services. Other Europeans are associated with diplomatic, military, and aid missions.

The Somali have long had strong prejudices against certain occupations, which they consider fit only for inferior persons; they regard working with one's hands as undignified. Guided by the principle that Allah will provide, many look upon the economic motive with obvious contempt. Other impediments include a very high rate of illiteracy, the small though increasing number of educational facilities in the country, the fact that Somali was not a written language until very recently (government employees are now required to learn written Somali and Italian and English will soon be dropped from among official languages); high disease rates; and the difficulty of providing health and educational facilities to the nomadic peoples.

Although the urban population of Somalia is still small, perhaps about 14 percent of the total, it has grown in recent decades with great rapidity. This produces the characteristic urban problems, but poverty makes them even more difficult of solution. Many city dwellers reside in either *aqal* or *arish*. The former, typical in the country, is a beehive hut of semicircular wooden struts covered with mats of grass and bark fiber, which can be carried by camel. Quite satisfactory, clean and healthy under nomadism, it quickly deteriorates in the city, creating unhygienic slum conditions. The *arish* is a wattle and daub structure, as unsatisfactory as the *aqal* for permanent dwelling. It houses ticks and bugs in its walls and mosquitoes in its rush roof, and it lacks light and ventilation. The cities in the north have a particularly high representation of these constructions, presenting a seemingly insuperable housing problem. The population of some towns varies enormously between the seasons, and is greatly reduced when the semi-nomads move away to summer pasture areas.

PHYSICAL BACKGROUND

Most of Somalia is a low plateau, averaging about 3,000 feet (900 m) in elevation, which rises fairly gradually from the Indian ocean.

In the north an extension of the Harar Massif runs in two ranges parallel with the Gulf of Aden coast to Cape Gardafui. Except in the north, landforms are seldom a limiting factor on economic use; climatic conditions are the controlling physical element.

Climatically, the whole area is hot and dry. Summer temperatures average about 80–95° F (26–35° C) but reach over 100° F (38° C) for many days. Rainfall is low and very irregular, varying from 1 to 5 inches (25–127 mm) along the coast and in the northern interior to a maximum average of about 24 inches (610 mm). In the dryland farming area between the two main rivers, the Juba (Giuba) and Shebeli (Scebeli) (Map 82), precipitation averages about 20 inches (508 mm). The heavier rains occur from April to June or July, lesser amounts are received in August and September, and almost none falls from December to March. Drought years are common; occasionally there is excessive precipitation, which leads to great destruction and an interruption of the normal cycle of farming and grazing. Much of the country is characterized by interior drainage and evaporite sediments, leading to a concentration of salts in both ground and surface water.

Vegetation on the plains is sparse except for thorn bush and other scrub growth. South of the Juba, however, the brush is often thick and mangrove is found along the coast. In the interior, "orchard bush" country is found—parched grassland with scattered thorny acacia trees. In the hills, incense and myrrh give rise to a gather industry of some significance.

GRAZING

Nomadic grazing is the most important activity, and live animal products, hides, and skins rank first among the country's exports, accounting for about two-thirds of their value. Climatic conditions restrict the number of cattle, hence sheep, goats, and camels are of greatest importance. Camels are used for milk and for transport, which has led one observer to dub the Somali nomad as "the parasite of the camel." Most nomads have an incredibly low standard of living; they suffer from a multitude of physical problems and are further restricted by cultural attitudes and practices. From the physical standpoint, there is inadequate and uncertain rainfall, which is the basic reason for the overwhelming dependence on pastoralism. About 40 percent of the area is practically useless because of low rainfall; the rest is periodically subject to severe droughts with consequent high stock losses. Low precipitation in turn means that fodder growth is inadequate and of poor quality. Overgrazing of the sparse growth contributes to soil erosion, which is prevalent and increasing throughout the country.

Stock and domestic water supply is often inadequate. Many of the older wells are poorly constructed, dangerous, and unsanitary, and often yield ground water of poor quality. The drilling of about a thousand wells in the last twenty years has improved this situation, but some difficulties resulted from the nomads' refusal to pay taxes on the wells and from clan jealousies, which required guarding wells to prevent their destruction.

Disease and insect pests are additional limiting factors. Some pastures are tsetse-ridden, and presence of the fly in the river valleys keeps graziers from the best watered areas. Two locust invasions may be expected each year and, although international control measures have reduced losses, some nomads have bitterly resisted using gammexane-treated bait in the belief that it caused livestock deaths.

Inadequate management, inter- and intratribal difficulties, and the usual attitudes associated with a cattle culture are further explanations for the generally low standard of Somali graziers. The government is attempting to improve conditions by a variety of measures, particularly through better veterinary facilities and the setting up, with World Bank assistance, of feedlots in the lower Juba.

TILLAGE AGRICULTURE

Tillage agriculture is severely restricted by the low precipitation and relatively meager surface and subsurface water resources. Cultivable land is estimated at about one-eighth of the total, but only about 5 percent of that or 0.6 percent of the total area is actually cultivated, and irrigated lands total only about 60,000 acres (24,300 ha) or 0.038 percent of the total.

The main areas for production of rain-grown crops are along and between the two main rivers where precipitation is somewhat more certain. Sorghum is the principal crop; sesame and peanuts are secondary. Yields fluctuate severely and output of food does not fulfill domestic needs; imports of foodstuffs and beverages represent about 40 to 50 percent of the total value of exports.

Bananas grown under irrigation in the lower Juba valley
Bananas account for about a quarter of Somali exports.

Irrigation agriculture in the north is confined to a number of date palm groves along the coast; in the south it is concentrated along the Shebeli and Juba, which drain large segments of southeastern Ethiopia. The Juba has the greater potential, with a flow about 2.5 times the Shebeli and some flow in all months, while the Shebeli is usually dry in February and March and its waters have a considerable alkaline content; both rivers have a high ratio between flood and low-period volumes, and construction of storage dams would be very expensive in relation to the return.

Cane sugar is the main irrigated crop on the Shebeli; it is produced on a modern plantation at Villabruzzi, about 100 miles (160 km) from Mogadishu, formerly owned by an Italian society. Production of about 30,000 to 35,000 tons of sugar at Jowhar meets most of the domestic needs, and sugar ranks second among marketed crops. About 170 small Italian farms produce bananas on the Shebeli near Genale.

The Juba supports most of the country's banana production; there are about 30 large growers in the lower valley. Bananas, which accounted for over two-thirds of Somalia's exports before independence, now account for about a fourth of the total value, though this represents a relative rather than an absolute decline. The precarious nature of the Somali economy is clearly revealed by the position with regard to banana sales. They were long indirectly protected and subsidized in the Italian market and their special status, which was supposed to end in 1967 in accordance with EEC requirements, was extended for two years because of the damaging impact of the Suez Canal closure. Somalia had already had a distance handicap as compared with some of its potential competitors. Italy again aided the Somali banana industry in the early 1970s by making a four-year grant to increase production. Other improvements include a switch from the Juba variety to the higher-yielding Poyo banana, shipping in cardboard crates, installation of modern handling facilities at Kismayu, and assignment of faster, better equipped vessels for the long run around the Cape. The tariff preference it enjoys in the EEC market may apply to new associate members in the years ahead, introducing yet another uncertainty. The marketing of bananas was taken over by the military government in 1970.

Other crops grown under irrigation include vegetables and cotton, but efforts to increase production of the latter, handled mainly by Somalis, have failed.

The present government has placed emphasis on state farms as the major way to increase production. In 1974 there were 14 such farms with a total of 26,750 acres (10,830 ha), most of which are in the Shebeli and Juba valleys. Some 9,300 workers are employed on the farms, which produce grains, cotton, oilseeds, and tobacco. Thus far they appear to be functioning well; the volunteer workers are reportedly quite enthusiastic.

NONAGRICULTURAL ACTIVITIES

Developments outside agriculture are of minor importance. The country is the world's leading exporter of frankincense, which, with myrrh, is collected from the woods of the north. A small amount of gum arabic is tapped, and the nut of the dom

Map 82. Economic map of Somalia

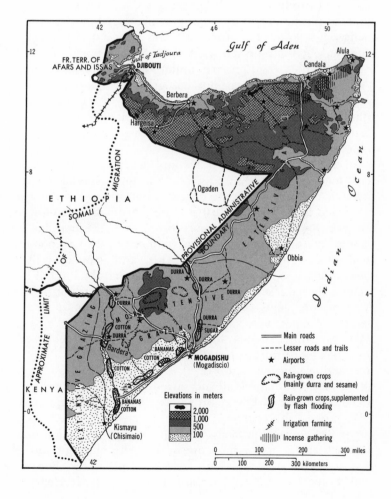

palm, called "vegetable ivory," is sold for the production of buttons.

There is a small but increasing production of rock lobsters and shrimp, sponges and pearls are collected along the coast, and guano is obtained from the island of Mait off the north coast. Though mainly a petty industry, fishing is the chief support of some coastal peoples. Industrial fishing is in the hands of a joint Russian-Somali fleet. The Russians have provided four trawlers and the Somalis conceded fishing rights beginning in 1973; the catch will be split between the two nations. Hunting yields a small commercial output of hides and skins; leopard skins from the area are considered the best in the world.

Mineral production is almost lacking. Deposits or indications of gypsum, phosphates, manganese, low-grade iron ore, and uranium have been reported, and methods of extracting the uranium are under study. The search for hydrocarbons has gone on for several decades with disappointing results, but gas finds in Ethiopia near the de facto border with Somalia have increased hopes for significant discoveries.

An acute shortage of artisans and an almost total lack of technicians, plus the tiny domestic market, inhibit the development of manufacturing. Most industries are small and are concerned with primary processing of raw materials; they include tanneries, food processing, ginning, fruit, fish, and meat canneries. An exception is the sugar factory, the largest industry in the country. Most of the plants operate below capacity because of inadequate supplies of raw materials. Consumer goods industries include soap-making, furniture, a state-owned textile mill, and production of soft drinks; there are also a variety of handicraft industries. While Somalia still has a liberal investment code, little interest has been shown by foreign companies and little is likely given the recent nationalizations of foreign enterprises. The cost of electricity is very high, and total installed capacity is only about 7,000 kw.

TRANSPORTATION

There are about 840 miles (1,350 km) of paved roads and 10,350 miles (16,650 km) of other roads and tracks in the country, but the flatness of much of the terrain and the long dry season make movement relatively easy over large stretches. Several major road-building programs are now in progress, and this will considerably extend the distance of paved and all-weather roads; most important is a 650-mile (1,045-km) north-south connection being constructed by the Chinese, which, it is estimated, will reduce the time between Mogadishu and Barao from two weeks to twenty hours. There are no railroads, except for the sugar plantation lines. Mogadishu, the capital, and Hargeisa have the two main airports, but Kismayu is slated to get a major international field, partly in the hopes of attracting tourists to the large game park in its hinterland. Somali Airlines serves ten internal points; it was organized in 1964 in conjunction with Alitalia.

Until recently none of the ports had deepwater quays. A small but modern port was constructed at Kismayu in the 1960s with special equipment to handle the banana trade; Mogadishu will have a 4-berth, 600,000-ton modern facility by 1976; and Berbera has been improved to accommodate two 10,000-ton vessels. With these facilities, Somalia will be well equipped in relation to the level of its seaborne traffic but not necessarily in relation to its extremely long coastline.

TRADE AND AID

The low level of trade, its characteristically unfavorable balance (exports have been only 66 to 75 percent of imports by value in recent years), and limited domestic generation of capital make Somalia very dependent on foreign aid for its development expenditures. For reasons that are not always entirely clear, numerous countries have been prepared to provide substantial grants and loans; receipts during the 1960s were the highest in Africa on a per capita or per-unit-area basis. Part of the aid appears to represent rivalry among donor nations, who may have assigned the country greater strategic significance than it deserves. Barring the discovery of oil or other minerals, it is difficult to see how very dramatic advances can be achieved for many years. But the Somalis are a proud and independent people who refuse to be pessimistic about the future.

THE FRENCH TERRITORY OF AFARS AND ISSAS

With a population variously estimated at from 137,000 to 200,000 on 8,494 square miles (22,000 sq km), the former French Somaliland is the only French territory remaining on the African continent. A major reason for the wide range of population estimates is a high level of illegal immigration; despite being ringed by barbed wire fences which are patrolled around the clock, the population of Djibouti is thought to have doubled in five years to 120,000.

The territory is essentially dependent on the port of Djibouti and the Franco-Ethiopian Railway. The port, situated on the deeply indented Gulf of Tadjoura, has two moles, one completed in 1970, providing deepwater accommodations for about eight vessels, plus special fueling facilities and a floating drydock. Before closure of the Suez Canal it had an important bunkerage traffic; in 1961, for example, about 85 percent of its 2.58 million tons was fuel oil figuring in both imports and exports. In 1970 traffic totalled 1.1 million tons of which about three-fifths was petroleum and one-tenth was water supplied to vessels. The increased favoring by Ethiopia of Assab over Djibouti was briefly noted in the previous chapter. France has attached considerable strategic value to Djibouti, recently increased by loss of the naval base at Diégo-Suarez on Madagascar. To aid its depressed economy a free port was es-

The city and port of Djibouti, French Territory of Afars and Issas The economy of the territory is dependent on the port of Djibouti and the transit traffic carried on the Franco-Ethiopian Railway.

tablished in 1971 and port extensions were made despite the lower traffic being handled. Reopening of the Canal could greatly aid Djibouti. And it should be noted that French presence there does have the benefit of keeping Ethiopia and Somalia from fighting over possession of the area and its port.

The indigenous population of F.T.A.I. consists mainly of Afars and Issas, rival but closely related groups which are in turn related to the Somalis, plus about 8,000 Arabs who own most of the buildings in Djibouti and control the vast bulk of its retail trade, several hundreds of Indo-Pakistanis, and a European population varying from 3,000 to 10,000 depending on how many French troops are stationed in the territory. Riots in 1966 preceding a referendum in 1967 designed to test the desire for continuing French presence led to the expulsion of many people whom the French claimed were aliens, a matter disputed by Somali nationalists. As borders mean nothing to the local people numerous "aliens" have returned in succeeding years.

Activities outside the port are extremely limited. Nomadic grazing is the most significant. There is some irrigated market gardening to supply Djibouti; it is almost entirely in the hands of foreigners as the Afars and Issas consider farming an inferior occupation.

The territory came under French authority in 1884 and is heavily dependent on direct budgetary support and development aid from France.

CHAPTER SEVENTEEN

East Africa

The area covered in this chapter includes Kenya, Uganda, and the United Republic of Tanzania, consisting of mainland Tanganyika and Zanzibar. These countries reached independence between late 1961 and late 1963, slightly after many other African countries. This delay was caused by the political importance of the European population in Kenya, and the difficulties of achieving internal agreement in Uganda (where the most advanced ethnic group, the Ganda, were reluctant to be included in a larger unified area), and in Zanzibar (where friction among its heterogenous peoples led to rioting in 1961). Shortly after Zanzibar became independent, in December 1963, the ruling Sultan was overthrown in a Shirazi-dominated revolt, and in April 1964 Zanzibar joined Tanganyika; but both countries continue to govern themselves and are united only in name and in the handling of official international relations.

While the whole area has a common heritage of British rule, the post-independence years have witnessed the adoption of strikingly different economic policies in each of the countries, which makes study of their present and planned developments a fascinating one. Their efforts to strengthen the preexisting joint services and customs union have also been most interesting, though the early intentions to form a federation have obviously been frustrated.

Kenya has opted for a mixed economy in which private enterprise has been strongly encouraged in many sectors, though state participation has been growing in recent years. Despite fears to the contrary, under President Jomo Kenyatta the country has followed moderate policies with respect to expatriates, but stronger moves toward Kenyanization are now being made.

Uganda was preoccupied in the early years with rivalries among the various ethnic groups. Prime Minister Milton Obote ousted the President (who was also *Kabaka* or King of the Baganda) in 1966, abolished the power of the traditional kingdoms in 1967, introduced a form of African socialism in 1969 stressing the Ugandanization of the economy, assumed a 60 percent takeover of many enterprises in the same year, and was himself ousted in a military coup in January 1971 by General Idi Amin. Amin's rule has been marked by internal dissension, suppression and elimination of dissidents, near warfare with Tanzania, expulsion of all noncitizen Asians as well as Israeli advisers and most British citizens, the nationalization of all British properties, and increased support from Arab nations, particularly Libya. These and other steps, including the closing of the borders for a fairly long period, have stifled much of the economy, including the growing tourist industry, though Amin's militant nationalism has won the plaudits of Africans in many countries.

Tanganyika (and it is necessary to divide the United Republic in examining policies) has adopted the most interesting policies under its idealistic leader, President Julius Nyerere, whose programs combine elements of Marxism and Christianity. Three basic elements are stressed under his program of *ujamaa* or familyhood: equality, cooperation among members of a community family, and a strong work ethic. There is much emphasis on self help even if this means eschewing some foreign aid and accepting a slower rate of economic growth. In line with these precepts, political leaders are prohibited from amassing wealth (they may not own company shares or houses to rent or receive more than one salary), many sectors of the economy have been nationalized (about three-fourths of large-scale economic activity is now publicly owned and managed by state corporations), attempts are made to encourage the formation of ujamaa villages as the avenue to rural socialism, and a considerable portion of government functions have been decentralized. Nonetheless, there has been increased foreign borrowing and the proportion of development expenditures from overseas sources has been rising rapidly, having been an estimated 56 percent in 1972–73.

Zanzibar was ruled from 1964 to 1972 by the whims of its leader, Sheik Karume, who repressed all dissent, closed the islands to most visitors, gave large sums for public housing on the mainland and for flood relief in Pakistan despite the poverty and poor housing on the islands and discrimination against Asian minority groups on Zanzibar, sought aid and advice from numerous Communist countries but developed an increasingly more enclosed economy, and, somewhat incongruously, built up large reserves in London banks. On the other hand, the Karume government instituted important land reforms, provided better health facilities, extended free education, and built rent-free apartments. Karume was assassinated in 1972 and his replacement, Aboud Jumbe, has followed more normal policies.

AGRICULTURAL POLICIES

All of the political units in East Africa are predominantly agricultural countries and the contrasts in their policies are probably best revealed in their approaches to this key sector.

KENYA

Kenya was the first African state to opt decisively for the western model of individual land tenure, and the bulk of its tillage farmers welcomed the switch from traditional systems, which had in a number of areas been distorted by increasing pressure on the land. Convinced that improvement in African farming areas could not be effective without changing the tenurial systems, the colonial government adopted in 1954 a five-year "Plan to Intensify the Development of African Agriculture in Kenya," usually called the Swynnerton Plan. Under this program, $29.4 million was to be expended on remaking the agricultural landscape in African areas; the United Kingdom and later the World Bank aided implementation of the program.

Involved in the plan were the consolidation of holdings, introduction of individual tenure, farm planning, expansion of cash crop production, provision of rural water supplies, a livestock program, and various reclamation schemes. Some consolidation had already taken place on African initiative, particularly among the Kipsigis but also in a few Kikuyu areas. But consolidation and registration of individual holdings, keystones of the whole plan, and farm planning on the scale envisaged made this probably the most important agricultural development scheme in Africa.

Consolidation and registration of title is an involved process from adjudication of existent holdings, through layout and assignment of new holdings, marking of new boundaries and delineating them by planted hedges, preparation of maps from aerial photographs recording the new farms, to final registration of individual properties. The rapidity with which it was accomplished in the Central Province may be partly attributed to the conditions existing during the Mau Mau emergency, particularly to the

Kikuyu land before land consolidation
Fragmentation of land holdings contributed to malpractices, which in turn were degrading the land or leading to outright soil erosion.

Kikuyu land near Kiambu after land consolidation
The multifaceted consolidation program has literally remolded the landscape over large areas of the highlands of Kenya.

practice of agglomerating villages for security reasons. But the program could not have moved forward so rapidly, both there and elsewhere, without the willing support of the local inhabitants, whose approval was and is required at each stage. Indeed the government, both before and after independence, has often been hard pressed to keep up with the demand for consolidation and registration.

Introduction of individual tenure is only one part of the plan. Of equal importance are the planning of farm layouts and a continuing effort to promote improved practices. Planned layouts require agricultural experts to arrange areas so that erosion con-

trol practices may be implemented, water will be available, farms will be divided as equitably as possible as far as soil and other factors of productivity are concerned, and access roads will be available. Since the goals are that livestock be integrated with crop farming and that cash crops be grown on each farm, agricultural education and information programs backed by continuing research are necessary. At the other end of the cycle, farmers must be assisted in marketing their output, while selection is required to maintain quality standards.

By mid-June 1962 some 293,000 farms had been enclosed, of which 186,000 with a total area of 1.1 million acres (445,000 ha) had been registered; by the end of 1972 some 631,600 holdings with an area of 6.0 million acres (2.4 million ha) were registered and an equal area was being processed, including some grazing lands. The whole program has literally remolded vast areas of the Kenya Highlands, brought about a near-revolution in smallholder practices, increased the production of cash crops and the share of these crops grown by Africans, made possible new initiatives in the application of farm credit and quality controls, and resulted in a much greater growth of processing and marketing cooperatives than in Tanzania or Uganda.

A second major program in Kenya, the transfer of land from Europeans to Africans in the Scheduled Areas, is discussed below. It, too, involved the assignment of individual tenure.

UGANDA

The strength of ethnic positions, which were enshrined in the first Uganda constitution, made government very wary of interfering with the existing and very diverse man-land systems. The ideal over most of the country is that of the yeoman cash-cropping farmer, and it is widely accepted that differences in income among rural Africans are a desirable consequence of individual initiative and ability. The government has attempted to aid the farmers, not by legalizing de facto systems of tenure, but through training centers, demonstration farms, extension services, and provision of loans.

An important exception to this general pattern was attempted in 1964, when the government decided to concentrate agricultural investment in group farms, most of which were in the north. The government cleared the land, subsidized tractor and technical services, and encouraged though did not require cooperative marketing. The group farming system failed because it was too costly, the tractors could not be adequately maintained, and, most important, the farmers neither understood nor showed any enthusiasm for the concept.

As a result, the overall approach in Uganda continues to be to work with the yeoman farmer. And it should be noted that Uganda had long had a much higher percentage of its marketed and export crops coming from individual African producers than Kenya or Tanzania, which meant that there was a better diffusion of the money economy.

TANGANYIKA

Government on the mainland part of Tanzania decided after independence to adopt a transformation rather than an improvement approach. The improvement approach called for working through farmers without radical changes in social and legal systems. It was seen as being ineffective because it would have required too many personal contacts between individuals and extension officers, who would be too difficult to supervise, and because peasants would be unwilling to adopt new techniques. The transformation approach, it was thought, would permit concentrating attention on the most important needs, ease the extension efforts by grouping farmers in settlement schemes, reduce resistance to innovation by

removing farmers from the traditional environment, and make it easier to provide social services.

Accordingly it was planned to set up 74 village schemes in the 1964–69 period, each including 250 families and involving an expenditure of about $420,000. The total allotment was about $31 million, or $1,680 per family. It became apparent by the end of 1965 that the program was not succeeding. The major explanations for failure were the unsuitability of some of the settlers (particularly the urban unemployed with little farming experience or interest), the perception of many settlers that it was a government job accepted only to receive rations and cash provided to each settler until the first harvest, poor initial planning, an excessive emphasis on social rather than productive aspects of the schemes, inability to use or maintain mechanical equipment effectively and economically, and severe shortages of skilled managers and bookkeepers. The per capita expenditures were also far out of line with average incomes. This approach was abandoned in 1966 in favor of modernizing the existing settlement patterns through the injection of capital and technical services. At that time fewer than 10 villages were viable.

The present program was formulated in a series of speeches and papers by President Nyerere. The famous Arusha Declaration, made in February 1967, stressed that all development must be based on the people's own efforts. His *Socialism and Rural Development,* issued in September of that year, stipulated that ujamaa villages would be the central route to rural socialism, that farmers would produce crops cooperatively, each being paid on the basis of the work contributed, but that formation of the villages must be voluntary and democratically achieved step by step. A year later *Freedom and Development* stressed that ujamaa villages can only be established by the will of the farmers, noted that the first step might be mutual help or some cooperative activity outside agriculture such as provision of water points, roads, housing, or schools, but stipulated that all members would be bound to decisions democratically taken.

The philosophy of ujamaa villages, whose eventual goal is communal holding and farming of the land, is based on the contentions that stable development can be achieved by spreading it as widely as possible, that communal work would contribute to achieving a classless society whereas focusing on progressive farmers would tend to create a middle class, and that it would tend to break down tribal considerations, which would likely become more rigidly entrenched under freehold tenure.

The very looseness of the present ujamaa village program makes assessment of its success difficult. Officially the number of such villages has increased from 1,200 with a population of 500,000 at the end of 1970 to over 5,500 containing more than 15 percent of the total population (14.3 million) by August 1973. But it is conceded that the vast bulk of the villages are in very early formative stages and one official estimates that less than 1 percent of the people involved live in a situation of communal ownership, work, or participatory decision-making.

While it is too early to judge the success of the ujamaa approach, certain questions may appropriately be posed:

(1) What incentive is there or can there be for farmers to join? Government exhortation and lengthy philosophical discussions may prove inadequate unless there is some incentive, the best being visible success. But there is little evidence that the villages are surpassing individual farmers, and the oldest schemes are not attracting new members.

(2) Can villagization spread development as widely as possible? Certainly the early schemes were far too capital intensive for that, and even the present ones fail to meet this objective because they are not that widely dispersed and have not been well accepted in those areas where individual farmers have been producing cash crops on individual holdings.

(3) Is the formation of a classless society likely to occur? The chances appear to be reduced because those villages formed in the early years will presumably be considerably ahead of those formed in ten or twenty years and because they are not being formed in just those areas where agriculture is best advanced—where the individual farmer is jealous of maintaining his farm and his achievements.

(4) Should the object of a classless society itself be questioned? Acquisitive and entrepreneurial instincts appear so prevalent and yet different among individuals as to make it extremely difficult to eliminate social stratification. Indeed these characteristics are important to development, including rural development, and numerous studies have shown that African farmers in any given area are far from being a homogenous group of individuals—there are significant differences in their skills, their willingness to innovate, their willingness and ability to work, etc. If these assets are suppressed to achieve a classless society, agricultural development will also be suppressed or slowed. This does not mean that it is undesirable to work toward more egalitarian attitudes.

(5) Will tribalism be better suppressed under a communal system? Only time will tell, but rejection of the more undesirable aspects of tribalism would appear in the long run to be more related to education than to landholding.

(6) Is the communal emphasis satisfactory from the technical standpoint? The

absence of fences in ujamaa villages, in strong contrast with the consolidated areas in Kenya, may make it more difficult for the individual to adopt soil conservation practices and almost impossible to adopt selective breeding of livestock and proper dairying practices. Despite these queries the ujamaa program holds great interest and one can only hope that the emphasis on self-help and hard work will succeed in modernizing peasant farming in Tanganyika.

So much attention has been focused on ujamaa villages that it is easy to forget that Tanganyika has also given attention to other approaches to agricultural development, including a plan to set up 29 State Farms from 1969 to 1974, establishment of about 10 ranches and 2 large dairy farms, and promotion of special crops through government and quasi-government plantations. The last are considered important to achieve commercial production of crops to replace sisal and possibly some coffee; plantation crops being promoted are tea, coconuts, sugar, seedbeans, cocoa, wheat, lime products, and essential oils.

THE EAST AFRICAN COMMUNITY

At the time of independence the three main countries inherited a customs union, a common currency, and a Common Services Organization. These activities involved more functional cooperation than any other African grouping and led to hopes that the move toward greater integration, including federation, might be facilitated. But the common currency was soon replaced by separate currencies, there were disputes over the costs and benefits derived from the various common services, and restrictions were placed by one or another country on the free movement of goods within the area. Indeed cooperation appeared to be in danger of collapsing completely until, in what was labeled as

an example of high statesmanship, the three heads of government signed a treaty, which took effect in December 1967, establishing the East African Community (EAC).

The single most difficult issue that had caused concern was the perception by Uganda and Tanganyika that Kenya was gaining the most from the customs union and common services, particularly because of the tendency for industry to concentrate in Kenya and for that country to have the lion's share of internal manufactured exports and total internal exports. The treaty tried to meet this issue directly and indirectly. Indirectly the two poorer partners gained by decentralization of certain of the common services: the East African Harbour Authority moved to Dar es Salaam, the Post and Telegraph Offices moved to Kampala, and the administrative headquarters of the Community were established at Arusha in Tanganyika.

More direct measures included establishment of an East African Development Bank (which was required to devote 77.5 percent of its funds equally to Uganda and Tanganyika and only 22.5 percent to Kenya), an agreement to harmonize fiscal incentives, temporarily continuing the previous licensing system which was supposed to favor the spreading of regional industries among the three, and, most important, provision for a transfer tax system whereby states in deficit in their total trade in specified manufactured goods may impose taxes on such goods up to the amount of the deficit. The transfer tax was designed to protect infant industries from older established internal industries as well as to achieve equilibrium in internal trade; it obviously represents an exception to the customs union provisions, but there is a complicated set of rules designed to restrict and finally eliminate its use.

The EAC has had strong appeal to other African countries, several of which have made applications to join. Discussions are further along with Zambia than with any other country and both that country and the community would appear to have the most to gain by its accession.

Unfortunately the EAC has been beset by a series of problems which have periodically threatened its existence. There have been disputes over various of the common services (including the inefficiencies of the Post and Telegraph) and over allocations of East Africa Railways; and there was the partial defection of Uganda from the East African Airways, which as a result almost split up in 1971. The headquarters staff at Arusha has suffered from the attractions of Nairobi, which the Ministers prefer, and from shortages in some specialized fields. And there has been a series of abrogations, including imposition of exchange controls by Uganda and Tanzania and controls on movement of goods not allowed by the treaty. The level of internal trade is discussed in the final section of this chapter.

Most threatening of all were the effects of the 1971 coup by Amin in Uganda. This led to disputes over the appointment of administrative officers, interruptions to telephone and air connections between Tanganyika and Uganda, renewed exchange controls and border closings, and an incursion into Uganda from Tanganyika in 1972 by supporters of former President Obote. Somehow, however, the community has managed to remain in being; very important in its survival have been the ability of Uganda and Tanganyika to talk via Kenya, mediation by Somali and Ethiopian leaders, and the strong ideological commitment of Presidents Kenyatta and Nyerere to cooperation. Any real optimism must be tempered, however, by the recollection of recurrent difficulties in the past and the realization that important ideological differences among the three governments persist.

POPULATION AND PEOPLES

East Africa is about a fifth the size of the United States. Tanganyika alone is about four times the size of Great Britain, Kenya is the size of France plus Belgium, and Uganda is a little smaller than the United Kingdom. The estimated populations and the areas of the countries are given in Table 16.

Africans make up about 98.5 percent of the total population. Numerous language and ethnic groups are represented, including some 120 tribes in Tanganyika. Some tribes, such as the Ganda and Nyoro in Uganda, had elaborate governmental hierarchies long before the colonial period; others had very poorly evolved political systems. The vast bulk of the tribes are concerned primarily with tillage agriculture, but there is a gradation from those with none or very minor pastoral interests to those with a primarily pastoral economy and little or no interest in farming. Some groups have made striking advances in recent decades; others have remained very little touched. Tribal boundaries tend to be rigid in East Africa, particularly in Uganda and Kenya, and stand as a deterrent in numerous areas to the most rational distribution of population and use of the land. Tribal rivalries and suspicions have colored most political events in Uganda. In Kenya the smaller tribes have feared domination by the Kikuyu and there have been periodic difficulties between the Luo and the Kikuyu. Kenyatta's prestige has dampened rivalries, which may arise again, however, over the choice of a successor. Tanganyika has no large, dominant tribe and has generally been spared the frictions experienced in other countries.

The African populations of East Africa are growing at high rates, about 2.7 percent per annum for Uganda and Tanzania and 3.3 percent for Kenya. A family planning program was begun in Kenya in the late 1950s, which in 1966 became the first country in mainland tropical Africa to adopt an interventionist population policy. It will be some years before any measurable impact may be expected, since there are deeply ingrained objections to the idea of restricting births. Nonetheless there is increasing appreciation of the difficulties of absorbing the

Table 16. Area and population of East African countries, 1962, 1974

	Area			Population	
	Sq. mi. '000	Sq. km. '000	% in water	1962 millions	1974 millions
Kenya	225	583	2.4	8.7	12.9
Uganda	91	236	17.7	7.0	9.5
United Rep. of Tanzania	365	945		9.9	14.7
(Tanganyika)	(364)	(943)	(6.0)	(9.6)	(14.3)
(Zanzibar)	(1)	(2)	—	(0.3)	(0.4)
Total	681	1,764	6.2	25.6	37.1

burgeoning numbers reaching working age and widespread concern regarding the already high level of unemployment in the urban areas. In the four years to 1974, for example, the need was to absorb about 716,000 people and it was calculated that about three-quarters of these would have to be absorbed in agriculture, one-fourth on present farms and estates and three-fourths on new farming units. Even more ambitious programs will be required in the future, presenting a frightening dilemma.

There were about 365,000 "Indians" in the area in 1961; the term includes Hindus, Pakistanis, and Goans, who are further divided by caste and strong religious and sectarian rivalries. Occupying a dominant role as traders, but also having a disproportionate representation in artisanal activities and many services, they have often been the subject of opprobrium and discrimination by both colonial and independent governments, though a realistic appraisal would reveal many significant contributions to the economy of all states. Occupying positions coveted by Africans, they have been increasingly restricted in recent years by various laws and edicts, and very large numbers were evicted from Uganda in 1972, including many who claimed to be citizens. Kenya has followed more moderate policies but has been serving

an increasing number of quit notices to Asians, leading to legal and illegal movements of capital out of the country and considerable emigration. The last estimate of the Asian population in Kenya was 139,000 in 1969—down 38 percent in two years. Tanganyika has attempted to build a nonracial society, but has nationalized Indian sisal estates and set up a National Trading Corporation designed in part to offset the near monopoly of Indians in the trading sector.

Arabs totaled about 122,000 before independence; most were in Zanzibar and along the coasts. They had contact with these areas long before the Portuguese explorers arrived, and they continue to operate dhows between the Arab countries and East Africa—moving back and forth on the alternating monsoon. They owned most of the clove and coconut trees on Zanzibar and occupied a special position because of the sultanate. Many were killed in the riots after independence and others fled the islands; their present numbers are not known.

Kenya had about two-thirds of the 100,000 Europeans in East Africa at the time of independence. The importance of European farming in that country was only a partial explanation of this concentration; also important were Kenya's position as the leading industrial area of East Africa, its handling· of almost all overseas trade for both Kenya and Uganda, and the position of Nairobi as the main center of the tourist trade. The European population of Kenya had declined to 40,000 in 1969. The European populations of the other countries have also declined, particularly in Zanzibar and in Uganda, where recent expulsions of Britons and Israelis led to a general exodus, including an estimated two-thirds of the doctors and many educators.

DISTRIBUTION OF THE POPULATION

The population pattern of East Africa (Map 83) displays some of the sharpest contrasts of any tropical African area. "Islands" of sometimes exceptionally high density abut regions of very low density. While historical factors are of considerable importance in explaining the present pattern, the correlations between population levels and physical factors are striking. The most important "control" is

Map 83. Population density of East Africa

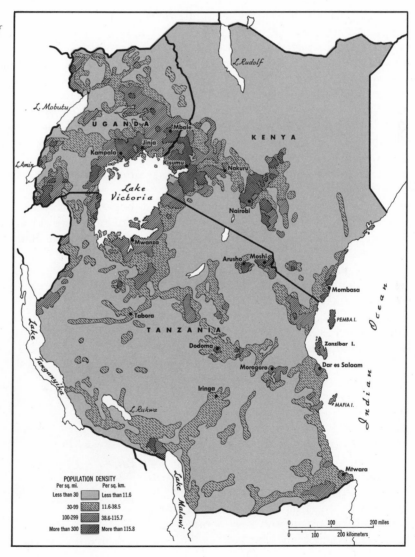

precipitation, intensive arable cultivation requiring a minimum rainfall of about 30 inches (760 mm) per year. Presence or absence of the tsetse fly is also of great importance, particularly in Tanganyika.

In Kenya about 88 percent of the total population resides on the one-sixth of the country with favorable precipitation. Densities in these areas average about 310 per square mile (120 per sq km), but reach over 600 per square mile (232 per sq km) in several districts. At the other extreme, about three-quarters of the area has only 8 percent of the population at densities averaging only 6.3 per square mile (2.4 per sq km).

In Uganda, which has an overall density on its land area of 127 per square mile (49 per sq km), the population is more evenly distributed because precipitation is better distributed, but over a third of the population lives at densities double that level. Kigezi in the southwest, parts of the higher lands of the West Nile area, and the slopes of Mt. Elgon have unusually high densities.

In Tanganyika, with a crude density of about 41 per square mile (16 per sq km) about half the population resides on less than a tenth of the total area. As in the other countries, the greatest densities exist on the superior-soiled, better-watered, and tsetse-free highlands. The lands around Lake Victoria also have densities well above the average.

The major cities have been mushrooming as in other parts of Africa, though urban percentages are still relatively low (Kenya, c.10 percent; Uganda, c.7 percent; Tanganyika, c.6 percent) except for Zanzibar with about 25 percent. The population of Nairobi, the leading city, grew from 119,000 in 1948 to 315,000 in 1962 and 509,000 in 1970. Dar es Salaam has grown from 69,000 in 1948 to somewhere around 400,000 in 1974. Kampala had only 24,000 in 1948 but is now believed to have over 300,000. Mombasa, with 246,000 in 1969, is the only other city over 100,000. The vast bulk of the non-African populations of East Africa reside in the major cities.

TRANSPORT IN EAST AFRICA

Before turning to the regions and economic sectors of the area it is appropriate to provide a summary of the transport patterns approached by looking at the major ports and their connecting links.

KENYA-UGANDA

The only ocean port of significance in Kenya is Mombasa, which handles almost all the overseas trade of Kenya and Uganda, a portion of trade for two sections of Tanganyika, the high-productive Kilimanjaro-Meru region and the Bukoba area west of Lake Victoria, most of the overseas trade for Rwanda, and some trade for Zambia which is not likely to continue after additional quays at Dar es Salaam and the Tazara Railway are completed. Mombasa is situated on an easily defensible island, which is now connected by causeways to the mainland. Access to its harbor is via a 1.2-mile (1.9 km) wide opening in the longshore coral shelf.

The original harbor is now used only by

Aerial view of Mombasa, Kenya
This major port handles almost all of the overseas traffic of both Kenya and Uganda.

dhows and small coastal craft. Seventeen deepwater berths are situated on Kilindini ("place of the deep waters") Harbor along the south side of Mombasa Island and at Kipevu on the mainland. Other facilities at Mombasa include a petroleum facility for 65,000-ton tankers, lighterage wharves, and small dockyards. Mombasa has experienced periodic severe congestion requiring several major extension programs. Some Kenyan officials have called for construction of a second major port, but there is room for expansion at Kilindini Harbor, which could be developed with a much lower investment. General cargo handled at Mombasa has been about 4 to 4.5 million tons, quite evenly divided between loadings and landings; this makes it the third ranking port on the Indian Ocean coast of Africa.

Mombasa is the seaward terminus of the Kenya-Uganda portion of the East African Railways (Map 84). Originally called the Uganda Railway and built in part to assist in the military and religious penetration of Uganda, the line played a significant role in the modern history of East Africa. It was first thought that Kenya was largely an impediment on the route to Uganda, but with the recognition that good, unoccupied lands were available in the Kenya Highlands and particularly with the desire to provide some kind of traffic that would make the railroad a paying proposition, the British and local governments decided to set aside land and promote European settlement in regions served by the railway. The repercussions of this original and rather innocent policy have obviously been enormous.

Construction of the line began in 1896, and by 1899 it had been carried inland 330 miles to Nairobi, which was nothing but a railway camp. By 1901 the line reached Kisumu on Lake Victoria, from which boat services ran to Entebbe and Port Bell, which was connected by a short line to Kampala. It was not until 1931 that the railway reached Kampala and in postwar years extensions

A combined passenger-freight train near Kikuyu, Kenya

have been made to southwest and northwest Uganda. Other extensions were made over the years in Kenya and a connection with the northern line in Tanganyika was built during World War I to facilitate military operations into what was then German East Africa. Freight tonnages on the Kenya-Uganda line have increased from 190,000 in 1913–14 to 898,000 in 1930, 3,345,000 in 1961, and over 5 million tons in 1973. There have been marked improvements in power equipment, rolling stock, signalization, and terminal facilities, and plans now call for complete dieselization by 1978. The line has been facing increasing competition from road services—particularly on the Mombasa-Nairobi stretch, which accounts for over a third of total revenue ton-miles. This reflects in part a rate structure that favors low-value agricultural produce, making it easier for trucks to skim off the higher-value, higher-rate traffic. Nonetheless, the railway still handles over half of Kenya's freight and the vast bulk of Uganda's overseas traffic.

Mombasa is also the eastern terminus of the Trans-Africa Highway, which is to extend 4,400 miles (7,080 km) to Lagos in Nigeria by 1978. Most of the 981 miles (1,578 km) in Kenya and Uganda are already paved. Kenya also plans to construct a pipeline from Mombasa to Nairobi where a new refinery would supplement output from the present refinery situated at Changamwe near Mombasa. Both Kenya and Uganda have relatively well developed road systems in the more developed portions of their countries.

TANGANYIKA

Mainland Tanzania has three major ports. Tanga serves the northeastern part of the country, which includes the leading sisal area and the productive coffee regions of Mounts Kilimanjaro and Meru. A lighterage port, it is situated on a low coral platform on the south side of Tanga Bay, which is protected by a headland that almost totally encloses the harbor. Tanga handles about 270,000 tons per annum. It is the terminus of a line that was constructed by the Germans to Moshi from 1893 to 1911 and extended by the British in 1927–29 to Arusha, 273 miles (439 km) from Tanga. As noted, this line is connected with the Kenya line; it was also tied with the Central line in 1962.

Dar es Salaam is the capital and the leading port of Tanganyika. It replaced the historic ports of Bagamoyo and Kilwa Kisiwani when Germany entered East Africa. Its trade increased only slowly, however, largely because its domestic hinterland is not particularly rich. In the interwar years, Tanganyika did not receive so much attention as many other African areas, partly because of its possible reversion to Germany. With the greatly increased interest focused upon it in postwar years, however, the traffic at Dar increased rapidly to far beyond the capacity of the existing facilities.

Possessing only a fair natural harbor, Dar es Salaam was a lighterage port until 1956, when the first deepwater quay provided space for three ships. Five more berths were finished in 1972, three in 1975. This large expansion was necessitated by the decision to use Dar as the major seaport for Zambia. Traffic at Dar was 743,000 tons in 1961 and increased to 3.7 million tons in 1974. With closure of the Zambian-Rhodesian border in 1973, Dar was called upon to handle over 1.5 million tons for Zambia, and when the railway is completed to the Copperbelt this level will doubtless be increased, which may require additional facilities.

Dar es Salaam is the terminus of the Central Line in Tanganyika; completed in 1914 it runs 780 miles (1,255 km) across the country to Kigoma, following the former Arab slave route most of the way. Despite considerable improvements this line is still not up to the standards of the Kenya-Uganda line. It handles most of the overseas freight for Burundi plus some traffic for Shaba and the Kivu Regions in Zaïre.

The story of Zambian efforts to reroute its traffic after the Unilateral Declaration of Independence (UDI) in Rhodesia in 1965 is given in the chapter on that country. Suffice it to point out here that a pipeline was laid on an emergency schedule from Dar to Zambia by 1968, that considerable tonnages were trucked on the Great North Road, and that Dar was also selected as the terminus of a new rail line, often called the Tan-Zam Railway, but officially the Tazara. Even before UDI, Tanzania and Zambia had appealed for financing from western sources to construct this line, but the reaction from national and international sources was negative because the existing line through Rhodesia was considered adequate, and after UDI few thought that Rhodesia could hold out for long against international sanctions. In 1966 Zambia and Tanzania decided to accept a Chinese offer to finance and construct the 1,160-mile (1,866-km) line from Dar to Kapiri Mposhi on the Zambian line. Financing is by a $400 million, interest-free loan with payments running for 30 years from 1983, shared equally by the two countries. China also supplied the rolling stock, including 102 locomotives. Local costs are paid by imports

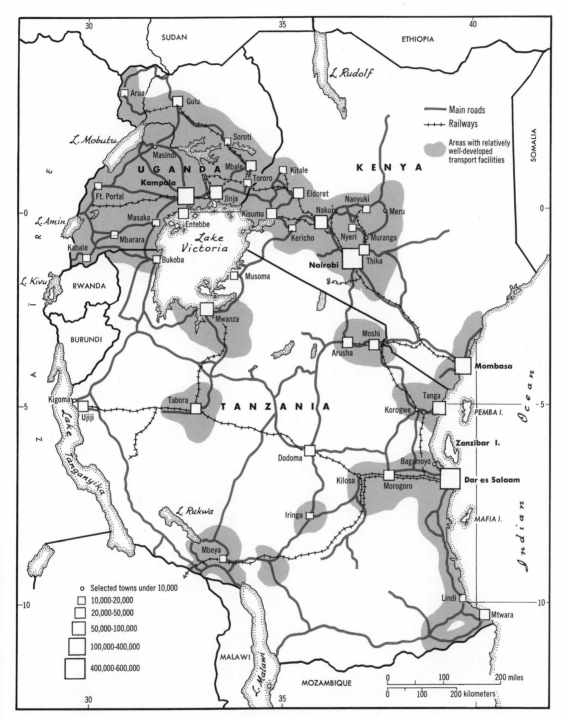

Map 84. Rail lines, main roads, and areas with relatively well developed transport in East Africa; cities and towns over 10,000

of Chinese goods for sale in the national markets.

Tazara has a particularly difficult route across the southern highlands, which has required numerous engineering works; the line will have a total of 5 miles (8 km) of tunnels and 31 miles (50 km) of bridges and viaducts. The Tazara and the Zambian Railway have a different gauge from the East African Railways, which explains the paralleling of the two lines in the section inland from Dar. Completed ahead of schedule in 1974, the line has a capacity of 2 million tons in each direction, which may be raised to 3.5 million tons up and down line with purchase of additional equipment. While the bulk of traffic will be Zambian, Tanzania hopes that Tazara will stimulate development in the potentially productive southern highlands.

In southern Tanganyika, Mtwara was equipped with a two-berth deepwater quay to handle expected large tonnages of peanuts from one of the Tanganyika Groundnut Scheme areas to which it was connected by rail. After collapse of that scheme and the low offering of traffic from the area the line was pulled up. Mtwara has been handling about 146,000 tons yearly, with some traffic brought by road from Zambia to relieve congestion at Dar.

Tanganyika's roads are not so well developed as those of the other countries, mainly because the productive zones are widely dispersed in a much larger area. The Great North Road and a road from Dar to Arusha and the Kenya border are now paved, replacing the formerly very difficult routes, and highways continue to receive substantial budget allotments.

Inland waterways, particularly Lake Victoria, are of some significance in East Africa. Kisumu in Kenya is the main port on that lake, but improvements at Mwanza have increased its handling of traffic for Tanganyikan areas bordering the lake.

East African Airways has normally been one of the more effective and profitable Af-

The M.S. Victoria leaving Kenya's Lake Victoria port of Kisumu

rican air lines but has run into financial difficulties in recent years, criticisms from Uganda and Tanzania that its routes favor Nairobi, increased competition from foreign lines in the carriage of international passengers and from charter services on internal routes, plus problems of management associated in part with the process of Africanization.

REGIONS OF EAST AFRICA

Diversity is a keynote of East Africa; it is apparent in the physical, cultural, political, and economic fields, and results in a remarkable heterogeneity of land-use patterns. This means that regionalizing the area, except by very small units, risks overgeneralization. Nonetheless, it is helpful to divide the area into large regions, within which subregions may be delineated when appropriate. The main focus is on commercial rather than subsistence economies. The major regions used are essentially delineated by topography except for the islands (Map 85).

ZANZIBAR

This islandic member of the United Republic of Tanzania consists of three islands: Zanzibar, about 53 miles (85 km) long and 24 miles (39 km) wide, lying about 22 miles (35 km) from the mainland; Pemba, 42 by 14 miles (67 by 23 km), situated 25 miles (39 km) northeast; and Latham Island, a small,

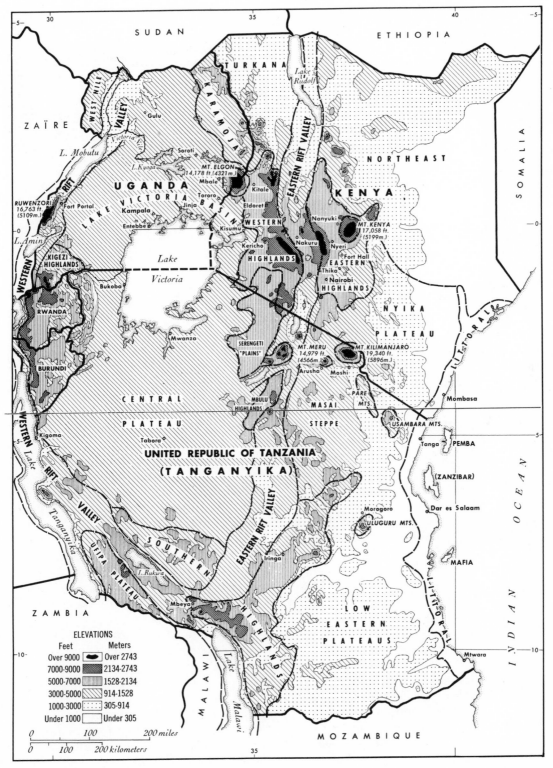

SUDAN

ETHIOPIA

ZAÏRE

Gulu

TURKANA

Lake Rudolf

WEST NILE VALLEY

KARAMOJA

EASTERN RIFT VALLEY

NORTHEAST

SOMALIA

RIFT

L. Mobutu

Victoria Nile

Soroti

L. Kyoga

MT. ELGON 14,178 ft.(4321m.)

Mbale

Kitale

Eldoret

KENYA

RUWENZORI 16,763 ft. (5109m.)

Fort Portal

UGANDA

Kampala

VICTORIA BASIN

Tororo

Jinja

WESTERN

Kisumu

HIGHLANDS

Kericho

Nanyuki

MT. KENYA 17,058 ft. (5199m.)

Nakuru

Nyeri

EASTERN

Fort Hall

Thika

Nairobi

HIGHLANDS

Entebbe

LAKE

WESTERN

L. Amin

KIGEZI HIGHLANDS

Lake Victoria

Bukoba

RWANDA

Mwanza

NYIKA

PLATEAU

LITTORAL

BURUNDI

SERENGETI "PLAINS"

MT. MERU 14,979 ft. (4566m.)

MT. KILIMANJARO 19,340 ft. (5896m.)

Arusha

Moshi

PARE MTS.

Mombasa

CENTRAL

MBULU HIGHLANDS

MASAI

STEPPE

USAMBARA MTS.

WESTERN

RIFT

Kigoma

PLATEAU

Tabora

Tanga

PEMBA

VALLEY

UNITED REPUBLIC OF TANZANIA

(TANGANYIKA)

(ZANZIBAR)

Lake Tanganyika

SOUTHERN

UFIPA PLATEAU

EASTERN RIFT VALLEY

Morogoro

ULUGURU MTS.

Dar es Salaam

L. Rukwa

Iringa

MAFIA

OCEAN

ZAMBIA

Mbeya

HIGHLANDS

LOW

EASTERN

PLATEAUS

INDIAN

Mtwara

ELEVATIONS

Feet	Meters
Over 9000	Over 2743
7000-9000	2134-2743
5000-7000	1528-2134
3000-5000	914-1528
1000-3000	305-914
Under 1000	Under 305

Lake Malawi

MALAWI

MOZAMBIQUE

0   100   200 miles

0   100   200 kilometers

Map 85. Generalized elevations and major regional divisions of East Africa

rocky, uninhabited island half covered with guano lying 40 miles (64 km) southeast of Zanzibar.

Zanzibar and Pemba are generally rather flat, though the latter has some small, steep hills and valleys. The bulk of the population of Zanzibar lives in the somewhat higher, better-soiled west; the east is a coral plain. Precipitation is relatively high and reliable, averaging about 60 inches (1,500 mm) on Zanzibar and 73 inches (1,850 mm) on Pemba. The climate permits a wide variety of crops, but little advantage has been taken of the opportunity for diversification.

The two main islands are inhabited by a great variety of peoples of whom the Shirazi, African peoples of mixed ancestry, are the largest group; others include mainland Africans, Arabs, Indo-Pakistanis, and a few Somalis, Seychellois, and Europeans. No accurate data are available by ethnic group, which has changed since independence because of the revolt against the Arabs and restrictions on the formerly substantial inmigration of Africans from Tanganyika.

Once notorious as a slave center, Zanzibar is most noted as the world's greatest producer of cloves. Cloves were introduced to Zanzibar in 1818 from Mauritius and, by order of the sultan, were planted in a ratio of three clove trees for every coconut tree. Today, about 6 million clove trees occupy about 80,000 acres (32,400 ha) on the two islands, permitting the country to export about 80 percent of the world's cloves. Nearly all the trees were owned by Arabs, but their estates were taken over and broken up into 3-acre (1.2 ha) plots after the 1964 coup. Most of the cloves are produced on Pemba, but Zanzibar Town is still the chief selling and shipping point and the site of the factory that distills oil of clove, used principally in the preparation of vanillin. Cloves are shipped mainly to South and East Asian countries, but also go to Europe and America.

In the 1950s the industry was seriously

A cluster of clove blossoms freshly plucked from a clove tree on Zanzibar

Zanzibar and Pemba lead the world in production of cloves.

threatened by two plant diseases, "die back" and "sudden death," which killed about half the trees on Zanzibar. The former was controlled by careful picking and the latter passed off in a mysterious cyclical pattern. In 1974 another severe disease attack wiped out nearly half the crop. Output normally follows a pattern of succeeding good and bad years, but exports are usually evened out. World prices have fluctuated markedly but have been very high in recent years, 5 to 7 times the prices prevailing in the early 1960s. Clove exports account for the bulk of Zanzibar's exports by value and are the main source of government revenue. Nonetheless, a significant part of the clove crop is wasted, largely because the government refused to permit the annual influx of pickers from the mainland.

The second major crop of Zanzibar is copra, processed in several oil factories. About 7 million coconut trees are planted on about 120,000 acres (48,600 ha) and are the dominant feature of most landscapes. A large number of nuts are consumed locally, while the fronds are used for thatching and baskets and the husk fiber for rope and twine. Exports of copra, coconut oil, oil cake, coconuts, and coir fibers normally account

A coconut nursery on Zanzibar
Coconut products are the second main export of Zanzibar.

for about a quarter of domestic exports, but their share has declined as the price of cloves has risen.

Manioc is the main food crop, but bananas and corn are also important. A wide variety of crops, including many of high value, could be grown; but the residents have preferred to rely on the two main cash crops and are likely to do so as long as prices remain high. Although there is a substantial animal population on the islands, the animals are not well cared for or properly integrated with tillage agriculture. Fishing employs about 9,000 people on an erratic basis with the catch averaging less than a thousand tons per man.

Zanzibar Town possesses one of the world's more romantic names, which could attract a large tourist trade. One of the oldest settlements south of the Sahara, it does have unusual charm and interest, but the government has restricted visitors and done little to promote the industry. It had a population of 68,400 at the 1967 census. Its lighterage port is capable of handling the relatively small overseas traffic; dhows also carry products between the islands and to mainland and Arabian ports.

THE COASTAL BELT

The true coastal belt or littoral zone is a relatively narrow plain except where it broadens in the Tana lowlands of Kenya and the Rufiji lowland in Tanganyika. Temperatures are warm and tropical; sensible temperatures are inclined to be high. Precipitation varies from conditions of semiaridity in the north to about 60 inches (1,524 mm) in southern Kenya and northern Tanganyika, and declines southward to about 35 inches (890 mm) on the southern Tanganyika coast. Monsoons play a prominent role in the climate of this region and other parts of East Africa; the northeast monsoon blows from December to March, the southeast monsoon from May to October. Much of the area has two rainy seasons.

Many of the coastal lands are formed of coral rocks, and light sandy soils are also characteristic. Coconut woodlands and bush are common vegetational types; mangrove swamps line the coastal creeks and river valleys; only occasionally is dense forest found.

Coastal peoples have had contact with the Middle East for over a thousand years. In the sixteenth and seventeenth centuries, the Portuguese challenged the position of the Arabs, erecting forts at various points including Fort Jesus at Mombasa, built in 1592. Portuguese influence has long since vanished, but the significance of Arab influence is now seen in the continuing dhow traffic and in the marked concentration of Arabs in the coastal cities. Much of the coastal zone has fairly high population densities, not only because of the cities, but also because this is a relatively favored and, for the most part, adequately watered zone.

AGRICULTURE. The most significant agricultural production of the coastal belt is that of sisal, which is particularly important in the area around Tanga and along the Central Railway inland from Dar es Salaam in Tanganyika (Map 88). Its production is not confined to the coastal zone, as it can be grown from sea level up to 6,500 feet (1,980 m) and most of Kenya's sisal is, in fact, concentrated fairly close to Nairobi.

Important factors in influencing the location of sisal production include:

(1) Its bulk, particularly before stripping. The usable fibers make up only 3 to 4 percent of the leaf weight, hence it is necessary to have the decorticating plant adjacent to the producing fields and a well-organized system of transporting the leaf to the factory. Even the dried and baled fiber is bulky in relation to its value, so there is a strong incentive to locate production near the coast and along rail routes.

(2) The relatively high capitalization of the decorticating plant, which makes it desirable to keep it in operation as continuously as possible, which in turn means that about 3,000 acres (1,200 ha) in various stages of planting, growing, and maturity are needed for support. While these factors make sisal

Worker on a sisal estate near Tanga
Sisal is the most important crop of the coastal belt but is also grown along the northern and central rail lines in Tanganyika and in the lower lands near Nairobi.

particularly appropriate for plantation-type operations, some is outgrown; and Tanganyika has attempted to promote production by smallholders on a sufficient scale to support a plant.

(3) Availability of water for decortication.

(4) Presence of lime soils, which give especially good yields.

Sisal was brought to Tanganyika by an agent of the East Africa Company in 1891 from Florida via Kew Gardens. Commercial production began in 1900 and sisal came to rank as the leading export of the country from 1925 to 1966. But increased competition from synthetics and from Brazilian sisal adversely affected the world price, causing a severe depression in the industry. Over 40 of the 170 plantations in Tanganyika and 42 of the 68 in Kenya were forced to close and sisal's share of exports declined from 30.7 percent of the total in 1962 to 10.5 percent in 1970 in Tanganyika and from 11.4 percent to 2.6 percent of the total for Kenya in the same years. In 1972, however, a dramatic increase in price occurred which has permitted the reopening of some plantations and even stimulated some new plantings. The tonnage of sisal exports from Tanganyika fluctuated around 200,000 tons during the 1960s but was only 145,000 tons in 1973; Kenyan sales have been about a quarter to a third of the Tanganyikan level. Employment in the industry has also declined sharply, not so much because of reduced volume but because much higher yielding stock and other improvements lowered the labor requirements to less than a quarter of the 1948 level per ton produced.

Production of cashew nuts is particularly important along the south coast of Tanganyika, but is increasing north of Dar and in the coast Province of Kenya as well. Exports from Tanganyika first achieved a significant level in 1945; they increased to about 40,000 tons in the early 1960s and 80,000 tons in the early 1970s, when that country accounted for over a quarter of world output. Kenya produced about 26,000 tons in 1972. Drought-resistant and not requiring very fertile soil, cashew nuts could be grown on a very large area along the coast. Until the recent development of special machines, all nuts were shipped to India for shelling, which was extremely labor intensive. Tanganyika now has two shelling factories, at Dar and Mtwara, with five more planned; a small plant at Kilifi in Kenya is being modernized.

Other cash crops produced in the coastal belt include coconuts, cotton, sugar, wattle, manioc, kapok, and a variety of fruits and vegetables for the large coastal cities. In the future, the delta of the Rufiji and other similar but smaller areas could become significant producers of paddy rice. Mafia Island is the site of a new 4,800-acre (1,940-ha) coconut plantation.

FISHING. Practiced all along the coast, fishing is with a few exceptions confined to inshore waters and characterized by primitive techniques. Among the improvements have been the introduction of nylon nets, outboard motors, and construction of new facilities at Mombasa. There appears to be a large potential for offshore fisheries and for catching high-value shellfish. Ocean fishing is less important in both countries than inland fishing.

THE LOW EASTERN PLATEAUS

The next major region consists of a series of low plateaus varying in elevation from 300 to 2,000 feet (90–600 m) and rising from the coastal belt by a series of scarps best marked at the Kenya-Tanganyika border and in the south. From the land-use standpoint Turkhana, in northwest Kenya and the northern part of the Rift Valley, may be associated with this region as may the somewhat higher Masai Steppe in Tanganyika. For most of the region the western bounds are the plateaus and mountains associated with the Eastern

Rift System, to which the outlying volcanic peaks of Mt. Kilimanjaro and Mt. Meru and the Pare, Usambara, and Uluguru Mountains more properly belong.

NORTHERN KENYA. Over half of Kenya is a sparsely populated, arid region containing about 750,000 people. Part of it is a volcanic desert, almost uninhabitable; most of it is an arid thornbush plain supporting only a small nomadic population. Except for the offtake from a few ranches and fish from Lake Rudolf, which has received 20 new vessels in recent years, almost nothing comes out of the region on a commercial basis. Although many areas show signs of excessive pressure on the range, the production per unit area is far below what it could be with improved pasture and livestock management, provision of water points, and sedentarization, but most pastoralists prefer their present way of life and the government has not been prepared to provide the funds and the force required to effect a change.

The northern border cuts across the tribal lands of the primitive, pagan Boran and of the Somalis. Intertribal skirmishes are not uncommon, while the desire of some Somali to unite with Somalia required police action for several years after independence. A few of the smaller game parks are located in the area and are attracting an increasing number of tourists. Construction of a good road link to Ethiopia will also result in more people seeing this otherwise little visited region.

The Tana River valley provides something of an exception, since it holds potential for fairly large scale irrigation, possibly as much as 300,000 acres (120,000 ha) supporting around 70,000 families. Encouraging success is reported from the Mwea rice scheme 20 miles southeast of Embu on a tributary of the Tana, where about 11,510 acres (4,660 ha) are irrigated on a run-of-river basis. A 1,200-acre (486-ha) pilot cotton scheme is underway at Galole on the lower Tana, where the bulk of the potential area is found. But irrigation projects are costly and the need to assure high productivity has led the government to adopt rigid controls on the farmers, to restrict crops to those that cannot be disposed of except through the managing agency, and to proceed slowly with the development of irrigation perimeters.

SOUTHEASTERN KENYA. This section, the Nyika Plateau, is somewhat better watered, but it is still moderately dry and precipitation is very unreliable. Again, it is sparsely inhabited except along the banks of the Tana and in the Teita Hills. Two large game reserves, Tsavo and Amboseli, are situated in this area; they have a considerable actual and potential value for the Kenya tourist industry (see below).

MASAILAND. Straddling the Kenya-Tanganyika border is a large region inhabited largely by about 230,000 pastoralist Masai. While their area is not confined entirely to the low eastern plateau, most of it is a vast, grass- and bush-covered plain receiving only about 25 inches of rain yearly. Left very much alone in the past, most of the

A Masai herding cattle in southern Kenya
The Masai, who practice a rigid cattle culture, are frequently cited as exemplifying the difficulties associated with achieving economic advance among African graziers.

tribespeople have preserved their cattle culture with few concessions to modernist forces. The Masai are among the richest of African tribes in cattle, and also have close to 2 million sheep and goats. The Masai are true pastoralists, depending on their stock for their livelihood (they drink milk and blood from the cattle but eat beef only on ceremonial occasions and use the smaller stock for meat, milk, wool, and skins). They are semi-nomads, remaining in the same general area most of the time unless forced to move by climate or disease infestation, though the livestock are moved away in the wet season to preserve the vegetation around the hamlets of low skin-covered huts. Government intervention with the Masai has not always been wise; provision of vaccine against rinderpest and of boreholes permitted a rapid increase in the cattle population, but without additional offtake the land became grossly overstocked and vegetation deteriorated. Nature provided a drastic, painful, but probably necessary solution in the form of a severe drought in 1961–62, which resulted in the loss of 50 to 70 percent of the cattle.

Kenya is now attempting to revolutionize the situation in Masailand by a land adjudication program designed to convert communal lands into private ranches held by groups that may be given corporate legal powers. The aims are to facilitate the economic development of the Masai and later other pastoralists without uprooting large numbers of people and at the same time conserving the rangelands. The government believes that the group ranches will permit economies of scale in provision of the physical and social infrastructure, improve grazing and health practices, and increase the sale of livestock, but it is prepared to enforce stocking limits if that proves necessary. Tanganyika is also giving attention to range development in its Masailand and has a pilot ranch at Kolomonik.

In the meantime, there have been signs of spontaneous change, with more Masai willing to sell cattle. Corned beef marketed by Liebigs for the Kenya Meat Commission comes from cattle in Masai country and from the north. Masai are also appearing in schools in large numbers, participating in the preservation of game parks in return for sharing in the benefits from tourist expenditures, and a substantial number of stockless Masai are turning to cultivation in the more arable parts, particularly in the Narok area where a major wheat and sheep scheme on some 20,000 acres (8,100 ha) could eventually be increased tenfold. Just how well the group ranches and other introductions will work remains to be seen. But certainly the days when the Masai could remain in haughty isolation have probably passed, because a growing population requires more intensive use of their lands, whose high potential areas will be developed by others if the Masai do not take steps to do so themselves.

SOUTHEASTERN TANGANYIKA. Much of the widening low plateau in southern Tanganyika is covered by *miombo* woodlands, baobabs, and thickets. Most of it is tsetse-ridden and sparsely populated and has only been the subject of development plans in recent years. The Makonde Plateau represents something of an exception. Covering about 1,200 square miles (3,100 sq km) in the southeast just inland from the coastal belt, it is densely populated by the industrious Makonde, who produce cassava and cashew nuts as cash crops.

Two of the three areas originally selected for the ill-fated Tanganyika Groundnut Scheme were situated in the eastern plateau region, one at Kongwa at the southern end of the Masai Steppe and one at Nachingwea in the far south. The scheme, which ran from 1947 to 1950, was eventually written off at a loss of about $73 million. It provided several expensive lessons, including the advisability of instituting pilot schemes before large-scale ventures are undertaken, and it demonstrated quite clearly the intransigent character of much of the African environment, where techniques applicable in the middle latitudes are not necessarily appropriate. Part of the Kongwa and Nachingwea areas were converted into ranches, while several tenant farms and larger "production farms" have continued to produce peanuts, soya beans, cashew nuts, and grain. Tobacco is grown in the southwest portion of the low plateau.

Tanganyika has also set aside several very large areas where settlements are not permitted; these will be developed as game parks, but thus far they are not included on many circuits, partly because the woodlands are not so suitable for game viewing as are the grasslands of the more popular parks.

There are many opportunities for extending irrigation on the rivers flowing across the low eastern plateaus. The Pangani, draining from the high-rainfall Kilimanjaro-Meru area, is well situated to relieve population pressure in that region, but its waters are somewhat saline and much of its plain is of doubtful suitability for irrigation. An area of 200,000 to 300,000 acres (88,000–120,000 ha) in the Pangani Basin is estimated to be subject to control in three floodplain areas. The Nyumba ya Mungu Dam, 140 feet (43 m) high and 1,300 feet (396 m) long, was constructed on the Pangani, 30 miles (48 km) south of Moshi, in 1966 to permit development of one area, provide flood control, and support an 8,000-kw hydroelectric station.

By all odds the greatest possibilities exist in the Rufiji Basin, which covers about a fifth of Tanganyika and which contains an estimated irrigable area of 1.5 million acres (600,000 ha). Achievement of such a level is regarded as a long-run goal; attention has thus far been focused on a sugar estate in the Kilombero Valley, which employs some 4,000 workers and produces about 40,000 tons per year. Plans call for more than doubling the output despite difficulties with plant diseases, soils not so productive as had

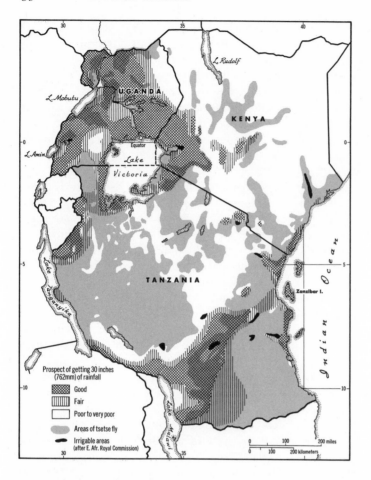

Map 86. Reliability of rainfall in East Africa; areas of tsetse fly

been thought, and rainfall interfering with the harvest. Other sugar estates are located at Mtibwa in the Morogoro region, where it is planned to produce 40,000 tons on a plantation and adjacent smallholdings and at Arusha Chini, south of Moshi, where 3,000 workers are employed on a 7,500-acre (3,000-ha) plantation. Other irrigation developments are taking place or are under study on the Wami, Mbarali, Ruvu, and Ruvuma Rivers, and at Stiegler's Gorge on the lower Rufiji.

GAME CROPPING. One possible solution for more effective use of the dry, tsetse-ridden low plateaus as well as for many other parts of Africa is game cropping. A systematic offtake of wild animals for supply of meat and skins has many possible advantages; it would (1) permit the use of otherwise very unattractive economic areas where rain-growing is difficult if not impossible and cattle-grazing is precluded by their susceptibility to sleeping sickness; (2) permit the use of resources not now being effectively tapped; (3) help the very valuable tourist industry by reversing the trend toward extermination of wildlife in many areas; (4) increase the protein intake in areas where there is a marked deficiency; (5) take advantage of the ability of wild animals to subsist on low-grade vegetation and of the fuller use of the grazing-browsing potential in which each animal of a mixed population has its own niche; (6) help in soil conservation because a variety of animals with differing feeding habits tends to keep the vegetation in balance and because there is no cultural impediment to keeping numbers in line with the carrying capacity.

These arguments appear very persuasive, but the key question remains—is it economic? There are many difficulties associated with the farming of wild animals, including the inaccessibility of many of the potentially interesting areas, the expense of hunting the animals or of directing them to convenient places for slaughtering, the problems of transportation, refrigeration, and preservation of game meat, the lack of acceptability of some species, and the conflict with other land uses in some areas. The low prices received for most of the meat in local markets appears to have been the most significant limiting factor to date. On the other hand, preparation of an area for game cropping may require only the construction of tracks and a few bridges as contrasted with requirements for tsetse eradication, fencing, water supplies, roads, and other infrastructural investments if the same area were to be developed for cattle ranching.

Lack of evidence makes it difficult to assess the claimed advantages and disadvantages of game cropping. There has been some experience in rotating game with cattle on a four-year cycle in eastern and southern Africa, a few individuals have raised specific wild animals on a regular ranching basis with apparent success, and a large game ranch has recently been set up in eastern Kenya by an American company which appears to be having particular success with the oryx. There has been a regular offtake of hippopotamuses and other animals in Ugandan game parks where excess numbers had led to soil erosion and destruction of vegetation upon which these and other animals depended. There has been no problem in disposing of the meat, which is delivered to local markets by truck within hours of slaughtering.

THE EASTERN HIGHLANDS AND RIFT VALLEY SYSTEM: KENYA

This region is a highly complex mountain, plateau, and rift valley belt. The Eastern Rift Valley may be considered as the axis of the system. About 30 to 40 but sometimes 60 miles (48–64–96 km) wide, it is well marked in Kenya and northern Tanganyika, less well to the south. The valley, which is often at a considerably high elevation itself, is typically bordered on both sides by mountains and blocks or horsts, while volcanic cones lie within it, are superimposed on the adjacent plateaus, or stand off at some distances as outliers of the system.

In southwestern Kenya are the Kenya Highlands. The floor of the Rift Valley, which lies at an elevation of 1,280 feet (390 m) at Lake Rudolf in the north of the country, rises to 7,000 feet (2,100 m) near Naivasha and falls again to about 2,000 feet (600 m) at Lakes Magadi and Natron. The valley is studded with inactive volcanoes. East of the Rift Valley in Kenya is the Aberdare Range, rising to about 13,000 feet (4,000 m); on the west the Mau Escarpment has elevations up to 10,000 feet (3,000 m). Lying to the northwest of the highland area and straddling the Uganda border is Mt. Elgon, 14,178 feet (4,321 m), while east of the Aberdares Mt. Kenya rises to 17,040 feet (5,194 m). The Rift System, with escarpments ranging up to 4,000 feet (1,220 m), has an important effect on transportation; its formidability is compounded by the fact that the plateaus bordering the valley are typically at their highest adjacent to the rift.

The involved topography of this whole highland region results in notable variety in climate, vegetation, and land use patterns; changes or successions often occur over very short distances. But this region contains much of the quality area of Kenya, much of its most densely populated districts, the major areas of former white settlement and of large plantations, the settlement schemes that have seen the subdivision of European farms and their allocation to Africans, and the vast bulk of the areas producing the high-value crop exports of the country. Factors favoring the Highlands include their greater and more certain precipitation, the general absence of the tsetse fly, the presence of better than average soils including some excellent volcanics, and the amelioration of temperatures and humidities. There are serious problems of population pressure in some sections, and soil erosion is also sometimes severe.

One of the main climatic attributes of the area is the reliability of precipitation (Map 86), while it is also favored with respect to the tsetse fly, which is not found over 6,000 feet (1,800 m) nor in the more arid regions. Because of the major distinctions between settlement patterns on the former "White Highlands" (now called the Scheduled Areas) and in the traditional African areas, it is appropriate to discuss them separately.

AGRICULTURE IN THE SCHEDULED AREAS. The areas of East Africa that were leased or alienated to non-Africans before independence are depicted in Map 87, which shows the large, irregular zone in Kenya formerly known as the "White Highlands." The whole question of reservation of lands on racial or for that matter tribal lines is fraught with serious moral, social, economic, and political problems. The existence of a small number of white settlers on a large block of land in Kenya was a root cause of the Mau Mau revolt and continued to exacerbate racial relations thereafter.

Alienated land in Kenya, 91.5 percent of which was in the Highlands, accounted for 5.71 percent of the total area but about 18.2 percent of the high potential lands and 24.9 percent of the available arable land in 1959. There were only 3,609 European agricultural holdings in 1960 employing about 6,900 full- and part-time Europeans, but they did employ 277,700 Africans, or 42

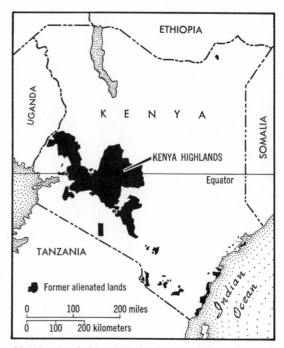

Map 87. Lands formerly leased to non-Africans in Kenya

percent of the total wage labor force, and about a million Africans resided on the farms. European holdings accounted for a gross production of $106 million in 1960 as compared to an estimated total of $134 million for African farms, and they accounted for about four-fifths of the total value of exports of the country. The White Highlands were, in fact, the center of economic activity, including many of the more important modern cities and much of the industrial development.

The former European holdings could be classified in several ways. The broadest division is into (1) comparatively large plantations producing much of the sisal, tea, and coffee, (2) the generally smaller mixed farms raising livestock, various field crops, and some of the high-value crops such as pyrethrum and coffee, and (3) ranches, which included most of the very large holdings. Crop products accounted for about three-fourths and livestock products for one-fourth of gross farm revenue. In November

1960 racial criteria were eliminated as a feature of the White Highlands, which thereafter became the Scheduled Areas. The very great importance of the European holdings presented a dilemma to the late colonial and early independence governments. Too drastic a change could almost literally have destroyed the economy, but the demands for land by landless and unemployed Africans required that some large redistributions be made. The compromise adopted was to retain the large plantations and estates producing tea and coffee (as these were particularly important in production of export crops and were as intensive in employment per unit area as smallholding operations would be), to allow most of the ranches to remain (as it was thought that Africans capable of running them were not yet available and most could not be converted to tillage agriculture), and to purchase most of the smaller, usually owner-managed, mixed farms for resettlement by Africans.

AFRICAN SETTLEMENT SCHEMES IN THE SCHEDULED AREAS. On January 1, 1961, a program to purchase European farms, subdivide, and resell them to Africans was inaugurated. It started slowly because of European fears that they would never receive full value for their farms, African inability to provide the down payment originally stipulated, and fears of conflicting tribal demands; but after some confusion, renegotiation, administrative overhaul, and other changes, the programs began to be intensively applied by the end of 1962. By 1966 some 29,096 families had been settled on 1.6 million acres (471,000 ha). There have been several distinct settlement schemes of which the Million Acre Scheme was most' important; it comprised 35 low-density schemes with farms averaging 37 acres (15 ha), 84 high-density schemes with farms of 27 acres (11 ha), and 16 large-scale cooperative farms and ranches. These settlement schemes involved the transfer and subdivision of about one-third of the mixed farming land in the former

An aerial view of European farming in the Rongai area of Nakuru District
Most of the area formerly held by Europeans has been subdivided and sold to Africans or has been taken over by African cooperatives.

European areas, while another third was transferred in large-scale units.

Large sums of money were expended on the various settlement schemes, including about $75 million on the Million Acre Scheme or around $2,000 for each farm. Over a third of this amount was in grants from the British Government and another two-fifths from overseas loans. The squatter schemes were far less capital intensive. By the mid-1960s the government began to slow down the transfer of lands, as it came to realize that it was using a very high proportion of available agricultural development funds for schemes whose economic benefits were marginal at best, while programs with higher returns were going begging. Costs per family were running about 15 times those for consolidation in traditional areas and only about one-fifteenth as many people were being aided. Furthermore, the financial transactions and reduced output of cash crops were having an adverse effect on the balance of payments.

In the period from 1967 to 1973, therefore, only 5 percent more land was added and the total number of families settled was increased to 36,118. While the policy of subdividing purchased farms has now been discontinued, the acquisition of European holdings continues in a new phase known as *shirika* settlement. Under this project a large farm is kept intact to be run as a single unit by members of the scheme. Each member gets a plot of about 2.5 acres (1 ha) for his own use and is paid for working on the rest of the farm, which is under the direction of a manager. By mid-1973 about 107,000 acres (43,300 ha) had been taken over by *shirika* schemes at a cost of $5.8 million, and 6,100 farmers had been settled; a major extension of the project was being planned.

The settlement schemes have faced a multitude of problems, including the following:

(1) *Unpaid loans.* Only about 55 percent of loans to African settlers are being paid on time and almost a quarter are over a year in arrears.

(2) *Low production levels.* Surveys have shown that, despite some improvement from the early years, less than a fifth of farmers were achieving target income levels and there was very high variability in the performances of individual farmers. A countrywide shortage of dairy cows decreased the ability to follow the plans as did a shortage of tractor contractors. The former shortage has been offset in part by keeping other livestock, while the latter problem has been reduced by extending loans to persons who make courses in operation and maintenance of tractors in preparation for becoming small-scale contractors. Other production problems include shortage of water points, inadequate availability of short-term credit to purchase seeds, fertilizer, and other imputs, and lack of farm-management skills.

(3) *Inadequate supervisory and extension personnel.* To be successful, the settlement schemes require careful layout of the subdivided farm, technical assistance to the new proprietor, a certain amount of farm training, especially for those with little or no previous experience, and maintenance of the necessary scientific and advisory services after settlement has been completed. These services require a substantial staff of trained agricultural workers and fairly heavy budgetary support, but despite comparatively impressive programs and considerable assistance from abroad there remains a shortage of adequately skilled advisory personnel. At first it was thought that supervision could be discontinued after 2½ years but a further 2½ years of "extended supervision" proved necessary, and even this was inadequate on all but the more successful schemes.

(4) *Selection of suitable cash crops.* A problem to some farmers both in settlement schemes and in the consolidated areas has been finding a suitable cash crop to permit a high return on the part of the farm devoted to other than subsistence production. Pyrethrum had been the most important cash crop on the settlement schemes but it faces continued competition from synthetic insecticides. Coffee would be a suitable crop in some areas but international agreements have precluded additional plantings.

(5) *Poor management of cooperative societies,* which market the bulk of products sold from the schemes. This problem is being alleviated by sending staff to courses in management and bookkeeping.

(6) *Inadequate access roads.* Despite the many difficulties affecting the settlement programs, they have brought about a peaceful conversion of very large areas, and sales through their co-

Picking pyrethrum in a settlement area
Pyrethrum has been the most important cash crop on these areas.

operatives increased fourfold in the period from 1965 to 1972, when they totaled $10.2 million.

FARMING IN THE TRADITIONAL AFRICAN AREAS. The highlands also contain the major areas of traditional African farming in Kenya—areas that accounted for about three-quarters of African agricultural production before independence. Considerable differences exist among the various ethnic groups and according to local physical conditions, but until the late 1950s most Africans were operating subsistence holdings and producing very little for commercial sale. Food supply in some areas was precarious; malpractices were leading to depletion of the soil and to outright erosion. Holdings in Kikuyuland and some other sections were grossly fragmented, the average number of separated plots in the Kikuyu areas having been eight for a 4-acre (1.6-ha) holding. Extreme cases include a 5-acre (2-ha) farm split into 39 noncontiguous plots and a parcel consisting of one banana tree. There was almost no integration of crop and livestock interests; crop yields were poor and grazing

practices destructive. Tenurial systems under which rights were always open to challenge made improvement difficult and precluded securing loans for modernization. No effective program to meet these problems was undertaken until adoption of the Swynnerton Plan in 1954. This Land Consolidation Program was summarized above.

In fact it is now more appropriate to think of this program as one of land adjudication and registration because many of the areas that entered in later years were not fragmented and did not need consolidation. In the mid-1960s new emphasis was placed on the registration program as the settlement schemes were slowed, and adjudication of rangelands began in 1974.

An interesting new spearhead program was adopted in the late 1960s. It was designed to increase job opportunities and raise rural incomes by coordinating development in selected areas, providing additional inputs of personnel and resources, and experimenting with new approaches. The areas chosen are intended to be representative of the varying ecological and human conditions of Kenya and thus permit the lessons that are learned to be applied in similar areas. Six areas in as many provinces were selected for the first phase with eight others listed for attention as soon as practicable.

It is apparent that the adjudication and settlement schemes have effected a revolution in Kenyan agricultural areas. While continuing evaluation is necessary, it appears that crop intensification has been successful in raising employment and the percentage of crops coming from African farms has been increasing satisfactorily. The share of small farms in marketed production increased from an estimated 40.7 percent in 1964 to 51.6 percent in 1972.

CASH CROPS FROM THE KENYA HIGHLANDS. The history of crop production in the highlands is replete with recurrent disasters. Many fortunes have been lost, regained, and lost again. But over the years a great deal has been learned about the vagaries of climate and the capacity of various soils, while new strains have been developed suited to the ecology of the area. Nonetheless, plant and animal diseases and insect pests are a problem that requires constant vigilance and continuing study in the various research and experimental stations. Crop and livestock emphases vary greatly from one area to another, depending upon precipitation, soils, elevation, aspect of slope, and other factors.

Coffee. Coffee is the most important cash crop in Kenya, contributing about 23 percent of the gross farm revenue and 27–29 percent of total exports by value. The heaviest concentration is found north of Nairobi between elevations of 5,000 and 6,000 feet (1,500–1,800 m), where coffee is the cash crop par excellence; but it is also found on the slopes of Mt. Kenya and Mt. Elgon and in parts of the Western Highlands. Until 1956, Europeans produced almost all Kenya

Picking coffee in Kenya
Coffee ranks as the most important cash crop of the country.

coffee exports; in fact, Africans were discouraged from producing coffee, ostensibly to ensure maintenance of quality standards. But by 1962 there were 110,000 registered African growers and by 1972 over 300,000 smallholders were accounting for about 45 percent of output while 306 estates produced the remainder. Yields on the estates are about double those on small farms, which have been using less mulch and fertilizer in recent years, and therefore have had a somewhat disturbing decline in quality.

African producers are served by some 155 cooperative societies, each operating one or more coffee factories. All coffee is marketed by the statutory Kenya Coffee Marketing Board. Output and earnings have fluctuated in recent years because of attacks of coffee berry disease and changing world prices. The International Coffee Organization also placed limits on the quantities sold in quota markets, but Kenya was successful in selling about 10 to 20 percent of its output to non-quota markets, albeit at considerably lower prices. Kenya coffee is high-grade arabica, commanding premium prices on the world market. In recent years production has been about 58,000–62,000 tons, valued at $53 to $65 million.

Tea. Ranking second to coffee among export crops, tea has seen a marked expansion in the last two decades. Kenya is Africa's leading producer of tea and Africa has increased its share of world exports to over 16 percent of the total. Tea needs an acid soil, well-distributed and heavy precipitation, and does best between 6,000 and 7,200 feet (1,800–2,200 m).

Like sisal, the considerable investment in factories requires a large supporting area producing as continuously as possible and it also needs a considerable degree of technical skill. These factors more or less dictated its introduction by large companies. In the early 1960s a tea factory for a 1,200-acre (486-ha) estate with 3,500 tea plants per acre (8,645 per ha) might cost about $340,000; by contrast, an acre in coffee requires only 540 plants (1,334 plants per ha) and an area only one quarter as large would justify erection of an $8,400 factory. Tea was, in fact, almost exclusively a plantation crop until the late 1950s, with the largest concentrations found around Kericho in the Western Highlands and Limuru, northwest of Nairobi.

Africans were encouraged to produce tea under a plan started in 1949, but the spectacular increase came after 1959 under one of the most successful smallholder schemes ever undertaken in Africa. Operated by the Kenya Tea Development Authority and aided by loans from abroad, the area in African tea holdings increased from 1,541 acres (624 ha) in 1959 to 12,600 acres (5,100 ha) in 1965 and 76,000 acres (30,800 ha) in 1974. The area in tea estates grew from 5,900 acres (14,600 ha) in 1959 to about 59,300 acres (24,000 ha) in 1973.

The share of production from smallholders was almost nil in 1959 and about 32 percent in 1973; their portion of the crop will increase because large acreages are not yet yielding, and it is planned to further enlarge their area in tea by almost half. Smallholder tea was first introduced on an outgrowing basis, but this is no longer considered necessary and it is planned to increase the number of factories processing smallholder tea from the present 15 to 37. Production has been extended on the Western Highlands and in the higher areas along the forest edge of Mt. Kenya and the Aberdares, where it is too cold to produce coffee, but the last phase has intensified plantings in existing areas to reduce transport and other costs. Africans tend to give individual attention to each bush and are aided by local agricultural supervisors and tea factory personnel, while the Tea Research Institute near

Kikuyu plucking tea leaves on an estate near Limuru, Kenya
Before 1959 all tea was produced on plantations, but a very successful smallholder scheme is rapidly increasing the share produced by individual African farmers.

Kericho is responsible for scientific research pertaining to the whole industry.

Exports of Kenya tea totaled 4,280 tons valued at $3.6 million in 1947, 10,900 tons and $9.3 million in 1962, and 49,200 tons worth $36.8 million in 1973. About 65 percent of exports are shipped to Britain, where they bring the highest prices on the market. Kenya and other African producers have lower production costs than traditional Asian sources, and, in some cases, the quality is higher; hence they have been reluctant to adopt quotas in proposed international agreements. Yet they are concerned with lower world prices and the danger of over-production. Tea exports were about two-thirds the value of coffee exports in 1972 but dropped to one-half of coffee's figure in 1973.

Pyrethrum. This daisy-like plant yields pyrethrins, which are used in insecticides; it has the advantages of being harmless to warm-blooded animals and of retaining its effectiveness when mixed in small ratios, while there is no evidence that insects have acquired an immunity to it. Kenya produces about 70 percent of the world's pyrethrum, output of dried flowers having increased from 1,500 tons in 1948 to 10,800 tons in 1962 and 13,800 tons in 1972. It is produced in widely scattered districts of the Highlands and 90 percent of output now comes from smallhold-ers as compared with 28 percent in 1958. As noted earlier, pyrethrum plays a very important role as the main cash crop on many settlement schemes.

Pyrethrum products rank third to fifth among Kenya's exports. Despite some concern with increased competition from other countries and from synthetics, the country has continued to push expansion and has tried to assure profitability by distributing high-yielding plants to all growers. Most exports are in extract or powdered form, and the dried flowers from the producers are processed at Nairobi and Nakuru.

Cereals. Corn is the most important staple in Kenya and occupies the largest area of any crop. Yields vary considerably depending on weather conditions. Marketed corn is handled by the Maize and Produce Board, which sets the producer price; any surplus after provision of the local market has normally been exported at a considerable loss. With new hybrid corn permitting much higher yields per acre, plans call for greatly increasing the production and exports, which can hopefully be purchased at prices permitting a profit to the Board. The marketing arrangements for corn have been severely criticized, first as being a subsidy to European producers and then as supporting production of a crop that could be partly replaced by higher value crops. Again the use of higher yielding varieties may permit a more rational price mechanism, while increased production may favor the introduction of feedlots and processing of part of the crop.

Wheat is the second major cash grain; it is produced mainly on large farms at elevations from 7,000 to 9,000 feet (2,100–2,700 m). Both the production of and the area planted to wheat have fallen off sharply in the last few years, which led the government to raise the guaranteed price in 1973. An area about 5.5 times the approximately 217,000 acres (88,000 ha) planted is suitable for wheat and, again, improved productivity may eventually permit export to world markets.

Sorghum and millet occupy about four times the area in wheat, but are mainly subsistence crops, particularly important in the drier areas. Small quantities of rice, barley, and oats are also produced.

Other cash crops. Additional cash crops produced in the highlands include tobacco, grown mostly for local consumption; pineapples, which support three canneries; other fruit, including passion fruit and mangoes; horticultural crops, some of which are flown to Europe; vegetables for the local city markets; and cinchona, which has recently been

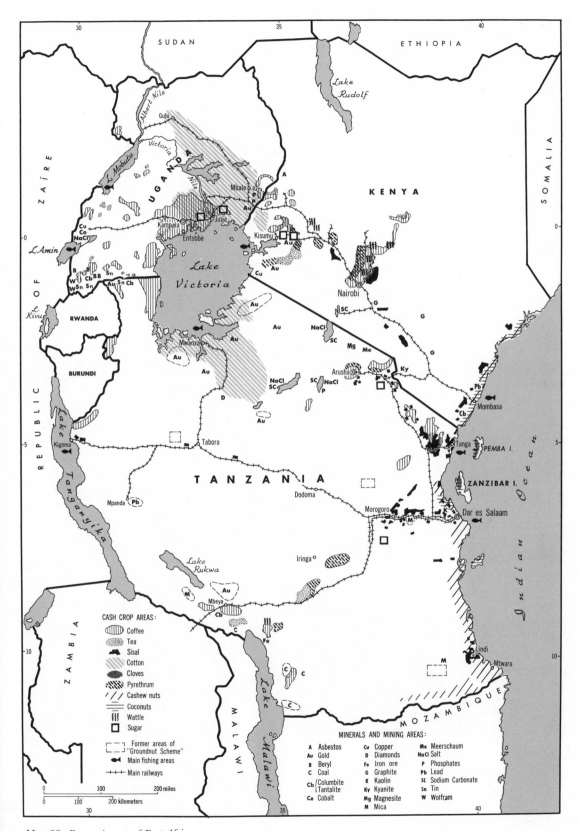

Map 88. Economic map of East Africa

planted on 200 acres (500 ha) in the Kericho area.

Livestock interests. Livestock and livestock products contribute approximately 30 percent of total gross farm revenue in Kenya. The highlands have a large number of the country's livestock and the potential for further development is very promising, particularly if standards can be raised to permit greater exports. European farmers succeeded in acclimating high-bred dairy cows to the highlands and, with a switch of many cows to African ownership, many smallholders are now successfully carrying on an integrated crop-livestock operation, though there has been some genetic deterioration caused by use of low-quality bulls. As noted earlier, a shortage of cows has adversely affected planned output on settlement schemes, but milk nonetheless accounted for five-eighths of their sales through cooperatives in 1972.

Seventeen commercial feedlots for beef cattle have recently been started in the Rift Valley, where a "disease free zone" has been established. The weight of range cattle is increased in these feedlots by 50 percent in 70 days, which will hopefully permit greatly increased exports of meat in the years ahead. A variety of other livestock is kept in the highlands, with a substantial part of the animal and product sale being marketed domestically.

Forest products. Kenya's main forests are also in the highlands above about 5,000 feet (1,500 m). There is a great variety of indigenous species, but most are slow-growing, so emphasis has been placed on plantings of conifers, eucalyptus, and wattle. Some 300,000 acres (125,000 ha) of forests had been planted by 1974; about a fifth is in the Eldoret and Turbo areas, part supporting a large new pulp and paper factory at Broderick Falls.

Wattle trees have been grown by Africans at higher elevations on the Aberdare slopes and on lower ones on Mt. Kenya, and by Europeans in the Uasin Gishu district where the main extract factory is found at Eldoret. In addition to providing bark and tanning extract, the trees supply useful wood. Competition from synthetic substitutes and from Argentinian quebracho reduced interest in the crop in the 1960s, but some new plantings were made under the 1970–74 plan.

THE EASTERN HIGHLANDS AND RIFT VALLEY SYSTEM: TANGANYIKA

The eastern mountain system of Tanganyika is more complex than that of Kenya and not so easily oriented to the Rift Valley. Its conformation has been likened to a huge figure 9, the loop encircling the Masai Steppe, the tail forming a great arc to the north of Lake Malawi. This includes the knot of mountains near that lake where the two rift systems coalesce, usually called the Southern Highlands. Moving north the figure includes the Iringa Highlands, the Rubeho Mts., the Uluguru Mts. (offset somewhat to the east), the Nguru Mts., the Handeni Hills, the Usambara Mts., the Pare Mts., and Mt. Kilimanjaro, at 19,340 feet (5,895 m) the highest point in Africa. Swinging around the loop we encounter Mt. Meru, the series of volcanic masses such as Loolmalasin and Ngorongoro, the Mbulu and Kondoa

Small stock on a ranch in the rift valley near Naivasha
While portions of the valley in Kenya are wet enough to support tillage agriculture, much is suitable only for extensive grazing.

Ranges, the Gogoland Hills, and the Mpwapwa Mts.

In Tanganyika the Rift Valley is well marked to about the Central railway; Lakes Natron and Manyara lie within it, with Lake Eyasi in a short westward bifurcation. It is well marked again in the south, particularly along the Lake Malawi axis. As in Kenya, the eastern highlands have a great variety of climate, vegetation, and land use patterns, and contain many of the most densely populated parts of the country and the most important producing areas for high-value export crops.

AGRICULTURE IN THE TANGANYIKAN HIGH-LANDS. As has already been suggested, many of the Tanganyikan highlands have an extremely dense African population, though there are always opportunities for intensification of land use, and in some cases for bringing additional lands under cultivation. The degree to which African agriculture has been commercialized varies widely according to tribal, climatic, soil, and other factors. Perhaps the most significant is the availability of transport and distance from the sea. The north, for example, is relatively well served and contributes significantly to Tanganyikan exports; the south, possessing some excellent soils, has been very remote until construction of Tazara and commercialization had been more modest and scattered.

Non-African estates, exclusive of those sisal estates located in this region, covered perhaps 200,000 acres (81,000 ha) in the various highlands, especially the Usambara Mountains and Mt. Meru in the north, but also in the Iringa Highlands and the mountains around Mbeya and Njombe in the south. Except for sisal and tea, however, non-Africans did not play the outstanding role they occupied in the agricultural production and exports of Kenya, though an estimated 40 percent of exports in the early 1960s came from the approximately 1 percent of the land they then held in lease. As noted earlier, the sisal plantations were nationalized; many of the other holdings have

also reverted to national ownership and Tanganyika is seeking aid from Britain to purchase some of the estates remaining in European hands.

The most important, and perhaps the most interesting, highland of Tanganyika is Mt. Kilimanjaro. The indigenous Chagga have developed an intensive agriculture on the slopes of this great volcanic peak between about 3,000 and 4,500 feet (900–1,400 m) and in places to about 8,000 feet (2,400 m). Its soils are light, moisture-holding, grey-brown volcanics of excellent fertility. Precipitation is heavy on the southwest to southeast slopes and increases with elevation; for example, the annual average at Kibosho, at an elevation of 4,850 feet (1,478 m), is 104 inches (2,642 mm), while it is only 18 inches (457 mm) at Arusha Chini, only 20 miles (32 km) to the south and east at 2,300 feet (700 m).

Some 400,000 Chagga live on Mt. Kilimanjaro. They reside on small plots not usually exceeding two acres (0.8 ha) on which the banana, eulesine for beer, and the cash crop, coffee, are grown. On the lower slopes, corn, onions, cotton, and manioc are produced. The Chagga had developed a remarkable system of irrigation on the moun-

A typical Chagga hut on Mt. Kilimanjaro
Coffee, the main cash crop, is grown under bananas, which are the most important subsistence crop.

tain long before the arrival of the Europeans; furrows, sometimes running for miles, bring water along the precipitous sides of ravines and then along the contours to the individual holdings where the water is distributed on a carefully organized basis. The system was apparently built primarily for support of eulesine and, while the Chagga propensity for consumption of beer has been criticized, it must be admitted that it led to an amazing agricultural development and indirectly to the promotion of a relatively advanced social system.

The Chagga also keep cattle, which are usually stall fed on banana stems and on grass laboriously brought up by the women each day from the lower slopes several miles away. Coffee was first grown on Mt. Kilimanjaro over 60 years ago by a Catholic mission. The industrious Chagga took to its production rather rapidly, and in 1925 formed the Kilimanjaro Native Coffee Union, which became the largest African cooperative society.

Mt. Meru has a land use pattern somewhat comparable to that of Mt. Kilimanjaro. Cattle, however, are usually grazed around the homesteads or, when possible, sent to outlying parts of the district. Coffee is again the major cash crop and the Meru Cooperative operates a central pulpery for the local coffees. The per capita GDP in the Moshi-Arusha districts, which include the two mountains, is among the highest in the country.

The Chagga and Meru peoples grow most of the country's arabica coffees, which account for about three-fourths of Tanganyikan coffee production by tonnage and a higher share by value. There are substantial opportunities for improving the standards of coffee production by reducing the excessive shade provided by banana trees under which the coffee is grown, through mulching with banana trash and grass, replacing old stock with selected seedlings, and improving the preparation of the bean.

The northwest slopes of Mt. Kilimanjaro and the Mt. Meru area comprise one of three leading pyrethrum areas in the country and the only extract factory is located at Arusha. Tanganyika ranks second to Kenya in output of pyrethrum products; most of the flowers are now grown by smallholders who receive licenses to protect against overproduction. Arusha is also the site of one of the meat-processing factories of Tanganyika Packers.

Between Mt. Kilimanjaro and Korogwe and forming the northeast portion of the topographic 9 are the North Pare, South Pare, and Usambara Mountains, all fairly alike. The Usambara Mountains form a compact block about 70 miles (113 km) long and from 20 to 40 miles (32–64 km) wide whose highest elevation is 7,550 feet (2,300 m). Eastern Usambara has a considerable number of large estates, with coffee and tea being the main crops. The other sections are densely populated. Corn is the chief subsistence crop, though bananas and sweet potatoes are important supplementary crops; there is also a dense livestock population. The main cash crop is rice, grown in the well-watered swamps along the foot of the mountain. Soil erosion reaches serious proportions in some areas on the predominantly red-earth soils.

West of Mt. Meru is a group of massifs called the "Winter Highlands," a volcanic region containing Mts. Loolmalasin and Ngorongoro and, within the Rift Valley, Mt. Lengai—still active. The huge caldera of Ngorongoro, with a tremendous number of wild animals on its floor, is one of the most spectacular tourist attractions in Africa. Commercial agriculture is not well developed in the area, except around Oldeani where a number of large farms and African smallholders specialize in wheat.

The Uluguru Mountains, south of Morogoro, have population densities exceeding 500 per square mile and cultivation has been carried onto excessively steep slopes. Efforts in the 1950s to introduce a variety of improvements—including bench terracing to reduce erosion, composting, stall feeding of cattle, and water control to maintain supplies for the sisal plantations in the Morogoro District—failed because of opposition from the residents. Similar schemes in other congested districts also had great difficulties, even though they were intended to relieve pressure on the land.

The Iringa Highlands, running south from the loop of the 9, support some 150,000 people on about 1,250 square miles (3,240 sq km), as well as a fairly dense livestock population. The friable soils of these highlands are easily eroded, and an effort has been made to provide grazing lands on the plains below the escarpment so that the overgrazed highlands could be given an opportunity to recuperate. Tea is grown in the Mufindi area; flue-cured tobacco comes mainly from Iringa; and pyrethrum is a third important cash crop.

In the south, the cluster of mountains at the head of Lake Malawi belongs to both the Eastern and Central Rift Valley systems. There are marked contrasts within this complex region, which is greatly disadvantaged by distance from the sea. Despite its isolation, commercial agriculture has developed to a degree, and the new railway is expected to stimulate further development. Tea estates are found in the Rungwe District south of Mbeya. Tanganyika produces about a fifth as much tea as Kenya; about three quarters still comes from estates, though their share of the total is decreasing.

Coffee comes from Mbozi, about 40 miles (64 km) west of Mbeya while smallholders are increasing their production of arabicas around Njombe and Mbeya. The area centered on Njombe has become the major Tanganyikan pyrethrum zone. Smallholder production of tobacco is important in the Iringa and Mbeya regions and is increasing in response to favorable prices. Near Njombe are the large CDC-supported Tanganyika Wattle Estates and the Ubena

Wattle Scheme. The government is setting up several large state farms to produce wheat in the area. And finally, about 700 acres (280 ha) roughly divided into estate and smallholdings are planted to lime trees and a lime extract factory sells the output to L. Rose and Company in London.

THE LAKE VICTORIA AND CENTRAL PLATEAU BASIN

A huge uplifted basin occupies the lands between the eastern highlands and the mountains and lakes of the Western Rift Zone, which Tanganyika shares with Zaïre, Rwanda, Burundi, and Zambia.

TANGANYIKA. In Tanganyika, the Central Plateau, with an average elevation of 4,000 feet (1,200 m), forms the largest region of the country. Much of it is covered with *miombo* woodlands in which *Bracystegia* is the predominant species. Most of this area is either sparsely occupied or completely empty, in part because it is infested with the tsetse fly, and human sleeping sickness is endemic. Annual precipitation is only 25 to 30 inches (635–760 mm) and is highly variable. Soils are generally poor except in the valley bottoms, where dark, heavy clays called *mbugas* are relatively fertile; they are not widely utilized, however, because it is difficult to work them with hand tools and because they are subject to seasonal flooding. A very large part of the Central Plateau is a peneplain whose monotony is relieved only by the protrusion of granitic inselberg.

That part of the Central Plateau lying south of the Central Railway has been called the "empty quarter." Around Tabora is a comparatively fertile but steppelike area, which supports somewhat greater densities; peanuts and castor beans are the main cash crops. Around Urambo, where the third area of the Groundnut Scheme was situated, the main cash crop is flue-cured tobacco grown both on non-African estates and on medium and small African farms. All of the tobacco is taken by the British-American Tobacco Co. (Tanzania) for the domestic market; that company and government agricultural staff are emphasizing extension of production by smallholders.

The woodlands of the Central Plateau produce some *mwinga* timber plus gums, honey, beeswax, and fibers; Tanganyika is the world's largest exporter of beeswax. It is not a richly forested country, however; only about 2 percent of its total area is in high forests. There is an increasing interest in softwood plantations, which are situated mainly in the upland areas; some hardwoods have also been planted, particularly teak.

Toward Lake Victoria, which lies at an elevation of 3,717 feet (1,133 m), is the "cultivation steppe"—a large, almost treeless area which has been cleared by the Sukuma. This area is better watered and is one of the more densely populated regions of the country. Overcrowding has worn out some of the lands and the amount of comparable new land available is limited. In recent decades the Sukuma have been increasing the production of cotton with some rapidity and they now produce about 90 percent of the country's total. Cotton ranks second only to coffee among Tanzanian exports. Families in the area cultivate about 6 to 9 acres (2.4–3.6 ha) each, about half in cotton and half in subsistence crops. Yields vary from about 300 to 700 pounds per acre (333–753 kg per ha) depending on soil fertility, weather conditions, and cultivation practices. Both yields and quality could be improved by more efficient handling and ginning. Early planting, tie-ridging to prevent erosion and conserve moisture, and application of manure and chemical fertilizers could double yields, but growers have been slow to adopt these practices despite their proved economic value.

Toward the west, along the Malagarasi and Ugalla River systems and near the border with Burundi, widespread swamps have repulsed settlement. But along the western side of Lake Victoria, the Bukoba District is a relatively favored part of Tanganyika. It may be divided into two major portions: a zone of rich lands along the lake shore which is well watered and supports about 80 percent of the population, and the more hilly hinterland interspersed with rock outcrops which has some good soils but unreliable water supplies. The shore zone is the main site for the staple banana, the main subsistence crop and raw material for beer for the Haya of Bukoba. Their main cash crop is coffee, mainly robusta but including some arabica. There is also a small sugar plantation in the area and some production of tea.

East of Lake Victoria, the farmers grow cotton, peanuts, rice, sesame, and corn, and carry on an extensive grazing. The huge uplifted Serengeti Plain is much drier; most of it is in the large game park of the same name, lush in the rainy season but sere and forbidding in the dry period, when vast stretches are charred by grass fires.

Within Lake Victoria are a large number of islands, some of which are of considerable interest. Ukerewe Island produces cotton and rice particularly well, in part because of its favorable rainfall. The smaller Ukara Island has such great population pressure that it is the scene of some remarkably evolved and intensive agricultural practices. Crops are grown in accordance with a carefully planned rotation, which includes the digging in of green manure crops and the provision of fodder crops for cattle. Livestock are hand fed and carefully bedded down in the huts, while their manure is composted and used on the fields. Stock numbers are rigidly controlled and when the livestock are taken through the fields they are muzzled to protect the crops. An experiment was once made to utilize these skilled farmers as demonstrators of their intensive practices on the adjacent mainland, but, divorced from the necessity for such a rigorous system, they soon fell into the ways used by the local farmers.

KENYA. The southwestern section of

Kenya, Nyanza Province and a part of Western Province, is part of the Lake Victoria Basin. Kisumu, on the Kavirondo Gulf, is the major lake port, handling traffic for all three East African countries. The area has an average elevation of about 4,000 feet (1,200 m) and receives from 40 to 70 inches (1,000–1,800 mm) of rain yearly. It has a dense population mainly concerned with subsistence production. Cotton is the main cash crop but production levels have rarely met the announced goals and are now about 30,000 tons a year. Efforts are being made to have farmers adopt improved practices, including timely planting, better weeding, the use of fertilizers, and proper control of pests; it is thought that only improved yields providing a better return to the farmers will lead to substantially increased production.

Most of Kenya's domestically produced sugar comes from this region. It is produced on large estates at Miwani, Muhoroni, Mumias, and Chemelil, and by several thousands of outgrowers. Production of sugar has increased from 37,000 tons in 1963 to 150,000 tons in 1971 but it fell off to only 90,300 tons in 1972; plans call for reaching an output of 235,000 tons in several years, which would permit elimination of imports.

There are several areas suitable for irrigation in the region, including 35,000 acres (14,200 ha) in the Yala Swamp, where a 500-acre (200 ha) pilot scheme has been completed, and an equal area in the Kano Plains, where the 2,000-acre (800-ha) Ahero pilot scheme is producing two crops of rice a year.

LAKE VICTORIA. An activity common to all three countries bordering the lake is fishing. It is generally confined to within a few miles of the shore and is characterized by the use of traditional techniques. Larger, powered vessels would permit fishing the deeper waters which are believed to have a substantial potential. About five-eighths of both Kenya's and Tanganyika's fish catch comes from Lake Victoria.

UGANDA. That portion of the Lake Victoria

Cutting sugar cane at Miwani, near Kisumu, Kenya
Kenya aims to achieve self sufficiency in sugar output with several large plantations in this area.

Basin located in Uganda includes some of the best quality lands in East Africa; indeed, it is one of the favored parts of the continent. Uganda is even more heavily dependent on agriculture than the other East African countries; agriculture accounts for about 54 percent of GDP, 85 percent of the active population, and over 90 percent of exports. About three-fifths of the tilled land is still used for subsistence production, but only one-sixth of the total available land is cultivated, and this does not yield anything like its full potential.

The richest part of the Uganda plateau is the crescent around the north end of Lake Victoria. Climatically it is favored by a relatively heavy and reliable precipitation (Map 86) and by the moderating influence of elevation. From the landform standpoint it is characterized by a seemingly endless succession of low, flat-topped hills, which gradually become less pronounced to the north. Local relief is about 500 feet (150 m); the peneplained hill tops are at elevations of about 4,300 to 4,400 feet (1,310–1,340 m). Soils are distinctly better than in most of tropical Africa; very typical is a catenal arrangement running from a thin covering of laterized, greyish soil on the hilltops to deep, rich, red and chocolate loam of excellent structure on the slopes, to alluvial clays that are virtually uncultivable on the valley floors, which are commonly swampy. Fortunately, it is the

A team of bullocks working in Teso, Uganda
Efforts have been made for some years to increase use of work animals in various parts of Uganda.

slope soils that cover by far the largest area. Where the land is not under cultivation the typical vegetation is a rank growth of grasses, especially the very high elephant grass, though considerable stretches of high forest remain. The high grasses grade into open woodlands merging into low tree–grasslands in the central part of the country.

The core area of the crescent is in East Mengo and West Mengo (formerly Buganda), which are the richest parts of Uganda. These two regions, which make up approximately 23 percent of Uganda's land area, have a population of some 2.85 million, about 30 percent of Uganda's total. The Ganda are the majority ethnic group, about a quarter of the population are Banyaruanda and Burundi, three-fourths of whom are migrant laborers who work on the farms of the Ganda.

Ganda are individualist farmers, operating family farms. In accordance with the Uganda agreement of 1900, about half of the lands of the Buganda Kingdom were allotted to notables of the tribe on a system of individual ownership. These areas, called "mailo" land from the square mile measurements employed, have since been subdivided into much smaller holdings, but many Ganda are still substantial landowners.

The typical land-use pattern over much of Ganda country is: (1) some grazing, but very little cultivation on the hilltops; (2) a close cultivation of the staple plantain banana, corn, some cassava, and sweet potatoes plus the cash crops—coffee or cotton—on the upper and middle slopes (this zone is the main site of the houses, which are usually surrounded by bananas); (3) some cultivation, particularly of sweet potatoes and grains on the lower slope where the soil is likely to be more acidic and sandy; and (4) papyrus swamps in the valley. There has been some drainage and planting of valley clays and swamps, but their potentialities are still inadequately understood.

A large number of livestock are kept in the south. While it has been typical for cattle to be herded and housed away from the cultivated area, a growing number of farmers realize the value of integrated crop and livestock farming. Few Ugandans had any appreciation of the value of manures, though the Ganda and others did mulch the bananas, but under prolonged prodding from agricultural officers manures are now better used, cattle paddocks are provided, and grade cattle are being kept, which requires a high standard of management and disease control. There would appear to be an excellent future for mixed farming, particularly for stall feeding with cut elephant grass, while rotation of crops and grass would be beneficial. There has been a notable increase in ox-drawn plowing, particularly in the east.

The general agricultural pattern of Ganda country is continued eastward to southern Busoga and Bukedi and southwestward into the Bukoba District of Tanganyika. But away from the lake, conditions are less favorable than in the crescent. In the short-grass zone, finger millet and sorghum replace bananas and corn, and cotton replaces coffee as the main cash crop. Cattle keeping is common, and while cattle are housed on the holding there is still little integration with tillage agriculture. North of Lakes Opeta and Kioga, which have large expanses of papyrus swamps both around and within them, precipitation drops off gradually, particularly to the northeast. Cotton production has increased north of Lake Kioga and has been aided by extension of the railway to Pakwach.

The Karamoja District in the northeast is a backward part of Uganda, mindful of the grazing areas of northern Kenya. The Karamojong and related groups are predominantly pastoralists but all produce some drought-resistant grains as a kind of evil necessity. Despite repeated efforts, they have remained reluctant to sell more cattle than is absolutely necessary, mainly because cattle are what they want and value most highly. The Karamojong and other pastoralist groups present dilemmas to many governments in Africa. One alternative is to leave such groups alone, in which case natural forces such as disease and drought would periodically adjust the livestock and human populations. It is questionable, however, whether the dynamics of the situation would permit such a museum-type approach. People may not want administrative authority and policing but they do want clinics, diptanks, and inoculations; sedentary neighbors will insist on protection and hence there must at least be control of the periphery; and the demographic dynamic may make it impossible to sustain the status quo. Population growth within a pastoral group will inevitably increase pressure on the land, which can only be met by periodic wars and famines, out-migration, adoption of more intensive practices, or the introduction of new activities. Population growth of more powerful outside groups may expedite the pressure through excision of more favored lands.

To the south and west of the fertile crescent, in Ankole and much of Toro, lands become higher and more rugged. They are mostly in open grasslands with scattered bush and trees and some riverine forest. Millet is the main crop; peas, beans, peanuts, and potatoes are also important. Cattle are relatively sparse except in part of Ankole, where cattle keeping is the major occupation. Uganda is fortunate in having less than

one-third of its area infested with the tsetse fly, but some otherwise attractive grazing lands in Ankole were infested. The Ankole Ranching Scheme has involved a massive eradication program and the establishment of 2,500-acre (1,000-ha) ranches operated under controlled conditions.

To the northwest, the Bunyoro District is fairly comparable to the Ganda areas. Occupied by the Nyoro, who have a kingdom similar to those of other groups in Uganda, it is a fertile country of small hills and swampy valleys with a dense population. Most Nyoro live in individual homesteads in rather closely settled areas separated from other such areas by wide stretches of uninhabited bush. Many years ago they owned great herds of cattle, but these were lost by war and disease and the Nyoro are now largely agriculturalists. Bunyoro accounts for about half of the tobacco produced in Uganda, most of it coming from the Waki Valley in the center.

Cash crops of the area. Cotton was long the prime commercial crop of Uganda. The main credit for its establishment goes to the Uganda Company, founded by wealthy supporters of the Church Missionary Society. Aided by the British Cotton Growing Association, organized in 1902 to develop alternative sources to the United States, the Uganda Company began distributing seeds to the local chiefs in 1904 and by 1911 production had grown to 2,500 tons and cotton was accounting for about half of total exports. Rapid progress continued after World War I and exports reached 73,000 tons in 1938. This level was not surpassed until 1966, when 80,000 tons were produced, and output has been below that in all but one of the last five years. Since the mid-1950s cotton has ranked second to coffee as the main cash crop and export of Uganda, and export value is 25–33 percent that of coffee.

Cotton is produced rather widely in Uganda. Busoga, Teso, Bukedi, and parts of Lango and Acholi are particularly important

Cotton brought to a ginnery in northern Uganda
Grown widely in the country, cotton ranks second to coffee among cash crops.

areas. The long-term trend has been downward in areas where coffee could be grown and upward in the lands north of Lake Kioga. Coffee is more popular than cotton because it requires only about 80 man-days of work per acre (200 man-days per ha) as compared to 140 for cotton and gives yields about two to three times as great. But cotton has the advantages that it can be produced in a much broader area of Uganda and that its production can be increased, whereas coffee sales have been restricted by agreement. Government campaigns to reach distinctly higher levels have, however, been frustrated.

Cotton is grown exclusively by Africans on plots averaging about three-quarters of an acre (0.3 ha). Yields compare very favorably with most rain-grown cottons in Africa, but could be considerably increased by proper spraying. The cotton is a good middling staple and is produced relatively cheaply. Ginning, which was long handled by Indians, causing periodic political difficulties, is now under African cooperatives.

Coffee was a traditional part of the Ganda farm long before it became significant as a cash crop. Its commercialization began in the 1920s and steady if not spectacular progress was made in prewar years. The big boom in

robusta production came in the 1950s and production averaged 150,000 tons in the first half of the 1960s, placing Uganda third among African producers with about 15 percent of the continent's total output. In more recent years production has averaged about 200,000 tons, about 3.5 times the Kenya level. But the value of Ugandan coffees per ton is only about two-thirds that of Kenyan coffees, mainly because over 90 percent is robusta. Improvement in quality, partly through the use of wet processing, could increase returns, while Uganda is also attempting to increase output of higher value arabicas. Coffee exports accounted for about two-thirds of total Uganda exports in 1973.

Other cash crops of the Lake Victoria Basin in Uganda include tobacco, sugar, tea, castor seed, peanuts, and soy beans. Sugar production doubled from 1953 to the 96,000-ton level of 1961 and now averages about 140,000 tons. About a third of production was formerly sold in Kenya, but efforts to achieve self-sufficiency in that country are closing the main extranational market. Most production has come from large Indian-owned estates in the Jinja area but increasing quantities came from outgrowers. These estates accounted for about two-fifths of the area in non-African holdings in Uganda, which was only about 98,000 acres (40,000 ha) in 1961. Takeover of the sugar estates led to management and maintenance problems which have not as yet been overcome.

Tobacco is grown almost entirely by African smallholders, and output runs around 4,000 tons for the whole country. Several tea estates are situated between Jinja and Kampala where tea was first produced in Uganda. This region is not so favorable ecologically as the higher lands of Mt. Elgon or the west, which are now the main producing areas.

THE WESTERN RIFT SYSTEM

As indicated earlier, this zone is shared with other nations. In Uganda it includes the

Lake Mobutu lowlands, very hot and dry; the remote region to the west of that with elevations of 4,000 to 5,000 feet (1,200–1,500 m) and precipitation of about 50 inches (1,270 mm), which produces about a quarter of the country's tobacco; Mt. Ruwenzori, a huge uplifted block lying within the rift and rising to 16,794 feet (5,119 m); Lake George and Lake Amin (Edward), along which is one of Uganda's two main game reserves; and the high land of Kigezi in the southwest corner, an upwarped shoulder of the Western Rift, culminating in the Mufumbiro volcanoes.

The area near Fort Portal is one of the leading tea-producing areas of Uganda and also accounts for a portion of the arabica coffee produced. Kigezi is a continuation of the highlands of Rwanda, a deeply incised upland which is densely populated and intensively farmed from the tops of the hills to the reclaimed swamps in the narrow valley floors, with remarkable terracing on its often extremely steep slopes. While handicapped by remoteness, which has been decreased by completion of a paved road from Kampala to Kabale, this beautiful region is contributing increasing quantities of arabica to Uganda's exports, while lower northern Kigezi has begun to produce tobacco.

Tea now ranks as the third export crop of Uganda; it is produced on Mt. Elgon, in the western highlands especially east of Fort Portal, and around Jinja. As in other African countries it was originally an estate crop; the maximum area in tea plantations was reached in 1964 with 24,000 acres (9,700 ha), but in 1971 it had declined to 18,000 acres (7,300 ha) as the area in smallholdings increased to 14,000 acres (5,700 ha). Most of the estates are now run by a subsidiary of the government-owned Uganda Development Corporation (UDC).

In Tanganyika a large part of the western border is formed by Lake Tanganyika, while a series of small highlands occurs along its eastern side until the large Ufipa Plateau, which lies between the Lake and Lake

A scene in the Kigezi District, southwest Uganda
This deeply incised upland area is greatly disadvantaged by remoteness, but is increasing the cash production of arabica and tobacco.

Rukwa, is reached. Very little commercial production comes from the western highlands of Tanganyika, their extreme remoteness having militated against development, but efforts are being made to promote high value crops in the sections west of Mbeya. This southwestern region should also gain from the Tazara line.

Fishing is of some importance in all the lakes of the western rift zone shared by Uganda, Tanganyika, and the other bordering nations. Uganda gets a catch from these lakes about equal to its Lake Victoria fisheries, with part being exported to Zaïre and Kenya. About a quarter of Tanganyika's catch comes from Lake Tanganyika, with a portion going to Zaïre, Burundi, and Zambia.

MINERALS OF EAST AFRICA

East Africa has been relatively unimportant as a mineral producer, and the search for new bodies has not as yet met great success.

TANGANYIKA

DIAMONDS. Now leading the mineral exports of East Africa, diamonds come from the Williamson mine near Shinyanga north of Tabora (Map 88). Discovered in 1940 by a Canadian geologist who spent six years exploring the country, the pipe at Mwadui is the largest ever found, an oval measuring about 3,500 by 5,000 feet (1,067–1,524m), which is over three times the size of the Premier pipe in South Africa. Numerous kimberlite pipes have been located in Tanganyika but no others have a paying diamond content.

The Williamson mine is now under Tanganyikan control and a decision has been made to reduce the level of production to sustain its life to 20 years instead of the 7 years previously planned. Diamonds account for over 90 percent of the total value of mineral output in Tanganyika and about 99 percent of the value of mineral exports, which were 7.6 percent of total exports in 1973.

OTHER MINERALS. Output from scattered localities of gold, which was long the main mineral produced in East Africa, has been declining steadily. A lead mine at Mpanda was operated from 1950 to 1960. There is a small output of mica, tin concentrates, salt, and meerschaum rock. The last comes from a deposit in Masailand close to the Kenya border and is used in the manufacture of high-value pipes in Arusha.

The soda ash factory at Lake Magadi, Kenya
Soda ash, used in making glass, is the main mineral export of Kenya. The raw material is renewed from springs in the lake.

Three deposits of coal are known, one east of and one near the northwestern end of Lake Malawi and the third east of Lake Tanganyika. The remoteness of these areas, the noncoking quality of the coal, and the necessity for underground mining militate against early exploitation, but there has recently been talk of Chinese support for development of an iron and steel industry based in part on these coals. Iron ore is found at Liganga only 30 miles (48 km) from coal but it is only of medium grade and difficult to smelt because of its titanium content.

Other minerals of potential interest include: a 10-million-ton deposit of phosphates apparently formed from guano accumulated on an island when Lake Manyara covered a much larger area; kaolin, found only 17 miles (27 km) west of Dar; soda ash from Lake Natron, which contains the world's largest known reserves; and magnesite, found near Longido on the Kenya border. The search for oil, which has gone on for over a decade, has been unsuccessful.

KENYA

SODA ASH. The main mineral export of Kenya is soda, produced from Lake Magadi by the Magadi Soda Company, formed in 1911 and now a subsidiary of the Imperial Chemical Industries. The raw material, known as "trona," is fairly evenly distributed over the lake and is renewed from springs that probably derive their heat from subterranean magma. Soda ash produced at the company's plant is used in glass production. Sales in recent years have been valued at about $5.4 million, which is about two-thirds of the value of total mineral output. Common salt is also obtained and marketed in the country.

OTHER MINERALS. There is a small amount of gold and gemstones and substantial production of limestone and other construction materials. Several deposits just being opened up will increase the value of mineral output:

a lead-silver mine at Kinangoni, 20 miles (32 km) north of Mombasa, is increasing its output from 210,000 tons in 1974 to double that level by 1978 with exports expected to yield $3 to $5 million; a fluorspar deposit in the Kerio Valley is scheduled to produce 72,000 tons from 1974 earning about $8.4 million in foreign exchange; and 500 tons a year of vermiculite are to be produced by a Japanese firm.

While exploration for hydrocarbons has not yet been successful, finds across the border in Ethiopia have raised hopes that discoveries will be made in northeastern Kenya. Nine companies are prospecting in eastern Kenya and in offshore waters.

Other minerals of possible future significance include diatomite, graphite, gypsum, kyanite, and asbestos.

UGANDA

COPPER AND COBALT. The most important mine in Uganda is the Kilembe copper mine in the foothills of the Ruwenzori, which provided the main reason for extending the railway west to Kasese. Mining began in 1948, but operations were not fully developed until 1956 when the rail line was completed and the copper was first smelted at Jinja. The ore has a 1.8 percent copper and an 0.14 percent cobalt content. Most of it is concentrated to a 27 percent product, which is railed with some of the oxides to Jinja, where smelting takes place. Low-cost power is supplied from the Owen Falls Dam, and the smelter is the most important consumer of electricity in Uganda.

Exports of 99.3 percent blister copper containing minor amounts of nickel, gold, and silver have been valued at about $15.5 million in recent years or about 5 to 6 percent of total exports. Cobalt concentrates have been stockpiled at Kasese, but plans now call for their being processed. Reserves of ore at Kilembe are somewhat uncertain and threats of nationalization have discouraged prospecting.

The Kilembe Copper Smelter at Jinja, Uganda
This plant is the most important consumer of electricity in East Africa. The mineral is railed from the Kilembe mine in the foothills of the Ruwenzori Mountains.

OTHER MINERALS. There is a small output of tin concentrates, wolfram, beryl, and salt in Uganda. Deposits of future interest include a very large reserve of phosphates in the Sukulu Hills near Tororo (too far from the coast to warrant export at present), iron ore in the same area, and vermiculite in Bugisu; prospecting continues apace for a variety of minerals, including uranium.

INDUSTRY IN EAST AFRICA

Kenya has enjoyed the greatest development of manufacturing industries in East Africa, particularly of market-oriented establishments; almost all of these industries have been introduced since World War II. While Tanzania and to a lesser extent Uganda have most of the industries found in Kenya, those in Kenya are often larger, are represented by multiple firms and plants, and frequently produce a wider range and more sophisticated products within a given branch. Many of the data on manufacturing in all the countries are less meaningful than they might be because of such practices as the inclusion, in tables of industrial activity, of mining, construction, handicraft activities, repairing, and the total employees of tea and sisal estates. The following summaries are nonetheless of some value in assessing the actual and comparative importance of manufacturing in East Africa.

In Kenya, the contribution of manufacturing and repair to the GDP was estimated at 11.5 percent in 1972. In that year 96,000 people were employed in manufacturing, and they produced goods valued at $784 million. Labor costs were calculated at 49.4 percent of gross production, down from 55.9 percent in 1967 because of improved productivity.

Tanganyika ranks second in industrial development. That country listed 6,500 registered factories employing 90,000, but these figures include many nonmanufacturing jobs, many establishments without power, and only 131 enterprises with over 100 employees. A more realistic figure is 55,000 employees in 1971, when manufacturing contributed an estimated 9.5 percent to the GDP.

In Uganda manufacturing accounted for about 10.0 percent of GDP in 1970. Expulsions of Asians, British, and others who played significant roles both in private industry and in the state holding company, Uganda Development Corporation, have seriously affected this sector, as has the nationalization of two large Indian companies. One of the latter, Madhvani, was the largest private industrial group in the country, with 20,000 employees, including agricultural workers. Some of the enterprises taken over were put under UDC, others under a special board pending their sale to Africans.

Zanzibar has very little industry other than the processing of raw materials.

There is some question whether all three countries may not have placed excessive emphasis on import-substitution industries. Because of inefficiency a number of these have required considerable protection, which has raised prices to the consumer and raised the costs of previously competitive industries insofar as these use protected goods and inputs. They may also divert resources away from more efficient and internationally competitive activities, which would have an adverse effect on foreign exchange earnings. Furthermore, some of the protected industries only add the "finishing touches," but may reap substantial returns for these small additions to the value of the product. Import substitution also tends to perpetuate inequalities, because wages in manufacturing rise more rapidly than those in other activities or than incomes in agriculture.

A good many industries, particularly in Tanzania and Uganda, operate well below capacity, which further reduces their efficiency. Some of the causes are the creation of excess capacity (partially explained by a "me-too" attitude whereby each country tries to copy the industrial make-up of the best developed, inadequate quality control, inability to compete with imports, irrational pricing, and loss of skilled managers and technicians; the last has adversely affected plants in Uganda.

There is now greater concern that manufacturing development be better related to other sectors of the economy. Kenya, for example, plans to give greater weight to the further processing of domestic raw materials, to those industries which have export potential, and to labor-intensive industries. It is commonly advised that a switch should be made from capital-intensive to intermediate technologies, but care must be taken not to create plants that require even greater protection or that forgo possible competitiveness in world markets by following this prescription. It may be wise, however, not to extend excessively generous investment al-

Inducements to train labor would also have a beneficial impact on employment.

INDUSTRIAL REPRESENTATION AND DISTRIBUTION

East Africa's industries are quite comparable to those found in other tropical African countries, and attention will therefore be focused only on aspects of special interest. In the field of agricultural processing, tea, sisal, and sugar factories are of importance; often of large scale and requiring substantial investment, they contrast strongly with the usually small-scale and more widely dispersed ginneries, oil seed presses, grain mills, and coffee washing plants. There is, however, a large coffee curing plant in Nairobi, while soluble coffee and tea factories are of special interest. Other raw material processing industries that are better represented in East Africa than elsewhere include processing of pyrethrum, canning of fruits and meat, and preparation of wild animal skins and trophies.

In the market-oriented sector, breweries,

Pineapples being canned at a factory in Kenya
The canning industry is better developed in Kenya than in other tropical African countries.

lowances for machinery and plant and to check a too-rapid rise in industrial wages, both of which encourage capital-intensive technology and reduce employment levels.

The Athi River Cement Works near Nairobi
Each of the East African countries has cement mills. A second mill in Kenya, at Changamwe near Mombasa, exports a portion of its output to other Indian Ocean countries.

cigarette factories, and cement mills may be found in all three countries and are, as usual, among the leaders. The textile industry has had a particularly vigorous growth in the last fifteen years and further expansion and diversification is planned. Bata, not unsurprisingly, has shoe factories in each of the countries. Fertilizer factories are larger than most such installations in tropical Africa; the one at Tororo in Uganda uses domestic phosphates while the new one at Tanga appears to be having difficulties despite considerable government assistance.

Uganda has one of the very few low-cost power oriented plants in Africa—the copper smelter at Jinja. Other plants at that city may have been attracted in part by power, but are not accurately designated as industries oriented toward low-cost energy. The meerschaum pipe factory at Arusha and the preparation of trophies at Nairobi and Kampala by internationally known companies are also exceptional; they manufacture and export products that require high skill.

There has been much talk, and in the EAC and previous agreements there was some provision for the promotion of regional industries—plants that might not be justified if they served only individual national markets but would be if assured the whole EAC market. Certain of these plants were to be alloted to each country, particularly to the weaker partners. In fact, little true cooperation has been forthcoming; there appears instead to have been retrogression in such industries as automobile tires, refining, fertilizers, and steel rolling mills in which unnecessary duplication threatens the viability of several plants.

The distributional pattern of industry in East Africa is roughly comparable to that of other tropical African countries—raw material processing is widely scattered and market-oriented industries are heavily concentrated in a limited number of cities. In Map 89 the pattern for Kenya is shown, based on the number of types of industry present in 1972, when there were 105 types. The dominance of Nairobi is greater than is suggested because there are multiple plants in many of the branches listed; of the 841 enterprises listed in the Kenya Ministry of Commerce and Industry's *Index to Manufacturers and Products 1972*, 60.8 percent were in Nairobi.

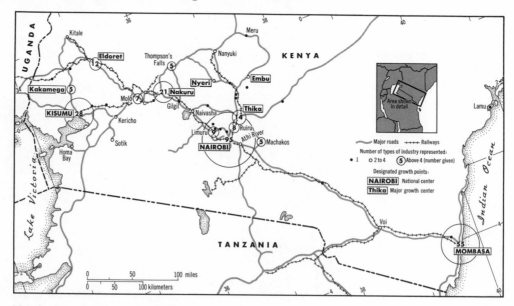

Map 89. Manufacturing centers in Kenya

Nairobi's advantages are excellent market location (60 percent of Kenya's population lives within a radius of 150 miles [240 km]), good road and rail connections, availability of the largest body of semiskilled and skilled labor, more industrial linkages and services than most African cities, nearness to government, and unusual attractiveness for foreign investors.

Mombasa also has an impressive development and accounts for about a sixth of employment in manufacturing and 18.5 percent of the 841 enterprises noted above. There is specialization in the processing of imported raw materials (e.g. refining, metal products, fertilizers) and in assembly industries. A portion of the cement and petroleum products produced at Mombasa is exported to other Indian Ocean ports.

Kenya is concerned about the dominance of Nairobi and Mombasa and aims to achieve in rural industries a rate of growth double that of the urban centers. Dispersion is also to be favored by the designation of growth points at which new plants will be encouraged to locate (shown in Map 89), and the establishment of industrial estates. A second major objective of these estates, and of new ones at Nairobi and Mombasa, is to encourage African entry into manufacturing, which is aided by provision of machinery and equipment on a 100 percent loan basis payable at 8 percent interest in eight to ten years and by free technical assistance.

In Tanganyika, Dar es Salaam is by all odds the leading industrial center, with over half of total modern manufacturing in the country. Arusha, Tanga, Moshi, and Mwanza are secondary centers. This country, too, has designated growth points, which are to be favored when appropriate: the four secondary centers plus Morogoro, Dodoma, Mtwara, Iringa, Mbeya, and Tabora. The selection of Dodoma as the new capital of the country will doubtless attract certain industries to this minor center.

Uganda has two cities occupying a dominant position in manufacturing: Jinja (where the Owen Falls Dam attracted a number of industries and which tends to have larger and more heavy industry) and Kampala (which has a large number of small, light, market-oriented industries).

FACTORS INFLUENCING FUTURE INDUSTRIAL DEVELOPMENT

RAW MATERIALS. In a brief look at the elements affecting the growth of manufacturing, it should first be noted that East Africa has a reasonably favorable raw material position, including the ability to produce an unusually wide range of food and industrial raw materials from the agricultural, fishery, and forestry sectors, all of which can be more intensively developed. While there is a reasonably diverse group of minerals, most will not likely require more than primary processing.

ENERGY. The energy position is somewhat spotty. Most of the coal is of only moderate grade and is located in remote areas of Tanganyika not otherwise attractive for manufacturing development. East Africa is well situated for the importation of petroleum from the Middle East and may, of course, itself become a producer. The 4-million-ton refinery at Mombasa serves not only Kenya but also Uganda, Mauritius, and the Seychelles. Study of geothermal resources in the Rift Valley in Kenya is continuing.

At present, hydropower resources are the main indigenous source of energy. Uganda has had the most significant development in the Owen Falls Dam at the outlet of Lake Victoria. Inaugurated in 1954, the plant at Jinja has a capacity that is fully developed at 150,000 kw. The dam and station were to some extent built on faith, with the expectation that large blocs of power would be taken by industry attracted to the site, but the installation has had surprisingly little direct effect on development. The only really large consumer is the copper smelter, but industrial and commercial sales on the grid supplied by Owen Falls take about half of the power produced. In 1958 Kenya agreed to take up to 30,000 kw of capacity over a fifty-year period. One advantage that to some extent offset the failure to attract industry has been the erection of a grid system that brings electricity to a very large portion of Uganda's towns and many rural areas, though construction of these lines has been costly.

With the normal rise in consumption of electricity, Uganda has had to give attention to installing new capacity. There is a very large potential in the country, mainly on the Victoria Nile, which drops a total of 1,690 feet (515 m) from Lake Victoria to Lake Mobutu. A decision to construct an 80,000 kw plant at Kabalega Falls ran into severe criticism because of its deleterious impact on one of Uganda's major parks and on the spectacular falls themselves, and has been withdrawn. An excellent site is found just 4.5 miles (7.2 km) below the Owen Falls Dam, where 180,000 kw may be developed. Uganda also has small hydro plants at Kikagati on the Kagera River, at Mobuku (where a 6,000 kw installation serves the Kilembe Mines), and at Kabale in Kigezi.

Kenya is not so favorably endowed in hydroelectric potential, and the characteristic great seasonal changes in flow would require construction of storage dams instead of merely the head-creating dams required on the Nile. The Tana and its tributaries have the greatest potential and have the advantage of permitting development in stages in line with increasing consumption. At present four installations in the so-called Seven Forks Scheme on the Tana system have a total capacity of 113,000 kw, while the total potential is estimated to be about 1 million kw.

Grids in Kenya have been developed in two major zones: the highlands are served by the Tana plants and by high tension lines from Uganda, with the Nairobi area taking about two-thirds of the total supply; the coastal zone is supplied by a 40,000-kw oil-fired plant at Mombasa, which consumes about a fourth of the country's total. The two grids have recently been connected to give greater flexibility and security.

Tanganyika has a series of relatively small hydroelectric stations: a 17,500-kw plant partially utilizing the Grand Pangani Falls on the Pangani River; a 21,500-kw plant at Hale, 5 miles (8 km) downstream, these two being connected by a 173-mile (278-km) line to Dar es Salaam; an 8,000-kw station serves the Moshi-Arusha area, and smaller plants serve Mbeya, Iringa, and Moshi. A 100,000-kw station at Kidatu on the Great Ruaha will soon be completed and will be doubled by 1985. Stiegler's Gorge on the Rufiji is being studied as a possible site for a 120,000-kw installation.

LABOR. The labor situation in East Africa is different from most of the areas thus far studied, because of the availability of Indian artisans and semiskilled workers. Their position has been resented by Africans and many Indians have been expelled from Uganda, while others have voluntarily left the other countries. There is no shortage of unskilled workers; indeed, unemployment in the urban centers is a major problem. Kenya on two occasions adopted a tripartite agreement, which called for government and employers of over ten people to hire additional workers, in return for which unions would forgo demands for higher wages. These agreements have been counterproductive. They have stimulated further immigration, which resulted in even greater unemployment. Wage levels have risen dramatically in some periods despite the availability of surplus workers; as noted, this has led to some distortions and inequities in the economy. There remains a serious shortage of skilled and managerial talent, which can only be improved over time. All three countries have restricted certain industrial activities, or trade in specific items, to locals as part of their overall Africanization programs.

MARKET. The market situation has important advantages and disadvantages. From the standpoint of numbers or incomes, none of the countries has a large market, but the combined market of the EAC is quite large by African standards. Kenya has the largest market and its per capita GNP, at $160 in 1971, was higher than those of Uganda ($130) and Tanganyika ($110). The high average incomes of the still fairly sizable European population in Kenya contributes to the market potential in that country, which also receives the greatest benefit from tourist expenditures, very important to the many producers of artistic handicrafts.

Kenya has some advantage in that its population and markets are concentrated in relatively small parts of the country. Uganda is advantaged by the concentrations south of Lake Kioga and its generally excellent road net. Tanganyika has a marked disadvantage in the wide dispersion and peripheral location of many of its richer nodes.

Map 90 is a kind of composite representation of a variety of assets and is intended to show the areas and nodes of relative strength for development.[1] It illustrates many of the points made above, but one may also note with respect to the EAC and cooperation in general that several of the areas delineated cross internal boundaries and that artificial barriers would disadvantageously distort the normal development of several nodal zones. This is particularly true for the major Kenya-Uganda zone of good-to-excellent potential, the Mombasa-Tanga zone, and the lands shared by all three countries around Lake Victoria. The Moshi-Arusha axis in Tanganyika, although not crossing a border, is reasonably well situated with respect to the whole East Africa area and would lose the advantage of centrality if barriers to the north were erected.

CAPITAL. East Africa remains heavily de-

[1] See William A. Hance, *African Economic Development*, 2nd ed. rev. (New York: Praeger for the Council on Foreign Relations), 1967, pp. 162–210 for a fuller development of these points.

pendent on outside sources for its industrial buildup, though many plant extensions have been locally financed and a number of wealthy Indians have invested in industry, disastrously in the case of Uganda. Governments have also been important in providing capital for industrial development, especially in Uganda, where the UDC participated in the introduction of several important industries, but also in Kenya, where there is considerable participation in investment through various government corporations. Tanganyika and Uganda have nationalized numerous enterprises but the former, at least, continues to welcome foreign investment in specific sectors. Political uncertainties in Uganda are likely to discourage new investment.

TOURISM

The tourist industry is of considerable significance to East Africa. Kenya has enjoyed the greatest boom, with the number of tourists rising from 61,000 in 1963 to 428,000 in 1972 when it earned more than $74 million in foreign exchange. Economically, investment in tourism is one of the most efficient ways of expanding employment, foreign exchange earnings, and national income. About 40,000 Kenyans are directly or indirectly employed in the industry, which is labor intensive, some three to four jobs being created for each "tourist year." The industry has been growing three times as fast as the economy as a whole and now ranks as the single most important revenue earner and contributes about 3 percent to the GNP.

Tanzania had 85,000 visitors in 1971, and tourism yielded $14 million in 1972. Uganda's tourist visits, which were 81,000 in 1971, had been developing very favorably until the Amin government sealed the borders; these were opened again in 1974, but the somewhat eccentric actions of Amin have seriously reduced the interest in visiting that country.

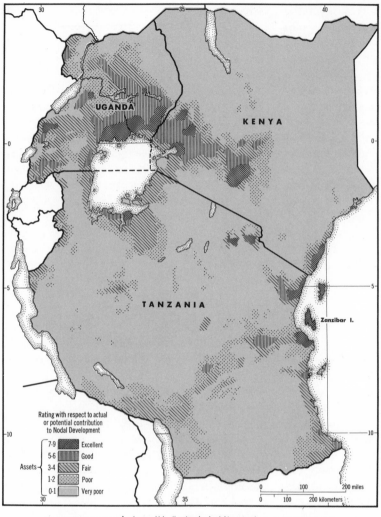

Map 90. Nodal developments in East Africa

Rating with respect to actual or potential contribution to Nodal Development

Assets

7-9	Excellent
5-6	Good
3-4	Fair
1-2	Poor
0-1	Very poor

Assets upon which ratings based and weights assigned

2	Good prospect of 30 in. (760mm) rain, no tsetse	1	Population density more than 100 per sq. mi. (38.6 per sq. km.)
1	Fair prospect of 30 in. (760mm) rain, no tsetse	½	Population density 31-100 per sq. mi. (12-38.6 per sq. km.)
1	Cash crop producing area	1	City or town over 10,000
1	Irrigable area	1	Electric power
½	Forest resources	1	In area with relatively good transport
½	Mineralized zone		

The prime attractions in East Africa are the magnificent game parks, but there are many other assets, including excellent beaches, a varied and often spectacular physical landscape, the very diverse and fascinating peoples, and all manner of outdoor activities from scuba diving to mountain climbing.

Tanzania and Uganda have long resented the fact that most tourists visiting East Africa use Nairobi as a prime base. Tanganyika, in particular, notes the very large number of tourists crossing from Kenya to visit its most favored areas, extending from Mt. Kilimanjaro westward to Serengeti and including Lake Manyara and Ngorongoro. The two main game parks in Uganda, Ruwenzori and Kabalega (Murchison) Falls, are more remote from Nairobi but were also often included on safaris originating in that city.

Both countries have attempted to stimulate use of their own services by various regulations regarding use of drivers and vehicles, etc. but they dare not institute too many restrictions because they might lose by doing so and this would also abrogate agreements under the EAC. Tanganyika built the large new Kilimanjaro airport in the hopes that jumbo jets full of tourists would arrive there and bypass Nairobi, but thus far the field is grossly underused and, in any case, the facilities to accommodate such large waves are not available. Uganda also lengthened its field at Entebbe for the same reason. Assuming the number of visitors continues to grow by about 20 percent per annum, which it did until 1973, however, it will not be many years before unused facilities will be sought out, little-visited parks will have to be developed, and new circuits will become popular.

In the meantime, it must be recognized that there are problems associated with tourism which may become serious if appropriate actions are not taken. One of these is the threat to the parks themselves. On the one hand, migration routes to them and reservoir areas outside their boundaries have been closed off, which increases the pressure from animals within them; on the other hand, the increased number of visitors, whose vehicles drive over the terrain and gradually destroy the vegetation, is also a danger. There is little doubt that restrictions must be placed fairly soon both on the number of visitors going to specific areas and on the free movement within them. There are also potential problems of culture conflict between the obviously wealthy and privileged visitors and the local residents, whose annual incomes are likely to be less than what the tourist spends in a day or two.

TRADE

The major foreign exports of East African countries in 1962 and 1973, the trade within the EAC, and the level of imports and ex-

ports for selected years from 1938 to 1973 are shown in Table 17. Major characteristics of the trade position include the rather broad range of exports, particularly for Kenya and Tanzania; the relatively high level of exports achieved by Uganda in relation to its size and population; the high dependence upon agricultural exports of all three countries; the markedly unfavorable balance in the external trade of Kenya, which is, however, offset by a favorable balance in intra-EAC trade, earnings from transit traffic, tourist receipts, and capital imports; the characteristically favorable though deteriorating balance of Uganda whose total balance of payments is now in serious deficit; and the unfavorable balance of Tanganyika, occasioned in part by imports for the Tazara.

Trade within the EAC was rising until 1970, but interference with free movement of goods across internal boundaries and political differences among the members caused some deterioration in succeeding years. Nonetheless, the level of intracommunity trade has compared very favorably with inter-African trade in general, a fact that should not be overlooked in assessing the EAC.

Much of the internal commerce has, until recently, been in the hands of resident Asians and foreign firms. All of the countries have taken steps to increase African participation in trading. Uganda adopted the most extreme policy—expulsion. Tanganyika attempted to place internal commerce under the State Trading Corporation, which suffered from inadequate managerial talent; the corporation has now been broken into three central and eighteen regional companies, but these still face shortages of qualified managers. In Kenya trade licenses in most rural areas and retail trading in specified commodities have been reserved for Africans, but the program of reserving retail trade for Kenya citizens has been pursued gradually.

Table 17. Trade data for East Africa, 1938–1973

A. EXPORTS BY COMMODITY, 1962, 1973 (*in million $U.S.*)

	Kenya		Uganda		Tanzania		Total	
	1962	*1973*	*1962*	*1973*	*1962* [a]	*1973*	*1962* [a]	*1973*
Coffee	29.7	101.9	56.5	202.8	18.4	70.5	104.6	375.2
Cotton	1.2	3.9	23.1	47.8	20.7	47.4	45.0	99.1
Tea	14.5	48.3	5.6	15.6	4.5	7.7	24.6	71.6
Sisal & products	12.1	15.3	—	—	44.1	42.4	56.2	57.7
Petroleum prod.	···	27.0	—	—	—	12.4	···	39.4
Cloves	—	—	—	—	—	33.2	—	33.2
Hides, skins, wool	5.3	18.0	3.3	4.7	4.2	6.8	12.8	29.5
Cashew nuts	···	2.0	—	—	6.5	24.8	6.5	26.8
Diamonds	—	—	—	—	15.2	24.2	15.2	24.2
Cereals	2.8	17.0	2.4	—	3.3	1.3	8.5	18.3
Animal feeds	···	2.7	···	6.3	···	7.6	···	16.6
Copper	1.3	—	10.1	15.6	—	—	11.4	15.6
Meat products	7.7	10.7	—	.3	6.5	2.6	14.2	13.6
Pyrethrum	8.9	10.4	—	—	.7	2.4	9.6	12.8
Ivory	···	9.8	···	.7	···	2.3	···	12.8
Tobacco	···	—	···	1.2	···	7.9	···	9.1
Soda ash	3.5	8.0	—	—	—	—	3.5	8.0
Cement	1.3	7.3	—	.4	—	.2	1.3	7.9
Beans, peas, pulses	—	3.3	—	—	2.2	3.2	2.2	6.5
Wattle products	2.4	3.8	—	—	1.2	2.4	3.6	6.2
Pineapples	1.9	4.2	—	—	—	—	1.9	4.2
Other	13.4	55.6	4.3	4.4	16.1	19.4	33.8	79.4
Total	106.0	349.2	105.3	299.8	143.6	318.7	354.9	967.7

[a] Excluding Zanzibar

B. IMPORTS AND EXPORTS, SELECTED YEARS, 1938–1973 (*in million $U.S.*)

	Kenya		Uganda		Tanzania [a]		Total	
	Imports	*Exports*	*Imports*	*Exports*	*Imports*	*Exports*	*Imports*	*Exports*
1938 [b]	35	40			14	15	49	55
1948 [b]	155	105			81	63	236	168
1958	170	93	76	130	94	121	340	344
1962	195	106	73	105	111	144	379	355
1972	497	254	114	259	363	284	974	797
1973	586	349	97	300	447	319	1,130	968

[a] Excluding Zanzibar except for 1972 and 1973

[b] Kenya and Uganda combined

C. INTRA-EAST AFRICAN TRADE, SELECTED YEARS, 1967–1973 (*in million $U.S.*)

	Kenya to		Uganda to		Tanzania to	
	Uganda	*Tanzania*	*Kenya*	*Tanzania*	*Kenya*	*Uganda*
1962	20.2	28.1	15.1	4.7	5.5	1.2
1967	41.4	31.9	28.6	6.7	9.2	2.2

	Kenya to		Uganda to		Tanzania to	
	Uganda	Tanzania	Kenya	Tanzania	Kenya	Uganda
1969	44.5	35.8	21.8	4.8	11.2	3.4
1971	53.8	41.2	22.4	2.2	22.1	5.3
1973	62.4	48.0	13.3	.3	21.7	2.5

··· not available
— nil or negligible

SOURCES: East Africa Common Services Organization, East African Statistical Department, *Economic and Statistical Review* (March, 1963); East African Community. East African Customs and Excise Department, *Annual Trade Report of Tanzania, Uganda and Kenya, 1973* (Mombasa, 1974).

CHAPTER EIGHTEEN

Rwanda and Burundi

From World War I until independence in 1962, Belgium attempted with considerable success to orient the economies of its trusteeship area of Ruanda-Urundi toward the Belgian Congo, with which it was joined in a customs union. It is, however, more appropriately a part of eastern Africa, as is recognized by the regional divisions of the ECA. In post-independence years trade has increasingly been directed eastward through Uganda or Tanzania, not only because the distances are less but also because the chaotic conditions in Zaïre precluded use of the former routes for extended periods and because the customs union and special arrangements for shipping via that country were discontinued.

Several keynotes may be given for the two countries: poverty, exceptional population pressure, serious ethnic rivalries (which have led to periodic conflicts with immense loss of lives and to political upheavals), and the predominance of high-based, heavily dissected hill lands in the topographic pattern of these tropical highland areas.

Poverty is evidenced by the continued predominance of traditional subsistence agriculture with relatively few improvements in techniques, low income levels (per capita GDP in both lands was estimated at $60 in 1971, the lowest in Africa), chronic imbalances in trade, payments, and government budgets, heavy dependence on foreign aid, and poorly developed social infrastructure. Nutritional levels are low, the incidence of the protein-deficiency disease, kwashiorkor, is high among small children, and the output of foodstuffs does not appear to be increasing with the rate of population growth. The rate of illiteracy is among the highest in Africa.

Pressure of population is suggested by the high average and local densities, by the serious degradation of lands over large areas, by a breakdown in the traditional agricultural system, by the food and nutritional situation, and by large-scale out-migration. Rwanda, with an area of 10,169 square miles (26,338 sq km) and about 4.0 million, had a crude density of 394 per square mile (152 per sq km) in 1974; Burundi, with 10,747 square miles (27,834 sq km) and an estimated 3.8 million had a density at the same time of 357 per square mile (138 per sq km). With a combined area only a forty-fifth of Zaïre's, the population of the two countries is about one-third as great. As elsewhere the population is unevenly distributed; the more densely populated sections have densities ranging from 600 to 800 per square mile (230–300 per sq km). About 40 percent of both countries are considered to be cultivable, which means that the density per arable square mile is about 932 (360 per sq km) in countries whose populations are 95 percent agricultural. These and the other keynotes are discussed below.

THE ETHNIC BACKGROUND

Three ethnic groups make up the indigenous population of Rwanda and Burundi; relations between the two major ones have a profound bearing on the economy and polity of the two countries.

About 88 percent of the population of Rwanda and 83 percent of that of Burundi are Hutu, a Bantu people who arrived in the area, presumably from the northwest, about 1,000 years ago. They are predominantly farmers and had developed some superior practices over the years, including rudimentary terracing and use of fertilizers. Tutsi, who reached the area some five centuries ago, are of Nilo-Ethiopic origin and are predominantly pastoralist adherents of a cattle culture. They comprise about 11 and 16 percent respectively of the total population of Rwanda and Burundi. The remaining group, the Twa, are pygmies and have been the longest established in the region; they retreated to the west and the mountain areas on arrival of the Hutu where they have a semi-gather economy and produce pottery.

Individuals of the three groups are distinguished by physical differences. The Tutsi, averaging about 4 inches (90 mm) taller than the Hutu and 8 inches (200 mm) taller than the Twa, are generally thin, have aquiline features, and are lighter skinned. The Hutu are more stockily built and darker. But the physical distinctions have been blurred by intermarriage and are far less significant than the social differences. When the Tutsi arrived in the area they established hegemony over the Hutu through possession of their long-horned cattle, conquest, and assertion of divine origin. A feudal system known as *ubuhake* developed, whereby the Hutu were given charge of the Zebu Ankole cattle, an allotment of land, and protection in return for performing a number of services and offering a number of gifts. The Hutu could use milk, the male calves, and meat and hides of dead animals; the clientship relation lasted as long as satisfactory relations were maintained with the overlord. The Tutsi themselves were divided into two kingdoms ruled by *mwami,* with well-developed hierarchical systems of chiefs and subchiefs, particularly in Rwanda.

SOCIOPOLITICAL CONFLICTS

The Germans maintained the feudal system when the area was part of German East Africa, as did the Belgians when it was mandated to that country under the League of Nations. It became increasingly clear, however, that *ubuhake* was an anachronism with dangerous political overtones and one that impeded rational development. In 1954, therefore, a decree was issued requiring that the chiefs surrender two-thirds of their cattle held under *ubuhake* to those who were actually looking after them. It was not possible to enforce the decree, however, and relations between the Hutu and Tutsi deteriorated rapidly, particularly as the Hutu became politically more conscious of the power of their larger numbers.

In Rwanda, following the death of the *mwami* in 1959, the Hutu rose up in a brief but destructive and, from their standpoint, successful revolution. It was more spontaneous than politically inspired: a large number of huts were burned; cattle and coffee plantations were destroyed; there were several hundred casualties and the new *mwami* and several thousands of Tutsi fled the country, setting a pattern which was repeated, reversed, and enlarged upon in succeeding years in one or another of the countries. At the time of independence on July 1, 1962, the *mwami* had been deposed in Rwanda, the Parmehutu Party (Party of Hutu Emancipation) had gained control of the government, and the Parmehutu leader was President of the new Republic. Tutsi refugees in adjacent countries tried to take the capital, Kigali, in 1963; but they were defeated and repressive measures against Tutsi living in the country resulted in perhaps 20,000 deaths (estimates of this and other bloody affairs in Rwanda and Burundi vary widely). In 1972 some 50,000 to 100,000 Hutu were reported killed, after having been charged with treason, while relations with the Tutsi deteriorated further when many were forced to resign their jobs or their places in the schools and in the national university at Butare. Widespread arson and the killing of several hundred to a thousand Tutsi led to an estimated 150,000 Tutsi fleeing the country. In mid-1973 President Kayibanda, who had ruled since independence, was ousted in a military coup led by Hutu officers. Discontent with the lack of economic progress and resentment over favoritism toward Hutu from the president's home area were given as explanations of the coup.

In the meantime there had also been rioting in Burundi before independence, though it did not reach the proportions of the Rwanda revolt and the *mwami* was retained in control with the understanding that his power would be limited. After independence (on the same day as Rwanda) a series of coups and countercoups, involving the murder of several prime ministers and the temporary exile of the *mwami,* were climaxed with the overthrow of Ntare V by his Prime Minister, Captain Michel Micombero, who declared Burundi a Republic in 1966. Tribal and regional friction continued and came to a head with the return of Ntare V to the country in April 1972 under an amnesty arrangement. He was arrested on arrival, which led to Royalist attempts to free him (he was later killed) and to an ill-planned and abortive Hutu uprising, which was savagely put down after 100,000 to 200,000 Hutu had been killed, including about a third of their literate numbers, in a four-month period. In May 1973 killings began again after Hutu refugees were reported to have conducted raids into the country. Again there was much burning of huts and over 150,000 Hutu fled, mainly to Tanzania.

The deplorable events in both countries have affected not only their internal relations but have also embroiled them with neighboring and distant countries. We need not detail the several changes made over the years; suffice it to note that relations between Rwanda and Uganda are now strained, as are those between Burundi and Tanzania. President Idi Amin of Uganda has promised aid to Burundi in the event of attack and Libya is reported to have provided military equipment to that country in recognition of its severing relations with Israel, which has provided aid to Rwanda. The Rwanda-Uganda and Burundi-Tanzania disputes are potentially damaging because they threaten severance of the major routeways for overseas traffic of the two small countries. Very obviously, the ethnic conflicts in both countries have had unfavorable effects on economic development. Nor is it possible to look with any optimism toward the future sociopolitical evolution, particularly in Burundi, where the minority Tutsi continue to retain power, holding most administrative jobs and about 85 percent of all important government posts.

THE POPULATION PATTERN

Certain features respecting population not yet mentioned are of interest in understanding the economic geography of the two countries. Demographic data are very incomplete, particularly estimates of mortality and fertility, while large-scale migrations, including refugee movements, are poorly recorded. Both populations are very young (according to the 1965 census, 46 percent of Burundi were under 15) and both are increasing rapidly, the UN estimates give 2.8 percent for Rwanda and 2.0 percent for Burundi. Infant and child mortality rates remain high.

It is not well understood why this part of Africa has such high population densities. Many diseases, including malaria, tuberculosis, and intestinal disorders, are as widespread as elsewhere; but the absence of the tsetse fly at higher elevations may provide a partial but indirectly important explanation. Probably more important was the existence of strong social and political institutions, which provided relative security—particularly when adjacent areas were subjected to slave raids. The fact that the ruling Tutsi elite were cattle keepers takes us back to the indirect significance of the tsetse-free highlands.

A distinctive aspect of the population pattern is its unusual dispersion, 98 percent of

Scene near Ruhengeri, Rwanda
The high population density of Rwanda and Burundi and their accentuated relief result in the farming of an unusually high percent of the total area and the use of some exceptionally steep slopes.

the total being rural. Bujumbura, the capital of Burundi and formerly of the whole trusteeship, is the only sizable city. Its population is perhaps 80,000, many of whom are from Zaïre and other countries. Rwanda's capital, Kigali, was just a village at the time of independence, but may now have 25,000 people. The rural landscape shows very few villages or even hamlets; the characteristic pattern is one of family compounds, with one or more beehive-shaped huts and courts in individual enclosures usually along the flanks of the endless series of hills on the highlands. This creates certain problems in providing amenities and schools—problems offset to some extent by the high densities over large areas. Only the lower areas of the two countries (the swampy valley of the Malagarazi River and the low Kumosi Plateau in Burundi, and the Bugesera and Kagera lowlands of southeast Rwanda) have relatively low densities.

In addition to the large number of refugees residing in adjacent countries there are as many as 300,000 Rwandans and Burundi who have settled definitively in Uganda over the past decade, plus 60,000 to 70,000 persons who migrate to Uganda and Zaïre seasonally to work on African farms in the former and estates in the latter. The Belgians attempted to resettle 100,000 inhabitants in Zaïre without achieving this goal; both governments now have programs designed to resettle peoples in the lowlands and in the higher areas, which also have lower than average densities.

PHYSICAL CONDITIONS

An unbroken chain of mountains forming the eastern side of the Central Rift system runs for about 174 miles (280 km) along the west of Rwanda and Burundi. Forming the watershed between the Zaïre and the Nile, it is nowhere less than 6,500 feet (1,980 m) high and has summits of over 9,800 feet (3,000 m) in the north. Eastward from these mountains is a highly dissected plateau sloping gradually toward Lake Victoria and having elevations varying from about 6,000 to 4,500 feet (2,430–1,370 m); this region forms the heartlands of both countries. In the north of Rwanda are the volcanic Virunga Mountains, still active in part, with Mt. Karisimbi rising to 14,788 feet (4,507 m). The vast bulk of both countries is in slopeland. A small portion of Burundi called Imbo lies in the downfaulted rift valley comprising the plain of the Ruzizi River and a narrow coastal plain along Lake Tanganyika. The eastern lowlands and low plateaus noted above are generally separated from the central region by a scarp about 650 feet (200 m) in height.

Except for the lower areas the precipitation ranges from about 40 to 60 inches (1,016–1,524 mm) a year, with two wet and two dry seasons permitting two crop cycles. There is considerable variability, which, because of the high population, led to recurrent famines in the past and food shortages in more recent years. An improved road system, requirements that manioc be planted as an insurance crop, and food storage programs have reduced the likelihood of famine. The natural forest vegetation has been almost completely removed in the last 500 to 600 years, leaving only about 2.9 percent of the combined areas forested. A rather remarkable program of afforestation with eucalyptus trees protects some of the steepest slopes and provides firewood, while plantings along the roads give a more closed appearance to the heartland than might be expected from the raw figure of forested area.

There are some excellent soils, including decomposed lavas in the northeast and alluvials in the valleys, but most soils are light, forest-derived soils overlying lateritic subsoils. The steep slopes, frequency of intense rains, and soil characteristics all favor rapid erosion, which is one of the most serious problems facing the country.

AGRICULTURE

About 40 percent of both countries is considered to be cultivable, while 40 percent of Rwanda and 30 percent of Burundi are suitable for pasturelands. In the heartland most of the hills are totally cultivated, including slopes of surprising steepness. Beans and corn are grown in the September–December wet season, sorghum in the January–July period. Bananas, manoic, and sweet potatoes are grown more or less independently of the seasons with bananas characteristically found in clumps around the compounds. Most of the bananas and part of the sorghum are used to make beers, which are consumed by the farmers' families and also provide a main item for local sale. The area under bananas has tended to increase in response to the demographic pressure.

Overcropping, overgrazing, and the use of excessively steep slopes has led to much loss of soil and some gullying. While rudimentary terracing was long practiced, the Belgians instituted programs to construct contour ridges and more effective terracing and to control torrential runoff. Since independence at least a third of these works have fallen into disrepair. Valley bottoms were previously unused or used for grazing cattle. Again the Belgians made a major effort to develop these valleys through drainage and water control, but the greatest progress has come as a spontaneous response to demographic pressure, and papyrus swamps are giving way to fields growing sweet potatoes and beans. Fragmentation of holdings and use of fields at growing distances from the individual compounds are common.

Livestock are more a sociopolitical than an economic matter in Rwanda and Burundi. They graze on about a third of the total area, some of which could profitably be put to crops or, probably better, an integrated crop-livestock agriculture, which would improve protein intake and yield enriching manures. At present and for the foreseeable future, however, cattle are likely to remain largely as a sign of wealth and prestige and to provide little subsistence and commercial value. Cows yield only about 40 gallons (150

Cattle near Kitega, Burundi
Valued as a measure of wealth and prestige, cattle are poorly commercialized in Rwanda and Burundi.

liters) of milk a year; the dressed weight of cattle is only about 250 pounds (113 kg), but cattle are rarely sold, and untanned hides provide only a very minor export. Cattle are not given adequate protection when corralled, sanitary practices are bad, selective breeding is impossible because communal pasturing is favored, the disease record is poor, and useless animals are retained to very old age. Conditions worsen as more land on the hills and in the valleys is put into cropping, which means that less fertilizer is available and overgrazing threatens further erosion. In addition to the cattle, of which there are at least double the proper number under the present system, there are perhaps 2 million sheep and goats. But meat supplies are estimated to be only about 5 percent of requirements.

COFFEE

The dominant cash crop is arabica coffee, first introduced by missionaries and by German officials. Production was not significant until the 1930s, however, when planting be-came quasi-obligatory; but the big increase occurred after World War II, and by 1959 about 560,000 peasants produced a record 29,300 tons, with 60–70 percent coming from Burundi. Growers usually have only 50 to 200 trees scattered among other crops on the slopes. After independence production fell drastically, partly because of its association with colonial authority and the loss of technical services, partly because of the need to grow more food, and partly because returns to the grower are low as transport from this remote region absorbs a considerable share of the market price.

Nonetheless coffee remains the prime export of both countries, accounting for 80 to 85 percent of the total value for Burundi and 50 to 60 percent for Rwanda. About 360,000 farmers in Rwanda and 400,000 in Burundi grow coffee, and output has recuperated since the mid-1960s to around 16,000 tons for the former and 25,000 tons for the latter, which has exceeded its quota under the ICO. While the production per tree remains low, improvement efforts have had some success.

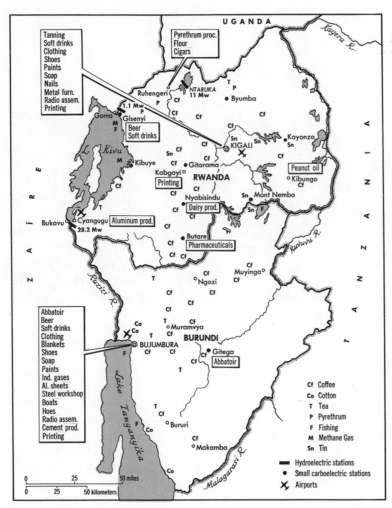

Map 91. Economic map of Rwanda and Burundi

OTHER CASH CROPS

Cash crops of lesser importance include cotton, tobacco, tea, and pyrethrum, but all cash crops are estimated to occupy only 3 percent of the total cultivated area (Map 91). Rwanda has had some success in diversification, particularly in tea and pyrethrum, while small tonnages of vegetables are air freighted to Europe.

Cotton ranks second among cash crops in Burundi. Produced in the Ruzizi Valley and along the shores of Lake Tanganyika on about 21,000 acres (8,500 ha), cotton accounts for about 9 percent of exports and fulfills local needs. While acreage has been

declining yields have almost doubled because of better seed selection and intensive spraying from the air. FED has been assisting in the formation of five tea plantations and outgrowing schemes; production began in 1967 and is expected to reach 520 tons in 1980, when it will rank second to coffee among exports.

RESETTLEMENT SCHEMES

Both under the Belgian and the independent governments considerable attention has been given to resettlement, primarily to relieve population pressure but also to develop little-used areas and to introduce new crops. The Ruzizi lowlands were one important region, where an area of 425,000 acres (170,000 ha) saw the beginning of development in 1949, including drainage of swamps for paddy rice, mechanical clearing, and provision of water points. Only about 9,000 families working about 86,000 acres (35,000 ha) were settled at the time of independence and production of cotton and rice have never reached the original goals. After 1960 a number of holdings were abandoned because of political difficulties, flooding, and neglect of the drainage system. The Burundi government has attempted to reinstitute development on a paysannat system, but farmers have not followed the set routine and only a fraction of the area is in use today.

Rwanda has resettled about 30,000 people on unused hills, again following a paysannat approach: a loop road, a community center, water points, contoured fields, forest reserves, and assigned pasture areas are laid out on each hill and individual holdings of 4.8 acres (2 ha) are assigned. It was planned that 60,000 settlers would be organized in such paysannats by 1975. Whether the results of these schemes justify the expense is questionable; spontaneous migration to new areas is probably far more significant in relieving pressure.

Both countries have also had tsetse eradication programs in the lower eastern regions, designed to open areas suitable for growing peanuts and corn and tobacco under irrigation as well as for the grazing of cattle. Other agricultural efforts include the provision of selected varieties of food crops and attempts to stimulate output of protein-rich foodstuffs; soybeans did not, however, appeal to the palates of the people. Belgium and FED have continued to aid in the reclamation of swamps.

NONAGRICULTURAL ECONOMIC ACTIVITIES

Fishing was long ignored, but efforts have been made since 1955 to rationalize the fisheries on Lake Tanganyika. Operations are carried on at night, with lamps used to attract the fish. About 13,000 tons are caught, mostly by traditional fishermen using pirogues, but increasingly from metal catamarans and by small boats owned by Greeks and worked by salaried fishermen. Tilapia is caught in the smaller lakes, both in Burundi and Rwanda, where about 1,500 persons are dependent on fishing.

There are a number of relatively small mining operations, particularly in Rwanda, where mineral exports account for about 35 to 40 percent of total exports. Tin concentrates provide the bulk of output; associated with tin are wolfram, tantalite, columbite, and beryl. There is also a small output of gold, most of which is smuggled out of the country. Burundi has a tiny output of tin and wolfram and one mine producing the rare earth Europium.

The Belgians sometimes talked of making a second Switzerland out of Rwanda-Burundi, basing industry on the readily available labor supply to produce goods for export. But manufacturing is very limited and confined largely to small primary processing plants. In Rwanda processing of cof-

A tin mine near Kigali, Rwanda
Tin and its associated minerals account for about 35 to 40 percent of Rwanda's exports. Most operations are small scale.

fee, tea, and pyrethrum is the main activity and there is a modern brewery at Gisenyi (Map 91); employment in manufacturing has declined since independence. Burundi has similar processing plants and cotton ginning plus, at Bujumbura, about 30 plants including a sizable brewery, clothing factories, a Bata shoe factory, two paint factories, and plants producing hoes, corrugated roofing, and pots. Plans to open a textile mill were frustrated when the company went into bankruptcy in 1970 before the plant was completed. Manufacturing accounts for 8 percent of GDP in Burundi and employs about 2,400 people plus perhaps 7,000 in handicraft occupations. It has been severely affected in recent years by the loss of Hutu workers, loss of the market in Rwanda, and depressed conditions in the city and country.

Most of the power in both countries is supplied by hydroelectric installations. Burundi receives electricity from the station near Bukavu on the Ruzizi and from two small carboelectric plants. It considers the Bukavu station insecure and is examining the possibility of installing a hydro plant on the Ruvubu near Gitega. Methane is dissolved in the waters of Lake Kivu below about 820 feet (250 m) and could conceivably be used in a nitrogen fertilizer factory. Rwanda is planning a joint hydroelectric development with Tanzania near the Rusomo Gorge of the Kagera River.

The countries could more appropriately emulate Switzerland in the development of tourism, but accommodations and transport are inadequate and the uncertain political conditions are scarcely conducive to attracting many visitors.

The road network of the area is unusually dense, but many of the roads are impassable in the rains. External contacts leave much to be desired and, as noted earlier, are strategically insecure because of disputes with neighboring countries. In recent years about 85 percent of Burundian traffic has moved to or from the lake port of Bujumbura via Kigoma, the Central Railway, and Dar es Salaam in Tanzania. Rwanda's trade has been largely reoriented through Kampala to Mombasa, but may be redirected more via road to Tabora in Tanzania and then on the railway to Dar.

Internal commerce in both countries is strongly dominated by aliens, mainly Asians, but also Arabs, Africans from the coast, and Europeans. The main export, coffee, goes mainly to the United States. Belgium remains the leading supplier but accounts for only 25 to 30 percent of the total for both countries. Burundi has applied for membership in the EAC but is not likely to be accepted under existing conditions. At present the future of these two excessively populated, ethnically torn, economically unviable countries appears extremely bleak.

Part Six
SOUTH-CENTRAL AFRICA

Long-hole percussion drilling at the Chibuluma mine in the Zambian copperbelt

CHAPTER NINETEEN

Rhodesia

Administered from 1889 until 1923 by the South Africa Company, then becoming a self-governing colony whose officials were elected from an almost-exclusively white registry and over which Britain exercised only nominal authority, a member from 1952 to 1963 of the Federation of Rhodesia and Nyasaland, this country, known as Southern Rhodesia until 1964, made on November 11, 1965, an illegal unilateral declaration of independence (UDI), which was immediately labeled by Britain as a rebellion. Since then, Rhodesia has been operating under international sanctions and has not been recognized by any nation, although South Africa and Portugal have aided the illegal government and other nations have been less than rigorous in applying sanctions. It is appropriate to approach Rhodesia as a case study in sanctions theory, but while doing so attention will be given to the economic-geographic elements examined for other countries.

SANCTIONS THEORY

According to sanctions theory, rarely expressed precisely, sanctions are expected to have an impact on the sanctioned power sufficient to cause it to do something it does not wish to do but that the sanctioning powers wish it to do, without recourse to war. Sanctions were first recognized as an international method of coercion in the League of Nations and they are included in chapter 7 of the United Nations Charter as the strongest economic measure short of a blockade, which is considered an "act of war." Inclusion of sanctions in these documents contributed to widespread acceptance of what is

perhaps a myth: that sanctions will indeed have the desired impact.

There is, however, another set of sanctions theory, perhaps best called a counter theory. It is that, except in times of war, sanctions are likely to have an effect opposite to that intended: the economy of the sanctioned power will be strengthened through import substitution and the willing acceptance of various restrictions on consumption and trading; the resolve of the sanctioned peoples, who rally to the defense of the threatened group, will be strengthened also, while the sanctioneers will fail to achieve unanimity of purpose, which will result in defections and a consequent breakdown or loosening of the sanctions. The weakening will of the sanctioneers will result from a desire not to lose trade and investments, evidence that sanctions are not working, sympathy for the sanctioned power, or greed on the part of individuals or nations who see the opportunity to make a profit by breaking sanctions. In the meantime, collusion between representatives of the sanctioned power and outsiders will gradually find and develop gaps in the sanction wall. This counter theory has been drawn from experience of various sanctioning efforts—against Italy, Spain, Japan, and others—none of which were successful.

Each sanctioning case, of course, has its own particularities, and there are at least three factors unique to the Rhodesian situation. First, those who speak for the Rhodesian nation represent less than 5 percent of the population, and the majority did not concur in UDI, formation of a republic in 1970, or many other measures adopted or imposed by the ruling Rhodesian Front government. Second, Rhodesia was bordered on two sides by states that opposed sanctions and on a third side by an independent black nation too weak and too dependent on its white-ruled neighbors to enforce sanctions. From the outset unprecedented gaps thus existed in the sanction wall. Third, there has been the rather remarkably long period over which sanctions have been nominally in force, including those imposed by the UN Security Council, which is supposed to give them the force of law for member nations. The context of the situation that led to sanctions is also notably different from those applying in previous cases. In a decade, a vast surge of independence moved across the African continent from the Maghreb to Malawi, freeing the vast bulk of its area and an even higher percentage of its people. When the wave reached Rhodesia, Angola, and Mozambique it met with an intransigence that stood in sharp contrast to the largely willing withdrawal of colonial powers from countries to the north. Weak Portugal and a small minority of white Rhodesians were able to stem the tide and insist that they had no intention of prematurely extending power to the majority of inhabitants in the lands they control. This position applies as well to South Africa and to the former trusteeship area it controls—Namibia (South-West Africa).

SUMMARY OF EVENTS SINCE UDI

Within a few weeks of UDI in November 1965 Britain applied a number of financial and trade sanctions, including a ban on imports from Rhodesia of tobacco (its main export), sugar, asbestos, copper, and meat, which together had accounted for about 95 percent of that country's exports; Britain also placed an embargo on shipments of oil to Rhodesia. The United States announced full support of Britain and instituted a boycott on imports from and a selective embargo on exports to Rhodesia. The United Nations General Assembly (UNGA), in the first of many resolutions (and it must be recalled that the General Assembly has power only to recommend), called on Britain to take all necessary steps to end the rebellion. On November 20,1965, the Security Council (UNSC) voted 10–0 (France abstained as it had in the UNGA vote on the grounds that it would be interference in a matter internal to Britain) to urge all states to impose an embargo on petroleum shipments to Rhodesia. The UNSC again discussed the Rhodesian situation in December 1965, when Britain succeeded in having its own primary responsibility in the area recognized.

In January 1966 the Commonwealth Prime Ministers—minus Presidents Nyerere and Nkrumah—met in Lagos and, after some vituperation, agreed to allow time for sanctions to work. While it is hard to imagine now, at the time it was widely believed that the sanctions would work; Prime Minister Wilson permitted himself to predict a Rhodesian capitulation "in a matter of weeks." In April 1966 Britain secured authorization from the UNSC to use force to intercept vessels carrying oil to Beira for Rhodesia. Exploratory talks between British and Rhodesian officials regarding a possible settlement were held over some months, culminating after a meeting of Wilson and Prime Minister Smith of Rhodesia on H.M.S. *Tiger* in December 1966 in a working document that was subsequently accepted by Britain but rejected by the Rhodesian government.

In December 1966 the UNSC adopted selected mandatory sanctions and in May 1968 it imposed a nearly complete economic embargo on Rhodesia. Later resolutions of the Council condemned Rhodesia and the states that had not enforced sanctions, called for the maintenance of sanctions, and, in 1973, asked Britain to take measures to end Rhodesia's economic blockade of Zambia and to secure the withdrawal of South African troops from Rhodesia. Efforts to condemn Britain in the UNSC for not using force and to extend sanctions to South Africa and the Portuguese territories were vetoed, including a March 1970 resolution which brought the United States' first veto in the UNSC. The UNGA, in the meantime, passed a series of recommendations calling for no recognition of Rhodesia until true majority rule was extended, extension of sanctions to South Africa and Portugal, and use of force by Britain.

Military actions in the years after UDI saw sporadic incursions of freedom fighters or guerrillas, mainly from Zambia but later from Mozambique. Since 1967 South African forces, military or paramilitary, have aided the Rhodesians in countering these incursions and patrolling the borders, while certain actions based in the Caprivi Strip of South-West Africa have been interpreted as threats to Zambia not to permit use of its territory by anti-Rhodesian forces. Immediately after UDI there was considerable concern that Rhodesia might curtail transmission of power from Kariba and thus seriously impair the Zambian economy. To forestall such a move, Britain sent small detachments of troops to Zambia and Botswana, but as the fear subsided Zambia requested removal of the troops and then embarked on construction of the north-bank plant at Kariba. While military actions played a minor role in the first six years and have never been on a large scale, beginning in 1972 attacks on isolated European farms, mining of roads, and some bombings caused increasing concern to white Rhodesians.

Following alleged attacks from Zambia the Rhodesian government closed the border with that country in January 1973 except for the movement of copper. This action was widely criticized within Rhodesia as well as by South Africa and Portugal and the Smith government reopened the border the following month. Zambia, however, announced that it would no longer permit shipments via Rhodesia and, with aid from the UN and numerous countries, proceeded as rapidly as possible to increase the capacity of shipping by alternative routes (see chapter 20).

Periodic efforts have been made both privately and officially to negotiate with Rhodesia and in December 1971 these led to agreed proposals whose acceptability by black Rhodesians was tested by the Pearce Commission in early 1972. The response was overwhelmingly negative and in November 1972 the British Parliament voted to extend sanctions again. In 1973–74 discussions were held within Rhodesia to determine if the blacks would be willing to support an agreement with Britain leaving the whites in control for a period of some years.

The 1974 coup in Portugal and the subsequent promise to extend independence to Mozambique in July 1975 drastically changed the geopolitical situation for Rhodesia, which could no longer count on using the external routeways via Beira and Lourenço Marques. In turn, this increased the pressure on South Africa, which could foresee growing insistence for extending sanctions to itself if it provided the only relief to UN measures against Rhodesia. South African Prime Minister Vorster, thereafter, played a key role in promoting negotiations between Smith and leaders of the black parties in Rhodesia. In November 1974 these leaders, released from detention by Prime Minister Smith, met in Lusaka and—with the help of the Presidents of Zambia, Tanzania, and Botswana—succeeded in agreeing, at least temporarily, on a united front. In December a ceasefire was announced between the leaders and Smith, but the major concessions made by each side had not been honored in early 1975 and it was apparent that continuing difficult negotiations would be necessary. Whatever the course of ensuing events might be, it was obvious that the independence of Mozambique would expedite the achievement of majority rule in Rhodesia.

POSITIONS OF THE SEVERAL PARTIES TO THE DISPUTE

RHODESIA

The early posture of the Rhodesian Front government reflected the belief of white Rhodesians that Britain would insist on too rapid an evolution to majority rule. The 1961 Rhodesian Constitution did endorse the principle of "unimpeded progress to majority rule" (UPMAR), but the wish was to extend the period of white dominance indefinitely. A gradual but seemingly inexorable shift toward the uncompromising, right wing of the Front and toward a kind of separate development increasingly akin to South African apartheid had begun even before UDI.

While a significant number of white Rhodesians opposed UDI and have favored a negotiated settlement, the vast bulk of the white populace has cooperated fully in making UDI work and in attempting to offset or circumvent sanctions. White Rhodesians long deluded themselves that they had strong support from the black population, revealing the remarkable lack of contact between the races that has prevailed.

Most of the more liberal opposition in the press, at the university, and in the churches has been squelched. Rhodesian blacks could have been presumed to oppose the Rhodesian Front, to favor ties to Britain solely as a brake on local whites, and to want majority rule within the foreseeable future. Their voices were, however, rather muted in the first six years, because their leaders were in detention, there was a split between the two major parties representing them, a good many chiefs dependent upon government salaries acquiesced in UDI, and the blacks perceived the impotence of their position vis-à-vis the rather effective security establishment of the rulers. Nonetheless, there was some guerrilla activity and much lack of cooperation in government programs; and when the visit of the Pearce Commission in early 1972 gave them the opportunity, blacks

Virginia flue-cured tobacco on a Rhodesian farm
Sanctions on purchase of tobacco from Rhodesia necessitated the reduction of production by more than half after the Unilateral Declaration of Independence in 1965.

registered a nearly unanimous "no" to the proposals for a settlement. Riots and unrest resulted in thirteen deaths.

BLACK AFRICA

The representatives of independent African governments have bitterly opposed UDI and the failure to correct the situation either by economic or military means. A variety of moves in the OAU and the UN have revealed their frustrated anger, but actions have not always been so firm as the pronouncements; a vote by the OAU Council of Ministers in December 1965 to break off diplomatic relations with Britain was repudiated by many heads of state, splitting the OAU in the following meeting, while contributions to the Liberation Fund have characteristically been in arrears. Efforts to extend the sanctions to Portugal and South Africa and to promote military action by the UK or the UN have been unsuccessful.

Among individual African countries Zambia has been most seriously affected by the Rhodesian situation. It had been dependent on Rhodesia for such items as transit of almost all of its overseas shipments, for coking coal used in smelting its minerals, for a substantial portion of its manufactured imports, and for the bulk of its electricity (which came from the jointly owned Kariba facilities whose plant was situated on the Rhodesian bank). Its efforts to offset these dependencies are covered in the next chapter.

Botswana occupied a position that precluded its taking actions against Rhodesia, which was fully understood by other black-ruled nations. The Malawi government also adopted the position that it was too poor and too dependent on migrant workers going to Rhodesia and South Africa to apply sanctions. Its posture was not accepted by other African nations, leading to a considerable degree of ostracization in African organizations.

PORTUGAL (FOR MOZAMBIQUE)

Portuguese policy appears to have been to try to be as unobtrusive as possible while earning as much as possible on transit traffic to any of the parties involved; its sympathies were clearly with Rhodesia. Failure to obey the mandatory sanctions of the UNSC meant that the two lines through Mozambique, which handled about 98 percent of Rhodesia's overseas trade before UDI, continued to be available to that country. While no oil destined for Rhodesia could be landed at Beira, the Limpopo Line from Lourenço Marques could be used, as could the line from that port via South Africa to the Rhodesian border. Because Lourenço Marques serves South Africa and Swaziland as well as Rhodesia the same measures adopted for Beira could not be applied there, as this would have been tantamount to sanctioning South

Africa—a proposal that would always have been vetoed in the UNSC, as it was in 1973. Portugal's attempt to maintain a low profile reflected its fear that sanctions might otherwise be applied to Mozambique or to Portugal itself.

Overthrow of the Portuguese government in 1974 and subsequent negotiations promising independence in 1975 to Mozambique naturally raised fears in Rhodesia that the routeways via that country might be closed under an independent government.

SOUTH AFRICA

The Republic early adopted an official policy of neutrality but refused to accept UN sanctions on the ground that they would be interference in the internal affairs of others and therefore contrary to its basic foreign policies. While pretending to be neutral and officially correct, South Africa has sent small detachments of "police" and several river boats to help the Rhodesians, its "business as usual" policy and "unofficial" support have provided a main prop to the rebel regime in Rhodesia, and it has colluded in the breaking of sanctions on imports and exports.

The South African government has, however, been critical of the Rhodesian Front on several occasions. It probably did not favor UDI, and it has both publicly and privately urged that a solution be negotiated. It was particularly angered by Rhodesia's closure of the Zambian border in 1973, which greatly reduced its growing sales to that country.

South Africa's willingness to aid Rhodesia was influenced by the desire not to expend large sums on subsidizing that country, unwillingness to damage its own industries by opening its markets to Rhodesian competition, the desire to avoid involvement in a conflict or encouragement of the extension of sanctions to itself, the knowledge that inclusion of Rhodesia in its sphere would worsen two factors of considerable significance to the Republic (the ratio of blacks to whites, and the defensive perimeter of the country), the realization of Rhodesia's lack of value as a defense ally (it is heavily dependent on South Africa for its existence), and the obvious conflict between support for Rhodesia and the Republic's so-called outward-looking policy involving the promotion of relations with black African nations. Despite these considerations, it is clear that South Africa permitted and connived at Rhodesia's staving off the worst impacts of sanctions. After it became clear that Mozambique was to become independent, however, its policy hardened and, as noted above, Vorster strongly insisted that Smith negotiate with the African leaders in his country.

THE UNITED KINGDOM

Britain's basic position has been that UDI was illegal. It had stated in advance that it would not use force, which some observers saw as a serious error; it did apply sanctions immediately and has enforced them more strictly than most countries. In 1966 it enunciated six principles that would have to be accepted before independence could be granted to Rhodesia (it has rarely been suggested that the UK would take over direct governing of the colony). In presenting these principles it will be noted how the 1971 proposals were intended to satisfy them because, while these proposals were rejected in 1972, it may be expected that renewed negotiations will follow them to a significant degree.

(1) Unimpeded progress toward majority rule (UPMAR) would have to be maintained and guaranteed. A new African higher roll with the same qualifications as the European roll, together with the existing lower roll, were to permit proceeding to parity in the Assembly, after which the Africans would presumably opt for direct elections on a common roll and achieve majority rule.

Various estimates have been made of the time required to achieve parity. The best estimate is probably not before 2035, which appears unrealistically far away. After failure of the negotiations on HMS *Tiger* UPMAR was replaced by NIBMAR (no independence before majority rule), but this principle was overlooked in succeeding negotiations and, while favored by Africans, was considered a "no-go" because of its complete unacceptability to white Rhodesians.

(2) Guarantees against retrogressive amendments to the constitution are to be supplied by a blocking vote in the hands of the directly elected Africans in the lower house.

(3) An immediate improvement in the political status of the Africans, to be satisfied by the new higher roll and a reduction in the qualifications for the lower roll. Release of detainees and increased expenditures on African education, partly through British aid, would also be expected to increase the number qualifying for the higher roll.

(4) Progress toward ending racial discrimination, to be enforced through a Declaration of Rights, recommendations from an independent Review Commission, and allocation of more land to the Africans.

(5) Acceptance of the proposals by the people of Rhodesia as a whole. The Pearce Commission was appointed by Britain to determine African attitudes, which were, as has been noted, negative.

(6) Assurance that there would not be oppression of the majority by the minority or vice versa. The proposals did not specifically deal with this point, but presumably the Declaration of Rights would be expected to be of value.

Britain's posture toward the Rhodesian situation was weakened by its decision not to use force, by a considerable sympathy for Rhodesian whites within both parties, and by a general lack of interest by most British people.

STRENGTHS AND WEAKNESSES IN RESISTING SANCTIONS

Sanctions have been in force, theoretically with increasing rigidity since UDI, or since 1965—a remarkable long period as compared with other sanctioning efforts. A variety of factors explain the ability of Rhodesia to withstand sanctions, some of which have already been suggested, including the crucial presence of two neighbors willing to disobey the sanctions mandated by the UNSC. A consideration that prevented the application of full pressure on Rhodesia was the desire not to damage the Zambian economy, which was very dependent on Rhodesia. Other factors are given in the following subsections.

THE POPULATION FACTOR AND EUROPEAN CONTROL

While the European population has always been a small proportion of the total (4.7 percent in 1965 and 4.6 percent in 1973), the government built up over the years a position of authority backed by small but effective military and paramilitary forces, which enabled it to retain control over the much larger African population. This control was maintained in part by suppression of dissidents and incarceration of African leaders who threatened white rule, and laws permitting punishment of offenders have become increasingly rigid. Disputes among major African groups, which persisted long after UDI, also weakened their position.

The demographic factor taken alone suggests an inevitability to the end of white rule, though one can only speculate about the date. A few comparisons will suffice to make the point: there were in 1972 only 255,000 Europeans compared to 5.3 million Africans; the estimated rates of natural increase are 1.1 percent for the Europeans and 3.6 percent for the Africans; the number of births of whites is less than 2 percent that of the blacks, while in another generation over 400,000 Africans will be born each year; and the percentage of population under 17 is about 53 percent for the blacks and 38 percent for the whites. In 1972 there were also some 27,300 Coloured and Asians in Rhodesia.

The Europeans keep close tabs on the dynamics of the white population, particularly on the figures for net migration. These reveal that the goals of attracting additional settlers have not been met, while in 1973 there was reduced immigration and some outflow of whites, related, at least in part, to the step-up in guerrilla activities and to the dislike by young Rhodesians of the commitment to serve in the security forces for a part of each year to age 40.

THE INFRASTRUCTURE

Rhodesia possesses a comparatively well-developed physical infrastructure, which has undoubtedly contributed to its capacity to resist sanctions.

TRANSPORT. The backbone of the transport system of both Rhodesia and Zambia prior to UDI was the Rhodesia Railways, whose older main line ran 1,270 miles (2,043 km) from the Mozambique border near Umtali to Salisbury, Bulawayo, Victoria Falls–Livingstone, Lusaka, the Copperbelt, and the Zaïre border. This somewhat devious route may be readily explained by a glance at a relief map; as far as possible the main line kept to the higher plateau levels and crossed the Zambezi just below but at the crest level of Victoria Falls, where the narrow gorge could readily be bridged.

For many years, Beira, to which the main line leads in Mozambique, was the gateway for the overseas traffic of Rhodesia and Zambia. Severe congestion at Beira and on the railway made it necessary to seek a second outlet, and in 1955 a line was opened to the southern Mozambique port of Lourenço Marques, which had the further attraction of available capacity. This Limpopo Line greatly relieved pressure on the railway and also opened up a hitherto undeveloped part of Rhodesia.

The first rail line to Rhodesia came northward from South Africa via Botswana to Bulawayo, and this line remained the only direct rail link with the Republic until 1974. The 398-mile (640-km) section in Botswana is owned and operated by Rhodesia Railways. The connection is important in traffic between Rhodesia and South Africa, but Republic ports were normally little used for overseas traffic because of their greater distance. Just what portion of overseas freight now moves via South Africa is not known, but it has presumably increased under sanctions if for no other reason than to conceal the source of destination of such shipments. Rhodesia long resisted effecting a cross-border rail link with South Africa at Beit Bridge for fear of increasing South African and particularly Afrikaner influence in that country, but a connection was finally completed in 1974.

Before UDI, substantial improvements had been made on the railway including relaying with 100-pound track, purchase of new diesel-electric and powerful Beyer-Garratt articulated steam locomotives and additional rolling stock, and installation of a centralized traffic control system, but burgeoning traffic, including the valuable metal exports from Zambia, were creating needs for further improvements. Total freight traffic on the Rhodesia Railways increased from 2.9 million tons in 1938–39 to 5.5 million tons in 1953 and 8.8 million tons in 1965. Coal and coke from Wankie and other minerals both from Zambia and Rhodesia normally accounted for over half of the total traffic.

After UDI the Zambian portion of the Rhodesia Railways was separated but, despite efforts to redirect Zambian traffic via alternate routes, large tonnages of both imports and exports continued to move via Rhodesia, with obvious advantage to the reduced Rhodesia Railway and to the earning of hard-currency foreign exchange by the sanctioned nation. This situation ended in 1973 with the decision by Zambia no longer to use the line through Rhodesia. This created a serious loss of revenues, roughly 20 percent, necessitating a 10 percent surcharge starting in April 1973. The Railway had also been permitted to run down because of the difficulty of importing replacements and the need to conserve foreign exchange, which required Rhodesia to plan for heavy expenditures beginning in 1973 to prevent further deterioration. Nonetheless, the line has been a major asset to Rhodesia and is particularly important for the carriage of mineral traffic and bulk agricultural produce from mines and farms in Rhodesia. To relieve pressure on the line plans were made to construct a 170-mile (274-km) pipeline from Somabula on the Limpopo line to Salisbury, but Mozambique's accession to independence in 1975 put an end to this project.

The trunk roads of Rhodesia parallel the rail line, with numerous feeders acting to some extent as branch lines. Competition with the railway is avoided by restrictions on haulage by road operators. An unusually high mileage of roads is paved, though many of the lesser roads are still precarious in the rainy season. The only waterway of significance is Lake Kariba, where recently installed ferries and hovercraft permit tourists to make a circuit of Lake Kariba, Victoria Falls, and the Wankie Game Park—three of the top attractions in the country—without considerable retracing of their steps.

The Central African Airways, one of the more successful of African air lines, has been subdivided into three airways and Rhodesian Airways has been circumscribed by limitations on landing rights outside of southern Africa. Rhodesia was advantaged by inheriting the headquarters and central maintenance facilities of both the CAA and the original Rhodesia Railways, at Salisbury and Bulawayo respectively.

Despite some efforts to discourage tourist visits to Rhodesia after UDI, a growing number of tourists went to Rhodesia until a shooting from Zambia and the stepped-up guerrilla activities of the last few years adversely affected the trade.

ENERGY. Rhodesia possesses large reserves of coal and hydroelectric potential. Until the 1960s, the major source of energy for both Rhodesia and Zambia was the Wankie Collieries, 200 miles (322 km) northwest of Bulawayo and 68 miles (109 km) southeast of Victoria Falls on the main line. Coal is mined at exceptionally low cost around Wankie, where proved reserves are given at about 890 million tons of low-ash, coking-quality coal and 330 million tons of poorer-quality coal; possible reserves total 412 million tons. Much of the coal lies close to the surface; the main seam varies in thickness from 11 to 34 feet (3.4–10.4 m) and is worked easily from inclined shafts at depths varying from 80 to 200 feet (24–60 m). Mining began at Wankie in 1903 and reached an output of 1 million tons in 1927, 2 million tons in 1949, and 3.5 million tons in 1965. Output declined after completion of the Kariba hydroelectric station and after UDI and has averaged about 3 million tons in recent years. But plans call for constructing a large carboelectric plant at Wankie to meet the increasing needs for electricity in Rhodesia; its capacity will be 240,000 kw in the first phase and 1.32 million kw by 1982.

Rhodesia also has coal reserves in the Sebungwe area northeast of Wankie and in the Sabi Valley. The latter area has about 5 billion tons of proved reserves, by far the largest in the country, and has seams averaging 46 feet in thickness; but its coals have the highest ash content of all the Rhodesian coals, averaging about 35 percent.

The Kariba Dam and plant on the Zambezi is now the main source of power for Rhodesia; it provides about 80 percent of total electricity consumed. The flow of the Zambezi River varied at Kariba from 15,000 to 200,000 cusecs before control, hence the need for a large capacity lake, met by construction of a 420-foot-high (128-m) dam, 1,900 feet (579 m) across at the top. Lake Kariba, averaging about 12 miles (19 km) in width, covers about 2,000 square miles (5,180 sq km). The lake itself has attracted several recreation communities and supports a modest fishing industry.

Phase one of the Kariba Project, inaugurated in 1960, involved construction of the dam and a 675,000-kw station on the Rhodesian side, involving a total expenditure of about $223 million. The output of this plant has been shared by Rhodesia and Zambia. Phase two entails placing a 900,000-kw plant on the Zambian side; delayed by financial and technical difficulties this plant is scheduled for completion in 1976. Kariba power has been sold at relatively high prices to pay off the debt promptly, but by 1981 its price will be lower and comparatively very favorable.

Kariba Dam on the Zambezi
Inaugurated in 1960 with an installed capacity of 675,000 kw, this Zambezi River station will have a capacity of 1.58 million kw upon completion of the north bank plant on the Zambian side.

It may be asked why the natural site of Victoria Falls was not developed first, before turning to a project requiring the construction of an expensive storage and head-creating dam. There were three objections to developing power at Victoria Falls: (1) the fear of destroying the beauty of a site with considerable value to the tourist industry; (2) the great fluctuation in flow over the Falls, the mean maximum flow being about ten times the mean minimum flow; and (3) the considerably greater distance of the Falls from major consuming centers. There are, however, two small installations on the Zambian side of the Falls. For the future as much as 520,000 kw could be developed on the Zambezi northeast of Wankie and about 1 million kw in the Marata Gorge downstream from Kariba.

There are no known deposits of petroleum in Rhodesia nor are the geological formations encouraging. The Rhodesians have considered construction of an oil-from-coal plant comparable to the Sasolburg plant in South Africa, but the capital expenditures would probably not be justified except as an emergency measure.

THE AGRICULTURAL SECTOR

THE PHYSICAL BASE. In Rhodesia a belt of generally level land above 4,000 feet (1,200 m) runs nearly across the country from northeast to southwest (Map 92). This area, about one-fifth of the country, contains most of the European population and the densest African settlement as well. It is the physical and economic backbone of the country. From this highveld region the land drops off through a broad, more heavily eroded, middle veld to the lowveld of the Zambezi Valley in the northwest and of the Limpopo and Sabi basins in the southeast. The margins of these troughs are everywhere marked by higher, rugged terrain, while the relief is gentle on the soft rocks of the troughs themselves. The Zambezi trough is deep and rather narrow, with very steep

Map 92. Physical map of southern Africa
Relief Map Copyright Aero Service Corporation

sides. Largely because of aridity and insect pests it has been little developed except for an irrigation scheme near Chirundu and several game parks. The southeastern lowveld is considerably broader and also largely undeveloped, but several sizable irrigation projects have been installed. The eastern border of Rhodesia is marked by mountain country, whose continuity is broken by a gap at Umtali.

The climatic and soil characteristics of Rhodesia are relatively unfavorable for intensive agricultural pursuits. The highly seasonal pattern of precipitation, with rainfall confined almost entirely to the summer, means that cropping is also confined to this season unless irrigation can be applied. Over most of the area rain begins about the end of November and ends in March, but the mountains receive more, and more extended, precipitation. The mean precipitation in Rhodesia is 26.1 inches (663 mm), though the highlands on the eastern border receive as much as 100 inches (2,590 mm). Extensive areas in the south and in the Limpopo Valley receive only 12 to 16 inches (300–400 mm). Rains are unreliable in their onset, duration, distribution within the season, and total amount. This is reflected in a wide variation in crop yields and in the carrying capacity of pastureland. Yields of the staple food crop, corn, fluctuate widely, and consequently foreign and internal movements vary sharply from year to year. Low yields result not only from low precipitation, but may be occasioned by excessively wet seasons as well.

Precipitation varies considerably with altitude, aspect, and latitudinal position. These differences now explain the localization of such crops as tea, tobacco, and sugar and they will lead in the future to more closely defined agricultural regions than now exist. The lower lands of the Zambezi and Sabi rivers, or the higher lands of eastern Rhodesia, will be looked to for specialized production to bring greater productivity and variety to the agricultural pattern.

Temperature characteristics are, in contrast to precipitation, unusually favorable. High temperatures usually associated with the tropics are considerably modified on the plateaus, and the light ground frosts, which are likely during the winter at higher altitudes, impose few limitations on the range of crops which may be grown. On the other hand, the prevailing high temperatures below 2,000 feet (600 m) permit the production of such tropical lowland crops as sugar.

Most of Rhodesia has mediocre soils, with a widespread deficiency of available phosphates, rapid depletion of nitrogen reserves when used, and structures that tend to deteriorate when subjected to cultivation. Soils vary, however, from fertile red clays to far more prevalent sandy soils of low basic fertility, which are quite suitable for flue-cured tobacco. European farmers—through better rotational practices, greater application of fertilizer, and more interest in mixed agri-

culture—have learned how to preserve and even improve their inherently poor soils.

About half of the country is suitable for semiextensive livestock production with supplementary production of drought-resistant crops on a third of that, and most of the rest is semiarid country suitable only for extensive grazing.

Insect and disease attacks are continuing problems. Rhodesia is, however, about nine-tenths tsetse free. Earlier, cotton was severely cut back because of plant disease, but renewed production in recent years has been successful. The existence of countless termitaria is another problem of some dimension.

THE DIVISION OF LANDS. The allocation of land between Europeans and Africans is a subject of intense concern, which may be expected to cause increasing dissension through the years. The latest division is as follows, by percent of total area (150,803 square miles, 390,581 sq km):

African Area		European Area	
Tribal Trust Land	41.3	General	40.1
Purchase Land	3.8	Other	6.5 [a]
Other	1.5 [a]	Total	46.6
Total	46.6	National Area	6.8

[a] Includes forest land, parks, and wildlife land

When it is recalled that about 62.8 percent of Africans live in the African areas, primarily dependent on the land, while only about 6,000 European farms and companies are engaged in agriculture in the European rural areas, the inequity of the situation becomes apparent. While it is true that about 20.6 percent of Africans reside in European rural areas and that European farming is far more productive than African farming, these considerations cannot override the long-run danger of the existing land distribution pattern. This becomes even more apparent when it is realized that most of the African lands are overworked and overcrowded while probably less than half of the potential arable area on European farms is actually being used.

The gross distribution of lands by racial group conceals disparities in quality of land held (Map 93). Europeans hold 76 percent of the highveld and highland areas, 40 percent of the middle veld, and 26 percent of the lowveld. An estimated 40 percent of the European area is in good quality lands whereas only a bit over 10 percent of African lands is in this category; Europeans also hold about 45 percent of the land with fair potential while 70 percent of African holdings are in areas with limited agricultural potential. The Africans are further disadvantaged by poorer roads and by greater distances from rail stations and towns.

EUROPEAN FARMING. White-owned farms vary greatly in size: about 30 percent have under 1,000 acres (400 ha) and occupy little more than 2 percent of the total area of European farms; 50 percent have between 1,000 and 5,000 acres (400–2,000 ha) and

occupy about 25 percent of the area; 16.6 percent have between 5,000 and 20,000 acres (2,000–8,000 ha) and 3.4 percent have over 20,000 acres (8,000 ha). These larger farms account for over 70 percent of the total area in European farms, with the bulk being in large ranches.

Most of the European holdings are on the highveld (Map 94), which is the most favorable land climatically and economically. The more important producing regions are around Salisbury and in the Eastern District, though the largest holdings are in the less favored parts of the country.

The European farm sector accounts for over 90 percent of the sales of crops and livestock, which were estimated at about $150 million in 1962 and $311 million in 1972. Output in 1973 was adversely affected by one of the more serious droughts of recent decades. The share of agriculture in the GDP has declined from 18.4 percent in 1965 to 17.4 percent in 1972 and 16.5 percent in

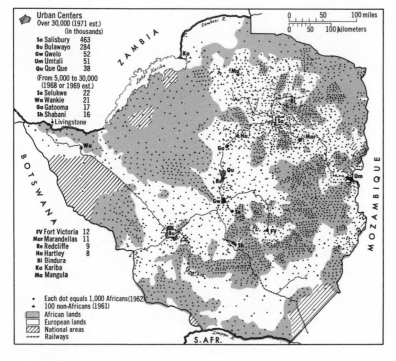

Map 93. Population and land distribution in Rhodesia

Map 94. *Agriculture in Rhodesia*

1973. Agriculture employs 4.2 percent of non-African employees and 38.4 percent of African employees.

AFRICAN FARMING. African agriculture is largely traditional and primarily concerned with self subsistence. Indeed the cash receipts per African in farming are estimated to have declined from about $7 in 1958 to $3.60 in 1970. Some indigenous systems display a keen sense of the relations between the character of the soil and the natural vegetation, but farming in the Tribal Trust Lands is marked by inadequate soil conservation practices, and rising pressure on the land threatens to destroy large areas. Even in the African Purchase Areas, where practices are somewhat more advanced and a higher share of output is commercialized, farming is not so well developed as had been hoped, and many farms are little better than those in the Trust Lands. The following data suggest the seriousness of the situation: an estimated 68 percent of Trust Land is unsuitable for cropping, 13 percent has been denuded, over 36 percent is considered to be heavily overgrazed, and the numbers dependent on the land may be expected to nearly double in a generation.

A major effort was made in the 1950s under the African Land Husbandry Act to transform the Trust Lands by introducing individual arable and grazing rights and fostering minimum standards of good husbandry. The program failed in part because the changes required were too drastic and too much in contrast with traditional attitudes and practices, and in part because land allocations would have resulted in large numbers of landless people. But failure can be attributed more importantly to distrust of the European leaders and resentment over the clearly visible inequalities in the distribution of lands by race.

CROPS. Corn occupies the largest acreage on both European and African farms in the area. Some of it is used as cattle feed, but most is used or sold to provide the main staple African food. In good years a substantial surplus is produced, much of which was formerly exported at a loss. Yields on European farms, which plant hybrid corn, have risen markedly, but the export of any surplus is now very difficult. The subsidization of corn prices played an important role in promoting European farming in Rhodesia, but it led to distortions in marketing arrangements, while the taxing of African corn sales has resulted in misunderstanding and resentment even though the proceeds were used to advance agriculture in the reserves.

Tobacco was by all odds the main export crop in the years preceding UDI. The very large increase in production in postwar years, one of the most spectacular in agricultural crops in Africa, was stimulated by ensured sales in the British market, where dollar shortages required sharp reductions of American tobacco imports, and also by the acceptability of the tobacco grown. Almost all of the flue-cured tobacco, the main type in Rhodesia, is grown on European farms, mostly in the northern highveld. Flue-cured tobacco grows best on light, sandy soils of relatively low fertility in areas with about 25 inches (635 mm) of rain and at elevations between about 3,000 and 4,000 feet (900–1,200 m).

Tobacco was the leading single export of Rhodesia before UDI. Difficulties in disposing of the crop under sanctions led the government to assign quotas to producers, reducing output from 300 million pounds (136

A barn for flue curing of tobacco
Tobacco farmers were subsidized after UDI to prevent collapse of the European farm sector.

million kg) in 1965 to 132 million pounds (60 million kg) from 1968 to 1971, but the quota was raised 11 percent in 1972 partly in expectation of reaching an agreement with Britain. While growers have been subsidized to prevent a collapse of the farming sector, the assured price has remained relatively low despite protestations, and about 1,300 tobacco farmers have surrendered most or all of their quotas, leaving some 1,600 growers to meet the quota.

Estimates of the amounts of tobacco held in stockpile vary widely, but they were probably between 200 and 300 million pounds (91–136 million kg) in 1972. There is evidence that much of the stockpile has been disposed of since then and a resurgence in demand permitted the government to raise the price about 15 percent in 1973–74. Growers had complained of a rising shortage of labor because wages had increased only 11 percent from 1963 to 1973 while the cost of living had gone up by 24 percent.

Other cash crops produced in Rhodesia include tea, cotton, peanuts, soya beans, sugar, wheat, and citrus fruit. Tea is grown in the Umtali and Chipinga districts and also at Inyanga; it is mainly marketed domestically. Cotton production has seen a remarkable expansion and it is now estimated that over 1,000 European and 2,000 African farmers in the African Purchase Area plus numerous smallholders in the Tribal Trust Lands are growing cotton and some 200,000 seasonal and permanent African workers have found employment in cotton production. About 85 percent of output is said to be exported.

Sugar is produced on irrigated estates; loss of foreign markets led to closing of the Chirundu estate and conversion to other crops, including wheat, which is now produced in quantities sufficient to meet about half of domestic needs. There appears to have been some recuperation in recent years and some success in selling sugar in foreign markets.

Huts and woodlands in an African reserve near Nyanyedzi
Agriculture in the reserves is largely traditional and primarily concerned with self-subsistence.

Several large estates produce citrus fruit, mainly for the domestic market. The Mazoe Estate and the Hippo Valley Estate, two of the largest privately owned citrus estates in the world, account for the bulk of output. An effort is being made to promote the growing of pyrethrum; in 1971 six tons of dried flowers were produced and an extract factory has recently been opened at Seki, near Salisbury.

The general shortage of water, particularly during the winter months, has encouraged giving attention to irrigation development and water conservation practices. Several thousands of small dams have been constructed on European farms and in African areas; thousands of boreholes and wells have been sunk. Large-scale irrigation schemes include the Kyle and Bangala Dams in the Fort Victoria district, which supply the Hippo Valley and Triangle Estates; the Sabi-Lundi Scheme, which involves gradual development of the considerable potential in that basin (a major project is now under preparation around Chiredzi); the Mazoe Dam; and the Chirundu Estate on the Zam-

The Nyanyedzi irrigation area in eastern Rhodesia
This is one of several schemes developed in the African areas of the Sabi Valley and Manicaland. The previous photo shows the nature of the area before irrigation water is applied.

bezi. A number of smaller schemes also exist, including nine for African farmers in the Sabi Valley and Manicaland. In 1973 the government announced plans to spend $6 million on development in the northeast border district, including construction of three dams to support irrigation schemes. It has been estimated that a total of 982,000 acres (397,000 ha) is irrigable in Rhodesia.

The need to improve soils places emphasis on the application of artificial fertilizers. Phosphatic fertilizers are produced from deposits at Dorowa, where the plant has recently been expanded. Nitrogenous fertilizers are produced as a by-product of coke oven operations at Redcliff and in a large new plant at Que Que. Crops and pastures respond very well to generous applications of nitrogenous fertilizer.

LIVESTOCK INTERESTS. The European-owned cattle herd rose from 1.6 million in 1965 to about 2.7 million in 1973; Africans own about 2.9 million head, but the value of livestock products marketed from European farms is over four times as great as from the African areas, where quality standards are poorer and where the cattle culture still inhibits commercialization. The hardy Africander cattle is preferred in many areas, though there has been increasingly successful cross-breeding with European cattle or maintenance of purely European breeds. There is considerable potential for expansion of livestock production. The major problems include a low average calving rate, which could be improved by supplying supplementary feed; intensification of bush vegetation, the control of which could increase capacity by 50 percent; and the poor quality of grasses in the dry period. Future emphases in much of the area suitable for cropping should probably be on mixed farming, because, with generally poor soils, any transformation to permanent cultivation will require composting with cattle manure. Furthermore, the climatic limitations on tillage agriculture suggest that livestock farming must eventually assume a much more important part in the agricultural economy.

FISHING AND FORESTRY

Fishing is of only minor importance in Rhodesia. Efforts have been made to stock Lake Kariba with bream and, more recently, with the Lake Tanganyika sardine or Kapenta, but results have thus far been somewhat disappointing. Some of the water-conservation projects are being used as fish ponds, but much more intensive development is possible. There is considerable sport fishing, especially for trout, in the colder waters of the eastern highlands. Rhodesian forestry output comes mainly from areas afforested with exotic conifers, eucalyptus, and wattle. Growth is very rapid; pine, for example, takes only 6 to 10 years to reach pulping size compared to 30 years in Sweden. Most of the requirements for wood, pulp, and paper are now met from domestic forests.

THE MINERAL INDUSTRY

Rhodesia produces an unusually wide range of minerals and its mining industry has ex-

High-bred Hereford cattle on a highveld farm
Such cattle are kept only in the favored areas. Large ranches in the drier areas must rely on hardy breeds such as the Africander.

panded considerably over the pre-UDI level. While the industry's share of the GDP declined from 7.0 percent in 1965 to 6.5 percent in 1973, the value of output increased about 87 percent in this period and in 1973 production reached a record level of R$135 million. Employment in mining and quarrying totaled 53,900 Africans and 3,670 non-Africans in 1971, an overall increase of 10,400 from 1965. Over half of African employees are aliens. In the previous year some 181 companies had 432 operating mines, but large companies such as Anglo-American and Rio Tinto Zinc account for a high proportion of the output of many minerals. Many data for the mining industry have been classified since 1965, but it is possible to discern the general trends.

Mining began in Rhodesia long before arrival of the Europeans; some 4,000 ancient workings have been found, from which an estimated 2 million ounces of gold were removed. The discovery of gold by Europeans in 1865 was the main reason for occupation of the colony; some 68,000 claims were recorded by 1894, and, with the exception of one year, gold remained the first interest in Rhodesian mining until 1948, when base minerals surpassed the value of its output. Today, about thirty minerals, metallic and nonmetallic, are produced, of which copper, nickel, gold, asbestos, and chrome are the leaders.

The central feature of mineral distribution in Rhodesia is the Great Dyke, an outcropping in a 350-mile (563 km) range of hills (Map 95). Vast chrome reserves, asbestos, nickel, and platinum are found within it, while the gold belts lie along either side and extend as far east as Umtali. North of Sinoia is a rich area of copper and mica fields; lithium and phosphates are found east of Fort Victoria; tin fields lie southeast of Wankie, and there are extensive iron ore deposits near Que Que and in the Bukwe Mountains. The coalfields occur in the lower-lying boundary areas.

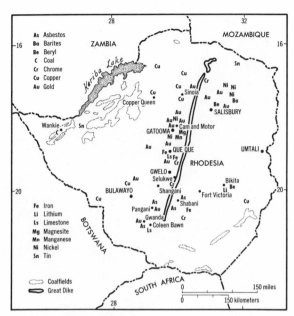

Map 95. Mining in Rhodesia

Copper has become the leading mineral in recent years and it is now won from 40 mines; it also occurs as a by-product of nickel. That metal has received notable attention in the last few years and may soon surpass copper; indeed, the most important feature of post-UDI mining developments has been the opening of several new nickel mines and completion of a smelting and refining complex at Bindura. A very large deposit may also be opened at Inyati in the near future.

Asbestos ranked first among mineral exports in 1965, accounting for a third of their total value. Rhodesia led in world production of top-grade chrysotile asbestos, but ranked third in total production. The largest workings are at Shabani and Pangani. Chrome ore ranked fourth in 1965 when about 19 mines were producing, of which the largest was Selukwe Peak. Ferrochrome plants exist at Gwelo and Que Que. Disruption of chrome supplies to the United States led to an abrogation of the UN sanctions when Congress passed Section 503, usually called the Byrd amendment, of the Military

Lumpy chrome ore ready to be hauled to Selukwe for treatment
Vast reserves of chrome, asbestos, nickel, and platinum occur within the Great Dyke.

Shaft headgear and mill at Shabani asbestos mine
Asbestos is one of the leading mineral exports of Rhodesia, which leads the world in top-grade chrysotile asbestos.

Procurement Act of 1971, permitting the import of chrome, ferrochrome, nickel, and asbestos from Rhodesia. In late 1973 this action was finally reversed, repealing what had been an embarrassment to the State Department and a thorn in the side of black Africans. Ironically, Section 503 did not reduce chrome ore imports from the USSR and a flood of ferrochrome imports from Rhodesia damaged the American ferrochrome industry.

Output of gold accounted for 51.7 percent of the total value of mineral production in Rhodesia from occupation of the colony to 1961, but a large number of small workings were closed after World War II and gold accounted for only about 22 percent of mineral exports by value in 1965. Sanctions provided an incentive to increase gold output, which could readily be exported to improve the balance of payments position. The marked increase in the sale price of gold in recent years has given further incentive and has raised the calculated reserves substantially. The 1971 value of gold production was estimated at $19.5 million and it is possible that gold could make a strong play to surpass copper and nickel as the most valuable mineral export. Most of the gold bodies are, however, small, heavily faulted, and widely scattered.

Resources of iron ore are of particular significance because of their impact on the possibilities of sustaining a domestic iron and steel industry. Occurrences are widespread and reserves are estimated to be at least a billion tons of 56 to 64 percent ore. The largest high-grade reserves are found at Bukwe near the rail line to Lourenço Marques; other large bodies are found at Chikurbi, near Salisbury, and at Redcliff and Beacon Tor, near Que Que. Ore at Redcliff supports the small steel mill there; ore, pig iron, and steel are exported, some under contract to Japan.

Other minerals produced include petalite (a lithium ore), beryl, tin, phosphates, corundum, magnesite, and platinum. Exploration has been stepped up in recent years for a variety of minerals.

MANUFACTURING

Rhodesia has unusual industrial diversity, with some representation in most important branches and an output of fairly sophisticated products in several. It held a predominant share of manufacturing in the Federation of Rhodesia and Nyasaland, with 72 percent of employment and 82 percent of the gross output in the three countries. An indication of the rapid progress made in Rhodesia can be seen in the rise of the gross value of production from $14.8 million in 1938 to $173 million in 1953 and $1,040 million in 1971. The index of manufacturing output reveals a growth from 84.0 in 1961 to 100 in the base year 1964 and 191.6 in 1973. Employment in manufacturing showed little growth from 1958 to 1965 but increased by 1971 to 105,800 Africans and 20,200 Europeans. In 1973 manufacturing accounted for an estimated 23.4 percent of the GDP.

With regard to the location of industry, primary processing of mineral and agricultural raw materials is typically close to the producing areas. Other manufacturing is heavily concentrated at Salisbury and Bulawayo, which account for about three-quarters of manufacturing output. The Midland towns of Gwelo, Que Que, Gatooma, and Hartley, and Umtali in the east, account for perhaps one-sixth.

A number of influences encouraged the rise of manufacturing in Rhodesia, including the high cost of importing goods to this landlocked country, the inability to obtain goods during World War II and the early postwar years, the desire to provide employment for the newly arrived immigrants, and the existence of sizable markets in Zambia and Malawi. Because agriculture could not be expected to absorb really large numbers of Europeans and the European population is highly urbanized, greater emphasis is

Interior of a large brewery at Bulawayo

placed on industrial development than might otherwise be expected. The government has participated directly in founding certain industries, including cotton spinning and iron and steel, and has attempted to stimulate industrial growth in other ways.

After UDI renewed efforts were made to promote import-substitute industries and it is claimed that 1,100 new industrial projects were started from UDI to 1971; these probably included many new products produced in old plants. New items introduced since 1965 include breakfast foods, foundation garments, cosmetics, aerosol sprays, deep freezers, balls, liners, and castings for the mining industry, record players, and toys.

The future growth of manufacturing in Rhodesia will depend upon the availability of markets, labor, power, raw materials, and capital, with political developments having a profound influence on most of these factors. The size of the market would appear to set definite limits on the size of the industrial es-

tablishment. Even if the European population, with a high average income, were to
double, it would still be too small to support
many types of industrial enterprises. This
provides a strong argument for working to
raise the African standard of living, for it is
only the indigenous populace that can provide a potentially large domestic market. But
the average annual wage of Africans, in fact,
increased only 28 percent from 1965 to 1971
and the ratio to the average European wage
fell from 1:10.5 in 1965 to 1:10.9 in 1972.

The advantages of the Rhodesian market
position have included the high purchasing
power of most Europeans, the demand stimulated by large and small development and
construction projects, the fact that about a
seventh of all Africans are wage earners, and
the considerable geographic concentration
of the richest market region on the highveld
along the line of rail. A major disadvantage
has been the loss of the Zambian market,
which is not likely to be opened until it is
clear that Rhodesia is on the road to majority
rule; the buildup of industries in that
country would, in any case, alter the nature
of the goods that could be marketed. It has
been possible to expand sales to South
Africa, but the strength of that country's industrial establishment makes competition

difficult in many sectors. A settlement of the
dispute would, on the other hand, provide
opportunities for the export of items semiprocessed from domestically produced agricultural and mineral raw materials.

Regarding labor, the rapidly growing
black population provides an adequate supply, but threatens increasingly serious unemployment problems. Foreign-born Africans,
who are willing to accept lower wages, continue to provide a significant part of the
labor force. At the time of the 1969 census
foreign Africans made up 7.5 percent of the
total African population and 18.5 percent of
those in European areas; 48.5 percent were
from Malawi, 32.5 percent from Mozambique, 13.0 percent from Zambia, and 6.0
percent were from other countries.

The system of land distribution in Rhodesia results in a heavy dependence on migratory labor. An indication of this position is
revealed by the presence of 53.7 percent of
African men in the European areas as compared with only 33.7 percent of the women
and 30.1 percent of the children in 1969.
This situation makes for high turnover and
low productivity. Labor employed in manufacturing is more stabilized than in agriculture and mining and this has contributed to
the considerable progress made in increasing
production per head of population, which
went up by about 69 percent from 1961 to
1971.

Shortages of artisans, skilled workers, and
managers are likely to be more restrictive in
the future than shortages of unskilled and
semiskilled labor, though Rhodesia does not
have the same limitations in this regard that
exist in most African countries. Further relaxation of the color bar would help this situation. It is possible, of course, that a settlement or increased guerrilla activity might
lead to a wholesale exodus of Europeans,
which would greatly alter the market, labor,
and management picture in Rhodesia.

The raw material and energy positions as
they affect manufacturing development are

The Rhodesian Iron and Steel Company plant at Redcliff
The steel mill has an exceptionally favorable raw material
position, but its small scale has made economic operation
difficult.

characterized by a wide variety of agricultural, forestal, and mineral products, plus substantial hydroelectric resources. They permit one to envision the development of a more rounded industrial economy than may be practicable in most African countries. The very low cost of coal must also be noted. One shortage which will require increasing attention is that of water supplies for domestic use on the highveld. Sanctions and exchange control have made it difficult for numerous manufacturers to obtain some necessary raw materials, spare parts, and new machinery in recent years.

From the standpoint of availability of capital, Rhodesia was one of the more attractive African areas for foreign investment in postwar years, though the greater part went to transportation, power, and mining rather than to secondary industry. Despite sanctions there has been a renewed interest in investment in mining, but industry has had to rely on domestically generated capital.

THE IMPACT OF SANCTIONS

It is apparent that Rhodesia has considerable strengths, which have permitted it to withstand sanctions: small but well-armed and well-trained military and paramilitary forces assisted by South African contingents; an economy noted for its relatively well-developed infrastructure, unusual diversity, and one of the best rounded manufacturing complexes among tropical African countries; a cushion to offset some strains in the form of foreign-born Africans who could be repatriated; considerable power over supply of coking coal, electricity, and transport for Zambia, with this leverage being used successfully until 1973; and the enormous gaps in the sanctions wall through Mozambique and South Africa.

It is not possible to measure the impact of sanctions with the desired precision, because the Rhodesian government has suspended publication of some important documents

and censors the press and radio. In addition, many countries have failed to report on trade as required by the UN, there has been collusion—particularly with South Africa—in hiding trade movements, and all sorts of subterfuge and false documentation have been employed by numerous parties.

Many nations were slow to apply sanctions and there seemed to be a tendency to assume that announcement of sanctions was tantamount to their application, whereas experience has shown that attention must be given from the start to the plugging of loopholes. In more recent years a growing number of countries appear to be ignoring sanctions and official actions against offenders are rare; in the period 1968–1972 there were 138 breaches reported to the UN, of which 127 were detailed by Britain; few were pursued beyond the investigatory stage.

Looking first at trade, it was thought in the beginning that a ban on petroleum and petroleum products would be the kingpin of sanctioning efforts and might be adequate in itself. And there were early favorable signs, including closing of the refinery near Umtali only two months after UDI, cessation of crude shipments via Beira, and increasingly restrictive gasoline rationing. However, alternative supplies and routes were gradually developed, and by the end of the first year Rhodesia was getting the same amount as before UDI, and by 1971 a considerable stockpile had been accumulated and rationing was discontinued. The Limpopo Railway and railroad haulage via South Africa and Beit Bridge have been the major new routes. In late 1973 the Arab nations announced a boycott on petroleum to South Africa, the Portuguese territories, and Rhodesia, but as of late 1974 there was little evidence that it was having any serious impact.

With respect to imports other than oil, it should be recalled that sanctions focused primarily on Rhodesian exports; the main impact on imports would thus be indirect, with lowered foreign exchange earnings reducing the ability to pay for imports. Certainly the gaps through Mozambique and South Africa have made it possible for Rhodesia to obtain physically such imports as were required. But financial stringencies have prevented that country from purchasing all the items needed and various serious shortages have gradually appeared, including especially railway locomotives and rolling stock and machinery for many industries, which has not been replaced as it normally would have been. Many industries and merchants have complained that government restrictions on imports have hurt their businesses in one way or another. That imports were adversely affected by sanctions is suggested from the data in Table 18, but the level has been more affected by Rhodesian restrictions than by the success of sanctions applied to shipments moving to that country. A battery of controls has been frustrating to many industries, while those which have been unable either to secure imports (e.g., petroleum refining, some branches of the chemical industry, and motor assemblers) or to export their prod-

ucts (e.g., tobacco products and, for some years, ferrochrome) were severely hurt. Some of these gradually found ways to circumvent sanctions; the vehicle assembly industry, for example, was able to secure kits from time to time, and in 1973 it was announced that a Rhodesian car was being built in Salisbury. And whatever the difficulties may have been, it is obvious that industrial expansion has cushioned much of the impact of sanctions and that many plants have been aided by import restrictions (e.g., foodstuffs, textiles and clothing, paper and printing, wood, and furniture).

With respect to exports, the impact of sanctions was quite pronounced in the early years despite the slowness in applying them, but a gradual increase brought them to the pre-UDI level in 1972, though the figures do not allow for inflation and a considerable adverse trend in the terms of trade until 1973. The recovery in trade was achieved not only by the relaxation of efforts of many countries, but also by false documentation, and by repackaging, remarking, and some partial processing in South Africa.

Table 18. Selected data on Rhodesia related to impact of sanctions, 1965–1973

| | | Foreign Trade | | | Average number of employees | | Net immigration of Europeans |
Year	GDP mil. R$	Exports,[a] mil. R$	Imports, mil. R$	Balance mil. R$	African '000	non-African '000	'000
1965	698	328	240	88	654	90	3.5
1966		207	169	38	642	90	−2.1
1967		202	187	15	657	92	3.3
1968		196	207	−11	690	96	6.2
1969	893	240	199	41	731	100	5.0
1970	992	265	235	30	743	104	6.3
1971	1,120	294	282	12	779	108	9.4
1972	1,289	345	274	71	848	113	8.8
1973	1,419	...[b]	...[b]	83	890	117	1.7

SOURCE: *Economic Survey of Rhodesia; Standard Bank Review,* May 1974.

[a] including gold.

[b] exports and imports no longer separately stated.

An aerial view of Salisbury, capital and main manufacturing center of Rhodesia

The most seriously affected export has been tobacco, whose controlled production has remained at less than half the 1965 level. And despite subsidies amounting to about $93,680 per year per producing farmer, many of the remaining growers are in difficulty. Although agricultural output has increased, diversification from tobacco has often brought much lower returns, and non-tobacco farmers are dissatisfied with their earnings. Switching from a labor-intensive crop such as tobacco had an adverse impact on African workers, though this was cushioned by repatriation of foreign Africans and the notable increase in cotton production. The cost to the government of maintaining productive capacity in tobacco has been heavy—for storage, for purchase at the support price, and in losses for those exports which are made (the trading loss in the period from UDI to 1972 has been estimated at about $122 million). Nonetheless, it appears that Rhodesia has found ways to dispose of a greater share of the crop in recent years and to have substantially reduced the stockpile. In 1974, in fact, the government withdrew limitations on the amount grown and the guaranteed basic price.

Sugar and corn have also been difficult to export, but the sale of beef and cotton has apparently presented no problem. Mineral sales abroad have been less affected than other commodities, as the 60 percent increase in output from 1965 to 1972 suggests. The construction sector, which was severely depressed at the time of UDI, recuperated in 1967 and reached a record level in 1972, when the value of building plans was over six times the 1966 level. The tourist industry has also prospered, with 343,000 visits recorded in 1971.

The railways have also suffered: the equipment, which is costly to replace and requires a considerable amount of foreign exchange, has run down; they have not been able to carry all of the traffic offered (which has affected from time to time, the delivery of cement, fertilizer, and oil) and they have lost the Zambian traffic (see chapter 20).

The interruption to the flow of capital has been one of the most severe impacts of sanctions, though it appears that in 1973 a net inflow occurred.

Measures of economic growth give further evidence of the impact of sanctions (Table 18). These data suggest that growth has not been very satisfactory. The GDP figures are misleading; they are not adjusted for inflation (which has, however, been reasonably well controlled), they need to be related to population growth, and they are inflated by government stockpiling of tobacco and some minerals. The increase in employment is also not adequate to meet the needs of an expanding population; in 1960 some 17 percent of the total African population was employed, whereas the figure was reduced to 14.5 by 1972. White prosperity has been well maintained and those Africans employed are apparently somewhat better off than in 1965; but there has been a failure to create adequate opportunities for the black popula-

tion, and provision of health, education, and housing for Africans has not kept pace with the rising needs.

The economy, it appears, has been going more sideways than up. There has been an unmeasurable but very high opportunity cost resulting from sanctions, and in the long run an economy such as Rhodesia's must remain stultified if it cannot expand exports or receive a greater inflow of capital.

The expected independence of Mozambique in July 1975, however, was a more severe blow to white Rhodesia than nine years of sanctions, revealing the very great significance in the gaps in the sanction wall that Portugal had provided.

CHAPTER TWENTY

Zambia

Economically one of the most important countries in Africa, primarily because of its position as the third-ranking producer of copper in the free world, Zambia achieved independence in October 1964. As Northern Rhodesia, it had been a member of the Federation of Rhodesia and Nyasaland from 1953 to 1963. Resentment over the favored position of white-ruled Southern Rhodesia in the Federation, a growing desire for independence and intense rivalry between the major political parties within the country led to much unrest in the years preceding independence.

Little more than a year later, UDI occurred in Southern Rhodesia, creating an immediate emergency because of the cessation of oil deliveries, and a pressing but longer-run problem of reducing or eliminating the previous very heavy dependence on Southern Rhodesia in a variety of spheres. Indeed, this dependence was so important that it partially explained the slowness in turning the screws on Rhodesia and the emphasis until mid-1968 on *selective* mandatory sanctions. It also created conflict in Zambia between those who wished to reorient traffic as rapidly as possible through independent black countries and those who took the more pragmatic view that Zambia should give highest priority to its own needs, irrespective of the sanctions against Rhodesia.

The fear that Zambia might be plunged into political chaos comparable to that experienced by Zaïre caused great concern among Western nations, who wished to preserve the flow of copper from Zambia. At the time of independence Zambia was little better off than Zaïre had been from the standpoint of the numbers of persons with higher education or with experience in administration. There were only 1,200 secondary school and 109 university graduates.

The country did, in fact, experience considerable internal political difficulties in the ensuing years, owing to continuing ethnic rivalries and personality clashes, but President Kenneth Kaunda succeeded in achieving order, although at the expense of detaining opposition leaders and terminating the two-party system—in 1972, when UNIP became the only legal party. Late that year a code of conduct was adopted similar to that enunciated by Tanzania in 1967, intended to prevent graft, nepotism, and the rise of a privileged class.

Domestic policies have focused upon attempts to broaden the base of the economy and speed Zambianization through the progressive replacement of expatriates and nationalization of several sectors, particularly trading, manufacturing, and mining. A large measure of state control is now exercised through a number of state holding corporations, which in turn control a series of parastatal companies in which Zambia has full or controlling, usually 51 percent, interests. Some of the companies are managed or provided with technical services by contract with minority shareholders.

Despite some difficulties and a number of very expensive takeovers, most of the parastatal companies have functioned with reasonable efficiency, and several minority holders have provided additional capital for expansion. Zambianization of employment, which has been rapid, must now be slowed because of the short supply of highly skilled

Zambians, particularly in mining and manufacturing. Indeed, one of the problems is securing the larger number of expatriates required to expedite localization. Dynamic training and educational programs will eventually meet this problem, but skilled geologists, mining engineers, and metallurgists, for example, require experience as well as training. In the meantime, the white population has declined from a peak of 77,000 in 1962 to about 40,000 in 1974.

REDUCING THE DEPENDENCE ON RHODESIA

A considerable effort has been necessary since 1965 to remove the very great dependence of Zambia on its southern neighbor, and a discussion of these efforts will serve to provide important background for understanding the economy of Zambia. As will be seen, the dependence was still very substantial for transport routes as late as the end of 1972; in January 1973, however, Rhodesia closed the border, ostensibly in retaliation for guerrilla incursions from Zambia, although it announced shortly thereafter that copper exports could continue to move by the Rhodesia Railways. But when the border was reopened in February, Zambia stated that it would no longer use the Rhodesian routeways, despite the difficulties and higher costs of using alternative routes. Thus while Rhodesia had successfully applied pressure in earlier years, and particularly when Zambia required large-scale supplies of corn because of a serious shortfall occasioned by drought, that country misjudged the Zambian resolve and found itself more hurt than helped, particularly through the loss of valuable transit traffic on the Rhodesia Railways. The decision by Zambia marked a notable turning point and will have significant results in a variety of spheres.

THE BASIC STRENGTHS OF ZAMBIA

The ability of Zambia to offset its dependence on Rhodesia is related, first and foremost, to its enormous reserves of copper, probably far above the stated 806 million tons of ore with an average metal content of about 3.5 percent as of the end of 1972. Indeed, it is this strength from which most other strengths derive. Copper, with cobalt mined as a by-product, accounts for 94 to 96 percent of the value of Zambian exports. It enables Zambia to sustain a high level of trade (second only to Nigeria among tropical African countries) and a generally favorable physical balance. This means, in turn, that the country can finance a very substantial development program largely from its own funds (the 1973 budget totaled $734 million of which about a quarter was for capital expenditures).

The mining industry provides about 24 to 40 percent of direct government revenues and additional indirect funds and accounts for over 40 percent of the GDP. It is the largest source of paid employment in the country, with 46,245 Africans and 4,600 expatriates being engaged in Copperbelt mining in 1972, earning a total of $156 million. The copper industry also supported the railway, although the line in Zambia was originally constructed to serve the copper industry across the border in Shaba, and the traffic it generates has justified the very

A raise borer reaming a 2-meter diameter ventilation raise in the Baluba copper mine
Production of copper and associated cobalt is the main strength of the Zambian economy.

heavy expenditures on improving alternate routes to the coast, including the Tazara line to Dar es Salaam.

Backward linkages account for large hydroelectric developments (the copper industry takes 85 percent of the total electricity consumed in the country) and for industries catering to the huge mining-metallurgical complex of the Copperbelt. Finally, location of the industry in a relatively small section of the country created an unusually concentrated and rich market justifying certain agricultural, manufacturing, and service developments that would otherwise have taken much longer to achieve. The six largest Copperbelt towns had an estimated 1972 population of 876,000, compared to 348,000 for the capital, Lusaka.

A second strength, obviously closely related to the first, has been the relatively solid overall economy. The GDP rose from $677 million in 1964 to $1,793 million in 1972. There has been considerable sectoral growth in construction, which has experienced boom conditions, especially in Lusaka, Kitwe, and Ndola, and in manufacturing, whose index of production with 1961 the base year stood at 297.8 in 1972.

MAJOR WEAKNESSES—
ACTUAL AND POTENTIAL

While Zambia has powerful strengths and very great potential for further growth there are certain problems which must be noted.

One of these is the unusually high one-product emphasis of the economy. This may appear fine when world prices for copper are rising as they were from 1964 to 1970 and again in 1973; indeed, the booming economy over the years to 1970 very largely reflected the rise in price rather than increased production (Table 19). It is likely to be traumatic when a price drop occurs, as it did for 1971–72. But even rising prices are not without problems; they may foster excessive demands for wage increases, encourage excessive development and recurrent expenditures (particularly for projects that are not self-sustaining, whose overhead may be difficult to maintain on lowered revenues), contribute to inflation, threaten the competitiveness of the copper produced, or stimulate efforts to substitute other metals for copper where that is possible (with the danger that some markets will be permanently lost). Price fluctuations are, of course, related not only to changes in demand but also to often unpredictable interruptions in supply. Examples include an eight-month strike in U.S. copper mines, a series of difficulties in Chile, and, with respect to Zambia itself, the inability to ship copper at desired levels owing to the Rhodesian situation, shortfalls in coking coal or rail wagons, and (in September 1970) a serious accident at the second-largest mine, Mufulira, which resulted in 89 deaths and the loss of a 150,000-ton productive capacity, which had still not been fully recovered in early 1974.

While high prices have their own set of problems the real crunch comes with lowered prices. Zambia, for example, experienced a 47 percent decline in the price of copper between March 1970 and the end of 1971. This resulted in markedly reduced foreign exchange earnings and government revenues, which in turn forced a considerable and quite painful retrenchment in development expenditures.

A second major weakness is the relatively poor development outside copper mining. This has been rapidly corrected in manufacturing and construction; but other mining activities are not very significant and the agricultural sector is particularly weak (see below), while tourism is much below the level in Rhodesia or East Africa.

Other problems include those related to employment. There are not now adequate opportunities for unskilled workers, a problem shared by most African countries. There has been a substantial increase in the numbers employed in transportation, construction, and government; but employment in mining has been rather static because of improving productivity, though this has flattened out or decreased slightly in recent years. Like other countries, Zambia has expelled foreign Africans to provide more jobs for nationals; many thousands were told to leave at the end of 1971. The generally booming economy has permitted a better record in job creation than in most African countries, but the goal for providing new jobs was only 60 percent filled in the first five-year plan and the second plan provides for only 100,000 new jobs in five years, whereas about 334,000 will be added to the labor force, including an estimated 186,000 in urban areas. An element of concern is the relatively very high wages for the employed labor force as contrasted with the rather meager improvements achieved by the bulk of the population resident in the rural areas and dependent on agriculture.

A different employment problem relates to the continuing shortage of skilled Zambians. The demise of the Federation and the independence of Zambia increased the shortage in that many functions which had been centered in Salisbury moved to or opened up in Zambia. While much attention has been given to education, including the opening of the University of Zambia, and there are an impressive number of training programs, particularly on the Copperbelt, the shortage cannot be eliminated in a brief period of years. The counterpart problem is the high dependence on expatriates in mining, government, and other sectors. This makes the government especially subject to pressure from dissidents or from the general population, many of whom do not really understand that it requires training and experience to run a computer or a copper refinery. The copper industry has reduced its expatriate employees from 7,641 in 1961 to 4,600 in 1972; it has considerable difficulty in attracting and holding these employees, the annual turnover having averaged 24.5 percent in the period 1968–1972. A somewhat delicate feature of the situation has been the inclusion of a considerable number of South Africans and Rhodesians among the expatriates, though it should not be assumed that they were unsympathetic to Zambia.

There are, as everywhere, political uncertainties which cause serious present and potential problems. Tribalism continues to be a strong force in Zambia and it has caused the government to move away from an open society and finally to adopt a one-party state, although elections among competing candidates are still held. Needless to say, many of the economic problems tend to contribute to political uncertainties.

Finally, among the weaknesses, has been the dependence of Zambia on Rhodesia. The

Table 19. Selected data on copper exports from Zambia, 1963–1973

Year	Exports		Indexes of copper exports		Unit value of copper $U.S. per 100 lbs.
	Total mil. $U.S.	% in copper	volume	price	
1963			100	100	
1966	690	93.2	102	189	48.96
1969	1,073	94.2	125	243	62.98
1970	1,001	95.3	117	244	63.25
1971	679	92.9	109	174	45.03
1972	758	90.6	122	169	43.86
1973	1,158	94.2	115	256	72.25

SOURCE: International Monetary Fund, *International Financial Statistics*, 27, no. 10 (October 1974): 28, 410–11.

ways in which it was dependent are outlined below, together with the steps that have been taken to offset them.

COAL AND COKE. Before UDI almost all of these supplies, amounting to about 1.2 million tons in 1965, came from the Wankie Collieries. Coke was required in the smelting of copper and other minerals. Coal deposits had been known to exist in Zambia, particularly along the Zambezi scarp, but no mining had taken place. A mine was opened in a crash program at Nkandabwe but was closed in 1968 after a better-grade deposit had been opened at nearby Maamba. Production reached 990,000 tons in 1973. Both of these operations had serious disadvantages: the considerable investment required, the high cost per ton produced (over twice that at Wankie), the poor quality of the coal (which has a 19.5 percent ash content and is not suitable for making lump coke), and the problem of transport. Over extended periods in 1966–1968 it was not possible to meet the needs of the smelters, which occasioned a reduction in output just when copper prices were very favorable.

For best results, Maamba coal must still be mixed with higher grade coal in the smelter, particularly in the Broken-Hill lead-zinc operations at Kabwe. At times coking coal was imported from South Africa via Lobito, which cost about five times the delivered cost of Wankie coal. Some gas-oil was flown in on occasion, and the Mufulira smelter was converted, at a cost of $490,000, to permit use of oil instead of coal; nevertheless, a decision was made to continue imports from Wankie. After definitive cessation of purchases from Rhodesia in early 1973, Zambia received some emergency supplies from the stockpile in Zaïre—which, somewhat ironically, will continue to receive Wankie coal via Zambia. With completion of the oil refinery near Ndola, heavy oil became available; this can

be used in place of at least half of the coal from Maamba. But increased prices for oil may have stimulated interest in special processing facilities to produce fine coke from Maamba coal, in seeking enlarged purchases of coal from Zaïre, and in a 110,000-kw thermoelectric plant proposed for siting at the colliery. The difficulties of moving adequate quantities of coal were greatly reduced by replacing a difficult road haul to railhead by a ropeway and rail extension, which were completed in late 1973.

ELECTRICITY. In 1965, 68 percent of Zambia's electricity came from the jointly owned Kariba installation, whose only plant was on the Rhodesian bank, which led to fears that transmission would be interrupted. There is a large consumption of electricity in the copper/cobalt refineries, which would have been crippled without power from Kariba. The Seke plant in Zaïre provided 9 percent and domestic sources provided 23 percent of total electricity consumed in 1965. It was recognized by all concerned that Kariba could not be replaced in a short period. The solutions to this problem have been:

(1) To continue to receive power from Kariba. And, over the years, the fears that Rhodesia would interrupt supplies diminished to the point that it was considered safe to construct the north bank plant at a cost of about $134 million. Construction began in 1970, but was delayed by unexpected rock troubles and bankruptcy of the contractor. The 600,000-kw Phase I is now scheduled for completion in 1976; an additional 300,000 kw will be developed in Phase II.

(2) To construct the Kafue Project. Phase I provided a 600,000 kw plant in 1972 where the Kafue River drops 1,964 feet (599 m) in 12.5 miles (20 km) into the Zambezi gorge. Phase II, under construction, is designed to even the seasonally fluctuating flow by erection

The Kariba Dam shortly after completion
Zambia is constructing the north bank plant to meet part of its growing needs for electricity.

of a 4,500-foot-long (1,370 m) dam with a maximum height of 185 feet (56 m) at Itezhitezhi near the Kafue Flats, which will permit the capacity to be raised to 900,000 kw. Cost of the Kafue Project is estimated at $312 million, up $89 million from the original estimates.

(3) To increase capacity of the installations at Victoria Falls from 8,000 kw to 108,000 kw by 1972; the output is fed into the grid system, which roughly parallels the line of rail.

(4) To increase, where possible, the capacity of smaller hydroelectric installations. The capacity of five such stations was increased from 40,100 kw in 1961 to 66,000 kw in 1973.

Plans for expanding the production of refined copper and continuing increases in general consumption mean that even these dynamic developments will need to be augmented in another decade or so. The most favorable site would be on the Zambezi, which would require cooperation with Rhodesia.

GOODS ORIGINATING IN RHODESIA. In 1964 about 40 percent of total imports by value came from Rhodesia, including not only coal and electricity, but also a considerable variety of manufactured produce and some foodstuffs. Zambia wished to replace these either from domestic sources or alternative sources, not only to decrease its dependence on that country, but as a contribution to the sanctioning efforts against it. A measure of the success is seen in the reduction of imports from Rhodesia to only 4.7 percent of the total by 1972. To Zambia's regret the South African share of Zambian trade increased, having been 14.6 percent of imports and 1.9 percent of exports in 1972.

TRANSIT TRAFFIC. Probably the most important dependency of all was Zambia's reliance on Rhodesian routes for the vast bulk of its overseas exports and imports. Before UDI almost 100 percent of exports moved through Rhodesia, with 99 percent moving on either to Beira or Lourenço Marques, and 1 percent to South African ports; 97.5 percent of overseas imports came through Rhodesia and about 2.5 percent from Lobito, Angola, through Zaïre. The dependence on Rhodesia for transport was heightened by joint ownership of the Rhodesia Railways and the Central African Airways, whose major facilities were located in Rhodesia.

Immediately after UDI it was found necessary to mount a rather intensive emergency airlift to supply Zambia with petroleum products, which Rhodesia refused to move so long as oil was refused to it. Britain, the United States, and Canada provided military and private planes and subsidized the operations; these were succeeded by lesser private airlifts, but movement of oil by air was extremely costly and, as surface routes managed to handle the necessary amounts, it was abandoned. The delivery of petroleum products was greatly improved by the construction of a 1,058-mile (1,702-km) pipeline from Dar es Salaam to Ndola, completed in August 1968 at a cost of $48 million. The ca-

The rail and road bridge linking Rhodesia and Zambia just below Victoria Falls
Before UDI over 95 percent of Zambian overseas traffic moved over this bridge, and after seven years of sanctions it still took over half of such movements. In January 1973, however, Zambia decided to stop any further shipments via Rhodesia. The line still moves coking coal and copper on behalf of Zaïre.

pacity of the line has since been increased to 1.1 million tons at a cost of $32 million. Products were moved through the line until completion of the Indeni refinery near Ndola, after which crude oil was piped.

Emergency flights were temporarily flown again after closure of the border in 1973, this time to move goods from Beira and supplies for the mining industry from South Africa.

With respect to rail and road links a number of alternatives were available, but each had drawbacks of greater or less significance. These routes, moving in a clockwise direction from the first, are as follows:

(1) Rail line thru Shaba (KDL) to the Benguela Railway, with the seaward terminus at Lobito. This route provided the only all-rail route other than the Rhodesia Railway. Among its disadvantages are that it transits Zaïre, where it is subject to heavy demands for the traffic of mineral-rich Shaba and where uncertain political conditions led to closing the line on several occasions; it crosses Angola, whose relations with Zaïre and Zambia threatened a cutoff more than once, where guerrillas closed the line for brief periods in 1966 and 1967, and which is unsatisfactory so long as it has a dependent relationship with Portugal; there have been shortages of rolling stock from time to time, while congestion at the port of Lobito has slowed movements and resulted in the imposition of surcharges on cargo moving through the port.

(2) Other rail-water routes through Zaïre. These are the *voie nationale* via rail, river, and rail to Matadi, and the rail line to Kalemie and the Lake Tanganyika Ferry to the Central Tanganyika Line. Both require expensive transshipments and have bottlenecks at one or more points.

(3) Road-rail routes via Tanganyika and Kenya connecting with the ports of Mtwara, Dar es Salaam, Tanga, and Mombasa. Connections included road to Mpulungu at the south end of Lake Tanganyika, ferry to Kigoma, and rail to Dar; road to Kidatu and rail to Dar; road to Mtwara; or road to Dar. Any of the routes leading to Dar could be extended by rail or road to Tanga or Mombasa, the latter having been used fairly intensively because of congestion at Dar. The road to Mtwara, which had spare port capacity, was so poor as to preclude more than a low level of traffic.

The most used of these alternatives has been the Great North Road to Dar, which came to be known as "Hell Run" in the hectic years after UDI. The quality of most of this 1,250-mile (2,011-km) road left much to be desired: bridge capacities were inadequate; sections were impassable in the rains; it was too narrow and twisting in the mountain sections. In the first three years over 100 drivers were killed on the run. Other problems included: the securing of adequate capacity; inadequate maintenance facilities, which resulted in a quarter to half of the vehicles being out of service in the worst period; handling the very varied types of cargo moving in and out (particularly petroleum products before the pipeline was completed); losses to damage and thievery; and congestion at Dar. Many of these problems have been ameliorated: the road had been fully paved and partly realigned by 1972; truck traffic was organized by adopting a convoy system, establishing spare-parts and repair depots, providing radio communications along the route, using containers, and requiring better crating. Dar remains a bottleneck despite installation of an offloading oil buoy,

greater use of lighters, and construction of additional quays.

(4) The Tazara Railway connecting Dar with the Zambian system at Kapiri Mposhi. This line has obviously been selected as the major future routeway for overseas movements. It was, until the independence of Mozambique, the only practicable and politically acceptable routeway available to Zambia, since all other routes passed through white-controlled countries or, in the case of the *voie nationale* in Zaïre, involved expensive trans-shipments. It is scheduled for completion in 1975, but there is doubt whether it will be used to full capacity because space is not likely to be available at Dar. One objection to Tazara is the fear of developing just as heavy dependence on one country as was true in earlier years for Rhodesia.

(5) Road and rail via Malawi and Mozambique ports. This involves truck movement on the Great East Road (previously dubbed the "Great Dust Road," this route was largely paved by 1975) and use of the Malawi and Mozambique rail lines to and from either Beira or Nacala.

(6) Road to the improved ferry connection to Botswana at Kazungula, and road from there to the rail line connecting with South Africa. This has obvious physical and political disadvantages, but it is the only practical routeway to the south not crossing Rhodesian territory. It is conceivable that some traffic with South Africa may follow this route but it may be cheaper to ship such goods via Lobito or Dar es Salaam.

To have completely boycotted the Rhodesian route would have severely damaged the Zambian economy and involved heavy additional expenditure. Despite the impressive efforts made to develop alternative routes the amount of traffic moving via Rhodesia never got below a third of the total in the period from UDI to 1973 and increased from 1969 to 1973 because of congestion at Dar, interruptions on the Lobito route, and the necessity to import corn during the period of shortage already noted. In 1972, 950,000 tons—or 63.8 percent of import traffic, exclusive of 510,000 tons of petroleum moving by pipeline—and 425,000 tons or 52.5 percent of export traffic were shipped through Rhodesia.

The decision taken in early 1973 to discontinue shipments through Rhodesia made it necessary to move an additional 1.3 million tons on other routes, an exercise that cost Zambia an estimated $66.5 million in capital expenditures in 1973–74 and $59 and $72 million in added recurrent expenditures in 1973 and 1974. The loss of receipts to Rhodesia was calculated at $57 million a year. It was planned that about half the total be moved via Dar or Mombasa, 35 percent by Lobito, and the rest via Malawi and Mozambique. In fact, the Lobito route appears to have handled a higher share, and by 1974 the problem had been reasonably well met through restrictions on imports, the arrival of many new trucks and trailers, and the training of 2,000 drivers and mechanics. Completion of the Tazara will provide further relief, whether or not it is possible to reroute all of the traffic through Dar es Salaam.

The problem of reorienting Zambia's traffic has, then, been a prolonged, expensive, and difficult one; but it now appears to be well on the way to solution. It is appropriate now to fill in the picture for Zambia by providing information on its physical background, population, and economic sectors.

THE PHYSICAL BACKGROUND

Average elevations in Zambia range between 3,000 and 5,000 feet (900–1,500 m); the greater part of the plateau lies between 3,000 and 4,000 feet (900–1,200 m). Shallow

swamps and lakes, such as Lake Bangweulu and the Kafue Flats, cover large areas of the flattish plateau surfaces. In many localities the level of the plateau is broken by hills, sometimes occurring as chains that develop into areas of more rugged country. The Muchinga Mountains, west of the Luangwa trough, form part of the great escarpment and have individual peaks rising to about 8,000 feet (2,400 m). The Luangwa Trough itself is believed to be an ancient depression and is covered with sedimentary formations in contrast to the bulk of the plateau area. In the northeast is a mountainous area associated with the Lake Malawi Rift Zone and isolated from the rest of the country by the low-lying unhealthy Luangwa Valley.

Zambia experiences most of the climatic and soil characteristics detailed for Rhodesia in the preceding chapter. Precipitation averages 40.3 inches (1,024 mm), and is higher toward the north and lower in the south, particularly in the Luangwa and Zambezi valleys. On the plateaus, soils are for the most part either highly leached and shallow or deep, loose Kalahari sands of low productivity. The most fertile soils are the "upper valley soils" found mainly in the lower Kafue basin between Pemba and Kabwe and the "lake basin soils" of the northeast, which carry a dense agricultural population. About five-eighths of Zambia is tsetse-ridden.

POPULATION PATTERNS

Kay suggests that five unequal sectors running approximately north-south may conveniently be used to delineate the population pattern of Zambia (Map 96).[1] Starting with the westernmost, the sectors are

(1) A moderately or sparsely populated area paralleling the upper Zambezi, within which are two "islands" of dense population: the heart of Barotseland, centered on the Barotse flood-

[1] George Kay. *A Social Geography of Zambia* (London: University of London Press, 1967), pp. 47f. Much of this section is derived from Kay.

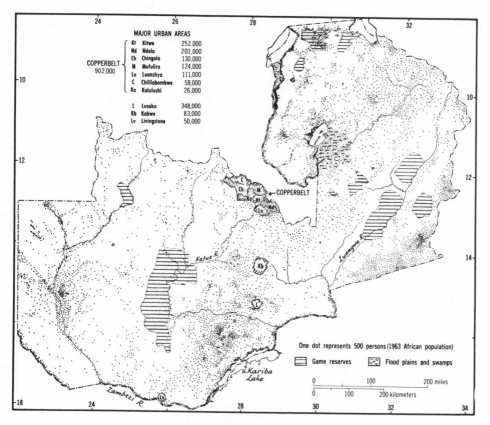

Map 96. Population map of Zambia

MAJOR URBAN AREAS

Kt	Kitwe	252,000
Nd	Ndola	201,000
Ch	Chingola	130,000
M	Mufulira	124,000
Lu	Luanshya	111,000
C	Chililabombwe	58,000
Ka	Kalulushi	26,000
L	Lusaka	348,000
Kb	Kabwe	83,000
Lv	Livingstone	50,000

COPPERBELT 902,000

One dot represents 500 persons (1963 African population)

Game reserves   Flood plains and swamps

The Zambezi River in the western part of Zambia
Because of special protectorate relations in the colonial period this part of Zambia is less developed.

plain, and the heart of Luvale country, astride the Zambezi upstream of Balovale. This is a tsetse-free area and the Barotse floodplain is considered the richest part of the province, though it is not capable under present land-use systems of supplying its densely populated portions with all the food it needs. The area is more isolated and less developed than other regions because of its remoteness from the line of rail, its relatively sparse road system, and the special status of Barotseland from 1890 to independence as a protectorate within the country, which meant that it received less attention. Very important in explaining the population concentration along the floodplain is the heritage of the powerful Barotse kingdom, one of several military states that existed in Zambia in the nineteenth century.

(2) A belt of very low densities paralleling the north-south section of the Kafue River and centered on the Kafue National Park. The Kafue Flats, once the floor of a lake, comprise over a million acres (400,000 ha) of grass-covered clay lands, much of whose area is flooded annually. Presence of the tsetse fly repulsed cattle-keeping tribes, while seasonal inundations of extensive stretches made tillage precarious.

(3) A belt of moderate to dense population bisected by the Shaba pedicle. The southern portion of this belt contains the main economically developed portions of the country: (a) the Copperbelt (which ranks as the single most important "island" of export production in tropical Africa), (b) Zambia's best-developed commercial agricultural zone, (c) the capital city, Lusaka, the most rapidly growing city in the country. The railroad, which parallels the country's main road, is a striking influence on the population pattern.

The northern portion of this belt, running northward of the Shaba pedicle, contains several "islands" of dense population: First are those associated with the fishing grounds of the lower Luapula, Lake Mweru, and the Bangweulu lakes and swamps; the average density on seven of the actual islands in the Bangweulu basin is now about 310 per square mile (120 per sq km), and concentration on manioc as a major subsistence crop has permitted higher densities than would otherwise have been possible. Second are a number of "islands" associated with important local settlements such as Kasama, Mansa, and Mpulungu. Existence of the Bemba kingdom contributed to some population concentra-

tion, though the Bemba characteristically did not settle so densely as peoples of the other major military states.

(4) A northeast-southwest belt of sparsely populated or empty lands associated with the Luangwa trough and the middle Zambezi Valley. These low-lying areas are unattractive for settlement. They are tsetse-ridden and hot and humid, and they have lower precipitation than the surrounding plateaus and a general scarcity of permanent surface-water supplies because most streams flowing from the north and west drain only the escarpment zones.

(5) A densely populated belt more or less paralleling the Luangwa Trough and lying on the eastern plateau. Two population cores are associated with Ngoni states, one of them centered on Chipata, the other on Lundazi. The present high densities of these core areas is explained in part by the inhibitions on expansion imposed by Pax Britannica, while the former no man's land surrounding them has gradually been filled in by neighboring groups who no longer needed to seek refuge from the Ngoni. Both this and the previous belt suffer considerably from remoteness and poorly developed surface transport.

AGRICULTURE

Agriculture accounts for only about 7.5 percent of Zambia's GDP. Less than 1.67 percent of the area is cropped in any one year, and about .1–.2 percent is under improved cultivation. A few years before independence about 1,185 European farmers holding 3.79 million acres (1.5 million ha)—though only a portion of this was actually farmed—accounted for about 70 percent of the marketed agricultural production of the country. By 1973 their number was down to 580 and this was reflected in lower production of tobacco, milk, and beef, though they still produced a major portion of marketed crops. There are, however, an increasing number of African commercial farmers and the government has determined to devote more attention to agriculture.

The main commercial farming areas are along the line of rail southward from Kabwe with the Mkushi block as an outlier (Map 97). Virtually all of the remaining large-scale European farms and most of the African cash production comes from these areas, with vacated European farms being used for cooperative, settlement, and state farms (see below). Former smaller European areas around Mbala (in the extreme north) and Chipata (near the Malawi border) have dwindled to extinction; in the latter area some thriving peasant schemes exist, which have been aided by paving of the Great East Road.

The traditional sector occupies about half of the total population—a relatively low proportion for tropical Africa—and produces about three-quarters of total argicultural output. Some indigenous systems show excellent empirical knowledge of the environment. For example, a "soil selection" system is practiced in Barotseland, where there are at least six distinct and specialized garden types on which the crops grown, the soil treatment, and the periods of cultivation and fallow vary according to the fertility and regenerative capacity of the soil.

A second example is *chitemene* cultivation, practiced over considerable areas of savanna woodlands. There are many varieties of chitemene, but the essential feature is the practice of lopping branches or felling trees over a wide area surrounding the actual garden and then burning them on the garden. Burning destroys weeds, makes clay soils more friable, and provides a fertilizing ash. Given static conditions, these and other systems can be considered as somewhat ingenious adaptations to the environment.

Modern practices and conditions have,

Chitemene *agriculture in Zambia*
Branches are cut from trees over a wide area, collected in piles, and burned to provide fertilization for a garden patch.

however, often led to the degeneration of the traditional systems, not all of which are good examples of proper land husbandry in any case. Partly because of the absence in some rural areas of from 40 to 70 percent of the able-bodied males, annual garden extensions have often been abandoned, even when there is more than ample land. Secondary crops are often neglected to the detriment of nutritional standards. In some areas, degeneration has resulted from the expansion of population and the consequent shortening of the period in bush or grass fallow, which should probably be close to 20 years on the old rock soils. These phenomena are by no means confined to Zambia; to some extent they are a largely inevitable effect of the introduction of a modern economy. But the degradation of large areas is a problem of major importance, which requires appropriate steps to offset its deleterious effects. An increased disharmony between rural and urban areas could otherwise prove to be a serious barrier to continued development in any sphere.

DEVELOPMENT AIMS AND PROGRAMS

In the colonial period a handful of European advisers worked mightily to encourage better practices but could not have been expected to bring about changes of the order required. By the time of independence a substantial area along the line of rail had been acquired by individual title and put partly to commercial production, a resettlement scheme had been opened in the heavily populated Chipata district of the Eastern Province involving the moving of 50 villages and 500 farmers, an Impact Scheme was launched in the north (which was overly ambitious and hence fell far short of its goals), and a beginning had been made on a program designed to modernize the practices of about 50,000 African peasant farmers. But most of the rural population remained relatively untouched and African agriculture was contributing very little to the money economy of the country.

Independent Zambia adopted the policy of promoting agricultural development to reduce the growing gap between the rural and urban/mining populace, to reduce the flow of migrants to the cities, to promote diversification, and to achieve self-sufficiency—a goal that was greatly strengthened by the shortfall of staple corn supplies, which placed the country at the mercy of Rhodesia in the late 1960s and 1970. The approach to rural development is patterned to a considerable degree on those applied to mining, industry, trade, and finance, with a series of parastatal corporations and subsidiary companies being responsible for overall aspects such as subsidy schemes for purchase of fertilizer and stimulation of corn production or provision of loans for extending and rationalizing operations. It was also proposed that the main institutional forms would either be cooperative or state farms. Many of the specific programs were poorly planned and initiated too abruptly, too massively, and without adequate technical advice and supervision, but, despite failures and some very substantial losses, the government continues to display a firm resolve to revolutionize indigenous agriculture as rapidly as possible.

Two basic approaches to rural development may be delineated: first, a somewhat piecemeal extension of services including shops, clinics, schools, roads, agricultural advisers, cooperative settlement schemes, and tractor mechanization. State farms are designed to increase rapidly the output of corn or special crops, which could replace imports, and by 1972 some 188,000 acres (76,000 ha) were in corn. The ranches are intended to build up large beef and dairy herds; the government has taken over some 300,000 acres (120,000 ha) for the establishment of 11 ranches which are intended to be stocked with 50,000 head in ten years and 6 dairy farms with a goal of 3,000 cows.

Cooperative settlement schemes have multiple objectives: to raise agricultural production, develop a spirit of communal enterprise (a keynote of the regime), resettle farmers from crowded areas, and encourage entrepreneurship. In about eleven settlement schemes 10–40 acre (4–16 ha) units were laid out and technical services were provided for the arriving farmers. In other cases a subsidy of $42 per acre ($104 per ha) was paid for land certified as having been stumped for cash cropping by a registered cooperative society with at least ten members. A major program of tractor mechanization was initiated in 1966; 200 tractors were delivered to 40 places. The cooperative and tractor programs have largely failed; land was opened to receive the subsidy and then uncropped or the crops went to waste, loans were taken apparently with no intention to repay them, tractor drivers were ill-trained and their machines lay idle after a brief period of use, and the mechanization schemes were reduced to 24 of which only 2 were successful. The government has not given up on the idea of cooperative farms; it continues to disfavor large private farms, which are to be limited to 25 acres (10 ha) in size.

In 1972 a new concept of Intensive Development Zones was introduced, which is intended to provide a wide range of services on a package basis to outlying areas; the program is expected to cost about $27 million over a five-year period. Zones marked for attention are around Chipata in Eastern Province, an area between Mbala, Nakonde and Kasama in Northern Province, and the Kabompo and Kasempa districts of Northwest Province.

CASH CROPS AND LIVESTOCK PRODUCTS

Corn occupies the largest acreage of any cash crop and is the main staple on subsistence holdings. Deliveries of corn to the National Agricultural Marketing Board declined, owing to low prices and to the exodus of European farmers. A low point was reached in 1969–1970, when a serious drought further reduced output, forcing Zambia to import over 30 percent of its commercial supplies. Determined that this should never happen again, the government raised the guaranteed price substantially which, together with better weather, raised deliveries from 1.49 million bags in 1971 to 4.4 million in 1972 and 6.76 million in 1973. This created problems of surplus supply; a million bags were sold to Zaïre in 1973 and large additional storage capacities were built at Ndola, Chisamba, and Kabwe. The 1974 crop was about the same as that of 1972.

Tobacco accounted for 69 percent of marketed agricultural production in 1969–70, but the output has declined seriously from the 1963–64 record 12,900 tons to 5,545 tons in 1972, when the area was down to 12,150 acres (4,920 ha). Most of the tobacco is Virginia flue-cured grown by expatriate and African farmers, while some Turkish and Burley is grown by Africans. Subsidies have been offered to tobacco growers since 1970, but these have neither prevented a continuing exodus of European growers nor stimulated African production satisfactorily. A program aided by the World Bank and the CDC involves the training and establishment of Zambians as tobacco growers, with a particular emphasis on one-acre schemes. More hopeful, too, has been an increase in yields per acre starting in 1971 and a notable increase in output from Zambian farmers, who now account for 52 percent of the crop.

Cotton is a relatively new crop whose production increased from 1,820 tons in 1968 to 11,920 tons in 1971 but then dropped to 8,123 tons in 1972 because of difficult planting conditions and a switch to corn to take advantage of the higher subsidy. But Zambia appears to be capable of meeting its goal of self-sufficiency in cotton. Two ginneries have been opened at Lusaka and Chipata. Peanuts of excellent quality are produced by the traditional sector, with about 80 percent of the marketed production being exported.

Efforts have also been made to stimulate the production of other import substitute crops, among them sugar, which was produced in record tonnages in 1973 from a large irrigated estate in the Kafue Flats; sunflower, to help meet the needs for vegetable oil; fruits and vegetables, whose output was inadequate and declining until 1972 (a cannery established at Mwinilunga has been packing pineapples, tomatoes, and carrots); coffee; and tea, which is being grown under irrigation at Kawambwe in Luapula Province where the first factory was installed in late 1973.

Interest in irrigation developments is likely to increase for the production of sugar, rice, wheat, tea, and other crops. On the plateau surface, attention will be given to the potentialities of the swampy and seasonally flooded areas, both the larger ones and the smaller *dambos* and *vleis*, areas of impeded drainage. The seven largest swamps and adjacent plains total 13,754 square miles (35,623 sq km), or 6 percent of the country. Just what the appropriate combination of uses should be in these areas will not be known until they have been much more thoroughly studied, but it is reasonable to believe that, with their apparently superior soils and with an excess of water, the effects of proper control will be highly rewarding. The Kafue Flats, where there is a total of 1.3 million acres (526,000) ha) subject to annual inundation, have been the scene of a 700-acre (280 ha) experimental farm, a large UN project, and now a West German supported scheme; sugar, wheat, barley, potatoes, and livestock have done well on the Flats, which also have considerable interest as a source of fish.

Output of animal products is well below needs, with commercial cattle slaughterings down about 20 percent from the pre-independence level and dairy products not being produced in line with increasing demands, despite various government schemes and subsidies. On the other hand, there has been a welcome increase in the output of dressed chicken, eggs, and pork, the Zambia Pork Production Co. being one of the few successful parastatal agricultural corporations.

FISHING AND FORESTRY

Fishing is carried on to some extent in all the larger lakes and swamps, but would appear capable of very great extension, which would be highly beneficial to a populace whose diet is generally deficient in proteins and calcium. The Kafue Flats are the most important producing area; Lakes Mweru, Bangweulu, Tanganyika, and Kariba are also important. Fishing is the mainstay of some districts; approximately 15,000 people are involved in commercial fishing while another 35,000 are concerned on a temporary or subsistence basis.

Over 50 percent of Zambia's area is covered with deciduous woodlands, but the quality of most trees leaves much to be desired. A large sawmill in the Livingstone area handles "black teak," mostly for railroad ties and mining timbers but also for parquet flooring and furniture. Afforestation has been carried out in several areas and it has been found that pine matures in 30 years

and eucalyptus in 8–10 years—in both instances more rapidly than local species. Because Zambia is a heavy net importer of wood and consumption is growing rapidly it has been proposed that plantings be increased to permit production of sawn wood, particle board, and pulp, with Kitwe to be the site of the new mills.

THE MINERAL INDUSTRY

The dominance of the copper industry in the Zambian economy has already been outlined. The Copperbelt mining and metallurgical operations rank second only to the South African gold industry among African metallic mineral producers.

Copper mining began on the present Copperbelt in 1921 at Nkana. The ores mined range in content from 2.92 to 4.84 percent copper and there is by-product cobalt, silver, selenium, and gold. Copper production has increased from 214,000 tons in 1948 and 375,000 tons in 1958 to 682,000 tons in 1973, while plans call for raising output to 900,000 by 1976. The proportion of electrolytic copper to total exports of the metal has grown from 46 percent in 1954 to 94 percent in 1973.

Location of the mines, smelters, and refineries on the Copperbelt is shown on Map 97. Two smaller copper mines are situated away from the belt, at Kansanshi and Kalengwa. The copperbelt contains about 22 percent of the total population of Zambia with over 90 percent of that in the urban centers (Maps 21 and 96). The 30 × 90 mile (48 × 145 km) area has the best developed infrastructure and services in the country and is a major market because of the relatively high incomes of many of the workers. Kitwe, the largest city, serves as a regional center because of its location and has important trade, service, and manufacturing functions as well as the mining and refining of copper. Ndola, the capital of the province, rivals Kitwe as a regional center, and has significant transport

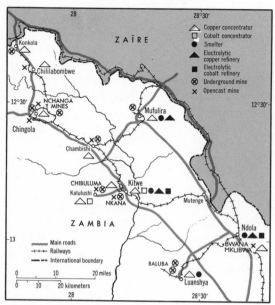

Map 97. *The Copperbelt, Zambia*

Large trackless diesel-powered front-end loader used in underground mining on the Copperbelt
Capable of moving many tons of material at a time, such machines suggest the highly mechanized and high-productive methods employed.

(rail yards, major airport), commercial, and manufacturing interests. The other towns are largely mining-metallurgical centers, though several have a considerable industrial representation (Map 99).

Aerial view of the Chambishi open pit copper mine
Separation and concentration facilities are seen in the foreground.

The Copperbelt has had development problems comparable to those of Shaba, some of which have already been noted. The provision of smelter fuel and electric energy presented numerous difficulties long before those occasioned by UDI in Rhodesia, as did the transportation of needed supplies and of the copper produced. Among labor-management problems have been the lengthy fight to remove the color bar and its racially based job specifications (finally won in 1960 when European miners voted to end it); the evolution from a largely migrant labor force to a more stabilized one (a declining problem as the average length of service has risen from 5.6 years in 1961 to 9.9 years in 1972 and the annual turnover rate has gone down to between 5 and 9 percent); how to respond to the sometimes exorbitant demands of militant labor unions, who have relatively little experience with or understanding of collective bargaining or of modern labor-management relations (a problem the independent government has met by imposing restrictions rather than further democratizing the relations); and the problem of Zambianizing the industry without affect-

Aerial view of the plant and works area at the Luanshya mine
A high-density township for workers is seen in the background.

ing its efficiency, requiring on the part of government a degree of pragmatic firmness that is not always politically palatable. Despite many tense situations and conflicting positions it is a tribute to the major parties concerned that the industry has continued to function effectively and to contribute increasingly to the dynamic if somewhat mercurial Zambian economy.

Other minerals produced in Zambia include:

(1) Coal, previously noted.

(2) Lead and zinc, mined by the Broken Hill Division at Kabwe, south of the Copperbelt. Mining began here in 1906 but the operations are now marginal (having required tax concessions to sustain production) and the main body is expected to be exhausted by 1983. A $31 million investment in special kilns completed in 1975 permits treatment of spoil banks, which will extend the life of the workings to 1990.

(3) Iron pyrites, coming from Nampundwe, 32 miles (51 km) west of Lusaka. This deposit, mined briefly in 1913–14, was reopened in 1970; it contains 16.1 percent sulfur and 0.63 percent copper; a sulfur concentrate is railed to Rokana for use in the smelter there.

(4) Gypsum, mined since 1973 in the Lochinvar game park at a rate of 24,000 tons a year and consumed domestically.

(5) Feldspar and fluorspar, from a new mine near Kariba and used by the glass factory at Kapiri Mposhi.

(6) Gold, from a small mine reopened near Mumbwa.

(7) Limestone, for the cement mills. In 1973 production was 978,000 tons.

Marginal deposits of manganese have been mined in Luapula province. Much exploration and prospecting is being carried on in various areas of Zambia and indications of numerous minerals have been

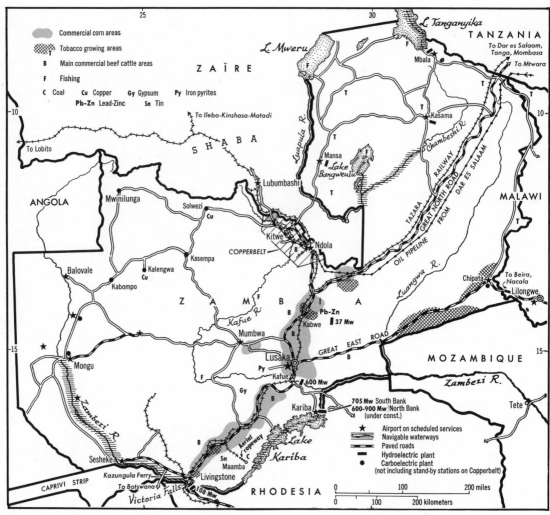

Map 98. Economic map of Zambia

found, including apatite, tin, tantalum, tungsten, and iron.

MANUFACTURING

Zambia has seen a dynamic growth of its manufacturing sector since breakup of the Federation. As is shown on Map 99, there is now a rather wide representation of consumer-oriented industries, many of which have only recently begun operations. The production index shows the several manufacturing sectors increasing output by factors of 2 to 10 from 1964 to 1972, while the combined index with 1969 as the base stood at 134.8 in 1972. The textile industry showed the most rapid increase as Zambia reduced its former reliance on Rhodesia and on other imports.

Lusaka, Kitwe, and Ndola are the leading manufacturing centers, and the vast bulk of the remaining establishments are situated either along the line of rail or in other Copperbelt towns. The reorientation of traffic may have an impact on industrial location, stimulating the rise of new centers in the

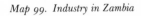
Map 99. Industry in Zambia

northeast, while Livingstone is hurt by finding itself at the end of the line instead of at the gateway to the country.

As the map shows, the types of industries represented are closely comparable to those of other tropical African countries except for a few factories catering to the huge mining industry, including those producing mining equipment and explosives.

While there are additional opportunities for new plants in Zambia, the growth of the manufacturing sector may not be so spectacular as it has been in the past decade because many of the larger installations are one-shot affairs, at least until considerable further growth in the economy has occurred. These include such industries as petroleum refin-

ing, sugar refining, and manufacturing of fertilizers and glass. Zambia is disadvantaged as far as export markets, for manufacturers are concerned by its landlocked position and by the high cost of labor as compared with neighboring states. At present only semi-manufactured copper products are exported.

CONCLUSION

Zambia has weathered an extremely difficult decade of independence with some remarkable achievements. The record in mining and industry has been impressive and efforts to prevent graft and to spread development to rural areas have been commendable. The country continues however, to face problems of actual and potential seriousness. Progress in the agricultural sector has been disappointing and the gains made have scarcely been commensurate with the expenditures. The rural-urban gap continues to widen and efforts have not been adequate to stem either the further concentration of activity on the line of rail or the massive inflow of migrants to the urban centers, many of whom live in shanty towns. An impressive road-building program and the success of the Intensive Development Zone schemes may ameliorate these problems somewhat, while completion of Tazara may stimulate new centers in the northeast. Unemployment is a serious and growing problem. Further Zambianization must be approached with care to make certain that locals are well qualified before they replace the skilled technicians, engineers, managers, and other specialists. Finally, continuing ethnic differences continue to make achievement of a Zambian nation a difficult problem.

CHAPTER TWENTY-ONE

Malawi

This small, often densely populated nation, formerly Nyasaland, became independent on July 6, 1964, a half year after dissolution of the Federation of which it was the weakest member. Considerable unrest occurred prior to independence and there were minor insurrections in 1965 and 1967, but President Hastings Banda has maintained firm control, suppressing any dissidence found to be developing. He has followed a conservative, pragmatic, paternalistic course both politically and economically. Malawi is one of the few African-led countries to develop a modus vivendi with white-ruled southern African countries, which has led to strong criticism from other African states and efforts by some to oust the country from the OAU. The President maintained, however, that Malawi could afford no other policy, because of the importance of migrant labor earnings and the dependence on Mozambique traffic. In 1974–75, however, South Africa was at least temporarily prohibited from recruiting mine workers in Malawi.

Malawi is one of the poorest of African countries with a per capita GDP estimated at $94 in 1972. Its physical and invisible trade balances have been increasingly unfavorable since 1964, and there are chronic budget deficits. The country's primary dependence is on agriculture; only about 150,000 persons are in paid employment within the country, while somewhere between 250,000 and 300,000 Malawians work abroad at any one time. The record of development since 1964 has, however, shown considerable improvement: the per capita GDP has grown by about 2.3 percent per annum despite a very high rate of population growth; the contribution of the modern agricultural sector to GDP increased from 15 to 35 percent between 1964 and 1970 while that of the subsistence sector declined from 43 to 16 percent; the annual budget deficits have been progressively reduced, lowering the reliance on Britain for direct subventions; and substantial progress has been made in the major economic sectors and in the provision of both the physical and social infrastructures. The country does remain dependent on foreign aid from an array of donors including South Africa for almost all of its development expenditures, and the problem of providing jobs for the increasing numbers reaching employment age is potentially serious.

POPULATION

Malawi's population was approximately 4.9 million in mid-1974, and is thought to be increasing by about 3.3 percent per annum. The total area of the country is 45,747 square miles (118,484 sq km), but the land area measures 36,324 square miles (94,079 sq km), giving a crude density of 135 per square miles (52 per sq km). Densities vary greatly, of course, ranging from about 470 per square mile (181 per sq km) in the south to 60 (23 per sq km) in the north. The average density per clear field pattern area is probably about 630 per square mile (243 per sq km). These obviously high densities help to explain the considerable land deterioration and the large number of applications for jobs (available on a contract basis) in other countries.

The Thyolo and Mulanje Highlands are

the most heavily populated parts of the country; their densities now probably exceed 900–1000 persons per square mile (347–386 per sq km). The region around Lilongwe on the central plateau is a second zone of high density. Only about 6 percent of the population is urbanized. Blantyre-Limbe, the chief commercial-industrial center, now has an estimated 146,000 people; Lilongwe, a fully planned city, which has been selected as the new capital because of its more central location, had only about 20,000 people in 1966, but may well have exceeded 50,000 in 1974; it is expected to reach 500,000 by the year 2000.

Roughly a quarter of the male labor force migrates, usually on 18- or 24-month contracts; about four-fifths go to South Africa and the rest to Rhodesia and Zambia. Remittances total between $4.8 and $9.4 million annually. While wages on foreign mines and farms are often in the lowest category, they are high by Malawian standards and workers often return to foreign jobs after less than a year at home.

Over 99.5 percent of the total population is African. Indians, who totalled 11,299 at the 1966 census, are employed in commercial, clerical, artisanal, and service positions, but have not been allowed to run stores in rural areas in recent years. About 7,000 Europeans reside in the country, playing important roles in estate agriculture, industry, education, missions, industry, and government services.

PHYSICAL BACKGROUND

A strip of land 560 miles (900 km) long and 50–100 miles (80–160 km) wide, Malawi is composed mainly of plateaus and highlands ranging in elevation from 3,300 to over 8,000 feet (1,006–2,439 m) plus a portion of the Great Rift Valley occupied by Lake Malawi in the north and by the Shire River in the south. The lake, 348 miles long (560 km) and 1,969 feet deep (600 m) stands at an elevation of about 1,553 feet (473 m), while the lower Shire Valley is only 120 to 300 feet (36–91 m) above sea level.

Malawi's precipitation levels, which vary greatly, are strongly influenced by topography; the highlands receive averages of about 30–60 inches (762–1,524 mm); but rainfall exceeds 80–100 inches (2,032–2,540 mm) on some steep slopes facing the rain-bearing winds, and is less than 30 inches (762 mm) in the lowlands. Year-to-year and within-season variations are considerable and greatly affect crop yields. In comparison to Rhodesia and Zambia, Malawi is favored by the percent of its area receiving adequate precipitation to support tillage agriculture.

AGRICULTURE

Over 90 percent of Malawi's population is dependent upon agriculture, which accounts for about 51 percent of the GDP and 94 percent of exports. About 13 million acres (5.3 million ha) or 56 percent of the total land area is considered to be cultivable, but less than a quarter of that is actually cultivated, despite the intense pressure in some regions. The government recognizes that agriculture must be looked to as the major sector for creation of new jobs, and feels that modernization of techniques can greatly increase production and raise rural standards.

About 82 percent of the land in Malawi is held under customary tenure, 16 percent is in public lands (including game parks, forest reserves, and town lands), and 2 percent is privately owned. For most of the population the man-land relationships involve the husband's moving to his wife's village upon marriage, matrilineal inheritance, and land allocation by village headmen. This system has broken down under the force of modern circumstances, but only to a limited degree.

Corn is by all odds the main crop; it occupies an estimated 78 percent of the total cropped area. Less than 5 percent of output is marketed, however. Other subsistence

crops include manioc, millet, sorghum, peanuts, pulses, and rice.

CASH CROPS

TOBACCO. A variety of tobaccos are grown in
Malawi and their exports account for about
40 percent of the total value of exports. Tobacco was introduced to the country in 1920,
most of the early production being flue-
cured leaf grown on European estates. Estates still account for over 40 percent of total
output and the vast bulk of flue-cured and
burley tobaccos, though some are farmed by
Malawi tenants under supervision of the
owners and Africans are being trained in the
more demanding methods required for
these higher-valued types. Malawians produce all of the dark fire-cured tobacco, used
for plug, twist, and roll tobaccos, which account for about 40 percent of total production by weight, and all of the sun/air-cured
crop, used mainly for pipe tobacco.

About 97,000 acres (39,000 ha) are
planted to tobacco, which is grown in many
parts of the country. The greatest concentration is in the Central Region stretching out
broadly to the west and northwest of
Lilongwe (Map 100); a portion of the flue-
cured crop is produced in the higher areas
of the Southern Region; and minor amounts
of Turkish tobacco are grown in the north.
All but the last are sold under the Tobacco
Control Commission at Limbe auction floors.
Output fluctuates considerably, but the
trend is distinctly upward.

TEA. First introduced to Malawi in 1878 by
missionaries, tea is now produced on about
40,000 acres (16,200 ha), mainly in the high-
rainfall Thyolo and Mulanje Highlands,
but also in the Nkhata Bay area, where tea
is irrigated by overhead sprinklers. Most
of the output comes from about 50 estates,
30 of which have their own factories, but
efforts have been made since 1966 to promote smallholder production; about 1,500
growers now account for about 6 percent
of the acreage and 2 percent of produc-

A tobacco field in the Southern Region
A variety of tobaccos are grown in Malawi with production
coming both from estates and smallholdings.

tion. The tea estates employ about 37,000
Africans, making them the largest employer
in the country. Tea accounts for about a
fifth of Malawi's exports; grades are lower
than those of East Africa and production
costs are adversely affected by the concentration of plucking in a relatively short period.
Output in 1973 was a record 23,500 tons.

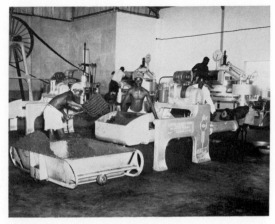

Workers in a tea factory at Mulanje
Tea estates are the largest employer of labor in Malawi.

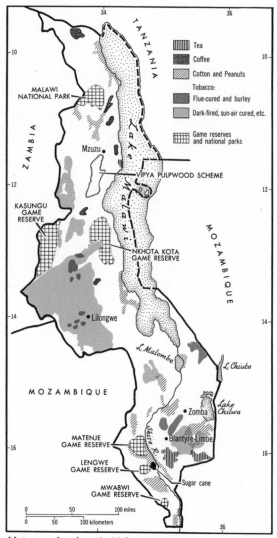

Map 100. Land use in Malawi

both to weather and insect attack, and the greatest need is the wider use of insecticides. The government aids the use of knapsack sprayers to reduce losses to the Red Boll-worm. Yields vary enormously among individual growers; a recent survey has revealed that 7.4 percent of the growers working 9.5 percent of the area in cotton produced 44.6 percent of the total crop.

Sugar is produced on a 12,000-acre (4,800-ha) irrigated plantation at Nchalo on the west bank of the Shire River; it employs about 4,300 Malawians and 73 expatriates. Most of the output has been absorbed domestically, but exports have gone to Zambia, and Malawi was assigned a 15,000-ton quota by the United States in 1973. Plans call for increasing the area in sugar to 21,000 acres (8,500 ha) by 1978.

In early postwar years it was hoped that Malawi would become an important exporter of tung oil, and estates were developed in four districts. Competition of other producing nations and from synthetics has led to the closing of all but one mill and at least temporary cessation of working the large CDC-supported estate on the Vipya Highlands. Rice production, sold on the domestic market, has increased rapidly in recent years.

AGRICULTURAL DEVELOPMENT PROJECTS

Major emphasis has been placed in recent years on several schemes designed to rationalize practices over large traditional farming areas. Most advanced of the schemes is the Lilongwe Land Development Program, largely financed by a $6 million IDA loan, begun in 1968 but based on earlier pilot schemes. When fully developed the scheme will involve about a half million people (c. 100,000 farm families) on 1.1 million acres (445,000 ha); at present about half of the area, divided into 20,000-acre (8,100 ha) blocks, has been brought under the scheme. The goals were to increase the marketable surplus of corn tenfold and to double pea-

OTHER CASH CROPS. Malawi exports a wide variety of other crops; peanuts, cotton, and pulses are the most important. Malawi peanuts are of excellent quality and a portion commands higher prices because they are sold as edible nuts rather than for oil; about three-fifths come from the Central Region but peanuts are grown in most parts of the country. Production of cotton is concentrated in the Lower Shire Valley and on the shores of Lake Malawi in the Central Region; output fluctuates widely in response

nut and fat cattle production; these goals had been exceeded by 1972 and tobacco sales had also doubled.

While the scheme involves a considerable number of changes, they are more evolutionary than revolutionary; farmers continue to grow crops and to use tools with which they are familiar. It has been shown that yields of corn can be at least doubled by use of synthetic varieties, fertilization, planting prior to the rains, proper spacing, and thorough weeding. Attention is also given to erosion control. These practices are promoted by extension workers on a ratio of 1:500 farmers, considerably below the original goal of 1:200. In addition, boreholes have been sunk, access roads constructed, market centers established, and a 161,000-acre (65,000-ha) ranch has been established. The scheme has been well received and is considered as a model for similar schemes in Malawi and perhaps other countries. Farmers were first somewhat suspicious, but the observed increases in yields were sufficiently convincing to gain increasing support. Payments on loans for purchase of seeds and other inputs have been made by 97 percent of the producers. Lands in the scheme are demarcated and farmers are invited to register titles, but this is on a voluntary basis because it was felt that compulsory individual tenure might discourage acceptance of the other aspects.

Two other comparable schemes are the Chikwawa Cotton Development Project, affecting 18,000 farmers on 530,000 acres (214,000 ha) in the south, and the Central Region Lakeshore Project with over 40,000 farmers on 640,000 acres (259,000 ha). Both of these have as their special objective an increased output of cotton, particularly by use of sprays.

Other programs include selection of *achikumbe* or progressive farmers for special assistance; training, by Israeli specialists, of Malawi Young Pioneers in agricultural techniques, home economics, and hygiene, with the expectation that they will either return to their home communities or help to develop settlement schemes; demonstration rice farms set up by Taiwanese advisers; the organization of eleven settlement schemes; and efforts to increase production of livestock products.

OTHER NONINDUSTRIAL SECTORS

Fishing is an activity of growing though minor importance, which contributes about $400,000 to exports. Of the total catch of about 25,000 to 35,000 tons, three-fifths comes from Lake Malawi, over a fourth from the lower Shire, and the rest from Lake Chilwa. Trawlers have been introduced on Lake Malawi, where modern facilities and a new cannery are found in the south; outboard motors and new nets are being used on Lake Chilwa, whose extent varies greatly from cycle to cycle.

About 91,000 acres (37,000 ha) have been afforested with exotic species, three-fifths in the Vipya Pulpwood Project. This scheme, started in 1964, is now supporting a pulp mill; its output is to increase to 100,000 tons by 1980.

Mining is of very minor importance in Malawi, though several bodies of potential interest are known, including bauxite on the Lichenya Plateau of Mt. Mulanje, rare earths near Balaka, and titanium sands along the shores of Lake Malawi. The bauxite deposit has been studied by several large companies but remoteness from the sea has discouraged exploitation; it has been suggested that aluminum might be produced, using electricity either from Cabora Bassa in Mozambique or from Shire River plants.

Tourism is in its infancy, but efforts are being made to attract more visitors by providing more lakeside resorts, improving certain roads, and adding facilities in the national parks (see Map 100). Much of the scenery is very attractive, but the country has not yet been discovered by large numbers or by agencies that could stimulate a greater influx. Most of the present visitors are Rhodesians or South Africans, but the need to proceed in military convoy across the Tete Road in Mozambique has discouraged automobile traffic.

INDUSTRY

Manufacturing is credited with 11.1 percent of the GDP, but it employs only about 25,000 workers (about 13 percent of all wage earners). Tea factories are the most important segment of the raw material processing establishments. Despite the small market, a considerable range of light industries is present and these permitted a 2.5–3.0-fold increase in industrial production from 1964 to 1973. The market-oriented establishments are heavily concentrated in Blantyre-Limbe, but Lilongwe may be expected to attract new factories (Map 101). The usual incentives have been offered, including participation by the Malawi Development Corporation.

The high cost of electricity was a deterrent until completion of a 24,000-kw hydroelectric plant at Nkula Falls on the Shire in the 1960s; a first-phase installation at Tedzani, four miles downstream, was completed in 1973 and the plant now has a capacity of 32,000 kw, which can be increased to 120,000 kw. The middle Shire drops 1,260 feet (384 m) in 50 miles (90 km) and at least 415,000 kw could eventually be developed.

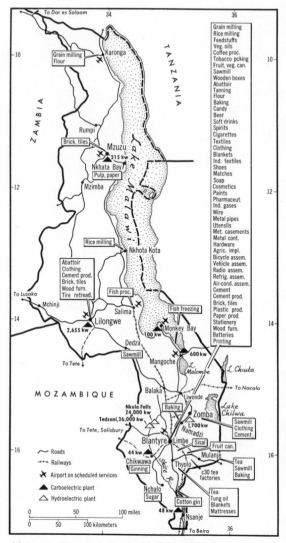

Map 101. Transportation and industry in Malawi

TRANSPORT AND TRADE

Malawi is served by two interconnected rail lines. The first runs from Salima Bay southward through Blantyre-Limbe to Beira in Mozambique; it handled almost all of overseas traffic until 1970, when a connection was made with the line in northern Mozambique terminating at Nacala. This line is advantaged by use of a fine harbor and avoid-

ance of congestion at Beira. Liwonde on the new line is to become the main terminal for lake traffic after the Upper Shire has been dredged to it, while an extension will probably be built from Salima to Lilongwe (Map 101).

Use of Lake Malawi as a routeway is restricted by the low level of traffic offered an inadequate port facilities; improvements are planned, including a shift of the main

Aerial view of Blantyre
Blantyre-Limbe is the leading commercial and industrial city in Malawi.

transfer point from Monkey Bay to Liwonde and construction of a protected harbor at Chinteche to serve the pulp factory. Output of that factory could raise lake traffic by about 100,000 tons; at present only about 17,000 to 23,000 tons are carried yearly (along with approximately 114,000 passengers) and the services run at a loss.

The road system has been quite inadequate, but it has been undergoing considerable improvements in recent years. The major roads in the south will soon be paved between the main centers and as far north as the new capital. The area north of that has been served by a single dorsal road, which crosses over into Zambia on one stretch, with a limited number of feeder roads leading to lake points. Now a lake road that will provide much better contact for several productive areas is nearing completion.

Air Malawi, which was spun off from the former federally owned Central African Airways, serves six internal centers and has international connections to seven African countries. Most migrant workers going to South Africa are flown in both directions. The airway normally runs at a profit.

Malawi's exports increased 2.4-fold from independence to 1972–73, when they totaled $85 million. Imports, however, increased 2.9-fold in the same period, thus enlarging the unfavorable balance. Reflected in these figures, however, is the substantial amount of development that has taken place. As already noted, Malawi gained after UDI in the carriage of transit traffic for Zambia, but this is not likely to remain at high levels after completion of the Tazara line.

Malawi is, in conclusion, a poor country and likely to remain so for some decades. Nonetheless its efforts to achieve higher levels have been commendable and the agricultural resource base permits some optimism. It is regrettable that very little effort has been made to reduce the burgeoning population growth, which tends to perpetuate the dependence on Rhodesia and South Africa to provide employment.

CHAPTER TWENTY-TWO

Angola and Mozambique

On April 25, 1974, the dictatorial regime that had ruled Portugal for close to 50 years was ousted in a military coup led by General António de Spínola, who had recently been the Deputy Chief of Staff and whose book *Portugal and the Future,*[1] published two months earlier, had caused a sensation by stating that the revolts that had been going on in the African territories since the early 1960s could not be suppressed by military means. The new government quickly began negotiations with the several liberation movements. Guinea-Bissau became independent on September 10, 1974, Mozambique was promised independence June 25, 1975, and Angola on November 11, 1975. Considerable unrest occurred both in Portugal and in the overseas areas, including several brief but bloody riots in Luanda and Lourenço Marques. At the time of writing many uncertainties existed regarding just how the transition would evolve, which could be reasonably peaceful or marked by bloody rebellions mindful of the early years of Zaïre.

The coup in Portugal was caused by weariness with the revolts in Africa. While the revolt in Angola appeared to have slackened in recent years, the liberation movement in Guinea-Bissau controlled large parts of that country, and Frelimo in Mozambique was widening actions in the north and in the Tete District. In addition to the personal sacrifices being demanded of the Portuguese, military expenditures occasioned by the revolts were taking up as much as 63 percent of the extraordinary budget and 23 percent

[1] António de Spínola, *Portugal e o Futuro* (Lisbon: Arcádia, 1974).

of the total budget, with additional contributions coming from the overseas budgets as well.

A brief account of Portuguese policies up to the 1974 coup will contribute to an understanding of the problems that the new countries are likely to face. Portugal had been in Africa for 500 years, which provided ample opportunity to develop several deeply ingrained myths that were explosively dissipated in the 1974 coup.

Portugal's African possessions were officially overseas states with local autonomy. There were, however, many characteristically colonial patterns that successive reforms instituted in response to the revolts had failed to erase. Portuguese policy long held that the overseas territories should pay for themselves, hopefully yield a profit on such investments as were made, provide raw materials and protected markets for metropolitan industries, and not compete with the metropole; investment was to be preserved primarily for Portuguese interests, partly to protect the political authority of the metropole. Policies with respect to the African population were paternalistic, the pater himself being poor and sometimes illiterate, devout, and a proponent of hard work as a moral and legal obligation. In postwar years two newly emphasized goals were the relief of metropolitan overpopulation by settlement of Portuguese in the states and the sale of territorial produce on world markets to contribute to exchange earnings for the escudo zone.

The uprisings in each of the territories led, on the one hand, to determined efforts to suppress them and, on the other, to a series of reforms without precedent in the long history of Portuguese colonialism. Labor laws abolished the requirements for cultivation and work in lieu of taxes, terminated the recruitment system through local administrators, and stipulated minimum wages, maximum hours, and improved conditions for rural workers.

Other reforms included a substantial extension of educational and health facilities, markedly enlarged development budgets, efforts to attract foreign investment (foreign investment had increased substantially, particularly in mining and water control projects, but a degree of caution and suspicion toward non-Portuguese interests persisted), the rationalizing of trade relations between Portugal and the overseas states, and the removal of restrictions on placement there of such industries as textiles, which might compete with those in the metropole.

In 1971 a considerably greater degree of autonomy was extended, and in 1973 elections were held to choose representatives to legislative councils; in Mozambique a nonwhite majority was elected and Africans made up a majority of the electorate in both Angola and Mozambique. While the states were theoretically empowered to govern themselves, make their own laws, negotiate contracts, and handle their own finances, Lisbon, in fact, maintained ultimate authority and direct control over defense and foreign policy. There was a difference of opinion as to just how large a step toward decolonization these moves were; liberals feared that they represented a move toward white-minority control à la Rhodesia, while conservatives contended that they represented an excessive granting of power to the territories, which they insisted must remain forever as integral parts of Portugal.

Special pride was taken by Portugal in the lack of racism in its overseas states, and it is true that there was a substantial number of persons of mixed race in them and that some Africans have been able to take advantage of the absence of legal barriers in employment. But Portugal's neglect of education—seen in the small number of black high school and university graduates—meant that whites continued to dominate the economies, while a substantial number of migrants from the lower strata of Portuguese society had taken jobs that in other countries would have normally been handled by Africans. And social discrimination, which all too readily translates into racial discrimination, had tended to increase in the burgeoning urban centers.

The direct impact of the revolts on existing economic production was not very large. Angola's coffee output rose even at the height of the revolt and, despite destruction of plantations that had yielded about 15 percent of the crop, it continued to expand in succeeding years. Many of the guerrilla activities in Mozambique were in areas of minor economic importance, though their discouragement to settlement in the potentially productive northern highlands and sporadic interruptions to traffic in the Tete area should be noted.

The political and financial status of Portugal helps to explain the failure to effect greater advances in Africa. One of the poorest countries in Europe, it is characterized by a considerable degree of rural underemployment. Politically, the country had seen little change since the 1920s; opposition was no more welcome or tolerated in the metropole than in its overseas states.

The cost of maintaining military forces in Angola, Mozambique, and Guinea-Bissau obviously prevented some more productive or socially desirable expenditures. On the other hand, the revolts awakened the Portuguese to the need for giving greater attention to the social infrastructure, expenditures of the military forces stimulated the local market, and quite a few physical infrastructural improvements—which had been neglected for decades and which are of continuing benefit—were undertaken with remarkable speed under the prodding of military authorities.

One of the goals not achieved was that of attracting large numbers of Portuguese settlers to the two states. The high cost—about $10,500 per family for those going to settlement schemes in the 1960s—prevented the achievement of any effective solution to Portugal's population problem, while Portuguese workers preferred to migrate to Western Europe instead of the little known, possibly unhealthy, less politically secure overseas states. Nonetheless, the European population of Angola increased from 44,000 in 1940, 79,000 in 1950, and 230,000 in 1965 to about 400,000 in 1974, giving that territory the second largest white population in Africa South of the Sahara. Mozambique's European population increased from 27,000 in 1940 to 102,000 in 1960, and possibly to 200,000 in 1974. In 1974–75 many whites left or were attempting to leave the two large territories; South Africa placed restrictions on their movement to that country.

Despite the relatively large European population in Angola and Mozambique there is a marked shortage of skilled labor and mana-

An alfalfa field at Cela, Angola
The goal of attracting large numbers of Portuguese to farming in Angola and Mozambique was never reached. Cela was one of the more successful of such schemes, however.

gerial talent in both countries. The migration from Portugal did not fill these needs satisfactorily and it will take some years before Africans can do so, despite a recent emphasis on technical education and the opening up in 1963 of universities in Luanda and Lourenço Marques.

While some have claimed that Portugal's major purpose in retaining control over its territories was to secure the economic benefits they provided, this position must be questioned in view of the heavy military expenditures, the deteriorating balances of trade and payments, and the loosening of metropolitan control over foreign exchange earnings, which previously benefited Portugal. Nor is Portugal's balance of trade with Angola favorable, as it once was. The precarious balance of payments of both Angola and Mozambique had required stringent control of imports, which contributed to inflation and created difficulties for some industries in securing equipment and raw materials from abroad.

Despite a record subject to criticism, in postwar years Angola and Mozambique did see notable advances in the development of their infrastructures, the output of goods and services, and the diversification of their economies.

Problems faced in the transition from colonial status to independence include potential conflicts among rival African groups, particularly in Angola; extraction of the Portuguese military forces; the future status of the relatively large Portuguese populations in Angola and Mozambique (some were strongly opposed to granting independence and would have favored a Rhodesian-type outcome; others will probably migrate to South Africa, Rhodesia, and Brazil; but many will wish to stay if it is at all feasible, and they can make important contributions to sustaining the economies of the two large countries); the extent of the continuing Portuguese presence in the areas; the reaction of South Africa, which has important interests in Mozambique; and the achievement of a successful transition from dictatorship to democracy in Portugal, which could lead to that country's joining the EEC. Angola and Mozambique are eligible upon independence to sign the Lomé Convention with EEC, as Guinea-Bissau has done.

ANGOLA

Angola is a roughly rectangular country with an area of 481,226 square miles (1,246,375 sq km) and a coastline about 1,031 miles (1,659 km) in length. It is approximately equivalent in size to Texas, California, and Arizona combined (or to Iberia, France, and Belgium) and it is fourteen times as large as the metropole. Although the Portuguese have been present along the coast for 500 years, there were only about 10,000 in the colony in the late nineteenth century, and sporadic efforts to settle the interior, which was not fully pacified until 1917, were largely unsuccessful. Real development did not begin until the 1930s, but has mainly been a product of postwar years.

PHYSICAL BACKGROUND

The vast bulk of Angola is an extension of the plateau in central Africa, which drops off to the Atlantic coast in two steps, widely spaced in the north, closer in the center, but almost joining in the south.

The coastal plain, with elevations to 1,300 feet (396 m), varies in width from about 12 to 100 miles (19–161 km), being greatest in the lower valley of the Cuanza just southeast of the capital. The presence of the cool, northward-flowing Benguela Current offshore moderates temperatures along the coast and reduces precipitation markedly. At Cabinda in the north the average rainfall is 25.4 inches (645 mm), at Luanda it is 13.3 inches (338 mm), at Lobito 10.5 inches (267 mm), and at Moçâmedes only 2 inches (51 mm). While these amounts limit rain-grown crops, there are opportunities for irrigation agriculture, which is further advantaged by the existence of some good soils along the river valleys.

The transitional lands and escarpments, ranging from 1,300 to about 3,300 feet (c. 400–1,000 m) in elevation, are varied in depth and aspect. In the north, the rise to the main plateau is gradual; in the central and southern sections, it is quite abrupt. Orographic precipitation contributes to the presence of a profuse vegetation in the north, which gradually evolves toward the south to savanna and steppe patterns.

The plateau itself has elevations from about 3,300 to 5,000 feet (c. 1,000–1,500 m) in the north and south except in the high Humpata Mountains in the south near Sa da Bandeira, but swells in the center to general elevations of 7,000 and 8,000 feet (c. 2,100–2,400 m) and over, especially along its western edge. The latitudinal position becomes dominant in setting the rainfall pattern, tropical rainy climate existing in the north and in the Cabinda Enclave, tropical highland savanna in the center, and highland steppe in the south. Annual precipitation is about 60 inches (c. 1,500 mm) in the north, 40–60 inches (c. 1,000–1,500 mm) in the center, and 25–40 inches (c. 600–1,000 mm) in the south. There are two well-marked seasons, the rainy season roughly from September to April, and the cooler, drier "Cacimbo" from May to September. The northern plateau is suitable for tropical crops such as coffee, oil palm, and cotton; inland from Lobito, the Benguela-Bié Plateau is appropriate for corn, peanuts, sisal, etc.; the southern Huila Plateau is mainly cattle country, whose southeastern portion consists of seemingly limitless stretches of sandy wastes interspersed with somewhat higher areas covered with thin forests. Temperatures on the plateau are everywhere moderated by elevation.

POPULATION

The population of Angola was estimated at 6.2 million in mid-1974, including about 400,000 Europeans and about 150,000 persons of mixed race. Over 90 Bantu-speaking ethnic groups make up the vast bulk of the Africans, though there are some Khoisan, believed to be descended from aboriginal races. In sharp contrast with most tropical African areas, many Europeans are small-holders, petty operators, and even unskilled laborers.

The overall density was only 12.9 per square mile (5.0 per sq km) in 1974 but it is, of course, variable from area to area. About a quarter of the country has under 2 per square mile (0.8 per sq km), the eastern and southern regions having the lowest densities. The density is about 32 per square mile (12.4 per sq km) on a third of the territory. Urbanization is not pronounced, but Luanda has seen a dynamic buildup in postwar years and its population rose from 230,000 in 1958 to 328,000 in 1970. Over 60 percent of the Europeans reside in the eight largest cities.

LAND USE REGIONS

Angola's economy is heavily dependent on extractive activities. About 85 percent of the African population is engaged in agriculture; 95 percent of exports are extractive products, of which 50 percent are agricultural, 42 percent mineral, and 7.7 percent fish and forest products. Only about 2 percent of the country is cultivated, however. A number of broad land-use regions, with differing economic and crop emphases, can be delineated.

THE CABINDA ENCLAVE. Measuring 2,794 square miles (7,236 sq km) in area, Cabinda may be considered as a distinct region not only because of its separation from the rest of Angola but also because of its significance in the production of tropical timber (Map 102). Almost all of the timber exported (c. 1.5 percent of total exports) from Angola comes from Cabinda, where forests cover the bulk of the interior and 84 percent of the whole enclave and where precipitation exceeds 50 to 60 inches (c. 1,250–1,500 mm)

a year. Some coffee and palm products are also exported, but Cabinda is now noted for having the main Angolan oil field and a promising phosphate deposit (see below).

NORTHERN ANGOLA. The northern section of Angola has a relatively undeveloped coastal lowland, a fairly broad and variable-surfaced subplateau, which drops into the Cuango Valley on the east, and a plateau which is less extensive than in central and southern Angola. The vegetation is steppe-like on the coast, luxuriant on the subplateau, and sparser on the somewhat monotonous plateau, which has only small pockets of workable timber.

The economic mainstay of the north is coffee, most of which is produced at elevations ranging from 1,300 to 5,000 feet (c. 400–1,500 m). Coffee has been the prime export of Angola, accounting for about 35 percent of the total, but petroleum moved into first place in 1973. It enjoyed a remarkable boom in postwar years; acreage in coffee rose from 88,000 (35,600 ha) in 1938 to over a million (400,000+ ha) after 1961; production increased from 19,000 tons in prewar years to an average of 48,900 tons in the period 1946 to 1951, and an average of about 211,000 tons in recent years, but further increases may be restricted by agreed quota.

Angola ranked during the 1950s as Africa's leading coffee producer, but it is now usually second to the Ivory Coast. European plantations and estates account for about three-fourths of production, with one very large plantation (148,200 acres, 60,000 ha) accounting for a quarter of the total. The emphasis on settlement of Portuguese brought several thousand small *fazendéiro* into coffee production on 125–250 acre (50–100 ha) plots to which title was secured if 10 percent had been developed in five years. Nearly a quarter of the total output comes from about 58,000 African smallholders, who were encouraged to join cooperatives in the coffee regions as a way of meet-

Coffee clearing on the Uige Plateau, northern Angola
Coffee is the main agricultural export of the country, which is the second-ranking producer in Africa.

ing the objections to practices that contributed to the revolt; some 130,000 Africans are also employed on European holdings.

The main coffee-producing areas in Angola are on the Uige Plateau, particularly around Carmona. A second concentration of estates is around Gabela inland from Amboím, while arabica coffee is produced on the highlands along the Benguela Railway. The vast bulk of Angolan coffee is robusta of good grade, much of which is used for the production of instant coffee.

Medium-staple cotton is grown both on European estates, which account for about 73 percent of the total output, and by perhaps 40,000 African smallholders. Up to 1961 Africans in the northern coastal area and along the Luanda railway grew coffee on a concession basis, which was considered exploitative, and sold the output at fixed prices below the world market level in order to favor the metropolitan textile industry. Strikes and uprisings in the Malange area resulted in abolition of compulsory plantings, the pricing system was changed, and after 1965 restrictions on the local textile industry were removed. These steps led, after a large-scale abandonment of production by

Africans in the early 1960s, to a renewed interest in the crop, whose output increased sevenfold from 1961–65 to 1971–73, when cotton came to rank second to coffee among Angolan agricultural exports.

Sugar is grown on a plantation basis along the lower courses of several rivers in Angola, both in the north and in the center; the main producing areas in the north are in the lower Dande and Cuanza valleys, relatively near Luanda. Until the early 1970s sugar was exported, but domestic consumption has increased with such rapidity that it is now imported from Mozambique despite higher production levels in Angola.

An important irrigation project is being developed in the lower Cuanza Valley about 18 miles (29 km) from Luanda, based upon the Cambambe Dam. The first phase involved drainage, flood control, and irrigation of 250,000 acres (100,000 ha), a considerable part of which is used to produce foodstuffs for the capital; eventually, about 750,000 acres (300,000 ha) may be controlled. The M'Bridge River Basin is the site of a large holding producing kenaf and abroma fibers plus coconuts and coffee.

The northern region also produces bananas, palm products, manioc, kapok, gums, wax, and other tropical products, mostly in small quantities. Banana exports have risen from 6,000 tons in 1964 to about 80,000 tons in 1973; saturation of the market in Portugal has led to efforts to find alternative outlets. Attention is also being given to improving the quality of the fruit and its handling, including the provision of special facilities at Luanda and at Lobito, near which much of the crop is produced. The area around Malange ships a considerable amount of market garden produce to Luanda, while tobacco is grown by Europeans in the same area and by Africans around Lucala, most of it going to factories in the capital.

MIDDLE ANGOLA. The coastal plain is quite narrow in this section of the country, but sugar plantations are found near Catumbela

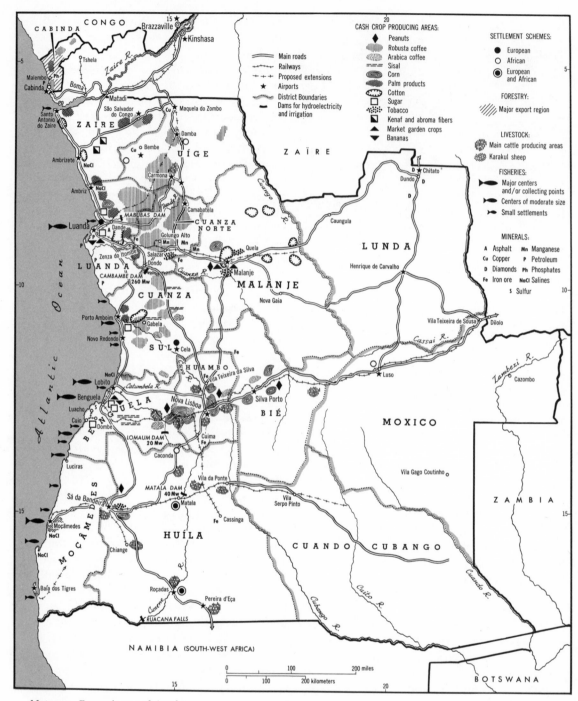

Map 102. Economic map of Angola

and in the Caporolo Valley, which drains into the Atlantic at Cuio. The central uplands, which contain the highest part of the country, form the Angolan watershed between the Zaïre and the Zambezi-Okavango River systems. Precipitation is quite high on the more elevated parts of the Bié Plateau, but gradually diminishes to the interior. The areas around Nova Lisboa and Silva Porto have been particularly favored for European settlement.

In the main producing belt, along the Benguela Railway, the land varies from rolling to extremely hilly with scattered inselberg and rough, infertile hillsides. Sisal, long the second most valuable export crop, is produced on European estates along the line (Cubal and Ganda are two of the principal centers). Sisal is also grown on the northern coastal plain. The marked decline in world prices for sisal forced many small growers out of business, but about 200 growers employing over 20,000 workers on some 300,000 acres (120,000 ha) continue to produce the fiber, which now ranks third among agricultural exports.

Corn, the fifth-ranking crop export, also comes mainly from the line of rail on the Benguela Plateau, though some comes from the Malange area in the north. The corn belt of the Benguela Plateau has moderately heavy red soil of low fertility, which is permitted to revert to bush after a few years of cultivation. About 95 percent comes from African farms. Yields are only half of the level in the Malange region, and total production fluctuates greatly, depending on seasonal conditions.

As noted earlier, arabica coffee is also grown along the rail belt. Peanuts, sesame, potatoes, wheat, beans, and peas are also produced in this section of the country. Beeswax is collected in the interior from crude, barrel-shaped, bark hives fastened in the trees. While three honey and wax crops can be collected annually, the crude methods employed cause considerable loss.

Livestock are relatively less important than in the south, but a factory at Nova Lisboa produces an unusual variety of animal products ranging from canned meat, soups, and soap, to pharmaceuticals and fertilizer.

SOUTHERN ANGOLA. The main agricultural concern of southern Angola is livestock raising, which is not yet well commercialized despite a threefold increase in cattle in the decade 1961–71 and big increases in the numbers of goats and sheep. The most important area is on the plateau in southwestern Angola in a zone extending from Vila Pereira d'Eca northward to the Benguela Railway. Some wells have been bored in the southern livestock area; a modern abattoir and deep-freeze plant is situated at Sá da Bandeira where a number of European ranches are located, a large German-owned ranch has been constructed near Vila da Ponte to raise cattle for corned beef, and cold stores have been built at Moçâmedes. Karakul sheep are raised in that area.

SETTLEMENT SCHEMES

A considerable effort, particularly beginning with the first six-year plan in 1953, was devoted to agricultural development schemes in various parts of the country. At first, the prime objective was to attract Portuguese colonization; this remained important, but Africans were later being trained to join the projects. A second important objective, arising after the revolt began in 1961, was to provide supervision of and security for African rural residents through gathering settlements around police posts, where it was also easier to provide social and commercial facilities. These agglomerated security villages, called *aldeamentos,* worked well in Uige, where 17 stable communities produced coffee successfully; their satisfactory functioning in the drier east, where production and sale of cash crops is more difficult, was less assured.

Details of the settlement projects varied considerably from scheme to scheme and

Santa Comba, a settlement scheme
Despite heavy subsidization, most of the schemes designed to attract Portuguese settlers were not successful.

evolved over time as the shortcomings of the early arrangements were revealed. Usually the Portuguese migrant family was provided with free transportation, a house, a farm holding that had been all or partly cleared and planted, work animals and cows, tools, and a cash subsidy for the first two years. Some settlers, after severe trials, achieved considerable success; many failed and left the projects; but, as noted earlier, the total number of Portuguese involved was much too small to have a noticeable impact on overcrowding in the metropole and the per capita cost to the government was totally out of proportion to any gains.

Settlement schemes are situated in the Loge Valley and at 31 de Janeiro (mainly for Africans) in the north, in the Moxico District near Luso and in the Bengo River Valley, just north of Luanda. The last provides foodstuffs and livestock for Luanda. One of the most important projects is located around Cela in a well-watered valley east of Novo Redondo. The Cela Scheme, which was started in 1952, has gone through sev-

eral phases; it nearly collapsed in 1963, but completion of a paved road to Luanda opened up new possibilities and the more successful farmers earned good incomes by selling potatoes, other vegetables and fruit, and dairy produce; this permitted an investment in tractors and other mechanical equipment and the hiring of four or more African workers per farm.

A very large scheme was planned in the southern Huila District, based upon control of the Cunene River, which flows southward across the district and then westward along the border with South-West Africa to the Atlantic. The first development was at Matala in 1954, where water from the dam is used to irrigate the alluvial flats on the right bank to produce tomatoes, tobacco, and wheat as cash crops and potatoes and corn as the main subsistence crops; a sixth of the cash crops goes to repay the capital costs of the prepared farms. There have been problems finding adequate markets for the produce and plant diseases have occurred. In 1969 a Portuguese–South African agree-

Aerial view of the Cela settlement scheme

Mending nets at Maçâmedes
While Angola ranks third among African countries in fish landings, the industry is less modern than those of South and South-West Africa.

ment was signed looking to the eventual control of the entire river, at an estimated cost of $640 million. The first phase involves construction of two dams, at Gove and Ruacana Falls, irrigation of about 50,000 acres (20,000 ha) at Matunto in Angola, and a large area in Ovamboland, South-West Africa, and installation of a 240,000–300,000 kw plant at Ruacana Falls, whose power will be used by the mineral industries at Tsumeb and Grootfontein and the fishing industry at Walvis Bay, all in South-West Africa. Later phases, whose schedules have not been set, would permit irrigation of 370,000 acres (150,000 ha), provide water to make 860,000 acres (350,000 ha) available for grazing, and support, at 15 dams above the Ruacana Falls, hydroelectric stations with a total capacity of about 297,500 kw. Whether an independent Angola will proceed with this scheme remains to be seen.

FISHING

Angola ranks third after South Africa and South-West Africa in fish landings by African countries, and fish products account for about 5 percent of total exports by value. Angola is advantaged by the Benguela Current which, like other cool ocean currents, has a large fish population, and by the existence of continental shelves considerably broader than for most of Africa.

While a large number of coastal settlements are engaged in fishing (Map 102), Moçâmedes district ports account for about 70 percent of the catch and contain the majority of the processing establishments. The Lobito/Benguela range ranks second with 20 percent of the catch, with Luanda landing the bulk of the remaining 10 percent. Mackerel and sardines make up most of the catch, over half of which is processed in fish meal factories; a third is dried or cured, and the remainder frozen or canned. Efforts to increase the percentage of frozen fish marketed, because they yield a substantially higher return, have been reasonably successful.

Angola's fishing industry does not compare in modernity to South Africa's or South-West Africa's. Over 6,000 craft participate, but only about 350 are reasonably large and modern and there are very few deep-sea trawlers, although Japanese, Russian, and Spanish trawlers get excellent catches of fish and shrimp operating about 50 miles (80 km) offshore. Recently a foreign-owned floating factory has been processing fish-oil and meal in Baía do Cesar in the south.

Except for diamonds, mining did not become important until after World War II. Large areas of the country have still not been well studied, but the geology of the country and finds to date suggest that minerals will become of increasing importance to the Angolan economy.

Diamonds were long the most valuable mineral. In the period 1920–40, they were the main financial support of the territory and they still rank third among exports with about a ninth of the total value. Angola ranks fourth among African diamond producers with about 5.5 percent of the total caratage; the value percentage, while not available, is higher because about 60 percent are gem diamonds. The only operation at present is conducted by the Angola Diamond Company (Diamang), the largest employer in the country, with about 600 Europeans and 28,000 Africans. Diamang's concession over most of the country expired in 1971, when it was reduced to 19,300 square miles (7,450 sq km) around Chitato in the northeastern Luanda district; De Beers took a 50 percent share in the same year. Since then two American Companies and one from South Africa have been granted concessions.

Petroleum was discovered in Angola in the mid-1950s on the coastal plain near Luanda and about 90 miles south. Oil from these fields is delivered to the refinery at Luanda, which has been expanded to a capacity of 1 million tons. But dynamic growth began with the discovery by a Gulf Oil subsidiary of a large deposit off the Cabinda Enclave; it is this field which has contributed to the rapid increase of production and exports in recent years, total output having risen from 114,000 tons in 1965 to 7.6 million tons in 1974, with exports exceeding coffee by value. In 1973, work began at Malembo on a

$6.4 million deepwater berth, linked to the mainland by a 5,578-foot (1,700-m) jetty and capable of loading 100,000-ton tankers. The Cabinda find stimulated prospecting, particularly off the mouth of the Zaïre, and led to a scramble for concessions along most of the coast.

Iron ore is the third ranking mineral and fourth ranking export of the country. First produced in 1957, output increased to 1.2 million tons in 1967 and 6.2 million tons in 1971. The Lobito Mineral Company mines iron ore at several sites in the Cassinga area, from which it is railed to Moçâmedes where mechanical loading facilities have been provided. High-grade (64 percent) ores at Cassinga are likely to be exhausted by 1980, but plans call for pelletizing or beneficiating about 5 million tons a year of medium-grade ore; two beneficiating plants with a combined capacity of 1.6 million tons are now operating at Cassinga South. About 70 percent of the exports have gone to Japan and the rest to Western Europe, but some fears have been expressed that increased shipments from Australia may reduce Japanese interest in the years ahead.

The Manganese Company of Angola mines and pelletizes iron ore at the Cassala-Quitungo deposits; although the high-grade deposits have largely been exhausted there, low-grade taconites are now worked and a contract calls for shipping 2 million tons a year to Japan in 150,000-ton vessels beginning in 1976. Further development there will probably result in upgrading and extending the Malange-Luanda line. Numerous other iron-ore bodies have been found in scattered areas of Angola.

The other minerals of present interest to Angola, none of which figure importantly in exports, include:

(1) Copper, found in several areas, including northern Mavoio.
(2) Gold, which comes from the M'Popo area south of Tchamutete. Deposits

have also been reported from other regions.

(3) Uranium, under study by the West German Urangesellschaft.

(4) Phosphate rock, found in a large body in the Cabinda Enclave.

(5) Sulfur, discovered in promising quantities in a gypsum strata near Benguela.

(6) Manganese, from North Cuanza and Malange districts.

(7) Asphalt rock, mined at Cacuaco, near Luanda.

(8) Salt, produced at many salines along the coast and marketed domestically.

Occurrences of many other minerals have been noted and geological and prospecting surveys are progressing in several parts of the territory.

INDUSTRY

The pattern of manufacturing development in Angola is quite comparable to that of other tropical African countries. The gross value of output increased an average of 18 percent a year in the decade 1961–71, when it totaled about $42 million. Raw material processing remains important and has increased with the buildup of extractive activities; but the real growth has been in import-substitute industries. The food and beverage sector accounts for about two-fifths of gross output, textiles for an eighth. A wide range of incentives is available to attract industry, and foreign enterprises have been more aggressively sought in recent years, particularly those that can bring know-how not now available. Red tape and import restrictions are negative factors. Luanda is by all odds the leading center, particularly for consumer-oriented establishments.

Hope for continuing growth of industry is based in part on hydroelectric developments. The largest installation is the Cambambe Dam on the Cuanza River about 110 miles (117 km) east of Luanda (Map 102). The river drops 3,300 feet (1,000 m) in a 60-mile (100 km) stretch and the dam represents the first stage of its harnessing. With a height of 279 feet (85 m) the dam permits the utilization of a 374-foot (114-m) head and now has a capacity of 260,000 kw. In the future, four dams upstream could provide an additional 2 million kw capacity. Power from Cambambe is transmitted to Luanda and to Malange.

The Lobito-Benguela and Nova Lisboa areas are served by the 20,000-kw Lomaum plant on the Catumbelo. The 72-foot-high (22 m) dam at Matala on the Cunene, with a capacity of 40,000 kw, provides power for Moçâmedes, Sá da Bandeira, and Porto Alexandre. Work was begun in 1973 on a $17 million installation at Chicapa to supply Diamang and Henrique de Carvalho, capital of the Lunda district.

The total hydroelectric potential of Angola is very large, estimated at a production level of 185 billion kwh. But large-scale installations would depend on attracting power-consuming industries such as aluminum smelting. Total and per capita consumption remain low, though they are in the upper third of tropical African countries.

TRANSPORTATION IN ANGOLA

One usually thinks of transportation as just a part of the domestic infrastructure of an African country, but in Angola—and even more so in Mozambique—railways and ports are also significant as earners of foreign exchange, because they handle transit shipments for a number of the most highly developed interior areas of the continent. Both territories are also unusually favored, especially by African standards, by the existence of a number of excellent natural harbors along their lengthy coasts. While Angola has only one rail line and one port now handling transit shipments, there are long-run possibilities for increasing such movements, both for Zaïre and for Zambia. The main features of Angolan transport may be summarized by noting the ports and their connecting routeways, starting in the north. The Cabinda Enclave is served by a small port at Cabinda and will soon have the Malembo pier for loading ore.

Angola has most of the southern bank of the maritime Zaïre River and maintains a small port establishment at Antonio do Zaïre, downstream from Matadi. North and south of Luanda, a number of small lighterage ports (e.g. Ambriz, Ambrizete, and Porto Amboím) tap a portion of the coffee trade of northern Angola. The last has an 80-mile (129-km) rail line extending to Gabela.

Luanda has a large natural harbor protected by a sandy spitlike island now joined to the mainland by a concrete causeway (Map 103). Deepwater facilities were first provided in 1945 when the present rectangular pier was completed, giving access to seven large and two small oceangoing vessels. With a traffic tonnage of 1.7 million in 1970, Luanda ranks first in the value and volume of general cargo. It exports over 80

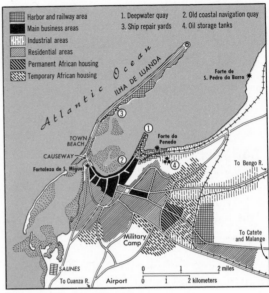

Map 103. The port of Luanda

percent of the country's coffee and a variety of other agricultural produce and is slated to become the second-ranking port for domestic mineral shipments, with special gear being installed for loading of iron from Cassala-Quitungo. The large European population of this capital city and its significance as the main manufacturing center account for the high proportion of imported goods landed at Luanda.

The railway serving Luanda was started in 1886 and reached its present terminus of Malange, 263 miles (423 km) inland, in 1909. For many years there was talk of extending the line into Zaïre, but this never came to fruition. The line has been operating well below capacity, in part because of increased competition from road haulers; it will have to be upgraded and extended to handle the iron ore.

Lobito's harbor, about midway along the coast, is protected by one of the sandspits that are among the most striking features of the Angolan coast. Its deepwater quays are arranged in an L-shape and are being extended to take 10 vessels; it handles about 1.7 million tons a year, but increased ship-

Part of the central business district and the main quay at Luanda
The capital, the main commercial and manufacturing center, and the residence of a large portion of the European population of the country, Luanda is also the leading general cargo port.

Aerial view of Lobito
The second main port of Angola, Lobito is the terminus of the Benguela Railway and handles large tonnages of metals and minerals for Zaïre and Zambia. Part of the quay space is along the protecting spit, which is such a common feature of Angolan harbors.

ments for Zambia may increase that level. The Benguela Railway permits Lobito to rank as the second port in tonnage traffic to and from Zaïre. Started in 1903, it was not connected with the line in Shaba until 1931. Extending 838 miles (1,348 km) from Lobito to the border, the railway follows a difficult route to the plateau which has, however, been improved with construction of the so-called Cubal variant. Wood, taken from extensive eucalyptus plantings along the line, is still used as a major fuel.

The potentialities of the Benguela line and of Lobito were largely unused until the 1950s because of the Belgian preference for the national route and the long-standing agreement of the Zambian copper companies to use the Rhodesia Railway. Congestion on the route to Matadi, plus the obvious advantage of eliminating rail-river transshipments, finally led to much heavier use of the direct line from Shaba to the coast. Low-value, bulk shipments of maganese ore from Shaba simply could not economically use the Zaïre routeway, though copper ingots could be shipped via Matadi.

Moçâmedes handles about 6.4 million tons, but 97.5 percent of that is exports, mostly iron ore. The city, situated in a fairly open bay with several inner indentations, was first provided with deepwater facilities in the mid-1950s. It has a substantial quayage for general cargo and fishing operations plus mechanical ore-loading facilities capable of loading 100,000-ton vessels at a rate of 3,000 tons per hour. The rail line from Moçâmedes was constructed to Sá da Bandeira between 1905 and 1923, and much later was extended to Serpa Pinto, a total distance of 469 miles (798 km), while a 56-mile (90-km) spur serves the Cassinga iron mines.

Other harbors of southern Angola are only partially developed, and the aridity of their hinterlands does not encourage much beyond fishing facilities. Porto Alexandre, which has an excellent harbor enclosed by a sandspit, is a leading fishing center; Luçiras also has a large deep harbor; while Baía dos Tigres in the extreme south has a vast harbor, again protected by a sandspit.

Inland waterways are of very minor importance in Angola. The Cuanza River is navigable for 120 miles (193 km) for boats drawing up to 8 feet (2.4 m), but the Luanda rail line and trucks have largely replaced it. The lower Dande is used on a 40-mile (64-km) stretch to transport raw sugar and palm products from plantations to the lighter port of Barra do Dande.

The roads of Angola, despite marked improvements in the last decade designed to provide paved connections to all district capitals, are still very inadequate. The revolt and the difficulties of agricultural schemes with poor linkages to market led to a greatly stepped-up road construction program, but many roads remain impassable in the rainy season. Angola has a modern and well developed air service, which was built up in part because of the poor surface connections in this large country.

MOZAMBIQUE

Despite contact with Portugal dating from discovery in 1489 and an established Portuguese presence since 1498, Mozambique is one of the less developed of African nations. It is, however, erroneous to equate presence with interest in the area. Except for extraction of gold and silver from the mines of Manica, development of Mozambique itself did not begin until four centuries after the first contacts; this lack of interest is explained by Portugal's greater concern with the lucrative Indian trade and with the colonization of Brazil, the presence of warlike tribes in Mozambique, the competitive Arab presence on the coast, and the lack of capital.

ECONOMIC GEOGRAPHIC KEYNOTES

RELATIVE LACK OF DEVELOPMENT. The relative underdevelopment of Mozambique is revealed in a variety of ways: by the existence of large, almost untouched areas, by a very inadequate infrastructure except for a few favored areas, by the low wage level and average income of the inhabitants, and by a low level of exports. Only in cashew nuts does Mozambique hold world rank.

SLOW NONAGRICULTURAL DEVELOPMENT. The dependence upon agriculture among domestic economc activities is very high. Agriculture contributes about 45 percent of the GDP and over 90 percent of exports, and about 90 percent of Africans within the province are engaged in agriculture. Other economic sectors have been relatively slow to develop, though accelerated progress has been made in manufacturing in the past decade, the huge Cabora Bassa hydroelectric installation will be the largest on the continent, and there are interesting prospects for mineral production.

Mining. Coal is now the only mineral produced in significant quantities, with output running about 325,000 tons a year. Mining of coal takes place in the Moatize field near Tete in the Zambezi Valley where easily accessible, good-quality reserves are estimated at 700 million tons (Map 104). Plans were announced in 1973 to open a new mine in the Chapunga area, which would increase output to about a million tons. Coal is also found near the Limpopo Railway and around Maniamba in the Nyasa Basin. Other minerals produced in small quantities include bauxite, beryllium, columbite-tantalite, copper, gold, rare earths, asbestos, gemstones, mica, and salt. Iron ore deposits are known in several districts and negotiations have been underway with a Japanese firm to produce 5 million tons of high-quality ore a year from an open pit near Namapa in Moçambique District. A $50 million investment would be required, including the cost of constructing a line to Nacala, and of docking facilities and a mechanical loader there. Another iron deposit may be opened in the Beira area when Cabora Bassa power becomes available. The search for oil has been underway since 1948, but thus far the only finds have been of gas, including a 600 billion cubic meter deposit at Pande, 60 miles (100 km) from Beira. As in Angola, prospecting for a variety of minerals has been stepped up in scattered parts of the territory.

Fishing. This industry has also been poorly represented, although resources in the Mozambique Channel are certainly adequate to sustain a much larger fleet. Except for export of some shrimp and lobster tails, the catch has almost exclusively been for local consumption, mainly in Lourenço Marques. But ten leading Portuguese fishing corporations have recently been investing in the industry, new canning factories have been installed along the coast, and a processing establishment has been started at Porto Amélia. An American company has also invested about $30 million, mostly for trawlers which will use Mozambique ports.

Forestry. Much of the better forested area is too remote to have encouraged exploitation. The main commercial area is situated along the line of rail north of Beira. A cellulose factory is planned for Chimoio by 1976, which will require extensive afforestation with eucalyptus and pine.

Manufacturing. The industrial sector is considerably smaller than Angola's, though it saw a fourfold increase in production in the decade 1962–72, at the end of which it accounted for 17 percent of the GDP. Important raw-material-processing plants include sugar mills, cashew nut decorticators, vegetable oil mills, sisal decorticators, sack and rope manufacturers. A considerable range of market-oriented industries is found in Lourenço Marques (L.M.) and to a lesser extent in Beira. Most establishments are quite small, the main exceptions being the sugar mills, cement factories and breweries, a textile mill at Vila Pery, the refinery at Matola near L.M. and a fertilizer factory under construction outside the capital.

THE ASSETS OF LOCATION. Mozambique's location with respect to its extranational hinterland is important. The territory capitalizes on this asset in three main ways: through transit traffic on its railways and at the ports of L.M., Beira, and Nacala, through receipts of Africans migrating to Rhodesia and South Africa for employment, and through the tourist trade which attracts thousands of people from these countries, particularly in the winter months. The number of tourist arrivals increased from 150,000 in 1964 to 583,000 in 1970, and in 1972 tourist earnings were rivaling transit traffic as a source of exchange. A fourth asset of location will soon be added when the bulk of electricity from Cabora Bassa is transmitted to South Africa. Mozambique's physical trade balance is very unfavorable, but earnings from these activities make very significant contributions to invisible earnings and permit, with capital inflows, a favorable balance of payments.

The railways and ports of Mozambique are the largest revenue earners in the country. In addition to its strategic location Mozambique is also advantaged by an irregular coastline with four harbors (L.M., Beira, Nacala, and Porto Amélia) fairly well spaced along its 1,700 miles (2,735 km), which can be entered by average deep-sea vessels.

While marked attention has been given to developing and modernizing the rail and road facilities handling transit traffic, the internal transport net has been relatively neglected. Until very recently, Mozambique to all intents and purposes has been divided into a series of east-west zones dependent on specific ports and on coastal or air movements for connections with the rest of the country. Roads are characteristically poor and there is only a sparse network, but the revolt led to a massive construction effort designed to add 3,850 miles (6,200 km) of new roads between 1971 and 1979, including a 620-mile (1,000 km) road between Beira and Mocimba da Praia in the extreme north, of obvious interest to the military. A 7,842-foot (2,390-m) bridge over the Zambezi has been constructed for this main national road.

Migrant labor earnings are another impor-

Aerial view of Lourenço Marques
The very much off-center capital of Mozambique owes most of its importance to the transit traffic it handles for South Africa and Rhodesia.

tant source of income for the country and an exchange earner for the government. It is difficult to estimate the total number of migrant laborers, because a considerable number move to adjacent countries clandestinely and others have become more or less permanent residents of South Africa or Rhodesia. About 85,000 Africans are recruited yearly for the South African mines under the Mozambique Convention (first signed in 1909 and continued under later modifications). Recruitment is handled by the Witwatersrand Native Labour Association (Wanela), which is responsible for the worker from the moment he signs his contract until his return home, though Portuguese labor officials maintained offices on the Rand to observe conditions. Under these arrangements, a portion of the pay of Mozambique Africans is withheld and paid to the Mozambique government in gold; that government then pays the returning worker in local currency. Mozambique gains not only in exchange but also through the expenditure of withheld earnings in Mozambique (about 83 percent of wages are received as deferred payments) and from several fees charged to the contract workers. In 1972 there were

132,000 officially registered Mozambiqui in South Africa plus an undetermined number of clandestine migrants probably totaling at least 30,000. The number of cross-border movements to Rhodesia has declined markedly since 1960. An act predating breakup of the Federation prohibited non-Federal Africans from working in the main urban centers with the exception of Umtali, close to the Mozambique border. The number of inmigrants from Mozambique to Rhodesia declined from 55,000 in 1957 before the act to 12,400 in 1962, and is now only about 4,000 a year. But 119,000 Mozambique Africans were counted in the 1969 Rhodesian census, suggesting that many were remaining there in order to retain employment. While migrant laborers make a significant contribution to Mozambique's income and currency exchange, the system has a somewhat unpalatable exploitative character and is objectionable in that it treats laborers as commodities subject to barter and because the absence of some 420,000 or more able-bodied males must reduce Mozambique's own domestic economic potential considerably. That earlier abuses have been progressively eliminated, that most laborers apparently look upon work on the Rand as an adventurous way of proving their manhood, and that employment abroad is favored over working in the country do not entirely offset the undesirable aspects of the system.

PHYSICAL OVERVIEW

With an area of 303,073 square miles (784,959 sq km), Mozambique is seven times the size of Portugal and about as large as Texas. It is about 1,220 miles (1,960 km) long and varies in width from 50 to 718 miles (80–1,155 km). Topographically, it may be broadly divided into three zones:

(1) The coastal belt, one of the largest in Africa, covers about 42 percent of the country. Narrow in the north, it widens constantly to the south, ex-

tends far inland in the Zambezi Valley, and includes most of the country south of Beira; the line of the great rift valley ends at the coast and is marked in Mozambique in the Shire Valley.

(2) A transitional zone of hills and low plateaus ranging from about 500 to 2,000 feet (c. 150–600 m) in elevation and covering about 29 percent of the country.

(3) Plateau and highland regions with average elevations of 3,000 feet (900 m), composed mainly of granitic and gneissic rocks. Most of the north between the lower Zambezi, Lake Malawi, and the coast is part of this zone; it is also represented in extensions of the Rhodesian highlands along the border with that country. In the extreme south, only a narrow fringe of hills exists along the border. A few isolated ranges rise above 4,500 feet (1,370 m) especially east of Lake Malawi.

Rainfall is heaviest toward the north and at higher elevations (Map 15). There are two well-marked seasons: the cool, dry period from April to September, and the hot, rainy period from October to March. Most of the country would be classified as tropical savanna, though conditions are more steppe-like in the low south and in the Zambezi Valley. The vegetation is characterized by grassy cover with scattered trees; only in some of the well-watered highlands and in the more humid back country of Beira are denser stands of timber found. Most of the soils of Mozambique are poor and sandy, but rich alluvials exist in several major river valleys and deltas, presenting attractive opportunities when river control has been effected.

POPULATION

The population of Mozambique was estimated to be 8.4 million in mid-1974, 97 percent of whom were Africans. Almost all of Bantu stock, the Africans are divided into 12 major and over 34 lesser tribes. The estimated growth rate is a relatively low 2.1 percent. The incidence of diseases is high, particularly of malaria, bilharziasis (second only to Egypt), tuberculosis, and helminthiasis, and life expectancy is only 41 years. Africans live mainly in small rural hamlets or villages, including in recent years fortified *aldeamentos* in the north designed to reduce contact with guerrillas.

The European population increased from 27,400 in 1940 and 103,000 in 1960 to about 200,000 in 1974. They are heavily concentrated in the south and in the main urban centers, with over half in the Lourenço Marques district. A considerable number of Europeans have migrated legally or illegally to South Africa in the last decade or so. Other non-Africans include about 45,000 persons of mixed race, over 20,000 Indians, and several thousand Chinese.

The most densely populated areas of the country are in the southern coastal stretch, parts of the Zambezi Valley, the pastoral Angonia Highlands, and the northern seaboard. Parts of the southern interior and the Save River area are practically empty, and much of the central and northern inland regions are also sparsely populated.

LAND-USE REGIONS

The physical attributes of Mozambique are such that approximately a third of the country is considered suitable for tillage agriculture. At present only about 6 percent is utilized for agriculture, two-thirds by Africans and one-third by Europeans, mainly on large estates and plantations. Most Africans practice a bush fallow agriculture in burnt clearings, typically own no cattle (approximately 70 percent of the country has tsetse fly), and farm only for subsistence. The main subsistence crops are corn, peas, manioc, sesame, and a variety of vegetables. Among cash crops, Africans account for all of the output of cotton and most of the rice. They also collect a variety of uncultivated products: cashew nuts, castor seeds, coconuts, jute, mufurra nut (an indigenous source of nonedible oil for soap), mangrove bark, and rubber. There are about 2.1 million cattle in the country, mostly in the tsetse-free south. The European or so-called organized agriculture is concentrated on a variety of cash crops: sugar, tea, coconuts, sisal, corn, tobacco, citrus fruits, bananas, vegetables, and flowers.

THE NORTH. The area north of the Zambezi Valley may be taken as one broad land-use region. Long undeveloped except for sisal estates along the coast and coconuts at Quelimane, it has more recently become the main agricultural region of the country (Map 104), accounting for three-fourths of cotton production, four-fifths of cashew nut and peanut sales, nearly all of the manioc and potatoes marketed, much of the rice, all of the tea, and smaller quantities of other crops including kapok, bananas, jute, sunflower, peas, tobacco, and wheat.

The extreme northern districts are less developed than the districts of Moçambique and Zambezia, except for scattered localities along the coast and production of cashew nuts on the Makonde Plateau inland from Porto Amélia. Eventually they could be comparably productive. This region possesses in Pemba Bay one of the finest natural harbors on the east African coast, but the low traffic moving through Porto Amélia on its southern shore is a measure of the generally low existing development. The port has a T-shaped quay capable of accepting one large and one coastal vessel.

It is the hinterland of Nacala that has seen the most dynamic developments in the north. The first center of Portuguese influence in the area was the town of Moçambique, founded in 1508; for centuries it was a military establishment and a revictualing and refitting center, and was the capital of Portuguese East Africa until 1898. Situated on a small coral island three miles from shore, its importance is now more historic

than economic. The regions lying in the interior of the mainland remained virtually untouched for over 400 years.

Nacala, on the eastern side of an extension of Fernão Veloso Bay, became the main port of the area in the 1950s, when a deepwater berth was constructed in the excellent harbor, protected even from cyclones that periodically strike the area. The port has recently received two new quays and ancillary equipment bringing its capacity to about 1.5 million tons; it handled 718,000 tons in 1972. Nacala has become the seaward terminus of the northern rail line, which originally extended from Lumbo, on the mainland opposite Moçambique, to Nampula, but which now traverses the country and connects with the Malawi line. As noted earlier, it handles a share of traffic for Zambia.

A more intensive development of the Nacala hinterland did not begin until after World War II. Four large concessionaires promoted the African production of cotton under a system comparable to that in Angola; the concession arrangements were ended in 1961 and a more favorable pricing system was introduced. Cotton is grown by about 600,000 families on about 750,000 acres (300,000 ha) in the country, mostly in the north; about 45,000 tons are produced annually. It ranked as the most important export crop until 1971, when sugar and cashew nut shipments had higher value. However, unlike the case in Angola no export occupies a dominant position.

Most of the country's cashew nuts, now the leading export, are produced in the north, particularly along the coast. Seven-eighths of the crop is picked by Africans, mostly from uncultivated trees. Some trees have begun to be cultivated, and the number can be expected to increase as the market expands, but many growers are reluctant to plant trees because they take 5 to 6 years to reach bearing age and 12 years to attain maximum fertility. The cashew tree was introduced to Mozambique from Brazil in the sixteenth

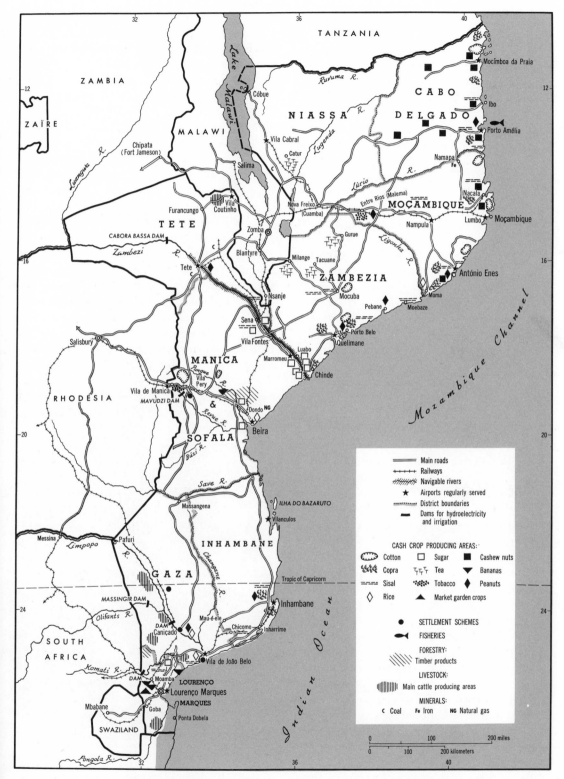

Map 104. Economic map of Mozambique

century, but commercial exploitation did not begin until the 1920s. Until the mid-1960s almost all of the nuts were bartered with storekeeper-collectors and shipped to India for the very difficult decortication, which required a large, low-cost, supply of labor. Machinery for processing first became available in 1962, and now the crop is processed in a score of factories, which sell not only the nutritious nut but the oil from between the two shells, which is used for a wide variety of industrial purposes; this has permitted the direct sale of output at higher returns, the United States being the largest purchaser. Output has increased in recent years from 120,000 tons in 1965 to about 180,000 tons in 1972, ranking Mozambique as the world's leading producer; plans call for an output of 300,000 tons by 1979.

European estates have long produced sisal; in the ealy 1960s some 22 plantations with about 125,000 acres (50,000 ha) in sisal were scattered along the coast. Output has remained at about 30,000 tons over the past decade despite lower world prices. A development of significance has been the colonization of several interior highland areas by Portuguese farmers, whose main cash crops are tea and tobacco. Tea comes from the Milange, Tacuane, and Gurue areas; about three-fifths is exported, permitting tea to rank fifth among exports with about 6 percent of the total value. Acreage under tea increased from 13,000 (5,260 ha) in 1956 to 40,000 (16,200 ha) in 1960 and about 60,000 (24,000 ha) in 1972, with production rising from 10,200 tons in 1961 to about 17,000 tons in recent years, when Mozambique has ranked fourth among African producers. About 55,000–60,000 Africans are employed on the tea estates, 90 percent of which are company owned and the remainder run by settlers. Large additional acreages are suitable for tea production, both in the present areas and in the high-lying, high-rainfall lands bordering Lake Malawi.

Tobacco is produced on European farms around Malema on the Nacala line. Output and exports have increased rapidly, but a portion of recorded exports may actually come from Rhodesia. Ecological conditions are similar to those of the Rhodesia highveld and the area planted to tobacco in Mozambique could be greatly increased.

The northern lands behind Quelimane may be considered as a subregion. Situated 12 miles (19 km) upstream from the mouth of the Rio dos Bons Sinais, Quelimane serves one of the most developed agricultural regions of Mozambique, including the tea area already noted. It is the terminus of an 89-mile (143-km) line to Mocuba, which is supplemented by extensive road services run by the railways. A coconut plantation near Quelimane, said to be the largest in the world, is part of the 67,000 acres (26,800 ha) in coconuts owned by the Boror Company, which is also the largest sisal producer.

A coconut plantation near Quelimane
Mozambique is the leading producer of coconut products in Africa.

About two-thirds of the estimated 12 million coconut palms in the country are found on European holdings, seven-eighths of which are on large company plantations and one-eighth on smaller, individually owned estates; together these account for about 70 percent of copra production. Some 300,000 Africans, many operating as outgrowers, have the remaining trees and 30 percent of copra output. Copra ranks sixth among exports, but the fourth-ranking vegetable oils include a considerable amount of coconut oil.

THE ZAMBEZI VALLEY. Most of this area is sparsely settled and little effort was made to develop it before 1948. The British-owned Sena sugar estates in the delta at Luabo and Marromeu, the largest in Mozambique, ship their output via Chinde. The company has about 45,000 acres (19,200 ha) in sugar and two of the four sugar factories; it also has a coconut plantation near Chinde and a cattle ranch near Luabo. Total output from the country is now about 326,000 tons, ranking Mozambique fourth among African producers; it is expected to increase to 600,000 tons by 1976.

As noted earlier, coal moves from the Tete area to Beira; there has been some discussion about moving coal and iron ore from that region via a 373-mile (600-km) pipeline to Chinde, where a 3.7 mile (6 km) pier would be used to load vessels of up to 150,000 tons. The Zambezi is navigable for nine months a year from 37 miles (60 km) above Tete to Chinde, but substantial control works and dredging would be necessary before large barges could use it.

Cabora Bassa and the Zambezi Valley Development Project. The Zambezi Valley will eventually become one of the most productive parts of Mozambique, and two distinct but interlinked projects situated in it merit special attention: the Cabora Bassa hydroelectric scheme, which is international in scope, and the Zambezi Valley Development Project (ZVDP), a multipurpose plan for the basin in

Mozambique, to be financed from other sources. The newly completed Cabora Bassa dam is situated in the Kebrabasa Gorge about 80 miles (128 km) upstream from Tete. Required to create a head, the dam also helps to even the flow of the river, whose volume is nearly double that at Kariba and has again acquired a marked seasonal rhythm. The arched-wall dam, measuring 550 feet (168 km) high and 984 feet (300 m) across at the crest, controls more water and can produce more electricity than either Kariba or the High Dam, though it is much less massive than the latter. Power will be developed in three stages: Phase I, begun in 1969 and completed in 1975, included construction of the dam and a south-bank 1.2 million-kw power station plus two 875-mile (1,400-km) transmission lines to the Apollo substation between Pretoria and Johannesburg in South Africa; the next two phases would bring the capacity first to 2.4 million and then to 3.6 million kw. Later still, additional dams might be placed on some of the tributaries. Thus Cabora Bassa will be one of

The Cabora Bassa Dam on the Zambezi in January 1974
One of the largest hydroelectric stations in the world, most of its power is slated to be sold to South Africa.

the largest hydroelectric installations in the world.

South Africa's Electricity Supply Commission (Escom) has contracted for a large block of the available power from Cabora Bassa—1.07 million kw a year from the third year of operation to expiration of the contract in 35 years—but a larger amount may be taken, because the lines can use 1.7 million kw of capacity. The Cabora Bassa project is basically concerned with electric power, and it is the South African agreement to purchase a large block that made it economically feasible. Other benefits, relating to flood control, irrigation, and transportation, are the concern of the separate ZVDP. Zamco, a consortium organized by Anglo-American Corporation of South Africa with French, German, and South African capital, arranged for financing and constructed the first phase of Cabora Bassa at a cost of $246 million; the three stages will cost an estimated $500 million.

South Africa's interest in Cabora Bassa is economic and political. It can readily absorb the power without becoming highly dependent on Mozambique; after the first years, when the price paid will be relatively high because of the need to repay the loans, the cost of power will be very attractive; and South Africa will avoid the problems of securing water for, and increasing the smog level from, an equivalent output in carboelectric plants, which provide the vast bulk of that country's needs. The Republic also saw investment in Mozambique as contributing to its outward-looking policy and as promoting stability in an important buffer area.

Portugal also saw political and economic benefits in Cabora Bassa. Political gains, it was presumed, would accrue from greater economic strength in Mozambique and from the interest that investors would have in sustaining the Portuguese presence there. From the economic standpoint, Portugal saw Cabora Bassa as the key to the development of a large and potentially rich part of Mozambique.

The greatest potential benefits to Mozambique come through the impact of Cabora Bassa on other developments in the Zambezi Basin, to be managed by ZVDP. A "starting plan" of indeterminate duration calls for expenditures of $176 million, of which about half would be for agriculture, a third for power, a ninth for transportation, and the remainder for social services and community development. In the agricultural sphere, Cabora Bassa will provide flood control for a large area and will permit eventual irrigation of as much as 3 million acres (1.2 million ha). Plans call for irrigating 200,000 acres (80,000 ha) of sugar, cotton, and kenaf fiber in the first phase and for developing dryland farming on 75,000 acres (30,000 ha) to produce food crops, citrus fruit, and beef. The ZVDP also calls for the exploitation of 500,000 acres (200,000 ha) of existing forest lands and the planting of an equal extent of exotic timbers, which might support a cellulose industry north of Tete. Transport benefits include construction of a new bridge at Tete, improved roads to serve the area, and better navigability on the Zambezi through regularization of the flow. The lake should provide some fishery resources.

Cabora Bassa and the participants in it were subjected to severe criticisms from various organizations opposed to the Portuguese presence in Africa and white domination in South Africa, and Frelimo said that it would disrupt operation of the project. Just what attitude will be adopted by an independent government toward the sale of electricity to South Africa remains to be seen. There is little question that Cabora Bassa is viable economically, but only if a large block of power is under contract. The ZVDP, which contains several elements unrelated to the dam and others unlikely to reach fruition for many years, cannot be judged for some time. There is no reason to believe that either will necessarily bring stability and happiness to the Zambezi Valley.

THE HINTERLAND OF BEIRA. Production from the Manica e Sofala district has been overshadowed by the importance of Beira and its railways in transit traffic for Rhodesia and Malawi and, before 1973, for Zambia. Portugal claims that Mozambique lost over $100 million in the period from UDI to 1972 because of sanctions on Rhodesia and estimates that Beira alone will lose an additional $11 million yearly because of the cessation of traffic with Zambia. There will, however, be an offsetting gain to Nacala and particularly to Lobito for some years.

Beira, located at the confluence of the Púnguè and Buzi Rivers, 15 miles (24 km) from the mouth of the former, has a mediocre physical site. Shipping is disadvantaged by shallow coastal waters, sand banks, a constricting bar in the river's mouth, restricted depths at low tide in the harbor, silting, and by an 18 to 23 foot (5.5–7 m) tidal range, which aids entry into the harbor but increases the problems of working at the piers. On land there is the additional difficulty of an unstable surface, which requires that some buildings be constructed on piles. Traffic at Beira totaled a record 4.5 million tons in 1965 but tended to weaken after UDI in Rhodesia and stood at 3.5 million tons in 1971. Nonetheless, its docking space is being extended.

The railway to Rhodesia was first constructed as a narrow-gauge line in 1893–97 but was relaid almost immediately to conform to the 3′6″-gauge line from the south to Salisbury. It has since been relaid with heavier track and realigned, particularly on the hilly section rising to Rhodesia. The Malawi link first went only from the north bank of the Zambezi to Blantyre, but a line from Dondo on the Beira-Rhodesia railway to the south bank was constructed after World War I and the two sections were connected by a long bridge across the Zambezi in 1935, after which Beira handled the vast bulk of overseas shipments for Malawi. A branch of the Beira line extends north of the Zambezi to Tete, and plans call for linking the Beira and Nacala lines within Mozambique.

The Beira area has had a considerable buildup of industry in postwar years, attracted in part by low-cost hydroelectric power from the Mavudzi plant on the Revue River 23 miles (37 km) from Vila Pery and 110 miles (177 km) from Beira (Map 103).

The domestic hinterland of Beira is the leading forestry and mining area of the country; the principal forest operations lie along the line north of Dondo. The main products are railway ties, parquet flooring, and hardwood logs, about two-fifths of which are exported—mainly to South Africa. The proposed cellulose factory at Chimoio near Vila Pery is expected to increase Mozambique's exports by about 15 percent when full capacity is reached in the 1980s.

The Buzi River Valley near Beira is the site of a Portuguese-owned sugar plantation, while a large new Swiss-Portuguese sugar estate is planned in the Púnguè basin about 43 miles (70 km) from that city. On the Manica-Sofala Plateau west of Beira only a thin veneer of agricultural colonists is found, despite the availability of an estimated 15 million acres (6 million ha) suitable for agriculture, especially in the Chimoio uplands. Corn is the outstanding crop from European farms now on the plateau, but some oilseeds, oil cake, fruit, vegetables, and sisal, plus small quantities of wheat and rice, are also shipped, while settlers at Zonue in the Revue Valley appear to be doing well with tobacco. Africans produce some cotton in the Gorongoza and Chimoio areas.

South of Beira to the Save River there is practically a no man's land except for coastal fishing villages. The improved road linking Beira and L.M. may open up new possibilities in this region.

THE SOUTH. That part of Mozambique south of the Save River is again dominated by a transport-industrial center, Lourenço Marques, the very much off-center capital of the country. Its port is one of the major gateways for the Transvaal, Rhodesia, Swaziland, and Botswana. Situated in the well-protected Delagoa Bay, it has had deepwater facilities since 1903. Its roughly 2 miles

(3.2 km) of quays can now accommodate 20 oceangoing vessels, while it has special facilities for loading coal and ore, large cold stores for fruit, grain silos, a new container depot, and, four miles (6.4 km) up the estuary at Matola, a petroleum refinery and storage tanks, lumber wharves, and a mechanical loader for iron ore from Swaziland.

A high percentage of L.M.'s traffic, which increased from 7.1 million tons in 1961 to 14.8 million tons in 1971, is in bulk commodities. But L.M. cannot take ore carriers above 65,000 tons and tankers above 30,000 tons and the increasing size of both led to a decision to develop a new terminal at Ponta Dobela, 25 miles (40 km) to the south. This new port will have a 5,578-foot (1,700-m) pier connecting to a highly automated jetty with a total capacity of 13.5 million tons for loading iron ore, coal, and phosphate rock. Dobela Point is in part a response to South Africa's new port at Richards Bay, 170 miles (275 km) south of Dobela Bay (see chapter 23). The Mozambique port has a superior physical site, can be developed in three years at an estimated cost about half that of Richards Bay, and will be able to take 250,000-ton vessels as compared to a maximum of 150,000 tons at the new South African port.

Lourenço Marques is important in the handling of South African traffic not only because it is the nearest port to the highly productive Transvaal, but also because the present version of the Mozambique Convention requires that at least 47.5 percent of the transit traffic destined for a region defined as the "competitive area" (that area enclosed by the stations of Pretoria, Springs, Germiston, Vereeniging, and Klerksdorp) go through L.M. In return, Wanela is permitted to recruit laborers from southern Mozambique to work in the gold and coal mines.

Four rail lines terminate at L.M.: the 57-mile (92-km) Ressano-Garcia line linked in 1894 to the South African Railways; the Limpopo line to Rhodesia completed in 1955 and having a length within Mozambique of 331 miles (533 km); a short line to Goba connecting with the Swaziland Railway, completed in 1962 primarily to move iron ore from that country; and a 19-mile (31-km) rail line serving banana plantations on the Incomati River.

The city of Lourenço Marques had a population of 384,000 in 1970. In addition to its great importance as a transport center, the state capital is also the country's main commercial, manufacturing, and cultural center, and the focus of an important tourist industry as well.

In southern Mozambique the most fertile agricultural zones are found in the valleys of the Limpopo, Incomati, and Umbeluzi Rivers. Dairying, market gardening, and cotton growing are carried on near L.M., while plantations of pine, cyprus, and eucalyptus are found in the north of that district. One of the country's large sugar plantations is situated at Xinavane on the Incomati River and Portuguese farmers produce bananas in the same valley. The Incomati-Movene Scheme calls for eventual irrigation of 120,000 acres (48,600 ha) by diversion of water from the Incomati to the Movene Valley for wheat growing.

The Limpopo Valley was the scene of the

Portuguese peasants working on a settlement scheme in the Limpopo Valley

country's most important settlement project, which was started in 1953. Control of the Limpopo began with construction in 1956 of a 2,280-foot-long (695 m), 42-foot-high (13 m) dam whose crest is used by the line to Rhodesia and which permits irrigation of 75,000 acres (30,000 ha). Each settler was provided with a three-room house, and a stable, storehouse, poultry yard, and large garden, plus allotments of irrigated and dry land, and loans for furniture, livestock, equipment, seeds, and one year's provisioning. In return for each holding, the state receives a sixth of the farm produce until the capital costs have been recovered. The success of the scheme was, like similar schemes in Angola, rather modest, with only a fraction of the expected numbers having joined. Difficulties encountered by the settlers included the high incidence of diseases, an oppressive climate, flooding, inadequate holdings, and problems in marketing the surplus production. In addition, the cost per family, over $10,000, was grossly out of line with general standards in Mozambique or returns from the scheme.

A new $67 million, 6-year development scheme for the Limpopo Basin is now underway, its major feature is the Massingir Dam on the Elephant River tributary, 17 miles (28 km) from the South African border. The dam is designed to control floods on the Limpopo, to irrigate 240,000 acres (97,000 ha) in the Limpopo Valley, and to produce hydroelectricity. The Limpopo Valley has one of the larger concentrations of cattle in the country, mostly Landim and Africander breeds held by Africans, but they are poorly commercialized at present.

Near Vila de João Belo, at the mouth of the Limpopo, is the Ihamissa Scheme, a government cooperative for Africans. This scheme has had much more favorable results than the other settlement projects of Mozambique. It will comprise some 20,000 acres (8,100 ha) of reclaimed marshland divided into 10-acre (4-ha) family farms producing rice, corn, beans, peanuts, manioc, and bananas. Vila de Joao Belo is the head of a short rail line, but its harbor has been deteriorating rapidly because of deposition of silt. Inhambane, farther north along the coast, is also the terminus of a short rail line and has a harbor equipped with a modern lighterage pier. Both small ports are tributary to L.M., through which their varied agricultural produce is transited.

One can only speculate about the future of the numerous programs begun by the Portuguese. Many of them are of undoubted value to an independent Mozambique, but Portugal may lose some interest in financing them unless its relations with such a government are reasonably cordial. One of the most difficult decisions facing an independent black-ruled Mozambique will be its relations with Rhodesia and South Africa. Members of the OAU, which have backed Frelimo, may expect a severance of relations; this would be an extremely costly choice because of the heavy reliance of the economy on transit traffic, tourism, labor remittances, imminent sale of electricity, and sale of a significant portion of its exports to its white-ruled neighbors. Closing the borders would also be highly damaging to Swaziland and Botswana, which depend on L.M. and Beira for most of their overseas shipments.

Part Seven
SOUTHERN AFRICA

Aerial view of Johannesburg with gold mine dumps in the background

CHAPTER TWENTY-THREE

The Republic of South Africa

South Africa is a land of contrast, a land of paradox and contradictions, a land of great problems and potentially great promise. Economically, it stands head and shoulders above other African countries; politically, it has brought upon itself the calumny of much of the world and appears to many observers to be determinedly set upon a course of self-destruction.

MEASURES OF ECONOMIC IMPORTANCE

The Republic's wealth is a tribute both to the great richness of its underground resources and to the ability of its people, particularly its dominant white minority. Its achievements have, however, required the contributions of all of its racial groups working in a kind of unequal partnership which, all too often, goes unrecognized. Its GNP in 1972 was $19.6 billion, about 23 percent of the African total, which is roughly comparable to Hungary or Denmark among European countries. The per capita GNP, $853, is the second best in Africa, where it is exceeded only by oil-rich Libya. However, in western Europe, only Portugal's per capita GNP is lower. South Africa's figure conceals wide differences among the racial groups; the whites have far higher incomes than the others, and many blacks are still living on a largely subsistence basis.

The economic position of the Republic may be further measured by looking briefly at the individual sectors of the economy. In mining, it produces about a quarter of the total value of African minerals. It ranks first in the world in the output of gold, antimony, and vanadium; second in gem diamonds, chrome, platinum, and manganese; and third in asbestos and uranium. It produces about 95 percent of the coal mined in Africa.

Agriculturally, despite numerous physical handicaps, South Africa leads the continent in the production of corn, sugar cane, and wool; it ranks second in cattle, tobacco, and wine, and third in citrus fruit and wheat. It also has first rank in Africa in fish landings.

South Africa has a much more diversified and developed industrial sector than any other African country. Manufacturing's share of the GDP has increased from 9.4 percent in 1919 to 23.1 percent in 1970. The Republic, which accounts for an estimated 55 percent of the total energy consumed and electricity produced in Africa, has a per capita energy consumption figure 8.5 times the continental average and the only one on the continent exceeding the world per capita average. It has the only well-developed iron and steel industry on the continent, and supplies a higher percentage of its own needs for manufactured goods than any other African country.

In the field of transport, South African Railways account for about five-eighths of the total rail net ton kilometers for the continent. The Republic, with about 27 percent of the rail mileage of Africa, possesses the only real rail network on the continent. Its road system of over 210,000 miles (338,000 km), including about 20,000 miles (32,000 km) of paved highway, is also superior, and the country has some 40 percent of the motor vehicles registered on the continent.

Table 20. Estimated population of South Africa by racial group, selected years 1904–1974

Year	Total thousands	African '000	African %	White '000	White %	Coloured '000	Coloured %	Asian '000	Asian %
1904	5,175	3,490	67.4	1,117	21.6	445	8.6	122	2.4
1936	9,588	6,596	68.8	2,003	20.9	769	8.0	220	2.3
1960	16,006	10,928	68.3	3,068	19.2	1,488	9.3	522	3.2
1970	21,448	15,058	70.2	3,751	17.5	2,018	9.4	620	2.9
1974	24,887	17,712	71.2	4,160	16.7	2,306	9.2	709	2.9

SOURCE: South Africa, *Union Statistics for Fifty Years, 1910–1960* (Pretoria, Government Printer, 1960), pp. A3–A5; *Bulletin of Statistics* (Pretoria, Government Printer, 1974).

N.B. African population undercounted in all censuses but probably less so in later censuses.

POPULATION AND PEOPLES

Despite the evident economic achievements of South Africa, it faces political and economic uncertainties of serious dimensions, most of which can be traced directly to its racial policies. The composition of the South African population is complex; four major groups are represented, and the cleavages or distinctions among the groups are constantly being sharpened legally. Table 20 gives the population by racial group and Table 21 presents information about the distribution of racial groups in the European area (c. 87 percent of the total) and the African reserves and Bantustans (c. 13 percent of the total area) according to the 1970 census.

The dominant white group, which now makes up about 16.7 percent of the total population, is itself split into two major groups: the Afrikaners—predominantly descended from Dutch and Huguenot immigrants whose first permanent settlement, at the Cape, dates from 1625—and those of British stock—who first settled in the early 1800s after Britain had taken over the Cape. About 58 percent of the Europeans speak Afrikaans at home, 37 percent speak English, and 5 percent speak other languages. Of the four provinces of South Africa, only Natal has a majority of English-speaking whites.

The fundamental racial policies of the exclusively white-controlled government reflect the minority position of the "European" population, which helps to explain the continuing efforts to attract white immigrants to the country. About 87 percent of the whites

Table 21. Distribution of the South African population by racial group in European and African areas, 1970

Area	Total '000	Total %	African '000	African %	White '000	White %	Coloured '000	Coloured %	Asian '000	Asian %
White	14,414	67.2	8,061	53.5	3,731	99.5	2,005	99.1	617	99.5
Urban	9,698	45.2	4,407	29.3	3,258	86.9	1,494	74.0	539	86.9
Rural	4,716	22.0	3,654	24.2	473	12.6	511	25.3	78	12.6
African	7,034	32.8	6,997	46.5	20	0.5	13	0.6	3	0.5
Urban			582	3.9						
Rural			6,415	42.6						

SOURCE: South Africa, *Bulletin of Statistics,* No. 4 (Pretoria, Government Printer, December 1971).

now live in cities and towns, there having been a particularly notable influx of Afrikaners in more recent years.

The Afrikaners are a fiercely nationalistic people, some would say more so than any other ethnic group on the continent. Afrikanerdom is not monolithic, but there is a highly complex structure of interconnecting links among their numerous political, religious, labor, intellectual, financial, and other special interest groups. Since 1948 they have been particularly strongly represented in government, where they hold over 90 percent of the key administrative, defense, and policy-making posts. The Boers, or farmers, still represent a kind of Afrikaner bulwark, more important than their numbers would suggest. The occupational pattern of the Afrikaners is, however, becoming increasingly similar to that of the English-speaking whites, though the latter continue to have a disproportionate number of the more highly skilled, entrepreneurial, and managerial positions in mining, manufacturing, and finance.

The largest group is the blacks, the vast majority of whom are Bantu, which is the term used for them in South Africa; they make up 70.2 percent of the population. About 53.5 percent of the total African population lives outside the Bantu areas, divided 55:45 between white urban and white rural areas. In the period 1969–70 the proportion of blacks in the European area declined because of the removal of perhaps 437,000 tenants and squatters from farms and "black spots," the endorsement out (ordering out) of urban areas of an estimated 400,000, the redrawing of boundaries to include some African townships in the Bantu areas, and more rigid control of influx. Nonetheless, the absolute number of blacks living in the European areas increased in the intercensal period by 1.2 million, and blacks outnumber whites there by about 2.2 to 1.

Nine major and several minor black ethnic groups, or "nations," are recognized, but

Part of Soweto, a very large African township outside Johannesburg
About 29 percent of the black population of South Africa lives in urban areas in the white portion of the country.

many of the long-term urban residents are thoroughly detribalized. The government was fond of noting that the white nation was the largest nation in the country until 1970, when the census revealed that they are now outnumbered by the Zulu and the Xhosa. Many foreign Africans are present in South Africa, most of whom are contract laborers on the mines. Estimates of their numbers vary from 562,000 in 1971 to the official figure of 441,000 in 1972, at which time they came from the following areas in the stated percentages: Lesotho, 29.9; Malawi, 29.7; Mozambique, 27.7; Botswana, 7.3; Swaziland, 2.3; and other African countries, 3.1 percent.

Third among the major racial groups are the "Coloureds," persons of mixed race, who represent 9.2 percent of the total population. About 87 percent reside in Cape Province and 75 percent are now urbanized. Last are the Asians, constituting 2.9 percent of the total. They are, as elsewhere in Africa, divided by racial and religious differences, with Hindus outnumbering Muslims by about 3 to 1. Asians were brought to South Africa between 1860 and 1911, when further immigration was prohibited, as field workers for the sugar plantations of Natal. About 85 percent live in Natal, with a heavy concentration around Durban, and 13 percent live in the Transvaal.

The discriminatory policies long pursued by South African whites have become increasingly rigidified under the predominantly Afrikaner Nationalist Party government, whose official goal is the separate development of the races. The basic aim of Apartheid, which is now called "separate development," is to secure white domination by dividing the racial groups according to living area, education, other facilities of all kinds, and even churches. Control is exercised through a battery of laws that give the government power to restrict residence, movements out of the Bantu areas, jobs, group meetings, etc. Such laws have rather completely eroded civil and individual liberties for all but the whites, and their liberties are infringed as a consequence. The adoption of apartheid may be explained by the isolation of the Afrikaners, especially after their great trek into the interior and the long period of cultural separation from changing Western thought; the lack of any unifying theme or ideal in the country; the force of Afrikaner nationalism; the long tradition of segregation through South African history; and the existence of a large dependent populace within the country.

To most South African whites, of whatever descent, it appears inconceivable that their way of life could be maintained if there were democratization; one man, one vote spells suicide to them. This explains why some people of unquestioned sincerity support separate development and strive to justify their position by working for greater economic and social advance for the African. Indeed, one of the stated goals of apartheid is to develop the African, and South African whites are quick to point out the relatively

high incomes (evidenced by the continuing willingness of hundreds of thousands of foreign Africans to migrate to the Republic for contract work) and educational standards of the Bantu, the substantial sums spent on African housing in the cities, and efforts to improve the economies of the reserves. While there is some justification to these claims, the amount spent on African education is far below that spent on the smaller number of white students. Furthermore, wage differentials are distressingly wide and African wages compare unfavorably with those in several African countries, the African reserves are characterized by increasing pressure, and there are the ever present bars to African advance in the European areas.

In recent years certain changes have occurred, which may alter the scene more rapidly. A series of strikes of African industrial and mining workers, occurred in 1972–74 despite their illegality, demonstrating to both blacks and whites the considerable de facto power held by the majority race. Several of the Bantustan leaders have spoken out regarding African desires for more land, more local autonomy—or even full independence—and removal of restrictive barriers; and the government can no longer follow the former practice of silencing dissidents by incarceration or house arrest because such moves would expose separate development as a meaningless proposition.

The government has responded to the increasing demands of the African leaders by granting wage increases, removing some elements of petty apartheid, and, more importantly, by proposing to expedite the devolution of power to the Bantustans. It has even been suggested that at least one of them, the Transkei, might be granted full independence within five years (see below).

The demographic dynamics of the Republic appear to make it impossible to maintain the present relations among the races over the long run. The *increase* of the black population in the last intercensal period was greater than the total white population at the end of the period, 1970. The rate of growth of the Bantu population is considerably higher than that of the whites. And there appears to be no possibility either of confining the bulk of the Africans to the reserves and Bantustans or of keeping the black-white ratio in the urban areas from widening in favor of the former. Of course, complete separation of the races was never contemplated, and the notion that all Africans belong to some "homeland," whether they ever lived there or not, is simply a device to justify treating them as foreigners in the European areas. But economic integration exists in the sense that the Bantu belong to the European economy while the Europeans rely upon them for the majority of workers in all sectors. Any real separation of the races would lead to a collapse of the whole economy. The millions now being spent on separate development and on the military and police forces represent a small part of the real cost of discrimination; the major economic cost stems from the brake on the whole economy resulting from restrictions on the advance of the nonwhite groups. The

moral and social cost of segregation and discrimination is incalculable.

THE AFRICAN AREAS

At present about 12.6 percent of South Africa's area is assigned to African reserves and Bantustans, divided into several hundred blocks ranging in size from 16,000 square miles (41,400 sq km) for the Transkei, larger than Switzerland, to small bits of isolated land. The Zulu, for example, have 29 main and 176 smaller units.

The African areas are considered by the government to be the homelands of all South African blacks, but only 46.5 percent of the black population was resident in them at the time of the 1970 census. The population density in them averaged 117.3 per square mile (45.3 per sq km), 3.3 times the density in the European area. Since the vast bulk of the population in the African areas is engaged in agriculture, the disparity is far more marked than these figures imply. Under the Land Act of 1936 the blacks are supposed to receive 13.7 percent of the total area, which is unacceptable to African

A Zulu kraal near Mahashini, northern Natal
African reserves and Bantustans cover about 13 percent of the country. Pressure on the land is severe in most areas.

leaders, and in late 1973 about 3.1 million acres (1.24 million ha) had still to be bought to reach this percentage.

At present plans are being drawn up to consolidate the African areas into eight homelands, each of which is to have one or more "viable" units. For example, KwaZulu would be consolidated from 29 large and 176 small units to 5 large units and 1 small unit. The consolidation plans involve both the excision of lands now held by blacks and the addition of state or private lands to their areas, which will involve some massive population movements. It has been estimated that creating eight fully consolidated homelands would have required the moving (in 1970) of 653,000 people.

The stated political goal for the homelands is eventual independence; in an intermediate step they are given the status of Bantustans with a degree of local rule, but the European government retains authority over external affairs, defense, certain aspects of the administration of justice, posts and telegraphs, railways and harbors, immigration, currency, public laws, external investment, and customs and excise. As of late 1974, six of the eight planned Bantustans had been created.

Physical conditions in the African areas vary enormously. A sizable percentage is in the better watered areas along the east, but most of the Transvaal reserves are in semi-arid regions along the north and northwest, and the eastern Bantustans are situated in some of the roughest topography in South Africa, with only 11–23 percent of their areas classified as flat to gently undulating (Map 105). Some notions of the limitations can be derived from the estimates that about 79 percent of the total is used for grazing and only 15 percent for tillage agriculture and that the areas are dependent on the European area for at least 40 percent of their food supplies.

From the standpoint of land use excessive pressure on the land and destructive prac-

tices have been keynotes. Holdings are too small to support a peasant farmer and his family. Other land use problems are comparable to those of other African areas: the deficiencies of communal tenure and grazing rights, excessive dependence on the work of women, and backward techniques contributing to serious land deterioration.

In 1961 the government launched the first five-year plan for upgrading the Bantu areas; expenditures grew at a slow but steady pace to a level of $237 million in 1971–72, not including amounts spent by the homeland governments. About a third of the Republic's allocations went to new townships, 27 percent to the social infrastructure, and much of the rest to administration. It is claimed that 60 percent of the total area has received attention with respect to erosion control, fencing, upgrading livestock practices, and other agricultural improvements. Some emphasis has been placed on irrigation and, where possible, on the promotion of labor-intensive crops such as tea, sugar, cotton, and pyrethrum. But further massive efforts will be required before farming in the Bantu areas reaches a modern level and in the meantime the pressure of population increases.

Nonagricultural sectors are very meagerly developed. About 3.6 percent of the total area is in woods and forests and there has been some afforestation, but the reserves are heavily dependent on imports for their

Cattle and goats in the Nqutu district of Zululand
Overgrazing and severe erosion are characteristic of large parts of the African areas.

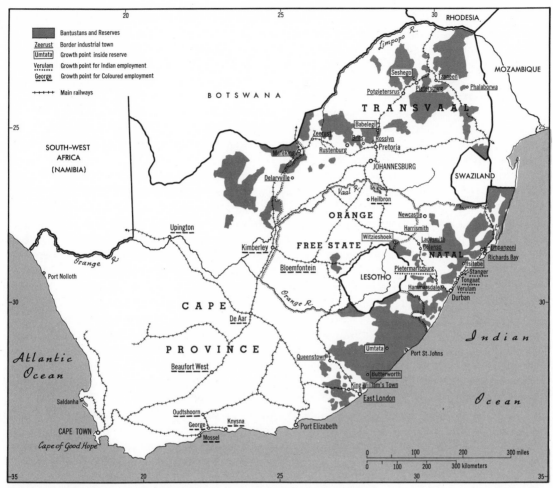

Map 105. African areas, border industrial areas, and growth points in South Africa

timber needs. The fishing industry is of minor importance. Tourism is restricted to the main roads traversing some of the areas plus a game park to be assigned to KwaZulu. Most of the mineral areas of the Republic are in the European areas, though there are mineralized zones in the Bantustans, particularly in the Transvaal and the northern Cape Province. Platinum is by far the most important mineral mined in the homelands; asbestos and chrome are also produced in large quantities, while a variety of other minerals are mined in lesser amounts. But the royalty paid to the Bantustan governments is usually only 10 percent and protective regulations enforced by white unions and acceded to by the government apply in the African areas as they do in the European areas.

Until recently there has been nearly a total absence of industry in the homelands. Removal of restrictions on foreign investment and introduction of agency arrangements have brought a number of industries; in the Transkei, for example, new plants are producing laboratory and medical office furniture, textiles, chemicals, and radios for sale throughout South Africa while a number of small factories cater to the Bantustan market. The number of industrial establishments in the homelands exclusive of the Transkei and the Ciskei is reported to have totaled 46 in 1973, most of which were recent introductions. It is conceivable that large-scale industries catering to the South African market could be established in the Bantustans if a sufficient number of foreign investors were interested, but the state authorities must approve such investment and their long-term policy is not yet clear. In 1974 the government was advertising in West Europe for potential investors in the Bantustans.

It is clear that the Bantu areas cannot now provide a living for their residents. While agricultural intensification is certainly possible, it is likely that half of the present population should be taken off the land if further destruction is to be avoided. The number of new jobs created in the African areas is put at 85,000 for the period 1962 to 1972, whereas at least that number is required each year. It has been estimated that 60 percent of cash receipts in the reserves are derived from migrant earnings and that their total GDP in the late 1960s was only about 2 percent of the South African total. Assuming the continuance of the present areas the only hope would appear to be in new industries within them or in the border areas (see below).

It is not surprising that the leaders of the several Bantustans have made increasingly urgent demands for rectification, and several have been very skillful in exposing the inconsistencies and immorality of separate de-

Training farm near Eschowe, Zululand
Efforts are being made to promote contour plowing and other erosion-control practices in the African areas.

velopment. They have sought rejection of the 1936 Land Act and addition of as much as 50 percent to the Bantustan areas, equal pay for equal work, an end to job discrimination and petty apartheid practices, education in English or Afrikaans at an early age, a halt to the transfer of blacks to the homelands unless the government is willing to provide jobs and housing for them, the delegation of more authority to the Bantustans, and the granting of full independence within a brief period of years.

The government has acceded to some of these requests. It has counseled patience in other instances—recommending, for example, that good farming practices be applied in the present areas before demands are made for more land. It has also toughened up the Terrorist Act, and, somewhat surprisingly, has intimated that at least one Bantustan, the Transkei, might indeed become independent within five years.

In concluding this section two points should be added: first, the homelands program does nothing for the Africans in the white urban areas, most of whom have been there for their whole lives and have no real ties in the Bantustans; second, there are no homelands for the Coloured and Asian peoples. Various areas have been allocated to them in the European areas, and they have not been permitted to live or work in other areas. In the early 1970s there was much public discussion of delineation of a Colouredstan, and, while the idea was rejected by the government as being impractical, the debate had value in exposing one of the many absurdities of apartheid. At the end of the debate the government said that the Coloured were "a nation in the making," which would get parallel, not separate, development.

MINING IN THE SOUTH AFRICAN ECONOMY

The Republic of South Africa, as has been suggested, is one of the leading mineral-

A view in the Transkei
This largest African area, now a Bantustan, may be given independence in a few years.

producing nations of the world. The sales of all minerals from the earliest date of record to 1973 total about $38 billion, with sales in the last 10 years exceeding those of the previous 45. In recent times, new records have been set nearly every year. The 1972 value of mineral sales was $2.7 billion and, with much higher prices being received for gold and other minerals, the 1974 figure was about $6 billion.

Mining's contribution to the GDP was given as 12 percent in 1972, when about 724,000 persons were employed, of whom 87.0 percent were African, 11.8 percent European, and 1.2 percent were Coloured or Asian. These figures only partially reveal the importance of mining to the South African economy—an importance more accurately portrayed by the facts that minerals and metals accounted for about 65 percent of exports—and gold alone for 36.7 percent—in 1972, that much of the transport and power services were constructed in response to the needs of the mining industry, that a large segment of industry is based upon and caters directly to the mining industry, that mining

indirectly permits the subsidization of a large part of South African agriculture, and that the progress of development depends significantly upon the level of mineral exports.

South Africa possesses a great variety of minerals, of which 26 have sales value exceeding $1 million, 17 have sales over $10 million, and 5 net more than $100 million. While South Africa has been more thoroughly prospected than most African countries, there are opportunities for initiating production in known bodies not now exploited and further discoveries may be confidently expected. The Republic is, indeed, a mineral storehouse, but the inevitable exhaustion that applies to any mineral body is a matter of continual concern to South Africans. This concern has led to a tendency to minimize the significance of mining to the economy and to the creation of programs designed to reduce the overall dependence on that activity.

GOLD

IMPORTANCE TO THE ECONOMY. Gold is by all odds the most important mineral produced in South Africa, measured by value of output, and historically it has accounted for over two-thirds of the total value of all minerals produced. In 1973, when the Republic mined over three-fourths of the free world's gold, revenues from the metal rose to an all-time record of $2.66 billion, up 77 percent from 1972 because of the large increase in bullion price (the average price was $51 per ounce in 1972 and $97 per ounce in 1973; it was expected to be at least $140 an ounce in 1974). The level of output, however, was the lowest since 1962—27.4 million ounces as compared to 29.2 million the previous year.

Gold mining forms the main basis for the economy of the Witwatersrand (the Rand) and the Orange Free State field centered upon Welkom. The large urban complex of the Rand, which has appropriately given its name to the South African monetary unit, has provided the stimulus for a considerable part of the Republic's agricultural and manufacturing development. With the closely associated developments at Pretoria and Vereeniging, the southern Transvaal industrial district is distinctly the most important in South Africa. While that area has other attributes, industry would scarcely have

started there had it not been for gold, which gives a locational orientation for a well-rounded industrial buildup that is unique in the world.

Gold mining also made it economically possible to construct railways up the difficult scarps from various coastal points, and the story of competition for the traffic of the Rand is a fascinating one in transport history. Durban and Lourenço Marques are particularly dependent on traffic to and from the Transvaal, but Port Elizabeth and East London also share in serving the area, as will Richards Bay after 1976.

Gold, like diamonds, was responsible for attracting large numbers of immigrants to the country and is a most important reason why South African whites have one of the highest income levels in the world. Gold mining has also provided ready employment for Africans from the Republic and other countries. While recruitment and handling of African laborers has had undesirable features (especially inadequate health measures in ealier years and the continued use of the compound system) it must be said that wages compare favorably with most but not all tropical African countries and that increasing numbers of Africans are prepared voluntarily to migrate to the gold fields for employment. Nonetheless wage levels are extremely low, even with recent high percentage increases, and the ratio between white and black wage levels is inordinately wide and not improving.

Finally, for many decades gold provided the South African economy with an extremely important stabilizing factor. A commodity that could be sold in international markets in whatever quantity produced at a price that did not vary in recessions obviously sustained other parts of the economy. The large increases in gold prices after its recent freeing from ties to the American dollar have provided a welcome boon to South Africa, though the element of stability is perforce reduced.

Gold mines in the Far West Rand field
Mines in this area are the deepest in the world.

The high dependence of the Republic's economy on gold, however, raises certain significant questions. How long can the economy continue to rely so heavily on the metal? What are the possibilities for developing a more rounded economy? The first question will be examined in the following section.

FACTORS IN THE PERMANENCE OF GOLD MINING. Gold is an exhaustible resource which must eventually give out. But the Republic's gold mines, which have led the world in output for 77 years, have had an unusually long life, and have already outlasted most competitors. This unusual endurance is caused by an uncommon array of fortuitous conditions.

Large reserves. Basically, of course, the most important factor is the vast extent of the reserves, which consist not of easily worked out alluvials or of thin, rich veins, but of low-grade and relatively uniform ores contained within a conglomerate matrix and extending over a large area. The gold-bearing conglomerates occur in thin tabular sheets called reefs, which occur in a sequence of sedimentary beds up to 5 miles (8 km) thick. The Rand, Orange Free State, and associated gold fields are by far the largest cognate group of metalliferous ore deposits being mined anywhere in the world. In 1974 their active mines stretched unevenly along an arc nearly 300 miles (483 km) in length. The fairly uniform nature of these deposits permits the industry to plan years in advance and to maintain rather even tonnages and grades.

Availability of coal. A second factor is the fortuitous presence of coal nearby. Originally, coal was obtained within the Rand itself, but the poor quality and difficult mining of this coal reduced it to a minor role, and the Witbank Collieries, some 90 miles (145 km) to the northeast, and those south of Vereeniging in the Orange Free State now supply the vast bulk of requirements.

Adequate water supplies. There are and have been certain problems associated with the supply of water in this semiarid area, but they have thus far been successively met. Control of the Vaal River has been particularly important.

Group administration. Under this organizational arrangement detailed comparative studies are made of every factor in costs that can be distinguished. Made available to each mine, these data permit the testing of individual operations against the average. New technological developments and scientific expertise are also shared, and cooperation is carried further by the provision of unified rail and power systems, central compressed air plants, and a central refinery at Germiston said to be the most efficient in the world. In addition, there is joint processing of by-products, especially uranium. The necessity for large amounts of capital and for making whatever savings in scale are possible helps to explain the highly developed cooperation in the industry. It may require $56 million to open a new mine.

Technological improvements. Technological development and innovation have successfully met many of the problems presented by the nature of the ore and ore bodies. Briefly, these problems include the low average grade of the ore (the average yield is about 1 ounce for each 2.97 tons of ore milled); extraction of as high a percentage of the gold as possible (the invention of the cyanide process in 1890 helped enormously); the great depth of working (few are less than 4,000 feet [1,220 m] below the surface and the deepest are 11,472 feet [3,497 m] down, which is deeper than any other mine in the world, and there are plans to take the Western Deep Levels mine to 12,500 feet [3,810 m]); difficult ventilation, occasioned by the large amount of dust, the great depths involved, and the increasing temperatures at lower levels (at 11,000 feet [3,353 m] the rock temperature is 123°F [50°C]); enormous ground pressure at depth (considered the most important physical problem today); and miscellaneous difficulties includ-

Gold mine dumps on the outskirts of Johannesburg

timber packs and roof bolting to support the overburden; underground refrigeration and improved ventilation techniques; pre-cementation of water-bearing strata; and a variety of improvements in the metallurgical processes.

Presence of by-products. Important by-products include silver, osmiridium, and platinum, and, most importantly, uranium oxide. When its extraction began in 1952, the gold mining industry benefited. Uranium in the gold reefs is the main part of South African reserves, which are estimated at 30 percent of the free world total at $6–$10 per pound ($13–$22 per kg). In 1972 the value of the 3,629 tons of uranium oxide produced was $34 million; prices have been low in recent years but are expected to increase along with the construction of more nuclear plants. South Africa is installing a $120 million full-scale prototype uranium-enrichment plant. It will use a new secret process developed there, which is expected to produce the lowest-cost enriched uranium in the world. Later, an $819 million plant would process 13,200 tons of uranium oxide to produce 2,600 tons of enriched product. This plant would require an amount of elec-

ing the narrowness of the reefs, the restricted ability to use gravity in removing the ore, heavy faulting in some of the mines, and the need to pump large quantities of water, especially from the Orange Free State mines.

A great many techniques have been developed to meet these problems: new methods, machines, and tools to speed sinking shafts, tunneling, drilling, loading, and hoisting;

Windbreaks placed on a mine dump in preparation for the planting of vegetation

tricity equal to about 23 percent of the 1973 generating capacity of the South African Electricity Supply Commission (Escom).

Availability of low-wage labor. The gold industry owes its continuance to the availability of large numbers of low-wage laborers. If the gold reefs were situated in the United States they probably would be of interest only to students of geology; they would not be worked. The average annual earnings of the 560,000 Bantu workers in mining and quarrying, of which the gold industry accounted for two-thirds of total employment, were $343 in 1972; the 60,000 white employees earned an average of $6,250.

Increase in gold price. The recent increases in gold prices will obviously have an important impact, lengthening the life of the industry, permitting the mining of lower grade ores, justifying larger employment, the opening of new mines and the reopening of abandoned mines. The price, which had been $35 per ounce from 1937 to 1968, climbed to as much as $196 an ounce in 1974. The increases have permitted long-delayed wage boosts, but they also increased profits and the value of gold shares dramatically.

After this brief account of the factors that explain the already long life of South Africa's gold mines, it is appropriate to ask again: How long may a high level of output be expected to continue? From the standpoint of physical reserves, one can only make a partially informed guess, because further exploration and the development of new techniques may well extend reserves considerably. We do know that production from the older fields has been declining since the 1950s, that quite a few mines have required subsidization, at least until recently, to sustain output, that only a very few new mines have been opened in the past decade, and that output has been rather static for about ten years. On the other hand, output from the Orange Free State, the Far West Rand, and the Klerksdorp Field has been in-

creasing. Mining began in the Orange Free State Field in 1948 and this field now accounts for over 35 percent of total output; the O.F.S. reefs have the advantages of appreciably shallower depths and higher gold content, but this field is relatively disadvantaged by greater faulting, a steeper geothermic gradient, and quantities of water which must be pumped to the surface.

Physically, it may be roughly estimated that a high level of production can be maintained for 40 years, but there is likely to be a gradual decline from the peak levels and perhaps a steeper one after about 1985. The president of the Chamber of Mines predicted in 1973 that gold output would decline by about 28 percent between 1972 and 1987. It is impossible to predict the effect the course of political events will have on the gold industry. Even though the industry has reasonable prospects of carrying on for 40 years or more, the heavy dependence of the country on this production provides an uneasy and finite base for its economy.

In examining the possibilities for diversification and for reducing the relative dependence on gold, it is appropriate to examine, first, the opportunities that exist through the production of other minerals. Even a cursory look reveals that the Republic is exceptionally favored in quantity and variety of mineral resources. It is, in fact, one of the greatest mineral storehouses in the world.

DIAMONDS

Since the output of diamonds of South Africa and South-West Africa is essentially integrated, the industry of the two countries will be discussed together. Their exports totaled $579 million from October 1973 to October 1974. South African production has been running around 7 million carats; they rank first among world producers of gem diamonds and second in output of industrial stones.

Diamond mining has played a very impor-

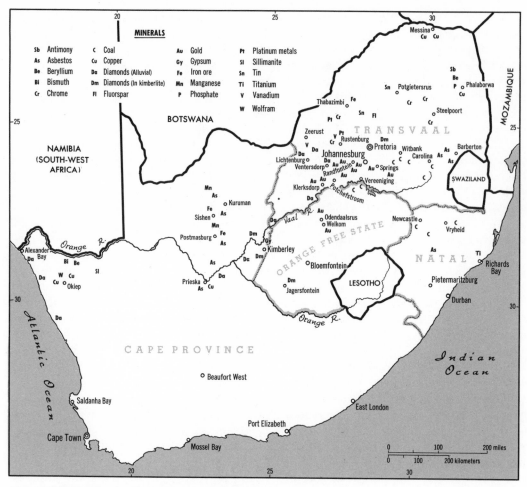

Map 106. Main mineral producing areas in South Africa

tant historical role in southern Africa. It attracted settlement and transport and provided large amounts of capital for other enterprises, including the production of gold, copper, and citrus fruit, and helped to open up the Rhodesias. In South-West Africa diamond mining is the leading economic activity, measured by value of output.

The circumstances under which diamonds are produced vary widely. The more important deposits in South Africa occur in volcanic kimberlite pipes, the main groups of which are centered on Kimberley, Pretoria, Jagersfontein, Koffiefontein, and Postmasburg. Only about a sixth of the more than

150 pipes thus far discovered contain diamonds in workable quantities and the bulk of production has come from eight pipes. The pipes vary greatly in size, the largest being the Premier Mine, 25 miles (40 km) east of Pretoria, which covered 78 acres (32 ha) at the surface; it is now mined from below the surface and has been worked to a depth of over 3,500 feet (1,067 m). This was the source of the 3,025.75 carat Cullinan Diamond.

Alluvial diamonds account for about 14 to 17 percent of the caratage of South African production but for 40 to 46 percent of value, because of the higher content of gemstones.

South-West Africa is notable for the high proportion of gemstones recovered—about 94 percent of the total as compared to a world average of 22 percent; it is the most important single source of quality gem diamonds in the world. Alluvial stones are won in the valleys of the Orange, Vaal, and Hartz Rivers between Prieska and Potchefstroom, in the West Transvaal, and on the coast of Namaqualand. In South-West Africa the major areas exploited are along the marine terraces at Oranjemund and Bogenfels, though the so-called Diamond Coast runs north as far as Conception Bay. Most of the output comes from onshore stripping, but foreshore mining has increased with the introduction of concrete prism walls permitting mining to the low-tide mark; the share from offshore suction dredging, which costs three to four times as much per carat recovered, has declined in recent years to only about 1 percent of the total. The average South African diamond mine moves about 30 million parts of ground to secure one part diamond; in the onshore South-West fields the ratio may be nearer 100 million to 1.

While the history of diamond mining is replete with difficult periods and lengthy closings of numerous mines, the years during and after World War II have generally been highly favorable, characterized by the frequent breaking of production records. The impact of periods of overproduction and lower prices is reduced by the Diamond Producers Association, which has apportioned production among its members, and by two international corporations, which control the bulk of international sales of gem and industrial stones.

COAL

While gold and diamonds are primarily important as earners of foreign exchange, coal is of enormous actual and potential significance as an energy base for the South African economy. Production of coal has risen from 48 million tons in 1961–62 to 62.4 million tons in 1973, and South Africa is now the world's seventh largest producer. By value of output, coal normally ranks second after gold, but in 1973 its sales value of $219 million was also exceeded by diamonds.

The known reserves of coal are given as 141 billion tons, rather evenly divided between the Natal and the Transvaal–O.F.S. fields, though large reserves are also believed to exist in the southeast of Cape Province. Seams are characteristically near the surface and are generally horizontal so that gently inclined shafts are usually used and coal is often removed by conveyor belt. Seams are thick, usually 4 to 15 feet (1.2–4.6 m), but up to 50 feet (15 m) thick in the Transvaal; they are thinner and more interrupted by faulting and volcanic intrusions in Natal. Underground mining is carried on largely by the room-and-pillar method, but the use of continuous mining machines is still not extensive. An increasing tonnage is being strip mined and a giant dragline with a 55-cubic-yard (42 cu m) bucket went into operation at the Optimum colliery in 1970.

While the mode of occurrence of South African coals is generally very favorable, this advantage is somewhat offset by the relatively poor quality of the bulk of the reserves. Most coal has a low calorific value and a high ash content and only 1.5 percent of the proved reserves are of coking quality. Attention is being given to mixing of coals and to the upgrading of noncoking coals by briquetting.

Coal is mined at exceptionally low cost, the average pithead price having been $1.93 per ton in 1973, about 28 percent of the U.S. level. This very low cost fuel is of great importance to the South African economy, particularly to the mining, iron and steel, oil-from-coal, cement, and transport industries.

The main producing areas are on the highveld around Witbank, Sasolburg-Vereeniging, Dannhauser, and Vryheid. Some coal has been exported (1.6 million tons in 1973) but large-scale shipments have been precluded by the shortage of transport and limited capacity at Durban. Several developments may alter this situation. One proposal calls for constructing a 500-mile (800-km) pipeline from a Transvaal mine to Richards Bay for shipment to Italy at a first-phase level of 5 million tons and a full-scale level of 15 million tons to be reached by 1983. A second plan would see construction of a pipe to carry coal to Saldanha Bay, part to be used for a 200,000-ton aluminum smelter and part to be exported. It is hoped that the value of coal exports will rise to about $ 110 million a year by 1980.

OTHER MINERALS

Some of the other minerals produced in South Africa are used primarily in the country, such as limestone, phosphates, and salt, and some are partly consumed domestically, such as iron ore, copper, asbestos, and manganese, but the most important function of most of these minerals is the earning of foreign exchange. In 1973 exports of minerals other than gold, uranium, and diamonds totaled $454 million; it is expected that their export value will double or treble in the coming decade. In many years, exports could have been greater had the railways been able to handle larger tonnages, but the bulk of a number of these minerals has been important in justifying the construction of a relatively superior rail system, and of special facilities in the main ports, as well as the opening of two new ports in the near future.

The most important mineral regions outside the gold reefs and diamond areas are:

(1) The Bushveld Complex. With the Great Dyke in Rhodesia this zone contains most of the world's chrome, the Rustenburg platinum belt, and deposits of iron, manganese, vanadium, titanium, nickel, copper, tin, bismuth, and asbestos. It appears in a 180-mile (290-km) crescent near Rustenburg (Map 106), reappears for 100 miles (160 km) near Lydenburg, and again for 40 miles (64 km) near Potgietersrust, and is one of the most remarkable mineralized areas in the world.

(2) The Transvaal Bushveld, situated around Phalaborwa in the northeast of the Transvaal Province, has iron, copper, phosphates, uranium, and other minerals.

(3) A zone from Prieska northward in the northwest Cape which contains iron, manganese, asbestos, diamonds, and copper.

(4) The area around Okiep in the western Cape, which contains copper and tungsten.

Very little mining now occurs south of about latitude 30°S.

METALLIC MINERALS. Among the precious metals produced in South Africa, osmiridium and silver come from the gold mines. The Republic is the leading Free World producer of platinum, the deposits being spread over a 9,266-square-mile (23,400-sq-km) area of the Bushveld Igneous Complex, particularly around Rustenburg. The use of platinum to reduce pollution from automobiles has greatly stimulated production in recent years and large-scale investments are underway for expanding present mines and opening new ones. Platinum is refined at Wadeville and at a new plant at Brakpan in the East Rand.

Copper ranks first among base metals. It is mined in a large opencast working at Loolekop near Phalaborwa, now the leading producer, at Messina near the Rhodesian border, around Okiep in Namaqualand, at a new mine near Preiska, and as a by-product of gold, platinum, and phosphate operations. The Phalaborwa mine is unique in Africa in the low copper content of the ores worked, only about 0.75 percent; in 1972 about 100,000 tons of copper were produced involving the moving of 42 million tons of ore and waste.

Manganese ranks after copper; South Africa is the second-ranking world producer of this most important additive to steel. The ores are mostly low grade, averaging about 40 to 48 percent metal content, but they have the advantage of being very hard, which minimizes disintegration in handling and storage. Production of manganese for carbon steel comes from the great Postmasburg deposits, 100 miles (160 km) west of Kimberley, and from west of Kuruman, both in the Cape Province; manganese dioxide used in the extraction of uranium oxide is supplied from scattered workings over a large area in the western Transvaal. Ferromanganese is produced in refineries at Newcastle and Cato Ridge in Natal and at Krugersdorp in Transvaal.

There are fairly large reserves of high-grade iron ore and enormous reserves of low-grade ore with a high silica content. The most important mines are at Thambazimbi ("mountain of iron") in the Transvaal and at Sishen in the northwest Cape. The cost of mining is exceptionally low. Ore from Sishen supplies the steel mill at Vanderbijlpark and has been exported since 1963 to Japan.

About a third of total production (c. 11 million tons a year) is now exported, but this share will change dramatically with the completion of a huge new investment program—the Sishen-Saldanha Bay project. The project involves opening a new surface mine at Sishen, laying a 232-mile (373-km) rail line to Saldanha Bay north of Cape Town on the Atlantic, enclosing the natural harbor there (which is four times larger than the present four main ports combined), and constructing storage and loading facilities to handle 250,000-ton ore carriers. Ore is to be exported at a 15-million-ton level by 1980 and double that level by 1990. These and ancillary needs will involve an investment of over $680 million, and shipments of ore are

expected to be valued somewhere between $190 and $230 million by 1980. A separate but related project involves construction of an $888–$1,036 million steel mill at Saldanha Bay to produce and export 3 million tons of semis beginning about 1978; it is being jointly financed by the South African Iron and Steel Corporation (Iscor) and the Austrian VOEST. Eventually the combined operations may be expected to bring in around a billion dollars in foreign exchange. Saldanha Bay, now only a fishing port, could also be developed to ship copper and manganese, while there is talk of constructing a huge drydock to service giant tankers and ore carriers using the Cape route.

Other metallic minerals of some significance include:

(1) Antimony, from the Letaba district, eastern Transvaal; South Africa is the leading non-Communist producer.
(2) Chromium, from various places in the Lydenburg and Rustenburg sections of Transvaal. South Africa is believed to have the world's largest reserves and is the leading non-Communist producer of this multipurpose mineral. Several plants, including a major one at Witbank, produce ferrochrome, mainly for export.
(3) Nickel, mainly produced as a by-product of the Rustenburg and Impala platinum mines.
(4) Tin, produced in the Transvaal Bushveld.
(5) Vanadium, which comes from very large reserves in the Bushveld complex, with mining centered at Roossenkal in the Eastern Transvaal and at Rustenburg.
(6) Zirconium, from Phalaborwa.

NONMETALLIC MINERALS. Except for diamonds and coal, asbestos is the most important nonmetallic mineral and one of the leading export minerals. It comes mainly from the Pietersburg and Lydenburg districts of the northern Transvaal and from near Kuruman in Cape Province. Phosphates come mainly from a mineral-rich volcanic pipe at Phalaborwa, where Foskor produces about 2 million tons of concentrates yearly for sale in the domestic market. Other nonmetallics include feldspar, fluorspar, gypsum, kyanite, limestone, magnesite, mica, pyrites, quartz, salt, and vermiculite.

It may be concluded that South Africa has an enviable position in the variety and quantity of its mineral resources. That country would certainly appear to have the greatest opportunities in Africa from the raw material standpoint to develop metal and metal-fabricating industries. Particularly significant are its reserves of coal and iron ore; while a shortage of coking coal may develop by 1990, it is possible that new processes will have been initiated by that time to permit using lower grade coals.

INDUSTRY IN THE REPUBLIC

In order to understand the potentialities for industry's reducing the dependence upon gold, one must examine the present state of development and analyze the attributes and deficiences of the physical and socioeconomic environment of South African industry. Manufacturing is credited with contributing 23.1 percent to the GDP as compared to 10.5 percent for mining and 9.0 percent for agriculture, fishing, and forestry. As suggested earlier these figures tend to understate the importance of mining. Employment in manufacturing increased from about 235,000 in 1938 to 1.16 million in 1970. The production index, with 1963 as the base year, stood at 199 in September 1973. Expansion has involved a diverse group of industries, which give South Africa a rather well rounded manufacturing sector. The most notable weaknesses are in the production of machine tools, machinery, vehicles, and newsprint, but advances are being made in several of these branches.

MAJOR BRANCHES OF INDUSTRY

If all phases of the metal and metal fabricating industry are combined, this branch ranks first in employment and gross value of output. The high importance and diversity of metal industries is unique on the continent.

The Republic is particularly favored by the possession of a number of modern steel mills, which produce pig iron and steel at very low cost. The two major plants are owned by the state corporation, Iscor. Its original plant at Pretoria has a capacity of 1.1 million tons. It is supplied with coal from Witbank, 66 miles (106 km) distant, and from its wholly owned mine at Dannhauser, 252 miles (405 km) away; iron ore comes from Thabazimbi, 150 miles (240 km) to the northwest; dolomite is quarried nearby and limestone is railed from northwestern Cape Province.

The Vanderbijlpark installation is newer and larger, having a capacity of 4.1 million tons. Located just west of the important steel-working center of Vereeniging and within the gold-bearing arc, it secures coking coal from Natal and lower-grade coal and electricity from the mines and pit-head stations in the immediate vicinity, iron ore mostly from Sishen, and fluxes from nearby and the northwest Cape. The Pretoria works

Electrolytic tinning line at the Vanderbijlpark steel works
This plant is the largest of three integrated steel works in South Africa.

specialize in bars, blooms, and billets, while the Vanderbijlpark mills produce plates, sheets, and strips. Both plants are well situated with respect to the country's leading industrial area and to the great markets of the mining industry.

A third integrated works, owned by a subsidiary of Anglo-American, is situated at Witbank; it produces about 300,000 tons of structural steels and rails, plus about 10,500 tons of vanadium pentoxide. A fourth plant at Newcastle, which produces pig iron for export, was taken over by Iscor in 1970 and is an integrated steelworks.

In addition to these mills there are several nonintegrated plants, including—at Vereeniging—the most important Klip Works, specializing in hollow drill steels, and the Vaal Works, producing a range of steels and castings—usually for the mining industry. Next there are a considerable number of ferroalloy plants producing a variety of alloys; important centers are Witbank, Kookfontein, Middleburg, and Cato Ridge.

Output of steel in South Africa increased from 344,000 tons in 1938 and 2.64 million tons in 1962 to 5.5 million tons in 1972–73. In 1973 a $4.4 billion expansion program was announced by Iscor calling for a production of 6.75 million tons by 1977 and 10.5 million tons by 1983 with capacities at that time to be 1.5 million tons at Pretoria, 3.0 million tons at Newcastle, and 6.0 million tons at Vanderbijlpark. Iscor will also hold a 51 percent interest in the Saldanha Bay plant mentioned earlier.

The metal products, electrical goods, and transport equipment industries are of growing importance to the Republic. The variety of products is far greater than for any other African country, with items of considerable complexity now being produced. The needs of the mining industry have stimulated output of tools, equipment, and machinery of increasing sophistication. The vehicle assembly sector is, with some seventeen plants, particularly well represented, and deliberate

efforts are made to increase the percentage of locally produced components, now about 52 percent of the total. A boom is occurring in the electronics industry with the introduction of television to the country in 1976.

The food, beverage, and tobacco sectors depend largely on South African raw materials and on the domestic market, but there is an important export of sugar, wine, canned fruit, and fishery products. There are about 300 primary mills and over 800 clothing and furnishing manufacturers in the textile and clothing branch. This sector is more heavily dependent on low wage labor than others and employs a higher percentage of nonwhites, including many Indian and Coloured women. It has faced a particular shortage of skilled labor, explained in part by the lower level of wages prevailing. The industry is protected and incentives have been extended in recent years to encourage movement to border locations (see below).

The chemical industry owes its importance partly to demand from the mines for explosives and chemicals, though a considerable range of products is produced for agriculture and for general consumption. Of special interest is the chemical complex at Sasol-

Part of the SASOL oil-from-coal plant
South Africa is expanding facilities at Sasolburg and constructing a new installation at Richards Bay.

burg centered on the government-owned SASOL oil-from-coal plant, the largest in the world. In 1971 a $98-million refinery using imported crude piped from Durban was added; the first inland oil refinery in South Africa, it is owned jointly by South Africa, the French Total, and the National Iranian Oil Company. At present SASOL produces 70 chemicals ranging from by-product nitrogen, used in a nearby fertilizer factory, to synthetic rubber stock, ethylene, and waxes. SASOL is also planning, in association with Shell/BP, a massive petrochemical complex at Richards Bay with double the capacity at Sasolburg, while the OAPEC boycott on oil shipments to the Republic has expedited plans to increase the oil-from-coal capacity at Sasolburg by 40 percent. Large crude oil refineries are also situated at Durban and Cape Town.

THE LOCATION OF SOUTH AFRICAN INDUSTRY

The distributional pattern of the Republic's industry reveals one major district in southern Transvaal and adjacent O.F.S. and a series of nodes associated with the major coastal cities. The southern Transvaal industrial district has a cruciform shape with each axis about 80 miles (130 km) in length and signs that it is evolving to a diamondlike pattern. Containing about 18 percent of the total population and 38 percent of the white population, the district has three major nuclei: Pretoria, Johannesburg and the inner zone of the Rand, and the Vereeniging-Vanderbijlpark-Sasolburg node. In addition to their importance in manufacturing these nuclei represent respectively the government administration, the national office, and the heavy industrial-chemical centers of the Republic. This district accounts for about 55 percent of the gross output in manufacturing, 37.5 percent of the country's industrial establishments, and around 45 percent of the labor force in secondary industry. Jo-

hannesburg alone is credited with 20 percent of the gross output of industry in the Republic.

There are numerous explanations for the dominant position of the southern Transvaal industrial district; foremost is its position with respect to the enormous market of the gold mining industry, but it is also well centered with respect to other mining communities, to the large conurbations of the area, and even to the national market. The availability of coal and cheap carboelectric power is a second major advantage, while of lesser significance are the availability of labor and the existence of a distance tariff in this inland industrial zone. The chief disadvantage of the district is its long distance from domestic ports, compensated in part by the relative nearness of Lourenço Marques, which is supposed under the Mozambique Convention to handle 40 percent of the merchandise imports for a defined area including most of the region. In the long run, the supply of water could become a more serious problem; it has already required rather substantial capital works programs. And the relatively declining position of the Rand in gold mining also raises interesting questions with respect to long-range industrial location. While this district is particularly important in the iron and steel, metal and engineering, and chemical industries, it also has sizable representation in the other major groups of South African industry.

Among the coastal centers, Cape Town and its satellite communities rank first in gross output, with roughly 15 percent of the total, but second in industrial employment. The western Cape district is important in the manufacture of foodstuffs, wines, tobacco products, textiles and clothing, and furniture. Durban and Pinetown rank just below the Cape district. The more important industries in Natal include sugar milling, rubber goods, chemicals, tanning, rayon textiles, paper, and oil refining. Richards Bay provides a new industrial node in Natal; it has a major aluminum refinery and is slated to receive a huge chemical complex.

Port Elizabeth and Uitenhage produce about 5 percent of the Republic's industrial goods. The assembly of motor vehicles and production of tires are of special importance, while textiles are of growing significance.

It has been government policy for some years that industry be encouraged to disperse more widely and, as an integral part of its separate development program, that industry be established in localities immediately adjacent to African reserves. This permits the use of African labor with their being in European urban areas only as commuters. Certain transportation rebates, tax, loan, and other incentives are offered to encourage such industrial placements, while the Physical Planning and Utilization of Resources Act of 1967 gave the government power to require the movement of specific plants, especially those with a high ratio of black to white employees.

The concepts and approaches of the border industry program have changed since its inception in 1960. In 1969 several of the successful nodes were removed as having achieved a self-sustaining level while a number of growth points were delineated for particular attention (Map 105). Advantages of the border industry program for the industrialist include the numerous inducements offered, the lack of necessity to provide housing (it is constructed by the government in townships within the reserves), and the possibility in many locations of paying lower wages and getting partial relief from job-reservation edicts. On the other hand, he may have to be satisfied with less-skilled labor, face difficulties in the recruitment of white employees, and be inefficiently located with respect to the market.

Advantages to the government are largely political, since the program tends to satisfy

Part of the central business district of Cape Town
Cape Town and its satellites comprise the second-ranking industrial district in South Africa.

the goal of decreasing the black to white ratio in existing cities, but economic gains accrue from the lower cost of constructing houses in the African areas and potentially from dispersion of industry. On the negative side, however, government may be promoting uneconomic location and is required to make expenditures that are either unnecessary or would otherwise be privately financed in part.

Advantages to the Africans are that the border program may create some new jobs, that housing is improved in the reserve townships, that more money will presumably be spent in the African areas, that migration may be reduced, and, perhaps most important, that the worker can live with his family. On the other hand, buildup of border industries is likely to make it more difficult to initiate industry within the reserves/Bantustans, to keep wages lower and hours of work longer, require those individuals moving to townships to relinquish claims to farming lands in return for a somewhat insecure tenure in the township, make the black a prisoner in his "homeland," and not provide real relief from job reservation discrimination.

It is difficult to assess the border industry achievements because of the lack of data, conflicting claims, and changing definitions. In the thirteen-year period from inception through 1973 it is claimed that 113,000 jobs were created in border industries, of which 91,000 were for Africans, and that $876 million has been invested in decentralized industries. This is not an overly impressive record, though it should be noted that there was a marked acceleration in later years, because the number of jobs created is nowhere near the number required to absorb new workers requiring employment or the number created in existing centers. Furthermore, there is in some cases a somewhat artificial character to the whole program in that most of the successful border industries have thus far been located in recognized manufacturing centers which happen to be adjacent to African areas, and there has even been some redrawing of boundaries, which distorts the success ratio.

It is entirely possible that additional boundary changes may be made and even that large African townships outside major cities may be given the status of reserves or Bantustans, thus markedly increasing the apparent success of the program.

In 1970–71 a new element in the dispersion of industry program was added with the delineation of fourteen "other economic development areas," ten of which are intended to provide employment for Coloured workers and four for Indians (Map 105), hopefully reducing their concentration in the major cities.

At present, there is very little industry within the African areas, but certain changes may result in some increase in the years ahead. Six growth points inside these areas have been named, and removal of earlier restrictions permits the starting of industries by white investors on an agency basis, with

white employment and ownership to end within a twenty-year period.

FACTORS INFLUENCING INDUSTRIAL GROWTH: THE PHYSICAL CONDITIONERS

The above brief review of South African industry and industrial location will permit a better understanding of the major factors that now affect and that will influence the further development of manufacturing in that country.

RAW MATERIALS. Obviously, the mineral position compares very favorably with most countries, although bauxite, fertilizer minerals, and sulfur must be imported in substantial tonnages and petroleum is lacking. Of greatest importance is the ability to produce low-cost steel, but there are numerous opportunities for the further processing of many minerals, both for domestic consumption and for export. The position with respect to agricultural raw materials is much less favorable, though such products as wool, corn, sugar, wine, and fruits do support sizable establishments. Fisheries supply a growing manufacturing sector, which can be further rationalized. The forest resource picture is not impressive, though plantings will doubtless support larger pulp, paper, and cellulose industries.

ENERGY. The energy position is characterized by the dominance of coal, available at exceptionally low cost. Coal provides an estimated 76 percent of total energy consumed; South Africa is unique in the persistently dominant status of coal in postwar years. With markedly increased prices for imported oil and the boycott on shipments from Arab countries announced in late 1973, the country may be expected to maintain or increase the role of coal, while many foreign countries have made inquiries about the availability of coal for export.

The government-owned Electricity Supply Commission (Escom) produces about 80 percent of electricty generated; the remainder

Taaibos carboelectric station near Vereeniging
Very low cost coal and large-capacity plants produce electricity at very favorable rates.

comes from municipal-, industrial-, and mine-owned plants. Total electricity generated in South Africa has increased from 9.3 billion kwh in 1948 and 24.6 billion in 1961 to about 65 billion kwh in 1972–73. Almost all electricity comes from carboelectric plants, some of which have very large capacities (the newest is a 3-million-kw installation). There is a notable concentration of large plants in east Transvaal, from which grids extend over much of the country including the 995 miles (1,600 km) to Cape Town. Escom's 21 stations produce the lowest cost steam-generated power in the world. The mines consume about 41 percent of Escom's sales. The low cost of electricity is attributable to the low cost of coal and labor, and to savings in scale because of the large demands of mining and industry.

Hydropower resources of the Republic are not very great and relatively few hydroelectric stations have been constructed. The small catchment areas of streams in the wetter east, the low and seasonally variable precipitation over the bulk of the country, and the general absence of attractive storage sites reduce the possibilities, despite the average high elevation of the area. Water control projects in South Africa are mainly concerned with urban-industrial supply and irrigation rather than with hydroelectric production. As noted in chapter 22 Escom has contracted to take a large block of power

from Cabora Bassa in Mozambique, but this is not likely to exceed 2 percent of total supplies in 1979.

Studies are underway regarding the erection of nuclear electric plants, the first of which is planned for Melkbosstrand, near Cape Town, with a capacity of 0.8–1.0 million kw by 1982, while it has been forecast that South Africa will be generating a third of its electricity from nuclear plants by the year 2000. Since 1964 an intensive search for oil and gas has been underway, both on and off shore, but only small offshore gas finds in Plettenberg Bay and off South-West Africa have been reported. The search is credited, however, with uncovering a huge uranium province in the central Karoo and a rich coal field in South-West Africa.

WATER SUPPLY. As suggested in the discussion of the southern Transvaal industrial district, water supply is a subject of some concern with respect to South African industry. In periods of severe drought the shortage in that area has nearly shut down industry and mining, partly because of unwise allocations of water to irrigation. In the future greater attention will have to be given to the treatment of sewage and industrial effluents to secure industrial water, to reduction of water for irrigation (which brings far lower benefits), to various schemes to direct water into the river systems best capable of serving the industrial communities, and to replacing some of the reservoirs (which are silting up at rates of 3 to 5 percent a year). The shortage of water may also influence certain industries to locate along the coast, especially in Natal.

Two diversion schemes that are likely to be completed in the near future are the Malibamatso Scheme in Lesotho (see chapter 24) and the Tugela-Vaal Project, which will pump water 1,640 feet (500 m) up from the eastward-flowing Tugela to the Sterkfontein Dam on the edge of the plateau on the fringe of the Vaal River catchment area. In addition to increasing the volume of the

Vaal, this will permit the storing of presently wasted flood waters in the existing Vaal dams, and yet keep the average level behind them lower to reduce evaporation losses.

OTHER PHYSICAL CONDITIONERS. An additional physical factor that has some influence on South African industry is the location of the country at the extremity of the continent in a position of partial isolation. The plateau configuration of much of the country also tends to increase the costs of transport, as does the scattered nature of industrial centers and the internal position of the southern Transvaal.

FACTORS INFLUENCING INDUSTRIAL
GROWTH: THE NONPHYSICAL
CONDITIONERS

MARKET. The restricted size of the domestic market is perhaps the most important limiting factor on South African industry. While the white population has a relatively very high standard, the 70 percent of the population which is Bantu has a total income somewhere between 25 and 30 percent of the national income, and individual African cash incomes are not adequate to stimulate production of a great variety of goods that might otherwise be justified. Furthermore, most South African political and economic policies work toward continuing the low standard of the African worker. An interesting example is presented by the television industry. Its introduction was postponed for several decades for fear that it would bring foreign influences which might undermine Afrikaans and the Afrikaner way of life; now with its start in 1976 the market is probably too small to justify manufacturing the tubes in South Africa.

Important in developing the domestic market for specific South African manufactures has been the government policy of tariff protection pursued since 1925.

The restricted size of the domestic market naturally leads to the examination of possibilities for catering to foreign markets. Within Africa, the largest available market would appear to be in Rhodesia, but that country has stringent exchange restrictions and pretensions toward developing an integrated industrial establishment. Many black African countries have placed embargoes on goods from South Africa, though a number do not enforce them. The Republic could assume a position of leadership in certain products, particularly of iron and steel and metal-fabricated items, but its racial policies restrict the realization of this potential. As for markets outside Africa, the greatest opportunities would appear to lie in the sale of semiprocessed metals, including steel.

HUMAN RESOURCES. Here there is a sharp artificial dichotomy between white and nonwhite. Whites receive excellent wages and fill practically all of the managerial, professional, and skilled labor jobs in industry, plus many of the semiskilled positions. Africans, Coloureds, and Indians provide the unskilled and many of the semiskilled jobs and receive wages that, irrespective of productivity, bear little relation to those of the whites. Although fewer than a quarter of persons employed in industry are white, they receive over three-fifths of wages, and the ratio does not appear to be improving despite recent increases in wages to Africans. The average African wage is below the Poverty Datum Line (PDL) calculated on government standards, and far below the "minimum effective level," put at 50 percent above the PDL. It is estimated that in the towns about 67 percent of Africans, 50 percent of the Coloureds, and 20 to 60 percent of Indians (depending on locale) live in poverty. Only 2 percent of the whites are at that level.

Advantages in the labor-management sphere as far as industrial development is concerned would include the low wages of Bantu, Asian, and Coloured labor, which permit exceptionally low-cost production of many items. A second advantage is that much of the nonwhite labor is more evolved and experienced than that of other African countries, hence the skill factor is relatively favorable—especially since a significant number are employed at levels that contravene the regulations. Third, the white population—by far the largest in sub-Saharan Africa—is familiar with the modern, machine civilization and capable of moving into industry with relative ease. The skill of many South African engineers, managers, and workers is unquestioned.

Unfortunately, the disadvantages in the labor-management area are equally if not more impressive; perhaps depressing is a better word. The most immediate bottleneck to industrial expansion is the shortage of managerial talent and of skilled labor. This shortage must be attributed in major degree to the policies, written and unwritten, that restrict the advance of the bulk of the population and thus markedly reduce the pool from which skill might be obtained. The shortages do result in encouraging the use of labor-saving equipment to improve productivity; they have forced the upgrading of nonwhites in the face of restrictive laws, and have created pressure to raise wages. But low wages for nonwhites combined with restrictions on their advance have a profound effect on the productivity of all races: whites do not need to produce as they are protected, others have little incentive to do so. It is not surprising that in the period 1952–74 improvements in productivity in South African industry were lower than in most Western developed nations.

In recent years considerable pressure has developed to rectify the situation to some degree. Many white industrialists have long called for removal of restrictive labor legislation, but white unions have joined the chorus, and even Prime Minister Vorster called, in late 1973, for "orderly change" to allow blacks to move into job categories previously reserved for whites. A series of illegal strikes by Africans in the past few years has played no small part in moves to increase wages and correct other conditions, not only in manufacturing.

The dependence on African labor in industry has been increasing over the years. They represented 44 percent of the total in the years 1924–30 and 54 percent in 1972, when whites made up 23 percent of the total employed, Coloured 17 percent, and Asians 6 percent. But the number of Europeans employed in secondary industry increased from 68,800 in 1929 to 283,000 in 1972, giving the lie to the notion that any gain by one group is at the expense of another.

The other disadvantages of the labor position are similar to those for most of Africa—impermanence, instability, and so forth. Despite some measures to reduce these deficiencies, many laws must have the effect of sustaining them. South African efforts to maintain the status quo in employment relations are not only directly contradictory to sound economics, but are fraught with the most serious long-run dangers for that country, while the effective contributions of all races could yield most promising rewards.

CAPITAL. South Africa is capable of providing a high portion of its own capital needs, but foreign investment has been substantial. Campaigns by church and other organizations calling for disinvestment from South Africa have thus far had relatively little impact, though criticisms of the policies of foreign firms operating in South Africa have resulted in certain improvements in the treatment of nonwhite employees. The government itself holds a considerable stake in manufacturing, including Iscor, Sasol, Escom, Foskor (Phalaborwa mining and fertilizer production), Saiccor (with participation in a large rayon pulp plant at Umkomaas, Natal), and the Bantu Investment Corporation (designed to promote industrial and other developments in the reserves and Bantustans).

To conclude this section on industry, it may be said that South Africa has very substantial opportunities in the industrial realm, but that it is not likely to realize them unless the bulk of its population is freed from the mass of restrictions curbing its advance and until its nonwhite population provides a more powerful market for South African manufactures.

AGRICULTURE

Agriculture appears to be a potential weak link in the South African economy. Development in some regions is impressive and there are potentialities for intensification and extension, but agriculture cannot be expected to replace gold or to provide more than a partial answer to the need for an expanding, more diversified economy. And, despite substantial improvements that have been made in the farming sector, its relative importance has declined and its index of production reveals a considerably slower rate of growth than that of other sectors. In 1970 agriculture, forestry, and fishing accounted for an estimated 9.0 percent of GDP. Agriculture remains, however, the most important single source of employment; about 91,000 white farmers and 950,000 African employees—including domestic servants—work on European farms, while there continue to be many persons engaged in agriculture in the reserves.

The labor situation on white farms has been the subject of considerable criticism; wages are too low to attract workers who can secure positions in other occupations, so farmers have depended in part on prison workers and clandestine migrants from Mozambique. Housing is also inadequate, and squatters do not follow soil conservation practices. In recent years a large number of squatters and tenant farmers have been removed from the white areas. White farms produce over 90 percent of the agricultural output in the country, including much of the staple corn consumed by Africans in the Bantu areas. The volume of agricultural production has kept pace with the population increase in the country and exports of farm products are valued at about 4.1 times imports.

THE PHYSICAL BASE

The most important explanation for the limited potentialities of South African farming is the difficulty of the physical environment (especially the deficiencies of climate and soils) but also the nature of landforms (particularly in the more favored climate regions).

TOPOGRAPHY. While it is possible to distinguish scores of landform regions in South Africa, the overall pattern consists of a narrow coastal belt on the west, south, and east, which rises sharply inland to the great interior plateau varying in elevation from about 2,000 to 6,000 feet (600–1,800 m). The plateau is characteristically bounded by a chain of mountains that form the outward-facing escarpments and that are highest on the east, where the Drakensberg Range has isolated peaks reaching 10,000 feet (3,048 m). At the base of the escarpment is the lowveld, generally below 1,800 feet (550 m), and in some places intermediate slopes provide a middle veld. The plateau surface is mostly a great plains area, varied in parts by broken country and by steep-sided, flat-topped masses.

Taken alone, the landform factor is not too unfavorable, and the vast plateau surface permits ease of movement over much of the country. But the form of South Africa as a great uplifted basin, which restricts the area in desirable coastal lowlands, confines the favorable Mediterranean climate to a very small section below the Karoo escarpment in the southwest (an area further limited by the folded mountains of that region), similarly restricts the potentially highly productive humid subtropical region to the east so that this most propitious climate occurs in large

Part of the Drakensberg Range
Extending along the east and rising often abruptly to elevations of as much as 10,000 feet (3,048 m), this range is a
formidable impediment on the routes between the coast and the interior plateau.

part in areas possessing landforms unsuitable for agricultural activity, and means that the best landform area is in the rain-shadow of the uplifted plateau edges. Other disadvantages of the topography, which influence all aspects of the economy, include the difficulty of building and operating transport lines from coastal points to the interior, and the lack of any navigable streams in the country.

CLIMATE. It is climate that presents the most severe limitations to agricultural intensity. Climate types vary from dry subtropical or Mediterranean in the southwest (a favored region for fruit, wheat, grapes, and tobacco) to humid subtropical along the east, suitable for a great variety of crops, to subtropical and tropical semiarid and arid regions over the bulk of the interior. Precipitation varies from 4 to 6 inches (100–150 mm) in the west to over 70 inches (1,780 mm) in very small portions on the eastern side of the mountains. Only 2.8 percent of the country has an average of over 40 inches (1,000 mm); only 10.2 percent has over 30

inches (760 mm), the amount needed for intensive farming with the rates of evaporation characteristic of the country; an additional 11.4 percent has between 25 and 30 inches (635–760 mm), which means that only 21.6 percent has annual rainfall adequate to support successful crop production. At the other extreme, 30.3 percent has less than 10 inches (250 mm), 64.7 percent less than 20 inches (500 mm) and 78.4 percent less than 25 inches (635 mm).

Other disadvantages of the climate include a generally high evaporation rate, high variability in rainfall, the occurrence of violent storms in the rainy season (which contribute to the serious soil erosion problem), and the marked seasonality of rainfall. Over 85 percent of precipitation occurs from October to April. Exceptions occur in the small Mediterranean region, which has rainfall maxima in winter, and on the south and east coasts, where rain falls year round (though with increasingly marked summer maxima toward the north).

There is a remarkable uniformity of mean

annual temperatures over the greater part of the country; the most striking differences are between those of the west coast—affected by the cold Benguela Current—and those of the east coast—influenced by the warm Mozambique Current. On the other hand, there are considerable differences in the annual temperature ranges, with greater extremes occurring in the inland locations. Sharp frosts occur in most of this area one or more times in the winter, while high summer temperatures reduce the effectiveness of the rainfall. On the highveld the winter cold and increased diurnal and annual ranges of temperature limit the production of tropical fruit, while winters are seldom sufficiently cold to permit the growth of middle-latitude deciduous fruit. Yields of corn on the plateau are reduced by the short growing season, by extremes of temperature and humidity, and by periodic droughts.

SOILS. In general, the soils of South Africa are relatively poor and difficult to manage. There is often a serious deficiency in phosphorus, humus content is low, soils mechanically weathered from ancient rocks are liable to compaction, there is much leaching in the rainy season, termites and grass fires have deleterious effects, and large areas are stony or mountainous. Elevation, steepness of slope, and rainfall characteristics lead to much soil erosion, which remains a serious problem despite a multipronged program to promote conservational practices. Despite the difficulties, progress has been made in scientific soil management, and the use of fertilizers has increased greatly.

THE IMPORTANCE OF PASTORALISM

The average low capacity of the land in the Republic explains the high importance of extensive grazing, the predominance of pastoralism, and the fact that wool is the most important surplus product from the land. From the early days, cattle and sheep raising have been the principal interests of the rural population. Even today, nearly 90 percent of the country's agricultural area is used for stock breeding, and livestock accounts for about 45 percent of the total earnings in agriculture, though this represents a reduction from 73 percent in 1918. Difficulties in livestock grazing include the low carrying capacities of the drier regions, heavy losses in drought years, the poor quality of many veld grasses, and the prevalence of animal diseases, said to be worse for sheep than in any other major producing country.

Wool is the most important pastoral product and ranks second or third among the country's exports. There are about 33,000 white and 12,000 nonwhite growers, but it is estimated that 750,000 people are supported by sheep farming. Average white sheep holdings are about 1,100 animals on 8,500–10,600 acre (3,440–4,290 ha) farms.

The present sheep population is about 34 million of which about 13 percent are non-woolled. The vast bulk of the woolled sheep are merinos. The average yield per sheep has gradually increased from 6.2 pounds (2.8 kg) in the early 1930s to 8.5–9 pounds (3.9–4.1 kg) today, while the total shorn wool production increased from an average of 238 million pounds (103 million kg) in the 1940s to 305 million pounds (138 million kg) in the 1960s. The 1973 clip was only 208

Merino sheep at Maclear, Cape Province
Aridity over most of South Africa limits the extension of tillage agriculture, and nearly 90 percent of the country is used for grazing. Wool is the leading agricultural export.

million pounds (108 million kg), because of drought conditions.

The arid Great Karoo is the homeland of the South African merino; Cape Province has about 66 percent of the sheep and O.F.S. has 23 percent. The value of wool sales fluctuates widely, having been about $231 million in 1973, up 165 percent from the previous year despite being 6 percent lower by weight. There is also a significant goat population and production of mohair.

The beef industry of South Africa is based primarily on Africander cattle, an indigenous breed well adjusted to local conditions, and is centered in the eastern half of the country. Bantu own about two-fifths of the approximately 12.3 million cattle, but production is normally not adequate to meet domestic needs. Output from dairy farms has markedly increased. South Africa produces about 28 percent of the cow's milk in Africa.

TILLAGE AGRICULTURE

Sixty years ago South Africa was unable to feed itself, although farming had been the traditional occupation for over 250 years. Today it produces most of its food requirements and about 30 percent of the agricultural raw materials consumed by secondary industry, and derives about 28 to 32 percent of nongold export earnings from crop and livestock products. Production of most such products is substantially above local demand, and as a result large quantities are exported, often at subsidized prices.

Practices in tillage farming remained poor for many years. But important changes have occurred in recent decades, including a great increase in mechanization (the number of tractors rose from 6,019 in 1937 and 22,292 in 1946 to over 170,000 today), a considerable replacement of oxen by beef and dairy cattle, greater attention to pasture improvement and rotational cropping, more regional specialization, increased application of fertilizer, and better erosion-control and moisture-conserving practices. Standards of Bantu farming remain very low and have not been given serious attention until recent years; average holdings in the African areas are extremely small.

A feature of South African agriculture is the controlled marketing of a variety of crops and livestock products. A score of control boards regulate some three-fourths of the value of agricultural produce marketed, sometimes at a substantial cost to the government and the consumer. Freight rates for corn and other products are heavily subsidized, as is the provision of irrigation water. This suggests again the dependence upon mining, from which much of the costs are indirectly extracted.

MAJOR CROPS. The principal field crops of

Africander bulls on a stud farm in the Orange Free State
The beef industry of South Africa is based primarily upon this indigenous breed.

South Africa are corn, wheat, oats and other cereals, potatoes, sugar, peanuts, sunflower, and tobacco; field crops account for about 45 percent of the gross value of agricultural production. Corn accounts for about 70 percent of the total cropped area (c. 10.5 million acres, 4.2 million ha), and 45 percent of the European tilled lands. It is the main staple food of the Bantu population and about a third of production is used as feed. The so-called "maize triangle"—with apexes at Smithfield, Vryburg, and Nelspruit—accounts for 95–99 percent of the marketed corn (Map 107). This region is rather sharply delineated on the west by the 20-inch (500 mm) isohyet, on the north by bro-

ken country and reduced effectiveness of precipitation, and on the east by the lower temperatures and moister summers as elevation increases toward the Drakensberg. Production increased about 5½ times from 1924 to 1973, which saw a record production; average yields have more than doubled with the use of hybrid seeds and better practices.

Wheat is a winter crop in the southwest Cape, the irrigated valleys of the Transvaal bushveld, and on the highveld of the Orange Free State. Production costs are high, yields low, and the country does not produce enough to satisfy domestic needs. Sorghum (usually called kaffir corn) is, like corn, produced in surplus; about half of the crop is

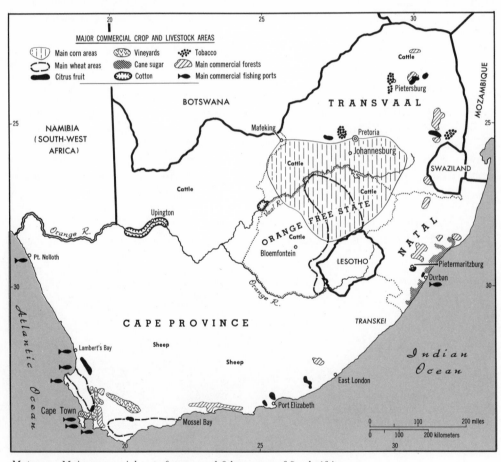

Map 107. Major commercial crop, forestry, and fishery areas of South Africa

grown by Africans and used mainly for brewing kaffirbeer, a nutritiously rich and alcoholically weak drink.

Sugar cane is grown in the coastal belt of Natal on about 800,000 acres (324,000 ha). The main belt stretches from 90 miles (145 km) south to 170 miles (274 km) north of Durban and inland usually no farther than about 8 miles (13 km) from the sea. Despite only marginally favorable climatic conditions, a two-year maturing period, and the practice of growing cane on slopes where machines cannot be used, South Africa ranks as the seventh world producer and exported more than a million tons valued at over $148 million in 1973 (a record). About 70 percent of the plantation acreage is owned by whites, though they represent only 25 percent of the 8,304 cane growers, Bantu and Asians accounting for about 53 and 22 percent of producers respectively. Part of the sugar is marketed in the United Kingdom under a bilateral agreement and part in the United States, where South Africa enjoyed a share of the market under the elaborate and politically manipulated quota system, which was

Sugar cane fields in Natal Province
About 800,000 acres (324,000 ha) in the coastal belt of Natal are planted to cane. The rugged terrain makes it difficult to introduce mechanization.

finally abandoned in 1974 when sugar prices had skyrocketed to unprecedented heights.

Horticultural products account for roughly a sixth of the gross value of agricultural production. Fruits are of considerable importance in specialized areas, and about 80 percent of total output is exported. Deciduous fruit, particularly apples, pears, peaches, and grapes, have gained rapidly

Vineyards near Worcester in the small dry-subtropical belt near Cape Town
About 250,000 acres (100,000 ha) are devoted to the vine.

and exports are now twice the value of citrus fruit; tropical fruits such as bananas, pineapples, mangoes, and papaya are also grown.

The major fruit-growing areas are in the western Cape, the coastal areas of the eastern Cape, around Durban, and in the margins of Transvaal. The Zebediela Estate in northern Transvaal is said to be the largest citrus estate in the world. South Africa is greatly advantaged by off-season production with respect to the markets of the northern hemisphere.

Viticulture is confined commercially to the southwest Cape where about 250,000 acres (100,000 ha) have been planted to the vine, which supports over 5,000 farmers and 100,000 people. Paarl is the main center and the headquarters for the Cooperative Winegrowers Association, which regulates the supply to the domestic market and has succeeded in enlarging the overseas market as well. The better quality wines are generally produced in the coastal belt on slightly acidic slope soils of moderate fertility and without irrigation. In the second main belt, along the Little Karoo, the grapes are irrigated on rich alluvial soils and yield a sweet wine, rather than the more delicate, dry wines of the coast. Wine is South Africa's oldest export, the first grapes having been pressed in 1859.

Tobacco, mainly a Virginia flue-cured type, is grown fairly widely, but the main producing regions are north and west of Pretoria; South Africa imports small quantities of tobacco for flavoring, but is also the second-ranking exporter on the continent. Cotton is grown in scattered areas, of which the most important is along the Orange River (Map 107), but output provides only about a third of domestic needs.

IRRIGATION DEVELOPMENTS. Since water is the most serious shortage in South African agriculture, it is not surprising that considerable attention has been given to irrigation projects. About 2.1 million acres (850,000 ha) are now irrigated; while this is less than 1 percent of the utilized agricultural area it ac-

counts for an estimated 17 percent of the value of agricultural output. Most schemes are scattered and rather small. The largest one is fed by the Vaal Haartz Dam and totals only 86,000 acres (34,800 ha). The long-term Orange River Project, started in 1963 and not due for completion before 1990, will irrigate 756,000 acres (306,000 ha), raise white farm output by an estimated 10 percent, and hopefully bring self-sufficiency in wheat and cotton production.

The project includes (1) construction of three major and nine lesser dams; (2) diversion by 51- and 32-mile (83- and 51-km) tunnels of some water from the Orange to the Fish and Sundays Rivers, whose valleys are better suited to irrigation; (3) piped water supplies to Bloemfontein, Kimberley, and Port Elizabeth; (4) installation of 20 hydroelectric stations with a total capacity of about 200,000 kw. The largest dam, named after Hendrik Verwoerd, was completed in 1972; it is 260 feet (79 m) high and 2,130 feet (649 m) long and can be raised later to give adequate regulatory capacity and offset the accumulation of silt.

Most of the more favorable irrigation schemes have already been developed and opportunities for extension are at best limited, because large quantities of the available water must be used to supply industrial, mining, and domestic requirements. It may, in fact, prove necessary to buy out some existing irrigation rights to secure additional quantities of water for these purposes. The problems inherent in water control schemes in South Africa are well illustrated by the Orange River Project; they include:

(1) The characteristically relatively small runoff of streams. The mean annual discharge at the Orange mouth is only 12.3 billion cubic meters from a catchment area of 328,000 square miles (850,000 sq km).

(2) The heavy losses from evaporation. Losses in the Orange River will be al-

most one-third more than that at the Vaal Dam, which is 6.7 feet (2 m) a year from an average depth of 11 feet (3.4 m).

(3) The aridity of large parts of the country. Over 98 percent of the Orange River runoff is contributed by about one-third of its catchment area.

(4) The problem of competing uses. It has been decided that all waters of the Vaal will be reserved for its catchment, that those of the Caledon will be allotted to the Orange Free State, and that part of the Orange River waters will be diverted for city use in other river basins.

(5) The relative poverty of the country in hydroelectric resources. The combined capacity of the 20 stations of the Orange River Project will be only 6.7 percent of the installed capacity of the largest carboelectric station.

(6) The high cost of most schemes. The cost-benefit ratio of most schemes, including the Orange River Project, compares quite unfavorably with schemes in other parts of Africa or on other continents.

OTHER ECONOMIC SECTORS

FORESTRY

The national timber resources of South Africa are very limited; the only indigenous high forests are confined to the constant rainfall area in the south and the seaward slopes of the mountains, occupying altogether only about 0.2 percent of the country. A vigorous afforestation program utilizing exotic species is designed to cover another 1.2 percent of the country by 1980, but opportunities for further extension are not great.

The most important tree economically is the black wattle, grown chiefly in the so-called mist belt of Natal at elevations ranging from 2,000 to 4,500 feet (610–1,370 m). Extract, bark, and chips are exported, while the remainder of the tree is used for pit props, as fuel, and for the manufacture of chipboard. Eucalyptus plantations are especially important for the mines and also supply rayon pulp and—with cuttings from coniferous plantings—paper pulp mills. Paper is also to be produced from bagasse in a large new mill in Natal. At present the local timber industry supplies less than half of the country's total requirements.

FISHERIES

South and South-West Africa, whose fishing industries are closely integrated, account for about 40 to 50 percent of total African fish landings and over half of the value of fish exports from the continent. Fishing is carried on all along the extended coast of the two countries, but the bulk of the catch comes from the Atlantic coast, where the cool Benguela current attracts large schools of fish. The industry is divided into the trawling section (which catches demersal or bottom-dwelling fish on a year-round basis) and the inshore section (which is further divided into pelagic fishing—operating in specific seasons when the pilchards have the maximum fat content—and the lobster fisheries).

Fishing jetty, Veldrift
South Africa leads the continent in fish landings.

The industry is highly organized, modern, and well equipped. Aircraft are used to spot the schools and electric fish finders are used on the boats. New harbor facilities, especially for trawlers, have been constructed at several ports in recent years. Overfishing threatened the industry in the 1960s, but strict quotas appear to have permitted recovery of the pilchards and lobsters. There is concern, however, over the large number of foreign trawlers operating off the coast, which catch more hake than is landed by the South African fleet. The pelagic fish, with a total catch of 1,059,000 tons in 1972, are used mainly for production of fish meal and oil, and also for canning, especially in South-West Africa. Collapse of the Peruvian fisheries in 1972 led to high world prices for fish meal, but the industry is required to sell a portion of its output at lower prices within the domestic market. Fresh and frozen fish are marketed throughout the area, which is well equipped with handling facilities. Lobsters are exported, mostly as frozen tails to the United States but also live and by air to France.

TOURISM

South Africa ranks after Morocco and Tunisia among African countries in the number of visitors entering the country, which increased from 276,000 in 1967 to about 600,000 in 1973 when their expenditures were estimated at $370 million. The white population of South Africa also contributes very heavily to support of the industry. Many parts of the country are of unquestioned beauty; facilities are better developed and more widely available than in any African country; and the somewhat overrated Kruger Park is a magnet for many visitors.

TRANSPORTATION

Reference has already been made to certain of the features of South African transport, so that only a brief summary is required at this point. The railways provide the domi-

nant form of inland transport, in large part because of the importance of bulk mineral and agricultural commodities and the considerable distances separating producing areas from ports and consuming regions, but also because they have been favored by rigid restrictions on highway competition. This policy may have turned out to be a wise one in view of the present situation with respect to oil. With about 14,000 route miles (22,500 km), South Africa has a rather well-developed network, connecting all of the major cities and towns and having links to all of the adjacent countries. Freight hauled has increased from 38 million tons in 1939 and 75 million in 1955–56 to 109 million tons in 1973. The state-owned South Africa Railways also maintains highway services on 33,000 miles (53,100 km), of roads; in 1973 these services handled 14 million passengers and 3.7 million tons of freight.

The railways have had persistent difficulties in handling the traffic offered, which has reduced the ability of the country to export minerals. Despite several massive investment programs, the need to purchase additional power and rolling stock, to realign difficult sections, double track some lines, and otherwise modernize the operation remain very great, and in late 1973 a new 10 to 12 year, $1.78 billion program was adopted to relieve pressure on the lines, on the rail yards, and at the harbors. New lines are under construction to Richards Bay and Saldanha Bay. Shortages of skilled and semiskilled staff have reached almost crisis proportions in recent years, necessitating some relaxation of job restrictions, but there is an unusually high ratio of whites in the service.

The Republic has over 210,000 miles (338,000 km) of roads, of which about 22 percent are all-weather and 9 percent are paved. A $1.85 billion ten-year program has recently been started to provide a national superhighway system. The total number of vehicles increased from 1.2 million in 1960 to 2.3 million in 1970.

An extensive network of domestic air services is operated by South African Airways. Intercontinental flights of S.A.A. are not permitted to land in most independent African countries, but routes via Portuguese territories have been used and nonstop service to Europe is now provided by 747s. Many other carriers continue to use the Jan Smuts airport, however, and these provide connections with many African countries as well as with other continents. Wanela runs services to pick up and return mineworkers from several foreign areas.

The ports of South Africa are fairly well spaced along the coast, but none of them are first-class natural harbors and all have had to be extensively improved to handle modern vessels and rapidly increasing tonnages. Because of its position as the domestic port nearest to the highly productive areas of the highveld, Durban ranks first among South African ports, handling over half of the cargo moving through them. Its total freight traffic has increased from 4.8 million tons in 1938 to 10.1 million in 1961 and 28 million tons in 1970–71. While its port facilities are generally modern and efficient it has been subjected to periodic extreme congestion—particularly after closure of the Suez Canal in 1967—and it is too shallow to take vessels of over 50,000 tons. An $89 million extension program is now underway and an offshore buoy was installed in the early 1970s, so that 200,000-ton tankers might be unloaded.

It became increasingly apparent, however, that none of the major South African ports could be adequately improved to handle the expected large-scale increases of coal and ores. Hence a decision was made to construct a new port at Richards Bay, about 100 miles (160 km) northeast of Durban. About $210 million will be spent in the first phase, which involves:

(1) Construction of a 3.3-mile (5.3-km) entrance channel protected by jetties;

Aerial view of Durban
The main part of the port is on the upper right, the Indian Ocean front on the upper left.

this length is required to provide the necessary stopping distance for large vessels even with the assistance of powerful tugs.

(2) Dredging a portion of the bay, whose average depth was only 3.3 feet (1 m), to take 150,000-ton vessels.

(3) Construction of a dike to protect this area against further deposition.

(4) Construction of three berths, two for coal and one for clean bulk cargo.

(5) Provision of ancillary services.

This port is expected to handle 12 million tons a year after it opens in 1976. It will then be extended as the need arises to as many as seven berths on the south side where dirty cargo will be handled and fourteen on the north, and probably be deepened to take 250,000-ton vessels. An aluminum smelter using imported alumina has already been constructed; large iron and steel, petroleum, and petrochemical industries are also likely to be situated at Richards Bay, which is being planned for a population of 800,000.

Cape Town now ranks second to Durban

among South African ports. It handled 4.7 million tons in 1961 and 8 million tons in 1970–71. Several extensions have been necessary here; material dredged from the harbor for the Duncan Dock was used to create a large foreshore area for port space and extension of the city's business district. An outer basin adjacent to Duncan Dock is part of a new $80-million extension program; seven additional berths, two of which will be equipped to handle containers, are to be completed in 1975 after which another basin is planned with 14 additional berths.

Port Elizabeth handles almost as much tonnage as Cape Town, but bulk manganese and iron ore account for a substantial part of the total. The only other ports of any significance are East London and Mossel Bay. Reference has already been made to the new iron ore port planned for Saldanha Bay.

South Africa has been building up its merchant marine for strategic reasons. The fleet rose from 120,000 GRT in 1953 to 538,000 in 1971. Safmarine has recently contracted with a Japanese shipyard for a 266,000-ton tanker. There are also plans to develop a container-ship fleet and to equip the main ports to handle containers in the traffic with Europe. The estimated cost is more than $575 million. Beginning in 1977 it is planned to have 14 container vessels on the northwest Europe and Mediterranean routes.

TRADE

Numerous factors influencing domestic and foreign trade have been touched upon in previous sections. Suffice it to note, in conclusion, that South Africa accounted for about 21 percent of total African imports and 22 percent of its exports in 1972. It is conceivable that political pressure to restrict trade with the Republic could seriously reduce the level, but no major trading partner has thus far imposed effective sanctions and many nations have been contracting for increased shipments of the wide variety of minerals available in the country.

If a number of Bantustans actually become independent it will be considerably more difficult to talk of applying sanctions to South Africa, which may be one reason for the apparent acceleration in delegating powers to the African areas.

CHAPTER TWENTY-FOUR

South-West Africa (Namibia)

South-West Africa, a German colony from 1892 until World War I, was then mandated to South Africa by the League of Nations, and has since been administered as almost an integral part of that country. The United Nations has renamed it Namibia and called upon the Republic to place it under trusteeship status, but the latter has refused to recognize UN authority over the matter. South-West Africa has been the subject of repeated UN resolutions, an abortive World Court case, and discussions between the Secretary-General and South African officials, but the Republic has given little ground; indeed, it is in the process of creating a series of Bantustans in the area. In December 1974 the UNSC unanimously adopted a resolution calling on South Africa to withdraw from South-West Africa with the aid of the UN. It also agreed to meet before mid-1975 and, if South Africa had not responded, to consider appropriate measures, an implied threat of the possible adoption of sanctions.

The most serious criticisms that may be made of the administration are comparable to those directed at South Africa itself: imposition of apartheid, and failure to give adequate attention to the advance of the non-white populace. The vast bulk of development investments have pertained to European mining, fishing, and agriculture, while the non-Europeans have played the usual roles of subsistence farmer and migrant worker. The South African Administration has, however, given considerably more intensive attention to the African homelands in the past decade—in part, one suspects, to indicate to its critics that it is a responsible ruler.

Conditions for the approximately 50,000 migrant workers have been worse than in South Africa, which led to a serious work stoppage in December 1971 and subsequent labor and political unrest more threatening than the later strikes in South Africa itself. The changes sought by the predominantly Ovambo strikers included an end to the contract labor system, a reduction in the term of work from the usual 18 months, higher pay, permission to bring one's family to the work area, freedom to select among employers (the diamond mines pay markedly higher wages), and removal of fences along the border with Angola, in which other Ovambo live. Some of these requests were met, at least in part, but the government also proceeded to proclaim laws in Ovambo to curb political activity and dispatched army units to ring the area.

PHYSICAL BACKGROUND

South-West Africa has an area of 318,259 square miles (824,292 sq km), about the size of France and Italy combined. This includes the Walvis Bay region, a 434-square-mile (1,124 sq km) exclave of the Cape Province administered as part of South-West Africa. Most of the territory is part of the great plateau of southern Africa, with an elevation of about 3,600 feet (1,100 m), interspersed with broken mountain masses ranging from 6,000 to 8,000 feet (1,800–2,400 m) and extending from the southwest to the Kaokoveld Mountains. The plateau swells toward the center of the country, has elevations of 4,000 to 5,000 feet (1,200–1,500 m) around the capital, Windhoek, and subsides into a vast alluvial plain in the north, about 2,000 to 3,000 feet (600–900 m) above sea level. A 1,000-mile-long (1,600 km) coastal plain extends the entire length of the country and inland from 60 to 100 miles (97–160 km).

The country is largely arid and semiarid. The Namib Desert along the coast is one of the driest in the world, with averages of less than 1 inch (25 mm) of precipitation yearly. Inland, rainfall increases from 3 to 16 inches (76–406 mm) in the center, and about 22 inches (559 mm) in the north. As is usual in dry areas, rainfall is highly unreliable and periodic droughts cause severe losses. Underground water is tapped where possible and ranchers construct relatively large dams to hold surface runoff, being careful where possible to conserve adequate stock water for more than one year in case of drought the following year. Water is also obtained from the Orange, Okavango, and Cunene Rivers, and by tapping the water flowing beneath the surface in the sandy beds of intermittent streams. Vegetation consists mainly of a variety of grasses and bushes, particularly the thornbush.

POPULATION AND PEOPLES

The population of the country as of the 1970 census was 746,000, consisting of 91,000 whites, 627,000 Africans, and 28,000 Coloureds; its mid-1974 population may be estimated at about 808,000. The largest number of whites are Afrikaans-speaking; the German-speaking population ranks second with about 22 percent, while the English-speaking whites are a relatively small group.

The nonwhite population consists of about a dozen very heterogenous groups who live in separate and often isolated segments of the country. The largest of the Bantu groups, the Ovambo, has about 46 percent of the total population; their "homeland" is a 20,650-square-mile (53,484 sq km) area in the north, which became the first Bantustan in South-West Africa. They provide about

80 percent of the total labor force, though two-fifths of the Ovambo are from Angola. Bantustans are proposed for at least five other ethnic groups whose populations range from only 18,000 to 64,000. The approximately 43,000 Hereros are probably the most advanced of the Bantu group; many are employed as clerks or storekeepers in the towns. They were a larger group until they were decimated in a revolt against the Germans in 1893. Today, the Herero consider themselves the natural leaders of the blacks, but the Ovambo's increased political awareness makes it unlikely that they would be so recognized in an independent Namibia.

The African population also includes about 21,000 primitive, nomadic Bushmen who hunt over vast stretches of the Kalahari and some 33,000 Nama or Hottentots, who live either on reserves in the south or on European farms and in the towns. Also contributing to the complex racial makeup are the Coloureds; the Reheboth community, descended from Dutch fathers and Hottentot mothers, who prefer the name Basters; and the Berg-Dama, a negroid people of unknown origin. Plans for the several Bantustans and reserves would involve a considerable redistribution of some groups, though the problem of consolidation is far less severe than in South Africa (Map 108).

Whites control approximately 70 percent of the total area, including the entire coastline (hence the fishing industry), all of the diamond and copper areas and most of the known mineral bodies, and the greater part of the cattle and Karakul sheep country.

The crude population density of South-West Africa was only 2.5 per square mile (1 per sq km) in 1974, but large parts are almost uninhabited, including the coastal Namib Desert (except for the ports and diamond mining centers), the steppelands of the east, and the Etosha Pan Game Reserve. The more heavily populated regions are the central uplands, containing a substantial portion of the European population and Wind-

Map 108. Economic map of South-West Africa

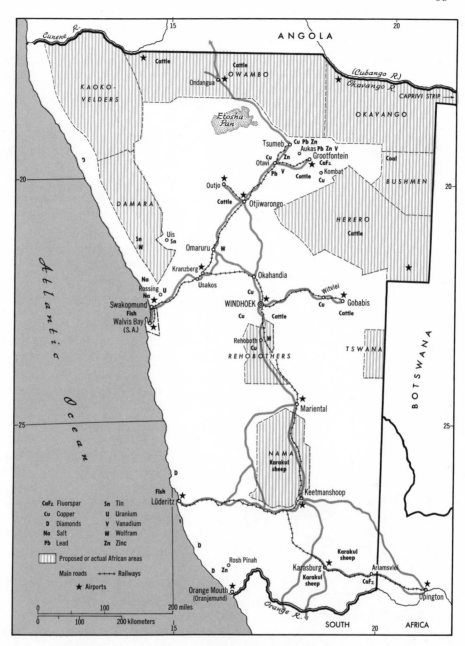

hoek with a population of about 70,000, and a strip in the north including Ovambo.

THE ECONOMIC SECTORS

Diamonds, copper, fish, and livestock are the main commercial concerns of South-West Africa. Many sectors are closely integrated with their counterparts in South Africa, trade figures are incorporated with those of the Republic, and output of minerals is not published officially. But company reports and other sources permit reasonably accurate assessments of most sectors, and it is clear that South-West Africa has one of the more dynamic economies in Africa. Its GNP was estimated at $487 million in 1970 and the average annual growth rate has been well above average. The budget in 1971 was $162 million, when exports totaled an estimated $280 million. Obviously, it is the white population, the investors, and South Africa who derive the main benefits, but many of the sectoral and infrastructural developments are of long-range value.

AGRICULTURE

Grazing is the most characteristic occupation of South-West Africa, and livestock products account for 98 percent of the value of farm output, which was about $78 million in 1972. Sheep and goats predominate in the south, with the karakul sheep being of particular importance. Introduced early in the century, karakul sheep provide one of the main exports of the country in the form of pelts used (mostly in the United States) for the production of Persian lamb "furs." The slaughtering of lambs when only a day old reduces the drain on the ewes and helps to explain the ability to maintain flocks in an area so arid.

Cattle assume importance in the center

Cattle at an artificial watering reservoir in the Ommetjette African Reserve
Aridity over almost all of the country makes extensive grazing the leading agricultural interest.

and north and are the main interest of several of the African tribes. About 500,000 to 600,000 head of stock are exported annually, mainly to South Africa and mainly from European ranches. Most of the roughly 5,000 white farmers are ranchers. Their holdings total 88.7 million acres (35.9 million ha), or about 44 percent of the total area, but about a fifth of the ranchers hold nearly half of the total in ranches of over 24,700 acres (10,000 ha). The carrying capacity is very low in much of the country. Only 1.1 percent of the land area is tilled, and the country is dependent on imports for most of its nonanimal foodstuffs. While considerable attention has been given to securing additional water for livestock and nonagricultural purposes, relatively little has been done to develop irrigation. The Ovambo Canal Scheme does provide water from the Cunene via a 59-mile (95-km) canal, which is used for improving pasturage and irrigating about 5,000 acres (2,000 ha); later phases will increase these areas substantially. The Hardap Dam on the Fish River, completed in 1962, controls about 7,400 acres (3,000 ha) developed by European farmers. The Okavango River could also be used for irrigation, but its extreme remoteness has discouraged anything but a small pilot operation.

Karakul sheep in arid grazing land
Pelts, used for Persian lamb coats, are one of the main exports of South-West Africa, which leads the world in their production.

FISHING

Fishing ranks third after mining and agriculture in economic importance. Walvis Bay and Luderitz are the main fishing centers. As noted in the preceding chapter, the industry is closely integrated with that of South Africa, with South-West Africa accounting for about 45 percent of the total catch.

MINING

Mining dominates the economy of South-West Africa in terms of value of production, exports, and number of wage employees. In 1972 it provided an estimated 60 percent of the GDP, 50 percent of government revenue, and over 60 percent of exports, while in 1971 some 12,800, or 30 percent of the African workers employed in the country, were in the mining industry.

Diamonds provide about 71 percent of the estimated value of mineral exports from South-West Africa. DeBeers' Consolidated Diamond Mines operates the richest diamond mine in the world, centered at Oranjemund. It currently produces about 1.6 million carats a year, averaging about 0.75 carats per stone. The south coast of the country is closed to almost all visitors and patrolled to protect the rich diamond deposits.

The second largest mining operation is at Tsumeb, where unusually rich ores with a metal content of about 21 to 25 percent are won; the main metals are copper, lead, and zinc, but by-product germanium, arsenic, cadmium, and silver are also present. Concentrates from smaller mines at Kombat, 65 miles (105 km) from Tsumeb, and the Matchless Mine, near Windhoek, are brought to the Tsumeb smelter for treatment. Based on present rates of output and known reserves the mines have a remaining life of only about 15 years.

A large, low-grade uranium body is under development at Rossing, 40 miles (64 km)

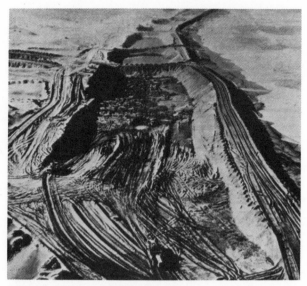

Diamond mining at Oranjemund on the coast
The retaining wall at right holds back the sea while the overburden is removed to reach the diamond bearing layers. About 95 percent of the diamonds recovered are gem diamonds in this, the world's largest opencast diamond-mining operation.

from Swokopmund, by a Rio Tinto Zinc subsidiary involving an investment of about $375 million. The British Energy Authority has contracted to take about 1,000 tons of concentrates a year beginning in the mid-1970s.

Most of the other mines in South-West Africa are small and widely scattered. They produce a variety of minerals, including copper, lead, zinc, vanadium, tungsten, tin, and fluorspar. A large, good-grade coal deposit has been found, but coal is now imported via rail or ship from South Africa. Intensive prospecting is taking place in many parts of the country, mainly for metallic minerals but a search for oil is underway along the coast and in the Etosha Pan area.

OTHER SECTORS

Such industry as exists in South-West Africa is associated mainly with mining, fishing, and agricultural activities. A few market-oriented plants produce such commodities as beer, clothing, soap, furniture, brushes, and bricks and tiles. Electricity has been provided mainly from coal-fired plants, but the cost of transporting coal makes the cost per kwh almost three times as high as at Johannesburg. An important aspect of the Cunene Project will be the transmission of electricity 441 miles (710 km) from the Matala and Ruacana hydroelectric stations to Tsumeb and centers to the south.

The country has attracted many South African tourists in recent years. There were 157,000 visitors in 1972. A major game park at Etosha Pan presents a considerable attraction, and additional parks are planned.

TRANSPORTATION

Considering the relative poverty of its surface, South-West Africa has a surprisingly good transport net. The main rail line, linked with Upington in the Republic, runs more or less north-south on the plateau (Map 108); western links connect with the ports of Luderitz and Walvis Bay, while branch lines run to mining communities and eastward to the ranching country centered on Gobabis. The 1,600 miles (2,574 km) of lines are operated, usually at a loss, as part of the South African system, which also runs regular motor services on about 5,600 route miles (9,000 km).

The road system consists of about 22,000 miles (35,400 km) of dirt and gravel roads and 1,713 miles (2,756 km) of paved roads, including a main north-south route extending the full length of the country.

Walvis Bay, a natural harbor, handles the bulk of territorial traffic. Its facilities include general cargo berths, mechanical ore loaders, and a tanker berth. Luderitz, with an artificial harbor protected by a jetty, is the only other port of importance. Only fishing vessels and shallow-draft coasters can be accommodated. It is the base of the rock lobster fleet.

CHAPTER TWENTY-FIVE

Botswana, Lesotho, and Swaziland

These three sharply contrasting countries are appropriately linked because of their common colonial experience as British High Commission Territories, their inclusion in the Southern African Customs Union and common currency area, and their heavy dependence on white-dominated southern African countries. Botswana became independent on September 30, 1966, Lesotho on October 4, 1966, and Swaziland on September 6, 1968.

The three countries were long neglected economically and politically and still do not have adequate administrative structures or statistical services, and they remain heavily dependent on expatriate advisers and administrators. Swaziland has surprising economic diversity and a per capita GNP considerably above the other countries; Botswana was one of the poorest nations in Africa until recently, but new mineral developments may soon raise the per capita GNP to one of the highest levels on the continent; Lesotho has few known resources, a serious problem of pressure on the land, and a continuing strong reliance on earnings from migrants working in South Africa.

Botswana, Lesotho, and Swaziland warrant attention in their own rights, but their location in southern Africa gives them wider interest for a variety of reasons: they complicate relations between the rest of the world and South Africa, particularly if sanctions were to be applied to the Republic; there is a desire to see them develop as examples of what the West would like to see evolve in the Republic; the damage to their economies that would result if South Africa cut back the number of foreign workers recruited; their relation to South Africa's efforts to develop a detente with black African states; the postulate that none of the three can have real political freedom unless they achieve greater economic freedom vis-à-vis their dominant neighbor; and the possibility that what happens in them may have a bearing on South African policies regarding its reserves and Bantustans.

The dependencies of the three countries on the Republic are many, while Swaziland also relies on Mozambique for transport and Botswana relies on Rhodesia and Mozambique for an important portion of its freight movements. They are dependent on South Africa for transport connections, employment of migrant workers and their remittances, customs receipts, markets, imports, advisers, investment, tourism, electricity, and for purchase of diverted water. It is impossible to remove certain of these dependencies, and development of the domestic economies of the three countries more often than not increases the reliance.

All of the countries have expressed their dislike of apartheid. The Swaziland monarch, King Sobhuza, has been most conservative in his relations with South Africa. President Sir Seretse Khama of Botswana has attempted within practical limits to follow an independent line. He has refused South African government aid and advisers while seeking them abroad, though he has accepted private investment from the Republic; and he has opposed dialogue between black African countries and South Africa and arms sales from Britain to the Republic. He has not felt it possible to implement sanctions on Rhodesia, a position accepted by other members of the OAU.

BOTSWANA

This country is a huge semiarid plateau with an average elevation of about 3,000 feet (900 m). It is higher and broken in the east and around Ghanzi in the west but gently undulating to flat over the bulk of its 231,804 square miles (600,372 sq km). The eastern region has the most favorable rainfall and agricultural potential and the bulk of the population; rainfall averages about 18 to 22 inches (457–559 mm) as compared to 12 inches (305 mm) or less in the drier west. About 84 percent of the country is covered with Kgalagadi sand, of aeolian origin and as much as 400 feet (120 m) thick, which supports a low savanna-type vegetation; much of the sand veld is uninhabited. Two areas in the west are of somewhat greater interest: the region around Ghanzi, where a number of European farms are situated on a limestone ridge relatively well supplied with water; and the area around Maun on the edge of the Okavango Swamps, which is primarily a cattle raising region.

The outstanding physical keynote of Botswana is aridity, which affects practically every aspect of its economic development. Four examples may be given as illustrations:

(1) Water supplies were so short in the two principal towns that the government seat was situated at Mafeking, 17 miles (27 km) inside South Africa, until a dam was constructed to provide water for the new capital at Gaborone.

(2) Large parts of the sand veld are described as so highly sensitive that they should probably be closed to all livestock development. The soil and vegetation have been compared "in their delicacy to the skin on the face of a beautiful woman: once destroyed, never replaceable."

(3) The difficulties faced by Bushmen in securing water, as described by Deben-

ham.[1] After finding a certain plant which indicates damp sand, a woman proceeds to scrape a hole in the sand to arm's length, then takes a grass stem or hollow reed, surrounds its end with fine roots or grass, and places that end in the bottom of the hole. The sand is carefully replaced around the reed and firmly tamped down. After an hour or two the woman begins to suck through the tube and in another hour or two the water begins to rise in the reed. With a second straw she expels it into one end of an ostrich egg. A number of these are filled, sealed with clay, and buried as reserves for dry weather.

(4) In the 1960s a severe and prolonged drought resulted in loss of 30 percent of the livestock, and at the time of independence the country was dependent on world famine relief for half of its food supply. In 1973 another drought reduced yields in the central districts to one-fifth of the "average" level.

POPULATION

The de facto population of Botswana at the 1971 census was recorded at 570,000, of whom 10,861 were foreigners; in addition 48,000 Tswana were listed as being abroad and there were an estimated 10,600 nomads, bringing the de jure population to about 629,000. The population is very youthful, 46 percent being under 15 and only 10 percent over 55 years of age, and the rate of increase was estimated at 3.08 percent.

Most of the population of Botswana is made up of eight main Bantu-speaking (Setswana) tribes, but there are many non-Tswana living within them on a socially inferior status and small numbers of Bushmen

[1] Frank Debenham, "The Kalahari Today," *The Geographical Journal* (March 1952), pp. 18–19.

(Sarwa) and Kgalagadi. About 80 percent of the population lives near the rail line which traverses the eastern edge, about 38 percent in large agro-villages with over 500 people.

Farming "lands" up to 30 miles (48 km) away are worked with these villages as bases, while beyond the "lands" are the cattle posts where herdboys tend the family livestock. From June to November only a small number occupy the cattle posts and "lands"; after the first heavy rains, usually in December, a part of nearly every family moves out to the "lands" to plow and plant, remaining there until after the harvest in March–April. In the northwest, people live in small settlements on islets in the swamp or along the main watercourses. The latest available estimates indicate that 67 percent of the population is engaged in traditional agriculture.

European farmers and ranchers are found in the Lobatse, Gaborone, Tuli, and Tati concession areas along the eastern border, along the Molopo River in the south, and on the Ghanzi Farms in the West. Gaborone, Francistown, and Lobatse are the main commercial towns; the capital's population increased from only 3,800 in 1964 to 18,900 in 1972 and is continuing to mushroom.

GRAZING

Owing partly to aridity but also to traditional interest, Botswana is predominantly a grazing country, animal products having accounted for about 80 percent of total exports until the recent diamond and copper/nickel developments. About 85 percent of the cattle are owned by Africans, but 21 percent of the rural population own no cattle, while 4 percent hold 30 percent of the total. The veld provides generally good grazing but at low carrying capacity, perhaps one head per 20 acres (8 ha) in the more favored southeast and one head per 50 acres (20 ha) in the Kalahari. Only about a sixth of the country is now used for grazing; those lands in use are overgrazed, particularly around

watering points. Disease problems are sometimes acute; periodic outbreaks of foot-and-mouth disease temporarily preclude foreign shipments, while adherence to the cattle culture contributes to overstocking. There has, however, been increasing willingness to sell cattle, a record of 209,000 head having been sold in 1973.

Several programs have been instituted in recent decades to improve the livestock economy, including the sinking of several thousand boreholes, construction of small dams, improving trek routes, and extending veterinary services. Opening of the Botswana Meat Commission slaughterhouse and canning factory at Lobatse in 1954 brought considerable improvement; it is now the largest nonmining industrial enterprise in the country. Under the present development plan each of the previous programs is to be extended and a new abattoir may be built in the north. But a major new emphasis is on the creation of 54 breeding and 15 fattening ranches varying from 16,000 to 20,000 acres (6,475–8,094 ha) except for one fattening ranch in the north of 190,267 acres (77,000 ha). Twenty-four of the breeding farms are to be for karakul sheep; they will be situated around Bokspits in the extreme southwest. Ninety-nine-year leases are also being granted for controlled grazing in parts of the State Lands which comprise about 100,000 square miles (259,000 sq km) of the undeveloped land. It is hoped that these and other improvements will bring the total cattle herd from 1.7 million in 1973 to over 2 million and almost double the offtake.

TILLAGE AGRICULTURE

It is estimated that about 8 million acres (3.2 million ha) of Botswana (about 5.7 percent of the whole) are potentially arable. Aerial photos reveal that the area actually cropped declined by half to about 500,000 acres (200,000 ha) in the 1960s, partly because of the extended drought, but also because of an increased interest in commercializing live-stock and possibly a shortage of oxen for the smaller farmers. The main subsistence crops are sorghum, corn, millet, beans, and cow-peas, but yields vary widely and the country has a substantial dependence on imported foodstuffs.

Efforts to upgrade African agricultural standards are centered on the Pupil Farmer Scheme and demonstration farms, about 50 demonstrators being graduated yearly. Farming improvements that are promoted include: winter plowing, early planting, cor-rect spacing, regular weeding, use of quick-maturing crops, manuring and fertilization, and proper rotations. An emphasis on mois-ture conservation is included in several of these practices. A system of land registration was initiated in 1970, though ownership in communal areas remains collectively with the tribe; it is hoped that this will encourage the adoption of improved techniques.

Most of the European farmers in Bo-tswana are interested primarily in ranching. Some, especially in the Ghanzi District, have a relatively important output of dairy prod-ucts. Europeans account for a dispropor-tionate share of the corn crop and for some citrus fruit, cotton, and peanuts usually mar-keted in South Africa. Most of the lands thus far brought under irrigation are on Euro-pean farms, but the largest scheme, in the Tuli Block, has only 4,000 acres (1,619 ha). The government has sponsored studies of the irrigation potential, which is fairly large, and instituted a demonstration scheme at Mahalapye, but indications are that available waters may be more effectively used for do-mestic, industrial, and livestock supplies than for irrigation, and that funds devoted to til-lage agriculture should be concentrated on dry land farming rather than capital-inten-sive irrigation schemes.

The largest amount of unused water in the country is available in the Okavango Swamps of the northwest. Possible control of this water has been the subject of speculation and periodic investigation for many years, including a current study being made by a UN team. The Okavango River, second only to the Zambezi in southern Africa, rises in Angola and brings a large seasonal flow to the apex of the delta, where it divides into two main streams (Map 109). The roughly triangular swamp totals about 4,000 square miles (10,360 sq km). Its upper part is per-manent and largely covered by papyrus; the remainder fills up annually, while varying amounts flow seasonally along the Boteti River and into the Makgadikgadi Pans. The vast bulk of water is lost through evapora-tion and transpiration in the swamps. Various plans have been suggested: direct-ing more water into the Thamalkane and Boteti Rivers for storage to feed irrigation perimeters; canalizing the outer rim and controlling the flow into part of the delta; piping water to the east of the country where the soils are better; and preserving the area as a natural park, since it has a great variety of game. Certainly, far more study is required before development is permitted if the delicate ecosystems of the area are not to be destroyed.

LABOR MIGRATION

Labor migration to South Africa began as early as 1871, when men went to work at the Kimberley diamond mines. In recent years 22,000 to 36,000 workers have contracted to work on the mines of the Republic, while about 20,000 others are employed there in agriculture and industry. The total number involved represents about 15 percent of Bo-tswana's employable population and is greater than the present number of wage earners within the country. Receipts in the form of deferred pay and remittances from those employed in mining alone contribute over $1.4 million yearly to the balance of payments, while migrants also bring a sub-stantial value of goods with them upon re-turn. Reliance upon this form of employ-ment would cause severe hardship should

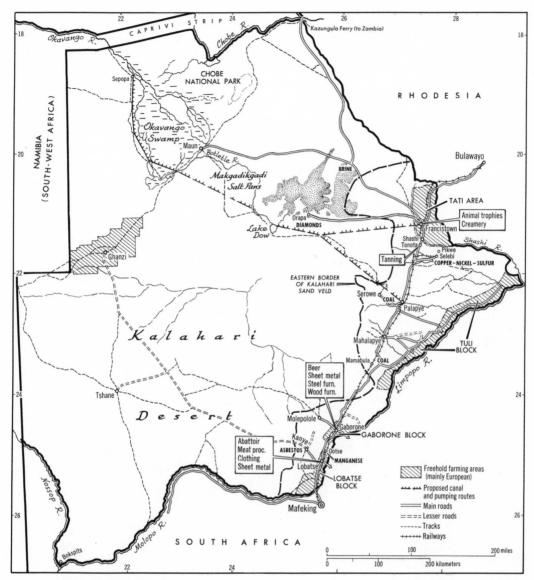

Map 109. Economic map of Botswana

South Africa decide to restrict the flow either for political or economic reasons.

MINING

Mineral output was of minor importance until the 1970s but it now promises to revolutionize the financial position of the country. Production during the 1960s was confined to a small output of asbestos, manganese, and kyanite, and output of these was interrupted or ended during the decade. Several bodies of manganese are known; Anglo-American mines the ore at Kanye near Lobatse and may open a mine at Ootse in the same area.

The first large-scale recent development was that of diamonds at Orapa, 143 miles (230 km) west of Francistown, to which it is connected by a new road. After much searching, De Beers found massive pipes in the area, the one at Orapa being the second-largest-known kimberlite pipe in the world. Opencast mining began in mid-1971 after an investment of about $30 million and production increased to 2.5 million carats in 1973, whose export value exceeded that of beef for the first time. While 93 percent of the output is of relatively low-value industrial stones bringing only $9.84 a carat in 1973, the Botswana Government received about $10.7 million in 1972–73. Beginning in 1975 its share of profits was increased from 50 to 75 percent. Negotiations have been underway for some months regarding opening a second pipe known as Dk 1, 25 miles (40 km) to the southeast, said to be smaller but containing better-quality stones.

The biggest development, however, is that of a copper-nickel deposit in the Selebi-Pikwe area, 55 miles (88 km) southeast of Francistown. It took 16 years after the original find to examine the body and then to arrange for financing. Capital costs totaled $170 million and provision of the necessary infrastructure cost $78 million for a total about three times the GDP of the country in 1968–69. Mining began in late 1973, the ores

The Selebi-Pikwe copper-nickel installation
Shown are the twin headframes at the right, the concentrator in the center, and the smelter, with its tall stack, in the background.

being smelted to produce a copper-nickel matte, which is then shipped to Braithwaite, Louisiana, where about 18,000 tons of refined nickel and 17,000 tons of refined copper a year will be produced, most of which will be marketed under contract in West Germany. About 127,500 tons of by-product sulfur will be sold in South Africa.

The infrastructure required included a dam on the Shashe River to provide water for the complex, a 45,000-kw electric plant (which involved opening a coal mine at Morupule), road and rail links, and a new township whose population is expected to rise to 25,000 by 1980. The production components include an underground mine, a small surface mine (which will last only a few years), a smelter whose 500-foot (152-m) stack is said to be the highest in Africa, and ancillary facilities. Botswana will benefit from the creation of several thousand jobs and a number of forward and backward linkages, but more importantly from a net increase in government revenues of about $6.7 million a year in the first three years. Several additional copper-nickel deposits have been under study, including ones at Maun, in the Tati Concession, and at the Bushman mine 75 miles (121 km) west of Francistown.

Other minerals of potential interest to Botswana include:

(1) Coal, of which there are very large reserves. There is some possibility that coal might be mined in sufficient volume to permit its being piped to South Africa or its supporting a large carboelectric station whose output would be transmitted to the Republic. Large-scale production would permit conversion of the present underground mine at Morupule, 6 miles (10 km) west of Palapaye, to a surface operation. A deposit at Mmamabula, 80 miles (129 km) south of Palapaye, has been under examination in recent years.

(2) Soda ash and salt, available in brine deposits at Sua Pan, part of the Makgadikgadi Salt Pans, may be opened up in the next few years.

(3) Gypsum, found south of Francistown.

Prospecting concessions have been granted over large parts of Botswana and a new taxing arrangement is intended to stimulate active searching. While developments in the mineral field are not yet exactly fantastic on a world basis they are nearly that by reference to pre-1970 Botswana. As is so often the case, however, their major contributions are in increased government revenues and exchange earnings. The numbers to be employed will have a relatively minor impact, as may be judged from the expectation that employment in the country will increase about 2,000 per year whereas the labor force is increasing by 10,000 a year. The government appears to be well aware of these facts and is striving to support the major employment sectors, farming and grazing, and to prevent the gap between rural and urban incomes from becoming excessively wide.

INDUSTRY AND TOURISM

Only about 40 manufacturing plants exist in Botswana, and no large development can be expected. As noted, the Botswana Meat Commission abattoir/cannery at Lobatse is the largest factory in the country; the Botswana Game Industries Co. at Francistown, which treats wild animal skins and mounts trophies, ranks second. Other industries with over 9 workers include six clothing makers, three establishments producing construction materials, two each of breweries, metalworking shops, and furniture factories, and one each of plants producing spirits, handbags, and paint, retreading tires, and assembling refrigerator and air-conditioning equipment. The increasing incomes in the country will justify additional import-substitute industries, which may be protected

within the customs agreement, but the small population restricts the total potential.

The tourist industry has a considerable potential, as Botswana has a large and unusually varied wild animal population. The 4,500-square-mile (11,655-sq-km) Chobe National Park is easily accessible from Victoria Falls, and the total area in game reserves has recently been raised to 30,000 square miles (77,700 sq km). A large investment in tourist infrastructure will, however, be required before this sector can expand significantly.

TRANSPORT

The major transport routes in the country are the 394-mile (634-km) rail line connecting South Africa and Rhodesia, owned and operated by Rhodesia Railways, and a partly paved, partly gravel, road paralleling it. An improved road is under construction to the ferry connection with Zambia at Kazungula, the only point at which Botswana has a boundary with an independent black African country. Should the OAPEC boycott on oil shipments to southern Africa restrict shipments via South Africa, Botswana could import oil via Tanzania and Zambia, but the delivered cost would be considerably above that on the present routing.

Botswana is undergoing one of the more dramatic transformations on the continent. At the time of independence there were only five Tswana in universities and 35 with degrees; the government relied on yearly grants-in-aid from Britain; incomes ranked among the lowest in the world. Evidences of poverty remain and will for some years. But mineral developments promise to reverse the traditionally unfavorable trade balance, to create new job opportunities and incentives for new industrial growth, and to provide greatly increased government revenues. The budget has, in fact, increased from $1.89 million in 1949–50 and $8.93 million in 1962–63 to $102 million in 1972–73. Nonetheless, dependence on a precarious precipitation remains heavy and the national greeting, *"Pula,"* "let there be rain," is not likely to be changed.

LESOTHO

Situated on the western watershed of the great Drakensberg Escarpment, Lesotho is entirely surrounded by Republic territory (Map 110). It may be divided into the following topographic regions, for which selected data for 1973 are provided:

	Percent of area	Percent of population	Percent of cultivated land	Population density Per sq. mile	Per sq km
Border lowlands	5.9	11	15	175	68
Lowlands	11.3	27	32	224	86
Orange River Valley	9.4	11	9	110	42
Foothills	15.6	21	17	126	49
Mountains	57.8	30	27	49	19
Actual totals	11,720 sq. mi. 30,355 sq. km.	1.1 mil.	1,560 sq. mi. 4,047 sq. km.	93.9	36.2

The border lowlands are characterized by impoverished soil and severe erosion. Extending eastward of this belt is a more fertile, less arid lowland belt between about 5,000 and 6,000 feet (1,500–1,800 m), interrupted by sandstone plateaus. These two areas, with about 38 percent of the total population, contain about 47 percent of the cultivated land. A third lowland area is a long wedge along the Orange Valley in the southern part of the mountain zone. The foothill zone is rolling country between 6,000 and 7,000 feet (1,800–2,100 m) interrupted by mountain spurs and river valleys. In the valleys and on the flatter areas, rich volcanic soils provide some of the best soils of the territory and erosion is not so advanced as in the west. Agriculture is, however, less developed than in the lowlands. The Maluti Mountain zone lies mostly above 8,000 feet (2,400 m) but elevations exceeding 10,000 feet (3,048 m) are found. Its deep valleys, separated by uplands and steep ridges, are fairly densely populated; the lands above

8,000 feet (2,400 m) are traditional "cattle post" country, formerly used exclusively for summer grazing. Population pressure has led to increasing settlement and arable farming in this zone. The short growing season and highly erodible thin soils are severe handicaps to agriculture, and in places the soils have been completely removed.

Temperature and precipitation vary greatly from region to region and within individual regions according to elevation and aspect of slope. Concentrated into a few months, yearly rainfall ranges from 75 inches (1,900 mm) along the scarp to 25–30 inches (635–760 mm) in the lowlands, but it varies considerably from year to year and within the season. The incidence of frost is erratic but may be disastrous and it may occur in any month in the mountains; hail is a summer hazard in all districts. Most of Lesotho is grass covered, and most of the trees that do exist have been planted in connection with gully control.

POPULATION AND LABOR SUPPLY

The vast bulk of the population of 1.1 million (mid-1973) are Sotho, a Bantu group, and all of the land of Lesotho is held in trust for the Sotho nation by the King and may not be alienated. There are about 1,600 Europeans and 800 Asians in the country. The 1969 labor force was estimated to be 430,000, of whom 270,000 were males. Estimates vary somewhat, but roughly 40 to 45 percent of the men are employed in South

Africa at any one time, 7 percent are in wage employment in Lesotho, and the remainder are engaged in agriculture in the country. Remittances and deferred pay contributed $8.5 million to the balance of payments in 1972, which was well above the total value of exports. In that year an average of 69,463 a month were employed in South African mines, but the total employment there was estimated to be about 130,000. In 1973 fighting took place between Sotho and other African employees at Carletonville in the Republic, resulting in the death of a number of Africans including 11 Sotho citizens; as a consequence 10,000 were repatriated, causing considerable concern with respect to possible alternative employment.

AGRICULTURE

The number one problem facing Lesotho is soil erosion which, despite efforts extending over many years, continues to destroy increasing acreages. Steep slopes, inadequate grass cover, thin soils, downpours, overgrazing, and poor farming practices all contribute to what is one of the worst erosion situations in Africa. Indeed, it has been predicted that at the present rate of erosion the land cannot support the Sotho nation for more than a half century. Control programs involve terracing, grass stripping, constructing dams, planting trees to arrest gullying, and limiting grazing rights. But control is much more than a physical problem; it is a social problem in that some required measures are restricted by the farmers and graziers, such as village relocation, closure of areas for recuperation, and restrictions on livestock numbers. It is also an economic problem because the government does not have adequate funds to mount a really effective control program.

Lesotho is mainly a grazing country, with animal products—chiefly live animals, wool, and mohair—usually providing 60 to 75 percent of exports. Most of the remaining exports are foodstuffs—mainly wheat, peas,

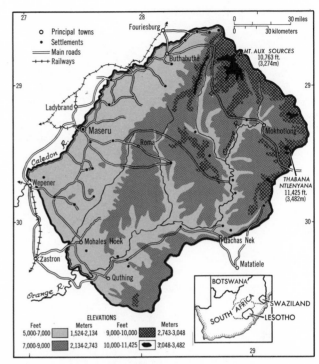

Map 110. Physical and transport map of Lesotho

and beans—but the import of foodstuffs, which make up only 20 to 26 percent of total imports, exceeds the total value of the country's exports. The livestock of Lesotho have been described as "a heterogenous collection of beasts showing signs of every conceivable strain." Yields are low and of poor quality; the birth rate for sheep and goats is very low, in large part because of overgrazing—particularly in the winter when the value of the grazings declines rapidly after the first frost. Improvement is sought in part by subsidizing the importation of first-class Merino and Angora rams from South Africa, but Sotho graziers periodically breed with low-class animals at the expense of quality to increase the hardiness of the animals. Improvements are also needed in grading, management of the clip, and marketing; but no real advance can be expected until an effective grazing control program has been adopted, which means abandonment of communal grazing and provision for some sup-

Corn fields and gully erosion in the western lowlands of Lesotho
Despite substantial efforts to protect the land, Lesotho continues to have one of the most serious soil erosion problems in Africa.

plementary feeding in the dry winter months.

About 13 percent of the country is used for tillage agriculture, including 29 percent of the lowlands and 8 percent of the highlands. The population density in relation to cultivated land is about 609 per square mile (158 per sq km), and it is thought that almost all of the cultivable land is actually in use. The main crop and foodstuff is corn, but it is often necessary to import additional tonnages to meet local requirements, and winter wheat is exported in exchange for corn. Corn occupies about 50 percent of the cropped area, sorghum and wheat 40 percent, and the remainder produces beans, peas, barley, oats, and miscellaneous fruits and vegetables. Farms are small and fragmented, yields are low but subject to considerable increase under proper practices.

Efforts to improve farming have centered on working with Progressive Farmers and on the Dryland Farming Development Project, intended eventually to rationalize mixed-farming operations on 120,000 acres (48,600 ha) of arable land and 180,000 acres (72,800 ha) of grazing land. An IDA- and AID-assisted pilot project under this program, involving an expenditure of $9.8 million, is under way at Thaba Bosiu.

NONAGRICULTURAL SECTORS

WATER RESOURCES. Lesotho is the source of a number of perennial rivers, including the Orange and the Caledon, which flow into the Republic. At present there is very little irrigation and the scope for its application is not large. Hydroelectric potentials are, however, very substantial, and diversion of water into the Elands River would increase the supply where it is most needed in the Republic. Many schemes have been proposed, and one has been under active negotiation in recent years: the Malibamatso Scheme, which would involve construction of a 250-foot-high (76 m) dam on the Malibamatso and a 36-mile (58-km) diversion tunnel to the Elands River at a cost of about $35 million. Later a power station could be added, but South Africa does not appear ready at the present time to provide the additional investment that would be required.

FORESTRY. Anglo-American and De Beers are investing some $960,000 over a twelve-year period to prepare 10–50 acre (4–20 ha) woodlots adjacent to all larger towns and villages. These are intended to combat erosion, and meet local needs for fuel and house poles.

MINING. Diamonds have been the only mineral of interest produced in Lesotho, but output has fluctuated widely from year to year and several interested companies have recently withdrawn. Diamonds were first produced by Sotho diggers from alluvial deposits, and in 1967 a digger found a 601-carat stone, the world's seventh largest, which sold for $302,400. (The finder's previous cash holdings were $4 and his first purchases were a new suit, three frying pans, and two new wives.) Later a score of pipes and five dikes were discovered and modern workings were opened. The yield is apparently not very attractive, but in 1974 the government reached an agreement in principle with De Beers regarding the opening of a mine at Letseng-la-Terai, which would give

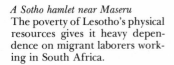

A Sotho hamlet near Maseru
The poverty of Lesotho's physical resources gives it heavy dependence on migrant laborers working in South Africa.

employment to about 400 Sotho and 60 expatriates and involve an investment of about $28 million. There is also a small output of semiprecious stones. Oil prospecting began in 1973 in the lowlands and the UN is supporting a large-scale mineral survey costing about $670,000.

MANUFACTURING. As would be expected, the industrial sector is very meagerly represented. Some emphasis has been placed on handcrafted items, of which mohair blankets and rugs have done best. Production of these started in 1969 at Teyateyaneng with only three workers, sales were so good that 400 workers are now employed and a factory is planned. Other plants, all small, produce flour, furniture, candles, beer, pottery, and clothing. Two industrial estates have been set up at Maseru and Ficksburg Bridge on land provided by the King to avoid tenurial problems. A corn mill, which will later produce oil and feedstuffs, and a tractor assembly plant have recently been placed at the latter, while the abattoir at Maseru is being modernized with Danish aid. Wages are lower in Lesotho than in South Africa, which might create opportunities for sale of manufactured goods in the Republic, but thus far only handicrafts and a new Italian-owned electric fixtures plant have catered to this market.

TOURISM. This is one sector which has shown dynamic growth in recent years. Before 1969 only about 3,600 visitors came yearly to Lesotho, but in 1971–72 60,000 were recorded and it is hoped to attract 200,000 to 300,000 by 1978. In 1970 a Holiday Inn and Casino was opened at Maseru and its capacity was soon doubled. It is now booked months in advance. Plans call for constructing two more hotels, rondavels in the mountains, and a ski resort in the Maluti Mountains, and for setting up a small national park near the capital.

TRANSPORT. With its very difficult topography, it is not surprising that Lesotho has a poorly developed transport network, but the emptiness of the road map as compared to surrounding areas is not entirely attributable to this factor. Maseru is the only town served by rail, a one-mile branch of the Bloemfontein-Durban line, though most lowland towns are only a few miles from a station in the Republic. The main north-south lowland road has been improved as have the road to Roma, main campus of the University of Botswana, Lesotho, and Swaziland, and a road into the mountains. Countless miles of bridle trails provide access to much of the highland areas; the Sotho using them ride oxen, ponies, donkeys, horses, and mules. The Sotho pony was once considered a valuable animal, but selective offtaking for sale and poor breeding practices have left a degraded animal today. Plans are being made to construct an international-class airport, which would permit overflying South Africa; there are local flights from Maseru to five towns and charter services are available to about 28 airstrips.

Lesotho is, then, an extremely poor country, heavily dependent on South Africa for its very existence. And each development undertaken appears to increase that dependence. The very serious soil erosion problem coupled with a rapidly rising population presents a frightening dilemma for the future.

SWAZILAND

Swaziland is the smallest (6,704 sq mi; 17,363 sq km; with maximum dimensions of 90 by 120 miles, 145 by 193 km), the richest, and the most complex of the three countries. Situated, in contrast to Lesotho, on the eastern watershed, it may be divided from the standpoint of topography and land use into four distinct regions, each running north-south (Map 111 and Table 22).

On the west the highveld is a northeasterly extension of the Drakensberg, a wide belt of granite-based, broken, heavily dissected country used for grazing sheep and some dryland farming and to support important forest plantations. The capital, Mbabane, and the two main mining centers are situated in this zone.

Second is the middle veld, which has an undulating to markedly rolling surface and is dissected at intervals by eastward-flowing streams. This is the main crop area of the country, producing mainly rain-grown crops but with an increasing area under irrigation. Soils are inherently not very fertile but yield well under proper use; rock is exposed on roughly a quarter of the area and very shallow lithosols occupy another quarter. With supplementary irrigation, a great range of tropical, subtropical, and middle-latitude crops can be grown in this zone. The zone's grazing potential is superior to the highveld's, but its "sourveld" grasses are not highly nutritious and about 6 to 8 acres (2.4–3.2 ha) are required to support one head of cattle.

Third is the lowveld, mostly gently undulating acacia bushveld, with rainfall too low to support more than grazing or hardy annuals such as sorghum and millet; temperatures are rather high. The presence of some very good alluvial and basaltic soils and of a number of streams provides attractive features for irrigation, and most of the area brought under irrigation is located on alluvials in this zone. Eradication of malaria has led to a rapid population increase, but bilharziasis is endemic. Overgrazing is serious in the lowveld.

The last region is the Lebombo Plateau, a narrow undulating zone that drops off abruptly into the lowveld and is deeply dissected by the three main streams leaving the country. Only about an eighth is good to fair arable land, while almost three-fifths is in outcrops or stony land.

POPULATION AND LAND DISTRIBUTION

The mid-1973 de facto population of Swaziland may be estimated at about 456,000, 97 percent of whom were Bantu-speaking Africans, mainly Swazis but also Zulu, Tonga, and Shangaan. About 75 percent are engaged in agriculture. There were approximately 48,000 wage employees in 1971, including about 6,500 Europeans, Eurafricans, and foreign Africans. Swaziland is much less dependent on migrant labor than Botswana or Lesotho and the numbers working in South Africa have tended to decline in recent years.

A complex landownership pattern is superimposed on a relatively simple topographic one. It stems from the granting of all manner of concessions in the period 1880–90 (including one for the sale of but-

Map 111. Economic map
of Swaziland

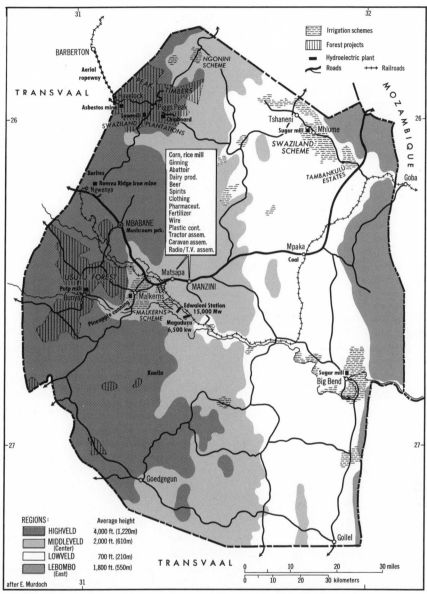

REGIONS: Average height
■ HIGHVELD   4,000 ft. (1,220m)
  MIDDLEVELD   2,000 ft. (610m)
  (Center)
  LOWVELD   700 ft. (210m)
  LEBOMBO   1,800 ft. (550m)
  (East)

after E. Murdoch

Corn, rice mill
Ginning
Abattoir
Dairy prod.
Beer
Spirits
Clothing
Pharmaceut.
Fertilizer
Wire
Plastic cont.
Tractor assem.
Caravan assem.
Radio/T.V. assem.

Irrigation schemes
Forest projects
Hydroelectric plant
Roads   Railroads

tons), some of which were three or four deep as far as land rights were concerned. These were fairly well sorted out in 1907 and additional lands were later credited to the Swazi, while Britain has set aside funds to assist in the eventual purchase of 960,000 acres (388,500 ha), providing that steps are taken to apply rational practices on the acquired lands. At present about 55 percent of the land belongs to the Swazi Nation; by law it is owned communally, but land-use rights are individually assigned and may be passed on. Most of the remaining land is held in freehold tenure by non-Africans.

The Swazi and European holdings are intricately interwoven, which results in land use patterns showing sharp contrasts from section to section and comparable dif-

Table 22. Selected data on the regions of Swaziland

Region	Area sq. mi.	Area sq. km.	% of total	General elevations feet	General elevations meters	Precipitation range inches	Precipitation range milli-meters	Approximate population, 1973 '000	%	Density per sq. mi.	Density per sq. km
High veld	2,000	5,180	29.8	2,700–6,100	823–1,859	40–90	1,016–2,286	128	28	64	25
Middle veld	1,800	4,662	26.9	1,100–3,500	335–1,067	30–45	762–1,143	191	42	106	41
Low veld	2,400	6,216	35.8	500–1,000	152–305	20–35	508–889	91	20	38	15
Lebombo Plateau	500	1,295	7.5	1,500–2,700	457–823	c.33	c.838	46	10	9	3
Totals	6,704	17,363	100.0					456	100	68	23

N.B. Areas of regions are rounded and do not add to the exact actual total.

ferences in population densities. The density on the separate blocks of African-held land is about five times as great as on the European areas, and overgrazing and other malpractices have led to greater deterioration on these lands than on the freehold land. The situation is further aggravated by the fact that about half of the freehold land is either unused or underutilized. About 15,000 families, possibly a fifth of the total, operate as squatters on freehold land.

The division of lands has led to repeated complaints by the Swazi, complaints that have been intensified because almost all of the main developments of recent decades in mining, forestry, irrigation, commerce, and industry have been focused in the European enclaves. In 1973, a number of partial nationalizations, added taxes on sugar and hotels, creation of an army, and assumption of full power by King Sobhuza II caused considerable concern in the expatriate community and a slowdown in new investment and construction.

AGRICULTURE

The Swazis are traditionally cattle culturalists, and despite an increasing acceptance of the commercial value of cattle, they tend to keep too many animals. This leads to soil erosion, which although less serious than Lesotho's still requires attention. Large parts of the highveld were formerly devoted to trek sheep from the Transvaal, but their numbers have now greatly decreased. Although over 75 percent of the Swazi Nation areas are used for communal grazing, livestock and livestock products account for only 3 to 5 percent of exports. The interest in cattle also reduces the attention that might be given to cropping, which occupies only about a seventh of the Swazi Nation land. Efforts to rationalize Swazi agriculture are focused on selected Rural Development Areas, where farmers and livestock are resettled on consolidated lands, grazing areas are fenced, contour terraces dug, small dams constructed, and the basic infrastructure is provided.

The modern sector of the dual agricultural economy consists of about 1,500 medium-sized and a few very large estates, 70 percent of the total having all or part of their holdings under irrigation. It accounts for most of the production of export crops and provides employment for about 21,500 Swazis, over 40 percent of the wage labor force.

A considerable variety of rain-grown crops

is found in Swaziland. Corn is the staple food and dominant crop, occupying about 70 to 85 percent of tilled lands except in the lowveld, where sorghum is more important. The country is approaching self-sufficiency in corn except in drought years. Millet, potatoes, legumes, peanuts, vegetables, cotton, and fruits are other dryland crops. The four most important rain-grown cash crops are corn, cotton, pineapples, and tobacco; 75 to 80 percent are grown on European farms, with an increasing percentage being put under irrigation.

The past two decades have seen a dynamic development of irrigation farming, and there are substantial opportunities for its further extension. About 60,000 acres (24,300 ha) or a fifth of the total estimated irrigable area is now under control. The largest project is the Swaziland Irrigation Scheme in the northern lowveld (Map 111), started by the CDC in 1950 and now having about 30,000 acres (12,100 ha) planted, mainly to sugar but also in rice and citrus fruit.

Other important irrigation schemes are the Tambankulu Estates in the Black Umbuluzi River area (the largest citrus estate), the Ngonini Estates in the lower elevations of the Pigg's Peak district (growing bananas and citrus fruit), and the Malkerns Irrigation Scheme watered by the Great Usutu in the middleveld (mainly devoted to citrus fruit and pineapples). African farmers have portions of the irrigated lands in smallholding schemes.

Sugar is the main irrigated crop and the country's leading export, foreign sales having increased from nil in 1958 to 54,500 tons in 1961 and 180,000 tons in 1971–72. It is processed in large mills at Mhlume and Big Bend, and discussions are underway about the construction of a third mill. Swaziland has quotas under the International Sugar Agreement and the Commonwealth Agreement, which should permit further expansion of production. The other major irrigated crops are citrus fruit and rice which, together with canned fruit, make up about 11 percent of exports.

FORESTRY

Three large plantations of pine and eucalyptus, covering about 215,000 acres (87,000 ha) or 5 percent of the total area of the country, have been developed in postwar years, all in the highveld, where pulpwood can be grown on a 12-year rotation as compared to 40 years in northern Europe. The most important is the Usutu Forests, said to be the second-largest man-made forest in the world, owned by Courtaulds, CDC, and the Swazi Nation. It produces unbleached sulfate pulp at Bunya, where the plant capacity was increased to 150,000 tons in 1972.

In the north, Peak Timbers and Swaziland

Landscape in the highveld of Swaziland
A considerable number of trek sheep are grazed in the highveld which also has important forest plantations. Part of the Havelock Asbestos Mine is seen in the middle distance.

Part of the Peak Timbers plantation in the Piggs Peak district of Swaziland Three large forest plantations in the highveld produce pulpwood, lumber, pit props, and other products.

Plantations have 76,000 acres (30,800 ha) in the Pigg's Peak district. There are five main sawmills producing lumber, pit props, telephone poles, and chipboards. Exports of pulpwood and wood products account for about 21.4 percent of total exports and the industry employs nearly a fifth of the wage labor force in the private sector.

MINERALS

Iron ore, asbestos, and coal are the main minerals produced in Swaziland. Asbestos has been mined at the Havelock Asbestos Mine since 1939; for some years it accounted for over 50 percent of government revenue and total exports, but its share is now only about 10 percent, because of developments in other sectors rather than reduced production. The mine is the world's fifth-largest producer of asbestos, but known reserves will be exhausted in about 20 years; another mine may be opened, however, near Emlembe. The rock is moved by a 12.5 mile (20 km) aerial ropeway to Barberton in the Republic and railed from there to Lourenço Marques.

Iron ore, developed at Ngwenya northwest of Mbabane, is now the leading mineral, accounting for about 21.4 percent of total exports. The high-grade ore will soon be exhausted, but medium-grade ores will extend the life of the mine to 1978; whether it will be possible to beneficiate and ship the lower grade ores from the area has not yet been determined. All shipments have gone to Japan under contract with two steel companies. The Ngwenya mine had an important additional benefit to the Swazi economy in that it justified construction of the Swaziland Railway across the country, serving the Usutu Forests, the coal mine, and the main commercial-industrial center of Manzini.

Swaziland also has deposits of good-quality coal, extending across the country on the eastern edge of the lowveld. About half of the present output of about 125,000 tons, mined at Mpaka, is used by the Swaziland Railway and half is exported. A study is being made of a proposal to erect a very large (1.5–2 million kw) electric plant near the mine, the bulk of whose power would be transmitted to South Africa, while another proposal is to use coal and iron ore in an $80 million plant to produce pig iron for export.

Other minerals produced are barites, kaolin, and pyrophyllite. Numerous mineral surveys have been undertaken in recent years and finds of various minerals have been reported.

INDUSTRY

The manufacturing sector employed 5,823 persons in 1970–71 or 12 percent of the total wage earners. The more important industrial developments in Swaziland are associated with the processing of extractive products, including the saw mills, pulp factory, sugar mills, fruit canning, cotton ginning, meat processing, and corn milling. But a beginning has been made on the installation of market-oriented establishments, which now produce beer, spirits, textiles, clothing, candy, plastic containers, wire products, and pharmaceuticals, and assemble tractors and caravans. An active effort has begun recently to attract plants that would look to the South African market and thus permit very much larger scale operations; thus far discussed are radio and TV assembly, zip fasteners, and production of refrigerator and deep freeze units. Many of the newer plants are located at the Matsapa Industrial Estate outside Manzini.

Swaziland has favorable conditions for small hydroelectric developments. Several of its important streams enter the territory in deep gorges at elevations of 3,300 to 4,500 feet (1,000–1,370 m) and flow across the country to leave at elevations of about 500 feet (150 m). A half-dozen small plants are now in being, including the 15,000 kw Edwaleni station on the Little Usutu River. For the future, about 30,000 kw could be developed by diverting water from the Great Usutu, while about 225,000 kw could be harnessed by 13 stations on that river. Many hydroelectric schemes will also increase or even out the water available for irrigation. In the meantime contracts have been concluded with Escom to take output from up to 35,000 kw of capacity to meet the rapidly rising needs for electricity.

Swaziland got an earlier start in the tourist industry than Botswana and Lesotho with the construction of a casino near Mbabane, which yields about $665,000 a year in government revenue; roulette wheels are not allowed in South Africa, which helps to explain the casino's popularity. Two new hotels were expected to increase employment in the tourist sector to about 2,000 and increase tourist visits from 300,000 in 1972 to 400,000 in 1975.

TRANSPORTATION

The 140-mile Swaziland Railway serves several of the most important regions of the country, including the Mbabane-Manzini axis. The road net totals about 1,510 miles (2,430 km) of which 114 miles (183 km) are paved, including the main highway across the country. Like Lesotho, Swaziland is considering the construction of an international airport. There are now four government airports, of which the Matsapa field is most important, and a score of private strips.

Swaziland is, then, a small but rich country with a wide range of economic activities. Its economic indicators reveal rapid growth: exports increased 4.4-fold from 1962 to 1972, government revenues grew from $10 million in 1961–62 to $78 million in 1972–73, and the estimated per capita GNP, growing by 4.7 percent per annum, is now about $200, removing Swaziland from among the 25 poorest nations in the world.

The country's greatest concern must be to narrow the gap between the traditional and modern sectors, to assure the effective and soil-conserving use of the land, to raise the incomes of the general populace to increase their well-being and expand the opportunities for further industrialization, and to spread development more evenly throughout the country.

A village in the Betsimitatatra Plain near Tananarive, Madagascar, surrounded by rice paddies

CHAPTER TWENTY-SIX

Madagascar

This "continent of the Indian Ocean" has many of the characteristics and problems of Africa, of which it is the largest appurtenance, yet it has distinctive physical, cultural, and economic features that give it peculiar interest. Madagascar owes its distinctiveness in considerable part to the Malayo-Polynesian origin of a substantial part of its populace, to the isolating influence of its location, and to its island character.

Likened to the imprint of a left foot in the sand, the "Grand Isle" stretches for almost 1,000 miles (1,610 km) as a giant breakwater off the coast of Mozambique, from which it is separated by the 250- to 500-mile (400–800 km) Mozambique Channel. Averaging about 230 miles (370 km) across and with a maximum width of 360 miles (580 km), this fourth-largest world island has an area of 226,657 square miles (587,041 sq km), larger than France and Belgium. Its population in mid-1974 was about 7.7 million.

Madagascar has had one of the poorer records of growth of African countries in the postwar years. Its balance of payments is characteristically in substantial deficit, and its per capita GNP showed almost no growth in the 1950s and grew by only 0.6 percent per annum from 1960 to 1971, when it was estimated to be $140. Only about 21 percent of the population is in the monetary sector of the economy and wage employees comprise only 8.3 percent of the working-age population. The value of exports increased between 1938 and 1970 at only one-third the rate for tropical Africa. The country's leading export since before World War I has been coffee,

but Madagascar did not share in the African coffee boom of the 1950s; its portion of the continent's production declined from 9.6 percent in 1947–49 to 4.3 percent in 1971. There have been repeated failures to meet goals announced for increased agricultural output and the purchasing power of farmers is estimated to have declined from 1967 to 1973. Per capita consumption of energy is about 21 percent of the African average, much of industry has been operating below capacity, and other sectors have not developed sufficiently to provide much relief to the gloomy situation.

Several features are disturbing when placed against this background: the heavy reliance on grants and loans to balance international payments and finance almost all development; a continual major dependence on expatriate enterprises (which produce almost all of the sugar, sisal, bananas, and cocoa and part of the coffee, rice, and other agricultural exports, and most of the mineral output, and which dominate the manufacturing, commercial, and financial sectors); an excessive concentration of activity on Tananarive (whose population has mushroomed to over 500,000); and an ever-increasing rate of population growth (births exceeded deaths by only 11,000 in 1946 but by about 157,000 in 1972). Adjustments to this dynamic change have thus far been seriously deficient.

It is no wonder that tensions have increased. Ethnic rivalries between the more evolved Merina, centered in the highlands, and various coastal groups, or *côtiers,* have led to clashes in each year since 1971 and were in considerable part responsible for the ousting of President Tsiranana by Army

Chief of Staff Ramanantsoa in May 1972. There has also been increasing friction in Franco-Malagasy relations, with Madagascar withdrawing from the franc zone and from OCAM, and arranging for the withdrawal of French troops. French willingness to continue grants has softened the resentments, but Madagascar appears determined to follow a strongly nationalistic line and a more rapid rate of localization. Even this causes internal dissent, however, because the côtiers resent the use of the Merina form of Malgache instead of one of their dialects.

THE PHYSICAL MILIEU

Madagascar's location helps to explain its isolation, the development of independent species of flora and fauna, the absence of other species common to Africa, the rather unusual crop emphases of the country, and some of the contrasts in cultural practices. Some islands are located in the stream of world movements and their interests are more universal than insular, but Madagascar's location has tended to minimize its contact with the outside world. Contributing to this isolation have been the inhospitable nature of much of the coast and the fact that the most evolved group, the Merina, deliberately isolated themselves in the center of the island and long discouraged contacts with surrounding peoples. While it is true that the island lay off the beaten track, particularly after the Suez Canal was opened, the degree of its isolation should not be overstressed. It is served by several liner and regular freight services and, if isolation were as severe as it is said to be, Mauritius, which is several days further away from Europe on the same services, should be equally underdeveloped, but, in fact, it has exports about half those of Madagascar from an area less than a third of 1 percent as great.

LANDFORMS

The island is divided into two main geologic zones, a highland region composed largely of Archean granites, and the western plain of almost undisturbed sedimentary rocks dipping gently towards the Mozambique Channel and occupying about a quarter of the total area. From the topographic standpoint, the dominant feature is the great central highland, which extends 720 miles (1,160 km) north and south with a maximum width of about 240 miles (386 km) (Map 112). This area is usually called the "central plateau," which conveys a highly erroneous impression of its surface features. Although the hills often appear to have a fairly uniform elevation and although they rise from the Archean base that is substantially elevated, the area is profoundly cut by valleys and the great bulk of it is in slope. It is more accurate to think of this important region as a high-based hill-land.

Rising above the average elevations of 3,900 to 4,600 feet (1,200–1,400 m) are three principal masses: Tsaratanana in the north, whose summit of volcanic rock rises to 9,436 feet (2,876 m); Ankaratra, an enormous volcanic massif with elevations of 7,000–8,600 feet (2,100–2,600 m) in the center of the country; and Andringitra, a vast denuded granite mass rising to 8,721 feet (2,658 m) south of Ambalavao. The highland area is generally higher in the east, where it drops off abruptly in one or two steps to the eastern coastal zone. Between the two scarps at about the middle of the island are some fairly extensive flattish areas, the Mangoro and Lake Alaotra basins.

From the standpoint of human use, the uplands are a more difficult region than the African plateau (inset, Map 113). There are the same difficulties of approach, the same fringing scarps, and the same lack of navigable streams that long repelled the opening up of Africa. But even when the heights are attained, the inhibiting influences of the landforms continue. Roads are constructed with difficulty; with the aim of reducing expenditure on bridges they meander endlessly along the fairly even crests of the dissected hills. Soil erosion occurs more readily

Map 112. Physical map of Madagascar
Photograph of relief map by permission Institut Nationale Géographique de France

and is more difficult to control; use of the sloping surfaces for agriculture is not attractive. On the other hand, there is perhaps an advantage in the more rapid accumulation of water in the valleys, which may then be controlled for the intensive cultivation of paddy rice.

The eastern coastal zone, only 10 to 50 miles (16–80 km) wide, is composed of downfaulted segments of the island's Archean massif, fringed by flat, low-lying swampland resting on Quaternary deposits. A series of lagoons lies behind the very narrow belt of sand, which extends in a remarkably straight line along the coast from Fenerive almost to Fort Dauphin, while one or more coral reefs, usually submerged, lie immediately off the coast. A final scarp or series of scarps, lying below the sea, leads to enormous depths very close to the island mass.

On the western side of the island a 60- to 120-mile (100–200 km) belt of plains and low plateaus dips gently toward the channel, while a series of more or less well-marked cuestas face the central highlands, from which the area is often separated by a profoundly eroded scarp zone, known in the center as the Bongo Lava.

The topographic features of Madagascar help to explain the division of the island into many economic subregions (Map 115) whose contact with one another is often very tenuous. One sees, as a consequence, the same sort of concentration of economic activity in coastal and highland "islands" that is discernible on the continent.

CLIMATE

Madagascar is entirely tropical, and its major climatic features are controlled most strongly by its position in the prevailing easterlies and by its relief. It may be divided into four main climate regions: a tropical rainy climate zone running almost the full length of the east coast and along the northwest coast; a tropical highland climate in the center; a tropical savanna region along the west and across the northern tip; and a tropical steppe climate zone in the southwest. The west and south suffer from irregularity of precipitation and from alternating excessive aridity and torrential rains. The climate regimes of Madagascar permit production of a great variety of tropical crops and the export list is an unusually long one; while this diversity cushions market swings it cannot be said to have given strength to the Malagasy economy.

Hurricanes strike the coasts one or more times each year. The east coast is most severely and regularly hit, but the west coast and highland areas are not immune. Periodically, east coast farms, cities, and transport facilities are severely affected. These disturbances account for the absence of Arab dhows in the coastal and international traffic of the east coast. while a considerable number engage in the carriage of freight along the west. And it is estimated that bad weather and heavy seas account for over half of the days lost by ships operating in Malagasy waters.

SOILS

Madagascar has been described as having "the color and fertility of a brick"; indeed it is sometimes called the Red Isle. Latosols, the curse of the tropics, cover not only the bulk of the rainy tropical east coast but also most of the highlands and part of the savanna area. They explain the bush fallow agriculture in the east, where it is known as *tavy*. Much of the sedimentary area is also disadvantaged by poor soils, being covered by an infertile skin of clayey sand. Nonetheless, the best regional soils are found in great concentric arcs in the drier western savanna and in the limestone Mahafaly Plateau in the south where, unfortunately, precipitation is most deficient and erratic. There are some good to excellent azonal soils, including those of volcanic origin in the Lake Itasy area west of Tananarive, the Antsirabe region to the south, and on the island of Nossi-Bé, as well as alluvials in the short valleys of the east and along the greater but more irregular streams of the west. In view of the generally deficient soils, it is perhaps fortunate that the prevailing agriculture of the highlands is confined to the usually small valley bottoms, and that it is centered upon cultivation of paddy rice.

Soil erosion is a problem of immense proportions on Madagascar, particularly on the central highlands, where bright red gashes scar the landscape over very wide areas. The major cause of erosion in the highlands is not farming of the slopelands, which is uncommon unless rice terraces are laboriously constructed, but the almost complete removal of the natural forest by deliberate firing, and subsequent overgrazing of the grasslands. There are few regions on the continent where soil erosion has been as severe as it is in central Madagascar.

THE FLORA AND FAUNA

The vegetation of Madagascar shows primary response to the climatic pattern, but misuse by man has degraded it over perhaps 70 percent of the island's surface. Most of the species are African, but those unique to Madagascar give some vegetational landscapes a fascinating appearance. The east coast and scarp zones and the northwest, except in the swampy littoral and where man

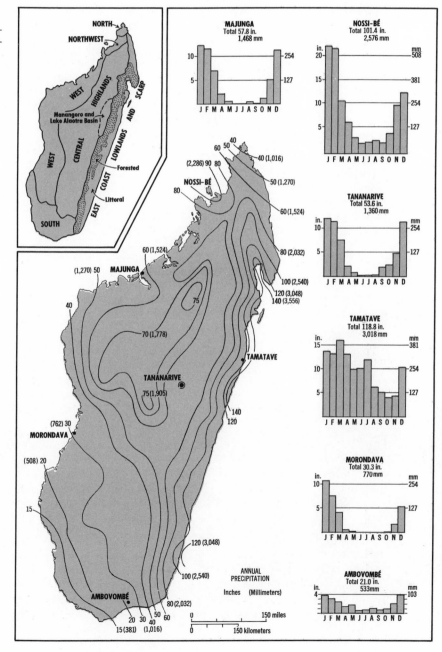

Map 113. Precipitation on Madagascar; selected climographs; physical regions

has cut the forest—either for commercial or subsistence reasons—have tropical rainforests with closely spaced trees of average height and with numerous species represented. These forests cover perhaps 10 percent of the island, but areas of *savoka* or secondary growth are greater than those of virgin forest.

Not many centuries ago the central highlands were covered with forest, but fires have destroyed all but a few remnants in inaccessible parts, and now this region is largely covered with grasses of only mediocre nutritive value. There is a reforestation program, but it has only begun to cover the huge denuded areas of the central region. The west and south have greatly variable vegetation types, sometimes low grasslands, often some type of bush country with surprising thickness despite the greater aridity. Finally, about 1,000 square miles (2,600 sq km) of mangrove forests fringe the west coast.

Government worker about to spray a hut near Fort Dauphin with DDT
A very successful antimalarial campaign in the 1940s and 1950s resulted in a dynamic increase in the rate of population growth.

Extensive gullying in the highlands near Lac Alaotra
Madagascar has extremely severe soil erosion problems caused by removal of the original forest and overgrazing.

Household articles removed before spraying
With the exception of a few pieces of furniture, these represented the total worldly possessions of the inhabitant.

As is true of Madagascar's flora, the fauna consists of numerous African species and others of domestic evolution. There is a complete absence of Africa's characteristic big game; the largest wild animals are the crocodile and the lemur, whose presence almost gave the name Lemuria to the island continent. Snakes are more common than in Africa, but none are poisonous.

Insects are of greater significance. The east coast long had an unsavory reputation as a highly malarious region, but in the 1950s a very successful campaign practically arrested malaria on the entire island. This was the major factor in the advancing growth rate of the population. Locusts constantly threaten crop production in the south. The absence of the tsetse fly is a blessing that has not been well utilized; it permits systems of agriculture not now applicable in large parts of the continent.

POPULATION

The estimated population of Madagascar in mid-1974 was about 7.7 million. It is very unevenly distributed (Map 114). About 20 percent of the island is virtually empty, notably the limestone areas of the west; areas with very low density include the region west of the central highlands and the steeper and higher parts of the elevated massifs and of the eastern scarps. At the other extreme, about 28 percent of the population lives at densities in excess of 100 per square mile (39 per sq km) on 2.3 percent of the area. The most populated portions of the island include the east-central part of the highlands, the borders of Lake Alaotra, and a series of nodes along the east coast from Andapa to Fort Dauphin.

The indigenous population comprises 18 ethnic groups. The ancestors of the Merina, Betsileo and Betsimisaraka, came across the Indian Ocean many centuries ago. They brought with them the language that developed into Malgache, as well as rice, and culture traits still common to Madagascar and Indonesia. Most of the other groups probably originated in Africa. Many of their customs—especially the attitudes of pastoralists toward their cattle—appear closely related to those of African graziers, and there are a substantial number of words in Malgache that have a Bantu or Swahili origin. There has, however, undoubtedly been a considerable mixture of various indigenous Malagasy.

There is considerable political rivalry between the 2 million highly evolved Merina, and the 3.8 million côtiers. Before the Europeans arrived the Merina had a detailed history, advanced governmental forms, and a fairly well developed exchange economy, and were practicing a highly organized, intensive agriculture not found in sub-Saharan Africa. An estimated 10,000 Merina could read and write in 1836; today the percent of their children attending school is well above the national average. Under the French the Merina came to hold most of the positions below the top levels in the government; they were also the clerks, the technicians, and the artisans of the island, and not just in their

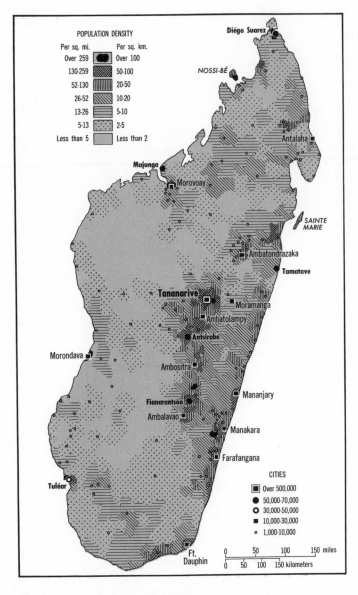

POPULATION DENSITY

Per sq. mi.	Per sq. km.
Over 259	Over 100
130-259	50-100
52-130	20-50
26-52	10-20
13-26	5-10
5-13	2-5
Less than 5	Less than 2

Diégo Suarez

NOSSI-BÉ

Antalaha

Majunga

Morovoay

SAINTE MARIE

Ambatondrazaka

Tamatave

Tananarive

Moramanga

Ambatolampy

Antsirabe

Morondava

Ambositra

Mananjary

Fianarantsoa

Ambalavao

Manakara

Farafangana

CITIES

- ■ Over 500,000
- ● 50,000-70,000
- ○ 30,000-50,000
- ▪ 10,000-30,000
- ○ 1,000-10,000

Tuléar

Ft. Dauphin

0	50	100	150 miles
0	50	100	150 kilometers

Map 114. Population density of Madagascar

traditional area. After independence the President and other high officials were côtiers, but the Merina continued to be disproportionately represented in government and other employment. As might be expected, this has aroused resentment among the other groups.

The non-Malagasy population consisted in 1971 of about 33,000 French, 37,000 Comorians, 17,000 Indians, 9,000 Chinese, and 2,500 others. A substantial number of the French are Creoles from Réunion and Mauritius, many of whom are small colons, foremen, operators of mechanical equipment, and small merchants. Metropolitan French are still present in considerable numbers though their representation in government has been dwindling rapidly and their dominance of the formerly influential Chambers of Commerce has been largely

A Merina village near Tananarive
The Merina are the most evolved ethnic group of Madagascar. Originating centuries ago in southeast Asia, they introduced paddy rice culture and the Malgache language to the island.

eliminated. The Indians and Chinese are often petty merchants and artisans. They have frequently been accused of sharp practices, but, since the Malagasy appear reluctant to enter the commercial field, they are almost indispensable intermediaries in some areas. The vast bulk of the nonindigenous population lives in the eight largest towns, with perhaps half in Tananarive. The capital—with its high concentration of cultural, commercial, industrial, and political institutions—is the Paris of Madagascar.

MAJOR LAND-USE REGIONS

Over 80 percent of the Malagasy population is dependent on agriculture, which accounts for just under a third of the GDP, and over 85 percent of exports. Most farmers have very little if any commercial production. Cultivated lands total only about 3 or 4 percent of the island's surface.

The following listing reveals Madagascar's unusual variety of both subsistence and cash crops:

(1) Food crops: rice, manioc, corn, beans, peanuts, potatoes, taro, sorghum, bananas, sugarcane, and other vegetables and fruits.
(2) Vegetable oils: coconut, castor, candlenut, peanut, tung.
(3) Fibers: sisal, raffia, paka, cotton.
(4) Stimulants: coffee, cocoa, tobacco.
(5) Perfumes, spices, etc.: ylang-ylang, lemon grass, cloves, pepper, vanilla.

The following crops are partially or entirely commercialized, classified on the basis of producer and destination:

(1) Grown almost exclusively on European holdings primarily for export: sisal, tung oil, pepper, ylang-ylang, lemon grass, cocoa, export bananas.

(2) Grown exclusively by Malagasy and having a considerable export: beans, raffia, oil-bearing seeds.

(3) Grown by Europeans and Malagasy and having a considerable export: coffee, tobacco, cloves, vanilla, sugar, manioc, rice, peanuts.

(4) Grown principally for food, mainly by Malagasy: rice, manioc, corn, peanuts, potatoes, market garden crops, taro, sorghum, bananas.

In 1950 about 3.7 million acres (1.5 million ha) were held by Europeans in large estates and small holdings of Creole colons, but only 10 to 15 percent were ever developed. Their holdings have now been reduced by 60 percent, with large companies having abandoned most of their lands in the west and the Creole farms in the east being on the way to extinction.

The island has a substantial and increasing livestock population though it is not anywhere nearly so well commercialized as it should be.

THE CENTRAL HIGHLANDS

The heart of the country is the central highland region (inset, Map 113) particularly the section stretching from the Tananarive region to around Fianarantsoa. Despite the difficult landforms and generally poor soils, this is the most populous part of the country, having about two-fifths of the total population. About 2.9 million are the more advanced Merina and Betsileo.

The two major concerns of the highland farmers are rice and beef cattle. Many of the valleys, often quite narrow, have been laboriously prepared for rice paddies to provide the staple food, the highlands having over half of the approximately 1.8 million irrigated acres on the island. Some terraces are reminiscent of those of Indonesia, the Betsileo having the reputation of constructing the finest ones on the steepest slopes. There are a few large plains, the two most

Rice terraces near Antsirabe
The Betsileo, closely related to the Merina, are credited with constructing the finest terraces on the island.

important being the Betsimitatatra Plain around Tananarive and the Lake Alaotra Basin on the intermediate level between the two eastern scarps. The former has been completely developed, though there are possibilities for better water control; the latter, which produces a large part of the high-quality export rice, still has extensive lands which could be put into paddies. Rice, the main subsistence crop of the island, is grown as a paddy crop wherever possible and paddy rice accounts for about 40 percent of the total value of agricultural output. Yields are low and practices poor; rice is all too rarely planted in rows, rotated, or fertilized—which could give considerably more satisfactory results—and losses to pests and diseases and in storage exceed 30 percent a year. Paddy and hill rice together occupy about 2.1 million acres (840,000 ha), or about a third of the total cultivated land. Despite considerable assistance from several international agencies, total production has not been increasing as it should and imports increased to 200,000 tons in 1973, when a severe shortage existed.

Although rice is particularly important on the highlands, a remarkable variety of other crops is produced, particularly around the capital, and the great Zoma at Tananarive is undoubtedly one of the finest markets in the tropics. There are a number of small specialized crop-production zones: the volcanic soils in the area around Lake Itasy yield peanuts, tobacco, arabica coffee, and tung oil; the high area around Antsirabe and Fianarantsoa produces coffee, tobacco, and white potatoes; the Lake Alaotra basin and the Mangoro Valley produce manioc, some of that portion produced on European estates being used to manufacture tapioca.

The variety of crops grown on the highlands should not permit one to lose sight of the importance of grazing. Over three-fifths of the island is used for grazing, but little care is taken of the cattle, most of which are Zebu, and there is very little selective breeding. Oxen are used in the highlands for drawing carts and for trampling the rice paddies preparatory to planting, but most are left to fend for themselves on the seasonally inadequate grass slopes of the region.

For the future, the two great needs of the highland area are to rationalize livestock practices and to intensify use of the hilltops and slopes without intensifying the already serious soil erosion. Experiments have

shown that by reforesting the steeper areas and by planting crops on the contour and in appropriate rotation, a type of mixed farming can be practiced that is far more intensive than the present nonintegrated system of rice in the valleys and cattle on the hills.

THE EAST COAST REGION

This region also has a dense population. Its wealth lies not in its soil, nor in the suitability of its landforms, but in the rainfall regime, which gives it a tropical rainy climate. The region may be divided into three north-south zones: the coast littoral, low and flat, with alluvials present to an important extent; a zone of foothills; and the scarp zone, a difficult barrier that is largely in forest. Paddy rice is produced in the lower valleys of the coastal zone, hill rice is grown in the foothills and slopes of the scarp zone as the major subsistence crop, but corn and manioc are grown on the drier soils.

Robusta coffee, which is the leading Malagasy export, is the principal cash crop. Its production provides a living for an es-

Zebu cattle in the hills west of Tananarive
Very little effort has been made to integrate crop and livestock farming, and Madagascar does not benefit as it might from absence of the tsetse fly.

Hill rice in an east coast valley near Tamatave
Small huts and platforms are used by children who attempt to chase birds from the fields.

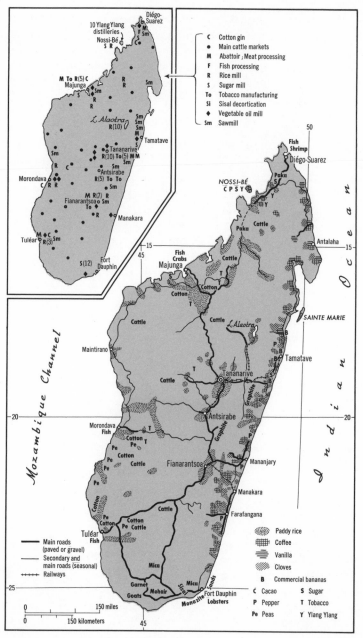

Map 115. Economic map of Madagascar

timated 80,000 indigenous planters plus a number of colons. As noted earlier, Madagascar did not keep up with other African producers when the big increase in output occurred; explanations for this include the failure to replace old bushes, destruction of plants by several hurricanes, and inflated costs of production. Despite several programs and much assistance, successive goals have not been met; and sales have fluctuated widely around an average of about 59,000 tons in recent years.

In addition to coffee, the east coast produces such cash crops as cloves, vanilla, sugar, bananas, and pepper. Madagascar produces about three-fourths of the world's vanilla, but competition from vanillin, made from cloves, tends to restrict sales; it ranks second to Zanzibar in output of cloves. These two crops alternate as second and third ranking crop exports of the island.

The systems of farming in the eastern region are backward and destructive and have led to soil degradation if not outright destruction. A more rational use of the area would involve reforesting the impoverished slopes, construction of numerous tilapia ponds in the smaller valleys, paddy cultivation and pasturing in the lower alluvial flats, and growing of coffee on the slightly higher, better drained lands along the rivers.

Communications are a difficult problem in much of this region. North-south roads are constantly interrupted by the many rivers flowing off the drenched eastern slopes; they generally do not penetrate far inland in the valleys, so many potentially productive areas are not open. The Pangalanes Canal (see below) does not satisfactorily replace the roads and there is only one modern port along the otherwise inhospitable coast.

Drying green vanilla beans
Madagascar is the world's leading producer of vanilla.

THE NORTH AND NORTHWEST REGIONS

In the extreme north there is a small region around Diégo-Suarez that has a savanna climate. Some corn, peanuts, and beef are produced and salines near Diégo provide most of the domestic requirements for salt. In the northwest Île Nossi-Bé and several rich valleys on the mainland are important economic "islands." The volcanic Nossi-Bé, long a center of colonization and now a budding tourist center, produces coffee, coconuts, sugar, pepper, vanilla, and perfume essences. Madagascar is Africa's leading producer of pepper, the vines typically being grown on shade trees in coffee and cocoa plantations.

On the mainland the fertile and extensive plain of the North Mahavavy is the site of a large, mechanized sugar development,

opened in 1954. Sugar output has been limited by quotas in French and OCAM markets; defections by purchasing nations in the latter have depressed exports in recent years but negotiations have been undertaken to renew contracts. The Sambirano Valley, the second rich alluvial zone, has soils derived from the high, volcanic mass of Tsaratanana. It produces coffee, cocoa, coconuts, sugar, essential oils, pepper, palm oil, manioc, and staple rice. The several parts of the north and northwest are more readily tied with each other by coastal vessel than by road.

THE WESTERN AND SOUTHERN REGIONS

The western region is a savanna area, while the southern region is steppelike with a long, rigorous dry season and occasional yet certain droughts which lead to great losses of livestock and human suffering. Composed of a long series of dry plains and low plateaus cut by large and small valleys, both areas contain the best regional soils on the island (still not very good, however), as well as some excellent alluvials. The major shortcoming of this extensive area is the meager and irregular precipitation.

Regionally, this is pastoral country, and cattle are the main interest of the tribes, which are grouped under the name Sakalava in the west, while the Mahafaly, Antanosy, and Bara are the major southern tribes. The area is the least important region of Madagascar, whose advance is impeded by the sparse network and poor quality of its roads.

Cattle raising suffers from physical difficulties, from malpractices of which grass firing is the worst, and from cultural attitudes. Many of the tribes are more concerned with their cattle as instruments of ceremony than as suppliers of food or sources of wealth. Cattle thievery is an honored custom, indeed almost a necessity if a young man is to be acceptable as a husband. Corralling the cattle at night to prevent their being stolen often takes place in unhealthy conditions leading to a high incidence of bovine tuberculosis. There has been strong disincentive, then, on the part of even the more evolved to attempt improvement of their practices.

Despite the difficulties, there is an export of animal products from all the west coast ports, but a recently installed meat-packing plant at Morandava lies empty because of inadequate supplies. Some goats are raised in the south, producing microdeserts around the settlements. At Ampanihy, an effort has

Goats near Ampanihy in southern Madagascar
Their grazing has created a microdesert around the village.

been made to develop a rug and drape industry in the government-subsidized Maison de Mohair.

Such intensive activity as there is in this region is found in the large and small valleys, where there are many additional opportunities for intensifying production, though control works will be required in some cases. In the valleys near Majunga there is an important output of rice, the island's largest and best-quality raffia production, and less important yields of tobacco and of sugar. The Betsiboka Valley is potentially one of the richest regions of the island, while the Morandava and Mangoky Basins are the sites of important multipurpose development schemes. Around Morandava and Tuléar, rice, lima beans, vegetables, and corn are the chief interests, while Virginia tobacco is grown in the high valley of the Tsiribihina. Efforts have been made for several decades to push cotton in several of the western valleys but with only modest success. The Mandrare Valley in the south is the main zone of sisal estates; in the mid-1950s about 25,000 people were employed in producing sisal, but the industry contracted considerably after the drop in prices of the 1960s. Some castor seed is produced around Ambovombé.

There is, then, great variety among the regions of Madagascar, far greater than can be conveyed in this brief discussion. Perhaps the most common characteristics are the intensity of use in some of the valleys, the very extensive use of surrounding areas, and the emptiness of enormous portions.

MINOR SECTORS

The fishing industry is poorly developed on Madagascar, except for the recent introduction of shrimp, lobster, and crab operations, which have contributed to rising exports. The Vezo engage in sea and lagoon fishing along the west, particularly from Tuléar, but they operate from pirogues and have only a small surplus production. Freshwater fish are caught in many of the rivers, in Lake Alaotra, and in the rice paddies, but mainly for local consumption. It is ironic that an island with many cattle and substantial opportunities for increasing the catch of fish should be plagued by a widespread protein deficiency.

Madagascar's forests are mainly found along the east and northwest zones, often on steep land and in areas which are not easily accessible. Some of its unique species have great beauty, but have been cut excessively and tend to be very slow growing. Afforestation began in the 1930s but has not been so intensive as it should have been, from either the standpoint of erosion control or supply of lumber and fuel, and Madagascar is a net importer of forest products.

The tourist sector is beginning to receive greater attention and the successes in the Seychelles and Mauritius suggest that considerable advances can be made. The island has a variety of attractions including Tananarive, one of the most attractive of African capital cities, the palm-fringed island of Nossi-Bé, excellent beaches on the west, and the southern area with its unique vegetation and presence of lemur. Consideration might well be given to setting up a game park in one of the empty areas and importing animals from East Africa. South Africa was planning an investment in tourist facilities at Nossi-Bé, but this and other joint projects came to an end with the ousting of President Tsiranana. A jet airport is, however, planned for that island.

Madagascar has a wide variety of minerals, but most occur in small and scattered bodies. The island is the leading world supplier of high-grade graphite, produced south of Tamatave and near Moramanga, and mica comes from a score of small exploitations north of Fort Dauphin. Production of both is limited by the size of the world market and increasing competition from substitutes. In 1969 a chromite deposit was opened near

Sorting garnets at a small mine near Ampanihy
Small-scale operations are characteristic of most mining on Madagascar.

Andriamena, 100 miles (160 km) north of Tananarive, involving an investment of $16 million. The concentrate is trucked 55 miles (88 km) to a short rail extension of the Lake Alaotra branch of the TCE rail line, and railed to Tamatave for export. These three minerals accounted for 4.2 percent of exports in 1972, a lower percentage than the first two enjoyed in 1938. There is also a small output of quartz, beryl, and garnets, while uranothorianite and monazite were mined from 1955 into the 1960s. Attention has been given to a bauxite deposit at Manantenina in the southeast, but it appears that the need to construct a wharf to take large ore carriers to sustain a yearly output of 5 million tons and the relatively small size and poor quality of the body have discouraged investment. Deposits of nickel, iron ore, and other minerals are under study, while prospecting for oil continues in the west, but a search extending over many years has so far yielded only minor evidence of natural gas. Madagascar has a small deposit of coal in the Sakoa basin in the southwest but most of the known reserves are mediocre, noncoking, high-ash-content coals.

THE MANUFACTURING SECTOR

About 45,000 persons are said to be employed in industry, but this includes artisans and an indeterminate number of non-manufacturing activities. The handicraft industry, with about 3,000 persons, is unusually well developed, producing a variety of products including raffia hats and handbags, some of which are exported. Not counting about 100 rice mills there are now about 200 manufacturing plants, of which about 30 process animal and vegetable products for export while 20 large and 150 small establishments produce goods for the internal market. There is excess capacity in many branches of industry.

As may be seen from the listing below and from Map 115, the range of manufacturing plants is fairly comparable to that of many tropical African countries. The list does not include those industries shown on Map 115.

The main limitations to increasing industry are the small size of the market, both in number and in purchasing power, and the highly fragmented character of the market. Other problems include the high cost of energy, inadequate capital, lack of skill and managerial ability, and local inflation. The National Investment Company attempts to promote manufacturing with various incentives, including the provision of minority-participation capital.

Over three-fourths of the market-oriented industrial establishments are located in Tananarive Province, though the two unusually large textile mills are situated at Antsirabe and Majunga which, with Tamatave and Diégo-Suarez, make up the more important secondary centers.

Although Madagascar rates low in the list of African countries in production and consumption of energy, it does have substantial resources. Nine hydroelectric plants provide about half of the electricity produced; the two largest stations are the 30,000-kw Mandraka plant 30 miles (48 km) east of Tananarive and the 11,050-kw Antelomita plant. About three-fifths of the total output of electricity is consumed in the capital city, but even there the bulk of houses are not served. The petroleum refinery at Tamatave does contribute to exports as it supplies the Comoros and Réunion as well as the Malagasy market.

TRANSPORT DEVELOPMENT

Madagascar's transport complex consists of a scanty road system, 540 miles (870 km) of railways in two lines, a rather unusual canal along much of the east coast, a remarkable air net, and 17 ports, only 7 of which handle more international than cabotage traffic.

INLAND TRANSPORT. The existing road net is rudimentary, marked by a somewhat distorted dorsal road with occasional ribs extending to the more important ports. Even parts of the dorsal road are in poor condition, while many of the secondary roads are seasonally impassable. The best-developed system radiates from Tananarive, and the

Manioc processing	Bags	Vehicle assembly
Fruit juice canning	Shoes	Radio assembly
Tanning	Matches	Sewing machine assembly
Shrimp packing	Soap	Cement
Flour milling	Paints	Bricks and tiles
Bakeries	Insecticides	Plastic products
Confectioneries	Pharmaceuticals	Bicycle tires
Dairy products	Industrial gases	Paper products
Brewing	Petroleum refining	Glass
Soft drinks	Metal construction	Wood furniture
Instant coffee	Household utensils	Batteries
Textiles	Metal furniture	Printing
Clothing	Metal containers	Ice

View of Tananarive
The capital was located on this large rocky hill in the Betsimitatatra Plain originally for protective reasons.

roads from the capital to the two main ports, Tamatave and Majunga, have recently been improved.

The main rail line of the island is the Tananarive–Côte Est Railway (TCE) connecting the capital and Tamatave, with branches serving the Lake Alaotra basin, shipments from the Andriamena chromite mine, and the secondary highland center of Antsirabe. The second line, the Fianarantsoa–Côte Est (FCE), connects the capital of the Betsileo country in the highlands with Manakara, an open roadstead port. Both railways are single-track, meter-gauge lines and have been dieselized since 1955. Carrying about a million tons a year, their traffic is heavily imbalanced in upline movements, especially on the more important TCE, which reflects the strong consumption power of Tananarive.

Both lines must surmount difficult scarps; minimum curve radii are only 262 feet (80 meters); maximum gradients are 2.5 percent on the TCE and 3.5 percent on the FCE; and numerous tunnels, culverts, and bridges were required in their construction. There has been talk for some years regarding the joining of the two lines by a 149-mile (240-km) link from Antsirabe to Fianarantsoa, which would result in some savings through consolidated maintenance in the excellent shops at Tananarive and through greater flexibility in use of equipment; but the small amount of traffic on the FCE and the difficult terrain that would have to be traversed have thus far discouraged the necessarily high expenditure that would be required.

Few of the rivers of Madagascar are of more than local significance in the carriage of goods. Those of the east coast are interrupted by rapids and those that flow into the Mozambique Channel are intermittent, seasonal, and plagued by shifting channels. Only the Betsiboka, which is navigable for 128 miles (206 km), is fairly heavily used. On the central highlands the irrigation and drainage canals of the Betsimitatatra Plain are used by pirogues, and a few light barges are employed on Lake Alaotra.

The most interesting inland waterway is the Pangalanes Canal, which runs for about 400 miles (640 km) along the east coast from Foulpointe to Farafangana, connecting a series of lagoons by cuts across the sills, or "pangalanes," that separate them. But the canal has never fulfilled its promise, in considerable part because only the section south of Tamatave is open to reasonably large barges.

Madagascar has an unusually well-developed air net, with 49 airports regularly served, the highest number for any African country. A close study might reveal that such an extensive system is somewhat of a luxury for a poor country, whose economy might be better aided by coming down to earth.

PORTS. The international traffic handled by Malagasy ports in 1972 was only 1.7 million tons, a rather feeble tonnage considering the size of the island and the fact that it has more ports than any other African country. Only Tamatave, which handles two-thirds of the total traffic of the seven ports serving international traffic, can be considered a modern port; only Tamatave, Diégo-Suarez, and Tuléar have some kind of deep-water quay. Many ports are no more than poor lighterage harbors with no protection for ships at anchor and with minimal shore and floating installations. This does not mean that there are harbors just waiting to be converted into modern ports; rather, the island is noted for the absence of indentations, especially on the east coast, while on the west coast the larger river mouths are unreliable because of rapid silting. It is ironic that Madagascar possesses one of the finest natural harbors in the world at Diégo-Suarez, situated where it is of only minor utility to the island.

Only a few ports have sizable productive hinterlands—Tamatave, Majunga, Manakara, Tuléar, and Fort Dauphin. Two others benefit considerably from coastal traffic though their hinterlands are small—Nossi-Bé and Diégo-Suarez. The remaining ports serve restricted areas only, and some would certainly not exist if better land communications were available. Finally, the dispersion of activity among 17 ports means that maintenance and development funds

must be divided and that only Tamatave gets adequate support to permit modernization. It is, in fact, now being enlarged and better equipped. A study is also underway regarding the possible improvement of Manakara. South African and French interests were planning to assist in the construction of drydock facilities at Narinda Bay in the northwest, for the supertankers and huge ore vessels rounding the Cape, but this scheme has been abandoned.

INTERNATIONAL TRADE

Perhaps the most important feature of the international trade position of Madagascar is the marked imbalance, imports being substantially greater than exports (1973 was an exception owing to high coffee prices). If this reflected a high level of investment for economic development it would not be disturbing, but it does not. Total international payments are balanced mainly by French subventions, upon which the country depends for practically all of the development budget.

Year	Exports (in million $)	Imports (in million $)	Exports as percent of imports
1938	23.6	17.4	135.7
1948	50.0	77.7	64.4
1958	82.6	108.4	76.2
1968	115.9	170.2	68.1
1972	163.7	202.0	81.0
1973	201.6	203.0	99.3

SOURCE: Madagascar, *Bulletin Mensuel de Statistiques.*

EXPLANATIONS FOR STAGNATION

Evidences of the unsatisfactory record made by Madagascar in the past decades are scattered throughout the preceding sections. Certainly the physical deficiencies cannot be blamed for the situation, because many of them are shared with other African countries and the island has certain advantages such as absence of the tsetse fly and large unused areas. Possibly the two most important explanations are the poor integration of the physical economy and certain inhibiting elements of Malagasy culture.

Economic sectionalism is a marked feature of Madagascar. The "islands" of economic activity are frequently discrete and poorly connected with others; many are dependent on slow and costly coastwise shipments. Heightening the disequilibrium is the location of the main consuming region in the central highlands focused on Tananarive. The inland transport net is not yet sufficiently well developed to integrate the numerous nodes, which contributes to the centrifugal effect. Thus the economy of Madagascar functions more like that of an archipelago of islands than of an island continent, which it is sometimes called.

Cultural features that may combine to retard growth are the exceptional insularity of the Malagasy, their conspicuous ethnocentricity, a high respect for custom, continuing feudalism (particularly among the Merina), and, most important, their "cult of the dead"—which embodies profound fatalism and concentration upon death. All of these traits help one to understand a certain shunning of innovation and a passive resistance to economic progress. The most promising way to change these attitudes would appear to be through education.

CHAPTER TWENTY-SEVEN

The Smaller Indian Ocean Islands

This chapter discusses three groups of small islands in the Indian Ocean which pertain to Africa—the Comoro Islands, the Seychelles, and the Mascarenes—and touches on the islands belonging to the British Indian Ocean Territory. Common denominators of the three main groups are high population densities, great and increasing pressure on the available resources, and very mixed populations of African, Asian, and European origin.

THE COMORO ISLANDS

Situated like stepping-stones between Madagascar and the mainland, the Comoros consist of four main volcanic islands, usually fringed with coral reefs. Their total area is 838 square miles (2,171 sq km) and their total population was about 336,000 in mid-1974, giving an average density of 401 per square mile (155 per sq km). From northwest to southeast they are (Map 116): Grand Comore, with slightly over half of the area and 47 percent of the population, which has two volcanic groups, the northern forming a vast plateau at elevations around 2,150 feet (655 m) interrupted with rounded hillocks, the southern having the still-active Karthala with an elevation of 8,120 feet (2,475 m) and a crater 2 miles (3.2 km) across; Mohéli, with 13 percent of the area and 4.2 percent of the population, whose central mountainous region rises to 2,820 feet (860 m); Anjouan, with 19 percent of the area and 36.0 percent of the population, whose ravine-scarred mountains culminate at 5,166 feet (1,575 m); and Mayotte, with 16.7 percent of the area and 12.8 percent of the pop-ulation, which has a mountainous north-south chain not exceeding 2,000 feet (610 m) with two sizable plains in the center and northeast.

The islands are tropical with a hot, rainy season from November to April and a moderately warm, dry period from May to October. The rich volcanic soils, most fertile on Mayotte and Anjouan, support a luxuriant vegetation below 1,300 feet (396 m) and forests containing mahogany and camphor above. Forests cover about a sixth of the total area and there has been some afforestation with teak and eucalyptus, while a variety of cabinet woods are used for furniture and house construction. There are few wild or domestic animals. Fish are quite plentiful offshore but landings barely meet local needs; a South African firm has shown interest in developing the lobster and shrimp catch for export.

Mayotte was ruled by France from 1843 and the other islands from 1886–1909. From 1912 to 1946 they were administered as part of Madagascar and increasing degrees of local autonomy were extended in 1956, 1961, and 1973, while in 1974 a referendum endorsed the wish to become independent by mid-1975, which the French Parliament was expected to approve.

The first arrivals on the islands were probably Bantu from East Africa. The Oimatsaha, an Indonesian people, arrived shortly before two Arab invasions, the last of which was in the fifteenth century, while Malagasy, best represented on Mayotte, arrived in a series of migrations and invasions beginning in the same century. Africans were later brought to the islands as slaves, giving rise to a feudal system whose heritage persists. A mixture of all groups, the Antalotes, are found on Anjouan and Grand Comore, which are the most Arabized of the islands. The vast bulk of the populace adheres to the Muslim faith; there are about 700 mosques on the islands. A few Indians and Europeans complete the heterogenous population.

The Comoros have a poorly developed social infrastructure and a precarious economy. Only a fifth of the school-age population is in schools; there are only 19 doctors, 2 dentists, and 1 pharmacy on the four islands; and there is serious malnutrition. The islands do not produce enough to feed themselves. Foodstuffs make up about 36 percent of imports; 25 percent of the working population is registered as unemployed; the government budget depends on direct subventions from France for 28 percent of receipts and on France, FED, and other agencies for all development expenditures; exports cover a decreasing percent of imports, only 55 percent in recent years compared to 85 percent in 1960.

Agriculture is by all odds the main activity, and about 53 percent of the total area is classified as useful. A great variety of subsistence

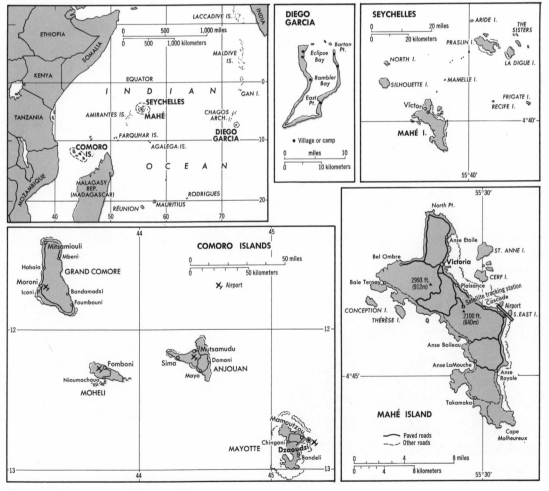

Map 116. Location map, Indian Ocean Islands

crops is grown; the main export crops are vanilla, perfume essences, cloves, and copra, which account for 90 percent of total exports. Lesser cash crops include pepper, coffee, sisal, and sugar. Most exports are grown on company plantations, 75 percent of which are French owned.

The other economic sectors are meagerly represented. Industry is limited to the processing of agricultural produce; plants include 38 perfume essence distilleries, copra drying facilities, sisal decortication, sugar milling, and oil expressing. Handicraft production is strong in wooden articles but appears to be declining. Electricity is produced in three small thermal and two hydropower stations. Tourism shows considerable promise but there are only a few small hotels. Whether Air France is prepared to run the kind of low-cost package tours that have brought a tourist boom to the Seychelles is a question, as it appears to favor the Antilles and Polynesia. There is some talk of South African interests, which were frustrated on Nossi-Bé, turning to the Comoros.

The transport infrastructure of the Comoros consists of about 600 miles (1,000 km) of roads, of which 85 percent are all-weather and about 30 percent are paved, four ports (only one of which, Dzaoudzi on Mayotte, has good protection for small ocean-going

Hand pollination of a vanilla orchid
One of the exports of the Comoros and of Réunion is vanilla.

vessels, the other islands having inhospitable coasts with steep cliffs, dangerous offshore coral reefs and islets, and heavy seas from December to March), and an airport on each island. Grand Comore's airport was improved in 1973 to accommodate large jets, and its location—on the slopes of an active volcano so that there is danger of lava flowing onto the field, and on a less attractive island from the standpoint of tourism—has been criticized as politically inspired. The Comoros depend on small aircraft, coastal vessels, and some dhows for interisland connections, and a considerable portion of international freight, which totals about 65,000 tons a year, moves via Malagasy ports. Exports go mainly to France, the United States, and Madagascar; imports, which have over three times the tonnage of exports, come mainly from France and Madagascar.

THE SEYCHELLES

The Crown Colony of the Seychelles is slated to become independent in late 1975. Located about 1,100 miles (1,770 km) east of Mombasa and 680 miles (1,094 km) northeast of Madagascar, the Seychelles are composed of two distinct collections of islands—the Mahé group, 32 granitic islands all within a radius of 35 miles (56 km) of the main island Mahé; and the outlying islands, 60 mostly coralline islands which are waterless and have no permanent population (Map 116). The area of granitic islands is 85 square miles, or 54.5 percent of the total, and Mahé's area accounts for 65 percent of the latter figure. This group has an equable climate despite its location near the equator, since temperatures are moderated by the marine influence. Rainfall on them varies from about 80 to 90 inches (2,032–2,286 mm) at sea level to 140 inches (3,556 mm) on the slopes, which culminate at 2,970 feet (905 m) on Mahé; they lie outside the cyclone belt. There is very little level land on the inhabited islands, and the soils, which are not inherently rich, require constant fertilization. Most of the original forest has long since been removed, but about 2,650 acres (1,072 ha) have been reforested to protect eroded hilltops and catchment areas and to provide for local needs. The islands are advantaged by the absence of numerous tropical diseases and of the tsetse fly.

The Seychelles were uninhabited to 1770 when the French colonized Mahé primarily to deny Britain a port of call on the route to India. In 1814 it became a British possession and was inhabited by an estimated 5,000 persons including at least 4,000 slaves from East Africa, Mauritius, and Madagascar. The abolition of slavery in 1834 destroyed the existing economy, which later came to depend on coconut and cinnamon products. French, British, Africans, Malagasy, Indians, Chinese, and others have contributed genes to the common pool, but, unlike the Mascarenes, Indians and Chinese make up little more than 1 percent of the population, though they provide the main mercantile class on the islands. The present population thus mainly represents natural growth from the early settlers plus about 2,400 Africans liberated by the British Navy from slave dhows between 1861 and 1872. Population growth was slowed at various periods by large-scale emigration, smallpox and influenza epidemics, and by the absence of a large number of young men serving in the Seychelles Pioneers in the 1940s. The 1971 census recorded 52,437 persons, 86 percent of whom were on Mahé, giving that island a density of 820 per square mile (317 per square km); most of the rest are on nearby Praslin and La Digue Islands. The population is believed to be increasing by about 1,400 per annum; the birthrate did show a welcome decrease from 40 per M in 1960 to 31 per M in 1971.

Until very recently the economy of the Seychelles was primarily agricultural. Holdings vary from plantations of 200 to 4,000 acres (80–1,600 ha) at one extreme to small-holdings of 1 to 10 acres (0.4–4 ha) at the other. The plantations account for most of the export production and employ about 5,000 people. Most of the cultivated land is in coconut palms, whose products account for about 50 to 60 percent of exports. Other cash crops are patchouli leaf, whose oil is used in perfumes; vanilla, once the leading export but now very minor; and tea, produced on a small plantation started in 1963 and on adjacent smallholdings in quantities to satisfy the local market and a small export. Various roots, fruits, and vegetables are grown for local consumption, but the islands are dependent for the vast bulk of their food on imports. Cattle are grazed in the coconut groves and pigs have been doing well since their introduction in 1967. The government agricultural program adopted in 1972 calls for the subdivision of larger holdings into small farms and an emphasis on production of foodstuffs for the indigenous and tourist population.

Nonagricultural productive sectors are poorly represented. The second-leading ex-

Victoria, capital of the Seychelles

port (c. 40 percent of the total), however, is cinnamon products from trees that grow wild all over the hillsides of Mahé. Traditional fishing is carried on close to shore; modern and larger boats would permit a greater catch and probably export of tuna and other pelagic species and justify long-planned cold storage facilities at Victoria, the capital and main port. The corraline islands, worked by labor under government contract, make a contribution to the economy through the collection guano, coconuts, birds' eggs, tortoises, and salted fish. Guano deposits have been exhausted on most of the islands, but 6,000 to 7,500 tons a year are still produced on two or three for shipment to Mauritius. Industrial representation is limited to the processing of copra, coir, cinnamon, and tea, an abattoir, and small plants producing soap, shoes, beer, and concrete products, plus a limited amount of artisanal workers making straw hats and baskets and articles from tortoise shells.

The really dynamic development, dating only from the early 1970s, has been in the tourist industry. While the Seychelles had long attracted a number of retired people from Britain, tourist arrivals totaled only 550 in 1968. With construction in 1971 of the islands' first airport and several hotels, plus scheduling of package tours by British airlines, the number rose to over 20,000 in 1973. Goals are to receive 75,000 visitors in 1977 and 150,000 in 1987, the last number now being considered as the maximum that can be absorbed without adversely affecting the beauty of the islands. Tourism has given a tremendous boost to the economy, bringing nearly full employment in 1972, substantial investment and foreign exchange earnings, and a potential market for local produce to replace high-cost imports. The rapid growth has presented new problems, however, including its stimulus to inflation, sometimes-violent labor disputes, the appearance of racial antagonism, and a drift from the land, which has brought a shortage

A tourist hotel at Anse aux Pins, Mahé Island in the Seychelles Construction of the first airport on the islands in 1971 sparked a boom in the tourist industry, which is now the main industry in the islands.

of labor to the plantations and inhibited the program to stimulate production of farm produce for the local market.

The physical infrastructure of the islands has also been undergoing rapid change in recent years. The airport, financed by Britain at a cost of about $12 million and constructed on made land about 10 miles southeast of Victoria, has obviated the necessity to go by sea in order to reach the islands. Roads are being extended and widened. A new port was opened with British aid of about $7 million in 1973 at Victoria, providing a 700-foot (213-m) pier with alongside depths of 35 feet (10.7 m). And two new reservoirs have been constructed to improve the water supply.

THE BRITISH INDIAN OCEAN TERRITORY

Four groups of islands were placed under the above heading in 1965 with their administration centered on Mahé, although they do not form part of the Seychelles. They are the Chagos Archipelago, south of India, and three island groups situated between the Seychelles and Madagascar—the Desroches Islands, the Farquhar Islands, and the Aldabras. There is no permanent population on them, though Aldabra Island is used as a fishing station and several have copra plantations producing a total of about 1,400 tons of copra yearly. The islands have strategic interests for both Britain and the United States in the Indian Ocean.

Considerable attention has been focused on Diego Garcia in the Chagos group in the last two years because it has a large lagoon suitable for naval facilities (Map 116). In 1973 the U.S. Navy established a modest communications center on the island and in 1974 it appealed to Congress for $32 million to expand the station to a full-scale naval and air support facility. American officials claim that this step is necessary to offset a "dramatic rise" in Soviet activity in the Indian Ocean, expected to intensify with the reopening of the Suez Canal; opposition both in the United States and abroad has contended that development of the base would stimulate an Indian Ocean arms race.

THE MASCARENES: MAURITIUS

Lying some 400 and 500 miles respectively (644, 804 km) east of Madagascar are Réunion and Mauritius (with several appurtenances, of which Rodriguez is the only one of any importance) collectively called the Mascarene Islands. They are volcanic in origin, and there is still activity on Réunion. Both are sugar islands and face problems of extreme population pressure.

Mauritius was discovered by Portuguese explorers in the early sixteenth century, but it had been used for shelter by Arabs and Malays for centuries before. Uninhabited and largely forest covered, it was first settled by the Dutch in the seventeenth century but was abandoned in 1710. Five years later the French claimed the island and renamed it Île de France; their first permanent settlement dates from 1722 and sugar soon became the major interest. By 1797 the population had increased to about 60,000, including 50,000

slaves from Africa and Madagascar. Seized by the British in the Napoleonic Wars, it was formally ceded to them in 1814 and was a colony until March 12, 1968, when it gained its independence, amidst a considerable amount of communal unrest accompanied by violence. A coalition government succeeded in the ensuing years in achieving a somewhat precarious cooperation among the parties, which tend to be racially oriented.

PHYSICAL CONDITIONS

With an area of 720 square miles (1,865 sq km), Mauritius is roughly oval in shape and composed of basaltic rock (Map 117). The land rises from a coastal belt fringed with lagoons and coral reefs to a central plateau with elevations of 900 to 1,900 feet (274–579 m). The central lands are bounded

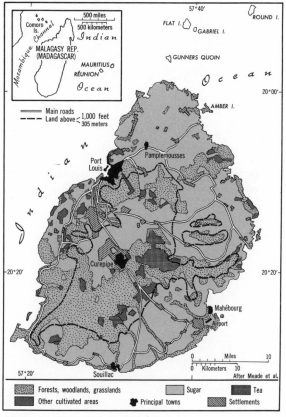

Map 117. Economic map of Mauritius

on all but the east by mountains, which were probably the rim of a great volcano and whose maximum elevations are about 2,600 feet (792 m).

Situated in the prevailing southeasterlies, Mauritius has temperatures moderated by its oceanic position and the trade winds. It may be divided into three climatic zones: (1) the subhumid coastal belt, covering about 22 percent of the island, with rainfall of 35 to 50 inches (889–1,270 mm) (the 10 percent of cane lands in this belt are dependent on irrigation); (2) the humid middle belt, about 43 percent of the island, rising to about 600 feet (183 m) on the windward side and 1,200 feet (366 m) on the leeward side, with 50 to 100 inches of rain (1,270–2,540 mm) (this belt has about 65 percent of the cane lands); and (3) the superhumid zones of the central plateau, lying above the second zone, having rainfall from 100 to 200 inches (2,540–5,080 mm), containing about 25 percent of the area planted to cane on about 35 percent of the total area of the island. Mauritius lies in the belt of maximum cyclonic disturbances and is periodically visited by severe storms. The worst in history, with wind velocities of 160 miles (257 km) per hour, struck in 1960, when well over half of the houses were damaged or destroyed and 60 percent of the sugar crop was lost.

The soils of Mauritius are somewhat like those of Hawaii but vary from clayey to gravelly in texture. Much of the area was originally strewn with large boulders; the lands from which these have been laboriously cleared are called "terre franche," but the process is not completed and large piles inhibit cultivation.

PEOPLES AND POPULATION

A great diversity and mixture of races inhabit the island. The descendants of the early French settlers and of Afro-Malagasy slaves, considerably mixed, are classified as the "general population"; they consist of about 28 percent of the total. The majority group is Indian, most of whom are descended from indentured laborers brought to Mauritius after the abolition of slavery in 1835. Totaling about 69 percent of the population, they are divided in the ratio of about 3.2 to 1 between Hindus and Muslims. Chinese make up about 3 percent of the population.

From the standpoint of economic activities, the owners and senior staff of the large sugar estates are mainly descendants of the original French settlers. "Creoles," of mixed origin, are largely urban dwellers, and are important in commerce and the civil service. Indians are mainly laborers or owners of smallholdings. The Chinese are mostly shopkeepers. English is the official language; French is spoken by most of the educated people; "Creole," a French patois, is the lingua franca; and there are several Indian languages and Chinese. Language requirements in the schools perpetuate the diversity, while the excessively traditional curriculum contributes to the prejudice against agricultural work by practically all school leavers.

The outstanding problem of Mauritius is overpopulation. The overall density was about 1,200 per square mile (465 per sq km) in mid-1974. Densities on the central plateau are about 2,100 per square mile (811 per sq km). Until postwar years the natural increase was quite small, but eradication of malaria and marked reductions in infective and parasitic diseases reduced the death rate 32 percent in one year and brought a population explosion to the island (Table 23). All races are aware of the problem and birth control campaigns have assisted in reducing the birthrate since 1950, a singular accomplishment among African countries.

Evidences of pressure include the growing unemployment rate, variously estimated at 16 to 25 percent of the labor force, the inability of development plans to absorb more than a portion of the persons coming on to the job market, inadequate dietal standards,

Table 23. Selected demographic statistics for Mauritius, 1938–1973

Year	Population (thousands)	Birth rate (per M)	Death rate (per M)	Crude increase (percent)	Infant mortality (per M)
1938	402	33.4	29.9	0.35	162.5
1943	417	33.1	25.9	0.72	141.5
1950	465	49.7	13.9	3.58	76.3
1960	639	39.6	11.3	2.83	69.5
1971	822	25.3	7.6	1.77	51.7
1972	834	24.6	7.8	1.68	...
1973 [a]	848	22.7	7.8	1.49	...

SOURCE: UN *Demographic Yearbook; Monthly Bulletin of Statistics.*

[a] Preliminary.

sometimes abominable housing conditions, and communal tension, which is never far below the surface.

AGRICULTURE

About 225,000 acres (91,000 ha) (49 percent) of the island is cultivated, all but about 16,000 acres (6,500 ha) to cane sugar. Food crops occupy less than 2 percent of the cultivated area, and the island is heavily dependent on imported foods, particularly rice.

The economy of Mauritius is dominated by sugar. About 40 percent of the active population is engaged in its production and, with by-product molasses, it accounts for 92 percent of total exports, and 30 percent of the GNP. About half of the acreage in cane is owned and cultivated by 21 companies, each having its own mill, 4 percent is owned by millers but cultivated by tenants, and the remaining is owned and cultivated by freehold planters, most of whom are Indians with average holdings under 4 acres (1.6 ha). About 40 percent of rural householders are landless. The large estates are highly organized, increasingly mechanized, and yields have increased over the years through selection, chemical weeding, fertilization, and irrigation.

Production of sugar has increased from an average 267,000 tons between 1929 and 1948 to 512,000 in 1953 and 715,000 tons in 1973. Mauritius has had quotas under the

Commonwealth Sugar Agreement (CSA), the U.S. Sugar Act, the International Sugar Agreement (ISA), and, beginning in 1973, the OCAM Sugar Agreement. Collapse of the ISA in 1973, expiration of the CSA and the U.S. quota in 1974, and French desires to favor beet sugar producers in the EEC market have caused anxiety, but high world prices and short supplies have reduced the fears, at least temporarily. China has recently offered to exchange rice for whatever amount of sugar Mauritius chose.

The only other cash crops of significance are tea, tobacco, and aloe. Tea is grown on

Sugar cane near Mahebourg, Mauritius
Sugar cane occupies about 45 percent of the island and provides over 95 percent of its exports.

about 11,000 acres (4,452 ha) in the cool, wet lands over 1,000 feet (300 m), an area which is expected to double by 1980 with the aid of a World Bank loan. About 78 percent of the output of about 4,000 tons is exported, mainly to South Africa, and tea exports account for about 3.7 percent of exports. Tobacco is sold to the two cigarette factories. Aloe, or Mauritius hemp, is used to produce about 1.5–2.5 million bags in six factories. Other crops are ginger, and a number of vegetables for local consumption.

The government has attempted to diversify production, but it has not proved easy to find crops that yield as well as sugar, provide equal employment, and have guaranteed markets. Some success has been achieved in increasing the number of dairy cattle; cows are typically stall-fed on grasses collected from wasteland or on cane tops. A few thousand acres are being added to the farmed area by terracing slopes of up to 50 percent grade and clearing brush to provide pasture.

OTHER ECONOMIC SECTORS

FISHING. Several thousand fishermen operating from small boats fish the lagoons and nearshore waters of Mauritius, but the island is a heavy importer. The Japanese have provided assistance to extend fishing to offshore areas, where the potential harvest is estimated at 300,000 tons.

INDUSTRY. The development of manufacturing has experienced two important phases in postwar years—the addition of market-oriented establishments to those processing for export, and more recently, the introduction of export-oriented plants based on the low cost of labor. Mauritius is one of the first African countries to have this type of industry, which the government hopes will provide the largest number of new jobs in coming years.

There is a surprisingly long list of market-oriented industries, very comparable to that of Madagascar or other African countries, but most are quite small and the size of the market limits expansion (Table 24). To attract export-oriented industries a variety of incentives is offered and 16 bonded processing zones have been delineated. At the end of 1973 some 33 industries employing about 5,000 workers had been set up in these zones, producing such goods as knitwear, shirts and other clothing, leather gloves, and wigs; another industry prepares jewels for Swiss watchmakers. It is hoped that electronic products, toys, and other goods will be added to this list. Several of the plants have been financed by Hong Kong entrepreneurs who can, by adding a minimum of 40 percent to the value of the processed item, escape quotas against goods of Hong Kong origin sold in the EEC market.

Electricity is provided from eight stations, including several small hydroelectric plants which are associated with reservoirs providing water for irrigation and domestic use.

TOURISM. This, too, is a rapidly growing sector. The number of visitors increased from 7,250 in 1963 to about 37,000 in 1971, and the goal is to accommodate 75,000 by 1975. Several new hotels have been constructed and twelve major projects are underway, which should provide jobs for an additional 3,000 employees.

TRANSPORTATION. Mauritius has a well developed road net, most of which is paved. The only port of significance is Port Louis, the capital, which has recently been improved with British and World Bank aid; it handled 1.5 million tons in 1972, rather evenly divided between loadings and landings. The Plaisance airport has numerous interconnections, including runs from Europe and Africa to Australia and south Asia, which contribute to the tourist industry.

DEPENDENCIES. Mauritius has three dependencies scattered rather widely in the Indian Ocean. Rodriguez, situated to the east, has an area of 42 square miles (109 sq km) and a population of about 24,000, 85 percent of whom are Creoles. Its economy, which has been described as one of survival rather than

Table 24. Industries of Mauritius classified by source of raw material and main market

PROCESSING LOCAL
PRODUCE:

Mainly for export

Sugar Mills
Tea factories
Ginger dehydration

*Mainly for local
sale*

Sugar refining
Spirits
Fiber decortication
Bags
Cigarettes
Bagasse boards
Plaster (shells)

MANUFACTURING FROM
IMPORTED MATERIALS

For island market

Tanning
Flour

Baking
Dairy products
Beer
Clothing
Shoes
Matches
Soap
Cosmetics
Perfume
Paints
Pharmaceuticals
Insecticides
Acids
Fertilizer
Petr. refining
Steel workshop
Utensils
Metal furniture
Metal casements
Hardware
Truck bodies
Ship repair

Cement products
Rubber products
Paper containers
Wood furniture
Glass products
Printing

For export

Knitwear
Children's clothing
Shirts
Leather gloves
Wigs
Hair pieces
Rattan furniture
Wire wool
Watch jewels

subsistence, is based entirely on agriculture and fishing. It produces enough corn on its hilly, volcanic terrain to meet its needs and it ships some livestock, fish, and vegetables to Mauritius, to which it is tied by the monthly service of a 2,000-ton vessel. The Agalega and Cargados Carajos shoals are uninhabited, but are visited twice yearly by fishermen from Rodriguez.

THE MASCARENES: RÉUNION

Réunion is like Mauritius in its volcanic origin, its population makeup, its heavy dependence on sugar, and its high population density. Politically it is an overseas department of France, with its own General Council acting on financial and budgetary matters. Discovered in the early sixteenth century by Portuguese navigators, it, too, was uninhabited. It received its first permanent settle-

ment in 1671, when its population was 76. In 1711 the populace had grown to 1,500 and it was about 46,000 in 1788, after many Africans and Malagasy had been brought to work on the sugar plantations. Following the abolition of slavery in 1848, most black workers quit and migrated to "Les Hauts," where they opened smallholdings, largely subsistent in character. A good many "petits blancs" had previously moved to the higher levels and their descendants continue to operate essentially self-subsistence units. Indians, Pakistanis, and Chinese were brought to the islands to replace the former slaves, mainly by the larger sugar-plantation holders, who were the only ones who could afford to pay their passage.

In mid-1974 the island was occupied by about 486,000 people divided approximately as follows: Afro-Malagasy, 58 percent; whites, 20 percent; Hindus, 20 percent; Chi-

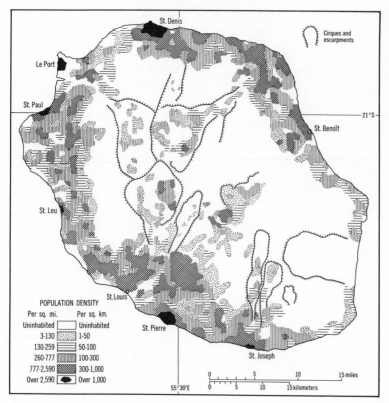

Map 118. Population map of Réunion

POPULATION DENSITY

Per sq. mi.		Per sq. km.
Uninhabited		Uninhabited
3-130		1-50
130-259		50-100
260-777		100-300
777-2,590		300-1,000
Over 2,590		Over 1,000

nese, 1 + percent; and Muslims, 0.5 percent. Its birth rate, 28.2 in 1973, remains higher than that on Mauritius partly because birth control, strongly rejected by the Catholic diocese, has only recently begun to be promoted. The death rate is only 7.2, however, so the crude rate of increase is still a high 2.1 percent, though it has declined from 3.27 percent in 1960. While one might have expected to see a substantial migration from the island to France, the annual quota of 5,000 is never filled, mainly because the cost is prohibitive for most of the population.

PHYSICAL BACKGROUND AND LAND USE

Réunion has an area of 969 square miles (2,510 sq km) and is roughly oval in shape. Two volcanic masses dominate the scene. The older and larger rises to 10,067 feet (3,068 m) in Piton des Neiges (Map 119) and covers about three-quarters of the island. It

is extremely dissected by erosion, three great pear-shaped cirques edged by sheer cliffs of as much as 3,000 to 4,500 feet (900–1,400 m) cut deeply into the dormant mass. The smaller massif, Piton de Fournaise, is active on an average of once every two years; it occupies the southeast quarter of the island and is joined to the main body by a broad, high saddle.

There is a sharp contrast climatically between the windward and leeward sides of the island. At Rivière de L'Est on the windward side, for example, precipitation is about 223 inches (5,664 mm), while at Le Port on the lee it is only 35 inches (889 mm).

The greatest part of the population is grouped along the main roads running along three bands on the mountain slopes: a coastal strip, almost uninterrupted except where recent lava flowed into the sea at the east end of the island and in the driest part

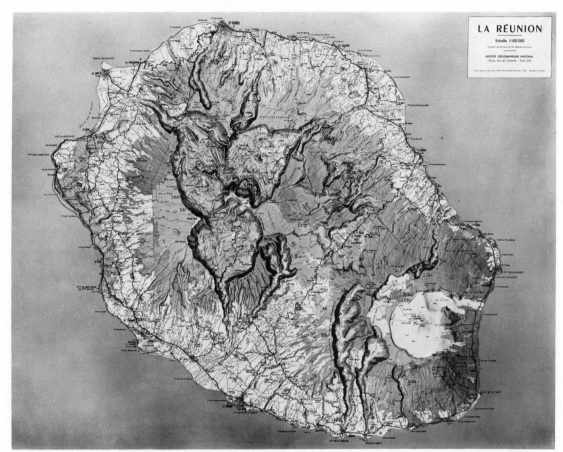

Map 119. Physical map of Réunion
Photograph of relief map by permission Institut Nationale Géographique de France

of the west; a band at about the 800–1,000 foot (240–300 m) line; and one at about 1,600–2,000 feet (490–600 m). A coastal strip about 4⅓ miles (7 km) wide contains about 85 percent of the total population (Map 118); settlements are characteristically on the undissected slopes rather than in the valleys. The pattern of population distribution means that actual densities are at least double the average of 502 per square mile (194 per sq km), and the man-to-arable land ratio is only about 1 to 0.31 acres (0.13 ha).

The major factors in restricting the population to little more than half of the island's surface are the high percentage of lands that are too high or too steep for effective use, exclusion from lands affected by recent volcanic flows, excessive precipitation on the windward side, excessive aridity on small portions of the leeward side, and the desire to avoid maximum exposure to the large number of cyclones which hit the island. The island supports its present population at a precarious level and does so only because of the richness of its soils, the ability to sell sugar at subsidized prices, and financial assistance from France and the EEC. Even then it is estimated that most of the population has less than two-thirds of the daily minimum calorie intake and, as on Mauritius, there are growing problems of unemployment and pressure on the land.

AGRICULTURE

Réunion's dependence on agriculture is even heavier than that of Mauritius, but its reliance on sugar is somewhat lower. Slightly less than a fifth of the island is cultivated, of which about two-thirds is in cane, which is grown up to the 1,300-foot (400 m) contour on the windward side and up to 2,950 feet (900 m) on the leeward side. About 22 percent is held by companies, the rest by a few individuals with large estates and by many smallholders. There is a total of some 25,000 planters of whom 18,000 are sharecroppers operating on company lands and allotted three-fourths of the crop they grow.

Production of sugar has increased more slowly than on Mauritius and is now about 230,000 to 270,000 tons per year. Sugar exports, which go mainly to France, account for about 88 to 90 percent of the total. It is milled in eight plants employing 2,800 persons.

Perfume essences make up the second-ranking export. Réunion is the leading world producer of geranium oil, produced by about 7,000 cultivators. Vetiver essence is also produced. Vanilla is the only other export of significance. Other cash crops grown on the island include tea and tobacco, which have been in decline, and market garden crops. But Réunion is highly dependent on imports for its food supply and recent inflated prices are adversely affecting the economy.

A scheme that may prove to be a prototype for others is the Bras de la Plaine project, involving the division of water from the stream of that name via a 3.5 mile (5.7-km) tunnel at the 1,300-foot (400-m) level to irrigate about 13,590 acres (5,500 ha) in a densely populated area on the leeward side. It is estimated that agricultural production on the area will be doubled.

OTHER ECONOMIC SECTORS

About 600 fishermen using 200+ small craft are engaged in fishing on the coasts. Their catch averages less than a ton per man per annum, but it is hoped to increase the catch substantially with a fleet being equipped by a Japanese company; in addition, 17 French vessels will soon be landing about 10,200 tons of tuna yearly.

Industry has seen a recent rapid increase with the number of plants rising from 12 in 1952 to 78 in 1974, but, with the exception of the well-equipped sugar mills and associated rum distilleries, most are small, and manufacturing provided employment to only 3,500 workers in 1972.

Electricity for the island is provided by three diesel stations and three hydroelectric plants, plus seasonal production from some of the sugar mills burning bagasse. The largest installation is the 17,400-kw Takamaka hydroelectric plant, opened in 1968. Its construction illustrates the difficulties of the topography; situated at a falls on the Marsouins River with a drop of 915 feet (279 m) in a region that receives 275–530 inches (7,000–14,000 mm) of rain, the site was not accessible by road, necessitating the use of an inclined plane and elevator to build the station, which was placed underground because of the danger of landslides. Réunion enjoys one of the higher per capita consumption levels for African countries.

Like its sister island, Réunion looks to tourism to provide a new and important sector. In 1972, however, there were only 216 rooms in island hotels. Even with the additional 250 planned it seems most unlikely that the goal of 150,000 visitors will soon be reached.

Réunion has a reasonably adequate road system, about half of which is paved. The road connecting the capital, Saint-Denis, with Le Port utilizes the right-of-way of the abandoned railway, replacing the extremely tortuous route previously used. The only port on the island is Le Port des Galets, which can take the largest vessels used on the service to Madagascar and the Mascarenes. It is an artificial basin cut in a low-lying gravel point on the northwest; until recently ships had to be roped alongside because of the narrow, right-angle entrance, but it is now undergoing extensive modernization. The airport of Réunion now receives daily flights by 707s and is being extended to accommodate jumbo jets; but charter flights will be required if the tourist industry is to flourish.

Part Nine
CONCLUSION

The University of Ghana at Legon

CHAPTER TWENTY-EIGHT

Development Problems and Potentials

The four introductory and 23 succeeding chapters devoted to individual countries or regions of Africa have illustrated the tremendous variety of the continent. An effort has been made to present the existing economic conditions and to point up the problems and potentials of each area. The limitations affecting development are physical, economic, political, and cultural. They impinge upon all aspects of the African scene, and the solution to some problems must await scientific research over an unpredictable time span.

The variety of Africa makes generalization a hazardous and often meaningless exercise. Yet there is a need for some comparative summarization, which can at least crudely measure the inhibiting influences and the potentialities of individual countries. Table 25 attempts such a summarization for 53 African countries.[1] The entries in this table are based on detailed assessments but are perforce subjective. They also represent estimates made at a given time; the situation for individual countries may alter abruptly as a result of changes in commodity prices and markets, plant or animal epidemics, abnormal climatic conditions, new mineral discoveries, or political events, and may alter more slowly with such factors as infrastructural developments, population growth, and rising levels of educational achievement.

The focus of Table 25 is upon economic

[1] This table represents an updating and extension of a table which first appeared in William A. Hance, "Development Limitations and Opportunities in Tropical Africa," *Africa Report,* January 1967, p. 45. See also Hance, *African Economic Development,* 2nd ed. rev. (New York: Praeger for the Council on Foreign Relations, 1967), pp. 242–297.

development. This is particularly important to note when examining the factors inhibiting development (columns 4–12). For example, political uncertainties and tribal conflicts may be inhibiting political and social development in many countries but may be relatively insignificant in impeding economic production.

FACTORS INHIBITING DEVELOPMENT

Columns 4 to 12 of the table list a number of handicaps to economic development. Since the data do not justify quantitative values here, only two types of entry are used in this segment: an X, denoting that the country has a more or less pervasive problem, and an R, indicating that the factor applies to one or more regions of the country. A space with no entry indicates that the country is relatively less influenced or not influenced at all by that specific factor. "Relatively" is a key word, because every country could be checked under some headings—for example, the need for roads, political uncertainties, or small size of the internal market. Other columns could also have been included if adequate information had been at hand: quality of soils, biotic disease problems, availability of skills, etc.

Twenty countries, with 33.7 percent of the area and 29.6 percent of the total population, are considered as having a more or less pervasive problem of population pressure. Fifteen other countries have serious problems of pressure in at least one region (see Map 22).

Aridity is a major inhibiting factor in 15

countries with almost 50 percent of Africa's area and 22.7 percent of its population. This includes Sudan and Senegal where the problem is regional. In 17 other countries, aridity severely limits regional advance.

Africa has the largest number of land-locked countries in the world, so it is not surprising that there are a substantial number of entries under the heading "remoteness" (column 6). Of the 33 countries checked in this column 13 are landlocked (Swaziland was the only landlocked state not included), 3 others were considered to have a pervasive problem, and 17 had substantial parts that suffer from remoteness (see Map 29).

With respect to transport needs, only five countries are indicated as needing new ports: the Sudan because of congestion at Port Sudan; Cameroon because of the inadequacy of the tiny ports in the west and of Douala; Zaïre because of the limited capacity of Matadi to meet national needs; Madagascar, which should probably have deepwater facilities at Majunga; and South Africa, whose needs will be met upon completion of new ports at Saldanha Bay and Richards Bay. A few years ago more countries would have been noted under this column, but new developments have eliminated the problem in such countries as Guinea (Port Kamsar), Ivory Coast (San Pedro), and Gabon (Libreville).

Eight countries are considered to be adversely affected by the absence of specific rail facilities. Upper Volta will need a line if its manganese ores are to be exploited. Niger could use a line to reduce the supply problems for its uranium mines. Guinea needs a line to the Mt. Nimba iron ores, but whether the line should run to Conakry instead of to the LAMCO line in Liberia is a question. I have reservations regarding the inclusion of Chad and Cameroon, but it is possible that an extension of the Transcameroon line into Chad would generate enough traffic to justify the investment. Zaïre is included be-cause of the desire for a national connection between the Matadi-Kinshasa and the Shaba lines. South Africa needs a new line to Saldanha Bay and double-tracking of other lines to handle larger mineral shipments.

Most African countries need new and improved roads, and the entries in column 9 indicate those countries which have particular needs.

Under the heading "political uncertainties" only those countries whose present and medium-range future is clouded by political problems are checked. The same is essentially true for those countries noted under "tribal rivalries." These items are subject to considerable volatility as has been illustrated in 1974 by events in Niger, Ethiopia, the Portuguese territories and other countries. It may be noted that many coups, attempted coups, and other forms of political tension have apparently had rather minimal effects

Nurse and technician at a hospital in Liberia
Improvements in the health field have been substantial, though very much remains to be done to achieve adequate levels.

on the economic sectors of the countries involved.

Column 12, which deals with the small size of internal markets, refers essentially to low purchasing capacity, which may relate to a small population or to low average incomes or both. Low purchasing capacity places severe limitations on many industries; it also creates difficulties in supporting an effective government. Thirty-three countries with 18.9 percent of the total population are checked under this heading. In fact, as was suggested in the examinations of the industrial sectors in the preceding chapters, every country could have been checked in this column, even South Africa, with the most developed economy on the continent, because of the low per capita incomes of its majority population.

OPPORTUNITIES FOR DEVELOPMENT

Columns 13 to 20 list six economic sectors plus irrigation and hydropower. Irrigation is of particular interest because of the possibility of increasing production under intensive conditions. An attempt has been made to rate each of the items for each of the countries as to whether the opportunities for development are considered excellent, good, fair, or poor (including nonexistent) in the short- to medium-run future. Only two sectors, agriculture and mining, appear capable at the present time of providing an adequate base for a satisfactory rate of growth. About the only exception to this generalization may be the Seychelles, which have a small enough population to permit tourism to rank as the leading economic sector. Important contributions can, of course, be made by the other sectors.

While it appears that the vast bulk of African countries must at present look to agriculture as their primary means of continuing growth, increasing production and productivity in this sector is not easy and the record

of many countries gives little reason for optimism about the near future. Despite much lip service paid to the necessity of improving agriculture, it has not in practice been assigned the priority or the funds it deserves in many countries—including Algeria, Nigeria, Zaïre, and Madagascar.

Valuable lessons may be learned from countries that have had relatively good success in this sector, including Ivory Coast, Cameroon, and Kenya. The efforts of Tanzania and Zambia have been commendable, but it is not clear that either has the answers to successful agricultural development. Other countries considered to be relatively favored for agricultural advance include Sudan, Guinea, Ethiopia, Uganda, Rhodesia, Angola, Mozambique, and Swaziland, though several of these must overcome cultural and political impediments if real progress is to be made.

Thirty countries, with 53.8 percent of the population and 41.0 percent of the area, are rated as having only a fair opportunity for agricultural development, while another ten with 3.7 percent of the population and 22.2 percent of the total area are considered to have a poor potential. In view of the heavy dependence on agriculture this is not an encouraging conclusion for African development. It must be reiterated, however, that general economic advance in most countries will be dependent on improvement in the agricultural sector. Whatever their quality, the soils remain the most valuable natural resource of Africa, and whatever the difficulty, their more effective use requires top priority in most countries.

Sudan is probably the only country for which irrigation projects can provide the main way of expanding agricultural output. Mozambique also has very substantial opportunities for extension of irrigation, but the need to do so is less pressing and the capital resources will not necessarily be readily available. Other countries where irrigation developments may make important if minor con-

Table 25. Development limitations and opportunities of individual countries in Africa

Key →

Factors Inhibiting Economic Development (columns 4–12): X—more or less pervasive; R—regional

Short- to Medium-run Opportunities for Development, by Sector (columns 13–20): 1–excellent; 2–good; 3–fair; 4–poor or non-existent

Difficulty in Achieving Economic Growth (columns 21–23): 1–not difficult; 2–not very diff.; 3–difficult; 4–very difficult

Country	Basic Measures		Factors Inhibiting Economic Development			Transport needs						Short- to Medium-run Opportunities for Development, by Sector								Difficulty in Achieving Economic Growth		
1	2 Percent of total area	3 Percent of total population, 1972	4 Population pressure	5 Aridity	6 Remoteness	7 Ports	8 Rail lines	9 Roads	10 Political uncertainties	11 Ethnic rivalries	12 Small internal market	13 Agriculture	14 Irrigation	15 Forestry	16 Fishing	17 Mining	18 Hydro-power	19 Manufacturing	20 Tourism	21 Money economy (Short run)	22 Income for average man (Short run)	23 Long run
Morocco	1.47	4.30	X	X	R							3	3	3	3	3	3	3	1	3	3	3
Algeria	7.86	4.14	X	X	R							3	3	4	3	1	4	1	1	1	3	3
Tunisia	.54	1.46	X	X								3	3	4	3	1	4	3	1	2	3	3
Spanish Sahara	.88	.01	X	X	X			X	X		X	4	4	4	3	2	4	4	4	2	4	4
Libya	5.80	.56		X	R			X	X		X	4	4	4	4	1	4	2	3	1	2	1
Egypt	3.30	9.45	X									3	3	4	4	2	2	3	2	2	3	3
Sudan	8.27	4.47		R	R	X		X				2	1	4	4	4	4	2	4	2	2	2
Mauritania	3.40	.33	X	X	R					X	X	4	3	4	2	2	2	3	4	2	4	4
Senegal	.65	1.12	R	R	R					X		3	3	4	2	4	4	3	2	3	3	3
Mali	4.09	1.43	X	X	X			X			X	3	3	4	4	3	4	4	3	4	4	3
Upper Volta	.90	1.52	X	X	X		R	X			X	3	3	4	4	3	4	4	4	4	4	4
Niger	4.18	1.14	X	X	X		X	X			X	4	3	3	3	2	3	4	4	2	4	3
Gambia	.04	.10									X	3	3	4	3	4	4	4	2	3	3	4
Guinea (Bissau)	.12	.13							X		X	3	4	3	3	1	4	3	4	4	3	3
Guinea	.81	1.12			R		R	X	X		X	2	3	3	3	1	2	2	2	2	4	1
Sierra Leone	.24	.71	R								X	3	3	4	3	3	2	3	3	3	4	3
Liberia	.37	.33						X			X	3	4	1	3	1	3	3	4	1	3	2

1	2	3	4	5	6	7	8	9	10	11	12	13	14	15	16	17	18	19	20	21	22	23
Ivory Coast	1.06	1.23		R	R							2	3	1	3	3	2	2	2	1	2	1
Ghana	.79	2.46	R	R	R							3	2	2	3	3	2	2	3	3	3	2
Togo	.18	.57	R	R								3	3	4	3	3	4	4	3	3	4	4
Dahomey	.37	.78	R	R					X		X	3	3	4	3	4	4	4	3	3	4	4
Nigeria	3.05	15.74	R	X	R			X		X	X	3	2	2	2	1	2	1	2	1	3	2
Chad	4.23	1.03	X	X	X		R	X		X	X	4	3	4	4	4	4	4	4	4	4	4
C.A.R.	2.05	.46		R	R	R		X	X			3	4	2	4	3	3	4	4	3	4	3
Cameroon	1.57	1.65	R	R	R	R	R	X				3	3	3	3	3	2	2	2	2	3	2
Congo	1.13	.27	R	R	R			X	X	X	X	3	3	1	3	3	1	3	4	3	3	3
Gabon	.88	.14			R		X	X			X	4	4	2	3	1	1	2	4	1	2	1
Equatorial Guinea	.09	.08						X	X	X	X	3	4	3	3	4	4	4	3	3	3	3
São Tomé e Príncipe	—	.02	X						X		X	3	4	2	3	1	1	2	2	1	4	1
Zaïre	7.73	6.20			R	X				X		2	2	3	3	3	2	3	2	2	3	2
Ethiopia	4.03	7.03	R	R	X			X				2	2	4	4	4	4	4	4	4	4	4
Somalia	2.10	.80	X	X				X			X	3	2	3	3	3	3	2	1	2	3	2
Kenya	1.92	3.27	R	R	X			X		X		2	3	3	3	4	4	3	1	3	3	3
Uganda	.78	2.84	R	R	R			X	X	X	X	1	2	3	3	4	2	3	1	2	3	3
Tanganyika	3.11	3.68	R	R	R							3	3	3	3	4	4	3	1	3	3	3
Zanzibar	.01	.11										2	4	4	4	4	4	4	1	2	3	3
Rwanda	.09	1.06	X		X				X	X	X	3	4	4	4	4	3	4	2	4	4	4
Burundi	.09	.92	X		X				X	X	X	3	4	4	4	4	4	4	2	4	4	4
Rhodesia	1.29	1.54	R		X				X	X		3	3	3	4	4	4	4	2	4	4	2
Zambia	2.48	1.20		R	X			X				2	1	3	3	2	3	2	3	3	3	2
Malawi	.39	1.27	X		X			X			X	3	3	3	3	3	3	4	3	1	3	3
Angola	4.11	1.58		R	R			X	X	X		3	2	3	3	2	1	2	3	2	3	2
Mozambique	2.58	2.31		R		R		X	X	X		2	1	3	2	2	1	3	3	2	3	2
South Africa	4.03	6.24	R	R	R			X	X	X	X	2	3	3	2	1	4	1	2	1	3	2
South-West Africa	2.72	.18	R	X	X		R		X	X		4	3	4	2	1	4	3	2	1	3	2
Botswana	1.98	.19		X	X		R	R			X	3	3	4	4	1	4	3	2	1	3	2
Lesotho	.10	.26	X					R			X	4	4	3	4	3	2	4	2	3	4	4
Swaziland	.06	.12						R			X	2	2	2	4	2	2	2	2	1	2	2
Madagascar	1.94	1.92		R	R	R		X	X			3	3	3	2	3	3	3	3	3	4	3
Comoros	.01	.08	X	X	X	R				X	X	4	4	4	2	4	4	4	3	4	4	3
Seychelles	—	.01	X								X	4	4	4	2	4	4	4	2	1	4	2
Mauritius	.01	.23	X							X	X	3	4	4	2	4	3	3	1	2	2	3
Réunion	.01	.13	X								X	3	3	4	2	4	3	3	2	3	3	3

tributions include Ghana, Nige::a, Ethiopia, Rhodesia, Zambia, and Swaziland.

Only fourteen countries would appear to have sufficient opportunities in new and expanded mining enterprises to make this sector the main hope for achieving an adequate growth rate: Algeria, Tunisia, Libya, Spanish Sahara, Niger, Guinea, Liberia, Nigeria, Gabon, Zaïre, Zambia, South Africa, South-West Africa, and Botswana. Other countries that should gain substantially from increased mineral production include Egypt, Rhodesia, Angola, Mozambique, Lesotho, and Swaziland.

Countries that have good to excellent opportunities for forestry development are Liberia, Ivory Coast, Ghana, Nigeria, Cameroon, Gabon, Congo, and Zaïre. Elsewhere afforestation will doubtless make significant contributions to domestic needs. Fishing can rarely be of more than minor importance, though many countries, including some of the landlocked nations, can greatly increase and rationalize existing operations. Care must be taken in a number of countries not to land excessive catches, particularly of the more recently exploited and high-value shellfish.

A school in Addis Ababa, Ethiopia
Improving educational achievement levels provide one of the most hopeful developments in Africa

Manufacturing expansion may be related to raw material (including energy) resources or to the domestic market; it is not likely to be skilled-labor oriented for some years, and only a few countries are beginning to see plants based on low-cost labor selling their output on the world market (Egypt and Mauritius) or within a customs union area (Swaziland and Lesotho vis-à-vis South Africa). Every country has some opportunities for development of local industry, but many are too resource-poor or have too small markets to warrant expectations of any substantial growth. In a number of cases import-substitution has been pushed somewhat too rapidly and the commonly heard prescription that agroprocess complexes are the answer does not appear to be valid for many countries because there already exists surplus capacity in many such industries. Countries that may be expected to make relatively rapid gains in manufacturing include Algeria, Egypt, Ivory Coast, Nigeria, Zaïre, Kenya, Zambia, and South Africa.

Despite the enormous resources of hydropower in Africa only a few developments have been successfully implemented as attractions to establishments oriented toward low-cost energy. There are now only three plants which may be said to have such an orientation—the Tema and Edéa aluminum smelters and the Jinja copper smelter. Hydroelectric installations that may stimulate further development include those in Guinea to permit the smelting of at least part of the domestic output of bauxite-alumina, plants in Gabon to support a pulp industry, the long-delayed Kouilou project in Congo, later stages of the Inga project in Zaïre, and stations in Angola. Just what impact Cabora Bassa will have on the Mozambique economy will depend on willingness of an independent government to sell large blocks to South Africa and the difficulty of attracting alternative large-scale consumers. In no case can hydropower developments be expected to play more than a supportive role in advancing the economies of African countries.

Tourism should make an increasing contribution to the GNPs of many countries. It is of particular value to the Seychelles, Kenya, Tanganyika, Morocco, Tunisia, and Mozambique. Its role could be profitably fostered in many others.

SUMMARY AND CONCLUSIONS

The last three columns attempt to assess the relative difficulties African countries face in attempting to achieve a satisfactory rate of economic growth. Columns 21 and 22 refer to the short run and distinguish between difficulties in achieving growth in the money economy and in income of the hypothetical average man. This distinction is made be-

The Kenyatta Conference Center, Nairobi
It may be hoped that the many meetings held at this center will contribute to an improved understanding of Africa and increasing cooperation among African nations.

cause there may be a sharp contrast in some countries in the two measures. Mining in Mauritania, for example, supported an unprecedented rate of growth in its exports, government revenues, and GNP, but the average citizen is very little affected and continues in his traditional pursuits. In Zaïre, as a second example, investment is being concentrated in a number of huge mining and infrastructural developments plus some manufacturing activity, while agriculture appears to be deliberately neglected, which means that a lower percentage of the population is involved in the money economy than before independence.

The estimates in the last three columns are derived in large part from the assessments of inhibiting factors and development opportunities in the previous columns. This sometimes makes it easier to arrive at the estimates; for other countries the uncertainty is compounded. Particular difficulty arises in trying to "guesstimate" what may happen in the short run in such countries as Ethiopia, Rhodesia, Angola, and Mozambique. Unusually high commodity prices may also improve the statistical achievement of countries that would otherwise have considerable difficulty in recording a satisfactory rate of growth. The impact of high petroleum prices on consuming countries, on the other hand, may slow advance in a number of countries. A world recession would, of course, adversely affect the economies of almost all African countries.

Twenty-nine countries with about 74 percent of both the population and area of the continent are assessed as being able to achieve with relative ease an adequate rate of growth in the money economy in the short run. In only seven countries with 8.1 percent of the total population is it considered that our hypothetical man will participate readily in this growth.

The last column rates each country with respect to the longer run with the assumption that inhibiting factors capable of amelioration will be diminished in significance. Achieving adequate growth in the long run is considered to be more difficult for some countries than achieving it in the short run. In some cases this is because population growth will make it very difficult to provide adequate employment opportunities (e.g., Algeria, Tunisia, Egypt, Nigeria, Zanzibar, Mauritius); in other cases it is because the sectoral development, which is stimulating current growth, cannot be expected to expand indefinitely (e.g., oil production in several countries, and mining in Spanish Sahara, Mauritania, Niger, Liberia, and other countries).

On the other hand, some countries have a more favorable outlook over the long run than in the immediate years ahead. This appears to be true for Guinea, Ghana, and Rhodesia. In Guinea the agricultural, manufacturing, and tourist sectors have been neglected and yet each has considerable potential. Ghana's economy has been adversely affected by early fiscal irresponsibility but its resources are reasonably comparable to those of the Ivory Coast and could sustain a good rate of growth. For Rhodesia the present political impediments will presumably be dissipated, although a massive exodus of whites could well shatter the economy for a period of years.

Twenty-two countries with 60.4 percent of the population and 56.3 percent of the area are rated as having a "not difficult" or "not very difficult" prospect of achieving satisfactory economic growth in the long run, while eleven with 7.4 percent of the population are

considered to face "very difficult" conditions in attaining an adequate long-term growth rate.

In conclusion, it seems appropriate to stress five points:

(1) The sector which must be given primary emphasis in most African countries is agriculture if for no other reason than that the bulk of the population of most countries is engaged in this sector and will be for at least several decades. Furthermore, no broad advance can be expected in manufacturing and services until greater progress has been made in agriculture.

(2) Contrasts among the countries as far as economic development is concerned are likely to widen in the years ahead. Some countries will move rapidly toward modernization and higher standards, while others will retain a predominantly subsistence economy for many years.

(3) Population pressure and a rapid rate of population growth will increasingly be seen as impediments to economic development.

(4) Economic development at a satisfactory rate is likely to be difficult for large parts of Africa.

(5) The most hopeful features of the entire African scene are the growing will and the broadening and deepening abilities of the Africans to participate in and achieve their own development.

Bibliography

GENERAL REFERENCES AND JOURNALS

Africa.

African Affairs.

African Development.

Africa Report.

Africa Research Bulletin.

African Studies Association. *Bulletin.*

African Urban Notes.

Barclays Bank. *Overseas Review.*

Economist (London).

Economist Intelligence Unit. *Quarterly Economic Reviews.*

Europa Publications. *Africa South of the Sahara 1974.* London, 1974.

——*The Middle East and North Africa 1973–74.* London, 1974.

Europe France Outremer.

Europe France Outremer. *L'Afrique d'Expression Française et Madagascar.* Paris, annual.

Industries et Travaux d'Outremer.

International Bank for Reconstruction and Development. *World Bank Atlas.* Washington, 1972.

International Monetary Fund. *International Financial Statistics.*

Jeune Afrique.

Jeune Afrique. *Africa 1973.*

——*The Atlas of Africa.* 1973.

Journal of Developing Areas.

The Journal of Modern African Studies.

Legum, Colin, ed. *Africa Contemporary Record: Annual Survey and Documents.*

Marchés Tropicaux et Méditerranéens.

Morrison, Donald G. et al. *Black Africa: A Comparative Handbook.* New York: The Free Press, 1972.

Rural Africana.

Savanna

Standard Bank Review.

United Nations. *Demographic Yearbook.*

——*Economic Bulletin for Africa.*

——*Housing in Africa.* New York, 1965.

——*Industrial Development in Africa.* New York, 1967.

——*Statistical Yearbook.*

——*Yearbook of National Accounts Statistics.*

——Economic Commission for Africa. *Survey of Economic Conditions in Africa, 1972.* New York, 1973.

——Food and Agriculture Organization. *African Agricultural Development,* New York, 1966.

——*Production Yearbook.*

United States. Bureau of Mines. *Minerals Yearbook.*

Wernstedt, Frederick L. *World Climatic Data, Part I, Africa.* Lemont, Pennsylvania: Climatic Data Press, 1972.

CHAPTERS I–IV: INTRODUCTION

Allan, W. *The African Husbandman.* Edinburgh: Oliver and Boyd, 1965.

Biebuyck, Daniel, ed. *African Agrarian Systems.* London: Oxford University Press, 1963.

Bohannon, Paul J., and George Dalton, eds. *Markets in Africa.* Evanston, Ill.: Northwestern University Press, 1962.

Brass, William et al. *The Demography of Tropical Africa.* Princeton: Princeton University Press, 1968.

Breese, Gerald. *Urbanization in Newly Developing Countries.* Englewood Cliffs, N.J.: Prentice-Hall, 1966.

Caldwell, John C. and Chukuka Okonjo, eds. *The Population of Tropical Africa.* New York: Columbia University Press, 1968.

Church, R.J. Harrison *et al. Africa and the Islands.* London: Longmans, 1967.

Cleave, John H. *African Farmers: Labor Use in the Development of Smallholder Agriculture.* New York: Praeger, 1974.

Commission for Technical Co-operation in Africa. Soils Map of Africa. Lagos, 1964.

de Wilde, John C. et al. *Experiences with Agricultural Development in Tropical Africa.* 2 vols. Baltimore: The Johns Hopkins Press for the I.B.R.D., 1967.

Dumont, René. *False Start in Africa.* New York: Praeger, 1966.

Eicher, Carl. "Employment Generation in African Agriculture." East Lansing, Michigan: Michigan State University, Institute of International Agriculture, Research Paper No. 9, July, 1970.

——"Tackling Africa's Employment Problems." *Africa Report.* January 1971, pp. 30–33.

Engberg, Holger L. "Sub-Saharan Africa," in J. Boddewyn, Holger L. Engberg et al., *World Business Systems and Environments.* Scranton, Penn.: Intext Educational Publishers, 1972, pp. 291–379.

Ewing, A. F. *Industry in Africa.* London: Oxford University Press, 1968.

Ford, John. *The Role of Trypanosomiasis in Agrarian Ecology: A Study of the Tsetsefly Problem.* London: Oxford University Press, 1970.

France, Institut National de la Statistique et des Etudes Economiques—Service de Cooperation, Institut National d'Etudes Démographiques. *Afrique Noire, Madagascar, Comores: Démographie Comparée.* 10 v. Paris: Délégation Générale à la Recherche Scientifique et Technique, 1967.

Gibbs, James L., ed. *Peoples of Africa.* New York: Holt, Rinehart and Winston, 1965.

Gourou, Pierre. *L'Afrique.* Paris: Hachette, 1970.

Grove, A. Thomas. *Africa South of the Sahara.* London: Oxford University Press, 1967.

Green, Leslie P. and T. J. Dennis Fair. *Development in Africa: A Study in Regional Analysis with Special Reference to Southern Africa.* Johannesburg: Witwatersrand University Press, 1962.

Green, Reginald H. and Ann Seidman. *Unity or Poverty? The Economics of Pan Africanism.* Hammondsworth, England: Penguin African Library, 1968.

Griffiths, J. F., ed. *Climates of Africa.* New York: Elsevier, 1972.

Gutkind, Peter C. W., comp. *Unemployment in Africa.* Montreal: Center for Developing Area Studies, 1973.

Hailey, Lord. *An African Survey.* London: Oxford University Press, 1957.

Hance, William A. *African Economic Development.* 2nd ed. New York: Praeger, for the Council on Foreign Relations, 1967.

——*Population, Migration, and Urbanization in Africa.* New York: Columbia University Press, 1970.

Hazlewood, A., ed. *African Integration and Disintegration.* London: Royal Institute of International Affairs, 1967.

Hoyle, B. S. and D. Hilling, eds. *Seaports and Development in Tropical Africa.* New York: Praeger, 1970.

Jackson, S. P. *Climatological Atlas of Africa.* Pretoria, 1961.

Johnson, E. A. J. *The Organization of Space in Developing Countries.* Cambridge, Mass.: Harvard University Press, 1970.

Jones, William O. *Manioc in Africa.* Stanford, Calif.: Food Research Institute, 1959.

——*Marketing Staple Food Crops in Tropical Africa.* Ithaca: Cornell University Press, 1972.

Kamarck, Andrew M. *The Economics of African Development.* 2nd ed. New York: Praeger, 1971.

Kuper, Hilda, ed. *Urbanization and Migration in West Africa.* Berkeley: University of California Press, 1965.

McLaughlin, Peter F. M., ed. *African Food Production Systems.* Baltimore: The Johns Hopkins Press, 1970.

Miner, Horace, ed. *The City in Modern Africa.* New York: Praeger, 1967.

Moss, R. P., ed. *The Soil Resources of Africa.* Cambridge, England: Cambridge University Press, 1968.

Mutharika, B. W. T. *Toward Multinational Economic Cooperation in Africa.* New York: Praeger, 1972.

O'Connor, A. M. *The Geography of Tropical African Development.* Oxford: Pergamon, 1971.

Oliver, Paul. *Shelter in Africa.* New York: Praeger, 1971.

Ominde, Simeon H. and C. N. Ejiogu, eds. *Population Growth and Economic Development in Africa.* London: Heinemann, 1972.

Oxford Regional Economic Atlas. Africa. Oxford, Clarendon Press, 1965.

Phillips, J. F. V. *The Development of Agriculture and Forestry in the Tropics: Patterns, Problems and Promise.* London: Faber, 1966.

Prothero, R. Mansell, ed. *A Geography of Africa.* New York: Praeger, 1969.

——*Migrants and Malaria.* London: Longmans, Green and Co., 1965.

Robson, P. *Economic Integration in Africa.* London: Allen and Unwin, 1968.

"Systèmes Agraires Africains," *Cahiers d'Etudes Africaines,* 12, no. 47 (1972), special number.

"Terroirs africaines et malgaches," *Etudes rurales,* No. 37–39 (January–September, 1970).

Thompson, B. W. *The Climate of Africa.* Nairobi: Oxford University Press, 1965.

Thomas, M. F. and G. W. Whittington, eds. *Environment and Land Use in Africa.* London: Methuen, 1969.

Zartman, I. William. *International Relations in the New Africa.* Englewood Cliffs, N.J.: Prentice-Hall, 1966.

CHAPTER V: THE MAGHREB AND NORTHWEST AFRICA

Adam, André. *Casablanca.* 2 vols. Paris: Centre National de Recherche Scientifique, 1968.

"L'Agriculture Socialiste Algérienne," *Annales Algériennes de Géographie,* 4, no. 8 (July–December, 1969): 1–226.

Area Handbook for Morocco. Washington, U.S. Government Printing Office, 1965.

Area Handbook for the Republic of Tunisia. Washington, U.S. Government Printing Office, 1970.

Barbour, Nevill. *Morocco.* London: Thames and Hudson, 1965.

——, ed. *A Survey of North West Africa* (The Maghrib). London: Oxford University Press, 1959.

Beguin, Hubert. *L'Organisation de l'Espace au Maroc.* Brussels: L'Académie Royale des Sciences d'Outre-Mer, 1973.

Benchetrit, Maurice. *L'Erosion Actuelle et ses Conséquences sur l'Aménagement en Algérie.* Paris: Presses Universitaires de France, 1972.

Capot-Rey, Robert. *Le Sahara Français.* Paris: Presses Universitaires de France, 1953.

Clarke, J. I. and W. B. Fisher. *Populations of the Middle East and North Africa*. Cambridge: Heffer, 1972.

Debbasch, Charles and Michel Camau. *La Tunisie*. Paris: Editions Berger-Levrault, 1973.

Despois, Jean. *L'Afrique du Nord*. 3rd ed. Paris: Presses Universitaires de France, 1964.

——*La Tunisie: Ses Régions*. Paris: Armand Colin, 1961.

——and René Raynal. *Géographie de l'Afrique du Nord-Ouest*. Paris: Payot, 1967.

Europa Publications Ltd. *The Middle East and North Africa 1973-74*. London, 1974.

Gallagher, Charles F. *The United States and North Africa: Morocco, Algeria and Tunisia*. Cambridge, Mass.: Harvard University Press, 1963.

Gautier, Emile-Felix. *Sahara: The Great Desert*. Translated by Dorothy F. Mayhew. New York: Columbia University Press, 1935.

Houston, J. M. *The Western Mediterranean World*. London: Longmans, Green and Co., 1964.

Institut Scientifique Chérifien. *Atlas du Maroc*. Rabat, various years, 1954–

International Bank for Reconstruction and Development. *The Economic Development of Morocco*. Baltimore: The Johns Hopkins Press, 1966.

Isnard, Hildebert. *Le Maghreb*. Paris: Presses Universitaires de France, 1966.

Le Coz, Jean. *Le Rharb: Fellahs et Colons: Etude de Géographie Régionale*. Rabat: Centre Universitaires de la Recherche Scientifique, 1964.

Maghreb (Paris).

Maghreb Digest (Los Angeles).

Mikesell, Marvin W. *Northern Morocco: A Cultural Geography*. Berkeley: University of California Press, 1961.

Noin, D. *La Population Rurale du Maroc*. 2 vols. Rouen: Publications de l'Université, 1971.

Osuna, Manuel G. *Geografía Urbana de Ceuta*. Madrid: Instituto de Estudios Africanos, 1972.

Oxford Regional Economic Atlas. The Middle East and North Africa. Oxford: Oxford University Press, 1960.

"Le Tourisme en Afrique du Nord," *Notes et Etudes Documentaires*, No. 3905-3906. Paris, 1972.

"Tunisia," *Focus*, 19, no. 5 (1969).

U. S. Army Area Handbook for Algeria. Washington, D.C.: U. S. Government Printing Office, January 1965.

Zartman, I. William, ed. *Man, State, and Society in the Contemporary Maghreb*. London: Pall Mall, 1973.

——"The Sahara—Bridge or Barrier," *International Conciliation*, no. 541, January 1963.

CHAPTER VI: LIBYA

Allan, J. A., K. S. McLachlan, and E. T. Penrose. *Libya: Agriculture and Economic Development*. London: Cass, 1972.

Area Handbook for Libya. Washington, D.C.: U. S. Government Printing Office, December 1969.

Farley, Rawle. *Planning for Development in Libya: The Exceptional Economy in the Developing World*. New York: Praeger, 1971.

International Bank for Reconstruction and Development. *The Economic Development of Libya*. Baltimore: The Johns Hopkins University Press, 1960.

"La République Arabe Libyenne," *Notes et Etudes Documentaires*, no. 3740-3741 (November 27, 1970).

Wright, John. *Libya*. New York: Praeger, 1969.

CHAPTER VII: EGYPT

Ayrout, Henry. *The Fellaheen*. Translated by John Alden Williams. Boston: Beacon Press, 1963.

Barbour, K. Michael. *The Growth, Location, and Structure of Industry in Egypt*. New York: Praeger, 1972.

Hurst, Harold E. *The Nile: A General Account of the River and the Utilization of Its Waters*. London: Constable, 1952.

Issawi, Charles. *Egypt in Revolution: An Economic Analysis*. New York: Oxford University Press, 1963.

Lacouture, Jean, and Simonne Lacouture. *Egypt in Transition*. New York: Criterion Books, 1959.

Little, Tom. *Modern Egypt*. New York: Praeger, 1967.

U.S. Army Area Handbook for the UAR (Egypt). 2nd ed. Washington, D.C.: U. S. Government Printing Office, 1964.

Wilber, Donald N. *United Arab Republic—Egypt: Its People, Its Society, Its Culture*. New Haven: Human Relations Area Files, 1969.

CHAPTER VIII: THE REPUBLIC OF THE SUDAN

Area Handbook for the Democratic Republic of Sudan. Washington, D.C.: U.S. Government Printing Office, 1973.

Barbour, K. Michael. *The Republic of the Sudan: A Regional Geography*. London: University of London Press, 1961.

Cunnison, Ian G. *Baggara Arabs*. Oxford: Clarendon Press, 1966.

Gaitskell, Arthur. *Gezira: A Story of Development in the Sudan*. London: Faber and Faber, 1959.

Hill, R. L. *Sudan Transport*. London: Oxford University Press, 1965.

Lebon, J. H. G. *Land Use in the Sudan*. Bude: Geographical Publications Ltd., 1965.

Oliver, J. *Problems of the Arid Lands: The Example of the Sudan*. Institute of British Geographers, Special Publication No. 1, 1968, pp. 219–34.

Reining, Conrad C. *The Zande Scheme: An Anthropological Case Study of Economic Development in Africa*. Evanston: Northwestern University Press, 1966.

Schlippe, Pierre de. *Shifting Agriculture in Africa: The Zande System of Agriculture*. London: Routledge and Kegan Paul, 1956.

Sudan Almanac.

Sudan Today. Nairobi: University Press of Africa, 1971.

Tothill, John D., ed. *Agriculture in the Sudan.* London: Oxford University Press, 1948.

Wynn, R. F. "Prospects for the Further Development of the Sudan's Share of the Water Resources of the Nile Basin," *East African Journal of Rural Development,* 4, no. 2 (1971): 37–66.

CHAPTERS IX–XII: WEST AFRICA

Ahn, P. M. *West African Soils.* London: Oxford University Press, 1970.

Area Handbook for Ghana. Washington, D.C.: U. S. Government Printing Office, 1971.

Area Handbook for the Ivory Coast. Washington, D.C.: U.S. Government Printing Office, 1973.

Area Handbook for Nigeria. Washington, D.C.: U. S. Government Printing Office, 1972.

Avenard, J. M. et al. *Le Milieu Naturel de la Côte-d'Ivoire,* 2v. Paris, O.R.S.T.O.M., 1971.

Bank of Sierra Leone. *Economic Review,* 7, nos. 2–3 (September–December 1972).

Banque Centrale des Etats de l'Afrique de l'Ouest. *Notes d'Information et Statistiques.*

Barbour, K. Michael, ed. *Planning for Nigeria: A Geographical Approach.* Ibadan: Ibadan University Press, 1972.

Barlet, P. "La Haute-Volta (Essai de presentation géographique)," *Etudes Voltaïques,* new series, no. 3, 1962 (1965), pp. 5–77.

Barral, H. *Tiogo: Géographique d'un Terrior Lela (Haute-Volta).* Paris: Mouton, 1968.

Beleky, Louis P. "The Development of Liberia," *Journal of Modern African Studies,* 11, no. 1 (March, 1973): 43–60.

Birmingham, Walter, et al. *A Study of Contemporary Ghana.* 2 vols. London: Allen and Unwin, 1966, 1967.

Boateng, E. A. *A Geography of Ghana.* 2nd ed. Cambridge, England: Cambridge University Press, 1966.

Buchanan, Keith M. and John C. Pugh. *Land and People in Nigeria.* London: University of London Press, 1955.

Caldwell, John C. *African Rural-Urban Migration: The Movement to Ghana's Towns.* New York: Columbia University Press, 1969.

——, ed. *Population Growth and Socio-economic Change in West Africa.* New York: Columbia University Press, 1975.

Chambers, Robert, ed. *The Volta Resettlement Experience.* New York: Praeger, 1970.

Church, R. J. Harrison. *West Africa.* 5th ed. London: Longmans, Green and Co., 1966.

Clarke, John I. *Sierra Leone in Maps.* London: University of London Press, 1966.

Clower, Robert W. *et al. Growth without Development: An Economic Survey of Liberia.* Evanston: Northwestern University Press, 1966.

Cornevin, Robert. *Le Dahomey.* Paris: Presses Universitaires de France, 1970.

——*Le Togo.* Paris, Presses Universitaires de France, 1967.

"Dakar," *Notes et Etudes Documentaires,* Nos. 3505–3506 (July 6, 1968).

Dickson, K. B. *A Historical Geography of Ghana.* Cambridge, England: Cambridge University Press, 1969.

Donaint, Pierre and François Lancrenon. *Le Niger.* Paris: Presses Universitaires de France, 1972.

Doute, Gilbert. "La République de Haute-Volta." *Notes et Etudes Documentaires.* No. 3818–3819, September, 27, 1971.

Eicher, Carl K. and C. Liedholm, eds. *Growth and Development of the Nigerian Economy.* East Lansing: Michigan State University Press, 1970.

Floyd, Barry. *Eastern Nigeria: A Geographical Review.* New York: Praeger, 1969.

Gallais, Jean. *Le Delta Intérieur de Niger: Etude de Géographie Régionale.* 2 vols. Dakar: I.F.A.N., 1967.

——*Le Delta Intérieur du Niger et ses Bordures.* Paris: Editions du Centre National de la Recherche Scientifique, 1967.

Garlick, Peter C. *African Traders and Economic Development in Ghana.* Oxford: Clarendon Press, 1971.

Gerteiny, Alfred. *Mauritania: A Survey of a New Nation.* New York: Praeger, 1967.

Ghana. Survey of Ghana. *Portfolio of Ghana Maps.* Accra, 1969.

Grayson, Leslie E. "A Conglomerate in Africa: Public-Sector Manufacturing Enterprises in Ghana, 1962–1971," *African Studies Review,* 16, no. 3 (December 1973): 315–46.

Hargreaves, John D. *West Africa: The Former French States.* Englewood Cliffs, N.J.: Prentice-Hall, 1967.

Helleiner, G. K. *Peasant Agriculture, Government, and Economic Growth in Nigeria.* Homewood, Illinois: Richard D. Irwin, 1966.

Hill, Polly. *Rural Hausa: A Village and a Setting.* London: Cambridge University Press, 1972.

——*Studies in Rural Capitalism in West Africa.* London: Cambridge University Press, 1970.

Hodder, B. W. and U. I. Ukwu. *Markets in West Africa: Studies of Markets and Trade among the Yoruba and Ibo.* Ibadan: Ibadan University Press, 1969.

International Bank for Reconstruction and Development. *Mauritania.* Washington, D.C., November 1968.

International Monetary Fund. *Surveys of African Economies.* Vol. 5. Washington, D.C., 1970.

Johnston, Bruce F. *The Staple Food Economies of Western Tropical Africa.* Stanford: Stanford University Press, 1958.

Kellerman, Jean. *Etude des Possibilités de Diversification de* la Production Rurale en Haute-Volta. Ouagadougou, 1967.

Lancrenon, François. "La République du Niger." Notes et Etudes Documentaires, no. 3994–3995. 1973.

Lawson, Rowena M. The Changing Economy of the Lower Volta. London: International African Institute, 1972.

Lericollais, André. Un Terroir Sérér du Sine (Sénégal): Sob. Dakar: O.R.S.T.O.M., January 1969.

Leroi-Gourhan, A. and J. Guiart. Communautés Villageoises Bwa, Mali-Haute Volta. Paris: Musée de l'Homme, 1973.

Mabogunje, Akin L. Regional Mobility and Resource Development in West Africa. Montreal: McGill-Queens, 1972.

Meillassoux, C. ed. The Development of Trade and Markets in West Africa. London: Oxford University Press, 1971.

Morgan, W. B. and J. C. Pugh. West Africa. London: Methuen, 1969.

Mortimore, M. J., ed. Zaria and Its Region. Zaria, Ahmadu Bello University, Department of Geography, 1970.

Ojo, G. J. Afolabi. Yoruba Culture: A Geographical Analysis. London: University of London Press, 1967.

Organization of African Unity, Scientific, Technical and Research Commission. International Atlas of West Africa.

Pallier, G. L'Artisanat et les Activités à Ouagadougou (Haute-Volta). Paris: Secrétariat d'Etat aux Affaires Etrangères, December 1970.

Poncet, Yveline. Cartes Ethno-démographiques du Niger. Niamey: Centre Nigérien de Recherches en Sciences Humaines, 1973.

Queant-Thierry, de Rouville C. Agriculteurs et Eleveurs de la Région du Gondo-Sourou. Ouagadougou: Centre Voltaïque de Recherche Scientifique, June, 1969.

Remy, Gérard. Les Migrations de Travail et les Mouvements de Colonisation Mossi. Bondy, France: O.R.S.T.O.M., 1973.

——Yobri, Etude Géographique du Terroir d'un Village Gourmanthché de Haute-Volta. Paris: Mouton, 1970.

République du Sénégal, Ministère du Plan. Cartes pour Servir à L'Aménagement du Territoire. Dakar: August 1965.

Rice, Berkeley. Enter Gambia: the Birth of an Improbable Nation. Boston: Houghton Mifflin, 1967.

Riddell, J. Barry. The Spatial Dynamics of Modernization in Sierra Leone. Evanston, Ill.: Northwestern University Press, 1970.

Rougerie, Gabriel. La Côte d'Ivoire. 3rd ed. Paris: Presses Universitaires de France, 1972.

Savonnet, Georges. Pina (Haute-Volta). Paris: Mouton, 1970.

Saylor, R. G. The Economic System of Sierra Leone. Durham: Duke University, 1967.

Schulze, Willi. A New Geography of Liberia. London: Longman, 1973.

Seck, Assane and A. Mondjannagni. L'Afrique Occidentale. Paris: Presses Universitaires de France, 1967.

Thompson, Virginia. West Africa's Council of the Entente. Ithaca: Cornell University Press, 1972.

——and Richard Adloff. French West Africa. London: Allen and Unwin, 1958.

Udo, Reuben K. Geographical Regions of Nigeria. London: Heinemann, 1970.

United Nations. Food and Agriculture Organization. Agricultural Development in Nigeria, 1965–1980. Rome, 1966.

——Crop Ecologic Survey in West Africa. Rome, 1965.

U. S. Army Area Handbook for Liberia. Washington, D.C.: U. S. Government Printing Office, 1965.

U. S. Army Area Handbook for Senegal. Washington, D.C.: U. S. Government Printing Office, August 1963.

United States. Department of Agriculture. The Agricultural Economy of Nigeria. Washington, D.C.: 1972.

University of Abidjan. Atlas de Côte d'Ivoire. Abidjan, 1971.

Vidaud, Pierre and Gilbert Doute. "La République de Côte-d'Ivoire," Notes et Etudes Documentaires, No. 3989–3990, May 21, 1973.

von Gnielinski, Stefan, ed. Liberia in Maps. London: University of London Press, 1972.

West Africa. (A weekly.)

Westebbe, Richard M. The Economy of Mauritania. New York: Praeger, 1971.

White, H. P. and M. B. Gleave. An Economic Geography of West Africa. London: G. Bell and Sons, 1971.

Wills, J. Brian, ed. Agriculture and Land Use in Ghana. London: Oxford University Press, 1962.

Woronoff, Jon. West African Wager: Houphouet Versus Nkrumah. Metuchen, N.J.: Scarecrow Press, 1972.

Wurta, Jacqueline. Adiamprikofikro-Douakankro: Etude Géographique d'un Terroir Baoulé de Côte-d'Ivoire. Paris: Mouton and Co., 1971.

CHAPTER XIII: MIDDLE AFRICA FROM CHAD TO CONGO

Ardener, Edwin, Shirley Ardener, and W. A. Warmington. Plantation and Village in the Cameroons: Some Economic and Social Studies. London: Oxford University Press, 1960.

Area Handbook for Cameroon. Washington, D.C.: U.S. Government Printing Office, 1974.

Area Handbook for Chad. Washington, D.C.: U. S. Government Printing Office, 1972.

Billard, Pierre. Le Cameroun Fédéral. 2 vols. Lyon: Imprimerie des Beaux-Arts, 1968.

Binet, Jacques. "La République Gabonnaise." Notes et Etudes Documentaires, No. 3703, June 27, 1970.

Bouquerel, Jacqueline. *Le Gabon*. Paris: Presses Universitaires de France, 1970.

Cabot, Jean. *Le Bassin de Moyen Logone*. Paris: O.R.S.T.O.M., 1965.

——and Christian Bouquet. *Le Tchad*. Paris: Presses Universitaires de France, 1973.

Cameroon. Ministry of Economic Affairs and Planning. *La Population du Pays Bamiléké et des Départements Limitrophes*. Paris: Société d'Etudes pour le Développement Economique et Social, 1966.

Capot-Rey, R., D. Audebert, and R. Owona. *Les Structures Agricoles de la Zone Littorale*. Cameroon: Secrétariat d'Etat au Developpement Rural, 1970.

Champaud, Jacques. *Mom, Terroir Bassa (Cameroun)*. Paris: O.R.S.T.O.M., 1973.

Diguimbaye, G. and R. Langue. *L'Essor du Tchad*. Paris: Presses Universitaires de France, 1969.

Guillot, Bernard. *Le Terre Enkou: Recherches sur les Structures Agraires du Plateau Koukouya (Congo)*. Paris: Mouton, 1973.

Hallaire, A. "Marchés et Commerce au Nord des Monts Mandara," *Cahiers de l'O.R.S.T.O.M., Série Sciences Humaines*, 9 no. 3 (1972).

Hugon, P. *Analyse du Sous-développement en Afrique Noire: Exemple de l'Economie du Cameroun*. Paris: Presses Universitaires de France, 1968.

Imbert, Jean. *Le Cameroun*. Paris: Presses Universitaires de France, 1973.

"L'Industrialisation des Etats de l'Union Douanière et Economique de l'Afrique Centrale (UDEAC). *Notes et Etudes Documentaires*, no. 3830, October 25, 1971.

International Monetary Fund. *Surveys of African Economies*, vol. 1, 1968; vol. 5, 1973.

Johnson, Willard R. *The Cameroon Federation*. Princeton: Princeton University Press, 1970.

Kalck, Pierre. *Central African Republic: A Failure in Decolonisation*. Translated by Barbara Thomson. New York: Praeger, 1971.

Le Rouvreur, A. *Sahéliens et Sahariens du Tchad*. Paris: Berger-Levrault, 1962.

Rubin, Neville. *Cameroun: An African Federation*. New York: Praeger, 1971.

Sautter, Gilles. *De l'Atlantique au Fleuve Congo: Un Géographie de Souspeuplement: Gabon et Congo (Brazzaville)*. 2 vols. Paris: Mouton, 1966.

Sikes, Sylvia. *Lake Chad*. London: Eyre Methuen, 1972.

Thompson, Virginia, and Richard Adloff. *The Emerging States of French Equatorial Africa*. Stanford: Stanford University Press, 1960.

Tissandier, Jean. *Zengoaga: Etude d'un Village Camerounais et de son Terroir au Contact Foret-savane*. Paris: Mouton, 1969.

Vennetier, Pierre. *Géographie du Congo—Brazzaville*. Paris: Gauthier-Villars, 1966.

——*L'Afrique Equatoriale*. Paris: Presses Universitaires de France, 1972.

——*Pointe-Noire et la Façade Maritime du Congo-Brazzaville*. Paris: O.R.S.T.O.M., 1968.

Weinstein, Brian. *Gabon: Nation-building on the Ogooué*. Cambridge, Mass.: M.I.T. Press, 1966.

CHAPTER XIV: ZAÏRE

Académie Royale des Sciences d'Outre-Mer. *Livre Blanc*. 3 vols. Brussels, 1963.

Area Handbook for the Democratic Republic of the Congo (Congo Kinshasa). Washington, D.C.: U.S. Government Printing Office, 1971.

Banque du Zaïre, *Rapport annuel*.

Belgian Congo and Ruanda-Urundi Information and Public Relations Office. *Belgian Congo*. 2 vols. Brussels: Inforcongo, 1960.

Belgium. Direction de l'Agriculture. *Volume Jubilaire du Bulletin Agricole du Congo Belge et du Ruanda-Urundi, 1910–1960*. Brussels, 1961.

Cornevin, Robert. *Le Zaïre*. Paris: Presses Universitaires de France, 1972.

Huybrechts, André. *Transports et Structures au Congo pendant la Décennie, 1959–1969*. Brussels: A.R.S.O.M., 1970.

International Monetary Fund. *Surveys of African Economies*. Vol. 4. Washington, D.C., 1971.

Jurion, F., and J. Henry. *De l'Agriculture Itinérante à l'Agriculture Intensifiée*, Brussels: I.N.E.A.C., 1967.

Robert, Maurice. *Le Congo Physique*. 3rd ed., revised. Liège: H. Vaillant-Carmanne, 1946.

Van den Abeele, Marcel, and René Vandenput. *Les Principales Cultures du Congo Belge*. 3d ed. Brussels, Ministère des Colonies, Direction de l'Agriculture, des Forets, et de l'Elevage, 1956.

Zaïre.

Zaïre, République du. Département de l'Economie Nationale. *Conjoncture Economique*. No. 13. November 1973.

CHAPTER XV: ETHIOPIA

Area Handbook for Ethiopia. Washington, D.C.: U. S. Government Printing Office, 1971.

Bequele, Assefa, and Eshetu Chole. *A Profile of the Ethiopian Economy*. London: Oxford University Press, 1969.

Cherian, K. *Ethiopia Today*. Addis Ababa: Central Printing Press, 1972.

Hoben, Allen. "Social Anthropology and Development Planning—A Case Study in Ethiopian Land Reform Policy," *Journal of Modern African Studies*, 10, no. 4 (December 1972): 561–582.

Huffnagel, H. P. *Agriculture in Ethiopia*. Rome: F. A. O., 1961.

Levine, Donald N. *Wax and Gold: Tradition and Innovation in Ethiopian Society*. Chicago: University of Chicago Press, 1965.

Lipsky, George A. *Ethiopia: Its People, Its Society, Its Culture.* New Haven: Human Relations Area Files, 1962.

Mariam, Mesfin W. *An Atlas of Ethiopia.* 2nd ed., revised. Addis Ababa, 1970.

National Bank of Ethiopia. *Quarterly Bulletin.*

Simoons, Frederick J. *North-West Ethiopia, Peoples and Economy.* Madison: University of Wisconsin Press, 1960.

Ullendorff, Edward. *The Ethiopians: An Introduction to Country and People,* 3rd ed., London: Oxford University Press, 1973.

United States. *Department of Agriculture. A Survey of Agriculture in Ethiopia.* Washington, D.C., March 1969.

Zerom, Kifle-Mariam. *The Resources and Economy of Ethiopia.* Menlo Park: Stanford Research Institute, 1969.

CHAPTER XVI: SOMALIA

Area Handbook for Somalia. Washington, D.C.: U. S. Government Printing Office, 1969.

International Monetary Fund. *Surveys of African Economies.* Vol. 2. Washington, D.C., 1969.

Karp, Mark. *The Economics of Trusteeship in Somalia.* Boston: Boston University Press, 1960.

Shilling, Nancy A. "Problems of Political Development in a Ministate: The French Territory of the Afars and the Issas," *The Journal of Developing Areas,* 7 (July 1973): 613–634.

Thompson, Virginia, and Richard Adloff. *Djibouti and the Horn of Africa.* Stanford: Stanford University Press, 1968.

CHAPTER XVII: EAST AFRICA

Agriculture in Kenya. Nairobi: University Press of Africa, 1968.

Area Handbook for Kenya. Washington, D.C.: U. S. Government Printing Office, 1967.

Area Handbook for Tanzania. Washington, D.C.: U. S. Government Printing Office, 1968.

Area Handbook for Uganda. Washington, D. C.: U. S. Government Printing Office, 1969.

Berry, Len, ed. *Tanzania in Maps: Graphic Perspectives of a Developing Country.* London: University of London Press, 1971.

Cone, L. Winston and J. F. Lipscomb. *The History of Kenya Agriculture.* Nairobi: University Press of Africa, 1972.

East Africa Journal.

East African Journal of Rural Development.

Ghai, Dharam. *Portrait of a Minority: Asians in East Africa.* London: Oxford University Press, 1966.

Good, Charles M., Jr. *Dimensions of East African Cultures.* East Lansing, Michigan State University: African Studies Center, 1966.

Harbeson, John W. *Nation Building in Kenya: The Role of Land Reform.* Evanston, Ill.: Northwestern University Press, 1973.

Hatch, John. *Tanzania, A Profile.* New York: Praeger, 1972.

Hazlewood, Arthur. *Rail and Road in East Africa.* New York: Augustus W. Kelley, 1965.

Helleiner, G. K., ed. *Agricultural Planning in East Africa.* Nairobi: East African Publishing House, 1968.

Ingle, Clyde R. *From Village to State in Tanzania: The Politics of Rural Development.* Ithaca: Cornell University Press, 1972.

International Bank for Reconstruction and Development. *The Economic Development of Kenya.* Baltimore: The Johns Hopkins Press, 1962.

——*The Economic Development of Tangayika.* Baltimore: The Johns Hopkins Press, 1960.

——*The Economic Development of Uganda.* Baltimore: The Johns Hopkins Press, 1962.

International Monetary Fund. *Surveys of African Economies.* Vol. 2. Washington, D.C., 1969.

Jameson, J. D. and D. Stevens, eds. *Agriculture in Uganda.* 2nd ed. revised. London: Oxford University Press, 1971.

Jätzold, R., and E. Baum. *The Kilombero Valley.* Munich: Ifo-Institut, African Studies No. 28, 1968.

Jensen, S. *Regional Economic Atlas, Mainland Tanzania.* Dar es Salaam: University College, Bureau of Resource Assessment and Land Use Planning, Research Paper 1, 1968.

Kenya. Survey of Kenya. *National Atlas of Kenya.* 3rd ed. Nairobi, 1970.

Langlands, B. W. *A Preliminary Review of Land Use in Uganda.* Kampala: Makerere University, Department of Geography. 1971.

McMaster, D. N. *A Subsistence Crop Geography of Uganda.* Bude, England: Geographical Publications, 1962.

Morgan, W. T. W. *East Africa.* London: Longman, 1973.

——*East Africa: Its People and Resources.* Nairobi: Oxford University Press, 1969.

——*Nairobi: City and Region.* Nairobi: Oxford University Press, 1967.

Nyerere, Julius. *Freedom and Development.* Dar es Salaam: Tanzania Publishing House, 1967.

——*Socialism and Rural Development.* Dar es Salaam: Tanzania Publishing House, 1968.

O'Connor, A. M. *An Economic Geography of East Africa.* 2nd ed. London: Bell, 1971.

——*Railways and Development in Uganda.* Nairobi: Oxford University Press, 1966.

Odingo, Richard S. *The Kenya Highlands: Land Use and Agricultural Development.* Nairobi: East Africa Publishing House, 1971.

Ominde, Simeon H., ed. *Studies in East African Geography and Development.* London: Heinemann, 1971.

Ouma, Joseph P. B. M. *Evolution of Tourism in East Africa.* Nairobi: East African Literature Bureau, 1970.

Parsons, D. J. *Systems of Agriculture Practised in Uganda.* 5 pts. Uganda: Department of Agriculture, 1960.

Pearson, Donald S. *Industrial Development in East Africa.* Nairobi: Oxford University Press, 1971.

Popovic, V. *Tourism in East Africa.* Munich: Ifo-Institut, African Studies No. 73, 1971.

Richards, Audrey and J. M. Fortt. *Subsistence to Commercial Farming in Present Day Buganda.* Cambridge, England: Cambridge University Press, 1973.

Rural Africana, no. 13 (Winter 1971). Special issue on East Africa.

Ruthenberg, Hans. *African Agricultural Production Development Policy in Kenya, 1952–1965.* Berlin: Springer-Verlag, 1966.

——*Smallholder Farming and Smallholder Development in Tanzania: Ten Case Studies.* Munich: Ifo-Institut, African Studies No. 24, 1968.

Rutman, Gilbert L. *The Economy of Tanganyika.* New York: Praeger, 1968.

Smith, Hadley E., ed. *Agricultural Development in Tanzania.* London: Oxford University Press, 1966.

Soja, Edward W. *The Geography of Modernization in Kenya.* Syracuse, N.Y.: Syracuse University Press, 1968.

Tanzania, Ministry of Lands, Settlement and Water Development. *Atlas of Tanzania.* Dar es Salaam, 1967.

Tanzania, University of Dar es Salaam, Bureau of Resource Assessment and Land Use Planning, *Research Notes.*

——*Research Papers.*

Uganda, Department of Lands and Surveys. *Atlas of Uganda.* 2nd ed., revised. Kampala, 1967.

von Haugwitz, H. W. *Some Experiences with Smallholder Settlement in Kenya.* Munich: Weltforum, 1972.

West, Henry W. *Land Policy in Buganda.* London: Cambridge University Press, 1972.

CHAPTER XVIII: RWANDA AND BURUNDI

Area Handbook for Burundi. Washington, D.C.: U. S. Government Printing Office, 1969.

Area Handbook for Rwanda. Washington, D.C.: U. S. Government Printing Office, 1969.

Baker, Randall. "Rwanda," *Focus,* vol. 23, no. 10 (June 1973).

"Le Burundi," *Notes et Etudes Documentaires,* no. 3364. Paris, 1967.

International Monetary Fund. *Surveys of African Economies.* Vol. 5. Washington, D.C., 1973.

Lemarchand, René. *Rwanda and Burundi.* New York: Praeger, 1970.

Mpozzagara, Gabriel. *Le République du Burundi.* Paris: Berger-Levrault, 1971.

Van der Velpen, Cl. *Manuel de Géographie du Burundi.* Brussels: Editions A. De Boeck, 1973.

CHAPTER XIX: RHODESIA

Anderson, R. et al. *An Agricultural Survey of Southern Rhodesia. Part II: Agro-Economic Survey.* Salisbury: Government Printer, 1961.

Brelsford, William V., ed. *Handbook to the Federation of Rhodesia and Nyasaland.* London: Cassell, 1960.

Federation of Rhodesia and Nyasaland. Federal Surveys Department. *Atlas of the Federation of Rhodesia and Nyasaland.* Salisbury, 1960–65.

Kay, George. *A Human Geography of Rhodesia.* London: University of London Press, 1970.

——*Distribution and Density of African Population in Rhodesia.* Hull, England: University of Hull, Geography Department, Miscellaneous Series 12, 1972.

Park, Stephen. *Business as Usual: Transactions Violating Rhodesian Sanctions.* New York: The Carnegie Endowment for International Peace, Special Rhodesia Project, 1973.

Roder, Wolf. *The Sabi Valley Irrigation Projects.* Chicago: University of Chicago Press, Department of Geography Research Paper 99, 1965.

Vincent, V. et al. *An Agricultural Survey of Southern Rhodesia. Part I: Agro-Ecological Survey.* Salisbury, Government Printer, 1960.

CHAPTER XX: ZAMBIA

Area Handbook for Zambia. Washington, D.C.: U. S. Government Printing Office, 1969.

Bostock, Mark and Charles Harvey, eds. *Economic Independence and Zambian Copper.* New York: Praeger, 1972.

Burawoy, Michael. *The Colour of Class on the Copper Mines; from African Advancement to Zambianization.* Lusaka: University of Zambia, Institute for African Studies, Zambian Papers No. 7, 1972.

Davies, D. Hywel, ed. *Zambia in Maps.* London: University of London Press, 1971.

Elliott, Charles, ed. *Constraints on the Economic Development of Zambia.* London: Oxford University Press, 1971.

Hall, Richard. *Zambia.* New York: Praeger, 1966.

Hellen, John A. *Rural Economic Development in Zambia, 1890–1964.* Munich: Ifo-Institut, 1968.

International Monetary Fund. *Surveys of African Economies.* Vol. 4. Washington, D.C., 1971.

Kay, George. *A Social Geography of Zambia.* London: University of London Press, 1967.

Rotberg, Robert J. *The Rise of Nationalism in Central Africa: The Making of Malawi and Zambia, 1873–1964.* Cambridge, Mass.: Harvard University Press, 1966.

Storrs, A. E. G. *A Study of Zambia's Natural Resources.* London: Oxford University Press, 1969.

Trapnell, Colin G. *The Soils, Vegetation and Agriculture of North-eastern Rhodesia.* Lusaka: Government Printer, 1953.

——and J. N. Clothier. *The Soils, Vegetation and Agricultural Systems of North-western Rhodesia*. Lusaka: Government Printer, 1937.

Young, A. *Industrial Diversification in Zambia*. London, 1974.

Zambia. Copper Industry Service Bureau. *Mindeco Mining Year Book 1972*. Ndola, 1973.

CHAPTER XXI: MALAWI

Agnew, Swanzie and M. Stubbs, eds. *Malawi in Maps*. London: University of London Press, 1972.

Brown, Peter and A. Young. *The Physical Environment of Central Malawi*. Zomba: Government Printer, 1965.

——*The Physical Environment of Northern Nyasaland*. Zomba: Government Printer, 1962.

Dequin, H. *Agricultural Development in Malawi*. New York: Humanities Press, 1969.

International Monetary Fund. *Surveys of African Economies*. Vol. 4. Washington, D.C., 1971.

"Land and Labor in Rural Malawi," *Rural Africana*, no. 20–22 (1973).

Nyasaland. Natural Resources Department. *The Natural Resources of Nyasaland*. Zomba: Government Printer, 1960.

Pike, J. G. and G. T. Rimington. *Malawi: A Geographical Study*. London: Oxford University Press, 1965.

CHAPTER XXII: ANGOLA AND MOZAMBIQUE

Abshire, D. M. and M. A. Samuels. *Portuguese Africa: A Handbook*. New York: Praeger, 1969.

Area Handbook for Angola. Washington, D.C.: U. S. Government Printing Office, 1967.

Area Handbook for Mozambique. Washington, D.C.: U. S. Government Printing Office, 1969.

Atlas de Moçambique. Lourenço Marques: Empresa Moderna, 1962.

Chilcote, Ronald H. *Portuguese Africa*. Englewood Cliffs, N.J.: Prentice Hall, 1967.

Duffy, James. *Portugal in Africa*. Cambridge, Mass.: Harvard University Press, 1962.

——*Portuguese Africa*. Cambridge, Mass.: Harvard University Press, 1958.

Spence, C. F. *Moçambique*. Cape Town: Howard Timmine, 1963.

Wheeler, Douglas and René Pelissier. *Angola*. New York: Praeger, 1971.

CHAPTER XXIII: THE REPUBLIC OF SOUTH AFRICA

Area Handbook for the Republic of South Africa. Washington, D.C.: U. S. Government Printing Office, 1971.

Backer, W. ed. *The Economic Development of the Transkei*. Fort Hare: University of Fort Hare, 1969.

Bell, Trevor. *Industrial Decentralization in South Africa*. Cape Town: Oxford University Press, 1973.

Board, C., R. J. Davies, and T. J. D. Fair. "The Structure of the South African Space Economy: An Integrated Approach," *Regional Studies*, 4 (1970): 367–392.

Cole, Monica M. *South Africa*. New York: E. P. Dutton and Co., Inc., 1961.

Duggan, William Redman. *A Socioeconomic Profile of South Africa*. New York: Praeger, 1973.

Grice, D. C. "The Approaching Crisis—Land and Population in the Transvaal and Natal." Johannesburg: South African Institute of Race Relations, 1973.

Grundy, Kenneth W. *Confrontation and Accommodation in Southern Africa: The Limits of Independence*. Berkeley: University of California Press, 1973.

Hance, William A., ed. *Southern Africa and the United States*. New York: Columbia University Press, 1968.

Hill, Christopher. *Bantustans*. London: Oxford University Press, 1964.

Horrell, Muriel, comp. *A Survey of Race Relations in South Africa* (annual). Johannesburg: South African Institute of Race Relations.

Houghton, D. Hobart. *The South African Economy*. 2nd ed. revised. Cape Town: Oxford University Press, 1967.

Olivier, Henry. "Water in Southern Africa," *Optima*, (June 6, 1970), pp. 58–67.

Pollock, N. C. and Swanzie Agnew. *An Historical Geography of South Africa*. London: Longmans, Green and Co., 1963.

South Africa. Department of Planning. *Development Atlas*. Pretoria: Government Printer, 1966.

Talbot, A. M. and W. J., eds. *Atlas of South Africa*. Pretoria: Government Printer, 1960.

Wellington, John H. *Southern Africa: A Geographical Study*. 2 vols. Cambridge: Cambridge University Press, 1955.

Wilson, Francis. *Labour in the South African Gold Mines, 1911–1969*. London: Cambridge University Press, 1972.

CHAPTER XXIV: SOUTH-WEST AFRICA (NAMIBIA)

Horrell, Muriel. *South-West Africa*. Johannesburg: South African Institute of Race Relations, 1967.

Kory, Marcelle. "The Contract Labour System and the Ovambo Crisis of 1971 in South West Africa," *African Studies Review*, 16, no. 1 (April 1973): 83–106.

Leistner, G. M. "South Africa's Economic Ties with South West Africa," *Africa Institute Bulletin*, 2 (April 1971): 111–121.

Mertens, A. *South West Africa and Its Indigenous Peoples.* New York: Taplinger, 1967.

Republic of South Africa. Department of Foreign Affairs. *South West Africa Survey 1967.* Pretoria: Government Printer, 1967.

Wellington, John H. *South West Africa and Its Human Issues.* London: Oxford University Press, 1967.

CHAPTER XXV: BOTSWANA, LESOTHO, SWAZILAND

Bawden, M. G. and A. R. Stobbs. *The Land Resources of Eastern Bechuanaland.* Tolworth: Directorate of Overseas Surveys, 1963.

Best, Alan C. G. *The Swaziland Railway: A Study in Politico-Economic Geography.* East Lansing: Michigan State University Press, 1966.

Botswana, Republic of. *National Development Plan 1973-78,* Part I, Policies and Objectives. Gaborone: Government Printer, 1973.

Fair, T. J. D., George Murdoch, and H. M. Jones. *Development in Swaziland: A Regional Analysis.* Johannesburg: Witwatersrand University Press, 1969.

Great Britain. *Basutoland, Bechuanaland Protectorate and Swaziland: Report of an Economic Survey Mission.* London: H.M.S.O., 1960.

International Monetary Fund. *Surveys of African Economies.* Vol. 5. Washington, D.C.: 1973.

Knight, David B. "Botswana," *Focus,* 20, no. 3 (November 1969).

Leistner, G. M. E. and P. Smit. *Swaziland: Resources and Development.* Pretoria: Africa Institute, 1969.

Smit, P. *Botswana: Resources and Development.* Pretoria: Africa Institute, 1970.

——"Mining Developments in Botswana," *South African Journal of African Affairs,* 3, no. 1 (1973): 44–55.

Potholm, Christian P. *Swaziland: The Dynamics of Political Modernization.* Berkeley: University of California Press, 1972.

Spence, J. E. *Lesotho.* London: Institute of Race Relations, 1967.

Stevens, Richard P. *Lesotho, Botswana, and Swaziland: The Former High Commission Territories in Southern Africa.* New York: Praeger, 1967.

CHAPTER XXVI: MADAGASCAR

L'Association des Géographes de Madagascar. *Atlas de Madagascar.* Tananarive, 1969–1971.

Bougeat, F. *Etude de la Basse Plaine de la Manambato.* Tananarive: O.R.S.T.O.M., 1964.

Bulletin de Madagascar.

Dandoy, Gérard. *Terroirs et Économies Villageoises de la Région de Vavatenina.* Paris: O.R.S.T.O.M., 1973.

Deschamps, Hubert. *Madagascar.* Paris: Presses Universitaires de France, 1968.

Donque, Gérald. *Madagascar, les Mascareignes et les Comores.* Paris: EDSCO Documents, November–December, 1965.

Guilcher, A. and R. Battistini. *Madagascar, Géographie Régionale.* Paris: Centre de Documentation Universitaire, 1968.

Heseltine, Nigel. *Madagascar.* New York: Praeger, 1971.

Isnard, Hildebert. *Madagascar.* Paris: Armand Colin, 1964.

Kent, R. K. *From Madagascar to the Malagasy Republic.* New York: Praeger, 1962.

Madagascar—Revue de Géographie.

Ottino, Paul. *Les Economies Paysannes Malgaches du Bas Mangoky.* Paris: Berger-Levrault, 1963.

Revue de Madagascar.

Thompson, Virginia and Richard Adloff. *The Malagasy Republic: Madagascar Today.* Stanford: Stanford University Press, 1965.

Trouchaud, J. P. *Contribution à l'Etude Géographique de Madagascar, Basse Plaine de Mangoky.* Paris: O.R.S.T.O.M., 1965.

CHAPTER XXVII: THE SMALLER INDIAN OCEAN ISLANDS

Area Handbook for the Indian Ocean Territories. Washington, U. S. Government Printing Office, 1971.

Barclays Bank. *Seychelles: A Barclays Bank International Economic Survey.* London, March, 1972.

Benedict, Burton. *Mauritius: Problems of a Plural Society.* London, Institute of Race Relations, 1965.

Charpentier, Jean. "Les Comores: Economie Agricole et Transports," *Les Cahiers d'Outre-mer,* 24, no. 94 (April–June 1971): 158–184.

Dallot, M. Louis. "Rodrigues, L'Oubliée des Mascareignes," *Afrique Contemporaine,* July–August, 1971, pp. 7–12.

International Monetary Fund. *Surveys of African Economies.* Vol. 4. Washington, 1971.

Lionnet, G. *The Seychelles.* Harrisburg, Pa.: Stackpole Books, 1972.

Meade, J. E. et al. *The Economic and Social Structure of Mauritius.* London: Cass, 1968.

Ramdin, T. *Mauritius: A Geographical Survey.* London: University Tutorial Press, 1970.

Scherer, André. "La Réunion," *Notes et Etudes Documentaires,* No. 3358, January 27, 1967.

Titmuss, Richard M. and B. Abel-Smith. *Social Policies and Population Growth in Mauritius.* London: Methuen, 1961.

Vellin, J. L. C. *Land Use in Mauritius.* London: London School of Economics, Geographical Papers, 1969.

Subject Index

Index of Geographical Names

Picture Credits

Algeria, Ministère de l'Information et de la Culture, 70
upper right, 107, 109, 110, 111, 113, 119, 121, 124,
125
Anglo-American Corporation, 560
Angola, Centro de Informação e Turismo, 63, 486, 489,
492, 493, 496, 497
Cameroon, Ministère de l'Information et de la Culture,
215, 286, 293, 295 lower, 301 upper, 302, 303, 308,
309
Chad, Direction de Documentation, 281
Compagnie des Mines d'Uranium de Franceville, 304
Compagnie des Potasses du Congo, 284, 305
Compagnie Togolaise des Mines du Benin, 39, 196
lower, 250
East African Railways and East African Railways and
Harbours, 1, 343, 378, 379, 382, 384, 395, 396, 397,
408, 411, 413
Egypt, Ministry of Information, 139, 148, 150
Ethiopia, Ministry of Information, 354 upper and lower
left
Firestone Tire and Rubber Company, 193, 220, 257,
612
French Territory of Afars and Issas, Service de l'Infor-
mation, 367
Gabon, Direction Générale du Tourisme, 277, 289
Ghana Information Services, 208, 216, 249, 263 left,
266 left
Kaiser Aluminum and Chemical Corporation, 266 right
Kenya Information Services, 371, 393, 405, 410
LAMCO Mining Company, 246, 247
Madagascar, Service Générale de l'Information, 591
Malawi, Ministry of Trade, Industry and Tourism, 10
lower, 80, 479, 483
MIFERMA, 189, 239
Morocco, Ministère de l'Information, 99
Mozambique, Centro de Informação e Turismo, 500,
504, 505, 508

Nigeria, Ministry of Research and Information, 261
upper
Olin Corporation, 242
Phillips Petroleum Company, 146, 181, 251
Roan C M, 431, 441, 458, 472, 473
Rhodesia, Ministry of Information, 51 upper right, 436,
450, 455, 463
Seychelles, Department of Tourism, 597, 598
Société des Mines de l'Aïr, 176, 241
South Africa Information Services, 511, 515, 517, 518,
520, 522, 524, 530, 531, 533, 535, 539, 540, 541, 543
upper, 545, 547, 552, 553
Standard Oil Company (N.J.), 91, 128, 129, 132
Sudan, Ministry of Information and Culture, 23, 157,
163, 165, 168
Tunisia, Office National du Tourisme et du Therma-
lisme, 116
Uganda, Department of Information, 406
The United Africa Company, Ltd., 212, 259, 263 right,
323
United Nations, 102, 166, 170, 175, 362, 364
United States, Agency for International Development,
229
Van der Borght, Dr. Henri, 59
Vanilla Information Bureau (USA), New York, 587, 596
Zaïre, Gécamines, 333, 334
Author, 10 upper, 27, 51 upper left and lower, 70
upper and lower left, 78, 81, 85, 98, 100, 101, 142,
143, 145, 158, 159, 161, 168 upper, 171, 195, 203,
204, 205, 206, 209, 210, 219, 228, 230, 231, 232, 236,
244, 255, 290, 295 upper, 296, 301 lower, 341, 346,
349 lower, 354 right, 355, 358, 385, 386, 388, 400,
401, 409, 429, 446, 447, 468, 521, 543 lower, 564,
565, 569, 570, 573, 580, 581, 583, 584, 585, 588, 589,
601, 609, 616, 617